DRIVER DISTRACTION

Theory, Effects, and Mitigation

DRIVER DISTRACTION

Theory, Effects, and Mitigation

Edited by

Michael A. Regan

John D. Lee

Kristie L. Young

CRC Press is an imprint of the
Taylor & Francis Group, an **informa** business

CRC Press
Taylor & Francis Group
6000 Broken Sound Parkway NW, Suite 300
Boca Raton, FL 33487-2742

Printed in the United States of America on acid-free paper
10 9 8 7 6 5 4 3 2 1

International Standard Book Number-13: 978-0-8493-7426-5 (Hardcover)

Library of Congress Cataloging-in-Publication Data

Driver distraction : theory, effects, and mitigation / edited by Michael A. Regan, John D. Lee, Kristie Young.
p. cm.
Includes bibliographical references and index.
ISBN-13: 978-0-8493-7426-5
ISBN-10: 0-8493-7426-X
1. Distracted driving. 2. Automobile driving. 3. Automobile drivers. 4. Traffic safety. I. Regan, Michael A. II. Lee, John D. III. Young, Kristie L. IV. Title.

HE5620.D59D75 2009
363.12'414--dc22 2008014178

Visit the Taylor & Francis Web site at
http://www.taylorandfrancis.com

and the CRC Press Web site at
http://www.crcpress.com

Contents

PART 8 Prevention and Mitigation Strategies

PART 9 Conclusions

Foreword

Driving a motor vehicle is, for many people, the most complex and potentially dangerous task they will perform during their lifetime. Despite this complexity, it is not uncommon for drivers to engage, willingly or unwittingly, in activities that divert their attention away from activities critical for safe driving. Distraction, it seems, is part of everyday driving. There is converging evidence, however, that it has potential to significantly degrade driving performance and safety. Sadly, the very group that has the greatest propensity to be distracted—young novice drivers—is the group that appears most vulnerable to its adverse effects.

The NRMA-ACT Road Safety Trust was formally established in 1992, with the principal objective of enhancing road safety for the benefit of the road-using community of the Australian Capital Territory (ACT), a jurisdiction located in southeast Australia. Since then, it has allocated around AUD16.3 million to some 260 innovative road safety projects, many of which have indirectly benefited communities elsewhere in Australia and overseas. The Trust recognizes the importance of driver distraction as an emerging road safety issue and is pleased to have been able to provide support for the writing of several chapters of this important book. The book brings together in a single volume much of the accumulated knowledge that exists on the topic of driver distraction. With contributions from more than 40 distinguished authors around the world, it provides the reader with a complete overview of the topic—the theory underlying distraction, its effect on driving performance and safety, strategies for mitigating its adverse effects, and directions for future research.

This book comes at a critical time, when road safety authorities around the world are just starting to recognize the importance of driver distraction as a road safety issue. Many are struggling to know how to deal with it. With perspectives from road safety authorities, vehicle manufacturers, equipment suppliers, injury prevention researchers, and academics, the book provides balanced and practical guidance on how to quantify the impact of distraction, manage it, and prevent it from escalating into a more significant problem than it already is. We are pleased to be part of this important, international, road safety initiative.

Professor Don Aitkin, AO
Chairman
NRMA-ACT Road Safety Trust

Acknowledgments

The editors wish to thank the following organizations and individuals for the important roles they played in enabling this book to be completed:

- The anonymous reviewers, recruited by CRC Press, for recommending that development of the book proceed.
- CRC Press for their professional guidance, trust, and patience.
- The French National Institute for Transport and Safety Research (INRETS) for providing funded time for Michael Regan to complete the book. We thank Dr. Corinne Brusque, in particular, for her support.
- The NRMA-ACT Road Safety Trust for a grant that provided Michael Regan and Kristie Young with some funded time to work on the book.
- The Monash University Accident Research Centre (MUARC) for providing Michael Regan and Kristie Young with some funded time to work on the book.
- The Monash University Accident Research Centre Foundation for providing John D. Lee with a grant that supported a sabbatical to MUARC in late 2006, which enabled him to work on the book.
- Laurie Sparke and Mike Hammer, of General Motors Holden, Australia, who sponsored research undertaken by MUARC that made the reporting of research in several chapters possible.
- Mike Perel, of the National Highway Traffic Safety Administration, for sponsoring and providing technical leadership on the SAVE-IT program, under which research represented in several chapters was conducted. Technical assistance and program monitoring was provided by Mary Stearns, of the Volpe Transportation Systems Center.
- The authors and coauthors, for their contributions, patience, and goodwill in adhering to the requirements of the editorial review process.
- Last, but not least, our partners and families for their untiring support during the long haul.

Michael A. Regan
John D. Lee
Kristie L. Young

Editors

Michael A. Regan is currently a research director with The French National Institute for Transport and Safety Research (INRETS), in Lyon, France, and is an adjunct professor in the Department of Applied Mechanics at the Chalmers University of Technology, in Gothenburg, Sweden. Before being seconded to INRETS, in March 2007, Mike was a senior research fellow, and program manager of Human Factors and Simulation, at the Monash University Accident Research Centre (MUARC) in Melbourne, Australia. Mike is an applied experimental psychologist and obtained his BSc (Hons) and PhD degrees from the Australian National University in Canberra, Australia. His PhD research focused on the human ability to divide attention between competing tasks. Mike's research in road safety has focused on driver distraction, human factors in the design and evaluation of intelligent transport systems, human-in-the-loop driving simulation, novice driver and passenger training, and human error in road transport. He is an ex-chairman of the Ergonomics Society of Australia and is affiliated with several professional organizations. He has authored and coauthored about 180 published articles, reports, and papers and sits on the editorial boards of the *Journal of the Australasian College of Road Safety, IET Intelligent Transport Systems* and *European Transport Research Review.* He sits on two Australian standards committees and is the Australian representative on International Organization for Standardization (ISO) Technical Committee 22, Sub-Committee 13-Ergonomics Applicable to Road Vehicles. Mike and his previous research team at MUARC have received several road safety–related awards, including the 2005 Australasian College of Road Safety Peter Vulcan Award. Mike will remain on secondment with INRETS until March 2010.

John D. Lee is a professor in the Department of Mechanical and Industrial Engineering at the University of Iowa, and is the director of human factors research at the National Advanced Driving Simulator. He is also affiliated with the Department of Neurology, Public Policy Center, Injury Prevention Research Center, National Advanced Driving Simulator, and Center for Computer-Aided Design. His research focuses on the safety and acceptance of complex human-machine systems by considering how technology mediates attention. Specific research interests include trust in technology, advanced driver assistance systems, and driver distraction. He has coauthored the book *An Introduction to Human Factors Engineering* and has authored and coauthored more than 120 articles. He received the Ely Award for best paper in the journal *Human Factors* (2002), the best paper award for the journal *Ergonomics* (2005), and is a Donald E. Bently Faculty Fellow. He is a member of the National Academy of Sciences Committee on Human Factors and has served on several other committees for the National Academy of Sciences. Dr. Lee serves on the editorial boards of *Cognitive Engineering and Decision Making; Cognition, Technology, and Work*; and *International Journal of Human Factors Modeling and Simulation*; and is the associate editor for the journal *Human Factors*.

Kristie L. Young is a research fellow with the Monash University Accident Research Centre (MUARC) in Melbourne, Australia. Kristie joined MUARC in January 2002 after receiving a bachelor of applied science (psychology) with first class Honours from Deakin University. Kristie has been involved in numerous projects at the center, including the TAC SafeCar project, effects of text messaging on young novice drivers, driver exposure to distracting activities, acceptability of in-vehicle technologies to road users, road safety education in schools, and intelligent speed adaptation for heavy vehicles. Her main research interests include driver adaptation to in-vehicle technology, driver distraction, human-machine interface (HMI) design, and intelligent transport systems. Kristie and other members of the TAC SafeCar team received the 2005 Australasian College of Road Safety Peter Vulcan Award for best paper at the 2005 Road Safety Research, Policing and Education Conference.

Contributors

Motoyuki Akamatsu
Institute of Human Science and Biomedical Engineering
National Institute of Advanced Industrial Science (AIST)
Tsukuba, Japan

Debbie L. Bakowski
Advanced Driver Support Systems
Delphi Electronics & Safety
Westfield, Indiana

Megan Bayly
Accident Research Centre
Monash University
Melbourne, Victoria, Australia

Linda Boyle
Department of Mechanical and Industrial Engineering
University of Iowa
Iowa City, Iowa

Peter C. Burns
Road Safety and Motor Vehicle Regulation
Transport Canada
Ottawa, Ontario, Canada

Judith L. Charlton
Accident Research Centre
Monash University
Melbourne, Victoria, Australia

Birsen Donmez
Department of Mechanical and Industrial Engineering
University of Iowa
Iowa City, Iowa

Frank A. Drews
Department of Psychology
University of Utah
Salt Lake City, Utah

Lutz Eckstein
Ergonomics and HMI
BMW Group
Munich, Germany

Jessica Edquist
Accident Research Centre
Monash University
Melbourne, Victoria, Australia

Johan Engström
Department of Humans, Systems and Structures
Volvo Technology Corporation
Gothenburg, Sweden

Brian Fildes
Accident Research Centre
Monash University
Melbourne, Victoria, Australia

James P. Foley
Intelligent Transportation Systems
Noblis
Falls Church, Virginia

Craig P. Gordon
Transport Research and Evaluation
Ministry of Transport
Wellington, New Zealand

Paul Green
Human Factors Division
Transportation Research Institute
University of Michigan
Ann Arbor, Michigan

Anders Hallén
Human Factors Engineering & Ergonomics
Volvo Car Corporation
Gothenburg, Sweden

Mike Hammer
General Motors Holden
Melbourne, Victoria, Australia

Peter A. Hancock
Department of Psychology, and Institute for Simulation and Training
University of Central Florida
Orlando, Florida

Joanne L. Harbluk
Road Safety and Motor Vehicle Regulation
Transport Canada
Ottawa, Ontario, Canada

Tim Horberry
Sustainable Minerals Institute
University of Queensland
Brisbane, Queensland, Australia

William J. Horrey
Center for Behavioral Sciences
Liberty Mutual Research Institute for Safety
Hopkinton, Massachusetts

Sjaanie Koppel
Accident Research Centre
Monash University
Melbourne, Victoria, Australia

Dave Leblanc
Engineering Research Division
University of Michigan Transportation Research Institute (UMTRI)
Ann Arbor, Michigan

John D. Lee
Department of Mechanical and Industrial Engineering
University of Iowa
Iowa City, Iowa

Stefan Mattes
Group Research & Advanced Development
DaimlerChrysler AG
Stuttgart, Germany

Suzanne P. McEvoy
The George Institute for International Health
The University of Sydney
Sydney, New South Wales, Australia

Mustapha Mouloua
Department of Psychology
University of Central Florida
Orlando, Florida

Michael A. Regan
Laboratory for Ergonomics and Cognitive Sciences Applied to Transport
French National Institute for Transport and Safety Research (INRETS)
Lyon, Rhône-Alpes, France
and
Accident Research Centre
Monash University
Melbourne, Victoria, Australia

Paul M. Salmon
Ergonomics Research Group
Brunel University
Uxbridge, Middlesex, United Kingdom

John W. Senders
Department of Mechanical and Industrial Engineering
University of Toronto
Toronto, Ontario, Canada

Matthew R.H. Smith
Advanced Driver Support Systems
Delphi Electronics & Safety
Carmel, Indiana

Alan Stevens
Transportation Division
Transport Research Laboratory
Wokingham, United Kingdom

Mark R. Stevenson
The George Institute for International Health
The University of Sydney
Sydney, New South Wales, Australia

David L. Strayer
Department of Psychology
University of Utah
Salt Lake City, Utah

Claes Tingvall
Swedish Road Administration
Borlänge, Sweden

Trent W. Victor
Volvo Technology Corporation/ Chalmers University of Technology
Gothenburg, Sweden

Christopher D. Wickens
Division of Human Factors
University of Illinois
Champaign, Illinois
and
Ma&D Operations
Alionscience Corporation
Boulder, Colorado

Ann Williamson
Department of Aviation
University of New South Wales
Sydney, New South Wales, Australia

Gerald J. Witt
Advanced Driver Support Systems
Delphi Electronics & Safety
Carmel, Indiana

Kristie L. Young
Accident Research Centre
Monash University
Melbourne, Victoria, Australia

Harry Zhang
Applied Research & Technology Center
Motorola, Inc.
Tempe, Arizona

Part 1

Introduction

1 Introduction

Michael A. Regan, Kristie L. Young, and John D. Lee

CONTENTS

Driving is a complex, multitask activity. Despite this, it is not unusual to see drivers engage simultaneously in other nondriving tasks. The potential consequences of doing so were borne out in a landmark incident that occurred on the morning of December 31, 2001, on Port Arlington Road, near Geelong, Australia. A 24-year-old female dentist was preparing to send an SMS message—"cu1"—using a mobile phone while driving along the road. She was writing the text message to a friend in Melbourne to let her know that she would be meeting her at one o'clock that afternoon. In doing so, she swerved into the adjacent bicycle lane and crashed into the back of a bicycle being ridden by a 36-year-old mechanical engineer, returning from a training ride. The rider was thrown against the windscreen and roof of the car, and landed on the side of the road. He died at the scene. The driver of the vehicle was brought to court in a landmark case—the first in which the use of a mobile phone had been blamed for a road fatality in the Australian State of Victoria. She pleaded guilty to culpable driving, was sentenced to 2 years' imprisonment (fully suspended), and was disqualified from driving for 2 years [1]. This incident, and the media attention that it attracted, was important in focusing attention on driver distraction as a road safety issue in Australia. The purpose of this book is to explain and avoid such tragedies.

Driver distraction can be defined as *the diversion of attention away from activities critical for safe driving toward a competing activity* (see Chapter 3). If the human brain were not limited in attending to multiple tasks at the same time, driver distraction would not be an issue. However, this is not so. Psychologists have known for more than a century that humans are fundamentally limited in their ability to divide attention between competing tasks [2] and that, under certain conditions (i.e., when the tasks are highly similar, highly demanding, and require continuous attention [3]), the performance of one or both will inevitably suffer. There has been much debate in the scientific literature about the locus of this limitation and the psychological mechanisms that give rise to it, and these issues are discussed in Part 2 of this book. Diversion of attention away from activities critical for safe driving toward a competing activity can occur willingly, such as when a driver initiates a mobile phone conversation, or it can occur involuntarily, such as when an item of information in the road environment (e.g., a moving billboard, an ambulance siren) compels the driver to attend to it. Indeed, the human mind is easily diverted from one activity to another, and there is good reason for this. From an evolutionary perspective, it is often advantageous. It is no accident of nature that certain objects, events, and activities are more diverting than others. There is biological advantage in having the human mind unwittingly orient itself toward objects, events, and activities that signify danger (such as a child running unexpectedly onto the roadway ahead, or a spider crawling on the windscreen) or to those that may be instrumental in perpetuating the species (such as other humans deemed to be attractive). Advertising material is designed to exploit this gift of nature; billboards are designed to attract attention.

Given that driving is a complex, multitask activity, some elements of the driving task itself, such as a flashing dashboard warning light or monitoring an unaccompanied child approaching the roadway, may also divert attention away from activities critical for safe driving. In some circumstances, the consequences of distraction may be negligible. In others, such as those that occurred in the text-messaging incident described earlier, the consequences can be tragic. The effects of distraction on driving performance and safety depend on many interrelated factors, which are discussed in later chapters of this book: the demands of the driving and nondriving tasks (see Chapter 19); the current state of drivers, for example, whether they are drowsy or inebriated (see Chapters 19 and 21); the personality of the individual (see Chapter 19); what is distracting them (see Chapters 11, 12 and 13); for how long they are distracted (see Chapters 17 and 18); when and where they are distracted (see Chapters 17 and 18); the momentary configuration of physical circumstances (see Chapters 2 through 4); the degree to which drivers, their vehicle, and the physical environment surrounding them is tolerant of the consequences of the distraction (see Chapters 13, 19, 26; and Part 8); and even a certain amount of luck. The prevention, mitigation, and management of distraction are, therefore, a complex undertaking.

There is converging evidence that driver distraction is a significant road safety issue worldwide. Findings from the analysis of police-reported crashes, reviewed in this book (see Chapter 16), suggest that driver distraction is a contributing factor in around 11% of crashes. Converging data, from a recent observational study in the United States involving 100 instrumented vehicles [4], suggest that distraction may be a contributing factor in up to 23% of crashes. While there is some disparity between these estimates, due to the different measurement methods and metrics used to derive them, there is good reason to believe that they underestimate the true scale of the problem (see Chapter 16). Not surprisingly, in many Organization for Economic Cooperation and Development (OECD) countries, driver distraction is increasingly becoming ranked, alongside speeding, drink-driving, and fatigue, as a significant contributing factor to road trauma [5].

The mobile phone (see Chapter 11), which is often denigrated as a source of distraction, is only one potential source of distraction. Scores of others exist (see Chapters 12 and 13). Not all are unrelated to the driving task, and not all degrade performance to the same degree (see Chapter 12). Some competing activities with the potential to distract drivers are neither new nor technical in origin. Indeed, many are considered to be part of everyday driving. These include talking to passengers, reaching for objects, operating and listening to the radio, smoking, eating, applying cosmetics, grooming, daydreaming, and attending to potential sources of distraction outside the vehicle cockpit (e.g., advertising and other nondriving-related material posted on billboards, the back of buses and taxis, and variable message signs).

Other activities with a potential to divert attention away from activities critical for safe driving have evolved as a consequence of the proliferation of new technologies finding their way into the vehicle cockpit. These include entertainment systems (e.g., compact disk players) and vehicle information and communication systems (e.g., traveler information systems and the Internet). Advanced driver assistance systems (e.g., route guidance systems and collision warning systems), although designed to support the driver, may distract the driver if poorly designed and located, or used inappropriately. Some of these existing and emerging technologies are hardwired into the vehicle during production, some are retrofitted as aftermarket products, and others, such as personal digital assistants (PDAs) and iPods, can be carried in and out of the vehicle. Driver interaction with these new technologies is becoming increasingly common and is likely to increase as more of these technologies find their way into the vehicle. While everyday activities such as talking with passengers and eating have the potential to distract drivers, governments have tended to focus more on the impact of distraction because of driver interaction with technologies, given that these are increasing in number and their functionality is expanding [6].

As a road safety issue, research on driver distraction is still in its infancy. Although people talk about distraction as if they know what it means, it is poorly defined, and theories and models of the mechanisms and features that characterize it are limited. There exists no universally agreed taxonomy of the sources of distraction, potential and actual, that exist, inside and outside the vehicle. Little is known about patterns of driver exposure to the various sources of distraction that exist or about the impact of these on driver performance, individually and in combination. The relationship between performance degradation, crash frequency, and crash risk is poorly understood. Programs and policies to address the issue are limited, and data on the effectiveness of existing initiatives are scant. Strategic cooperation between key stakeholders in dealing with the issue is lacking.

It is noteworthy that the study of distraction has been confined almost entirely to the road transport domain, although related work has been going on for some time in the computing and aviation domains, under the guise of "interruptions" and the closely related topic of mental workload. Even within the road transport domain, the focus of distraction efforts to date has been on drivers: distracted walking and riding, whether on bicycles or motorcycles, are potential areas of concern almost totally unexplored and researched. There is little practical guidance on how to align and prioritize research to support the development of effective distraction prevention and mitigation strategies, even for car drivers, and the design, development, and evaluation of community-based prevention and mitigation strategies are still in their infancy.

There are many sectors of the community with a vested interest in preventing and mitigating the potential effects of distracted driving: the motoring public, vehicle manufacturers, road and transport authorities, the police, the media, motoring clubs, equipment manufacturers and suppliers, standards organizations, road safety bodies, driver trainers, academics, and others (see Chapters 2, 14, 25, and 33). Typically, each party is concerned with only a subset of the issues relevant to the causes, effects, and management of distracted driving. Because the study of distraction is still in its infancy, there is relatively little information available on the topic, and the information that is available is contained within a variety of disparate sources.

This book brings together a substantial portion of the body of knowledge that currently exists on the topic of driver distraction. It provides the reader with a broad overview of the topic: the theory underlying distraction, its effects on performance and safety, strategies for preventing and mitigating its effects, and directions for future research.

Driver distraction has the potential to escalate into a more serious problem as more and more objects, events, and activities, inside and outside the vehicle, compete for the driver's attention. To date, regulation has been the main tool of governments in dealing with the issue, without much evidence-based research to support this decision making [7]. However, simply banning the installation and use of certain devices while driving, such as mobile phones, and banning drivers from engaging in other potentially distracting activities, is not a constructive way forward in dealing with a complex road safety problem. Distraction is an inevitable consequence of being human, and it is incumbent on governments to accept that driver distraction cannot be eliminated. Attention should be given to designing a road transport system that minimizes driver exposure to avoidable sources of distraction. Such a system would mitigate the effects of distraction and tolerate the consequences of distraction through better road and vehicle design. In doing so, the system must be designed so that it does not place demands on drivers, which are inconsistent with their limited capacity to attend simultaneously to competing activities. To do so will require a sea change in current thinking (see Chapters 2 and 30 through 33). In the meantime, there is sufficient knowledge about driver distraction to support the development and implementation of a wide range of potentially effective countermeasures.

This book is divided into nine parts that are sequenced according to the title of the book, *Driver Distraction: Theory, Effects, and Mitigation*.

Part 1 introduces the reader to the topic. Part 2 defines distraction, discusses philosophical issues pertaining to it, and presents theories and models to explain the mechanisms that give rise to it. Part 3 reviews the various tools, techniques, and methods that have been developed to measure and quantify the effects of distraction on driving performance and behavior. Part 4 reviews the actual effects of distraction on driving performance. Part 5 reviews what is known about the relationship between distraction, crashes, and crash risk, presents a taxonomy for classifying known sources of distraction, and reviews what is known about driver exposure to different sources of distraction. Part 6 of the book reviews factors that mediate the effects of distraction on driving performance and safety, such as driver's age, experience, and state of mind. Part 7 reviews the various principles, guidelines, and checklists that have been developed to support the human-centered design of systems and artifacts with which drivers interact within the vehicle and on the roadway, to minimize distraction. Standards that have been developed to standardize the design and measurement of distraction are also reviewed. Part 8 reviews the range of countermeasures that have been, and are being, developed to prevent and mitigate the effects of distraction. Recommendations are made for further countermeasure development. Finally, Part 9 draws conclusions derived from the material reviewed and makes some general recommendations for further research.

This book provides a balanced treatment of the topic of driver distraction, with contributions from more than 40 distinguished authors from around the world, representing a broad range of stakeholders: vehicle manufacturers, road and transport safety authorities, equipment manufacturers and suppliers, standards organizations, road safety bodies, academics, and human factors and injury-prevention researchers. All chapters were peer-reviewed. It is recognized that human operators in domains other than road transport, such as rail, process control, and aviation, are vulnerable to the effects of distraction. While the focus of the book is on driver distraction, the information in it is also relevant to the understanding and management of distraction in these and other domains.

There is much that can be done to manage driver distraction and to prevent it from escalating into a more serious road safety problem than it already is. It is hoped that the information, insights, and advice contained in this book will help to inform and guide this process.

REFERENCES

1. Adams, D., Text message driver who killed cyclist goes free, *The Age*, November 11, 2003, available at http://www.theage.com.au/articles/2003/11/10/1068329487085.html.
2. James, W., *The Principles of Psychology*, Holt, New York, 1890.
3. Gladstones, W.H., Regan, M.A., and Lee, R.B., Division of attention: The single-channel hypothesis revisited, *Quarterly Journal of Experimental Psychology*, 41A, 1–17, 1989.
4. Klauer, S., Dingus, T., Neale, V., Sudweeks, J., and Ramsey, D., *The Impact of Driver Inattention on Near-Crash/Crash Risk: An Analysis Using the 100-Car Naturalistic Driving Study Data.* DOT Technical Report HS 810-594, National Highway Traffic Safety Administration, Washington, D.C., 2006.
5. Report of the Parliament of Victoria Road Safety Committee Inquiry into Driver Distraction. Parliamentary Paper No. 209, Session 2003–2006, Parliament of Victoria, Melbourne, Australia, 2006.
6. *Strategies for Reducing Driver Distraction from In-Vehicle Telematics Devices: A Discussion Document*, Transport Canada, Toronto, Canada, 2003.
7. Hedlund, J., Simpson, H., and Mayhew, D., *International Conference on Distracted Driving: Summary of Proceedings and Recommendations*, The Traffic Injury Research Foundation and the Canadian Automobile Association, Toronto, Canada, 2006.

Part 2

Definitions, Theories, and Models of Driver Distraction

2 On the Philosophical Foundations of the Distracted Driver and Driving Distraction

Peter A. Hancock, Mustapha Mouloua, and John W. Senders

CONTENTS

2.1 OVERVIEW OF THE CHAPTER

The philosophical and scientific lenses through which we view safety act to constrain the ways by which we seek to understand and improve it. The differentiation among identifiable causal factors and the parsing of collision events into their various physical configurations have led to conceptually contingent approaches to the possible solutions we have explored and adopted. Thus, rear-end collisions beget rear-end collision warning technologies, and alcohol-related fatalities give rise to social, technical, and legal initiatives to curb drunk-driving. This chapter addresses the conceptual and philosophical underpinnings of a commonly identified and progressively more evident safety hazard—driver distraction and the corollary in the distracted driver. As is a logical extension of this proposition, we argue that there are two fundamental forms of this hazard. "Driver distraction" occurs when circumstances act to displace the primacy of the social role "driver" in the person's on-road behavior. Thus, a woman turning around to reseat her unrestrained infant is now "attentive" to her role as mother but "distracted" from her role as driver. The second form of distraction is "distracted driving," in which the individual retains the primary role as the "driver" but circumstances act to divert attention from the appropriate course of action to other momentarily inappropriate components of the driving task or the external environment. We argue that the very nature of driving as a "satisficing" task encourages and reinforces the first form of distraction, while the specific nature of vehicle control in dynamic environments means that, on occasion, this second form is virtually unavoidable. Adverse consequences, in the form of collisions, result from a sequence of interdependent events of which this momentary lapse in attention is certainly one. Blame for such collisions is a viscerally satisfying but scientifically stultifying correlate of these adverse outcomes. We offer the beginnings of an alternative interpretation of distraction effects, which suggest potential progress beyond the unhelpful propensity to attribute monocausal blame.

2.2 A PUZZLE TO BEGIN

Imagine that a man is in his car and proceeding down a busy street. Suddenly he becomes aware that coiled on the passenger seat beside him is a poisonous snake that is about to strike. He takes his hand from the wheel to parry the reptile's attack and naturally directs his attention toward the snake and away from the roadway. At that very moment a child following its lost ball runs out into the street in front of his oncoming vehicle. Fortunately, on this day, the child stopped in time and the snake's attack was thwarted. But suppose the outcome had been different. The ensuing collision indeed raises highly problematic philosophical, logical, moral, and scientific issues concerning the nature and interpretation of behavior of both the man and the child during such a sequence of events. Suppose the child had been killed and the individual in the driver's seat survived. What would be the legal ramifications? Would you consider the individual in the vehicle responsible? What sort of punishment (if any) would be appropriate in such a case? What if he saw the child and managed to swerve at the very last moment but if doing that resulted in the snake's striking and causing his death? What, if any, are the sequelae of this eventuality?

Although the facile moral seems to be "Don't drive with snakes," we can imagine any number of situations in which the attention of the individual in charge of the vehicle is drawn to circumstances other than its momentary control. The example of a mother struggling with a young infant that has somehow managed to extricate itself from the safety seat is one close to our collective experience. It may be sufficient for authorities to utter relatively meaningless platitudes such as, "Pull over safely to the side of the road, turn on the hazard lights, and deal with the situation while stopped." However, in science we have to explore beyond the shallow assumption that everything, both inside and outside the vehicle, is in control of the individual seated behind the wheel. However, before we begin our examination, perhaps it is best to provide brief consideration to the current perspectives on driver distraction.

2.3 THE BACKGROUND ON DRIVER DISTRACTION

In recent years, we have come to hear more and more regarding the problem of driver distraction (see, e.g., Refs 1–3) and the possible ways this concern may be addressed.[4] Although distraction has been a category used in collision etiology and epidemiological evaluation for some years,[5] there is no doubt that the recent popularity of portable cell phones and other such devices has put the issue of driver distraction prominently to the fore of both social[6–10] and scientific consciousness.[11–17] It has brought forward important ergonomic issues such as whether regular cell phones are more or less distracting as compared with hands-free technologies (see Ref. 18). These evaluations have featured methodologies derived from human factors and cognitive science to answer what have become pressing social issues. Paradoxically, it does not appear that cell phones themselves rank as the greatest source of distraction during driving. For example, eating and drinking have been observed to be a problem of much higher frequency, whereas adjustments of the in-vehicle entertainment system appear to occur at almost an order of magnitude greater than cell phone use.[19,20] It must, of course, be remembered that these respective trends do not remain stable over time. They certainly vary with the popularity and familiarity of particular handheld devices.[21–23] For example, even the most modern surveys have yet to include, in their analysis information, rapidly emerging technologies such as the iPod (e.g., Ref. 24), although we might well suspect that such devices are liable to be implicated in many collisions at this moment in time (see Ref. 25). Parenthetically, within a decade the iPod itself will be replaced by some newer system and this chapter will appear updated accordingly. However, although the sources of distraction are in a state of constant change, the issue of distraction is itself a persistent concern. This persistent concern is not only emphasized in the present volume but also reflected in special issues of journals such as *Human Factors*, which published a whole volume specifically on driver distraction (see Ref. 3). It was in this special issue that Sheridan[26] provided a control theoretic interpretation of the distraction issue, and the model that he presented is used in our present arguments here. Similarly, the time horizon for action specified in the companion paper by Lee and Strayer[3] proves to be the most useful source of inspiration. These respective advances point us toward the crucial insights to be derived from quantitative modeling efforts, as well as companion empirical evaluations (see Ref. 27) and in-depth, on-road studies.[28,29] The most recent and most extensive of these evaluations of on-road naturalistic driving have identified inattention and distraction as major contributors to vehicle collisions and therefore absolutely central to future improvements of transport safety (see Ref. 30). However, in addition to these important empirical and modeling advances, we still need to explore the fundamental philosophical foundations of distraction. Indeed, the very term "distraction" itself suggests some form of indictment. It implies that there is necessarily a source of "attraction" toward which the individuals are required or obliged to direct their attention. The directing of attention is primarily a voluntary and intentional act in which the individuals must also actively exclude other sources of stimulation that seek to "capture" their attention (see Ref. 31). We shall return to this conundrum of voluntary/involuntary attention later. However, since the precise phrase under consideration is "driver distraction," let us deal first with the problems raised by the designation "driver."

2.4 WHEN IS A DRIVER NOT A DRIVER?

We are going to begin with the optimistic but unfounded assumption that you are not reading this while seated in the driver's seat of a moving vehicle. However, we are going to assume that you most probably possess a valid driving license. Are you then, at this moment in time, a driver? We suspect that the general belief would be that one is not a driver while sitting in an office or at home and that the role of driver begins with the intent to use the vehicle. Of course, in law there are exceptions even to this definition. For example, in some countries intoxicated individuals can be arrested for merely sitting behind the wheel with the keys in the ignition even if they have no intention to drive. Alternatively, if we asked whether you are a licensed driver on a sit-down survey right now, you would probably check the box saying yes. This would confirm that you have the right to assume the socially designated role of driver, but you are not exercising that right at this particular moment. But this assertion is true of many roles. You may also be a spouse, a parent, a sibling, or one of a number of other roles, but since you are at work, you are not acting in these roles now, at precisely at this moment. Similarly, you might be a scientist, a psychologist, an engineer, or one of many other types of professionals and you may be pursuing this role right now, but while going home, this professional role diminishes in importance; it does not disappear altogether. The point here is that our social existence means that multiple roles are attributed to us at different times and through various agencies, and the role of a driver is simply a member of this overall set. Although the active boundaries on the role "driver" are more constrained than those of others such as "family member," it is still the case that the person who sits behind the wheel, starts the vehicle, and causes it to move is not a sterile, disembodied, driving entity but is rather a multifaceted individual, whose confluence of social responsibilities means that critical role conflicts are, at times, almost inevitable.

If the driving task required our full attention at all times, and any minor slip automatically led to certain death, then we would be justified in our assumption that the role of a driver was one that excluded both the conscious consideration of, and the active exercise of, any other possible role. Although no driving situation is quite fraught with such peril (although certain motor-racing situations can indeed get very close), we do make an evident distinction in required levels of responsibility, for example, between professional drivers and the general driving public. We hold professionals to a higher standard of certification and higher levels of social disapprobation if they fail in their duty, since their role as a professional driver is viewed somehow as more binding than that of an average motorist. In contrast to this all-consuming, draconian perspective, driving is not a task that requires continual optimal driver response.[32] Actually, it is a task that is much more accurately described as being "satisficed" for the majority of the time. The term "satisficed" was coined by the Nobel Prize winner Herbert Simon to indicate behavioral situations in which individuals do not have to do their very best but rather, only have to do well enough to get the job done.[33] Indeed, if driving were constituted so as to demand our optimal performance, there would be a strong human factor and human-centered design imperative to reduce such inordinate demands and return the task to one amenable for a "satisficing" level of response. Given this situation, it may well be possible to create a theory of driving behavior, which is based on the notion of "periodic optimality." Yet herein resides one of the first great issues in driver distraction. When we create a performance task that can, under the vast majority of conditions, be completed without the full attention of the individual involved, what happens to their residual attention that is then left untapped? The answer, of course, is that it is used to servicing one of many other in the hierarchy of goals created by each of the other roles that occupy the person who is driving. Thus, in the great tradition of perpetuating stereotypes, the salespeople are thinking of their sales, the drug users their next fix, the scientists of their next experiment, and the drunk of not much at all! Truly, we create almost all technology with an eye to this ease of use and in doing so we very rarely account for the residual attention of the user, which is not employed by the current task. Consequently, with an adaptive system such as a human being, we should not be surprised that such available attention-requiring resources are then actively employed elsewhere.

If, during the vast majority of time, the individual can control the vehicle with relatively little effort and thus have a considerable degree of attention that may be directed to other tasks, what might these specific tasks be? It is true that some tasks such as navigation can help support the role of "driver." However, as we have noted, in a large percentage of situations they support other, competing goals. In terms of control theory, the excess of energy available in the "activation" component interacts with the ordering of functions of the "intention" component (see Ref. 26; Figure 2.1). Consequently, when we talk of "driver" distraction, society associates all the actions

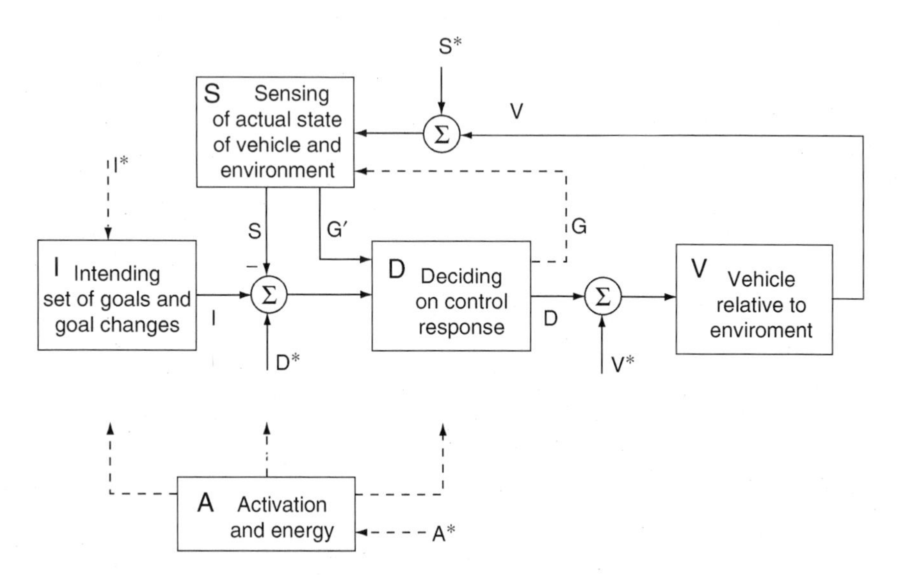

FIGURE 2.1 Control theoretic approach to driver performance. (After Sheridan, T.B., *Human Factors*, 46(4), 587, 2004.)

of the individual that are undertaken in the vehicle with the role of the driver. This assumption is false. For example, in the case of the "in-car snake," we are actually more justified in referring to the role of that individual as a "snake fighter," and according to Hobbes,[34] the moral justification for personal survival has to be judged against the potential for increased momentary risk to other individuals in society. The relative level of immediate risk from the inflamed serpent is obviously much higher than the putative risk to others on the open roadway ahead. Thus, in this specific situation, it is the "driver *qua* driver" who is distracted. However the "person *qua* person" is not; their attention is directed exactly where it should be. It is just that the ascendant role of an aspiring snake-attack survivor has overtaken the socially imposed role of driver. In retrospect, when we undertake collision investigations, we are often justified in attaching the appellation "distracted driver" to a specific event like the one described. However, are we morally justified in attributing blame in such circumstances? In seeking to specify blame, society has to make a retrospective judgment as to the value of the alternative task that was competing with the driving task. Obviously some competing tasks are judged to be trivial, such as shaving or applying makeup. In such circumstances, the individual who voluntarily engages in these competing activities is readily judged culpable because the imperative to shave or put on makeup is negligible when set against the requirement to control half a ton of metal going at 60 mph. In contrast, other competing actions such as paying attention to a pregnant wife are seen as much more justifiable. There are also cases, for example, when an individual is experiencing a heart attack, which is viewed as involuntary and therefore not culpable at all. But where does this leave exceptionally important but still voluntary activities such as "snake fighting" or more commonplace actions such as a police officer using his radio to direct colleagues to the site of an emergency or a surgeon consulting on the phone about a life-saving procedure? What of culpability in regard to distraction in these cases if the individuals are then involved in a collision? The setting of any criterion of culpability is a societal decision. It is very much analogous to setting the beta level in signal detection theory[35] and can and indeed should be modeled as such. It requires a collective agreement on the level of acceptable risk associated with the transport systems we create. We shall discuss the emergent issue of blame later since it is a very thorny problem and requires us to understand our shared, moral obligation to our own personal survival and that of our fellow members of society. Discussion of the nuances of the blame game is to come.

The present set of observations can lead to some perverse outcomes if we are not very careful. Consider that a driver might sue a manufacturer because the operation of the vehicle that they purchased did *not* demand all of their attention. Imagine the various forms of legal defense against culpability that might be constructed from the multiple social roles interpretation presented earlier. For instance, a mother might well deny responsibility for causing a collision on the grounds that she was acting first as a mother and not as a driver. These dangerous social possibilities arise from the ambiguity of the fuzzy boundaries to the roles we voluntarily assume or which are imposed on us, and the fact that drivers who are not fully occupied can cogitate on issues other than momentary vehicle control. In large part, this is a function of the system of transportation we collectively choose to create and these issues related to distraction are therefore equally applicable to the operation of a horse, a skateboard, a set of roller blades, or a commercial jet. The fundamental issues remain common across all these modes of assisted motion, that is, the outcome of failure is a brisance of uncontrolled kinetic energy—a collision.[36,37]

What we have talked of so far pertains largely to the idea of the social role of the driver and the outcomes that can occur when an individual implicitly or explicitly deviates from it while in charge of a vehicle. We have noted that such deviation can be more or less justified, depending on the specifics of what is happening at that point in time. In general, we view this "driver distraction" as a "between social role" effect—an effect at one level above that of actual vehicle control. However, there is an equally important component of the argument, which we choose to call "distracted driving" in which the individuals involved never vary in their primacy of driver as a social role but rather choose to engage in one of the ostensibly inappropriate behaviors within the driving task. We call this latter typology "distracted driving," and in contrast to "driver distraction," it is more aptly viewed as a "within social role" effect. It is to this facet of the argument that we now turn.

2.5 WHAT IS DISTRACTION AND WHAT IS ATTRACTION?

Since the seemingly simple issue of distraction appears to be leading us into some very complicated territory, perhaps it is useful to start with a straightforward case away from the context of driving. Imagine that a teacher wants a child to focus on a particular picture in a book. If the child appears to be directing its gaze and attention to this picture, and especially if it is not directing gaze or attention to something else, the teacher may be content that, to the degree that it can be observed, the child is accomplishing the goal that was set. To go beyond purely individual and subjective experience, this act of attraction necessarily requires an external arbiter (in this case the teacher) to specify the goal. It also requires a further arbiter (who can also, but need not necessarily, be the teacher) to say whether that goal is being achieved or not. When no such arbiter is immediately present, things threaten to just become a matter of individual subjectivity. Indeed, this was the fundamental conundrum of introspectionism that bedeviled psychology at the turn of the twentieth century, and found its anodyne in radical behaviorism.[38] To be meaningful then, the very idea of attraction requires some externally mandated goal and some form of referee to arbitrate as to whether this act of attraction has been successful. But who arbitrates in driving? If we are to understand distracted driving, we must understand what the drivers are being distracted from by knowing what they should be attracted to. Simply asserting that society requires safe vehicle operations underspecifies this goal. Underspecification almost inevitably leads to vacuous and circular reasoning in which distraction is defined in terms of collisions that are proposed as being caused by distraction. Although this level of conceptualization may be sufficient for the blame game, science demands more.

In the example of the child and the teacher, the task itself can be described very specifically. Thus, the teacher might ask the child to point to the picture of the giraffe on a page. Assuming there is only one such image and the child points directly to it, then the teacher may be satisfied that the goal of the task (the child's comprehension of giraffes) has been achieved. In this situation, there is an explicit objective, one single solution and very tight spatial and temporal boundaries on what connotes success. In fact, the task has all the hallmarks of optimization. Unfortunately, as we have seen, driving is not like this. There is no ever-present authority figure looking over our shoulders adjudicating the correctness of our actions. After we have passed through driving school and successfully obtained a license, this authority devolves into the periodic presence of law enforcement officers, or in Britain, diffuse and rather generally hated, speed cameras. The absence of an external arbiter is one concern but judging what attention is actually required to be directed to (i.e., providing a closed-end description of what should be done) is even more difficult. Thus, objects and entities that prove to be relevant at one moment can well become sources of distraction the next moment. The sources of stimulation to which the driver should be attracted always change in a very dynamic manner in the course of any journey. Unfortunately, society does not express what the driver should be paying attention to at this level of dynamic detail. Traffic control devices and resident traffic laws provide support for the accomplishment of this dynamic task, but because it necessarily involves interaction with other drivers, such control devices and regulations cannot deterministically specify what *should* occur in each and every situation. Although different countries use traffic control devices and law enforcement to different degrees, collisions remain relatively rare across national boundaries, and to a surprising extent, insensitive of such local variations.[39,40] In the world of legislation, distraction is often judged against the imperatives implied by these devices and regulations. Thus, people receive citations for failing to stop at a stop sign, etc. However, such conventions have little to say about the psychology of driver attention—despite their ubiquitous employment of vacuous phrases such as "driving without due care and attention." This process eventually devolves into the circularity of reasoning we have already discussed, and merely serves to maintain the status quo by satisfying a naïve sense of causation. It does not lead to improved safety.

Before we proceed to a more detailed search for a scientific definition of distraction, we do need to add some caveats to what has been said. Driving, being predominantly performed in a social setting, means that normal, everyday motorists are aware of the other drivers, who also occupy the roadways in their immediate vicinity. However, there are some occasions when driving can be a singular experience. For example, if you are on a closed track with no other drivers present and no possibility of damaging yourself or other people, then many of the foregoing observations do not apply. Perhaps it is the freedom from such constraints that makes these conditions so attractive and enjoyable, and indeed some facets of the sport of motor racing so popular. Many of the observations that we have made do not apply under such circumstances of free expression. Second, our discussion is really confined to situations in which drivers are consciously endeavoring to avoid collision. What we have said largely does not apply to drivers in demolition derbies. This restriction also rules out situations in which people explicitly seek suicide or homicide through the use of a vehicle, although such events do occur at a nontrivial rate (see Ref. 41). It has also strained acquaintance with joyriding thrill seekers, who do understand the social constraints, but intentionally seek to fracture them. In essence, we are not considering the case in which individuals purposively and consciously act contrarily.

The case of driver impairment (whether is permanent impairment through circumstances such as brain damage, or temporary impairment through disturbances such as alcohol and fatigue effects) is an interesting intermediary situation, which shall be discussed later. However, having expressed dissatisfaction with the general definition of distraction, we are impelled to examine the definition question in more depth to see whether further resolution is possible.

2.6 LOOKING FOR A DEFINITION OF DISTRACTION

Defining distraction is akin to defining a negative, since distraction not only has the connotation of a negative activity but also clearly implies a more important positive state of attraction. If we can specify what drivers should be attracted to, then at least we can specify when and where distraction occurs, even if we are not immediately able to identify the motivation behind it. This *a priori* specification also circumvents the problem of the *post hoc, ergo propter hoc* (after this, therefore because of this) type of reasoning so beloved of sports commentators and legal personnel alike. Unfortunately, there is currently no assured method of specifying, *a priori*, what any particular driver in any particular situation should necessarily be paying attention to. In case you suspect that this is not an especially complex endeavor, think of specifying what each driver should do in relation to the actions of other drivers who fracture either the explicit rules of static traffic control devices or the more ubiquitous strictures embedded in the omnipresent traffic laws. One cannot presume perfect compliance on behalf of all other members of the driving community. Laws, traffic devices, physical constraints, vehicle capabilities, roadway geometries, etc., represent our conscious intent to impose order on a partially predictable but complex and dynamic world. But however well intentioned these efforts are, they cannot dictate a deterministically "correct" course of action for each and every driver under each and every driving circumstance. Therefore, since we cannot "know" exactly what it is that each driver should be "attracted" to at each and every moment in time, our driving actions are propositions set before an indeterminate world. Consequently, either we, or our neighbors, will periodically be unlucky enough to be caught in a web of unfortunate circumstances. This is what Haddon referred to as, "the loosing of the tiger".[36] A more formal, physics-based approach to this general issue of inherent unpredictability, which uses the Minkowskian "space–time" framework can be found in Hancock[42] and Moray and Hancock.[43] If our efforts to impose order on a partially chaotic world appear to be doomed to failure, can we ever hope for improvement? We believe that one possible line of progress lies in the development of a general theory of driving that is prescriptive and positive in nature. Although we cannot report any immediate solution, there are signs that such a goal is feasible based on a number of insights published over the several past decades.

2.7 A THEORY OF DRIVING

The most influential of all reports concerning driving is the early classic by Gibson and Crooks.[44] In this seminal work, the authors took the first steps toward a field-based theory of driving (see Ref. 45). Gibson[46] developed this into an apparently complete ecological theory of perception that represents one crucial foundation to a full understanding of purposive driving behavior. As we have seen, although it is not possible to say what each and every driver should be doing at each and every moment, it is possible to specify some principles, which can form the basis of a foundational theory. At the highest level of abstraction, Gibson and Crooks identified driving as a general form of need reduction enacted through locomotion. The function of movement velocity here is a reflection of the need to traverse from point A to point B. For example, emergency vehicles have a great imperative to achieve this transition as quickly as possible, whereas many touring drivers actually enjoy the process of driving itself, and thus have little imperative to complete their journey at top speed. Deceleration, or braking, also serves the same purpose of travel, but largely by providing a mechanism through which the driver can avoid incipient collision, either with fixed or moving objects. Thus, in case of an emergency vehicle, the act of braking to avoid collision retards the overall goal of fastest possible transit but failure in this latter collision-avoidance task may well mean the journey is never completed. Braking and speed also deal with the case of obstructions in front of the vehicle and the threat of collision from a faster-moving vehicle coming from behind. Velocity and braking are directly mapped to one specific control in the vehicle.

To deal with lateral sources of threat, Gibson and Crooks proposed an "envelope" around the vehicle that they termed the "field of safe travel." Any potential threat of collision modifies this dynamic field and the task of the driver is to assure that the physical dynamics of the vehicle, that is, its minimal stopping distance, always remains within this field of safe travel. Logically, rationally, and semantically, the "field of safe travel" is the exact equivalent of what today we refer to as "situation awareness" (SA; see Refs. 47 and 48), although the illustration of the "field of safe travel" is very visual and immediate whereas the contemporary conception of SA is to some extent more projective in nature. Since the imperative in driving is preservation of the acceptable ratio of the vehicle's stopping distance to the boundary of the field of safe travel, we can equate this with the driver's situation awareness with respect to the capacities of their vehicle at hand, or more generally the global human-machine status.[42] Attraction can now be defined as the active search for incursions into the field of safe travel and distractions are the direction of attention away from this required process. In more contemporary parlance, this might be termed maintenance of appropriate situation awareness (see Ref. 26; Figure 2.2). However, here we return to problematic philosophical grounds, the phrase because "situation awareness" has little more explanatory value than the accident-distraction circularity we criticized previously (see Refs. 49 and 50).

Although the contribution of Gibson and Crooks[44] is an important beginning, it is not a complete theory of driving, because the focus is overwhelmingly on a spatial perspective. The equally important temporal aspect of driving remains insufficiently emphasized and is masked by the static illustrations of the identified "field of safe travel." We have previously noted that the requirement for an "optimal" response in driving is a very rare event indeed. Most driving demands can be easily reconciled with a "satisficing" strategy. When exceptional circumstances do arise, they can quickly exceed even the driver's very best efforts to resolve (e.g., an unanticipated incursion into the path of progress, which means the drivers would have to respond

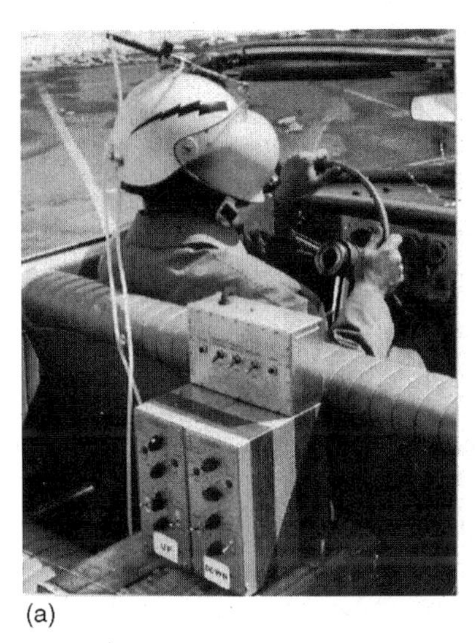
(a)

(b)

FIGURE 2.2 Showing the two states of the Senders et al. experiments with visual occlusion showing the helmet (a) closed and (b) open. The experiment demonstrated that the road drove the driver as much as, if not more, the driver drove the road. (From Senders, J.W., Kristofferson, A.B., Levison, W.H., Dietrich, C.W., and Ward, J.L., *Highway Research Record*, 195, 15–33, 1966.)

in an interval lower than their best possible response time). Understanding this temporal portrait of the driving task (see Ref. 51) can lead to an evident recognition as to where human, human–machine, and purely machine-automated responses are required.[52] This resolution of the temporal demands of driving is crucial since it will begin to dictate what sort of augmented control support is provided by vehicles in the future. We can easily imagine a situation in which vehicles are moving at such a speed that all human control is removed from the momentary (inner) control loop (see Ref. 26). This issue goes beyond ground vehicles because the issues pertain to every form of powered transportation. Resolution of these adaptive allocation questions thus appears to be a goal, which might be achieved in the immediate future (see Ref. 53). In light of these steps toward the beginning of a theory of driving, couched in the more general terms of human-machine control, let us look at the proposition: is perfect performance possible?

2.8 IS PERFECT PERFORMANCE POSSIBLE?

Let us consider for a moment you are driving down the road and approaching an intersection having a signal light. The scene in front of you is a dynamic one and is constantly evolving. The light is at present green but of course this may change. The intersection is also currently clear, but of course this may also change. Now let us run the scenario forward. As you approach the actual crossing, the visual angle between the traffic light and the forward view across the junction begins to increase. There comes a point in time when one cannot see both together. This is not a divided attention issue or one of foveal versus peripheral field of view; it is a simple question of structural interference, the driver's eye cannot see both at the same time. The design efforts of traffic engineers in terms of sight lines, seek to reduce any such occurrences of ambiguity, and on most occasions they are very successful. However, let us consider this example as one of inherent ambiguity. Where is the driver to look? If you look at the light to see a possible change, you cannot look at the intersection and vice versa. The pragmatist will say that by the time the light changes, the intersection should be free. However, the simple fact is that driving presents many such ambiguous situations in which, whatever "correct" action one is actually accomplishing, there is another equally "correct" action that one must neglect. What of distraction in such circumstances? Can we say the driver involved in a collision in such circumstances is distracted and is not driving with due care and attention?

There is a way of thinking that demonstrates that the notion of inherent ambivalence is not mere assertion. Senders[54] has recently sought to weld together a sampling model, which was purpose-developed for driving[51] with Estes'[55] stimulus sampling theory and more modern developments in the change blindness paradigm.[56] In an early study, Senders and his coworkers[51] looked to examine the obligatory sampling rate of the driver in relation to the demands of the roadway. The primary insight derived from this investigation was that "it was not that drivers decided when to look at the road but rather that the road demanded, from time to time, that the driver looks at it." Under controlled field conditions, it was possible to adjust the velocity of the vehicle in relation to the steering demand of the road ahead to provide either a comfortable or distressing sample rate. In a technical sense, the closure of the sampling helmet (as shown in Figure 2.2; see also Chapter 9 of this book) acted to contract the Gibson and Crooks[44] "field of safe travel." A threshold is crossed when the ratio of the field of safe travel to the minimum stopping zone (which is a character of the vehicle's momentary physical dynamics) drops below unity. It may well be that this transition is equivalent to the sense of distress noted by Senders et al.,[51] which occurs when the driver perceives that too long a time has passed since their last glimpse at the roadway. Most importantly, Senders and his coworkers emphasized just how much a dynamic process this is, with the information profile changing on a moment-by-moment basis.

Now we can see that it is primarily the demand/affordance structure of the external environment that generates the imperative for driver attentional direction. The putative "attraction" discussed earlier is represented in these demand/affordance conditions much more than it is the voluntary action of the driver. The problem now is that not all such demand/affordance conditions are deterministically predictable. Random, low-frequency events, referred to in Gibson and Crooks[44] as incursions, do occur—the child does run onto the road, the drunk driver does swerve unexpectedly, and tires do blow out. Although we are able to calibrate the obligatory rate of looking when we can determine the whole of the demand, as is possible in certain controlled conditions, we cannot yet do this on the road because of the unpredictability of these low frequency, yet crucial incursions. The recently completed studies from the 100-car naturalistic driving evaluation is an evident step forward in this matter in that we can now begin to approach the quantification of such incursions with some degree of numerical accuracy (see Ref. 28). Thus, although we still cannot say with certainty what the driver must be attracted to and logically we cannot determine all conditions in which the driver is thus "distracted," fortunately, progress is being made on this front.

Despite this inherent uncertainty and logical shortcoming in expressing exactly what distraction is, all is not doom and gloom. In practice, very few instances of either manifest distraction (i.e., the distraction of a driver that is evident to an outside observer) or intrinsic distraction (i.e., the distraction of a driver that occurs as a result of change of their internal state but is not externally observable) are actually punished by collision. Herein is another important behavioral issue. Because manifest driver distraction is so rarely punished, the learning process is vastly suppressed by both the frequency and the pertinence of the feedback that the driver receives. Even when collision occurs, as in the case of infrequent and unpredictable disruption, the personal and legal tendency is to wrap such events together with all the forms and nuances of distraction that we have discoursed on here. This strategy may be satisfying to a naïve sense of causality but it is fatal to the progress of a science of safety. It is what we have called the futility of "the blame game."

2.9 THE BLAME GAME

There is an anecdotal story about blame that is to some extent comical and instructive. Following a collision, a man was filling out a form concerning the event and came to the question of cause. In answer, he put "Act of God." Later, in billing him, the insurance company sent a letter that began "Dear Mr. God." Although we agree this is not a wonderful joke, it does speak about our belief regarding causation and the sources of causation in our world. It is evident here that members of the insurance community are ready, willing, and able to recognize other individuals as centers of causation (see Ref. 57). Financially, it would be unwise for them to do otherwise. We also recognize the physical forces of nature as sources of causation and like the man involved in the aforementioned collision, some people trace all these sources of causation back to one "first" or original cause. In the Greek world, momentary passions and abnormal actions were often attributed to the visitation of a demigod.[58] In the Middle Ages, demonic possession was a legitimate excuse for certain transient behaviors (e.g., Ref. 59). The modern materialistic outlook is much more likely to eschew such supernatural explanations of behavior. In case you are interested in testing this latter proposition, try citing "demonic possession" to account for your next speeding ticket! In the previous discussion, we have made a strict division between drivers who proceed consciously to distract themselves from their responsibility for vehicle control, and the vast majority of individuals for whom driving distraction was inadvertent and in a sense involuntary. This is an important distinction since the former always implies an intentional act, whereas the latter can result from numerous influences, some of which supersede even the most assiduous driver's voluntary efforts. In seeking to fix responsibility, we must continue to recognize this difference between voluntary and involuntary activity.

Our driving environments prove to be replete with "thieves of attention." These are analogous to what James Reason has referred to as "thematic vagabonds" (see Ref. 60). Indeed, why would organizations pay enormous amounts of money to place advertisements alongside roadways if they did not think drivers ever paid attention to them? Are such organizations to blame for "driver distraction," when the drivers suddenly and momentarily change their role to become a "consumer," or indeed for driving distraction when attention is drawn away from the road toward a conspicuous sign? This question also applies to the almost obligatory act of orienting to crucial stimuli, an effect which advertisers also take great advantage of. The act of perception is a symbiotic one between the perceiver and what is perceived. Therefore, are not the individuals or groups who manifestly seek to pervert driver attention away from the road as culpable as the drivers themselves who apparently let this happen? Indeed, we should always remember that other people are themselves almost always a major source of distraction. And what of the situation, as shown in Figure 2.3,[61,62] where there are multiple traffic devices present at one time? Absence of traffic signs would be taken as culpable inaction on behalf of local authorities, but what is the order of importance of traffic control devices when so many are present at one time? As Gilbert and Sullivan had it in one of their productions "if everybody's somebody, then no one's anybody!" The same thing applies to this "Tower of Babel" of signage. Is such a local authority responsible for creating driver confusion and associated distraction in such circumstances? Blame is thus a game played largely by those in the legal profession to identify a single causational source in environments that manifestly have multiple causes. In this sense, as one of us[63] has noted, what is "lawgical" often bears no necessary relationship to what is "logical." If, for one moment, we suspend the idea of transferring blame to one single, all-powerful entity, then the ultimate arbiter in these situations is society itself. The attribution of blame to a driver who is accused of being distracted is a societal decision embodied in the

FIGURE 2.3 Some driving environments are overcrowded with signage, some of which are obligatory traffic control devices, others commercial advertising hoardings. Understanding that the illustrated scene is experienced dynamically by the driver, where is the obligatory point of attraction? (From Hastings, M., *Daily Mail*, September 13th, London, 2006; Massey, R., *Daily Mail*, September 12th, London, 2006.)

rules and regulations of that society. Our role as scientists and safety professionals is to inform society about these various strictures and adjust the law to accord with this understanding.

Driving, as we have seen, is a satisficing task. It is one in which all drivers frequently, and on certain occasions necessarily, fail to maintain their attention toward the "correct" source of attraction. Infrequently and unpredictably, these momentary failures encounter the precise environmental circumstances that induce collision. In Haddon's[36] terms, we "meet the tiger." Society is content, in general, to chastise those unlucky drivers who find themselves involved in these rare collisions. This does not, of course, exempt those drivers who consciously make the decision to neglect their responsibilities. However, if collisions became more frequent by several orders of magnitude, society would not single out these "bad" individuals but would seek to make corrections at a systemic level (see Ref. 64). However, we have been generally content to ratify our collective, institutional schizophrenia, which "blames" the "bad" drivers while encouraging the production of ever greater numbers of technologies that inevitably redirect drivers' attention from the ever-more satisficed task of vehicle control. As Churchill once trenchantly remarked, "if you really want to know the answer, follow the money." In the present circumstances, driver distraction is currently being used in the same convenient way that "pilot error" was formerly employed in aviation. It satisfies the legal mind and its requirement for requisite blame, but such usage will not lead to a safer transportation system. This latter goal requires us to take a much more rationale approach to driver distraction and it is toward this goal that we have offered the present observations.

2.10 SUMMARY AND CONCLUSIONS

We have asserted that there are two primary forms of the distracted driver. One is contingent on social role, where the driver has been distracted away from the primary function of driving. In the terms that Sheridan[26] has elaborated in his control theoretic approach, this form of distraction is a disruption to the goal ordering of the "intention" function. There are understandable causes of such deviation since attending to the position of the vehicle on the road does not take primacy in every situation. We feel that such deviations occur because driving itself does not require continuously perfect behavior and total attention. The problem with such behavior is that unexpected events continually arise in a more or less random way. Less than complete attention to the driving task, due to overwhelming and competing demands, is often considered somewhat benignly by those considering any subsequent collision that may have occurred. However, the deliberate failure to attend to driving is still considered blameworthy by society and is treated as such in the courts of law.

There is a second order of distraction that we have called "distracted driving." Here, the driver has not been distracted from the driving role *per se* but is simply engaged in what is considered the "wrong" aspect of the driving task at the time in question. Thus, a driver consulting an in-vehicle device or responding to a dashboard warning light is not "paying attention to the road" but may still be responding to an important driving demand. This is in contrast to a driver whose attention has been captured by an advertisement for goods or services unrelated to the task of vehicle control. Again, in terms of the control theory model, this interruption is primarily due to the sensing function (Sheridan,[26] Figure 2.1). A driver who runs a red light (i.e., pays careful attention to the road but misses the traffic control device) is now not driving without "due care and attention." Traffic control devices, traffic laws, and law enforcement personnel in addition to other agencies seek to specify what a driver *ought* to be doing at any one time. However, their coverage is necessarily incomplete and there are many situations in driving in which a deterministic statement as to what the driver should be doing is not possible. There are worse situations in which laws, devices, enforcement agencies, etc. necessarily contradict themselves so that a driver under such circumstance cannot "do the right thing." These indeterminate and contradictory situations arise frequently since attention is fundamentally a single-channel system and demands for attention themselves are statistically distributed in time. It is inevitable that occasional failure to attend at the moment of need will occur.[51] However, since a collision can be conceptualized as a Markov process in which driver state is only one of several prepotentiating factors, and since the nuances of the proliferation of the adverse sequence of events depends on the momentary configuration of physical circumstances (often involving the presence and action of another driver), relatively few of these situations result in collision. When they do, it is easier to point to the proximal human agent and consider the issue essentially "solved." As Reason[64] has so eruditely demonstrated, this naïve conceptualization of causation is, in reality, both antiscientific and counterproductive.

Are we then to abandon the idea of the distracted drivers or driving distractions? We do not think so. Rather, what is needed for progress is two great steps, which perhaps the present volume can help take. First, we need a renaissance in driver studies, which looks to develop a central theory of driver behavior (see Ref. 65). Although Gibson and Crooks[44] is a very useful beginning, it is only that—a beginning. Senders has produced qualitative models that can articulate and integrate with this initial conception and, in showing the primacy of task demand, point us in the direction of trying to understand the nature of rare events. The recent quantitative approaches expressed by Sheridan[26] and Lee and Strayer[3] provide most useful vistas for further modeling efforts but our purview must go beyond this one limited realm of driver activity. In essence, we are asking for a theory of purposive human behavior in context and, in the long run, nothing else will suffice. Perhaps a marriage of ecological and quantitative behavioral science can achieve this.[48] If driving research can lead the way, it will have proved its value (see Ref. 66). The second step concerns the use of accidents and collisions as marker and basic research data. Collisions are often shocking, damaging, and life-changing events. It is somewhat natural that road-safety researchers focus on them. However, they are unpredictable, nonlinear, and infrequent events, which make for a confusing guide to understanding the nature of the behavior that creates them. In essence, they are the tails of the distribution where the intervehicle distance is zero or less. Perhaps we cannot eliminate all errors, and our attempt to seek rational explanations for all collision events induces a false sense of confidence. In reality, collisions are an almost inevitable result of the form of transportation system that we have created.[67] Our collective goal must be nothing less than the aspiration for the elimination of all transportation accidents, however chimerical this goal might appear in an uncertain world (see Ref. 68).

The Swedish idea of "vision zero" is a laudable one, even if it is practically unattainable. To undergo the sort of sea change in safety envisaged by such programs, we have to question the very way in which we view the driving task. Although human beings often see more through tears than through telescopes, it is a sad and unscientific way to approach life. It is the task and responsibility of the safety researcher to minimize those tears. We hope that the initial thoughts that we have offered here can help in this process.

ACKNOWLEDGMENTS

This research was facilitated by the Department of Defence Multidisciplinary University Research Initiative (MURI) program, P.A. Hancock, principal investigator, and administered by the Army Research Office under Grant DAAD19-01-1-0621. The authors wish to thank Drs. Sherry Tove, Elmar Schmeisser, and Mike Drillings for providing administrative and technical direction for the grant. We would very much like to thank Drs. John Lee, Neville Stanton, Gareth Conway, and Geoff Underwood for their helpful comments on an earlier version of this work, and Gabriella Hancock for her editorial assistance. The comments of the editors and a number of other unknown reviewers are also very much appreciated by the authors. The views expressed here are those of the authors and do not necessarily reflect the official policy of the U.S. Army.

REFERENCES

1. Caird, J.K. and Dewar, R.E., Driver distraction, In *Human Factors in Traffic Safety*, 2nd edition, Dewar, R.E. and Olson, R., (Eds.), Lawyers & Judges Publishing, Tucson, AZ, 2007.
2. Karlsson, R., Evaluating driver distraction countermeasures. Master's Thesis in Cognitive Science, Linkoping University, Linkoping, Sweden, 2004.
3. Lee, J.D. and Strayer, D.L., Preface to the special section on driver distraction, *Human Factors*, 46, 583, 2004.
4. Donmez, B., Boyle, L.N., and Lee, J.D., The impact of distraction mitigation strategies on driving performance, *Human Factors*, 48(4), 785, 2006.
5. Treat, J.R., A study of the pre-crash factors involved in traffic accidents, *The HSRI Review*, 10(1), 1, 1980.
6. Cottle, M., My roving barcalounger. *Time*, August 1st, New York, 2005.
7. Drucker, J. and Lundegaard, K., As industry pushes headsets in cars, U.S. agency sees danger, *The Wall Street Journal*, July 19th, New York, 2004.
8. Massey, R., Now drivers who use mobile phones face three penalty points, *Daily Mail*, July 19th, London, 2004.
9. Mimran, E., Integrating the control of in-vehicle devices, *Autobeat Daily*, March 28th, 2006.
10. Wald, M., 'Hands free' cell-phones may still be road risk, *The New York Times*, April 2nd, 2002.
11. Ålm, H. and Nilsson, L., The effects of mobile telephone task on driver behaviour in a car following situation, *Accident Analysis and Prevention*, 27(5), 707, 1995.
12. Curry, D.G., In-vehicle cell phones: Smoke, but where's the fire, *IEEE Spectrum*, August 16–18, 2001.

13. Goodman, M.J., Bents, F.D., Tijerina, L., Wierwille, W., Lerner, N., and Benel, D., *An Investigation of the Safety Implications of Wireless Communications in Vehicles* (DOT HS 806-635), National Highway Transportation Safety Administration, Washington, 1997.
14. Kantowotz, B.H., Using micro-worlds to design intelligent interfaces that minimize driver distraction, *Proceedings of the International Symposium on Human Factors in Driver Assessment, Training, and Vehicle Design*, 1, 42, 2001.
15. Laberge-Nadeau, C., Maag, U., Bellvance, F., Lapierre, S.D., Desjardins, D., Meissier, S., and Saidi, A., Wireless telephones and the risk of road crashes, *Accident Analysis and Prevention*, 35, 649, 2003.
16. Lamble, D., Kauranen, T., Laasko, M., and Summala, H., Cognitive load and detection thresholds in car following situations: Safety implications for using mobile (cellular) telephones while driving, *Accident Analysis and Prevention*, 31, 617, 1999.
17. Redelmeier, D.A. and Tibshirani, R.J., Association between cellular telephone calls and motor vehicle collisions, *The New England Journal of Medicine*, 336(7), 453, 1997.
18. Cooper, P.J., Zheng, Y., Richard, C., Vavrik, J., Heinrichs, B., and Siegmund, G., The impact of hands-free message reception/response on driving task performance, *Accident Analysis and Prevention*, 35, 23, 2003.
19. Stutts, J.C., Reinfurt, D.W., Staplin, L., and Rodgman, E.A., *The Role of Driver Distraction in Traffic Crashes*, AAA Foundation for Traffic Safety, Washington, 2001.
20. Stutts, J.C., Faeganes, J., Rodgman, E., Hamlett, C., Meadows, T., and Reinfurt, D., *Distractions in Everyday Driving*, AAA Foundation for Traffic Safety, Washington, 2003.
21. Evans, L., *Traffic Safety and the Driver*, Van Nostrand Reinhold, New York, 1991.
22. Evans, L., Transportation safety, In *Handbook of Transportation Science*, Hall, R.W., (Ed.), Kluwer Academic Publishers, Norwell, MA, 2003.
23. Evans, L., *Traffic Safety*, Science Serving Society, Bloomfield Hills, MI, 2004.
24. Horrey, W. and Wickens, C.D., Examining the impact of cell phone conversations on driving using meta-analytic techniques, *Human Factors*, 48(1), 196, 2006.
25. Landsdown, T.C., Brook-Carter, N., and Kersloot, T., Distraction from multiple in-vehicle secondary tasks: Vehicle performance and mental workload implications, *Ergonomics*, 47(1), 91, 2004.
26. Sheridan, T.B., Driver distraction from a control theory perspective, *Human Factors*, 46(4), 587, 2004.
27. Horrey, W., Wickens, C.D., and Consalus, K.P., Modeling drivers' visual attention allocation while interacting with in-vehicle technologies, *Journal of Experimental Psychology: Applied*, 12(2), 67, 2006.
28. Dingus, T.A., Klauer, S.G., Neale, V.L., Petersen, A., Lee, S.E., Sudweeks, J., Perez, M.A., Hankey, J., Ramsey, D., Gupta, S., Bucher, C., Doerzaph, Z.R., Jeremeland, J., and Knipling, R.R., *The 100-Car Naturalistic Driving Study, Phase II: Results of the 100-Car Field Experiment, DOT HS 810 593-April*, Virginia Tech Transportation Institute, Blacksburg, VA, 2006.
29. Hancock, P.A., Lesch, M., and Simmons, L., The distraction effects of phone use during a crucial driving maneuver, *Accident Analysis and Prevention*, 35, 510, 2003.
30. Klauer, S.G., Dingus, T.A., Neale, V.L., Sudweeks, J.D., and Ramsey, D.J., *The Impact of Driver Inattention on Near-Crash/Crash Risk: An Analysis Using the 100-Car Naturalistic Driving Study Data, DOT HS 810 594-April*, Virginia Tech Transportation Institute, Blacksburg, VA, 2006.
31. Recarte, M.A. and Nunes, L.M., Mental workload while driving: Effects on visual search, discrimination, and decision making, *Journal of Experimental Psychology: Applied*, 9(2), 119, 2003.
32. Hancock, P.A. and Scallen, S.F., The driving question, *Transportation Human Factors*, 1(1), 47, 1999.

33. Simon, H.A., *The Sciences of the Artificial*, MIT Press, Cambridge, MA, 1969.
34. Hobbes, T., *Leviathan*, London, Crooke, 1651.
35. Swets, J.A., *Signal Detection Theory and ROC Analysis in Psychology and Diagnostics*, Erlbaum, Mahwah, NJ, 1996.
36. Haddon, W., On the escape of tigers: An ecologic note, *Technology Review*, 72, 44–47, 1970.
37. Hancock, P.A., The tale of a two-faced tiger, *Ergonomics in Design*, 13(3), 23–29, 2005.
38. Watson, J.B., Psychology as a behaviourist views it, *Psychological Review*, 20, 158–177, 1913.
39. Sivak, M., Motor vehicle safety in Europe and the USA: A public health perspective, *Journal of Safety Research*, 27(4), 225–231, 1996.
40. Sivak, M., How common sense fails us on the road: Contribution of bounded rationality to the annual worldwide toll of one million traffic fatalities, *Transportation Research: Part F*, 5, 259–269, 2002.
41. Porterfield, A.L., Traffic fatalities, suicide, and homicide, *American Sociological Review*, 25(6), 897–901, 1960.
42. Hancock, P.A., *Essays on the Future of Human-Machine Systems*, Eden Prairie, Banta, 1997.
43. Moray, N.P. and Hancock, P.A., Minkowski spaces as models of human-machine communication, *Theoretical Issues in Ergonomic Science*, 2008, in press.
44. Gibson, J.J. and Crooks, L.E., A theoretical field analysis of automobile-driving, *American Journal of Psychology*, 51, 453–471, 1938.
45. Lewin, K., *Principles of Topological Psychology*, Heider, E. and Heider, G.M., (Trans.), McGraw-Hill, New York, 1936.
46. Gibson, J.J., *The Ecological Approach to Visual Perception*, Erlbaum, Mahwah, NJ, 1979/1986.
47. Endsley, M., Measurement of situation awareness in dynamic systems, *Human Factors*, 37(1), 65–84, 1995.
48. Hancock, P.A. and Diaz, D.D., Ergonomics as a science of purpose, *Theoretical Issues in Ergonomic Science*, 3(2), 115–123, 2001.
49. Flach, J., Situation awareness: Proceed with caution, *Human Factors*, 37(1), 149–157, 1995.
50. Smith, K. and Hancock, P.A., Situation awareness is adaptive, externally directed consciousness, *Human Factors*, 37(1), 137–148, 1995.
51. Senders, J.W., Kristofferson, A.B., Levison, W.H., Dietrich, C.W., and Ward, J.L., The attentional demand of automobile driving, *Highway Research Record*, 195, 15–33, 1966.
52. Hancock, P.A., Evaluating in-vehicle collision avoidance warning systems for IVHS, In *Concurrent Engineering: Tools and Technologies for Mechanical System Design*, Haug, E.J., (Ed.), Springer, Berlin, 1993, pp. 947–958.
53. Hancock, P.A., On the process of automation transition in multi-task human-machine systems, *Transaction of the IEEE on Systems, Man, and Cybernetics, Part A: Humans and Systems*, 37(4), 586–598, 2007.
54. Senders, J.W., Human Control of Simultaneous Variables, Paper presented in Linkoping, Sweden, March, 2006.
55. Estes, W.K., Toward a statistical theory of learning, *Psychological Review*, 57, 94–107, 1950.
56. Simons, D.J. and Ambinder, M.S., Change blindness: Theory and consequences, *Current Directions in Psychological Science*, 14(1), 44–48, 2005.
57. Olson, P.L. and Faber, E., *Forensic Aspects of Driver Perception and Response*, Lawyers and Judges Publishing Co., Tucson, AZ, 2003.

58. Calasso, R., *The Marriage of Cadmus and Harmony*, Cape, London, 1993.
59. Kemp, S. and Williams, K., Demonic possession and mental disorder in medieval and early modern Europe, *Psychological Medicine*, 17(1), 21–29, 1987.
60. Wallace, B., *External-to-Vehicle Driver Distraction*, Scottish Executive Social Research: Stationery Office Bookshop, Edinburgh, Scotland, 2003.
61. Hastings, M., A plague on this clutter, *Daily Mail*, September 13th, London, 2006.
62. Massey, R., Danger! Rural blight ahead, *Daily Mail*, September 12th, London, 2006.
63. Senders, J.W., Analysis of an intersection, *Ergonomics in Design*, 6(2), 4–6, 1998.
64. Reason, J., *Human Error*, Cambridge University Press, Cambridge, 1990.
65. Dewar, R.E. and Olson, R., (Eds.), *Human Factors in Traffic Safety*, Lawyers and Judges Publishing, Tucson, AZ, 2002.
66. Hole, G., The Psychology of Driving, Erlbaum Associates, Mahwah, NJ, 2007.
67. Peden, M., Scurfield, R., Sleet, D., Mohan, D., Hyder, A.A., Jarawan, E., and Methers, C., *World Report on Road Traffic Injury Prevention*, World Health Organization, Geneva, 2004.
68. Lundegaard, K., Naik, G., Molinski, D., Wonacott, P., and Onsanit, R., New world health goal: Halting the rise in traffic deaths, *The Wall Street Journal*, April 7th, New York, 2004.

3 Defining Driver Distraction

John D. Lee, Kristie L. Young, and Michael A. Regan

CONTENTS

Distraction-related crashes impose a substantial cost on society. By one estimate, cell phone–related crashes are responsible for $43 billion in costs each year in the United States alone.[1] Some of these costs are balanced by the benefits associated with the distracting activities. As an example, in-vehicle information systems (IVISs) can reduce the monotony of driving and enable people to accomplish other tasks while driving. IVIS includes a wide variety of entertainment and information systems that range from MP3 players and cell phones to navigation systems and Internet content.[2,3] Consequently, the benefits and the distraction potential vary widely. An important challenge that the designers and policy makers are facing is how to maximize the benefits of this technology and minimize the costs associated with the distraction potential of IVIS. This challenge is compounded by many other activities that compete for the attention of the drivers, such as eating, grooming, and talking with passengers.

Managing the trade-off associated with distraction requires a clear definition of distraction and an understanding of how the design features of IVIS and the characteristics of other competing activities influence distraction. Distraction is a complex phenomenon, and some argue that understanding distraction requires a comprehensive theory of the behavior of the driver (see Chapter 2). Such a theory is beyond the scope of this chapter, not to mention beyond the current understanding of human cognition. Although a comprehensive theory may not be possible, considering several complementary theoretical perspectives can help designers and policy makers to manage distraction. Each theoretical perspective implies a different cause of distraction and identifies different design and policy considerations to reduce distraction-related crashes.

Complementary theories also suggest different independent and dependent variables and different paradigms for system evaluation. For example, considering distraction in terms of concurrent task demands might involve the theoretical constructs of multiple resource theory (see Chapter 5) and experimental protocols that are used to assess dual-task interference, whereas considering distraction in terms of task timing would lead to a very different experimental approach. Likewise, a theory of distraction that considers only the individual driver will neglect traffic and societal contributions to distraction. Addressing the challenge of distraction demands a more comprehensive description of the behavior of the driver than that afforded by a simple information processing description of dual-task performance.

Chapter 4 and this chapter present an initial attempt at combining several theoretical perspectives to guide design and policy considerations regarding distraction. This chapter begins with a brief discussion of the definitions of distraction and the underlying issues, and suggests our definition. It concludes by describing the role of distraction as a cause of crashes.

3.1 DEFINING DISTRACTION

Distraction, and more generally inattention, has been a concern for road safety professionals and other stakeholders for many years,[4–6] but the recent controversy regarding cell phones has prompted a surge of research in this area.[7–9] Many definitions of distraction and related phenomena have accompanied this increased interest in distraction. These definitions have important implications for assessing the magnitude and cause of the distraction problem. A recent naturalistic driving study found that nearly 80% of crashes and 65% of near-crashes included inattention as a contributing cause.[10] Importantly, the authors defined inattention as including general inattention to the road, fatigue, and secondary task demand. Inattention occurs in a broad class of situations in which the driver fails to attend to the demands of driving, such as when sleep overcomes a drowsy driver. Inattention represents diminished attention to activities that are critical for safe driving in the absence of a competing activity. One way to distinguish between inattention and distraction is that distraction involves an explicit activity (e.g., dialing a cell phone or daydreaming) that competes for the attention of the driver, as compared with a cognitive state (e.g., drowsiness or fatigue) that leads to diminished capacity to attend to the roadway. The following sample of distraction definitions, which has emerged over the last 20 years, is consistent with considering distraction as a subset of inattention:

1. Diversion of attention from the driving task that is compelled by an activity or event inside the vehicle.[11]
2. A shift in attention away from stimuli that is critical for safe driving toward stimuli that are not related to safe driving.[12]
3. Any activity that takes the attention of a driver away from the task of driving.[13]
4. Driver distraction occurs when a driver is delayed in the recognition of information needed to safely accomplish the driving task because some event, activity, object, or person within or outside the vehicle compelled or tended to induce the driver's shifting attention away from the driving task.[14]

5. The involvement of drivers that takes his or her attention away from their intended driving task (Chapter 13)[15].
6. Driver distracters include those objects or events both inside and outside the vehicle that serve to redirect attention away from the task of driving or capture enough of the attention of the driver such that there are not enough attentional resources for the task of driving.[16]
7. A disturbance imposed within a lateral or longitudinal control vehicle loop.[17]
8. Driver distraction implies that drivers do things that are not primarily relevant to the driving task (driving safely), and this disturbs attention needed when driving safely.[18]
9. Distraction occurs when attention is withdrawn from the driving task, which results in delayed responses to driving events, increased perceptions of workload, and, in some cases, disruptions of speed and lane maintenance.[19]
10. Distraction can be defined as misallocated attention.[20]
11. Distraction occurs when a triggering event induces an attentional shift away from the task, in this case driving.[21]
12. A diversion of attention from driving, because the driver is temporarily focusing on an object, person, task, or event not related to driving, which reduces the awareness, decision making, or performance of the driver, leading to an increased risk of corrective actions, near-crashes, or crashes.[22]
13. Any event or activity that negatively affects the ability of a driver to process information that is necessary to safely operate a vehicle (Chapter 11).
14. A form of inattention that shifts attention away from the task at hand (Chapter 21).

Some definitions consider distraction in terms of its effect on driving performance, others describe it in terms of the activities or objects that lead to distraction, and most describe distraction as something that disrupts *the driving task*. Table 3.1 presents five elements of distraction (labeled *source*, *location of source*, *intentionality*, *process*, and *outcome*) and some examples of each element.

The first column of Table 3.1 shows that although distraction may be derived from one or more sources, ultimately, some activity directly contributes to distraction. Events

TABLE 3.1
Common Elements of Distraction Definitions and Examples of Each Element

Source	Location of Source	Intentionality	Process	Outcome
Object	Internal activity (e.g., daydreaming)	Compelled by source	Disturbance of control	Delayed response
Person	Inside vehicle	Driver's choice	Diversion of attention	Degraded longitudinal and lateral control
Event	Outside vehicle		Misallocation of attention	Diminished situation awareness
Activity				Degraded decision making
				Increased crash risk

can initiate activities that then involve a person or object. In the case of internal activity, the object might only be a mental abstraction, such as when the driver is distracted by thinking about vacation plans. Distracting activities can involve sources either inside or outside the vehicle, and, for the purpose of a general definition of distraction, it does not matter where the source is located. Furthermore, distraction can occur when extremely salient perceptual cues compel drivers to attend to a source against their will or when drivers willingly divert their attention from the road or from the primary driving task. Both willing and unwilling engagements in activities can be distracting.

Many definitions invoke the construct of attention, as well as the distribution of attention, in describing the underlying process by which distractions affect drivers. Distraction is fundamentally related to the process associated with distributing attention. Being independent of the psychological theories that describe attention, an important consideration in some of these definitions is that distraction is associated with the breakdown of a dynamic control process.[23] Many definitions also consider the outcome of distraction, such as the effect of distraction on reaction time, lane position, crash risk, or driving safety. Defining distraction in terms of specific outcomes is problematic because the presence or absence of distraction then depends on a somewhat arbitrary selection of measures and combination of roadway events. This is important because some studies have shown that cognitive distraction actually enhances lateral control performance, and, at the same time, it diminishes visual scanning behavior.[24]

The use of different, and sometimes inconsistent, definitions of driver distraction can create a number of problems for researchers and road safety professionals. First, the lack of consistent definitions across studies can make the comparison of research findings difficult or even impossible, as even seemingly similar studies can be examining slightly different concepts and measuring different outcomes. Inconsistent definitions can also lead to different interpretations of crash data and, ultimately, to different estimates of the role of distraction in crashes. For example, the decision to include poorly timed driving activities as forms of distraction can change the estimate of the involvement of distraction in crashes by up to one-third (see Chapters 15 and 16 for further discussion of how different definitions can affect crash involvement estimates). These issues highlight the need to develop a common, generally accepted definition of driver distraction. To this end, the following definition has been developed, which builds on the considerations discussed earlier and previous definitions of distraction:

> Driver distraction is a diversion of attention away from activities critical for safe driving toward a competing activity.

3.2 DISTRACTION AS A CAUSE AND CONTRIBUTOR TO CRASHES

This and other definitions of distraction face a challenge similar to the definitions of human error.[25,26] There is a danger of *post hoc* attribution of distraction as a cause of a crash. Only after the fact is it obvious what particular activities were critical for safe driving and how the drivers should have distributed their attention. Because driving is varied and complex, it is not feasible to define a normative model of how drivers should attend to the roadway, and therefore it is difficult to assess the degree to which there is a distraction-related inappropriate distribution of attention. Not defining the distraction in terms of a specific outcome (e.g., delayed accelerator release reaction time to a braking lead vehicle) also leaves open the question of how to operationalize a diversion of attention away from critical driving activities.

Related to the issue of defining an appropriate distribution of attention, defining distraction as a diversion of attention from activities critical for safe driving suggests that some driving activities can be a potential source of distraction. Driving is a complex, multitask activity, making it likely that the demands of one element of driving will interfere with another element. Considering *driving* as a single activity in defining distraction oversimplifies a complex activity and neglects important driving-related distractions that drivers must manage.

The potential for driving tasks to pose a distraction becomes more important as advanced driver-assistance systems (ADASs) become more prevalent and powerful. Interaction with these devices is one of the many activities that constitutes driving and so can represent an additional source of driving-related distraction. Even mundane driving-assistance technology can distract. For example, a poorly timed glance to a rear-view mirror could delay a driver's response to a braking lead vehicle. Poorly designed collision warning systems may be even more likely to distract drivers, particularly if several warnings occur at the same time or the warning is poorly timed. Navigation represents a driving-related task with substantial potential to distract, whether it is the challenge of reading a paper map spread out over the steering wheel or entering a destination into a navigation system while the vehicle is in motion.[27–29] Considering driving tasks as potential distractions is important because ADAS will become an increasingly powerful force in directing the attention of the drivers to what the ADAS estimates to be critical elements of the roadway environment. To the extent that the ADAS correctly identifies critical elements, it will mitigate distractions but could also distract drivers if it directs attention away from these critical elements (see Chapters 15 and 28).

Distraction reflects a mismatch between the attention demanded by the road environment and the attention devoted to it. More precisely, it reflects a mismatch between the attention devoted to a safety-critical driving activity and the attention demanded by it. For simplicity, the following discussion refers to the demand of a safety-critical driving activity as critical roadway demand or simply roadway demand. Drivers can fail to devote sufficient attention to the road when the combined demands of the roadway and a competing activity, such as interaction with an IVIS, exceed drivers' capacity to perceive and respond to roadway events. The degree to which the driver's engagement in a competing activity poses a threat of distraction depends on the combined demands of the roadway and the competing activity relative to the available capacity of the driver. Figure 3.1 plots the variation over time of the driving demand and the demand of a competing activity (interaction with an IVIS). The gray area of Figure 3.1 shows that high levels of each demand do not necessarily lead to distraction; distraction occurs when high levels of demand for both the roadway and the competing activity coincide and jeopardize drivers'

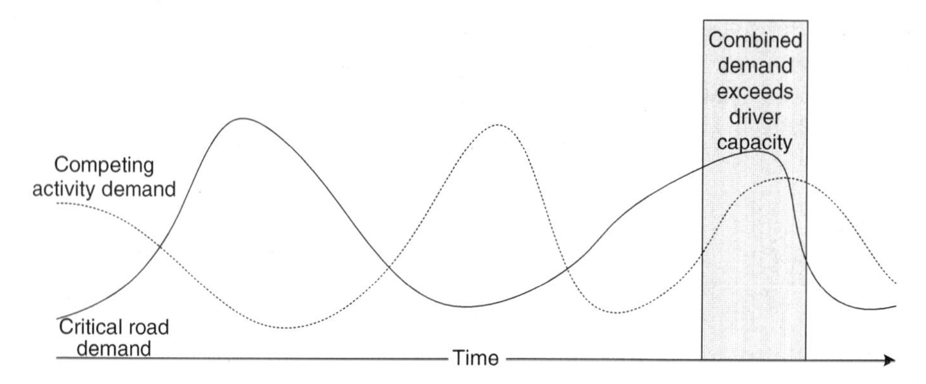

FIGURE 3.1 The demands of the roadway and the competing activity lead to distraction-related mishaps when both combine to exceed the drivers' capacity to respond.

capacity to respond[30] (see also Chapter 27 for a detailed discussion of roadway demand). For example, even if an IVIS is particularly demanding, this may not lead to distraction if the roadway demand is low. However, if roadway demand peaks suddenly and unexpectedly, such as when a child runs onto the road, then the driver may not have sufficient attentional capacity to respond to this increased demand. Distraction, therefore, is a property of the joint demands of the IVIS and roadway and the driver's distribution of attention to meet those demands.

Importantly, distraction-related mishaps *tend* to occur when IVIS and roadway demands combine to threaten drivers' capacity to respond. Because of the inherent variability of roadway and IVIS demands, identifying distraction as a cause of a crash is problematic. Even if a driver devotes relatively little attention to the competing activity, the driving demand could be surprisingly large and overwhelm the driver's ability to respond appropriately. The *post hoc* analysis associated with crash data often focuses on the crash event, which ignores the processes that generated the event and limits consideration of causes to those immediately surrounding the event, rather than the pattern of activities that preceded it.[31] Such *post hoc* analysis fails to distinguish between situations in which a driver was devoting a nominally appropriate degree of attention to the critical driving tasks and situations in which a driver was not.

Figure 3.2 plots the distributions of attention demanded by the roadway and that devoted by the driver as one way to place distraction-related mishaps in the context of the overall process that generates them. The distribution of attention demanded by the roadway reflects the variation in roadway demand and that devoted by the driver reflects the variation of the driver's attention to the roadway. Drivers avoid distraction-related mishaps when the tails of the distributions do not overlap—when there are no instances where roadway demands exceed the drivers' attention to these demands. According to this perspective, distraction represents a diminished safety margin as defined by the degree to which the tails of the distribution overlap; the less the overlap, the less likely the roadway demands will exceed a driver's capacity to respond.

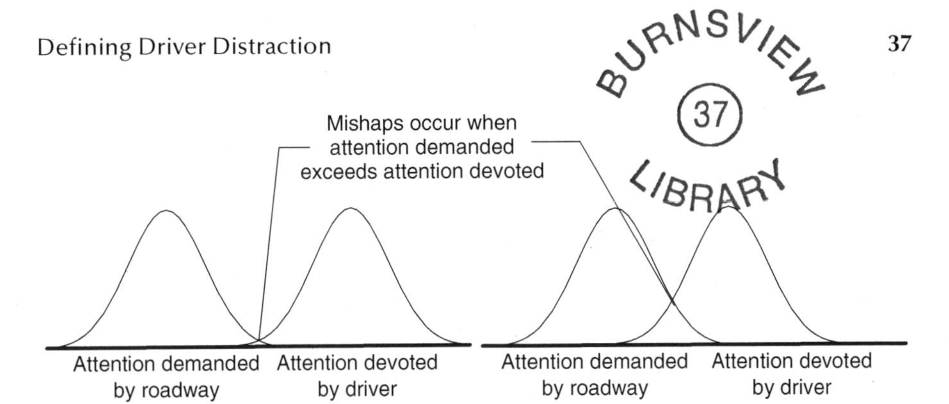

FIGURE 3.2 The distributions of attention devoted by the driver and attention demanded by the roadway for an attentive driver on the left and a distracted driver on the right.

The distributions on the left side of the figure represent a nominally attentive driver, and the ones on the right represent a distracted driver. Even with nominally appropriate attention to driving, the distributions of attention demanded and that devoted will overlap, such that there will be instances where attention demanded exceeds attention devoted and mishaps might occur. Thus, even nominally sufficient attention can result in "distraction"-related incidents when, for example, a driver interacting with the IVIS is unlucky to look away from the road at the wrong moment. If driving is considered a monitoring task, such instances are unavoidable in any situation except the one in which the driver focuses exclusively on the driving task[32,33] (see also Chapter 2). As a consequence, a definition of distraction describes inappropriate *distributions* of attention, where a distracted driver inappropriately distributes attention over time relative to the demands of the roadway and, as a consequence, has a higher probability of a mishap. Distraction is a property of the distributions and not the individual events. Because such an inappropriate distribution does not guarantee a mishap, or even a decline in performance, drivers may not realize they are distracted; as a consequence of this imperfect link between distraction and safety consequences, it might not be obvious to designers that there are flaws in their products.

The timeline in Figure 3.1 has no units and this omission reflects an important consideration of distraction. IVIS and driving demands vary over time horizons that range from seconds to days. Distraction reflects an inappropriate distribution of attention between the roadway and the competing activities, such as IVIS interactions, at any of these time horizons. The overlap of these distributions depends on the separation of the means of these distributions as well as their spread. Distraction occurs if the separation of the distributions diminishes or if the distributions widen.

The time horizon affects the shape and rate of change of the distributions. A short time horizon results in narrow distributions where there is little uncertainty in the current demand, but the mean of these distributions could change quickly. At a long time horizon (hours or days) the distribution of demand is broad, but the mean of the distribution changes slowly, reflecting the typical demands of the roadway. The short timescale can lead to distraction-related mishaps when the driver is not able to respond to rapid changes in demands, whereas the long timescale can lead to mishaps because drivers fail to maintain adequate separation of the distributions, and events from the tails of the distributions surprise drivers. Avoiding distraction-related mishaps depends on controlling attention to the road at long time horizons (to minimize situations in which demands change so quickly that the driver cannot adjust) and at short time horizons (to compensate for the differences between the typical and the actual roadway demands). Distraction represents a breakdown of control of the distribution of attention at one or more time horizons.

3.3 CONCLUSION

In summary, distraction shares some of the definitional and causal attribution challenges of human error. Defining distraction in terms of behavioral outcomes makes for a clear, concrete definition but results in a somewhat arbitrary pairing of dependent variables and driving situations. Defining distraction in terms of a delay in the accelerator release in response to a braking lead vehicle also provides a clear definition but may fail to capture important aspects of distraction. Such a restrictive definition limits its generalizability and also our understanding of distraction. An alternative is to define distraction in terms of a diminished safety margin associated with the overlap of the distribution of attention demanded by the road and that devoted to the road. This more abstract definition suffers from a substantial challenge of operationalizing these distributions, but it leads to a broader consideration of the factors affecting distraction.

Considering distraction as a problem of controlling the safety margin associated with the overlap of the distribution of attention demanded by the critical driving activities and attention devoted to these activities emphasizes the factors that contribute to overlaps in these distributions. These factors vary according to the time horizon, with the overlap of broad tails dominating at the long time horizon and rapid changes in the separation of the distributions dominating at the short time horizon. With a short time horizon, distraction depends on the competition for processing resources associated with the concurrent demands of the road and the competing activity.[34,35] In contrast, with a long time horizon, distraction depends on a random walk process in which productivity pressures, impoverished feedback, and indistinct safety boundaries lead drivers to adopt unsafe practices that increase the chance for roadway demands to exceed the attention devoted to the roadway.[36] Between these two extremes, distraction depends on task timing and the ability of drivers to avoid overlapping attention-demanding activities.[37,38] The process of distributing attention between the road and the competing activity at each of these time horizons is quite different and so implies different theoretical perspectives and models. Each of these perspectives helps to define design approaches, evaluation criteria, and mitigation strategies; each perspective also gives a slightly different meaning to the general definition of distraction: *Driver distraction is a diversion of attention away from activities critical for safe driving toward a competing activity.*

ACKNOWLEDGMENT

This chapter was substantially influenced by conversations with Professor Thomas Triggs of Monash University Accident Research Center.

REFERENCES

1. Cohen, J.T. and Graham, J.D., A revised economic analysis of restrictions on the use of cell phones while driving, *Risk Analysis* 23(1), 1–14, 2003.
2. Lee, J.D. and Kantowitz, B.K., Network analysis of information flows to integrate in-vehicle information systems, *International Journal of Vehicle Information and Communication Systems* 1(1/2), 24–43, 2005.
3. Walker, G.H., Stanton, N.A., and Young, M.S., Where is computing driving cars?, *International Journal of Human-Computer Interaction* 13(2), 203–229, 2001.
4. Brown, I.D., Tickner, A.H., and Simmonds, D.C.V., Interference between concurrent tasks of driving and telephoning, *Journal of Applied Psychology* 53 (5), 419–424, 1969.
5. Sussman, E.D., Bishop, H., Madnick, B., and Walter, R., Driver inattention and highway safety, *Transportation Research Record* 1047, 40–48, 1985.
6. Treat, J.R., Tumbas, N.S., McDonald, S.T., Shinar, D., Hume, R.D., and Mayer, R.E., Stansifer, R.L., and Castellan, N.J., *Tri-level Study of the Causes of Traffic Accidents: Executive Summary*, NHTSA, U.S. Department of Transportation, Washington, D.C., 1979.
7. McCartt, A.T., Hellinga, L.A., and Bratiman, K.A., Cell phones and driving: Review of research, *Traffic Injury Prevention* 7, 89–106, 2006.
8. Horrey, W.J. and Wickens, C.D., Examining the impact of cell phone conversations on driving using meta-analytic techniques, *Human Factors* 48(1), 196–205, 2006.
9. Haigney, D. and Westerman, S.J., Mobile (cellular) phone use and driving: a critical review of research methodology, *Ergonomics* 44(2), 132–143, 2001.
10. Klauer, S.G., Dingus, T.A., Neale, V.L., Sudweeks, J.D., and Ramsey, D.J., *The Impact of Driver Inattention on Near-Crash/Crash Risk: An Analysis Using the 100-Car Naturalistic Driving Study Data*, Report No. DOT HS 810 594, National Highway Traffic Safety Administration, Washington, D.C., 2006.
11. Treat, J.R., *A Study of Precrash Factors Involved in Traffic Accidents*, Centre for Automative Safety Research, Adelaide, 1980, p. 21.
12. Steff, F.M. and Spradlin, H.K., *Driver Distraction, Aggression and Fatigue: A Synthesis of the Literature and Guidelines for Michigan Planning*, Report No. UMTRI-2000-10, The University of Michigan Transport Research Institute, Ann Arbor, MI, 2000.
13. Ranney, T.A., Mazzae, E., Garrott, R., and Goodman, M.J., NHTSA driver distraction research: Past, present, and future, available at http://www-nrd.nhtsa.dot.gov/departments/nrd-13/driver-distraction/Welcome.htm, Accessed on January 28, 2003.
14. Stutts, J.C., Reinfurt, D.W., Staplin, L., and Rodgman, E.A., *The Role of Driver Distraction in Traffic Crashes*, AAA Foundation for Traffic Safety, Washington, D.C., 2001.
15. McAllister, D., Dowsett, R., and Rice, L., Driver Inattention and Driver Distraction (No. 15), Virginia Commonwealth University Transportation Safety Training Center (Crash Investigation Team), 2001.
16. Manser, M.P., Ward, N.J., Kuge, N., and Boer, E.R., Influence of a driver support system on situation awareness and information processing in response to lead vehicle braking, *Human Factors and Ergonomics Society 48th Annual Meeting*, New Orleans, Louisiana, 2359–2363, 2004.
17. Sheridan, T.B., Driver distraction form a control theory perspective, *Human Factors* 46(4), 587–599, 2004.
18. Patten, C.J.D., Kircher, A., Ostlund, J., and Nilsson, L., Using mobile telephones: cognitive workload and attention resource allocation, *Accident Analysis & Prevention* 36(3), 341–350, 2004.
19. Laberge, J., Scialfa, C., White, C., and Caird, J., Effects of passenger and cellular phone conversations on driver distraction, *Transportation Research Record: Journal of the Transportation Research Board* 1899, 109–116, 2004.

20. Smiley, A., What is distraction?, *International Conference on Distracted Driving*, Toronto, Ontario, Canada, 2005.
21. Horberry, T., Anderson, J., Regan, M. A., Triggs, T. J., and Brown, J., Driver distraction: The effects of concurrent in-vehicle tasks, road environment complexity and age on driving performance, *Accident Analysis & Prevention* 38(1), 185–191, 2006.
22. Hedlund, J., Simpsom, H., and Mayhew, D., International Conference on Distracted Driving. Summary of Proceedings and Recommendations, *International Conference on Distracted Driving*, Toronto, Ontario, Canada, 2006.
23. Sheridan, T.B., Driver distraction from a control theoretic perspective, *Human Factors* 46(4), 587–599, 2004.
24. Engstrom, J., Johansson, E., and Ostlund, J., Effects of visual and cognitive load in real and simulated motorway driving, *Transportation Research Part F-Traffic Psychology and Behaviour* 8(2), 97–120, 2005.
25. Rasmussen, J., Human error and the problem of causality in analysis of accidents., *Philosophical Transactions of the Royal Society Series B* 327(1241), 449–462, 1990.
26. Senders, J.W. and Moray, N.P., *Human Error: Cause, Prediction, and Reduction*, Lawrence Erlbaum Associates, Hillsdale, New Jersey, 1991.
27. Dingus, T.A., Antin, J.F., Hulse, M.C., and Wierwille, W.W., Attention demand requirements of an automobile moving-map navigation system, *Transportation Research* 23A(4), 301–315, 1989.
28. Antin, J.F., Dingus, T.A., Hulse, M.C., and Wierwille, W.W., An evaluation of the effectiveness and efficiency of an automobile moving-map navigational display, *International Journal of Man-Machine Studies* 33, 581–594, 1990.
29. Dingus, T., Hulse, M., Mollenhauer, M.A., Fleishman, R.N., McGehee, D.V., and Manakkal, N., Effects of age, system experience, and navigation technique on driving with an advanced traveler information system, *Human Factors* 39(2), 177–199, 1997.
30. Verwey, W.B., On-line driver workload estimation. Effects of road situation and age on secondary task measures, *Ergonomics* 43(2), 187–209, 2000.
31. Leveson, N., A new accident model for engineering systems, *Safety Science* 42(4), 237–270, 2004.
32. Moray, N., Monitoring, complacency, scepticism and eutactic behaviour, *International Journal of Industrial Ergonomics* 31(3), 175–178, 2003.
33. Moray, N., Are Observers Really Complacent When Monitoring Automated Systems? *Proceedings of the IEA2000/HFES 2000 Congress*, I-592–I-595, 2000.
34. Strayer, D.L., Drews, F.A., and Johnston, W.A., Cell phone-induced failures of visual attention during simulated driving, *Journal of Experimental Psychology-Applied* 9(1), 23–32, 2003.
35. McCarley, J.S., Vais, M., Pringle, H.L., Kramer, A.F., Irwin, D.E., and Strayer, D.L., Conversation disrupts visual scanning of traffic scenes, *Human Factors* 3, 424–436, 2004.
36. Rasmussen, J., Risk management in a dynamic society: A modelling problem, *Safety Science* 27(2–3), 183–213, 1997.
37. Carbonell, J.R., A queueing model of many-instrument visual sampling, *IEEE Transactions on Human Factors in Electronics* HFE-7(4), 157–164, 1966.
38. Senders, J.W., Kristofferson, A.B., Levison, W.H., Dietrich, C.W., and Ward, J.L., The attentional demand of automobile driving, *Highway Research Record* 195, 15–33, 1967.

4 What Drives Distraction? Distraction as a Breakdown of Multilevel Control

John D. Lee, Michael A. Regan, and Kristie L. Young

CONTENTS

Chapter 3 offered a definition of distraction and pointed toward some of the challenges in identifying distraction as a cause of motor vehicle crashes. As with most safety-related mishaps, distraction-related incidents and crashes are complex events with multiple causes. Definitions of distraction, related models of performance, and crash analyses can pursue an arbitrarily long causal sequence in explaining the situation.[1] Chapter 3 also characterized distraction as a failure to maintain an appropriate distribution of attention relative to the demands of activities critical for safe driving. This chapter considers the processes that underlie distraction in an effort to provide a useful causal explanation of distraction-related crashes and to describe why drivers fail to maintain an appropriate distribution of attention. This explanation of how attention is diverted away from activities critical for safe driving describes distraction as a breakdown in a multilevel control process, with a different timescale characterizing each level.

4.1 DISTRACTION AS A BREAKDOWN OF MULTILEVEL CONTROL

A common perspective regarding distraction is that drivers passively respond to the demands of driving and competing activities and that doing two things at once compromises performance of one or both of the activities (i.e., dual-task interference). This perspective captures important cognitive constraints regarding the degree of interference between concurrent tasks, but it does not account for how drivers distribute their attention between such tasks, distribute tasks over time, or choose to engage in tasks. Considering drivers as active controllers provides a useful perspective on distraction.

Driving, as a control process, has been described in terms of three levels: operational, tactical, and strategic.[2–5] The operational level concerns the lateral and longitudinal control of the vehicle and occurs at a timescale of milliseconds to seconds. Tactical control concerns the choice of lanes and speeds and occurs at a timescale of seconds to minutes. Strategic control concerns decisions regarding routes and travel patterns and occurs at a timescale of minutes to weeks. Each of these three levels of control applies to activities critical to safe driving and to competing activities.[6]

These three levels can also be used to describe the control of attention to competing activities. At the operational level, drivers control resource investment; at the tactical level, they control task timing; and at the strategic level, they control exposure to potentially demanding situations. Distraction-related mishaps result from a breakdown of control at any one level, and from the accumulation of control problems that compound as they propagate across levels. A consequence of this perspective is that distraction-related crashes result not only from dual-task interference but from drivers' inability to control potentially distracting interactions.[7]

The following definition of distraction guides this discussion: *Driver distraction is a diversion of attention away from activities critical for safe driving toward a competing activity.* Specifically, "competing activity" refers to interactions with in-vehicle technology, passengers, food, thoughts, and noncritical driving activities. These noncritical driving activities can involve in-vehicle technology, as with distraction associated with a navigation system, or a conversation with a passenger regarding the choice of speed. For the purposes of this chapter, the discussion of distraction and distribution of attention is grounded in a concrete scenario involving in-vehicle technology, but the general process applies to other competing activities.

Consider the following scenario. Anticipating a long drive, Sam decides to insert an MP3 player into his car's audio system and select a playlist as he begins to drive out of the city. After inserting the player, he glances down to view the catalog of playlists and begins to scroll through the list. Meanwhile, a car abruptly merges in front of Sam's car to exit the freeway, suddenly shrinking the gap between his car and the car ahead. Sam begins to slow to widen the gap, then looks back to the playlist to continue scrolling down the list. Overshooting the desired playlist, Sam continues to look at the MP3 player as the car ahead brakes suddenly to accommodate other vehicles entering the highway. Sam looks up to find himself crashing into the vehicle ahead.

Figure 4.1 shows some causes of distraction-related mishaps in terms of a multilevel control framework and highlights several contributors to the distraction-related

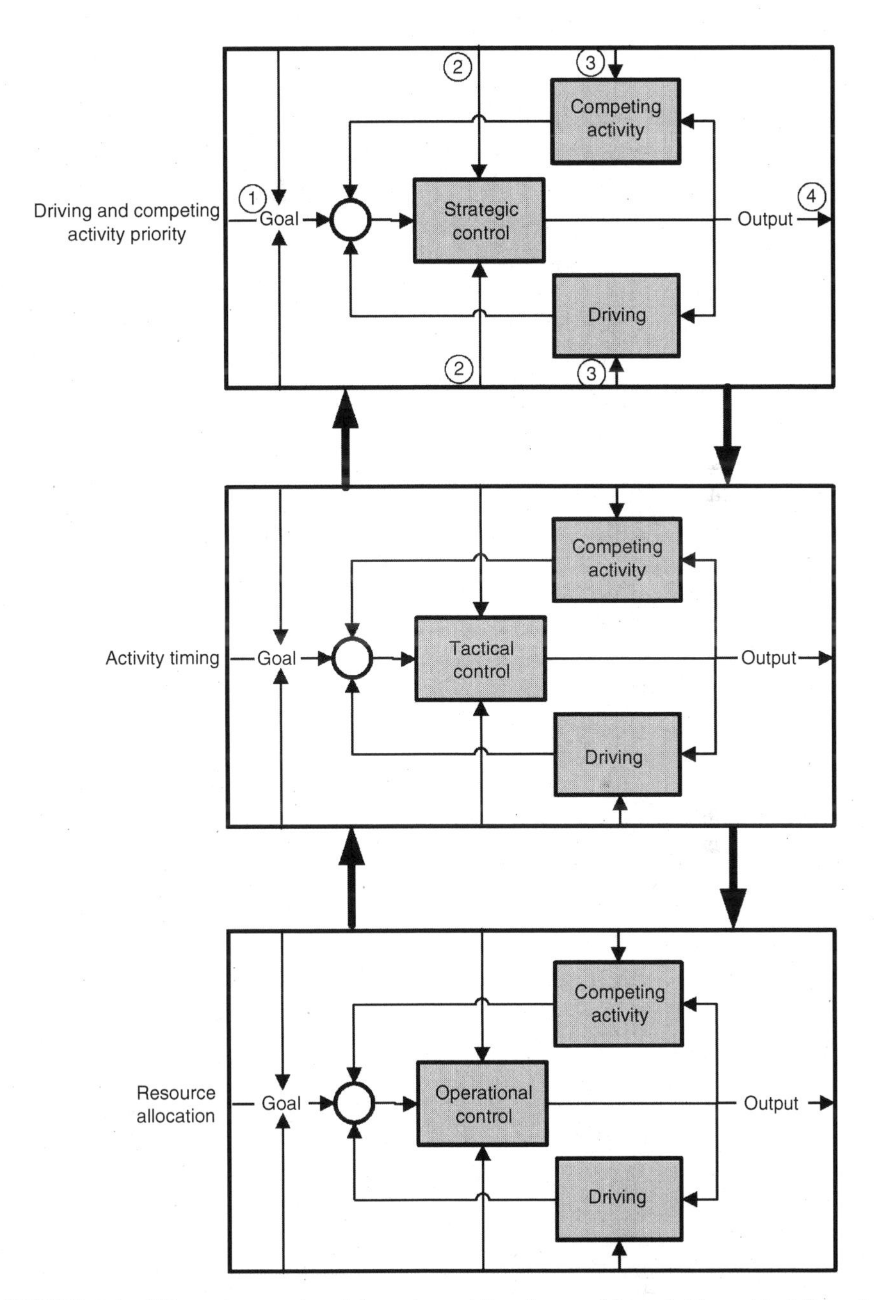

FIGURE 4.1 Distraction as a breakdown in multilevel control in activities critical for safe driving and competing activities. Numerals indicate interactions between levels: (1) adaptive control, in which the output of one level affects the goal of another level; (2) feedforward control, in which the output of one level affects expectations and appropriate response schema at another level; (3) cascade effects, in which the output of one level influences the control dynamics of another level; (4) the output supports feedback control for a given level and adaptive control for other levels. The numerals at the strategic level apply to the tactical and operational levels. The heavy lines between levels encapsulate these interactions.

crash in this scenario. One contributing factor is the breakdown of control at the operational level, where the visual and cognitive demands of selecting the playlist interfered with Sam's attention to the road. Another contributor was the cascade effect set in motion when Sam overshot the playlist at the operational level, which led to a longer-than-expected task duration. This overshoot compelled continued interaction and undermined control at the tactical level by inhibiting Sam from adapting his speed and headway to the changing traffic conditions. At the strategic level, the decision to interact with the playlist in a relatively demanding roadway environment also contributed to the crash. Each of these causes represents equally valid explanations of the crash, and each demands a different set of theoretical considerations. Importantly, each also points to different design and policy strategies to prevent distraction-related crashes.

Three types of control are active in each level in Figure 4.1. Breakdowns in any one of these three types contribute to driver distraction. Each type of control has important limits, as shown by the columns on the right of Table 4.1. Feedback control uses the difference between a goal state and the current state to guide behavior. To be successful, feedback control depends on timely, precise information regarding the difference between the current state of affairs and the goal state. In driving, such feedback is often delayed and noisy. Furthermore, for feedback control to be effective, the time constant of drivers' response to the error signal must be fast relative to the system needing to be controlled so that responses can be made before the system diverges by an unacceptable degree from the desired state.

The second type of control, feedforward control, uses the anticipated future state of the system to guide behavior. Feedforward control is critical for safe

TABLE 4.1
Challenges for Each Type of Control for Each Time Horizon

Control Type	Operational: Control Attention to Tasks (milliseconds to seconds)	Tactical: Control Task Timing (seconds to minutes)	Strategic: Control Exposure to Tasks (minutes to days)
Feedback—reactive control based on past outcomes	Time constant of driver response is slower than that of driving demands	Feedback is too delayed or noisy to guide behavior	Poor choices might not affect performance
Feedforward—proactive control based on anticipated situation	Task demands are unpredictable or unknown	Task timing is unpredictable or unknown	Potential demands are unpredictable or unknown
Adaptive—metacontrol based on adjusting expectations, goal state, and task characteristics	Tasks that lack a graded effort/accuracy trade-off	Biological and social imperatives not calibrated to task importance	Poor calibration regarding interaction between driving and IVIS goals

driving, enabling experienced drivers to anticipate, detect, and respond to hazards in a proactive manner.[8,9] Expectations associated with feedforward control have a powerful effect on drivers' reaction time to events, reducing reaction time by 750 ms compared with unexpected events.[10] Feedforward control can compensate for the limits of feedback control, but it suffers from other problems. Feedforward control requires an accurate internal model of the future state of the system and an absence of any major unanticipated disturbances. The uncertainty associated with poor mental models of driving situations and competing activities, coupled with the inherent variability of driving demands, limits the effectiveness of feedforward control.

The third type of control, adaptive control, reduces the difference between the goal state and the current state by redefining the goal state.[11] As such, adaptive control represents a type of metacontrol that is critical for accommodating change in the operating environment. While the traditional focus has been on feedback and feedforward control, adaptive control represents an important option for drivers. With adaptive control, drivers adjust their response to task demands according to their performance criteria and capacity. For example, drivers adjust their tolerance for errors in lateral control and tolerate the occasional lane deviation as they focus on an in-vehicle interaction. The success of adaptive control depends on task flexibility over time, and task performance depends on the effort invested. If a task cannot be delayed, or if task performance declines abruptly with diminished effort, then adaptive control may be ineffective. Another challenge for adaptive control is potentially poor calibration regarding the expected driving performance achievement and engagement in a competing activity, such as using an in-vehicle information system (IVIS). Drivers may not realize the consequences of adopting less ambitious goals for performing the driving task. Table 4.1 summarizes some of the reasons why each control type might fail at each time horizon.

In addition to the control challenges outlined in Table 4.1, control can suffer from interactions across the three time horizons. Cascade effects occur when the outcome at one level of control affects control at another level or when control breakdowns at one time horizon undermine control at the others. Such effects are often nonlinear in that small perturbations at one level can lead to catastrophic effects at another. Poor choices at the strategic level can propagate downward and make control at the tactical and operational levels more difficult. In the previous example, the choice to start using the MP3 player in an urban environment imposed greater demands on the tactical control of task timing. Likewise, errors at the operational level can propagate upward to tactical control, as when overshooting in selecting a playlist compelled an unexpectedly long interaction. Control at the longer time horizons constrains that of the shorter time horizons by specifying the goals and tasks at shorter time horizons. Control at the shorter time horizons creates disturbances that undermine control at the longer time horizons. Distraction-related incidents occur when the demands of driving and competing activities combine to undermine control at any one time horizon or initiate cascade effects across the horizons.

Beyond cascade effects, saturation effects represent another important factor that can undermine control.[12] Saturation effects occur when control limits and safety margins are reduced, making effective control vulnerable to small perturbations. At the operational level saturation can be defined in terms of spare capacity, and at the tactical level it reflects the utilization rate, which is the percentage of time the driver is busy responding to the IVIS or driving demands. At the strategic level, saturation reflects the overlap of the demand distributions discussed in Chapter 3—the probability that a driving situation will occur in which the demands will exceed the attention devoted to the roadway. As saturation increases, the potential for breakdown at a given level increases. Such a breakdown might then lead to cascade effects across levels. If the other levels are also highly saturated, they may have little capacity to accommodate these perturbations, prompting further cascade effects that undermine driving safety.

In summary, contrary to many views of distraction, drivers are not the passive recipients of IVIS and other competing demands, as well as driving demands. Instead, it is argued here that drivers actively control the mechanisms that give rise to the distraction they experience. This control occurs at three interacting time horizons (operational, tactical, and strategic) and is achieved by three types of control (feedback, feedforward, and adaptive). The limits of control at each time horizon, and the cascade effects across levels, cause distraction-related mishaps. Saturation effects exacerbate cascade effects because a highly saturated controller is vulnerable, so that small perturbations can cause catastrophic failures. Labeling distraction as simply a problem of information overload neglects the failures that lead to temporally constrained driving and competing demands that overwhelm the driver's capacity. Defining distraction as dual-task interference neglects the failures that occur at the tactical and strategic levels and how these failures interact.

The following sections describe control at the three time horizons. Each of these descriptions provides a different vocabulary for discussing distraction, different tools for assessing it, and different approaches to reducing it. The discussion begins at the operational level, with a description of resource competition between IVIS and driving demands. Following that, the tactical level is considered, with a description of task timing of IVIS and driving demands. The section finishes at the strategic level, with a description of exposure to IVIS and driving demands.

4.2 OPERATIONAL CONTROL: DISTRACTION AS RESOURCE COMPETITION

In the scenario involving selecting an MP3 playlist, distraction arises from resource competition associated with the task of playlist selection and the demands of event detection and vehicle control. Distraction is most likely to occur when the resource demands of the driving task overlap with those of the competing task.[13]

Multiple resource theory provides a useful theoretical perspective for describing driving and IVIS task demands.[14] This approach describes attentional resources in terms of modes, codes, and stages (see Chapter 5 of this book for more detail). When both driving and IVIS tasks demand the same type of resources, performance on one or both tasks suffers. To the extent that the tasks involve different resources, performance will be relatively unaffected. The task of dialing a cell phone with a voice recognition system while negotiating a curve, compared with dialing using a standard keypad, illustrates the benefit. Voice dialing demands resources associated with the oral/auditory mode of interaction, whereas the standard keypad involves resources associated with visual/manual mode of interaction. The stage (perception, cognition, and response selection) seems to be a particularly important consideration. Both basic and applied research has shown a processing bottleneck with response selection.[15,16] Two concurrent tasks involving response selection causes performance to suffer, even when they require different modes and codes of processing.[17]

Not surprisingly, drivers' ability to keep the car in the lane and respond to braking lead vehicles diminishes when they look away from the road.[18,19] Substantial research shows that even when two tasks engage separate resources, as when driving while holding a conversation with a hands-free cell phone, they increase the reaction time to events such as a braking lead vehicle by approximately 300 ms[20–22] and even degrade perceptual judgments.[23] Performance suffers because both tasks require a certain degree of central processing resources, but these effects are less than for tasks that compete for visual/manual resources.[24–26] The recent Crash Avoidance Metrics Partnership (CAMP) Driver Workload metrics project, for example, revealed that although auditory-vocal tasks degraded driving performance, their effects were not as pronounced as those of the visual-manual tasks.[24]

Multiple resource theory can be operationalized using a demand vector associated with driving and IVIS tasks. The degree to which resource requirements overlap defines the decrement of performance in the two tasks (see Chapters 5 and 15 for details). Drivers use adaptive control to modulate this performance decrement by choosing an allocation policy that distributes resources between the competing tasks. The performance of one task could be preserved by preferentially allocating resources to that task.

Several factors undermine drivers' control of distraction at the operational level. Feedback control is problematic because drivers receive misleading feedback regarding the success of their allocation policy. The capacity of ambient (or peripheral) vision to support lane position makes it possible for drivers to look away from the road and receive positive feedback regarding their ability to drive safely. Because hard-braking events are rare, drivers fail to receive regular feedback regarding their impaired performance on event detection, which requires focal vision. Feedforward control is also compromised by a limited ability to anticipate the confluence of IVIS and driving demands. The effectiveness of adaptive control depends on the degree to which performance of the IVIS task declines as resources are allocated toward driving tasks. If IVIS performance declines abruptly with a small decline in resource allocation, then drivers may be unable to adopt an allocation policy in which they modulate their IVIS performance goals to accommodate driving demands. When it is impossible for the driver to change the demand associated with the performance of a task, the task is said to be *unadjustable* (see Chapter 15 for more detail).

Considering distraction as resource competition at the level of operational control leads to three general design considerations. Most obviously, minimizing the overall demands associated with IVIS interactions or the demands of driving will tend to diminish distraction. Advanced driver support systems, such as collision warning systems and intelligent speed adaptation, can reduce driving demands and diminish distraction. More subtly, creating IVIS interactions that avoid direct competition for common resources associated with driving tasks will also diminish distraction. This suggests that IVIS designs that rely on verbal/auditory interactions, such as voice-based control, rather than visual/manual interactions will reduce distraction. However, this approach may not always be effective because reducing resource competition does not eliminate dual-task decrements, and voice control of some tasks may involve more effort than their visual/manual counterparts. For example, a continuous control task, such as raising a window, may be more difficult and time consuming with voice control. A third consideration of this approach has not been widely discussed: that is, to support a graded rather than "brittle" resource allocation and performance trade-off for IVIS interactions.[27] For example, analog displays make it possible for drivers to extract approximate information with little effort and more precise information with greater effort. In contrast, digital displays provide only highly precise information, but only with focused effort. Analog displays tend to support adaptive control, so that drivers can easily modulate their attention to the device, extracting approximate information with a single brief glance and more precise information with a longer glance. This helps drivers succeed with their interactions even if they devote most of their attention to the road.

4.3 TACTICAL CONTROL: DISTRACTION AS FAILURE OF TASK TIMING

In the scenario of searching for an MP3 playlist, distraction depends as much on the timing of the interaction as on the resource demand of the interaction. It is the timing of the scrolling activity relative to the lead vehicle braking that is particularly problematic. In this example, the IVIS interaction associated with overshooting the playlist provides a particularly compelling incentive for the driver to extend the interaction, increasing the delay in responding to roadway events. Distraction is likely when a breakdown of tactical control leads to a situation where multiple tasks must be performed at the same time.

Queuing theory provides a useful theoretical perspective to consider distraction at the tactical level. According to a queuing theory representation of driving, drivers act as a server, processing tasks sequentially and causing tasks to wait if the server is occupied processing another task. Tasks awaiting the server accumulate in a queue. Such a representation has a long history in human performance modeling and offers a perspective for addressing distraction that focuses on task timing rather than competition for multiple resources.[28–31]

In contrast to the multiple resource perspective of distraction, a queuing theory perspective describes the demands on the driver in terms of the policy for queuing tasks, task timing, and how easily the tasks can be interrupted. As such, it provides a framework for considering how drivers plan and manage their interactions with critical driving tasks and competing activities.

A common measure in queuing theory is the utilization rate of the server, which corresponds to the time spent in processing tasks divided by the total time available to process tasks. In terms of distraction, utilization rate can be considered as the time spent responding to critical driving tasks and competing activities. An important insight from queuing theory concerning distraction is that any nonzero utilization rate will delay some percentage of incoming tasks. This means that even an IVIS that requires drivers to respond to relatively infrequent, short tasks will delay response to driving demands. Specifically, the expected delay increases with the utilization rate (ρ) and decreases with the rate at which tasks are processed (μ). Assuming new tasks arrive according to a Poisson distribution, Equation 4.1 predicts the delay.

$$\text{Delay} = \frac{\rho}{\mu(1-\rho)} \tag{4.1}$$

Although Equation 4.1 and queuing theory provide an elegant description of how a driver might process driving and competing tasks, it fails to consider the active role drivers play in determining the timing of activities. Drivers are not passive servers who respond to tasks as they appear. Drivers can reduce delays if task demands can be predicted and if tasks can be interrupted. Characteristics of the driving situation and the IVIS interact to determine how successfully drivers time tasks to avoid the delays predicted by Equation 4.1. Three characteristics of competing tasks can undermine control at the tactical level (see also Chapters 15 and 19):

- A task is said to be *unignorable* when it is so compelling or demanding that the driver cannot delay engagement.
- A task is *unpredictable* when its onset is unexpected or its duration and demand cannot be foreseen by the driver.
- A task is *uninterruptible* when it cannot be easily disengaged or cannot be resumed after interruption.

Some tasks are not easily ignored. The characteristics that make it difficult for drivers to ignore competing activities include a combination of internal and external forces that determine when the driver initiates or delays a task. Biological and social imperatives affect the degree to which driving and competing activities demand attention to initiate or continue an interaction. In the case of initiating a cell phone conversation, these factors might range from a general need to call or a need to call at a specific time, to an external reminder to call (e.g., a PDA reminder) or a cell phone ring. Preliminary research suggests that drivers tend to neglect future driving demands and focus on the current demands of competing activities. As an example, drivers tend to answer ringing cell phones independent of the upcoming driving demands.[32] The social imperatives that induce drivers to respond to these demands are exacerbated by a tendency for drivers, particularly young drivers, to fail to anticipate hazardous situations.[33] The tendency to neglect upcoming driving demands and respond to social and biological imperatives regarding competing interactions may play an important role in driver distraction. As mentioned in Chapter 2 of this book, in responding to some categories of distraction (e.g., children), drivers assume different social roles (e.g., parent), which make it difficult for them to ignore the competing activity.

Drivers can also avoid delays in responding to the road by interrupting competing activities and returning their attention to the activities critical for safe driving. Interruptions that prevent goal rehearsal or that occur in the middle of the task result in longer resumption times.[34] These results are consistent with the goal activation model[35] and suggest that increasing the duration of a competing task may make drivers less able to interrupt that task and return to driving tasks. The goal activation model also predicts that the distraction posed by a competing activity may persist even after it has ended, because drivers continue to think about certain aspects of the competing activity.

Going beyond the ability to manage task timing through their decision to initiate or interrupt tasks, drivers can also actively negotiate the timing and nature of the tasks. Drivers actively negotiate with other drivers on the road to widen their safety margins. Likewise, a driver might adjust the pace of a conversation or delay an interaction based on road demands. In these situations, drivers actively adapt demands of driving and competing activities to arrive at an acceptable combination.[36]

Human–human communication provides a useful metaphor for how drivers adapt the demands of the competing tasks.[37] Human communication is a collaborative process supported by back-channel communication.[38–40] Back-channel responses[38,41] refer to the hearer's use of peripheral utterances, such as "uh-huh" or "yeah," to provide feedback that the utterance is being understood and to coordinate turn taking.[42] Back-channel utterances represent a large proportion of conversations—19% by one estimate.[43] Although many speech theorists focus on back-channel communication as speech acts (e.g., "uh-huh" or "hm"), back-channel communication can also take the form of pauses, intonation, gestures, and facial expressions. Back-channel responses support grounding. Grounding is the development of a shared context that supports joint understanding and the timing of interactions. Without back-channel communication and the grounding that it supports, the goals of communication are unlikely to be met and direct communication will likely fail.

As with conversation, back-channel cues support drivers' understanding of the driving context, help coordinate the timing of interactions, and guide the adaptation of demands. Back-channel communication is already a critical component in driving. For example, drivers respond to the slippery feel of tires on an icy road to moderate their driving behavior—not just relying on the information provided by weather reports or even on focused observation of the roadway. For those who drive manual cars, it is not necessarily the position of the tachometer that lets the driver know when to shift. The sound and vibration of the motor are also essential, even though few people focus their attention on these cues. Drivers would lose a critical component of how they sense and perceive the driving environment if they did not have such back-channel cues. Although the ideas of back-channel communication were initially developed to describe communication between people, the concepts seem relevant to any situation that demands dynamic coordination between multiple entities.[44] For example, an IVIS might use the pauses between voice commands of the driver to identify situations where the driver might be engaged in a demanding driving situation.

Feedforward control is difficult because driving and some competing activities, such as IVIS demands, are unpredictable. Another challenge to effective feedforward control is that breakdowns in control at the operational level can lead to unexpected demands and poor management of the IVIS and driving demands at the tactical level. Speech recognition systems, particularly in the context of a noisy car, will likely induce errors. Such errors can lead to an unanticipated and increasing spiral of demand. Inexperience also undermines feedforward control in a way that can be particularly devastating.[45] Interaction with IVIS will likely exacerbate problems of feedforward control and the difficulty drivers have in anticipating and responding to upcoming demands. Concerning adaptive control, face-to-face conversations allow participants to accommodate some of the demands on the speaker by adjusting their engagement on the basis of back-channel cues that support common ground and efficient turn taking—passengers will suspend a conversation when drivers encounter high-demand situations. This also occurs at a much more limited level for drivers talking on a cell phone. Interactions with IVIS devices currently lack this ability to establish common ground, making adaptive control somewhat difficult, particularly if IVIS interactions are not ignorable or interruptible.

Considering the driver in terms of queuing theory leads to several important design considerations. Queuing theory suggests that minimizing distraction should focus on reducing the likelihood that tasks will simultaneously demand the driver's attention. This implies that, first, to the extent possible, IVIS demands placed on the driver should be timed to avoid conflict with upcoming traffic demands—for example, notifications of low-priority information concerning an upcoming restaurant could be delayed if the driver is approaching an intersection.[46] Workload managers work on this principle by prioritizing, delaying, or canceling IVIS interactions and system information if the concurrent demands of driving and the IVIS are too high (see Chapters 26 and 28 of this book). Second, IVIS devices should avoid interactions that are hard to ignore, interrupt, or resume. Finally, future systems could benefit by considering the benefits of back-channel communication to make the behavior of the IVIS more predictable to the driver and facilitate adaptive control. These design considerations and those associated with multiple resource theory at the level of operational control are not mutually exclusive. Multiple resource theory can define the degree to which a performance declines if tasks are performed simultaneously and queuing theory defines how likely tasks are to be performed simultaneously.

4.4 STRATEGIC CONTROL: DISTRACTION AS INAPPROPRIATE PRIORITY CALIBRATION

In the scenario of Sam searching for an MP3 playlist, distraction depends as much on the decision to use the device in a challenging driving environment as the specific timing of the interaction or the resource demand of the interaction. The decision to use a device in a challenging driving environment can depend on many factors:

- Drivers' awareness of the demands associated with using the device in that environment
- Drivers' appreciation for own ability to handle the demands associated with using the device in that situation
- Propensity of the individual to take risks
- Existence or absence of laws that permit its use in this context
- Productivity and other pressures
- Driving culture and societal norms

Had Sam delayed the interaction, initiating the search in a less demanding driving situation, the crash would not have occurred. Distraction is likely when drivers engage in demanding competing activities in situations in which the critical driving task is also likely to be demanding. Queuing theory describes the delays associated with a particular utilization rate and how these delays can be mitigated by supporting the driver in managing task timing. Control at the strategic level concerns how drivers select a target utilization rate and determines an acceptable demand profile of driving and competing activities.

The concepts underlying diffusion may provide a useful theoretical perspective to consider distraction at the strategic level. According to this perspective, drivers' choice of behaviors follows a random walk process that is influenced by productivity pressures and safety concerns. Drivers do not seek risky situations and rarely consciously balance risk and productivity. Instead drivers engage in different behaviors in a somewhat random fashion, with productivity pressures pushing this random process toward behaviors that involve increasing attention to tasks that are not critical to safe driving. In this situation, productivity pressures can push behavior toward and over safety boundaries. Such a representation has been developed to describe how safety practices erode in other complex sociotechnical systems.[47]

In controlling the distribution of attention at the strategic level, a fundamental challenge is that in driving, the safety boundaries are not very salient and the production pressures will gradually influence driver behavior to migrate into increasingly unsafe situations. The drift toward safety boundaries reflects a breakdown in feedback control. One reason for such behavior is that driving provides poor feedback, particularly concerning the inappropriate engagement in competing tasks. Because driving is often forgiving, drivers can neglect the driving task to a dangerous degree and suffer no immediate consequences. Even when drivers receive feedback in the form of a crash, it seldom results in a lasting change in behavior.[48] This poor feedback also causes drivers to overestimate their driving ability. In one study, half the drivers judged themselves to be among the safest 20%, and 88% believed themselves to be safer than the median driver.[49] Providing better feedback might lead drivers to adopt safer behavior.[50–52] Similarly, a well-designed device that reduces distraction at the operational level may actually undermine driving safety if it encourages drivers to use the device more frequently while driving. This *usability paradox* occurs when increased ease of use reduces the risk of any particular interaction, but increases overall risk by encouraging drivers to use the device more frequently. This tendency for drivers to adapt to improvements and undermine the expected safety benefit is a common phenomenon. For example, when roadway improvements are made (lanes widened, shoulders added, lighting improved), speeds often increase and undermine the potential safety benefit.[53]

Feedforward control at the strategic level suffers from the inherent variability of the driving environment. Even relatively demanding situations, such as driving on congested freeways, do not always pose a critical demand for the driver. Likewise, even situations that are typically not demanding can occasionally require drivers' full attention. The inherent variability of driving demands and the challenge of estimating the typical demand of a driving situation undermine the effectiveness of feedforward control, particularly for novice drivers who have not developed an appreciation for the demands associated with various driving environments and the demands of various competing tasks. Not only have novice drivers not developed an appreciation of the demands associated with driving and of competing activities, but they have not developed an appreciation of their own capacity to meet those demands.[54]

At the strategic level, adaptive control depends to some extent on a societal judgment of what constitutes acceptable risk and safe driving. The strategic decision to carry a cell phone into a car and generally intend to answer incoming calls depends on the driving culture and associated social norms concerning acceptable driving behavior. Such social norms may be the most powerful factors governing distraction, but may be the most difficult to quantify and shape. Subtle design modifications that reduce distraction at the operational level of behavior may have a much smaller effect on driving safety compared with changes in societal norms that influence the strategic level and make the use of a device while driving taboo. Societal response to traffic deaths illustrates this tendency. Recent high-profile catastrophes (e.g., the Oklahoma City bombing, shootings at Columbine High School, terrorist attacks on September 11, 2001, and Hurricane Katrina) caused less than 5,000 fatalities. In contrast, 42,636 lives were lost in 2004 alone as a result of vehicle crashes in the United States. Compared with motor vehicle fatalities, fatalities associated with disasters have had a much greater influence on American public policy and individuals' behavior. The American public seems to consider the loss of an average of 116 lives each day in crashes as an acceptable risk of transportation. Calibrating the public to the human toll of attending to competing tasks could lead to cultural changes that promote substantially safer driving.[55]

4.5 CONCLUSIONS

Defining distraction as a diversion of attention away from activities critical for safe driving toward a competing activity implies a breakdown in the control of attention. Considering distraction as a problem of control at three different timescales describes the causes of distraction differently than many other accounts of distraction. These differences lead to a range of design considerations that may help reduce distraction-related crashes. At the operational level, cognitive and motor constraints that govern driver performance tend to influence distraction. At the tactical level, attitudes and intentions that govern driver behavior tend to influence distraction. At the strategic level, societal norms and culture influence the roles that drivers adopt and these roles, in turn, govern the likelihood of distraction. These distinctions point toward fundamentally different and complementary contributions to distractions, which each require different design and policy strategies to mitigate. At each of these levels, a control-theoretic approach that considers the limits of feedback, feedforward, and adaptive control leads to novel design considerations. These considerations focus on improving feedback regarding the effects of neglecting the driving task, supporting more accurate expectations regarding the demands that compete for drivers' attention, and promoting more appropriate safety boundaries and performance criteria that guide control.

ACKNOWLEDGMENTS

Some of the ideas in this chapter emerged as part of the SAVE-IT program under contract by Delphi Corporation and sponsored by the U.S. DOT National Highway Traffic Safety Administration, Office of Vehicle Safety Research. The authors acknowledge the technical support provided by Mike Perel of NHTSA and Mary Stearns of the U.S. DOT Volpe Center. The chapter was also substantially influenced by conversations with Professor Thomas Triggs of Monash University Accident Research Center. We are also grateful for the contributions of Teresa Lopes of the University of Iowa Public Policy Center for assistance in preparing this manuscript.

REFERENCES

1. Rasmussen, J., Human error and the problem of causality in analysis of accidents, *Philosophical Transactions of the Royal Society Series B* 327 (1241), 449–462, 1990.
2. Sheridan, T.B., Big brother as driver: New demands and problems for the man at the wheel, *Human Factors* 12 (1), 95–101, 1970.
3. Michon, J.A., Explanatory pitfalls and rule-based driver models, *Accident Analysis and Prevention* 21 (4), 341–353, 1989.
4. Ranney, T.A., Models of driving behavior—a review of their evolution, *Accident Analysis and Prevention* 26 (6), 733–750, 1994.
5. Michon, J.A., A critical view of driver behavior models: What do we know, what should we do? In *Human Behavior and Traffic Safety*, Evans, L. and Schwing, R.C. (Eds.), Plenum Press, New York, 1985, pp. 485–520.
6. Lee, J.D. and Strayer, D.L., Preface to a special section on driver distraction, *Human Factors* 46, 583–586, 2004.
7. Sheridan, T.B., Driver distraction from a control theoretic perspective, *Human Factors* 46 (4), 587–599, 2004.
8. Underwood, G., Crundall, D., and Chapman, P., Selective searching while driving: The role of experience in hazard detection and general surveillance, *Ergonomics* 45 (1), 1–12, 2002.
9. McKenna, F.P., Horswill, M.S., and Alexander, J.L., Does anticipation training affect drivers' risk taking? *Journal of Experimental Psychology-Applied* 12 (1), 1–10, 2006.
10. Green, M., How long does it take to stop? Methodological analysis of driver perception-response times, *Transportation Human Factors* 2 (3), 195–216, 2000.
11. Bellman, R. and Kalaba, R., On adaptive control processes, *IRE Transactions on Automatic Control* 4 (2), 1–9, 1959.
12. Leveson, N., A new accident model for engineering systems, *Safety Science* 42 (4), 237–270, 2004.
13. Wickens, C.D., Multiple resources and performance prediction, *Theoretical Issues in Ergonomics Science* 3 (2), 159–177, 2002.
14. Wickens, C.D., Processing resources and attention, In *Varieties of Attention*, Parasuraman, R. and Davies, R. (Eds.), Academic Press, New York, 1984, pp. 63–102.
15. Strayer, D.L. and Johnston, W.A., Driven to distraction: Dual-task studies of simulated driving and conversing on a cellular telephone, *Psychological Science* 12 (6), 462–466, 2001.
16. Pashler, H., Dual-task interference in simple tasks—data and theory, *Psychological Bulletin* 116 (2), 220–244, 1994.
17. Gladstones, W.H., Regan, M.A., and Lee, R.B., Division of attention: The single-channel hypothesis revisited, *Quarterly Journal of Experimental Psychology* 41A, 1–17, 1989.

18. Lamble, D., Laakso, M., and Summala, H., Detection thresholds in car following situations and peripheral vision: Implications for positioning of visually demanding in-car displays, *Ergonomics* 42 (6), 807–815, 1999.
19. Senders, J.W., Kristofferson, A.B., Levison, W.H., Dietrich, C.W., and Ward, J.L., The attentional demand of automobile driving, *Highway Research Record* 195, 15–33, 1967.
20. Horrey, W.J. and Wickens, C.D., Examining the impact of cell phone conversations on driving using meta-analytic techniques, *Human Factors* 48 (1), 196–205, 2006.
21. Alm, H. and Nilsson, L., Changes in driver behaviour as a function of handsfree mobile phones: A simulator study, *Accident Analysis and Prevention* 26 (4), 441–451, 1994.
22. Alm, H. and Nilsson, L., The effects of a mobile telephone task on driver behavior in a car following situation, *Accident Analysis and Prevention* 27 (5), 707–715, 1995.
23. Brown, I.D., Tickner, A.H., and Simmonds, D.C.V., Interference between concurrent tasks of driving and telephoning, *Journal of Applied Psychology* 53 (5), 419–424, 1969.
24. Angell, L.S., Auflick, J., Austria, P.A., Kochar, D., Tijerina, L., Biever, W.J., Diptiman, T., Hogsett, J., and Kiger, S., *Driver Workload Metrics: Task 2 Final Report*, National Highway Traffic Safety Administration, Washington, D.C., 2006.
25. Hurwitz, J.B. and Wheately, D.J., Using driver performance measures to estimate workload, In *Proceedings of the Human Factors and Ergonomics Society 46th Annual Meeting*, Human Factors and Ergonomics Society, Santa Monica, CA, 2002.
26. Serafin, C., Wen, C., Paelke, G., and Green, P., Car phone usability: A human factors laboratory test, In *Proceedings of the Human Factors and Ergonomics Society 37th Annual Meeting*, Human Factors and Ergonomics Society, Santa Monica, CA, 1993, pp. 220–224.
27. Smith, P.J., McCoy, E., and Layton, C., Brittleness in the design of cooperative problem-solving systems: The effects on user performance, *IEEE Transactions on Systems Man and Cybernetics Part A-Systems and Humans* 27 (3), 360–371, 1997.
28. Liu, Y., Feyen, R., and Tsimhoni, O., Queueing network-model human processor (QN-MHP): A computational architecture for multitask performance in human-machine systems, *ACM Transactions on Computer-Human Interaction* 13 (1), 37–70, 2006.
29. Carbonell, J.R., A queueing model of many-instrument visual sampling, *IEEE Transactions on Human Factors in Electronics* HFE-7 (4), 157–164, 1966.
30. Rouse, W.B., Human-computer interaction in multitask situations, *IEEE Transactions on Systems, Man, and Cybernetics* SMC-7 (5), 384–392, 1977.
31. Horrey, W.J., Wickens, C.D., and Consalus, K.P., Modeling drivers' visual attention allocation while interacting with in-vehicle technologies, *Journal of Experimental Psychology-Applied* 12 (2), 67–78, 2006.
32. Nowakowski, C., Friedman, D., and Green, P.A., An experimental evaluation of using automotive HUDs to reduce driver distraction while answering cell phones, In *Proceedings of the Human Factors and Ergonomics Society 46th Annual Meeting*, Human Factors and Ergonomics Society, Santa Monica, CA, 2002.
33. Fisher, D.L., Laurie, N.E., Glaser, R., Connerney, K., Pollatsek, A., Duffy, S.A., and Brock, J., Use of a fixed-base driving simulator to evaluate the effects of experience and PC-based risk awareness training on drivers' decisions, *Human Factors* 44 (2), 287–302, 2002.
34. Monk, C.A., Boehm-Davis, D.A., and Trafton, J.G., Recovering from interruptions: Implications for driver distraction research, *Human Factors* 46 (4), 650–663, 2004.
35. Altmann, E.M. and Trafton, J.G., Memory for goals: An activation-based model, *Cognitive Science: A Multidisciplinary Journal* 26 (1), 39–83, 2002.
36. Esbjornsson, M., Juhlin, O., and Weilenmann, A., Drivers using mobile phones in traffic: An ethnographic study of interactional adaptation, *International Journal of Human-Computer Interaction* 22 (1–2), 37–58, 2007.

37. Wiese, E.E. and Lee, J.D., Attention grounding: A new approach to IVIS implementation, *Theoretical Issues in Ergonomics Science* 8 (3), 255–276, 2007.
38. Clark, H.H. and Wilkes-Gibbs, D., Referring as a collaborative process, In *Intentions in Communication*, Cohen, P.R., Morgan, J., and Pollack, M.E. (Eds.), MIT Press, Cambridge, MA, 1990, pp. 463–493.
39. Cohen, P.R. and Levesque, H.J., Preliminaries to a collaborative model of dialogue, *Speech Communication* 15, 265–274, 1994.
40. Goodwin, C., Between and within: Alternative sequential treatments of continuers and assessments, *Human Studies* 9, 205–217, 1986.
41. Schegloff, E.A., Discourse as an interactional achievement: Some uses of 'Uh Huh' and other things that come between sentences, In *Georgetown University Roundtable on Languages and Linguistics*, Tannen, D., Georgetown University Prcss, Washington, D.C., 1982.
42. Clark, H. and Brennan, S., Grounding in communication, In *Socially Shared Cognition*, Resnick, L., Levine, J., and Teasley, S. (Eds.), American Psychological Association, Washington, D.C., 1991, pp. 127–149.
43. Jurafsky, D., Bates, R., Coccaro, N., Martin, R., Meteer, M., Ries, K., Shriberg, E., Stolcke, A., Taylor, P., and Ess-Dykema, C.V., Automatic detection of discourse structure for speech recognition and understanding, In *Proceedings IEEE Workshop on Speech Recognition and Understanding*, Santa Barbara, CA, 1997, pp. 88–95.
44. Brennan, S.E., The grounding problem in conversations with and through computers, In *Social and Cognitive Approaches to Interpersonal Communication*, Kruez, R.J. (Ed.), Lawrence Erlbaum Associates, Mahwah, New Jersey, 1998, pp. 201–225.
45. Fisher, D.L., Pollatsek, A.P., and Pradhan, A., Can novice drivers be trained to scan for information that will reduce their likelihood of a crash? *Injury Prevention* 12, 25–29, 2006.
46. Sohn, H., Lee, J.D., Bricker, D.L., and Hoffman, J.D., A dynamic programming model for scheduling in-vehicle message display, *IEEE Transactions on Intelligent Transportation Systems*, 9, 226–234, 2008.
47. Rasmussen, J., Risk management in a dynamic society: A modelling problem, *Safety Science* 27 (2–3), 183–213, 1997.
48. Rajalin, S. and Summala, H., What surviving drivers learn from a fatal road accident, *Accident Analysis and Prevention* 29 (3), 277–283, 1997.
49. Svenson, O., Are we all less risky and more skillful than our fellow drivers? *Acta Psychologica* 47, 143–148, 1981.
50. Donmez, B., Boyle, L.N., and Lee, J.D., The impact of driver distraction mitigation strategies on driving performance, *Human Factors* 48 (4), 785–804, 2006.
51. Donmez, B., Boyle, L.N., and Lee, J.D., Drivers' attitudes toward imperfect distraction mitigation strategies, *Transportation Research Part F: Traffic Psychology* 9 (6), 387–398, 2006.
52. Donmez, B., Boyle, L.N., Lee, J.D., and Scott, G., Safety implications of providing real-time feedback to distracted drivers, *Accident Analysis and Prevention* 39 (3), 581–590, 2007.
53. Evans, L., *Traffic Safety and the Driver*, Van Nostrand Reinhold, New York, 1991.
54. Mitsopoulos, E., Triggs, T., and Regan, M., Examining novice driver calibration through novel use of a driving simulator, In *Proceedings of the SimTect 2006 Conference*, Melbourne, Australia, 2006.
55. Moeckli, J. and Lee, J.D., The making of driving cultures, In *Improving Traffic Safety Culture in the US: The Journey Forward*, AAA Foundation for Traffic Safety, Washington, 2006.

5 Models of Attention, Distraction, and Highway Hazard Avoidance

Christopher D. Wickens and William J. Horrey

CONTENTS

5.1 OVERVIEW

Driver distraction is characterized by the diversion of attention away from the driving task in favor of a secondary activity. Cell phone conversations or interactions with in-vehicle information systems are prime examples of distracting activities. Driver distraction is clearly a major highway safety problem. For example, an estimated 10–50% of motor vehicle crashes involve some form of driver distraction or inattention.[1–4] A recent naturalistic on-road study suggests that more than 78% of crashes and near-crashes are attributable to driver inattention.[5] Furthermore, the marketing push for greater e-commerce and entertainment in automobiles, which may be particularly appealing for drivers who are often stuck in long traffic jams, threatens to impose even more distractions on drivers.[6]

The safety issue of driver distraction can be aligned fairly closely with the failure to respond to unexpected hazards in the roadway (e.g., a lead vehicle braking sharply for no apparent reason; pedestrians, animals, vehicles, or other obstacles that move suddenly into the driver's path). Thus, this chapter will focus on the successes and failures of noticing and attending to critical discrete hazards and events, in the multitask scenario, that are characteristic of much of driving.[7] It may be argued that the major component of the failure to respond to these events is the initial failure to notice them in a timely fashion (i.e., an attentional failure). That is, drivers may not notice the events at all or may not notice them until too late, at which time a safe braking or steering maneuver would be impossible or ineffective.

In Section 5.2, we describe two psychological phenomena (inattentional blindness and change blindness) and one selective attentional model that, collectively, underlie the success and failure of noticing events in the traffic environment. We then consider how this attentional model, along with models of multiple resources, belongs to a class of models that can underlie proposed solutions for the distracted driver.

5.2 THE ATTENTIONAL PHENOMENA OF "BLINDNESS"

The traffic environment is highly complex, with many dynamic and changing elements. The correct detection, identification, and assessment of these elements are crucial for safe driving. Unfortunately, human observers are limited in their ability to process information, which often leaves them susceptible to missed objects, events, or features. Two attentional phenomena underscore these limitations: *inattentional blindness* and *change blindness*. Both involve a failure to detect an object or event (change), or notice features of an object that should be quite noticeable under other circumstances. Inattentional blindness (IB) describes a failure to see unattended objects.[8,9] IB can be applied most directly to the "looked but did not see" phenomenon, since this can clearly be associated with a failure of attention. For example, a driver may fail to notice an oncoming vehicle while trying to negotiate a left-hand turn. Change blindness (CB) describes a failure to see unattended changes.[10–12] For example, drivers may fail to notice the sudden appearance of a pedestrian from behind a parked vehicle because their attention was diverted away momentarily.

A representation for understanding these phenomena in the context of visual scanning is outlined in Table 5.1. Here the population of events, features, or changes that go unnoticed by an observer can be divided into two mutually exclusive classes, those that are in foveal vision and those that are not. Furthermore, the latter class can be arrayed along a continuum defining the eccentricity from foveal vision at the time the change (event) occurred, or defining the location of the object whose attribute (or existence) is to be later recalled. In general, we consider foveal vision to be a proxy for attention in this applied setting.[13]

As shown in Table 5.1, failures of attention can result in missed objects or events in the traffic environment. Cell 1 characterizes a failure of divided attention wherein an observer is directly fixating the relevant object yet fails to recognize it because attention is being diverted to a concurrent distracting activity. Strayer et al.[15] have documented an increase in IB to previously fixated roadside targets for drivers engaged in a cell phone conversation. In Cell 2, drivers are less likely to miss the target object because they are paying attention to the driving task, looking at the object of interest, and not currently dividing attention between two activities. (Recall that these two cells do not encompass situations where the viewing of a changing element is disrupted by an eye blink or the like. These cases would fall into the corresponding Cells 3 and 4.)

The right-hand column in Table 5.1 deals with situations in which the object or event occurs in the peripheral visual field or when visual information is temporarily

TABLE 5.1
Possible Outcomes in the Detection of Objects and Changes, Based on Location Relative to the Direction of Eye Gaze

	Location of Object or Event			
		In Foveal Vision		**Outside Foveal Vision**[a]
Distraction present	1	Failure of divided attention—inattentional blindness (looked but did not see)	3	Failure of focused attention (inappropriate scanning)—change or inattentional blindness
No distraction present (on task)	2	Correct detection	4	Failure of optimal scanning[14]—change or inattentional blindness[b]

Note: Distraction refers to instances where attention is redirected toward another task (visual or nonvisual) or where drivers are subject to internal distractions (e.g., daydreaming). Such distractions can involve enduring engagement in the task (cell phone conversation) or the momentary switch of attention to another event (the momentary distraction of the cell phone ring).

[a] A range of peripheral eccentricities as well as instances where no peripheral information is available (e.g., during blinks).

[b] Such failures could occur when the gaze is redirected to another location but serving the task of hazard monitoring.

unavailable (as in the case of saccades, blinks, or fixations so far away from the changing object [>40–50°] that the latter is essentially invisible). When drivers are engaged in a distracting task (Cell 3), they may be diverting their visual attention away from driving-relevant information and toward areas that support the distracting activity. For example, they may be looking at their cell phone while keying in a phone number. Thus, a failure in focused (or selective) attention can result in change blindness (for events) or inattentional blindness (for objects).

Unfortunately, observers who are not subject to distracting tasks may also fail to notice important features of the traffic environment. That is, even when scanning different parts of the visual scene appropriately, there is a risk that important features are missed in unattended areas (Cell 4).[14] For example, this cell describes the driver who is actively scanning the busy roadway for hazard information, but whose eyes are temporarily removed from the location of a pedestrian in the roadway, because the driver is looking at a car in the opposite lane that is poised to make a left turn across the path. Although these failures can (and do) occur in nondistraction situations, we emphasize the fact that drivers engaged in a distracting task (Cell 3) are slower to detect or are more prone to miss elements of the traffic environment than drivers who are not distracted.[16,17] Thus, distracting tasks can exacerbate the risk of missing relevant road hazards.

Within this representation, it is clear that models of visual scanning most directly apply to the Cells 3 and 4. As demonstrated, the absence of focused attention at the appropriate location can lead to inattentional blindness (to objects and features) and change blindness (to changing elements). In the naturalistic roadway environment, relevant traffic information spans a much larger visual area than the 4° of foveal vision, suggesting that at any given moment, a great majority (over 95%) of the visual field or traffic environment is susceptible to such "blindness."

When applied to driving, three features of the research on these two cognitive failures are important. First, studies of object properties that are correctly or erroneously recognized (as for IB) are only partially relevant because crashes result from a failure to notice, not a failure of more enduring memory (tested by recognition), the paradigm used by Strayer et al.[15] Second, many studies of change blindness employ the "flicker paradigm" to ensure that a changed item is not visible at the time it is changed.[16,18,19] Such studies lack the full ecological validity relative to naturalistic circumstances when the eye remains open, but is fixated elsewhere in the visual field. Furthermore, the task of explicit change detection (where changes are anticipated) may not fully generalize to one that examines the actual collision avoidance maneuver in response to introduction of an environmental hazard, as an implicit measure from which the failure to notice can be inferred (although see Ref. 19). Third, some studies that have examined more directly hazard collisions under distraction[20,21] have not measured scanning, so the distinction between the two columns of the table regarding naturalistic hazard detection cannot be assessed. Horrey et al.[22] have, however, intersected these two, as will be discussed later. Although we note the lack of full relevance of any of these categories of studies, the partial relevance of all of them converges to highlight the importance of these phenomena in crash prevention.

In considering the robust phenomena of inattentional and change blindness, it is worth reiterating a key modifier: *unexpected events*. In general, the prevalence of these attentional failures is reduced when the object in question is anticipated by the observer.[8,23,24] Thus, unexpected events are the most problematic. In the traffic context, these may be somewhat harder to define quantitatively because these events can take on many different forms. Intuition informs us that the likelihood of a "hard brake" by the car in front of us on a fast-moving freeway is small, as is the sudden emergence of a car at a blind intersection or one through which unobstructed passage is invited by a green light. Moreover, pedestrians usually have enough good sense not to run out into traffic from behind a parked vehicle, so these events too are rare and hence unexpected.

According to the arguments given earlier, considerable gains in safety can be made by minimizing failures of focused attention in Cells 3 and 4. Section 5.3 describes a model of scanning (SEEV) that predicts the amount and distribution of time that vision is away from the forward roadway (or any other visual area). Such a model predicts the vulnerability to change or inattentional blindness. Of course, such a scan model cannot account for the remaining attention failures that occur within Cell 1 (IB and "looked but did not see") and so we describe a second model, based on multiple-resource theory (MRT), that can help predict the degree of success of detection in Cell 1, where attentional failures may be less strongly related to visual scanning behavior.

5.3 THE SEEV MODEL

The SEEV model predicts how observers will allocate their visual attention to different areas of interest in various operational environments, such as the vehicle.

According to this model of selective visual attention, the purpose of visual scanning is to bring task-critical information into foveal vision. Scanning is guided by the influence of four factors: Salience & Effort (both bottom-up influences on attention) and Expectancy & Value (both top-down influences).

Salience, the most intuitive of the four components, reflects the fact that attention is normally captured by salient events in the visual field (or more broadly, in the sensory array).[25] Flashing lights are salient, as are objects that stand out against their background (i.e., high relative contrast), such as a brightly clad pedestrian standing against a dark backdrop or the onset of the brake lights of a leading vehicle. It is important to note that the salience of visual objects or events varies as a function of their location on the retina. That is, events that are salient in or near the fovea may not be so when located in the periphery, and thus sudden increases in illumination that would normally capture attention near the fovea may be subject to change blindness when positioned at greater eccentricity.[26,27] There are many psychophysical studies that have documented detection thresholds for different types of stimuli at different degrees of eccentricity.[28,29]

Effort is an inhibitory component that discourages observers from scanning between two locations that are far apart. In general, as the physical distance between two information sources or the time needed to access the information increases, the likelihood of scanning decreases.[30] This includes longer scans, such as moving attention between the roadway and the radio console, as well as those supported by head movements, such as checking the rearview mirrors or a blind spot before changing lanes. Furthermore, studies have shown that the breadth of visual scanning tends to decrease under conditions of mental workload,[31] suggesting that information access effort is a limited resource and subject to competition from concurrent tasks.[32]

Expectancy characterizes the tendency for observers to look at or sample sources of higher information bandwidth (event rate) more frequently. That is, observers tend to look where they expect to find a lot of task-relevant information. Expectancy is an optimal driver of visual scanning, provided the information sources are relevant to the task at hand.[33] In the driving context, drivers often sample the roadway because they expect to see more information relevant to safe driving. However, visual in-vehicle tasks compete for this visual attention—the degree of which will vary as a function of the amount and frequency of information that is conveyed there, even if it is not relevant to the driving task itself.

Value accounts for the fact that observers tend to sample sources of information that are of more value to a task (or more costly if we fail to look).[30] Hence, the higher cost of missing events in the forward roadway (e.g., braking vehicles) is more of a determinant of visual scanning (higher value) than is the cost of missing an event on the instrument panel or the value of processing an in-vehicle task (lower value).[22]

The collective influences of these four factors, each of which can be quantified and defined independently, can be expressed in an additive model, shown in the following equation:

$$P(A) = s(S) - ef(EF) + ex(EX) + v(V) \tag{5.1}$$

where $P(A)$ represents the probability of attending to a given location or area of interest (AOI). The capitalized terms, corresponding to salience and expectancy (S, EX), represent properties of the particular AOI within the visual field. EF defines the distance between any two AOIs, and V defines the value or importance of the to-be-performed task served by that AOI. The lowercase terms (s, ef, ex, v) represent the relative strengths of the different components in guiding attention. The two top-down, knowledge-driven factors (expectancy and value) can be thought of as influences on how to *optimally* allocate attention, according to an expected value model,[34] and hence can be said to represent components of the operator's *mental model*. In contrast, the two bottom-up factors (salience and effort) can be thought of as "nuisance influences" that should, optimally, have no influence on scanning *unless they are directly correlated with expectancy and value*, an issue discussed in what follows.

The SEEV model has been validated several times in predicting the allocation of visual attention (eye movements) in the context of both aviation[34,35] and driving.[22] In the latter context, Horrey et al.[22] measured visual scanning behavior, lane-keeping ability, and responses to unexpected hazards as drivers drove a high-fidelity simulator along traffic-filled roads, with different levels of lateral turbulence. Drivers performed a concurrent visual (distracting) task presented on a dashboard-mounted display. They found a high correlation ($r = .92$ and $.97$ in two experiments) between SEEV model predictions and actual scanning of the different areas of interest as driving conditions were varied.

It is important to consider, in a bit more detail, the meaning and implications for driving safety of the two top-down, knowledge-driven effects of expectancy and value on driver safety. First, with regard to expectancy, we argue that the *well-calibrated* driver will have expectancy of hazard appearance directly corresponding to the bandwidth or event rate of hazards (see also Refs 30, 33, and 36). This will, of course, be context dependent. The bandwidth of relevant traffic information will be low when the driver is driving slowly on an empty freeway, but high when he or she is driving fast along a crowded roadway through a shopping district. Models exist to objectively assess the bandwidth or "hazard exposure" rate of different driving conditions.[37,38]

Second, with regard to value, it is important to note that while the value of safe driving (e.g., noticing and avoiding hazards) exceeds that of performing in-vehicle tasks, there may be considerable modulation in the latter, particularly as other roles and responsibilities of the driver are considered. For example, returning a cell phone call concerning a family emergency may be assumed to be highly valuable. Third, our intention in creating an additive rather than multiplicative relationship between expectancy and value (in spite of the fact that the multiplicative one is often chosen in decision theory) is to signal the independent contributions of each. Thus, a source of information should be looked at, possibly frequently, even if no events are expected there, if the cost of *not* noticing such an event, should it ever occur, is quite high. For example, the driver on an empty straight freeway should look outside more than the extremely low hazard rate would dictate,[14] and not allow the low event expectancy to trigger long glances to in-vehicle information.

Finally, it is important to ask how the relative weights of the four components, s, ef, ex, and v, are determined, and this is indeed a considerable challenge, given that the units of measurement of the four measurable components (*S*, *EF*, *EX*, and *V*) are all quite distinct from each other. Space prevents a detailed discussion of this issue here (see Ref. 34). However, two approaches can be taken: (1) it is possible to scale all variables on an ordinal scale (e.g., lowest to highest bandwidth of three channels, 1, 2, 3; least to most valuable, 1, 2, 3; closest [least effortful] to most separated [most effortful], 1, 2, 3), and this approach works well in predicting scanning; and (2) the model can be of use as a guide for optimizing display layout, as discussed in the following.

SEEV and Hazard Vulnerability. The SEEV model can be run in a dynamic Monte Carlo simulation, which will allow it to create a series of scan trajectories to various areas of interest within the driving environment. Importantly, these trajectories can reveal a critical statistic in hazard research called the "mean first passage time,"[36] the period in which the eyes are not looking forward and, hence, the driver is vulnerable to roadway hazards. Because it is a stochastic simulation model, SEEV can produce a distribution of such times that, in turn, can reveal statistics such as the longest 5% of in-vehicle glances and estimates of how long these glances will be. Furthermore, characteristics of the driving environment and the in-vehicle task can be altered in the model to assess their impact on hazard vulnerability. In this regard, data from elsewhere appear to suggest a discontinuous increase in hazard vulnerability (as revealed by crash and near-crash rates) when this time is above around 1.5 s.[5]

Focal versus Ambient Vision. Although the SEEV model can be used to characterize the likelihood of scanning task-relevant information (and hence success in hazard detection), it appears to have relatively little power in predicting lane-keeping ability.[22] The reason for this lies in the distinction between the two visual systems: *focal*, served primarily by the fovea and employed for object recognition; and *ambient*, served by processors all across the retina, and therefore well served by the periphery.[39] Lane keeping is well supported by ambient processing of the visual flow field, without necessarily requiring focal vision (predicted by the SEEV model). However, the existence of this dissociation still allows us to concentrate on SEEV as the key to understanding the most important aspect of distraction—failing to detect and recognize those discrete roadway hazards.

There is, however, an important aspect of the focal-ambient distinction that warrants consideration in this context. That is, in many circumstances, particularly night driving, ambient vision provides a continuous feedback signal that "I am driving safely." As Leibowitz et al.[40] have pointed out, such feedback can lead to a sense of overconfidence and, consequently, result in a faster driving speed than is appropriate, given the reduced ability of drivers to detect hazards using focal vision in these conditions. Similarly, we may consider that we are driving safely because lane keeping can be preserved using the upper visual field, even as focal vision is diverted for in-vehicle tasks, thereby providing unwarranted confidence that unexpected hazards can also be noticed in this manner.[22]

In this sense, we can consider the influence of ambient vision on SEEV as an indirect one, mediated by the sense of (overconfidence in) driving proficiency when focal vision is away from hazard areas, or when the driving is too fast to allow focal vision to accomplish timely hazard recognition. Thus, the driver's assessment of hazard vulnerability is not well calibrated with actual vulnerability, a phenomenon well documented more broadly in metacognitive research.[41]

Summary. Safe driving requires the correct detection, identification, and assessment of a wide range of dynamic traffic elements. In summarizing this section, we note that both inattentional and change blindness describe vulnerability in noticing such changes in hazard expression. SEEV predicts when such changes may be out of foveal vision (and therefore more likely to be completed changes and less detectable). Finally, when changes do occur within the fovea, task distraction (multitasking with inappropriate prioritization) may still compromise change detection, as this is now represented, in Cell 1 of Table 5.1, by inattentional blindness.

5.4 THEORY-BASED SOLUTIONS TO THE PROBLEM OF HAZARD UNAWARENESS

The SEEV model and the phenomena of "blindness" (inattentional and change) indicate serious vulnerabilities in monitoring for hazards in the driving context. In response, we can propose several theory-based remedies, articulated within the broad category of design.

5.4.1 Design Solutions Imposed by the SEEV Model

Here we discuss two types of solutions, addressing issues of visual scanning based on the SEEV model:

Designers should configure displays so that information access effort is negatively correlated with expectancy. Therefore, information sources that are accessed relatively frequently (high expectancy) are located proximally (reduced effort). Such a solution dictates a head-up display (HUD) location for frequently used sources of in-vehicle information[42,43]—a solution that has indeed been found to improve unexpected hazard response.[20] However, HUDs can inadvertently produce clutter, which can make it difficult to extract relevant information. As such, capitalizing on the nonlinear relation between effort and visual separation, it is possible to present HUD information just below the forward view, so that it neither requires much effort to access nor obscures roadway hazards or other relevant driving information.[20,44,45] For example, several studies suggest that the optimal placement of such a display would be at around 5° head down, great enough to avoid the clutter of overlapping images, but small enough to allow adequate focal vision on the highway, while display information is fixated.[20]

Designers should make salience positively correlated with value. In other words, more important (valuable, safety related) information will also be highly salient or attention grabbing. Such a solution translates to development of visual cues or auditory alarms. Imposing this solution sometimes involves some degree of automation, in which an intelligent vehicle "decides" what is important to the driver (e.g., a suddenly decelerating car ahead). Such automation will be bound to lead to mistakes in its inference.[46,47] Hence, it is appropriate to consider how low this reliability can be, to still be useful in collision warning or hazard awareness, as well as whether any imperfect automation should be biased to produce more false alerts or more late alerts (or misses). These issues are beyond the scope of the current chapter; however, a recent review of the literature suggests that this kind of automation should have a reliability of at least 0.80 (i.e., a false alert rate of no more than 0.20) to be useful.[47]

5.4.2 Design Solutions Imposed by the Multiple-Resource Model

Thus far, our discussion has focused on the allocation and limits of selective attention. The multiple-resource model is one that addresses the limits of *divided attention* as people (here drivers) engage in multitasking, and, in particular, is concerned with the magnitude of interference from nonvisual tasks within Cell 1 of Table 5.1. Details of this model, as applied to driving, can be found elsewhere.[48–51] We provide a quick overview here.

According to the model, the degree of time-sharing success between any two tasks can be predicted by the joint difficulty of the two tasks (demand level) and the degree to which they overlap in the demand for common resources. The resources vary along multiple dimensions: processing stage (perceptual-cognitive versus action), processing code (verbal versus spatial), perceptual modality (auditory versus visual), and visual channel (focal versus ambient, as described earlier). Since driving (both hazard monitoring and lane keeping) is primarily a visual-spatial-motor task, it is predicted (and observed) to be fairly efficiently time shared with tasks that are auditory and language based (both in perception, i.e., hearing, and in action, i.e., speaking). Furthermore, because ambient and focal vision use separate resources, lane keeping and hazard monitoring can be well time shared, as long as the latter has foveal vision available. The multiple-resource model allows prediction of the distinct advantage to be gained from off-loading visual tasks to auditory delivery,[20] and voice response,[52,53] as well as off-loading route guidance information from spatial maps to voice-based commands.[54,55] A computational version of the multiple-resource model[48] has been used to successfully predict *how much better* interfaces using separate resources will be compared with those that demand common resources.

It is important to note that whereas the demand level and resource overlap predict how much mutual interference there will be between driving and a distracting task, there is a third component of the multiple-resource model—resource allocation policy—that determines the extent to which the time-shared components will suffer. Thus, it is this allocation policy that is responsible for occurrences in which more resources are given to an "engaging" cell phone conversation than to roadway hazard monitoring. Hence, the perceptual-cognitive resources required by both activities may be misallocated—to the detriment of the hazard-monitoring task. The extent to which different factors control this allocation policy—engagement, priority, difficulty—remains one of the greatest challenges to the study of attention management.

5.5 IMPLICATIONS FOR METHODOLOGY

The SEEV model is a probabilistic, stochastic model of information acquisition. As such, it can characterize not only the average scan profiles but also the extremes. This point brings to relevance a critical issue regarding the research methodology that is often used to examine safety issues in the laboratory, an issue we discuss briefly as follows.

While the probability of distraction (D) given an accident (Ac)—*P*(D|Ac)—is relatively high (10–60%), the converse probability of an accident given a distraction—*P*(Ac|D)—is quite low.[5] This low value reflects the fact that most drivers have presumably optimized their scan strategy between the traffic environment and inside the vehicle to minimize vulnerability, and have maintained sufficiently short downward dwells (usually <1.6 s; but not always[38]) so that hazards can be fixated on before it is too late. Yet *P*(Ac|D), while low, is not zero. This fact creates a challenge for researchers in their study of the "psychology of rare events."[56] Whereas most human factors research and analysis is dominated by the statistics of the mean (*t* tests, ANOVAs), what really counts in crash investigation is the statistics of the tail ends of the distribution. It is the slowest-responding driver or that driver with the longest head-down glances who may contribute to the rare crash. These low N (sample size) circumstances do not lend themselves well to traditional statistics. For example, Horrey and Wickens[38] found that substantially different statistical conclusions emerged regarding the relationship between scanning and crashes in a simulator when the data at the tails of the distribution were analyzed, rather than those pertaining to the more traditional mean values.

5.6 CONCLUSIONS

In conclusion, using psychological theory and data to address issues in highway safety involves at least three steps, two of which are illustrated here.

First, it is necessary to document that a psychological phenomenon is a real safety concern outside the laboratory where it is typically studied. The many chapters of this book, and statistics cited, certainly do this for distraction in general, and we believe that we have done so for the two "blindness" phenomena (inattention and change) discussed in this chapter.

Second, it is necessary to demonstrate that manipulations, validated in the laboratory, will reduce undesirable phenomena. Here again, we believe that we have demonstrated the strong influences of SEEV parameters on scanning behavior that can assure that the roadway sources of hazard information can remain longer in or near foveal vision, so that the two blindness phenomena can be reduced, but not eliminated.[57] It is important to note, however, that solutions described here do not address the "looked but did not see" phenomenon, characteristic of much of IB research. Multiple-resource theory, characterizing the interference between perceptual and cognitive processes, provides a consistent accounting for this, but does not directly prescribe solutions, which are probably best embodied in training.

Third, it is necessary to demonstrate that valid manipulations, in the laboratory or simulator, can generalize to improving actual driving safety on the highway. This most challenging step has not, to our knowledge, taken place regarding attentional phenomena, but surely such research lies within the near future.

ACKNOWLEDGMENTS

We are grateful to John Lee, Michael Regan, Kristie Young, Dave Melton, and William Shaw for their helpful and insightful comments regarding earlier versions of this chapter.

REFERENCES

1. Wang, J.-S., Knipling, R.R., and Goodman, M.J., The role of driver inattention in crashes: New statistics from the 1995 Crashworthiness Data System, *40th Annual Proceedings of the Association for the Advancement of Automotive Medicine* (pp. 377–392), Vancouver, BC: AAAM, 1996.
2. National Highway Traffic Safety Administration, *An Investigation of the Safety Implications of Wireless Communications in Vehicles* (Report DOT HS 808–635), Washington, D.C.: NHSTA, 1997.
3. Ranney, T.A., Mazzae, E., Garrott, R., and Goodman, M.J., NHTSA driver distraction research: past, present, and future. Available online at http://www-nrd.nhtsa.dot.gov/departments/nrd-13/driver-distraction/Papers.htm (accessed July 10, 2006), 2000.
4. Weise, E.E. and Lee, J.D., Attentional grounding: A new approach to in-vehicle information systems implementation, *Theoretical Issues in Ergonomics Science* 8(3), 255–276, 2007.
5. Dingus, T.A., Klauer, S.G., Neale, V.L., Petersen, A., Lee, S.E., Sudweeks, J., *The 100-Car Naturalistic Driving Study, Phase II—Results of the 100-Car Field Experiment* (Report No. DOT HS 810 593), Washington, D.C.: National Highway Traffic Safety Administration, 2006.
6. Ashley, S., Driving the info highway, *Scientific American*, 285, 52–58, 2001.
7. Wickens, C.D., Lee, J., Liu, Y.D., and Gordon-Becker, S., *An Introduction to Human Factors Engineering* (2nd ed.), New York: Addison Wesley Longman, 2003.
8. Mack, A. and Rock, I., *Inattentional Blindness*, Cambridge, MA: MIT Press, 1998.
9. Carpenter, S., Sights unseen, *APA Monitor* 32, 54–57, 2002.
10. Rensink, R.A., Change detection, *Annual Review of Psychology* 53, 245–277, 2002.
11. Simons, D.J., Current approaches to change blindness, *Visual Cognition* 7, 1–15, 2000.
12. Simons, D.J. and Ambinder, M.A., Change blindness: Theory and consequences, *Current Directions in Psychological Science* 14, 44–48, 2005.
13. Wickens, C.D. and McCarley, J.S., *Applied Attention Theory*, Boca Raton, FL: CRC Press, 2008.
14. Moray, N., Monitoring, complacency, scepticism and eutectic behaviour, *International Journal of Industrial Ergonomics* 31(3), 175–178, 2003.
15. Strayer, D.L., Drews, F.A., and Johnston, W.A., Cell phone-induced failures of visual attention during simulated driving, *Journal of Experimental Psychology: Applied* 9, 23–32, 2003.
16. McCarley, J.S., Vais, M.J., Pringle, H., Kramer, A.F., Irwin, D.E., and Strayer, D.L., Conversation disrupts change detection in complex traffic scenes, *Human Factors* 46, 424–436, 2004.
17. Richard, C.M., Wright, R.D., Ee, C., Prime, S.L., Shimizu, Y., and Vavrik, J., Effect of a concurrent auditory task on visual search performance in a driving-related image-flicker task, *Human Factors* 44(1), 108–119, 2002.
18. Pringle, H.L., Irwin, D.E., Kramer, A.F., and Atchley, P., The role of attentional breadth in perceptual change detection change detection, *Psychonomic Bulletin & Review* 8(1), 89–95, 2001.
19. Caird, J.K., Edwards, C.J., Creaser, J.I., and Horrey, W.J., Older driver failures of attention at intersections: Using change blindness methods to assess turn decision accuracy, *Human Factors* 47(2), 235–249, 2005.
20. Horrey, W.J. and Wickens, C.D., Driving and side task performance: The effects of display clutter, separation, and modality, *Human Factors* 46(4), 611–624, 2004.
21. Strayer, D.L. and Drews, F.A., Multitasking in the automobile, In *Attention: from Theory to Practice*, Kramer, A., Wiegmann, D., and Kirlik, A. (Eds.), Oxford, UK: Oxford University Press, 2007.

22. Horrey, W.J., Wickens, C.D., and Consalus, K.P., Modeling drivers' visual attention allocation while interacting with in-vehicle technologies, *Journal of Experimental Psychology: Applied* 12(2), 67–78, 2006.
23. Fadden, S., Ververs, P.M., and Wickens, C.D., Costs and benefits of head-up display use—a meta-analytic approach, *Proceedings of the 42nd Annual Meeting of the Human Factors and Ergonomics Society* (pp. 16–20), Santa Monica, CA: HFES, 1998.
24. Rensink, R.A., When good observers go bad: Change blindness, inattentional blindness, and visual experience. *Psyche*, 6. Available online at http://psyche.cs.monash.edu.au/v6/psyche-6-09-rensink.html (accessed December 18, 2006), 2000.
25. Itti, L. and Koch, C., A saliency-based search mechanism for overt and covert shifts of visual attention, *Vision Research* 40(10–12), 1489–1506, 2000.
26. Nikolic, M.I., Orr, J.M., and Sarter, N.B., Why pilots miss the green box: How display context undermines attention capture, *International Journal of Aviation Psychology* 14(1), 39–52, 2004.
27. Sarter, N.B., Mumaw, R., and Wickens, C.D., Pilots' monitoring strategies and performance on highly automated glass cockpit aircraft, *Human Factors* 49, 564–74, 2007.
28. Anstis, S., Motion perception in the frontal plane: Sensory aspects. In *Handbook of Perception and Human Performance*, Vol. 1, Boff, K.R., Kaufman, L., and Thomas, J.P. (Eds.), New York: Wiley, 1986, pp. 16.1–16.27.
29. McKee, S.P. and Nakayama, K., The detection of motion in the peripheral visual field, *Vision Research* 24(1), 25–32,1984.
30. Sheridan, T., On how often the supervisor should sample, *IEEE Transactions on Systems Science and Cybernetics SSC*-6(2), 140–145, 1970.
31. Recarte, M.A. and Nunes, L.M., Effects of verbal and spatial-imagery tasks on eye fixations while driving, *Journal of Experimental Psychology: Applied* 6, 31–43, 2000.
32. Liu, Y. and Wickens, C.D., Visual scanning with or without spatial uncertainty and divided and selective attention, *Acta Psychologica* 79, 131–153, 1992.
33. Senders, J.W., The human operator as a monitor and controller of multidegree of freedom systems, *IEEE Transactions on Human Factors in Electronics* Hfe-5, 2–6, 1964.
34. Wickens, C.D., McCarley, J.S., Alexander, A., Thomas, L., Ambinder, M., and Zheng, S., Attention-situation awareness (A-SA) model of pilot error. *In Human Performance Models in Aviation*, Foyle D., and Hooey, B. (Eds.), Boca Raton, FL: CRC Press, 2008.
35. Wickens, C.D., Goh, J., Helleburg, J., Horrey, W.J., and Talleur, D.A., Attentional models of multi-task pilot performance using advanced display technology, *Human Factors* 45(3), 360–380, 2003.
36. Moray, N., Monitoring behavior and supervisory control, In *Handbook of Perception and Human Performance*, Vol. 2, Boff, K.R., Kaufman, L., and Thomas, J.P. (Eds.), New York: Wiley, 1986, pp. 40.1–40.51.
37. Horrey, W.J. and Wickens, C.D., Focal and ambient visual contributions and driver visual scanning in lane keeping and hazard detection, *Proceedings of the Human Factors and Ergonomics Society 48th Annual Meeting* (pp. 2325–2329), Santa Monica, CA: Human Factors and Ergonomics Society, 2004.
38. Horrey, W.J. and Wickens, C.D., In-vehicle glance duration: Distributions, tails and a model of crash risk, *Transportation Research Record* 2018, 22–28, 2007.
39. Previc, F.H., The neuropsychology of 3-D space, *Psychological Bulletin* 124(2), 123–164, 1998.
40. Leibowitz, H.W., Owens, D.A., and Post, R.B., *Nighttime Driving and Visual Degradation* (SAE Technical Paper Series #820414), Warrendale, PA: Society for Automotive Engineering, 1982.

41. Bjork, R.A., Assessing our own competence: Heuristics and illusions, In *Attention and Performance XVII. Cognitive Regulation of Performance: Interaction of Theory and Application*, Gopher, D. and Koriat, A. (Eds.), Cambridge, MA: MIT Press, 1999, pp. 435–459.
42. Kiefer, R.J., Defining the "HUD benefit time window," In *Vision in Vehicles-VI*, Gale, A.G., et al. (Eds.), Amsterdam: Elsevier, 1995, pp. 133–142.
43. Kiefer, R.J., Quantifying head-up display (HUD) pedestrian detection benefits for older drivers, *Proceedings of the 16th International Conference on the Enhanced Safety of Vehicles* (pp. 428–437), Washington, D.C.: National Highway Traffic Safety Administration, 1998.
44. Wickens, C.D., Dixon, S.R., and Seppelt, B., Auditory preemption versus multiple resources: Who wins in interruption management? *Proceedings of the 49th Annual Meeting of the Human Factors and Ergonomics Society*, Santa Monica, CA: HFES, 2005.
45. Gish, K.W. and Staplin, L., *Human factors aspects of using head-up displays in automobiles: A review of the literature*, Interim Rep. DOT HS 808 320, U.S. Department of Transportation, Federal Highway Administration, Washington, D.C., 1995.
46. Parasuraman, R., Sheridan, T.B., and Wickens, C.D., A model for types and levels of human interaction with automation, *IEEE Transactions on Systems, Man and Cybernetics, Part A* 30(3), 286–297, 2000.
47. Wickens, C.D. and Dixon, S.R., The benefits of imperfect diagnostic automation: A synthesis of the literature, *Theoretical Issues in Ergonomics Science* 8(3), 201–212, 2007.
48. Horrey, W.J. and Wickens, C.D., Multiple resource modeling of task interference in vehicle control, hazard awareness and in-vehicle task performance, *Proceedings of the Second International Driving Symposium on Human Factors in Driver Assessment, Training, and Vehicle Design* (pp. 7–12), Park City, Utah: University of Iowa, 2003.
49. Wickens, C.D., Multiple resources and performance prediction, *Theoretical Issues in Ergonomics Science* 3(2), 159–177, 2002.
50. Wickens, C.D., Multiple resource time sharing model, In *Handbook of Human Factors and Ergonomics Methods*, Stanton, N.A., Salas, E., Hendrick, H.W., Hedge, A., and Brookhuis, K. (Eds.), London, UK: Taylor & Francis, 2005, pp. 40-1/40-7.
51. Wickens, C.D., Attention to attention and its applications: A concluding view, In *Attention: From Theory to Practice*, Kramer, A.F., Wiegmann, D.A., and Kirlik, A. (Eds.), Oxford, UK: Oxford University Press, 2007.
52. Tsimhoni, O., Smith, D., and Green, P., Address entry while driving: Speech recognition versus a touch-screen keyboard, *Human Factors* 46(4), 600–610, 2004.
53. Ranney, T.A., Harbluk, J.L., and Noy, Y.I., Effects of voice technology on test track driving performance: Implications for driver distraction, *Human Factors* 47, 439–454, 2005.
54. Streeter, L.A., Vitello, D., and Wonsiewicz, S.A., How to tell people where to go: Comparing navigational aids, *International Journal of Man-Machine Studies* 22(5), 549–562, 1985.
55. Srinivasan, R. and Jovanis, P.R., Effect of in-vehicle route guidance systems on driver workload and choice of vehicle speed: Findings from a driving simulator experiment, In *Ergonomics and Safety of Intelligent Driver Interfaces*, Noy, Y.I. (Ed.), Mahwah, NJ: Lawrence Erlbaum, 1997, pp. 97–114.
56. Wickens, C.D., Attention to safety and the psychology of surprise, *Proceedings of the 2001 Symposium on Aviation Psychology*, Columbus, OH: The Ohio State University, 2001.
57. Wickens, C.D. and Alexander, A.L., Attentional tunneling and task management in synthetic vision displays, *International Journal of Aviation Psychology*, in press.

Part 3

Measurement of Driver Distraction

6 Measuring Exposure to Driver Distraction

Suzanne P. McEvoy and Mark R. Stevenson

CONTENTS

6.1 INTRODUCTION

Researchers undertake simulator and on-road studies to quantify the degree to which distraction impacts different aspects of driving performance by examining factors such as lane keeping, reaction time, speed variability, following distance, and situational awareness. However, such findings do not tell us whether this degree of degradation will actually compromise road safety. To this end, information on exposure is needed.

The frequency with which a driver distraction will contribute to a crash is a function of two factors: how often drivers engage in the activity while driving and the risk of crash conferred by the activity. Accordingly, the extent to which drivers are exposed to a distracting activity while driving is an important consideration in assessing whether or not that exposure results in adverse outcomes such as road crashes. However, determining drivers' level of exposure to distracting activities while driving is not straightforward, and research in the field is in its infancy.

There are four components that need to be considered when measuring exposure: two relate to time (person-time exposed), namely, the cumulative exposure while driving and the duration of the current encounter; the other two relate to the magnitude of the threat, namely, the hazard intensity during the current encounter and the driver's ability to mitigate the danger. Importantly, the magnitude of the threat will be influenced by extrinsic and intrinsic factors. Extrinsic factors may include the driving task, traffic density, speed and weather conditions, and intrinsic factors may include an individual's risk-taking propensity, driving experience, and the effects of alcohol or fatigue. Many of these factors may change between episodes of exposure.

There are several reasons why measuring exposure to distracting activities while driving is difficult. First, distracting activities while driving tend to be short and episodic. Second, the effect of practice on risk may not be known and may modify the link between the exposure and the outcome. Third, the effect of the exposure may change as a function of the exposure itself: for example, the intensity of mobile phone conversations while driving can vary. Fourth, decisions about when to engage in a distracting activity are often, though not always, volitional and will vary among drivers.

Epidemiologists use a number of methods to measure exposure and estimate risk under real-world driving conditions. In this chapter, we review the types of study designs that have been used to measure drivers' exposure to distracting activities, including the strengths and weaknesses of each of the designs.

6.2 MEASURING EXPOSURE TO DISTRACTIONS WHILE DRIVING

Researchers must incorporate appropriate measures of exposure when designing studies. To date, much of the data on exposure to distracting activities while driving have been limited to self-reported behavior, which may be subject to bias, particularly, recall bias. In some instances, it has been possible to validate self-report with other data, for example, participants' phone activity records, in examining phone use while driving. Researchers have also used roadside observations to collect data on mobile phone use and passenger carriage at a point in time. Most recently, video footage over longer periods of time has been used to monitor a range of distractions while driving in selected and small groups of drivers.

6.3 CROSS-SECTIONAL SURVEYS

Cross-sectional surveys, often referred to as prevalence surveys, can be used to ascertain information about exposures to driver distraction, as well as outcomes, and are elicited from individuals within a population at the same point or period in time. Cross-sectional surveys can also be used to obtain information on population attitudes and beliefs toward various distractions, which can be important in planning road safety initiatives. When the participants resemble the population from which they were sourced, such samples are said to be representative, and can provide very useful data to guide policy. Characteristically, these surveys are quick to perform, relatively inexpensive, and the analysis is generally straightforward using standard statistical methods.

There have been a number of road user cross-sectional surveys undertaken to quantify the prevalence of distractions, in particular mobile phone use, while driving. These surveys have been conducted in countries such as Australia [1,2], Canada [3], Finland [4], New Zealand [5], and the United States [6]. There have been fewer surveys that have collected specific data on other distracting activities that can occur while driving [1,6]. Generally, information has been collected using telephone interviews, handouts, or mailouts.

In addition to quantifying mobile phone use or other distracting activities while driving, cross-sectional surveys have documented which groups are more likely to engage in these activities while driving, drivers' attitudes about the risk of crash associated with common distracting activities, and the adverse consequences resulting from distracting activities, including the proportion of crashes in which driver distraction was cited as a contributing factor. In relation to phone use, information about the current pattern and frequency of phone use while driving and the extent to which drivers use handheld phones while driving has also been obtained.

In most surveys, over 50% of drivers have reported phone use while driving at least occasionally [2,4,5,7,8]. Nonetheless, the proportion of drivers who report phone use while driving does vary by jurisdiction and over time. In Finland, following the introduction of a ban on handheld mobile phone use while driving, use of handheld mobile phones while driving fell sharply ($p < .001$), then rose suggesting that the effect of the law on phone use while driving was difficult to sustain [8]. However, use of handheld mobile phones while driving is generally, though not always [7], lower in jurisdictions with restrictions than in those without, for example, Rajalin et al. [8] versus Sullman and Baas [5].

Another specific type of cross-sectional survey is the personal transportation survey. These have been conducted in many jurisdictions to quantify travel behavior. In the United States, such surveys have been conducted periodically. For these surveys, a stratified random sample of households is selected and every member of the household aged 14 years and over is interviewed about trips made on a preassigned day for which a travel diary has been provided. For example, in 1995, over 40,000 households were approached for the survey [9]. In 1988, a travel survey was also conducted in Ontario, Canada. Over 2000 drivers provided demographic details and kept trip logs for periods of 1 or 3 days, including information on travel distances and passenger carriage. The data from these surveys have been used as baseline exposure data in studies assessing the risks associated with passenger carriage (see Chapter 17) [10,11]. While both the surveys used a stratified random sampling strategy, the response rates* in the Canadian survey were low (as low as 13% in one stratum) [12].

The potential limitations of cross-sectional surveys can include a nonrandom participant selection process, a low response rate (as raised previously), and bias resulting from self-report. If the participant selection process is inadequately designed or the response rate is low, the sample may not be representative of the source population because the drivers who are not involved may differ from the participants in relation to the factor(s) of interest, for example, in the frequency and types of distracting activities that they engage in. Moreover, participants themselves may moderate their responses to provide socially desirable answers. Importantly, any resultant bias from these limitations would tend to underestimate the extent of the problem of driver distraction rather than overestimate it. Although cross-sectional surveys provide information on

* The response rate is the number of respondents divided by the number of people who were approached and eligible to participate in the study (respondents and refusals).

exposure, they cannot capture all the components of exposure discussed in the introduction. A driver's cumulative exposure to a distracting activity while driving, the role of effect modifiers,* and the driver's individual ability to mitigate the risk attributable to a distracting activity are not easy to quantify using this method. Finally, cross-sectional surveys *per se* cannot be used to estimate the risk or likelihood of having a crash associated with the occurrence of a distracting activity. Such studies are primarily descriptive and cannot determine with certainty whether the exposure preceded the outcome. Accordingly, cross-sectional studies can suggest hypotheses but other epidemiologic designs, including quasi-experimental studies, such as the case-control study, are needed to test them. Relevant quasi-experimental studies are discussed later in the chapter (see also Chapter 17).

6.4 ROADSIDE SURVEYS

Although there are a number of roadside surveys that have measured driver distraction, the focus of these studies has been primarily on handheld mobile phone use while driving [13–18]. These have commonly, though not always, been performed in the context of assessing the effects of legislative changes banning the use of handheld devices while driving. These surveys have generally involved daytime and weekday collection of data by observers on the prevalence of handheld mobile phone use while driving at selected road sites, most commonly intersections. Some studies have required that drivers be stopped to assess additional details, including sex, race, estimated age group, and vehicle type. The U.S. National Occupant Protection Use Survey also collects information on passenger carriage [14].

The prevalence of handheld mobile phone use while driving, as measured by roadside observational studies, varies by jurisdiction but is generally between 2 and 6%. Moreover, some studies have shown that handheld mobile phone use while driving has risen over time [13,14]. This coincides temporally with increasing phone ownership. The introduction of restrictions to handheld mobile phone use while driving is the exception. In these jurisdictions, handheld mobile phone use characteristically falls significantly after the introduction of the ban [15,19,20]. However, there is evidence that the falls are not maintained over the longer term [8,16].

Hands-free phone use while driving cannot be adequately quantified using roadside observations, with the likelihood that the observations could underestimate or overestimate exposure because it is often not possible to tell if a driver is using a hands-free device or not; alternative explanations could include a driver singing, talking to oneself, or speaking to a passenger, or on a Citizens' Band (CB) radio. Another drawback of this method is that the observations of handheld mobile phone use tend to occur during daylight hours, on weekdays, and under low-speed conditions for visibility or logistical reasons. The overall effect of this may be to overestimate exposure to handheld phone use while driving because the usage pattern may

* An effect modifier is a factor that modifies the effect of a causal factor under study. In other words, effect modification is said to be present when the association between the exposure and outcome of interest varies by the level of a third factor (the effect modifier). Last J.M. *A Dictionary of Epidemiology*. 3rd Edition. Oxford University Press, New York, 1995; Hennekens C.H., Buring J.E., and Mayrent S.L. (eds). *Epidemiology in Medicine*. Little, Brown and Company, Boston, MA, 1987.

vary by time of day, day of the week, and road type. There is some evidence of this being the case [21].

To address the concerns about lighting and speed and to estimate the prevalence of some other types of distracting activities, Johnson et al. [21] examined ~40,000 photographs of drivers on the New Jersey turnpike taken at all hours and at different sites along the turnpike. A panel of three trained coders examined each photograph and a rating was deemed reliable if at least two of the three coders agreed. In addition to handheld mobile phone use, the authors reported on other in-vehicle distracting activities including adjusting controls, smoking, drinking, eating, reading, and interacting with passengers. However, this type of study can provide no information on certain types of distracting activities including thinking about other things and distractions outside the vehicle—although we recognize that there are few valid methods for measuring such distractions. Moreover, as with other roadside surveys, the capacity to assess hands-free mobile phone use is limited.

Similar to cross-sectional data, roadside surveys cannot capture some important aspects of exposure including a driver's cumulative exposure to a distracting activity while driving, the role of effect modifiers, and the driver's individual ability to deal with the risk attributable to a distracting activity. Indeed, roadside surveys represent a snapshot in time only.

6.5 QUASI-EXPERIMENTAL STUDIES

6.5.1 CASE-CONTROL STUDIES

In a case-control study, groups of individuals, defined by whether or not they have sustained a given outcome, are compared with respect to their exposure histories. For example, data from a case-control study of car crashes resulting in occupant injury in Auckland, New Zealand, were used to investigate the effects of passenger carriage on the risk of car crash injury [22] (for the results see Chapter 17). In that study, cases were drivers who were involved in crashes in which at least one occupant was hospitalized or killed, and controls were a cluster sample of drivers drawn from randomly selected sites on the road network and not involved in a crash.

Case-control studies can provide information on the odds (or risk) of an outcome (such as a motor vehicle crash) associated with an exposure of interest (e.g., a specific type of driver distraction).* In the preceding illustration, passenger carriage was the exposure of interest. Expanding on that, if carrying passengers occurs equally or less commonly during a trip in which a crash occurs compared with a trip in which no crash occurs, then passenger carriage will not be associated with the adverse outcome or may be protective. However, if passenger carriage occurs significantly more frequently during trips in which a crash occurs than during uneventful trips, then passenger carriage is said to be associated with the adverse outcome, in this case, a crash.

* For further discussion about epidemiologic studies, see Hennekens C.H., Buring J.E., and Mayrent S.L. (eds). *Epidemiology in Medicine*. Little, Brown and Company, Boston, MA, 1987; Rothman K.J. and Greenland S. (eds). *Modern Epidemiology*. 2nd Edition. Lippincott, Williams and Wilkins, Philadelphia, PA, 1998.

Verifying exposure is an important consideration in these studies. When assessing an exposure, such as the presence of passengers (or for that matter, any driver distraction), investigators may rely on self-reported data. However, ideally, other methods will be employed to validate self-report. For example, in the case of controls, if observations are being taken at the roadside, investigators could record the presence and number of passengers in the vehicle by direct observation. For cases, other records could be used to verify self-reported information, for example, from police reports taken at the scene of the crash. One other concern is the response rate for cases and controls. If passenger carriage influences the participation of cases or controls, then a low response rate in either group may bias the risk estimates. For example, if controls without passengers are willing to participate more frequently than those with passengers, then there would be a low proportion of controls carrying passengers and this could overestimate the risk of crash associated with passenger carriage. To assess this, investigators would need to record the relative proportions of nonresponders (i.e., eligible cases and controls who decline to participate) who were or were not carrying passengers.

In a case-control study, there is a need to control for potential confounders because cases and controls may differ in relation to factors other than the exposure of interest. These factors could include driving experience, alcohol use, and road and weather conditions. In planning the study, a strategy needs to be developed to take this into account in the design or analysis phase of the study. One of the drawbacks of a case-control study is that there may be unknown factors which consequently cannot be adjusted for. Finally, the study needs to have an adequate sample size (requiring *a priori* sample size calculations) to explore the associations of interest. For example, if a researcher is interested in examining whether age modifies the effect between crash outcome and passenger carriage, then a study has to have a sample size with the statistical power to enable an analysis by age group.

6.5.2 Case-Crossover Studies

A case-crossover study is a variation of a case-control design in which cases act as their own controls (in contrast to a case-control study in which the controls are not the same individuals as the cases). Thus, the design controls for characteristics of the driver that may affect the risk of a crash but do not change over a short period of time. For example, driving inexperience can result in crashes; however, because drivers act as their own controls, driving experience is taken into account. By way of comparison, in a case-control study, one would need to adjust for the differences between cases and controls either in the recruitment phase (e.g., by matching cases and controls on the basis of certain characteristics) or in the analysis phase (e.g., by adjusting for potential confounders using multiple logistic regression techniques). The case-crossover design is useful when a brief exposure causes a transient rise in the risk of a rare outcome such as a crash. As a distraction is transient in most situations, this is an opportune design for determining the role distraction plays in crash propensity.

Two case-crossover studies [23,24] have compared drivers' use of a mobile phone at (and just prior to) the estimated time of a crash with the same driver's phone use during another suitable time period (for the results see Chapter 17). Thus, these studies have been able to estimate the odds of a crash associated with mobile phone use while driving. Because it is important that risks during control periods and crash trips are similar, phone activity during the hazard interval (time immediately before the crash) is compared with phone activity during control intervals (equivalent times in the recent past during which participants were driving but did not crash, e.g., McEvoy et al. [23]).

As discussed earlier, validating exposure to driver distractions is important. In the studies by McEvoy et al. [23] and Redelmeier and Tibshirani [24], phone activity records from the participants' telecommunication companies were used to ascertain mobile phone use while driving during the hazard and control intervals. In the study by McEvoy et al. [23], phone company data were reconciled with self-reported information on phone use. Unfortunately, the use of a case-crossover design to examine other distracting activities has been limited by the lack of available data sources to validate self-report about such activities prior to a crash and during equivalent control intervals, in contrast to mobile phone use.

In assessing exposure to mobile phone use during the hazard interval, misclassification bias, which in this case is defined as a systematic error arising from incorrectly assigning exposure, may be a problem because the precise time of a crash may not be known with certainty. Given that phone use in the aftermath of a crash is common, calls made after the crash could be incorrectly classified as having been made before the crash. Doing so would overestimate the risk associated with phone use. Thus, researchers have used a number of techniques to reduce the likelihood of such an error. These have included using several sources to ascertain the time of collision [23,24], using the earliest reported time in the event of a discrepancy [23,24], and assessing the level of agreement between self-report and phone activity records and reconciling inconsistencies [23].

Another potential limitation of epidemiological studies, such as case-crossover studies, is a low response rate. While the response rates were high in the two case-crossover studies presented here (70% in Redelmeier and Tibshirani [24] and 91% in McEvoy et al. [23]), it is possible that drivers who refused to take part differed from the participants. For example, an individual may have refused to participate because of concern about the potential liability of admitting to phone use prior to a crash. The overall effect of such a bias would be to underestimate the risk of a crash associated with phone use. By necessity, in each of these studies, phone use during an interval of up to 10 min (and 5 min) prior to the crash was examined. Thus, not all drivers were actually using the phone at the time of the crash. Accordingly, while the case-crossover studies performed to date represent good epidemiological evidence about the role of mobile phone use in crashes, the findings do not confirm a causal association.

6.5.3 Cohort Studies

In a cohort study, groups of individuals defined on the basis of their exposure status are followed to assess the occurrence of injury. While this study design is useful in road safety research, for example, to examine the association between a history of drunk-driving motor vehicle crashes and subsequent alcohol-related diseases [25], exposure to distracting activities while driving is, by nature, generally short, episodic, and difficult to record accurately (often relying on self-report) and this has limited the use of the cohort design for this purpose.

6.6 NATURALISTIC DRIVING STUDIES

Naturalistic driving studies have been conducted to assess drivers' exposure to, and the adverse consequences of, distracting activities [26,27]. These studies combine experimental and epidemiological techniques. Video cameras and other sensors are used to monitor the behavior of drivers in their everyday driving. The study by Stutts et al. [27] examined 3 h of driving footage collected over a 1 week period from each of 70 drivers. The presence and types of distracting activities undertaken by the drivers and their driving performance were quantified and assessed. In the 100-car naturalistic driving study [26,28], drivers were followed for 1 year to measure their exposure to distracting activities while driving and to determine the antecedents of any critical events, namely, crashes, near-crashes, and other incidents requiring an evasive maneuver.

In the 100-car study by Dingus et al. [26], five camera views were used. These were the driver's face and driver's side view, the passenger's side view of the road, the forward and rear roadways, and a view of the driver's hands, steering wheel, and instrument panel. For privacy reasons, there were no cameras to assess the presence of passengers in the vehicle and no audio was recorded. Sensors were used to measure other aspects of driving, including longitudinal and lateral acceleration and distance to lead and following vehicles. Provided the footage was available and of good quality, trained coders were able to quantify drivers' exposure to a range of in-vehicle distracting activities including handheld mobile phone use, adjusting in-vehicle equipment, drinking and eating, operating personal digital assistants (PDAs), reading, and personal grooming.

Advantages of the study performed by Dingus et al. [26] are that the data could be used not only to quantify exposure but also to calculate the population-attributable risk percentage* and to estimate the odds (or risk) of an at-fault crash or near-crash attributable to categories of driver distraction (simple, moderate, or complex) and certain types of driver distraction, for example, dialing a handheld device [28]. For the purposes of the study, complex secondary tasks were defined as tasks requiring multiple steps, multiple eye glances away from the forward roadway, or multiple button presses. Moderate secondary tasks were defined as those requiring, at most, two glances away from the roadway or at most two button presses. Simple secondary tasks were those that required one or no button presses or one glance away from the forward roadway. Using a case-control methodology, the presence of driver distraction during critical events, namely, at-fault crashes or at-fault near-crashes requiring an evasive maneuver, was compared with the presence of driver distraction during

* The population-attributable risk percentage is the proportion of disease or injury within a population that can be ascribed to a risk factor, for example, percentage of road deaths attributable to drink driving.

randomly selected control or baseline periods during which no critical incidents occurred, to calculate an odds ratio. The results are described in Chapter 17.

In this study, the population-attributable risk percentage described the percentage of all crashes and near-crashes occurring in the population that were attributable to secondary task distraction and certain distracting activities of interest. Calculation of a population-attributable risk percentage requires the population exposure estimate and the odds ratio, both of which were available from the data that were collected and analyzed. Moreover, through use of video recording of a driver over a period of time, theoretically, data on all four components of exposure discussed in the introduction can be gleaned, namely, the cumulative exposure to distracting activities while driving, the duration of each encounter, the hazard intensity during an encounter, and the driver's ability or skill to mitigate the danger.

The naturalistic driving studies performed to date have limitations [26,27]. Some of these were common to both studies [26,27]: small, nonrepresentative, volunteer samples; the difficulty in reliably capturing some types of secondary distracting tasks, such as the drivers' level of cognitive attention, the presence of passengers, and some outside distractions; the difficulty in identifying distracting activities during periods of reduced visibility, for example, during nighttime driving; and a lack of data on the role of driver distractions in serious crashes resulting in driver injury. Only five of the 15 police-reported crashes in Dingus et al. [26] involved air-bag deployment and injury, and there were no crashes documented in the study by Stutts et al. [27].

Other limitations related to one or other of the studies. In the study by Stutts et al. [27], only a limited number of hours of driving were recorded, a relatively low interrater reliability in coding distracting activities was noted (65–70%), and the outcome was driving impairment rather than crash risk. In the study by Dingus et al. [26], the cost of running the study was high (as drivers were followed for a 1-year period); and several logistical issues were encountered including the types of vehicles that could be instrumented, dealing with equipment failure, legal and confidentiality concerns, and the complexity of analyzing the vast data output (thousands of driving hours).

6.7 FUTURE DIRECTIONS

With the development of the naturalistic driving design, remarkable progress has been made in the capacity to assess drivers' exposure to many distracting activities while driving. However, much work remains to be done. Establishing risk estimates for the gamut of distracting activities that occur while driving remains an important area for research. Further investigation into the circumstances during which distracting activities present the greatest risk is also warranted. Though all the studies we have discussed have limitations, naturalistic driving studies have the advantage that on-road behavior is captured using cameras. Accordingly, exposure to many types of distracting activities can be quantified. Moreover, naturalistic driving studies can also be used to estimate the risk associated with certain types of driver distraction and can shed light on the circumstances during which such activities present the greatest hazard. Thus, further studies of this kind, employing sound epidemiological methods and larger, more representative, driving populations are warranted.

REFERENCES

1. McEvoy S.P., Stevenson M.R., and Woodward M. The impact of driver distraction on road safety: Results from a representative survey in two Australian states, *Inj Prev*, 12, 242–247, 2006.
2. McEvoy S.P., Stevenson M.R., and Woodward M. Phone use and crashes while driving: A representative survey of drivers in two Australian states, *Med J Aust*, 185, 630–634, 2006.
3. Beirness D.J., Simpson H.M., and Desmond K. *The Road Safety Monitor 2002. Risky Driving*. The Traffic Injury Research Foundation, Ottawa, Ontario, Canada, 2002.
4. Lamble D., Rajalin S., and Summala H. Mobile phone use while driving: Public opinions on restrictions, *Transportation*, 29, 223–236, 2002.
5. Sullman M.J.M. and Baas P.H. Mobile phone use amongst New Zealand drivers, *Transport Res F*, 7, 95–105, 2004.
6. Royal D. National survey of distracted and drowsy driving attitudes and behaviours, 2002, The Gallup Organization, Volumes I and III. Prepared for the National Highway Traffic Safety Administration, March 2003.
7. Gras M.E., Cunill M., Sullman M.J., Planes M., Aymerich M., and Font-Mayolas S. Mobile phone use while driving in a sample of Spanish university workers, *Accid Anal Prev*, 39, 347–355, 2007.
8. Rajalin S., Summala H., Poysti L., Anteroinen P., and Porter B.E. In-car cell phone use and hazards following hands free legislation, *Traffic Inj Prev*, 6, 225–229, 2005.
9. Research Triangle Institute and Federal Highway Administration. User's guide for the public use data files, 1995 Nationwide Personal Transportation Survey. October 1997. Available at http://nhts.ornl.gov/1995/Doc/UserGuide.pdf; accessed July 20, 2006.
10. Chen L.H., Baker S.P., Braver E.R., and Li G. Carrying passengers as a risk factor for crashes fatal to 16- and 17-year-old drivers, *JAMA*, 283, 1578–1582, 2000.
11. Doherty S.T., Andrey J.C., and MacGregor C. The situational risks of young drivers: The influence of passengers, time of day and day of week on accident rates. *Accid Anal Prev*, 30, 45–52, 1998.
12. Chipman M.L., MacGregor C., Smiley A.M., and Lee-Gosselin M. Time vs. distance as measures of exposure in driving surveys, *Accid Anal Prev*, 24, 679–684, 1992.
13. Eby D.W., Vivoda J.M., and St Louis R.M. Driver hand-held cellular phone use: A four-year analysis, *J Safety Res*, 37, 261–265, 2006.
14. Glassbrenner D. Driver Cell Phone Use in 2004, Overall Results (DOT HS-809-847), Washington, D.C.: US National Highway Traffic Safety Administration, 2005.
15. Johal S., Napier F., Britt-Compton J., and Marshall T. Mobile phones and driving. *J Public Health Oxf*, 27, 112–113, 2005.
16. McCartt A.T. and Geary L.L. Longer term effects of New York State's law on drivers' hand-held cell phone use, *Inj Prev*, 10, 11–15, 2004.
17. Taylor D.McD., Bennett D.M., Carter M., and Garewal D. Mobile telephone use among Melbourne drivers: A preventable exposure to injury risk, *Med J Aust*, 179, 140–142, 2003.
18. Horberry T., Bubnich C., Hartley L., and Lamble D. Drivers' use of hand-held mobile phones in Western Australia, *Transport Res F*, 4, 213–218, 2001.
19. McCartt A.T., Hellinga L.A., and Geary L.L. Effects of Washington, D.C. law on drivers' hand-held cell phone use, *Traffic Inj Prev*, 7, 1–5, 2006.
20. McCartt A.T., Braver E.R., and Geary L.L. Drivers' use of handheld cell phones before and after New York State's cell phone law, *Prev Med*, 36, 629–635, 2003.
21. Johnson M.B., Voas R.B., Lacey J.H., McKnight A.S., and Lange J.E. Living dangerously: Driver distraction at high speed, *Traffic Inj Prev*, 5, 1–7, 2004.

22. Lam L.T., Norton R., Woodward M., Connor J., and Ameratunga S. Passenger carriage and car crash injury: A comparison between younger and older drivers, *Accid Anal Prev*, 35, 861–867, 2003.
23. McEvoy S.P., Stevenson M.R., McCartt A.T., Woodward M. et al. Role of mobile phones in motor vehicle crashes resulting in hospital attendance: A case-crossover study. *Brit Med J*, 331, 428–430, 2005; doi:10.1136/bmj.38537.397512.55.
24. Redelmeier D.A. and Tibshirani R.J. Association between cellular telephone calls and motor vehicle collisions, *New Engl J Med*, 336, 453–458, 1997.
25. Stevenson M.R., D'Alessandro P., Bourke J., Legge M., and Lee A. A cohort study of drink driving motor vehicle crashes and alcohol-related diseases, *Aust NZ J Public Health*, 27, 328–332, 2003.
26. Dingus T.A., Klauer S.G., Neale V.L., Petersen A. et al. *The 100-Car Naturalistic Driving Study, Phase II – Results of the 100-Car Field Experiment.* (Technical Report No. DOT HS 810 593.) National Highway Traffic Safety Administration, Washington, D.C., April 2006.
27. Stutts J.C., Feaganes J., Reinfurt D., Rodgman E. et al. Driver's exposure to distraction in their natural driving environment, *Accid Anal Prev*, 37, 1093–1101, 2005.
28. Klauer S.G., Dingus T.A., Neale V.L., Sudweeks J.D., and Ramsey D.J. *The Impact of Driver Inattention on Near Crash/Crash Risk: An Analysis Using the 100-Car Naturalistic Driving Study Data.* (Technical Report No. DOT HS 810 594.) National Highway Traffic Safety Administration, Washington, D.C., April 2006.

7 Measuring the Effects of Driver Distraction: Direct Driving Performance Methods and Measures

Kristie L. Young, Michael A. Regan, and John D. Lee

CONTENTS

7.1 INTRODUCTION

In modern vehicle cockpits, drivers have access to a range of entertainment, information, communication, and advanced driver assistance systems (e.g., navigation systems). Whether or not these technologies distract drivers depends in large part on the way they are designed and used. Assessment methods and metrics that are sensitive to the effects of different in-vehicle technologies on driving performance are needed to inform the safe design, deployment, and use of these devices.

Many methods and metrics are available for evaluating the impact on driving performance of driver interaction with secondary tasks. However, the selection of measurement methods for driver distraction research, as in other areas of research, should be guided by a number of general rules related to the nature of the task under study and the properties of the method itself. To be considered an appropriate measurement technique, a method must be *valid* (i.e., it measures what it claims to measure) and *reliable* (i.e., the results obtained are consistent across administrations) and have high *sensitivity* (i.e., be likely to detect the effect of an activity or technology on driving performance even when that effect is small). The results obtained must also have *external validity*; that is, they should be able to be generalized to real-world situations or to situations and individuals beyond the scope of a specific study.[1]

This chapter reviews a range of assessment methods and metrics that have been used to assess the impact of distraction on driving performance. The available evidence regarding the reliability, validity, sensitivity, and generalizability of these is discussed. The focus of the chapter is on "direct" driving performance methods and metrics; that is, methods and metrics that directly assess objective measures of driving performance. Surrogate measurement methods, which simulate and assess particular aspects of the driving task (e.g., the lane change test and the visual occlusion technique), are discussed in Chapters 9 through 11 of this book.

7.2 DRIVING PERFORMANCE MEASUREMENT METHODS

7.2.1 ON-ROAD AND TEST-TRACK STUDIES

One of the most realistic, and ecologically valid, ways of measuring the potentially distracting effects of tasks that compete for the driver's attention is to conduct an on-road study. Using this method, drivers are required to drive an instrumented vehicle on real roads for a specified period while various driving parameters are recorded using data loggers. Driving performance while engaging in a secondary task is then compared against a baseline or reference condition, such as driving when not interacting with any device.[2] On-road studies include naturalistic driving studies, field operational tests (FOTs), and more highly controlled on-road experiments.

FOTs and naturalistic driving studies are conducted over a period of weeks, months, or years and involve drivers driving an instrumented vehicle (sometimes their own vehicle) as part of their normal, everyday driving activity. These studies are less obtrusive than on-road experiments as they do not involve the presence of an experimenter and drivers are free to drive whenever and wherever they wish. The main difference between FOTs and naturalistic driving studies is the level of experimental manipulation involved. FOTs are designed to measure, under natural driving conditions, drivers' interaction with one or more in-vehicle systems of interest and usually involve these systems being activated and deactivated over the course of the study. The Australian TAC SafeCar project is an example of a FOT, where drivers interacted with a suite of in-vehicle intelligent transport systems (ITS) during their everyday driving.[3] In FOT studies, the effectiveness of the systems in optimizing driving performance and safety (relative to nonuse of the systems by the same drivers or relative to a control group driving vehicles not equipped with ITS technologies) is usually the main focus of research. However, naturalistic driving studies involve no experimental manipulation and simply gather, using video cameras and other on-board sensors, data on drivers' everyday driving behavior and activities, which may or may not include interaction with ITS and other technologies Original Equipment Manufacturer (OEM)-fitted, retrofitted, or nomadic). In the latter studies, the impact of everyday driver interaction with various sources of distraction on driving performance and crash risk (e.g., a mobile phone) is more the focus of research and can be observed and determined *post hoc* from analysis of the logged driving and video data. The largest naturalistic driving study conducted to date is the 100-car study (for more detail on this and other naturalistic driving studies, see Chapters 5 and 17 of this book).

On-road experiments involve relatively shorter time periods, of minutes or hours, and a higher level of experimental manipulation. The drivers follow a specified route (usually with the experimenter present) and are required to perform one or more tasks during particular segments of the drive, the timing of which is usually controlled by the experimenter.

On-road studies yield vast amounts of driving performance data under conditions that are relatively representative of actual driving. However, on-road studies, and particularly FOTs and naturalistic driving studies, are time consuming (often taking months or years to complete), are expensive to conduct, and require the storage and analysis of enormous amounts of data. In FOTs and naturalistic studies, researchers also have little or no experimental control over factors such as weather and traffic conditions. On-road experiments afford a higher level of experimental control, but the presence of the experimenter can also cause drivers to behave differently than they normally would.

Test-track studies, conducted on closed roads or dedicated test circuit, closely approximate real-world driving and have been used extensively to examine the distracting effects of secondary tasks that compete for the driver's attention.[4–6] Here, participants are required to drive an instrumented vehicle along the test track. Information on participants' driving performance while engaging in secondary tasks is collected by a data logger or an observer, or even instrumentation embedded in the roadway. These data are compared with a baseline condition to determine the level of distraction imposed by the secondary task(s). This method approximates real driving conditions while affording the experimenter greater control over factors such as traffic signal timing (see, e.g., Ref. 7). Driving under the more controlled conditions of a test track also minimizes the safety risks associated with conducting distraction research on real roads.[8] However, the nature of the test-track course can affect the data collected. If the course is relatively short and there is little or no traffic or obstacles on the track, then drivers may not assign as much priority to the driving task (and greater priority to the secondary task) as they would on real roads.

Drivers may, for example, spend a greater amount of time glancing away from the road than they would under real driving conditions, and this could alter the influence of the secondary task on driving performance. In addition, safety considerations make it impossible to expose drivers to demanding situations, where the consequences of driver distraction may be most critical.

7.2.2 Driving Simulators

Driver distraction research often makes use of driving simulators, as they allow for the examination of a range of driving performance measures in a controlled, relatively realistic, and safe driving environment. Driving simulators, however, vary substantially in their characteristics, and this can affect their realism (or fidelity) and the validity of the results obtained. A distinction is often made between low-level, mid-level, and high-level driving simulators. High-level, or high-fidelity, simulators offer a realistic driving environment, complete with a vehicle cab with realistic components and layout, advanced graphics with close to a 360° field of view, and a sophisticated motion base. Mid-level simulators have a realistic cab or vehicle, large projection screens, and sometimes a simple motion base. Low-level, or low-fidelity, simulators offer less realistic driving environments and usually consist of a PC or desktop workstation and a simple buck with controls.[9–11]

Driving simulators have a number of advantages over on-road and test-track studies. First, they provide a safe environment in which to conduct research that might be otherwise too dangerous to be conducted on-road.[8] Although test tracks can be used to evaluate the distracting effects of in-vehicle systems on driving using single-vehicle scenarios, using multiple-vehicle scenarios in such situations is potentially hazardous. Driving simulators, in contrast, provide a safe environment for the examination of distraction issues using multiple-vehicle scenarios, where the driver can negotiate very demanding roadway situations while engaging in secondary tasks.[8] Second, greater experimental control can be applied in driving simulators compared with on-road studies, as they allow for the type and difficulty of driving tasks to be precisely specified and any potentially confounding variables, such as weather, to be eliminated or controlled for.[11] Third, the cost of modifying the cockpit of a simulator to allow for the evaluation of new in-vehicle systems may be significantly less than modifying an actual vehicle.[11] Finally, a large range of test conditions (e.g., night and day, different weather conditions, or road environments) can be implemented in the simulator with relative ease, and these conditions can include hazardous or risky driving situations that would be too difficult or dangerous to generate under real driving conditions.[11–13]

The use of driving simulators as research tools does, however, have a number of disadvantages. First, data collected from a driving simulator generally include the effects of learning to use the simulator and may also include the effects of being directly monitored by the experimenter.[2] Second, driving simulators, particularly high-fidelity simulators, can be very expensive to install and manage and require a higher level of expertise to operate than other equipment used to measure driver distraction (e.g., surrogate measures such as visual occlusion goggles or the lane change test), particularly because they also often require the experimenter to operate peripheral equipment such as eye tracking systems.[11] Simulator discomfort is another problem encountered with simulators and is particularly common among older drivers and females, who experience higher dropout rates than younger, male drivers.[8,14,15] In addition, simulator discomfort can undermine driving performance and confound the measurement of distraction-related performance decrements.[16] However, one of the most problematic aspects of driving simulator research that has major implications for driver distraction research is the effect of the simulator on drivers' priorities in relation to the performance of primary (i.e., driving) and secondary tasks. The cognitive resources that drivers devote to the primary and secondary tasks while in the simulator may differ significantly from those deployed on actual roads because driving errors in the simulator do not have serious consequences.[8] Thus, drivers may glance away from the road for a greater length of time when dialing a phone in the simulator than they would in the real world because they know their safety will not be compromised. This is a contentious issue in driving simulator research and raises the question as to how valid are driving simulators as tools for conducting driver distraction research.

7.3 SIMULATOR FIDELITY AND VALIDITY

7.3.1 FIDELITY

As discussed earlier, a simple description of driving simulators places them on a continuum that ranges from low to high fidelity. Fidelity refers to the level of realism inherent in the virtual world. The closer a simulator approximates real-world driving, in terms of the design and layout of controls, the realism of the visual scene, and its physical response characteristics, the greater fidelity it is reported to have.[9,13] Numerous dimensions of fidelity have been proposed, many of which relate to the simulator's technical or physical characteristics, but these characteristics may not necessarily correspond to the degree to which the simulator replicates the driving experience.

Rehmann et al.[17] proposed that there are four interrelated dimensions of simulator fidelity: equipment fidelity, environmental fidelity, objective fidelity, and perceptual/psychological fidelity. Equipment fidelity refers to the degree to which the simulator replicates the appearance and feel of the real-world system, in terms of the layout of the vehicle cockpit and the size, shape, color, and position of the vehicle/system controls. Environmental fidelity concerns the extent to which the simulator replicates motion and visual cues, and other sensory information from the real-world environment. Objective fidelity refers to the degree to which a simulator replicates its real-world counterpart in terms of dynamic cue timing and synchronization (e.g., timing of the visual cues matching steering inputs). The fourth dimension, perceptual or psychological fidelity, is concerned with the degree to which the driver perceives the simulation to be a believable reproduction of the real driving task, and the degree to which the driver's pattern of interaction with the driving environment and system controls corresponds to real-world driving.[17] Simulator fidelity is a complicated issue with many factors to consider, and simulators cannot always be accurately described using a three-tier classification system of low, medium, and high fidelity.

The level and type of fidelity required by a simulator depends on the type of research being conducted. It has been suggested that higher fidelity levels are required for research where the results of the simulation are used to draw conclusions about real-world driving performance, as when assessing whether interaction with an in-vehicle device distracts drivers.[13] In terms of the specific aspects of simulator fidelity that are most important for distraction research, little research exists that can be used to guide this decision. However, knowledge regarding what driving performance measures are affected by distraction can provide some useful insights into what aspects of simulator fidelity might be important. For example, distraction, particularly visual distraction, has been shown to affect drivers' ability to maintain lateral position.[18,19] In turn, a lack of motion and visual cues has been shown to affect the precision of lateral position control to a greater extent in simulators than actual vehicles, because the absence of visual and kinesthetic feedback leads to a decreased ability to select appropriate steering corrections.[11,20] Thus, it appears that environmental fidelity, and the precise replication of motion and visual cues in particular, is important for the accurate measurement of the effects of distraction on lateral control. Distraction has also been shown to affect drivers' visual scanning patterns and their ability to detect events occurring in the periphery,[18,21] suggesting that a display screen with a wide field of view is important to be able to capture the effects of distraction on the detection of objects or events occurring in the driver's peripheral field of view. A simulator's fidelity can thus affect how sensitive it is to the effects of distraction. This issue is explored further in Sections 7.3.3, 7.4, and 7.5.

The location of the in-vehicle system under evaluation, relative to the driver and the roadway, and the type and layout of its controls are also important. The location of the system in the simulated vehicle and its visual angle from the road should match precisely its placement in real vehicles because its distance from the forward view directly contributes to the degree of distraction it imposes on drivers. For example, a study on monitor location within the vehicle revealed that as the downward viewing angle of the display increased, the drivers' ability to detect that they were closing in on a lead vehicle decreased.[22] In addition, the types of controls used and their layout should be consistent across the simulated and real systems. Discrepancies in the location and design of the in-vehicle system between simulated and real vehicles may lead drivers to interact with the system differently in the simulator and, thus, lead to driving performance being differentially affected across the simulated and real-world environments.

These are but a few examples of simulator fidelity dimensions that may be important for distraction research. It is likely that there are many more aspects that are important for this type of research. Identifying the specific aspects and levels of simulator fidelity that are required for distraction assessment should be a focus of further research.

7.3.2 Validity

Simulator validity typically refers to the degree to which behavior in a simulator corresponds to behavior in real-world environments under the same conditions.[10,20] The validity of a simulator can be affected by its level of fidelity. This, and the following, section will explore the issue of simulator validity and its relationship to fidelity.

The best method for determining the validity of a simulator is to compare driving performance in the simulator to driving performance in real vehicles using the same driving tasks.[20] A number of studies have examined driving simulator validity and have generally found good correlations between simulated driving performance and driving performance on real roads.[10,18]

There are two types of validity: absolute validity and relative validity. If the numerical values for certain tasks obtained from the simulator and actual vehicles are identical or near identical, absolute validity is said to have been achieved.[9,23] Relative validity is achieved when driving tasks have a similar affect (e.g., similar magnitude and direction of change) on driving performance in both the simulator and real vehicles.[23] Although limited, research has generally found that simulators demonstrate good *relative* behavioral validity for many driving performance measures, although *absolute* validity has rarely been demonstrated.[9,11,20,23–25]

7.3.3 Relationship between Simulator Fidelity and Validity

High-fidelity simulators offer a more realistic driving environment and generally support greater validity than lower-fidelity simulators; however, they can be much more expensive to build and operate than lower-fidelity simulators and can be less sensitive to the effects of distraction.[18] An important consideration for simulator users is whether the increased costs of high-fidelity simulators are offset by their greater validity and whether the decreased validity of low-fidelity simulators is offset by their greater sensitivity.

Several driving simulation studies have examined how the level of fidelity of a simulator affects driving performance and the validity of the results obtained. Early research suggested that the presence of a moving base and higher image resolution may increase the absolute validity of driving simulators.[10,20,25] For example, the performance of certain driving tasks, such as speed control and lane-keeping performance, are less precise in low-fidelity, fix-based simulators than in high-fidelity, motion-based simulators or real vehicles, because of the absence of haptic and motion cues.[11,20,25] However, some recent research has found that lower-fidelity simulators demonstrate levels of validity comparable to those of their higher-fidelity counterparts.[11,18]

A study conducted by Reed and Green[11] assessed the validity of a low-cost driving simulator, in high- and low-fidelity mode, for use in measuring the distracting effects of using a mobile phone. The simulator used was the Driver Interface Research Simulator located at the University of Michigan Transportation Research Institute (UMTRI). This comprised a 1985 Chrysler Laser with an instrumented steering wheel and brake and accelerator pedals. The visual scene was projected onto a screen that gave a 22 × 33 degree field of view. Participants were tested under two fidelity conditions: high and low. In the high-fidelity mode, the visual scene included a colored, textured background and roadside objects, whereas in low-fidelity mode, the scene was black with white road edge and center lines.

Twelve participants drove an instrumented car along a freeway route and a simulated highway route while periodically dialing a mobile phone. Measures of lane position, speed, steering wheel angle, and throttle position were recorded and compared for the simulator and actual driving conditions. Results revealed that only one significant difference was found between fidelity conditions: the effect of visual fidelity on the standard deviation of steering wheel angle was in the opposite direction for males and females. Given the lack of differences observed between the two fidelity conditions, the validity of the simulator was evaluated for the high-fidelity condition only. When the simulator and on-road results were compared, it was found that mean speeds were similar in the simulator and instrumented vehicle; however, lane keeping was less precise in the simulator than in the instrumented vehicle. More specifically, the variation in lane position was twice as large in the simulator as in the instrumented vehicle under the phone task condition, which may reflect drivers' tendency to be less cautious about making driving errors in the simulator because the consequences for doing so are far less severe than those in actual vehicles. Reed and Green concluded that the simulator demonstrated good absolute validity for speed measurements and good relative validity for the effects of the phone task on lane keeping.

More recently, a study by Engström et al.,[18] conducted as part of the HASTE project, compared the effects of visual and cognitive demand on driving performance in a range of test environments, including static and moving-base simulators. The fixed-base simulator was the Volvo Technology (VTEC) simulator located at Volvo Technology Corporation in Sweden. This simulator comprises a Volvo S80, with a 135° horizontal field of view without rear projection. The moving-base simulator was the Swedish National Road and Transport Research Institute (VTI) simulator, located in Lidköping, Sweden, which is a high-fidelity, dynamic simulator. The VTI simulator consists of a Volvo 850 vehicle and a 120° visual view with no rear projection. Drivers drove along a motorway in either the fixed- or moving-base simulator, once while performing no secondary task and once while performing either a surrogate visual IVIS task (the arrows task) or a surrogate cognitive-auditory IVIS task (the Auditory Continuous Memory Task). The difficulty level of the surrogate tasks was varied during the drive. Results revealed that the effects of visual and cognitive load on driving were largely consistent across the static and moving-base (dynamic) simulators. One point of difference was that the effect of the surrogate tasks on lateral vehicle control was greater in the fixed-base than in the moving-base simulator, which suggests that the lower-fidelity simulator may have been less valid but more sensitive.

It is important to note that although the aforementioned research is promising in terms of the ability of a low-fidelity simulator to accurately measure the effects of driver distraction on driver performance, demonstrating that one simulator's being valid for the assessment of a particular driving task does not assure that all simulators will be equally valid. In general, the validity of simulator results will depend on the degree to which the simulator replicates the confluence of driving and in-vehicle task demands that occur in actual driving situations. The absolute degree of simulator fidelity may be less important than its specific characteristics. It is recommended that the validity of individual simulators be established separately for each driving situation they are used for.[13]

7.4 THE RELATIONSHIP BETWEEN SENSITIVITY AND TEST METHODS

A general finding that has emerged from driver distraction and workload research is that the different assessment methods or test environments demonstrate varying levels of sensitivity to the effects of secondary tasks. In particular, a number of studies have shown that as the realism of assessment methods or test environments increases, the less sensitive they become to the effects of secondary tasks on driving performance. Sensitivity refers to the ability to detect even small changes in driving performance due to, in this case, the performance of a secondary activity. Blaauw[20] demonstrated that a fixed-base driving simulator was able to differentiate between experienced and inexperienced drivers on secondary tasks with greater sensitivity than an instrumented vehicle during an on-road test. Reed and Green[11] also found that the effects of dialing a mobile phone on lane-keeping performance were more pronounced in a low-cost driving simulator than on the road. Age effects on lane-keeping performance while performing the concurrent dialing task were also of a greater magnitude in the simulator than in the on-road test.

More recently, the results of the CAMP Driver Workload Metrics project revealed that the laboratory tests generally generated larger or more discernable effects than the test-track or on-road tests, and for some driving measures, effects were observed in the laboratory tests that were not observed in the on-road and test-track studies. For example, a number of event detection measures in the laboratory were capable of discriminating between the effects of high- and low-workload auditory-vocal tasks on driving, whereas these measures were not able to distinguish between the effects of high- and low-workload tasks on driving on the test track. The effect sizes found in the on-road and test-track studies were largely comparable across tasks and driving performance measures.[26]

The studies conducted as part of the HASTE project also revealed differences in the discriminability of test environments; however, the results were more diverse. These studies examined the effects of surrogate and actual in-vehicle tasks on driving performance in the laboratory, in a range of driving simulators and on real roads. Overall, the results revealed that, in line with the CAMP results, the effects of the surrogate visual and cognitive in-vehicle tasks were generally larger in the laboratory than in the driving simulators or the on-road tests. The effect sizes found in the driving simulator and on-road studies were, however, largely comparable across the driving performance measures examined. One difference that was observed across the field and simulator trials was that physiological workload and steering activity was higher in the field than in the simulators, which the authors believed reflected an increase in effort and a lower error tolerance in real traffic, possibly due to the increased risk associated with this environment.[18]

The relative sensitivity or discriminability of testing environments was more varied when drivers interacted with actual in-vehicle systems (navigation and traffic information systems). In contrast to the CAMP project results, both the on-road tests and the driving simulators demonstrated larger effect sizes than the laboratory results. However, the effects of the actual in-vehicle systems on driving performance were larger in the simulator compared with the on-road tests.[27] It is not clear why these discrepancies across the studies occurred, but they may relate to the different secondary tasks used.

A number of explanations exist for why less realistic test environments, such as low-fidelity simulators, provide larger secondary task effects than on-road or test-track tests. First, drivers' priorities and attention/effort allocation in laboratory tasks and simulators may differ from real-world driving environments because the risk of injury or property damage is absent in these environments. Thus, drivers may be less concerned about missing hazardous events or drifting from their lane in laboratory and simulator tests because the consequences for doing so are far less severe than they are on real roads, or even test tracks. Second, task effects may not be as apparent in on-road and test-track environments because of the greater amount of "measurement noise" present (e.g., uncontrolled variables such as traffic density or weather conditions). Finally, the lack of vestibular and tactile cues or feedback in laboratory and simulator tests may make driving more demanding, leading to less spare capacity and, consequently, larger effects of secondary tasks. In particular, the lack of feedback may reduce drivers' ability to choose appropriate steering corrections when they deviate from their correct lane position, resulting in the observed larger lane-deviation effects.[11,26]

The ability of laboratory and simulator tests to detect small effects is further enhanced by the relatively low cost of testing. The relatively low cost of low-fidelity and laboratory tests makes it possible to collect data from many more participants than would be possible in a high-fidelity simulator. As a consequence the statistical power of comparisons is much greater.

Until the exact nature of this trend is better understood, system designers should not rely solely on laboratory tests to evaluate the safety implications of in-vehicle systems. Rather, it is advisable to include them to initially highlight potential problems with systems, or as part of a batch of evaluation tests that also include on-road or simulator trials.

7.5 FIDELITY, VALIDITY, SENSITIVITY, AND COST TRADE-OFFS

Choosing which method to use when evaluating an in-vehicle system is often a trade-off between level of fidelity, validity, sensitivity, and cost. On-road studies, for example, offer the greatest level of fidelity and external validity. They are, however, expensive and time consuming to conduct and do not offer the same level of experimental control as a driving simulator or laboratory tests, and thus may be less sensitive. One promising trend to emerge from the literature is that low-fidelity simulators offer a similar level of sensitivity and validity as high-fidelity simulators for evaluating the effects of secondary tasks on driving performance. Low-fidelity simulators can, thus, be used by in-vehicle system designers to evaluate the impact of in-vehicle systems on driving performance, without sacrificing any sensitivity to system effects and without incurring the high setup and operating costs associated with high-fidelity simulators. However, no simulator, regardless of its level of fidelity, can completely replicate driving in the real world. In distraction research, the different attention allocation policies and patterns of secondary task interaction used by drivers in simulated and real driving environments is of particular concern. Thus, although simulators offer a relatively inexpensive test environment for initial assessment of the distracting effects of a device, in-vehicle devices should also be evaluated in an on-road setting before final decisions are made regarding its suitability for use while driving.

7.6 MEASURES OF DRIVING PERFORMANCE

Choosing appropriate driving performance measures or variables, which are compatible with the measurement technique(s) being used and sensitive to the particular aspect of distraction being evaluated, is just as important as using appropriate measurement methods. Ideally, the selection of driving performance measures should be based on a consideration of the type of system being evaluated and theories or previous research findings regarding how certain tasks influence driver behavior. In turn, the selection of assessment methods will often be informed by the particular driving performance measures being examined. A deliverable from the European Aptive Integrated Driver-vehicle Interface (AIDE) project (Deliverable 2.2.1) provides a comprehensive review of driving performance measures that can be used in the evaluation of in-vehicle systems.[28] A brief review of some of the most common driving measures that have been used in distraction research is provided in the following section along with example research findings.

7.6.1 LONGITUDINAL CONTROL

A range of longitudinal control measures can be examined in distraction research. Two of the most common include measures of speed and following distance. These are discussed in Sections 7.6.1.1 and 7.6.1.2.

7.6.1.1 Speed

The relationship between speed and crashes is widely recognized in the road safety community, and as such, speed is a commonly used dependent variable in transportation human factors research, including driver distraction research. A number of speed-related measures can be calculated, including mean and 85th percentile speed, maximum speed, and the standard deviation, or variability, of speed. Several on-road and simulator studies have found that drivers display greater variation in driving speed and throttle control when using a mobile phone, and this has been demonstrated for hands-free as well as handheld phones.[5,11,29] It has also been shown that drivers display a tendency to reduce their speed when talking on a mobile phone, which is believed to be a form of behavioral adaptation to reduce primary task demand or increase the safety margin.[29,30] In a driving simulator, Srinivasan and Jovanis[12] found that mean speeds were lower when drivers manually operated a route navigation system. Operating a CD player while driving can also result in reduced simulated driving speeds[31]; however, the use of voice inputs to operate CD players has been shown to reduce the likelihood of traveling at speeds that are considered to be too low.[32]

7.6.1.2 Vehicle Following (Headway)

Vehicle following, or headway, measures are also commonly employed in driver distraction research. Several specific vehicle following measures have been commonly used, including mean headway (distance or time based), minimum headway, and standard deviation of headway. Headway is an indication of the safety margin that drivers are willing to accept, and thus, short headways are often interpreted as being indicative of degraded driving performance and a measure of high secondary task load.

A number of studies have, however, found that drivers tend to adopt longer headways when interacting with secondary tasks, particularly visual tasks.[19,33,34] In a simulator study, Greenberg et al.[19] found that drivers increased their headway when engaging in a visual secondary task and that this effect was particularly pronounced for older drivers. Also, as part of the HASTE project, Östlund et al.[34] found in a series of laboratory, simulator, and on-road studies that both time and distance headway increased when drivers were performing a visual surrogate IVIS task, but not during cognitive IVIS operation.

7.6.2 Lateral Control

Lateral control measures commonly examined in distraction research include lane keeping and steering measures. These are discussed in Sections 7.6.2.1 and 7.6.2.2.

7.6.2.1 Lane Keeping

Lane keeping, or lateral position, refers to the position of a vehicle on the road in relation to the center of the lane in which the vehicle is traveling. Decrements in lateral position control are used as a measure of secondary task load when evaluating the effects of in-vehicle systems on driving performance. The most commonly used lateral position metrics are mean lane position, standard deviation of lane position, and number of lane exceedences (LANEX).

Research suggests that drivers' ability to maintain their lateral position on real or simulated roads is adversely affected when performing secondary tasks, particularly tasks requiring large amounts of visual attention.[18] Drivers make a greater number of lane position deviations and exceedences while dialing or talking on either a handheld or a hands-free mobile phone, even when driving on straight roads with little traffic.[5,11] Research also suggests that drivers make a greater number of lane deviations and exceedences when manually entering details into a route guidance system or when following navigation instructions presented visually, rather than through voice guidance.[35,36] Tuning the radio, interacting with a CD player, or listening to radio broadcasts can also degrade driving performance, as measured by lane position deviation.[31,37,38]

An interesting finding with respect to lateral control is that moderate levels of cognitive load have been shown to lead to more precise lateral control, by reducing lane-keeping variation. Visual load, in contrast, has been shown to increase lane-keeping variation.[18,19]

7.6.2.2 Steering Wheel Metrics

Measures of steering wheel movement have been used extensively in many forms of driving research. These include standard deviation of steering wheel angle, steering wheel reversal rate, steering wheel angle high-frequency component (HFC), steering wheel action rate, and steering entropy (less predictable steering behavior).[34] In driver distraction and workload research, steering wheel movements are considered to be an indicator of secondary task load. When driving under normal conditions (i.e., when not performing a secondary task), drivers will make a number of small corrective steering wheel movements to maintain lateral position. When engaging in a secondary task, however, particularly a visual-manual task, drivers will often make a number of large and abrupt steering wheel movements to correct heading errors.

7.6.3 Event Detection and Reaction Time Measures

Event detection and reaction time metrics have become increasingly popular in in-vehicle system research and evaluation, primarily because of the relationship between these measures and risk of crash involvement. A range of event detection and response time measures can be examined, including number of missed/detected events, number of incorrect responses made, and response time and distance (e.g., distance from event when detected).

Drivers' ability to detect and react to external events or objects has been shown to be impaired by the use of in-vehicle devices, particularly when these devices are complex. A number of studies have found that using either a handheld or hands-free phone can increase drivers' reactions to hazards and common road events (e.g., traffic light changes) by up to 30%.[29,34,39–41] In a driving simulator study, Srinivasan and Jovanis[12] found that drivers' reaction times to vehicles crossing their path or to traffic light changes increased when they were listening to navigation instructions from a route guidance system that issued turn-by-turn navigation instructions. Accessing and reading e-mails using a voice-based in-vehicle e-mail system while driving has also been found to increase drivers' reaction times to a braking lead vehicle by up to 30% in a simulated driving environment.[42]

7.6.4 Gap Acceptance

Negotiating gaps in traffic is a complex task requiring considerable visual guidance and attention. Gap acceptance measures that have been used in distraction research include number of collisions initiated and size of gaps accepted. Research shows that when using in-vehicle devices such as a mobile phone, drivers tend to accept shorter gaps in traffic when turning than when not using a phone.[43] A test-track study by Cooper and Zheng[4] also found that when using a mobile phone, drivers do not consider weather or road surface conditions when making a decision to turn across oncoming traffic. In particular, when using the phone and the road surface was wet, drivers initiated twice as many collisions as when not using the phone.

7.6.5 Subjective Mental Workload

Subjective, or self-reported, workload measures require the participant to rate his/her perceived level of workload shortly after completing a task. Several simple subjective mental workload scales have been developed to measure an individual's perceived workload. Some of the main scales used in the driving domain include the NASA Task Load Index (NASA TLX), the Subjective Workload Assessment Technique (SWAT), the Modified Cooper Harper Scale (MCH), and the Rating Scale Mental Effort (RSME). The NASA TLX and SWAT scales are multidimensional scales designed to address several different dimensions of workload, such as performance and mental effort. However, ratings from the subscales are frequently combined in an equally weighted average.[44] Another multidimensional workload scale that has recently been developed to assess the level of workload associated with in-vehicle tasks is the Driving Activity Load Index (DALI). The DALI is a modified version of the NASA TLX that has been specifically tailored to the assessment of in-vehicle systems/tasks in the automotive environment and has been validated as part of the AIDE project.[45] The MCH and RSME scales, in contrast, are unidimensional scales that rely on just one dimension (e.g., invested effort) to assess workload. Subjective mental workload measures are appealing because of their low cost, ease and speed of administration, and the fact that they are nonintrusive. Subjective assessment techniques do have a number of drawbacks, however, including participants forgetting aspects of their performance posttrial and difficulties in determining whether participants are reporting overall workload levels averaged over the entire task or to specific peaks in workload.[46] The reader is referred to Stanton et al.[46] for more detail on subjective workload measures.

Research has demonstrated that using a mobile phone of any type to talk, dial, or answer while driving on real roads results in increased workload and greater levels of frustration, particularly when the conversation is complex or highly emotional.[6,47] Entering destination details into a route guidance system while driving also increases drivers' subjective workload, particularly if the system is operated manually rather than through voice activation.[12,36] Finally, accessing and reading e-mail using an in-car e-mail system, even when it is voice activated, has been found to increase drivers' subjective workload, and this increase is further heightened as the system becomes more complex.[42]

7.6.6 Choosing Driving Performance Measures

Driver distraction is a multidimensional construct, which means that no single driving performance measure will capture all the effects of distraction. A large number of driving measures exist, making it difficult to know which ones to include in an evaluation. Of course, the decision regarding which set of measures to use should be guided by the specific research question under examination. However, recent research offers insights into what measures are most appropriate for particular evaluations.

A number of on-road and simulator studies, for example, have found that visual and cognitive distraction differentially affect different driving performance measures.[18,19,21] Specifically, visual distraction has a greater effect on lateral control measures, whereas cognitive distraction affects visual scanning behavior to a greater degree than visual distraction. Thus, the type of competing task being assessed should guide measurement selection. The HASTE and CAMP projects have also attempted to identify a set of valid and reliable driving measures that should be used when assessing distraction.[24,26] Although this research offers some guidance regarding what driving performance measures to use, there does not exist a universally agreed set of driving performance measures to use in distraction evaluations. All researchers and system evaluators can do is use a range of driving measures that are valid, reliable, and sensitive to the type of distraction being evaluated.

Another challenge in relation to driving performance measures is the absolute versus relative interpretation of evaluation results. Driving performance measures do not always have a monotonic relationship with safety or crash risk, and this makes absolute interpretation of the effects of distraction on driving performance difficult (see Chapter 4 for further discussion of the nonlinear relationship between driving performance and crashes). For example, if engagement in a secondary activity increases the standard deviation of lateral position by a certain amount, this does not mean that crash risk will be increased; crash risk may increase, remain unchanged, or decrease, depending on the prevailing driving conditions at the time. An absolute interpretation of driving performance results requires a clear relationship between the driving performance measure in the test environment and the consequences for safety outcomes *and* a clear criterion for determining what a safe system is. The challenges associated with defining an "acceptable" level of driving performance and establishing performance criteria are discussed in Section 7.7.

7.7 REFERENCE TASKS AND PERFORMANCE CRITERIA

When evaluating the effect of in-vehicle systems on driving performance, researchers sometimes employ a reference task. A reference task is a task that is used as a benchmark for defining the maximum level of secondary task demand that is deemed acceptable for a driver to cope with when driving.[48] Driving performance when performing the reference task is compared with driving performance while interacting with the in-vehicle device under examination. If the level of driving performance when concurrently performing the secondary task under evaluation is poorer than that associated with concurrent performance of the reference task, then the secondary task is deemed to be unsafe to perform while driving. Currently, there is no single agreed "best" or standardized reference task to use when evaluating in-vehicle systems. Nor is it always clear what constitutes an "acceptable" level of driving performance degradation. The development of standard reference tasks that can be used in the evaluation of in-vehicle systems is the focus of a Preliminary Work Item currently being developed by Sub-Committee 13 of Technical Committee 22 of the International Organization for Standardization (ISO).

A number of different reference or criterion tasks have been used previously in driver distraction research. One approach is to assume that the criterion task is no task. Here, any driving performance degradation deriving from interaction with a competing secondary task is regarded as an unacceptable level of degradation. Several studies examining mobile phone use have compared phone use while driving with simply driving while engaging in no secondary task (e.g., Ref. 8; see also Ref. 49). However, a number of researchers have questioned whether driving on its own is a suitable reference activity when comparing secondary tasks. It is possible that just driving is too stringent a benchmark given that most secondary tasks can be expected to impose at least some additional demand on drivers. This is a particularly important argument when the IVIS task replaces one that is currently performed in the car through a different means, such as navigation with a paper map.[35]

The Alliance of Automobile Manufacturers (AAM) in the United States has included, in their guidelines, time-based criteria stating that visual-manual tasks performed while the vehicle is in motion should require no more than 20 s total glance time and that single-glance durations should not exceed 2 s. In-vehicle tasks that comply with these criteria are considered safe to perform while driving. Data from the 100-car study provide support for the 2 s criterion. Klauer et al.[50] found that single-glance durations of greater than 2 s increased near-crash/crash risk by at least two times, whereas glance durations of less than 2 s were not associated with increased crash risk. However, there is a lack of evidence to support the criterion that tasks requiring less than 20 s total glance time can be safely performed while driving.[51]

Some studies examining the distracting effects of mobile phones have used manual radio tuning (e.g., tuning into a specified radio station using buttons, dial, or toggle switch) as a reference task, claiming that if using a mobile phone degrades driving performance to a comparable or lower degree than tuning a radio, then it is an acceptable task to perform while driving. The AAM also advocates the use of manual radio tuning as a criterion task against which to compare telematics device use. The adoption of the radio tuning task as a benchmark is based on the assumption that tuning the radio imposes only a moderate and socially accepted level of risk, and its impact on driving is reasonably well understood. Research suggests, however, that even radio tuning can degrade driving performance to an extent that safety may be compromised.[38,52] The type of host radio device used when performing the radio-tuning task can have a profound influence on the degree of workload deriving from performance of the task. Several studies, for example, have examined the effects of the radio-tuning task on driving performance using a continuous tuning dial rather than the preset buttons found on many modern radio systems. Use of this type of interface is likely to require a greater amount of time to tune a station than simply pressing a preset button. Similarly, radio control position and display size can have substantial effects on the demands of radio tuning, making it surprisingly difficult to ensure uniform implementation of this reference task. This may have the effect of increasing the distracting effects of the radio-tuning task and, thus, may lead to erroneous conclusions regarding the acceptability of performing certain secondary tasks while driving. More generally, a visual-manual reference task may affect driving performance in a quite different manner than a cognitively demanding task. A cognitive task, for example, might not degrade lane position maintenance to the same degree as a given reference task, but it might undermine event detection to a much greater degree. As a consequence, comparisons to a single-reference task can underestimate the safety consequences of distracting tasks.

Other studies have compared the effects of using mobile phones with driving under the influence of alcohol. Burns et al.[29] compared the driving impairment caused by using handheld and hands-free mobile phones to that caused by having a blood alcohol concentration (BAC) over the U.K. legal alcohol limit (80 mg/100 mL or 0.08). Twenty participants were tested using the Transport Research Laboratory (TRL) advanced driving simulator. Drivers' reaction times to hazards were, on average, 30% slower when conversing on a handheld mobile phone than when driving under the influence of alcohol. Drivers also demonstrated reduced speed control when using a mobile phone, but not when under the influence of alcohol. However, drivers' standard deviation of lane position when they were under the influence of alcohol was significantly higher than when they were talking on the phone or when driving with no phone or alcohol. The authors concluded that certain aspects of driving are impaired to a greater extent when using a mobile phone than when driving while intoxicated.[29]

Strayer et al.[53] also compared, using a driving simulator, the driving performance of mobile phone users with drivers who were intoxicated by alcohol (BAC of 0.08) to establish a benchmark for assessing the relative risks of phone use while driving. They found that although the impairments associated with mobile phone use may be as great as driving with a BAC of 0.08, the exact nature of these impairments differed across the two conditions. When using the mobile phone, drivers were involved in more rear-end collisions, their reaction times to a lead braking vehicle were reduced, and their following distance variability increased by 24%. When intoxicated, drivers displayed a more aggressive driving style. They followed the pace car at a closer distance than they did in the baseline condition, and they braked with 23% more force than they did in either the mobile phone or baseline conditions.

A number of problems have been identified with using alcohol intoxication as a benchmark for establishing the risks associated with various distracting tasks. First, although both alcohol and mobile phone use have been shown to impair driving performance, they have been shown to affect different driving performance measures—or to affect the same driving performance measures to different degrees or even in the opposite direction—suggesting that the mechanisms underlying degraded driving performance differ between the two forms of impairment. Second, distraction tends to be relatively transient (i.e., lasting only as long as the driver is engaging in the distracting activity), whereas alcohol impairment persists over prolonged periods of time (i.e., usually over the entire length of a drive). Consequently, the time frame for exposure to risk is relatively greater for alcohol impairment than for impairment deriving from distraction. It is important to note that this issue is relevant to all reference tasks, not just alcohol. Selecting a reference task that is as similar as possible to the task under evaluation in terms of its duration and the mechanisms by which it affects driving is fundamental for obtaining valid results.

Some researchers have suggested that a more appropriate reference task may be an alternative application of the particular secondary function under evaluation. For example, finding a point of interest on a paper map may be used as a reference task against which to compare the effect of an in-vehicle route guidance task.[26] One problem with this approach is that the alternative application of the task may itself impose a high level of demand on the driver and, as such, would not be deemed an "acceptable" benchmark against which the task of interest is compared. Another problem is that for some new tasks, such as iPod interactions, there is no alternative application of the task because the device is so novel.

It is clear that the development of distraction reference tasks is in its infancy, and as such, there is no universally agreed "best" reference task against which to compare the effects of new in-vehicle systems. It is likely that the most appropriate reference task will differ depending on the type of in-vehicle system and, hence, the specific type of distraction being evaluated (e.g., a reference task that imposes visual-manual load on the driver when evaluating a visual-manual task). Ideally, a reference task is one that is unambiguously defined, is repeatable across different test environments, and induces distraction in a manner similar to the task under evaluation.

7.8 CONCLUSION

This chapter has provided a review of the various methods, driving performance measures, and reference tasks that can be used to directly assess the effects of secondary tasks on driving performance. The performance of secondary tasks has a wide range of effects on driving performance, and a number of methods and driving performance measures exist to quantify these effects. Each of the assessment methods has advantages and disadvantages. Deciding which method to use is often a trade-off between cost, validity, and experimental control. Driving simulators, for example, provide a safe and controlled environment for conducting distraction research and evaluation; however, they lack the realism that on-road evaluations offer, particularly if they are lower-fidelity devices.

One promising finding that has emerged from driving simulation research is that lower-fidelity simulators offer similar levels of sensitivity and validity as high-fidelity simulators for evaluating the impact of distraction on driving performance. This suggests that system designers can, therefore, use lower-fidelity simulators to evaluate the impact of in-vehicle systems on driving performance, without sacrificing sensitivity and without incurring the high costs associated with high-fidelity simulators.

There are still a number of areas of distraction assessment that require further development or understanding, particularly in relation to the relative sensitivity of test environments and reference tasks. First, a number of studies have found that, paradoxically, the more realistic is a test method or environment, the less sensitive it is to the effects of secondary tasks on driving performance. The mechanisms underlying this trend are not currently understood, and as such, system designers should not rely solely on one assessment method to evaluate the safety implications of in-vehicle systems. Second, it is important that appropriate reference tasks be developed to provide a benchmark against which the impact on driving performance of in-vehicle tasks can be established. Appropriate reference tasks are ones that are unambiguously defined, are repeatable across different test environments, and impose a similar type of demand or distraction as the task under evaluation.

Finally, there is currently little consensus regarding which assessment method or driving performance measures should be used for the evaluation of particular tasks. As such, in the past, evaluation studies tended to use a large range of methods and driving measures. Efforts are under way in a number of projects (e.g., HASTE, AIDE, and CAMP), however, to define a set of assessment methods and driving measures that can be used to assess different categories of in-vehicle systems. The outputs of these projects should be closely monitored.

Given the high rate at which technologies are proliferating the vehicle market, it is important that a set of standardized assessment methods and reference tasks, which are valid, reliable, inexpensive, and sensitive, are identified to inform the safe design and use of these systems. The remaining chapters in this section discuss the various surrogate methods that are being used to evaluate the potentially distracting effects of in-vehicle systems.

REFERENCES

1. Kantowitz, B.H., Selecting measures for human factors research, *Human Factors* 34, 387–398, 1992.
2. NHTSA, *Internet Forum on Driver Distraction* [On-line], available at www.nrd.nhtsa.dot.gov/departments/nrd-13/driver-distraction/welcome.htm, accessed on November 10, 2006.
3. Regan, M.A., Young, K.L., Triggs, T.J., Tomasevic, N., Mitsopoulos, E., Tierney, P., Healy, D., Tingvall, C., and Stephan, K., Impact on driving performance of intelligent speed adaptation, following distance warning and seatbelt reminder systems: key findings from the TAC SafeCar Project, *IEE Proceedings Intelligent Transport Systems* 153 (1), 51–62, 2006.
4. Cooper, P.J. and Zheng, Y., Turning gap acceptance decision-making: the impact of driver distraction, *Journal of Safety Research* 33 (3), 321–335, 2002.
5. Green, P., Hoekstra, E., and Williams, M., *On-the-Road Tests of Driver Interfaces: Examination of a Route Guidance System and a Car Phone*, University of Michigan Transport Research Institute, Washington, D.C., 1993.
6. Harbluk, J.L., Noy, Y.I., and Eizenman, M., *The Impact of Cognitive Distraction on Driver Visual Behaviour and Vehicle Control*, Report No. TP No. 13889 E, Road Safety Directorate and Motor Vehicle Regulation Directorate, Ottawa, Canada, 2002.
7. Hancock, P.A., Simmons, L., Hashemi, L., Howarth, H., and Ranney, T., The effects of in-vehicle distraction on driver response during a crucial driving maneuver, *Transportation Human Factors* 1 (4), 295–309, 1999.
8. Goodman, M.J., Bents, F.D., Tijerina, L., Wierwille, W., Lerner, N., and Benel, D., *An Investigation of the Safety Implications of Wireless Communication in Vehicles*, Report No. Report No. DOT HS 808-635, U.S. Department of Transportation, National Highway Traffic Safety Administration (NHTSA), Washington, D.C., 1997.
9. Godley, S.T., Triggs, T.J., and Fildes, B.N., Driving simulator validation for speed research, *Accident Analysis and Prevention* 34 (5), 589–600, 2002.
10. Kaptein, N.A., Theeuwes, J., and van der Horst, R., Driving simulator validity: some considerations, *Transportation Research Record* 1550, 30, 1996.
11. Reed, M.P. and Green, P.A., Comparison of driving performance on-road and in a low-cost simulator using a concurrent telephone dialling task, *Ergonomics* 42 (8), 1015–1037, 1999.
12. Srinivasan, R. and Jovanis, P.P., Effect of in-vehicle route guidance systems on driver workload and choice of vehicle speed: Findings from a driving simulator experiment, In *Ergonomics and Safety of Intelligent Driver Interfaces*, Noy I.A. (Ed.) Lawrence Erlbaum Associates, Inc., Mahwah, NJ, 1997, pp. 97–114.
13. Triggs, T.J., Driving simulation for railway crossing research, In *Seventh International Symposium on Railroad-Highway Grade Crossing Research and Safety—Getting Active at Passive Crossings*, Monash University, Clayton, Australia, 1996.
14. Hein, C.M., Driving simulators: six years of hands-on experience at Hughes Aircraft Company, In *Proceedings of the Human Factors and Ergonomics Society 37th Annual Meeting*, Seattle, Washington, 1993, pp. 607–611.
15. Johnson, D.M., *Introduction to and Review of Simulator Sickness Research*, Report No. Research Report 1832, U.S. Army Research Institute-Rotary Wing Aviation Research Unit, Fort Rucker, AL, 2005.
16. Bittner, A.C., Gore, B.F., and Hooey, B.L., Meaningful assessments of simulator performance and sickness: can't have one without the other, In *Proceedings of the Human Factors and Ergonomics Society 41st Annual Meeting*, Santa Monica, CA, 1997.
17. Rehmann, A.J., Mitman, R.D., and Reynolds, M.C., *A Handbook of Flight Simulation Fidelity Requirements for Human Factors Research*, Report No. DOT/FFF/CT-TN95/46,

U.S. Department of Transportation, Federal Aviation Administration, Atlantic City, NJ, 1995.

18. Engström, J., Johansson, E., and Ostlund, J., Effects of visual and cognitive load in real and simulated motorway driving, *Transportation Research Part F: Traffic Psychology and Behaviour* 8 (2), 97–120, 2005.
19. Greenberg, J., Tijerina, L., Curry, R., Artz, B., Cathey, L., Grant, P., Kochlar, D., Kozak, K., and Blommer, M., Evaluation of driver distraction using an event detection paradigm, *Journal of the Transportation Research Board* No 1843, 1–9, 2003.
20. Blaauw, G.J., Driving experience and task demands in simulator and instrumented car: a validation study, *Human Factors* 24, 473–486, 1982.
21. Recarte, M.A. and Nunes, L.M., Mental workload while driving: effects on visual search, discrimination and decision making, *Journal of Experimental Psychology: Applied* 9, 119–137, 2003.
22. Asoh, T., Kimura, K., and Ito, T., JAMA's Study on the location of in-vehicle displays. Paper 2000-01-C010, In *SAE Conference Proceedings*, 2000, pp. 37–42.
23. Harms, L., Experimental studies of dual-task performance in a driving simulator—the relationship between task demands and subjects' general performance, *IATSS Research* 16 (1), 35–41, 1992.
24. Carsten, O.M.J., Groeger, J.A., Blana, E., and Jamson, A.H., *Driver Performance in the EPSRC Driving Simulator (LADS): A Validation Study.* Final report for EPSRC project GR/K56162, Leeds University, Leeds, UK, 1997.
25. McLane, R.C. and Wierwille, W.W., The influence of motion and audio cues on driver performance in an automobile simulator, *Human Factors* 17, 488–501, 1975.
26. Angell, L.S., Auflick, J.L., Austria, P.A., Kochhar, D.S., Tijerina, L., Biever, W., Diptiman, D., Hogsett, J., and Kiger, S., *Driver Workload Metrics Project: Task 2 Final Report*, National Highway Traffic Safety Administration, Washington, D.C., 2006.
27. Johansson, E., Carsten, O., Janssen, W., Jamson, S., Jamson, H., Merat, N., Östlund, J., Brouwer, R., Mouta, S., Harbluk, J., Anntila, V., Sandberg, H., and Luoma, J., *Validation of the HASTE Protocol Specification.* HASTE Deliverable 3, available at http://www.its.leeds.ac.uk/projects/haste/downloads/Haste_D3.pdf, accessed on October 31, 2006.
28. Johansson, E., Engström, J., Cherri, C., Nodari, E., Tofferi, A., Schindhelm, R., and Gelau, C., *Review of Existing Techniques and Metrics for IVIS and ADAS Systems.* AIDE Deliverable D2.2.1, available at http://www.aide-eu.org/res_sp2.html, accessed on March 24, 2006.
29. Burns, P.C., Parkes, A., Burton, S., Smith, R.K., and Burch, D., *How Dangerous is Driving with a Mobile Phone? Benchmarking the Impairment to Alcohol*, TRL Report 547, TRL Limited, 2002.
30. Haigney, D., Taylor, R.G., and Westerman, S.J., Concurrent mobile (cellular) phone use and driving performance: task demand characteristics and compensatory processes, *Transportation Research Part F* 3, 113–121, 2000.
31. Jenness, J.W., Lattanzio, R.J., O'Toole, M., and Taylor, N., Voice-activated dialing or eating a cheeseburger: which is more distracting during simulated driving? In *Human Factors and Ergonomics 46th Annual Meeting*, 2002, pp. 592–596.
32. Gartner, U., Konig, W., Wittig, T., and Bosch, R., *Evaluation of Manual vs. Speech Input when Using a Driver Information System in Real Traffic*, available at http://ppc.uiowa.edu/driving-asse...nt%20Papers/02_Gatner_Wittig.html, accessed on February 10, 2003.
33. Hjälmdahl, M. and Varhelyi, A., Speed regulation by in-car active accelerator pedal—effects on driver behaviour, *Transportation Research Part F* 7 (2), 77–94, 2004.
34. Östlund, J., Carsten, O., Merat, N., Jamson, S., Janssen, W., and Brouwer, R., *Deliverable 2—HMI and Safety-Related Driver Performance.* HASTE Project. Report No. GRD1/2000/25361 S12.319626, Institute for Transport Studies, Leeds, UK, 2004.

35. Dingus, T., McGehee, D., Hulse, M., Manakkal, N., Mollenbauer, M., and Fleischman, R., *Travtek Evaluation Task C3—Camera Car Study*, Performance and Safety Sciences, Inc., Iowa City, 1995.
36. Tijerina, L., Parmer, E., and Goodman, M., Driver workload assessment of route guidance system destination entry while driving: a test track study, In *Proceedings of 5th ITS World Congress*, Seoul, Korea, 1998, p. 8.
37. Jancke, L., Musial, F., Vogt, J., and Kalveram, K.T., Monitoring radio programs and time of day effect simulated car-driving performance, *Perceptual and Motor Skills* 79, 484–486, 1994.
38. Wikman, A., Nieminen, T., and Summala, H., Driving experience and time-sharing during in-car tasks on roads of different width, *Ergonomics* 41 (3), 358–372, 1998.
39. Brookhuis, K.A., de Vries, G., and de Waard, D., The effects of mobile telephoning on driving performance, *Accident Analysis and Prevention* 23 (4), 309–316, 1991.
40. Strayer, D.L. and Johnston, W.A., Driven to distraction: dual-task studies of simulated driving and conversing on a cellular telephone, *Psychological Science* 12 (6), 462–466, 2001.
41. Tokunaga, R.A., Hagiwara, T., Kagaya, S., and Onodera, Y., Effects of conversation through a cellular telephone while driving on driver reaction time and subjective mental workload, In *Transportation Research Board 79th Annual Meeting*, Washington, D.C., USA, 2000, p. 14.
42. Lee, J.D., Caven, B., Haake, S., and Brown, T.L., Speech-based interaction with in-vehicle computers: the effect of speech-based e-mail on drivers' attention to the roadway, *Human Factors* 43, 631–640, 2001.
43. RoSPA, *Mobile Phones and Driving: A Literature Review*, The Royal Society for the Prevention of Accidents, Birmingham, UK, 1997.
44. Nygren, T.E., Psychometric properties of subjective workload measurement techniques: implications for their use in the assessment of perceived mental workload, *Human Factors* 33 (1), 17–33, 1991.
45. Pauzié, A., Manzano, J., and Dapzol, N., Driver's behavior and workload assessment for new in-vehicle technologies design, In *Proceedings of the 4th International Driving Symposium on Human Factors in Driver Assessment, Training, and Vehicle Design*, Stevenson, Washington, 2007.
46. Stanton, N.A., Salmon, P.M., Walker, G.H., Baber, C., and Jenkins, D.P., *Human Factors Methods. A Practical Guide for Engineering and Design*, Ashgate, Hampshire, UK, 2005.
47. Matthews, R., Legg, S., and Charlton, S., The effect of cell phone type on drivers subjective workload during concurrent driving and conversing, *Accident Analysis and Prevention* 35, 441–450, 2003.
48. ISO/WD 26022, Road vehicles—ergonomic aspects of transport information and control systems—Simulated lane change test to assess in-vehicle secondary task demand, International Organization for Standardization (ISO), 2007.
49. Young, K., Regan, M., and Hammer, M., *Driver Distraction: A Review of the Literature*, Report No. 206, Monash University Accident Research Centre, Clayton, Australia, 2003.
50. Klauer, S.G., Dingus, T.A., Neale, V.L., Sudweeks, J.D., and Ramsey, D.J., *The Impact of Driver Inattention on Near-Crash/Crash Risk: An Analysis Using the 100-Car Naturalistic Driving Study Data*, Virginia Tech Transportation Institute, Blacksburg, VA, 2006.
51. Transport Canada, *Strategies for Reducing Driver Distraction from In-Vehicle Telematics Devices: A Discussion Document*, Transport Canada, Ottawa, Canada, 2003.
52. Briem, V. and Hedman, L.R., Behavioural effects of mobile telephone use during simulated driving, *Ergonomics* 38, 2536–2562, 1995.
53. Strayer, D.L., Drews, F.A., and Crouch, D.J., A comparison of the cell phone driver and the drunk driver, *Human Factors* 48 (2), 381, 2006.

8 Surrogate Distraction Measurement Techniques: The Lane Change Test

Stefan Mattes and Anders Hallén

CONTENTS

8.1 INTRODUCTION

More and more technologies such as telephones, navigation systems, and entertainment systems are becoming available for use in passenger cars. This development inevitably forces car manufacturers, after-sales equipment providers, and also authorities to consider the unwanted potential of such systems to distract drivers from the primary task of driving. Methods suggested to investigate this question range from simple laboratory tests to extensive on-road studies, each of which has advantages and drawbacks. The lane change test (LCT) is a relatively new dual-task approach for evaluating the impact on driving performance while simultaneously performing a secondary task. It combines the advantages of classic reaction time measurement approaches with those of driving simulator techniques. It is meant to

FIGURE 8.1 LCT driving scene with signs. The signs indicate that a lane change from the current lane (here center lane) to the right lane is required.

provide a simple, cost-efficient, reliable, and valid tool for evaluating the demand of in-vehicle information and communication systems.

The basic idea of the LCT is to measure and quantify a degradation in driving performance that occurs when a test participant performs secondary tasks simultaneously when under test. The primary task consists of a simple, but highly structured, driving simulation. Test participants have to drive along a straight, three-lane road at a fixed speed of 60 km/h and are requested to repeatedly perform lane changes according to instructions on signs that are placed in pairs along the roadside (see Figure 8.1). What is measured is performance degradation on this primary LCT task due to the demand of performing at the same time a specific secondary task under test. The method can be applied to a large variety of secondary tasks as long as these are not in conflict with the LCT itself, such as tasks that require speed variations to be performed or tasks that require the driver to follow certain navigation instructions. Among the applicable tasks that can be assessed are all kinds of interactions with in-vehicle information, communication, entertainment, and control systems: manual, visual, haptic and auditory/vocal, and combinations thereof. The LCT applies to the testing of original equipment and aftermarket systems, as well as to integrated and portable systems, and can be implemented on a personal computer or in a driving simulator.

8.2 DEVELOPMENT OF LANE CHANGE TEST

The LCT was developed in the research project Advanced Driver Attention Metrics (ADAM; 2002–2004), which was a joint project of the car manufacturers BMW AG and DaimlerChrysler AG. One of the project goals was to find a surrogate technique for measuring driver distraction that is not only as valid and reliable as possible but also has a strong focus on efficiency, which means that it should not require sophisticated and expensive equipment. Further, it should be simple enough to be performed by a majority of companies manufacturing in-vehicle devices. Finally, such a relatively simple surrogate measurement technique should be designed to "distinguish black from white instead of detecting all shades of gray." In other words, it should mainly be intended to provide an objective means for filtering out extremely demanding secondary in-vehicle tasks, thereby allowing more precise questions regarding the details of system design to be addressed using more laborious methods such as eye glance measurement or tailored on-road tests.

The basic approach in the development of the LCT was to combine features of driving simulations with features of reaction time paradigms. Driving simulations are frequently used to evaluate in-vehicle tasks due to the numerous advantages of this research approach (see Chapter 7). It is obvious that there is a relatively high equivalence to real driving (e.g., with respect to posture and structure of visual-manual control effort required by the primary task). These important features provide driving simulations with face validity. However, realism in the scenarios might make data analysis and interpretation ambiguous; typically, variables of longitudinal and lateral performance (among others) are analyzed, and changes in one variable must be interpreted in light of the other variable. For example, changes in lateral performance must be interpreted differently for a driver who decides to reduce speed when asked to make an input on an in-vehicle system as compared with a driver who prefers to keep speed constant. Such complications make data interpretation a challenge.

In typical reaction time paradigms, compared with driving simulations, the options for test participants are largely reduced. One popular example of a reaction time task used in the domain of traffic research is the peripheral detection task (PDT; see Chapter 10). The PDT is one specific example of the so-called probe reaction task. In such tasks, a simple stimulus (e.g., a sound or a small light) is presented repeatedly and a test participant typically has to respond with a simple key press as fast as possible while performing a secondary task simultaneously. It is assumed that reaction time in this paradigm reflects limited cognitive capacity,[1] and probe reaction time is therefore an index for the cognitive complexity of the concurrent task. When the probe paradigm is used in automotive research to study driver demand, it is sometimes used in a dual-task approach (secondary task under test versus probe reaction task) and sometimes in a tertiary task situation with the following components: (1) driving (real or simulated), (2) secondary task under test, and (3) probe reaction task. The advantages of such reaction time paradigms are good experimental control (at least for the dual-task approach), the high number of measurements one can get within limited time, and, resulting from this, the high reliability that is typical of reaction time experiments. The disadvantage of the dual-task approach is that realism is quite low. Instead of driving, participants do nothing between two stimuli, and this might well lead to different strategies in terms of attention allocation as compared with a more realistic driving scenario. The ternary-task approach (which incorporates real or simulated driving) can be questioned in terms of intrusiveness since the third task (detect the probe stimulus and respond to it) is neither part of the driving task in a real situation nor of the investigated secondary task.

The LCT has features of both the driving simulation approach and the probe reaction paradigm. It can be regarded as a simple driving simulation with stimuli (signs) and responses (initiating lane change maneuvers) embedded in the driving task. This driving task, although largely artificial, nevertheless features aspects of real driving such as perception, reaction, maneuvering, and lane keeping. From another point of view, the LCT can be seen as a stimulus–response paradigm with complex stimuli: signs with "X" and arrows (see Figure 8.1), complex responses (steering maneuver), and a tracking task between two consecutive trials (lane keeping). Viewed in this light, it is clear that the LCT is not meant to represent a realistic driving simulation. It should rather be regarded as a surrogate driving task, which shares certain critical similarities with real driving. It is noteworthy that even in the most advanced driving simulations, validity is limited. Hoedemaker distinguishes between absolute validity (the absolute value of a certain metric, e.g., number of lane exceedences, corresponds to real driving) and relative validity (absolute values of metrics may be wrong, but relations between conditions, e.g., different secondary tasks, correspond to real driving) for driving simulators. Hoedemaker argues that meaningful conclusions can only be drawn on the relative size of effects (i.e., comparison of various independent variables) but not on the absolute values obtained in a driving simulator.

8.2.1 Lane Change Test Driving Task

The LCT comprises a simple driving simulation that requires a test participant to drive along a straight, three-lane road at a constant, system-controlled speed of 60 km/h. Participants are instructed, by pairs of signs that appear simultaneously at variable intervals on both sides of the road, in which of the lanes to drive. Each lane is 3.85 m wide and track length is 3000 m (3 min at the system-controlled speed), which is sufficiently long to collect about 2 min of data on secondary-task performance. The signs are balanced such that each of the six possible lane changes (from left to middle, from left to right, and so on) is often performed equally (three times per combination for a track length of 3000 m). The mean distance from sign to sign is 150 m (a minimum of 140 m plus an exponentially distributed random variable with a mean of 10 m and an upper limit of 50 m). This leads to intervals of 8.4 to 11.4 s between two lane changes. The lane change signs are always visible but remain blank until the lane instructions on the signs appear instantaneously (i.e., pop up) at a distance of 40 m before the signs. This was done to avoid dependency on presentation quality or visual acuity of the test participants as would be the case if the information on the sign was always visible. Each sign contains two "X"s and one downward arrow, indicating the appropriate travel lane into which to change (see Figure 8.1). Each sign prompts a lane change, which means that there are no "catch trials" where two consecutive signs point to the same lane. The participants are instructed to change lanes quickly and efficiently as soon as the information on the sign becomes visible. Preferably, the lane change should be completed before the sign is passed. Furthermore, they are encouraged to stay in the middle of the lane between the lane changes. More technical details will be available in draft International Organization for Standardization (ISO) standard 26022 (in progress).

8.2.2 Lane Change Test Procedure

Test participants practice the lane change task as a single task first. The mean deviation values are analyzed and testing is continued only after a criterion of a lane change task score no larger than 1.2 m is met. Typically, this is achieved by most of the participants in the second or third practice run. Then the dual-task situation is practiced to learn the simultaneous handling of the primary and secondary tasks. No instructions are offered to participants on how to prioritize between the primary and secondary tasks. Participants are instructed to perform the primary and secondary tasks simultaneously to the best of their ability.

Practice is typically followed by a baseline run (only LCT without secondary task). Such baseline runs are embedded symmetrically in the design; for example, at the beginning and at the end, and if time allows for a third baseline run, additionally in the middle of the experiment. The LCT is applied in a within-subjects design. This allows for each participant to experience all relevant conditions and facilitates statistical comparisons. The various types of secondary tasks can either be performed blockwise (the same secondary task within one track, i.e., blocked runs) or in a mixed design (different secondary tasks within one track, i.e., mixed runs). If a mixed design is conducted, the type of secondary task is marked down by the experimenter during data collection. Each of the designs (blocked versus mixed) has advantages and drawbacks. The drawback of a blocked design is that it might appear unnatural or artificial (e.g., changing the sound setting of the radio over and over again). However, a blocked design makes instructions easier to administer and comprehend, and the participant has only one type of secondary task to keep in mind. In a mixed design, the participant has to memorize all possible secondary tasks. Therefore, the learning and training effort for a mixed design is higher than that for a blocked design. However, one can state that a mixture of secondary tasks within one track appears more natural. In either case, each participant should be given a clear explanation of the system operation and the secondary task(s) of interest. Before testing starts with the LCT, each participant should have at least two practice trials and up to five trials for each type of secondary task being investigated; fewer practice trials can be used if the participant is adequately prepared for the task.

It is important that the overall measured epoch of data recording for each secondary task is sufficiently long. Therefore, short tasks (e.g., sound adjustment) have to be performed more often than longer tasks (e.g., destination input to a navigation system). Overall time on task for each type of secondary task should be no less than 2 min. If it is obvious that the participant refrains from operating the secondary task (due to low motivation, inability, or adopting an unwanted dual-task strategy), this part of the data should be excluded from data analysis. During testing, the experimenter typically gives brief verbal instructions to the participant, such as navigation targets to be entered in a system or targets to locate on a paper map.

8.2.3 Lane Change Test Analysis

The aforementioned structural similarity of the LCT with a classical reaction time experiment enables examination of some data equivalent to reaction time, for example, the time between information onset (i.e., sign information appearing) and the initiation of the lane change maneuver. One can also calculate measures of lane keeping quality between the signs, as it is usually done in driving simulation studies. Also, the quality of the lane changing maneuvers could be evaluated as an index of how much a driver under test is distracted by a secondary task. Finally, the number of missed signs can be used as a further index of demand due to a secondary task.

To keep the data analysis simple, but at the same time to allow for the complexity of driving, a single measure was developed, which covers all of these features: the deviation from a normative path model (see Figure 8.2). This deviation measure covers important aspects of the drivers' performance; namely, their perception (late perception of the sign or missing a sign), quality of the maneuver (slow lane change results in larger deviation), and lane keeping quality, which all result in an increased deviation. Average deviation is the total area between a "normative" model and the driving course divided by distance driven.

As outlined earlier, the lane change signs contain no information (i.e., they are blank) until 40 m before the sign. At this point, the instruction content of the sign becomes visible. Assuming a reaction time of 600 ms to initiate the lane change, participants cover a distance of 10 m within this time 16.67 m/s (0.6 s = 10 m). This reaction time was estimated to be a minimum for such a choice reaction task. Therefore, the normative lane change starts 30 m before a "lane change" sign to take reaction time to the signs into account. The lane change has a length of 10 m regardless of whether a single- or a double-lane change is to be performed. This feature of the normative model is due to empirical observations: drivers appear to make more pronounced steering maneuvers when they have to change over two lanes as compared with one lane. It is noteworthy that it should not be possible for the participant to perform better than the normative model since that would equally lead to a measured deviation from the normative model, which would, in this case, be invalid.

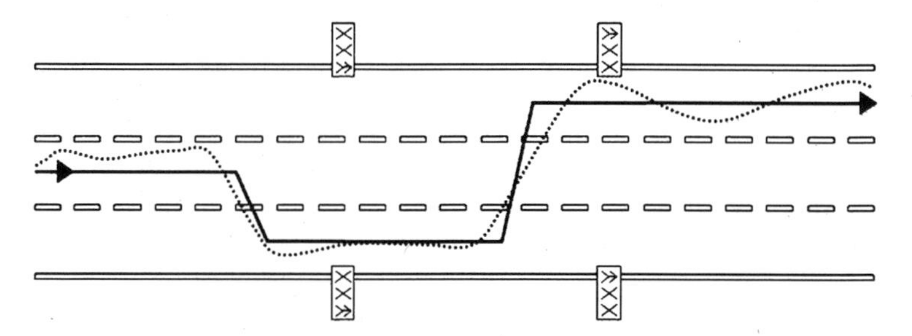

FIGURE 8.2 Normative path. According to the sign, the driver has to change from the center to the right lane and then to the left lane (driving direction is from left to right). The driving course (dotted line) is compared with the normative model (solid line). Average deviation from the normative model is the total area between the normative model and driving course divided by distance driven. (Note that the figure is not to scale.)

The mean deviation between the normative model and the actual driving course is calculated as the performance parameter of major interest. Only the section(s) of a track in which the participant is performing the respective secondary task is used for analysis (i.e., those sections between tasks or where the experimenter gave instructions are excluded from the analysis). To facilitate this, the relevant sections will typically be marked by the experimenter with appropriate markers in the recordings during the session. Epochs with secondary tasks that were not completed (due to time limits), however, can be included in the analysis as long as the participants worked on the secondary task to the best of their ability, as instructed.

The resulting mean deviation values are calculated for each experimental condition of interest; for example, for baseline driving (primary task only), for each secondary task under test, and for reference tasks (i.e., tasks that are not of interest *per se*, but are used for comparison). Outlier detection and correction is applied on the basis of experimental conditions. For each experimental condition, mean deviation values differing by more than two standard deviations from the average of that condition are set to the value that corresponds to the average plus or minus two standard deviations. Since the results will be typically positively skewed (more extreme values above average than below average), this procedure will more often make corrections to high outliers as compared with low outliers. The mean deviation values are then analyzed using an appropriate statistical test (e.g., analysis of variance or *t* test, depending on the specific design and research question).

8.2.4 Development of the LCT in the Advanced Driver Attention Metrics Project

The ADAM project developed and tested a systematically designed set of surrogate driving methods, which should be simple and efficient but at the same time should go beyond simple task durations measurement (e.g., 15 s rule; see Chapter 25). Several surrogate methods were considered:

1. *Occlusion method.* This technique was employed as the only non-dual-task method in the project (see Chapter 9 for a detailed description).
2. *Simple reaction time measurement.* A test participant has to respond via a key press to a simple visual stimulus on an empty screen. Reaction time to this probe stimulus and errors are measured while the participant performs a secondary task. This method can be regarded as one variant of the PDT (see section 8.2 and Chapter 10).
3. *Choice reaction time measurement.* Since real driving comprises multiple stimuli with multiple responses, a step toward this direction was made. Therefore, the surrogate primary task was now a three-choice reaction task. Simple arrows (left, right, down) were presented on a computer screen, and participants had to respond by turning a game steering wheel to the left or to the right or by pressing a foot pedal correspondingly. Note that the general setup of this method was very similar to the setup shown in Figure 8.3, with the driving scene on the screen replaced by the arrows.

FIGURE 8.3 LCT desktop setup. In a typical desktop setup, a game steering wheel with foot pedals and a cathode ray tube (CRT) monitor are used. The device under test is placed at a position that is close to the position intended for usage in a vehicle.

4. *LCT.* As described earlier, the LCT was designed as a next step toward greater similarity to the driving task (driving simulation), although retaining most of the features of a choice reaction task.
5. *Fixed-base driving simulation.* A simple driving simulation with a car-following scenario on a motorway with surrounding traffic was used.

A set of 12 secondary tasks was tested with each of these surrogate methods in separate studies. The secondary tasks included those that were performed with an advanced in-vehicle information system (e.g., changing radio stations, input of a street address to a navigation system) and other tasks that people do while driving, such as searching for coins or looking up cities on a paper map. The same 12 secondary tasks were tested extensively in a high-end driving simulator with a 360° field of view and an advanced motion system. The driving measures included lateral and longitudinal control, eye glance behavior, and subjective ratings. The summarized results from 85 test participants for the 12 secondary tasks were then compared with the results obtained with each of the surrogate methods. The results revealed that the LCT showed the highest correlation with the results of the high-end driving simulator. In a subsequent validation study, a slightly different set of secondary tasks (adjusted to on-road testing) was tested in a field study with an instrumented vehicle on a German highway and with the same sample of participants in the laboratory with the LCT. Over the eight secondary tasks, the correlation between lateral performance in the field study (standard deviation of lane position) and the LCT value was $r = .715$.

8.3 EXAMPLE STUDY: EXPLORING TWO NAVIGATION SYSTEMS

A prototypical experiment with the LCT was carried out at the Waseda University in Tokyo. The purpose of this study was to test several tasks on two aftermarket navigation systems and to test the sensitivity of the LCT to both cognitive distraction (i.e., distraction due to tasks, which requires mainly cognitive operations but little or no visual or manual effort) and visual-manual distraction. One navigation system had a remote control with several buttons and switches, which are all intended to be operated with the thumb, and the other was operated via a touch screen. A total of 20 participants (10 male, 10 female; aged 21–59 years) were tested in individual sessions of about 90 min. Each was tested on nine tasks for about 150 s each and additionally in three baseline runs without any secondary tasks (i.e., driving only). The nine tasks were

Cognitive easy. Continuously add 2 from a given starting number of 282.

Cognitive hard. Continuously subtract 7 from a given starting number of 581.

Visual easy. Here, a visual-manual search task was employed. In this task, the participant had to find the one larger circle in an array of circles that were displayed on a screen at a position similar to the typical position of an in-car navigation system. They then had to move a cursor over this target by means of a small keyboard and confirm this selection. Immediately after that, the next array of circles were presented.

Visual hard. This task differed in two ways from the visual easy task. First, the difference in size between the small and large circles was much smaller than in the visual easy task. Second, the marker (gray area in Figure 8.4) was narrower, that is, it covered only one-tenth of the screen so that several keystrokes were necessary to move the marker over the target. Figure 8.4 shows screen examples for the hard condition.

Remote control—map scale. This task was considered to be quite easy. The experimenter named a target scale setting (e.g., "500 m") for the navigation

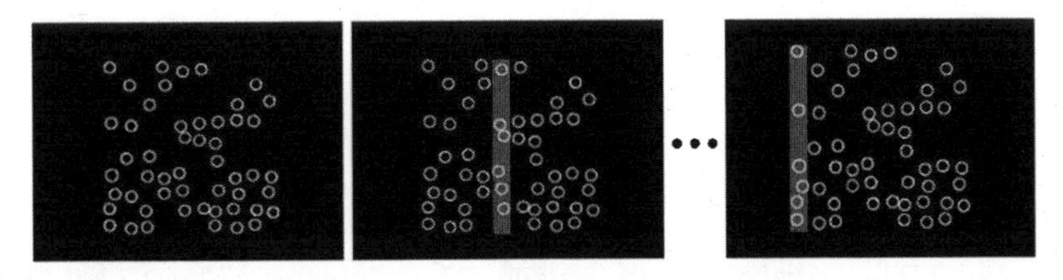

FIGURE 8.4 Example screens of visual search task (difficult condition; sequence from left to right). The participant is required to detect the larger circle on the screen. Three fingers of the hand that are not used for steering rest comfortably on a keyboard at a position comparable to the typical position of a gearshift. The gray bar has to be moved over the target with the left and right keys (here five steps from center of the screen to the left of the array). The selection is then confirmed with the key in the middle. In the easy condition, the distractors (nontarget circles) are much smaller than the target, and the gray bar covers half of the screen so that only one key press (left or right) is required to mark it.

map, and the participant then adjusted the scale with a two-way rocker switch on the remote control.

Remote control—show points of interest. The experimenter gave a target category such as "hospital" or "gas station," and the participant's task was to navigate through three screens in the system menu, select the target category from a list, and confirm this selection so that the targets were displayed on the navigation map. This task was solved using an eight-way control with a push function for confirmation.

Remote control—map scrolling. The participant's task was to move the target cross on the navigation screen to one of five possible targets (eight-way control). As soon as the participant reached the target, the experimenter named the next target.

Remote control—navigation input. The participants had to navigate to the speller of the address input (three steps in the menu) and enter a target city name (three hiragana characters) with the eight-way control.

Touch screen—navigation input. Here, the participants had to push a hard key first and then two keys on the touch screen until the target city name could be spelled (three hiragana characters). Note that this task was largely comparable to the navigation input on the remote control system.

Baseline. At the beginning, in the middle, and at the end of the experiment, the participants made one LCT run without a secondary task.

All tasks were repeatedly performed for one LCT run (e.g., blocked design). At the end of each run, the participants were asked to rate the difficulty of that condition on a scale of 1 (very easy) to 5 (very hard).

8.3.1 Results for LCT Mean Deviation

Figure 8.5 summarizes the main results of this study. The comparisons were analyzed with *t* tests and one-way ANOVAs. As can be seen, LCT values (mean deviation) differ significantly for the two cognitive tasks, the two visual-manual tasks, and for the five tasks on the remote control system. The comparison of the navigation input tasks on the two different systems (remote control versus touch screen), however, revealed no significant effect. For the analysis shown in Figure 8.6, the sample has been split into a group of younger participants ($n = 11$, 21–39 years) and older participants ($n = 9$, 40–59 years). The secondary tasks are ordered according to the LCT mean deviation value of the younger subsample for illustrative purposes. As can be seen, the order of the secondary tasks is mostly the same for both subsamples with the map-scaling task being the only exception from this picture. There is some spreading of the two curves to the right, which was statistically confirmed by a significant interaction in an ANOVA with the factors task and age groups, $F(11,198) = 5.1$, $p < .001$. This interaction can be interpreted such that difficulties with the more complex tasks are more pronounced for older participants. However, it is also noteworthy that most practical comparisons (e.g., comparing the two levels of the cognitive tasks) would lead to almost the same decisions for the two age groups.

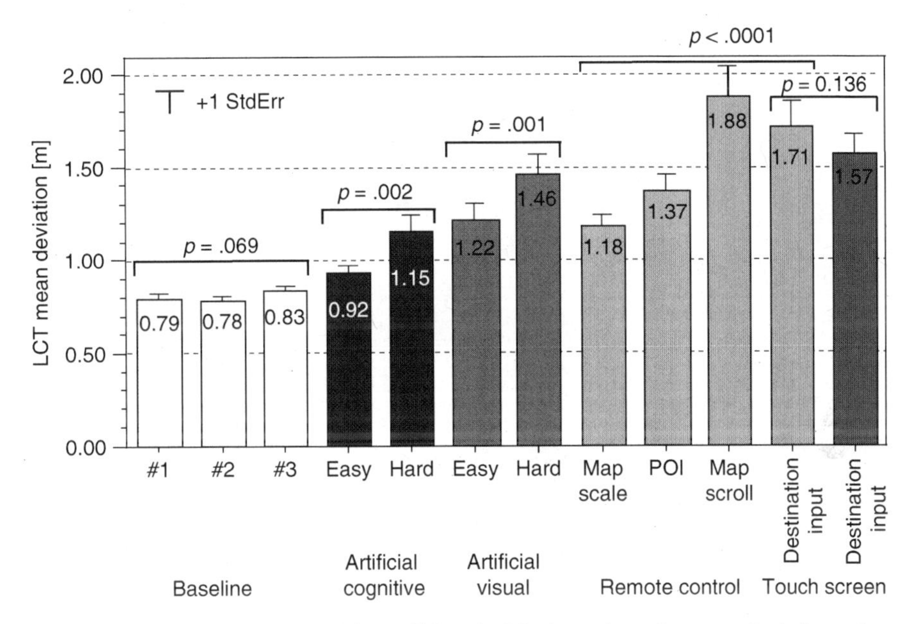

FIGURE 8.5 Results of an empirical LCT study. The bars show the mean deviation values for several secondary tasks tested in this study. The *p* values show the statistical significance level for selected comparisons.

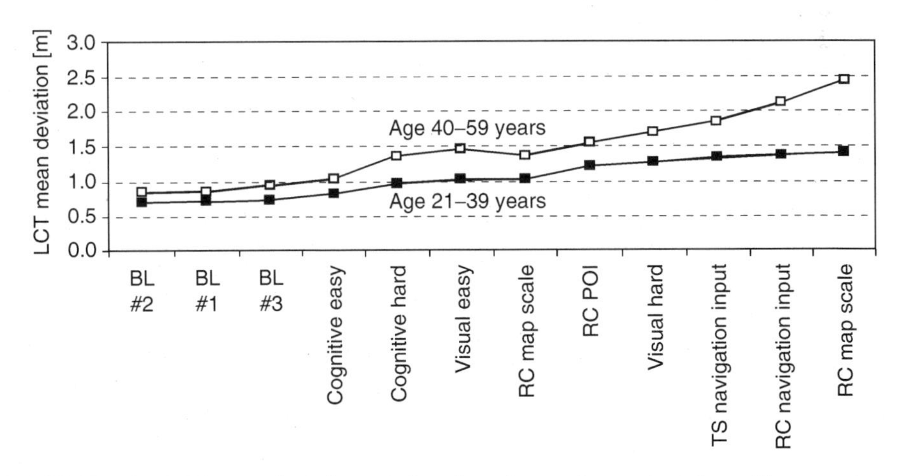

FIGURE 8.6 LCT mean deviation as a function of secondary task and age. The secondary tasks are ordered according to the mean deviation values of the younger subgroup.

8.3.2 Results for Task Duration

Task durations were measured for completed tasks (i.e., when the participant was able to solve the task before the end of the LCT track). For the artificial cognitive tasks or the artificial visual-manual task, the duration for a single item was calculated (one number or one target detection, respectively), although the participants

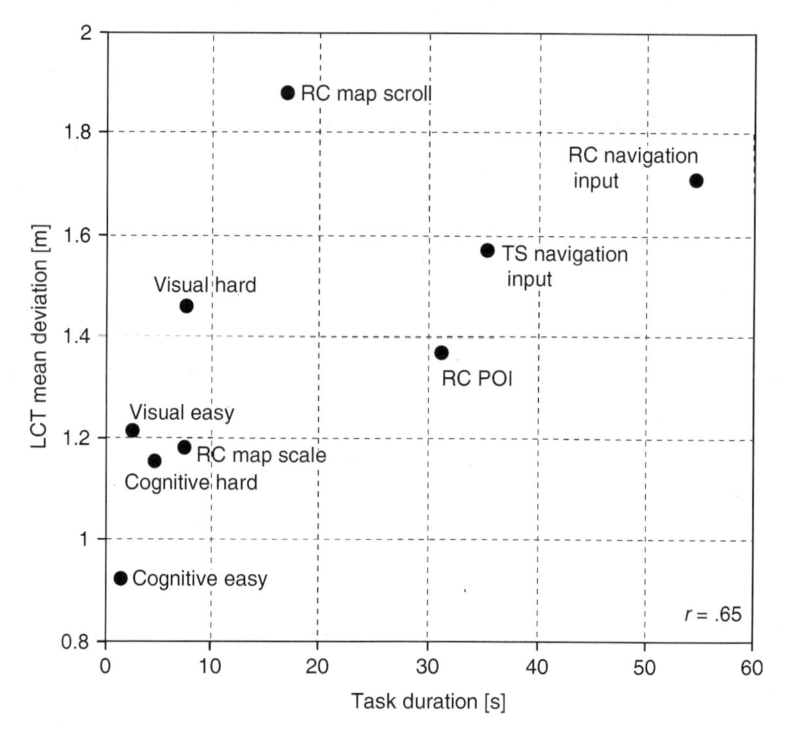

FIGURE 8.7 Correlation of LCT mean deviation and task duration. (LCT mean deviation values are the same as those depicted in Figure 8.5.)

worked in a more or less continuous stream on these tasks. Figure 8.7 shows the correlation of task duration and LCT mean deviation values. The correlation of $r = .654$ provides some support for the frequently proposed assumption that task duration is a predictor of driver distraction. However, a closer look at the data exemplifies the general claim that dual-task methods such as LCT can measure task intensity beyond task duration. Map scrolling is identified convincingly by the LCT and by subjective ratings on task difficulty as being the most challenging of the tasks. The duration of this task, however, is relatively short when compared with the other tasks.

8.3.3 Subjective Ratings

The participants were required to give an immediate rating on subjective task difficulty for each secondary task (i.e., after each test drive with a secondary task). Ratings ranged from 1 (easy) to 5 (difficult). The lowest average ratings were found for the easy cognitive task (3.30) and the highest ratings for the navigation input using the remote control system (4.85). The correlation between subjective ratings and the LCT deviation value over the nine secondary tasks was $r = .834$, whereas the correlation of subjective ratings with task duration was $r = .622$. The high correlation between the LCT deviation value and the subjective ratings is in line with previous findings. For example, Mattes[3] reported a correlation of $r = .897$ over 12 secondary tasks.

When the difficult cognitive task is eliminated, the correlation of LCT and subjective ratings increases to $r = .961$. The mean subjective rating for the cognitive task was 4.45, whereas a linear regression of the other secondary tasks would predict a rating of only 3.65. In other words, the difficulty of the hard cognitive task was subjectively overestimated by the participants relative to the moderate LCT mean deviation it produced (1.15 m). An alternative explanation could be that the participants estimated the impact of the hard cognitive task correctly, whereas the LCT value obtained underestimated the level of difficulty. However, the very high linear correlation for the eight remaining tasks when the cognitive hard condition is eliminated speaks against this explanation. This is because one may well assume that there is some variation in cognitive complexity between the eight tasks, and if cognitive complexity would disturb the correlation (with subjective ratings too high as compared with low LCT values), the overall correlation would hardly be $r = .961$.

In brief, this experimental study with the LCT demonstrated that secondary tasks of a visual-manual nature (circle detection) can be distinguished by the LCT as well as secondary tasks on the basis of cognitive operations (counting tasks). Most importantly, in-vehicle system tasks of different complexity could be distinguished by the LCT (remote control system). An "outlier" in the comparison of the subjective ratings with LCT results (the difficult cognitive task) exemplifies the necessity of an objective, experimental method, when compared with a subjective evaluation of task difficulties.

8.4 STANDARDIZATION OF THE LCT IN ISO 26022

Since 2003, attempts have been made in the ISO Technical Committee 22 (Subcommittee 13, Working Group 8) to standardize the LCT. There has been general agreement that a valid, reliable, and sensitive method to estimate driver demand brought about by concurrent performance of a secondary task is needed. Standardizing such a task should facilitate comparisons among test sites considerably. The general approach in these standardization efforts is to describe the LCT as precisely as necessary, although keeping restrictions with respect to hardware requirements as low as possible. For example, application is not limited to the desktop setup as depicted in Figure 8.3. Instead, the method can also be implemented in a seating buck, driving simulator, or production vehicle—either with a desktop display or with beamer projection. The aforementioned ISO standard cites the appropriate viewing angles for these conditions and also provides a description of how to set up the LCT in a real vehicle. An advantage of the LCT compared with "traditional" driving simulators is the ease with which a production vehicle can be set up for testing because only the steering wheel movements need to be transferred to the LCT software.

8.5 FURTHER DEVELOPMENT AND OTHER RESEARCH

An obvious improvement of the LCT might be seen in the refinement of the analysis. During the development of the LCT in the project ADAM, tests were conducted to determine if individual normative models rather than a unified one could improve the sensitivity of the method. This idea seems to be straightforward: parameters for the normative model can be calculated from the baseline runs for each subject.

This takes into consideration individual differences, and error variance might well be reduced. Parameters that can be adjusted are the preferred subjective middle of the lane, the onset of the lane change, and also the duration of the lane change. However, during the ADAM studies, the advantages of such an additional analysis step had been considered too small when compared with the increased effort required. Recently, Tattegrain Veste et al.[4] reported advantages for such an individual analysis for their own data. A reanalysis of the data from the Waseda University study, however, again brought no advantage of such an individual analysis (i.e., effect strength did not increase). In contrast, individual adjustments brought a clear improvement when applied to another data set that came from a study of three different cognitive tasks. That study had provided hardly any significant results. Statistical significance of the main effect (comparison of three secondary tasks) changed from $F(2,34) = 1.01$, $p = .374$ for the regular analysis with the standard normative model to $F(2,34) = 2.97$, $p = .065$ when individual normative models were applied. More research is needed to find out whether such adjustments should be generally recommended. If so, the procedure for individual adjustment needs to be carefully defined; for example, which parameters of the normative model should be adjusted (lane offset for each of three lanes, start of lane change, or duration of lane change) and which epochs of the data should be used for parameter fitting.

Meanwhile, there have been attempts to expand the diagnostic power of the LCT. The aforementioned idea to calculate more detailed parameters instead of the overall mean deviation measure has been tested within the European adaptive integrated driver-vehicle interface (AIDE) project.[5] Here, parameters have been estimated for response delay, lane change duration, and lane keeping variation between signs. This analysis had been stimulated by the findings from the HMI and the safety of traffic in Europe (HASTE) project—visual secondary tasks affect both event detection performance (e.g., increased PDT response time) and lateral control (e.g., increased standard deviation of lane position), whereas cognitive secondary tasks lead only to reduced event detection performance although lateral control is unaffected or even improved.[6] In the AIDE project, however, a reanalysis of LCT data for cognitive secondary and visual secondary tasks did not confirm this pattern. The parameters for response delay and lane keeping quality did not show the expected sensitivity to distinguish cognitive from visual-manual distraction. However, in a recent study, Engström and Markkula[7] tackled the same basic question with a different research approach. The authors focused on frequency statistics of extreme incidents in the LCT data. In their reanalysis of the same data as reported in the AIDE study,[5] they convincingly demonstrated that cognitive workload results mainly in an increased number of detection or interpretation errors (changing to wrong lanes or disregarding lane change signs), whereas visual-manual tasks mainly resulted in control errors such that the driver produced large overshoots or completely came off the track. This was confirmed both in qualitative and quantitative analyses. These findings suggest that the LCT might well be developed further into a research tool that allows for deeper analyses that go beyond the aforementioned claim to only distinguish black from white.

Beside such attempts to extend the analysis, variations of the driving task itself have been tested. Numerous variations were investigated during LCT development in the ADAM project, such as variations in speed, length of intervals between the signs, number of lanes (five instead of three), or even the use of a leading vehicle that drives along the normative path. In particular, variations of the presentation mode have been investigated (e.g., permanently visible information on the sign or lanes without signs where the sign or other markers pop up at a certain distance). This finally led to the setup with permanently visible signs, but information that appears at a distance of 40 m only. Another research idea was followed by Schwalm,[8] in which he replaced the visual lane change information with auditory instructions to test effects of spatial congruency for auditory information.

8.6 DISCUSSION

In the short time that the LCT has been used in automotive research, it has received considerable interest. The exemplary study reported in this chapter demonstrates, well, how the LCT can serve as an objective means to categorize in-vehicle secondary tasks. It was shown that (a) the LCT can distinguish among secondary tasks of different complexity, (b) that it can be applied to both visual-manual and cognitive tasks, and (c) that the LCT provides an objective measure of the level of distraction, which cannot be replaced by measuring only task duration or by subjective ratings of task difficulty by drivers.

Beyond this practical use of the LCT, the method has stimulated significant research effort. The attempt by ISO Committee TC22/SC13/WG8 to standardize the method is extremely helpful here, since a standardized procedure will clearly facilitate comparison of research results coming from different laboratories. Future research should show whether the LCT can be developed further into a tool for more in-depth analysis of secondary tasks. First steps in this direction have been reported, as demonstrated by those studies trying to identify cognitive and visual-manual distraction. Further exploration of ancillary conditions will help to improve the sensitivity and interpretation of the method. A study reported by the Japanese Automotive Research Institute (JARI),[9] for example, points to the fact that experimental effects obtained with the LCT might be diminished with increasing LCT experience of the participants. It needs to be determined whether such findings can be replicated, and whether this observation is specific to the LCT or it is a general feature of this kind of dual-task method.[10] Variations to the LCT for research purposes will also show whether the field of application of the LCT can be extended.

Despite the possibility for improvement and modification, the main value of the LCT is the application of the standard version as a simple, cost-efficient, yet reliable and valid tool for in-vehicle task comparison. Early experiences with the LCT suggest that it might well meet the claim of being the helpful link between synthetic laboratory measurement and driving stimulation it was designed to be.

REFERENCES

1. Kahneman, D., *Attention and Effort*, Prentice-Hall, Englewood Cliffs, NJ, 1973.
2. Hoedemaker, M., *Driving with Intelligent Vehicles*, Delft University Press, Delft, 1999.
3. Mattes, S., The lane-change-task as a tool for driver distraction evaluation, In *Quality of Work and Products in Enterprises of the Future*, Strasser, H., Kluth, K., Rausch, H., and Bubb, H. (Eds.), Erognomia, Stuttgart, Germany, 2003, p. 57.

4. Tattegrain Veste, H., Bruyas, M.P., Letisserand, D., and Blanchet, V., Lane change test: Results with adapted trajectories, LCT Task Force meeting of ISO TC22/SC13/WG8, Munich, May 8, 2006.
5. Östlund, J., Peters, B., Thorslund, B., Engström, J., Markkula, G., Keinath, A., Horst, D., Mattes, S., and Föhl, U., Driving performance assessment—methods and metrics. EU project AIDE, IST-1-507674-IP, D2.2.5, 2005.
6. Östlund, J., Nilsson, L., Carsten, O., Merat, N., Jamson, H., Jamson, S., et al., Deliverable 2—HMI and safety-related driver performance (No. GRD1/2000/25361 S12.319626). Human Machine Interface And the Safety of Traffic in Europe (HASTE) Project, 2005.
7. Engström, J. and Markkula, G., Effects of visual and cognitive load on the lane change test—preliminary results, LCT Task Force meeting of ISO TC22/SC13/WG8, Munich, May 8, 2006.
8. Schwalm, M., Die Schonung kognitver Ressourcen durch die Nutzung auditiv-raeumlicher Kommunikation in Navigationssystemen (Saving cognitive resources by using auditory-spatial communication in navigation systems), Diploma Thesis, University of Saarland, Saarbrücken, Germany, 2005.
9. Japanese Automotive Research Institute, Application of LCT to auditory-oral tasks and these combined visual-manual tasks, LCT Task Force Meeting of ISO TC22/SC13/WG8, Munich, May 8, 2006.
10. Shinar, D., Tractinsky, N., and Compton, R., Effects of practice, age, and task demands, on interference from a phone task while driving, *Accident Analysis and Prevention*, 37, 315–326, 2005.

9 Now You See It, Now You Don't: Visual Occlusion as a Surrogate Distraction Measurement Technique

James P. Foley

CONTENTS

As can be seen throughout this book, driver distraction, regardless of the source, is a potentially serious safety problem. If a driver is not attending to the environment in which the vehicle is being operated, and if something changes in that environment, then the driver may have to act to avoid a crash. Driving has been described as years of boredom interrupted by seconds of terror. Driver distraction increases the probability that the driver will experience those seconds of terror.

To devise effective countermeasures to driver distraction, there is a need to measure and quantify (if possible) the level of distraction. Unfortunately, driver distraction is a multivariate phenomenon that no single measure can capture.[1] This chapter will address one surrogate measurement method, "visual occlusion," which has been widely adopted as a measure of the visual demand of a task performed concurrently with driving. As the visual demand of a secondary task increases, the likelihood that it will divert attention away from tasks critical to safe driving also increases.

The visual occlusion method is described in national standards (Society of Automotive Engineers [SAE] J2364[2]), international (International Organization for Standardization [ISO] 16673[3]), and voluntary (The Alliance of Automotive Manufacturers [The Alliance][4] and Japanese Automobile Manufacturer Association [JAMA][5]) guidelines and has been extensively researched (see Ref. 6 for a summary of the literature, and Chapter 24 of this book).

9.1 BACKGROUND

Driving is primarily a visual–manual task. The driver visually assesses the roadway, the intended path of the vehicle, and any obstacles or threats in that path and makes decisions on steering, acceleration, and braking to safely navigate the vehicle to its intended destination. Senders et al.,[7] the originators of the occlusion technique, called this the demand of the roadway. For an experienced driver, this primary driving task becomes a "brain stem" activity, where for most of the drive, little conscious effort is needed or expended. The driver continually looks to the roadway, checks mirrors, looks in different directions to assess traffic, determines whether an intended maneuver is possible, assesses risk, makes decisions, and activates vehicle controls. Some of the driver's eye glances will not be directly related to driving: for example, when looking at a passenger, reading a billboard, tuning the radio, finding and inserting a CD, or interacting with a navigation system. Thus, during the course of normal driving, there typically is a pattern of eye glances to and from the roadway (i.e., between the primary driving task and in-vehicle secondary tasks).

As more capabilities and features are added to the vehicle, be it for entertainment (e.g., MP3 player), information and communication (e.g., Internet connectivity, real-time traffic information), or advanced driver assistance (e.g., crash warning and mitigation systems), there will be more opportunities for the driver to be distracted by in-vehicle devices. A key determinant of distraction is where the driver is looking at any given point in time, and for how long.[8] When the driver is looking at the forward roadway, it is likely (but not guaranteed) that attention is being paid to the primary driving task. However, if the driver is looking at an in-vehicle device (e.g., radio or navigation system), it is less likely that the driver will notice and respond to an event in the forward roadway.

An observational study of "naturalistic driving" has found that engaging in a visually or manually complex secondary task increases the driver's risk of a crash or near-crash by three times compared with engaging while attentive in only the primary task of driving (i.e., baseline driving[8]; see also Chapters 16 and 17 of this book).

An important question is how to measure the impact of a given in-vehicle device on the driver's ability to drive safely. Is there a reliable method to determine the visual demand of an in-vehicle secondary task without the time and expense of naturalistic observation or direct measurement of eye glance behavior? Direct eye glance measures are regarded as the "gold standard" of measurement, but they are difficult to collect and time-consuming to analyze. Eye glance measures require human observers to reduce and code the data from video recordings, a tedious and time-consuming process,[9] or to utilize automated eye tracker systems that require dedicated hardware and software as well as expertise in their application.

Because of these factors, direct eye glance measurement is difficult to incorporate in a development process for a new product. An effective surrogate measurement method, especially one that can be used early in the product design cycle, is desirable and useful. The main goal when developing and designing in-vehicle information systems (IVISs) is to support drivers' acquisition of information quickly and safely without distracting them from the primary driving task.[10] Visual occlusion (when vision is obscured) provides one such a method. The occlusion method consists of systematically obscuring the participant's (driver's) vision and then removing the obscuration. This can be accomplished by turning a display on and off or by physically blocking the participant's vision intermittently with a screen or similar device such as eye goggles that contain a shutter device, which opens and shuts. Section 9.2 provides details of the occlusion procedure.

The pattern of back-and-forth eye glances (to and from the roadway) by the driver in the dual-task condition forms the basis for the occlusion technique (see Figure 9.1). The occlusion method estimates visual demand associated with performance of a secondary task using a protocol for intermittent viewing of the in-vehicle system. The periods of occlusion are considered to represent the time spent looking at the roadway.

The occlusion technique was pioneered by Senders et al.[7] to model driver behavior based on information theory. The development of the technique was prompted by self-observation of driving behavior during heavy rain and relating vehicle control (speed) to the limitation of the sweep of the windshield wipers intermittently clearing only a small portion of the windshield. Senders and Ward[11] developed and then refined a simple model that hypothesizes that a driver's attention is, in general, not continuously but only intermittently directed to the road. Between observations of the road, the driver develops an uncertainty about both the position of the vehicle and the presence of other vehicles or obstacles. When that uncertainty exceeds an internal threshold, the driver again looks at the road. As noted, Senders's experiments were designed mainly to investigate the primary driving task, and not to determine the maximal acceptable distraction from an IVIS. In this original context, occlusion measures the voluntary attention required as the driver responds to the demand of the roadway, which encompasses the characteristics of the car, the characteristics of the road, the conflicts in the traffic stream, movement of pedestrians, and other related factors.

More recently, and for the purposes of this chapter, the occlusion technique has been used as a method for quantifying the visual demand of IVISs. The Alliance principle of human-machine interaction[4] states that "systems with visual displays should be designed such that the driver can complete the desired (secondary) task with sequential glances that are brief enough not to adversely affect driving" (p. 22). Occlusion has been widely studied by automotive human factors experts and has been

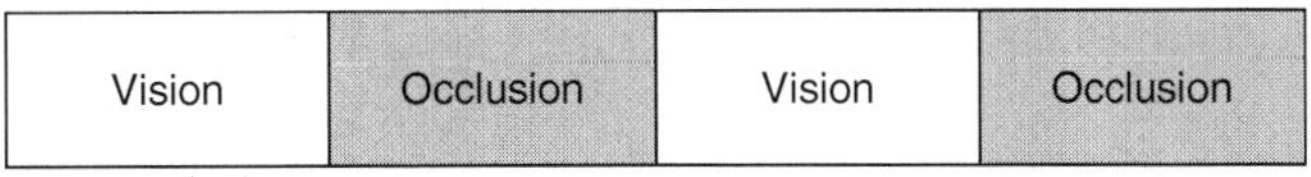

Vision	Occlusion	Vision	Occlusion

FIGURE 9.1 Vision and occlusion intervals.

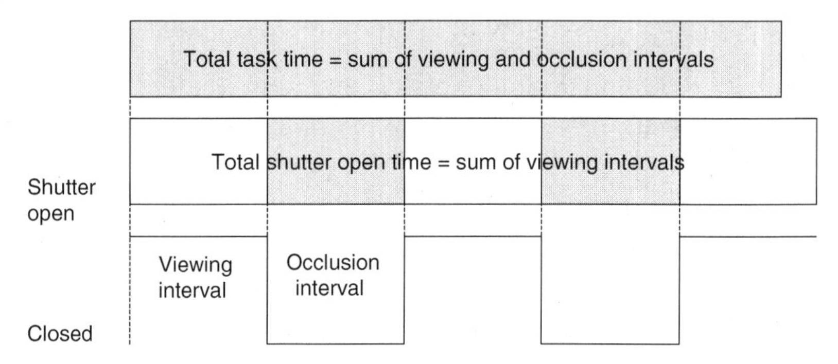

FIGURE 9.2 Illustration of occlusion and timing measures.

the subject of two major standardization efforts by the SAE and ISO (see Chapter 25). Figure 9.2 provides an illustration of use of the occlusion technique and the related timing measures.

9.2 PROCEDURE

For full details on how to apply the occlusion method and calculate occlusion parameters and measures, the reader is referred to the ISO occlusion technical standard 16673.[3] The key elements of the technique are described in Table 9.1.

The occlusion technique measures the performance of a participant on a secondary task both when there is no occlusion during the performance of the task (total task time unoccluded [TTTUnoccl]) and when the task is periodically occluded (total task time occluded [TTTOccl]), as described in the following text. Task durations are timed from the end of the instructions until the participant says "done." During the occlusion interval, neither the interface display nor controls are to be visible, but operation of the controls is permitted (note that most input occurs typically when vision is available, and usually one or two control actions are completed per open interval).[12] This protocol simulates drivers looking at the road but continuing to enter information via manual control. Any task that is a visual or visual-manual task is appropriate for measurement of visual demand using the occlusion technique. A task with a TTTUnoccl of less than 5 s is excluded because only one full occlusion cycle of shutter closed and shutter open is completed within 5 s, and this resolution is insufficient for the procedure to be meaningful. TTTUnoccl is calculated only if the resumability ratio, R (see the following text), is to be calculated; otherwise it is not needed.

Experimental investigations have included both system control of occlusion, where there is a systematic opening and closing of the device at predefined time intervals, and participant control of the device, where the shutter is opened on demand (see Ref. 6, p. 12). During the development of ISO 16673, an examination of the literature indicated that system control provided equivalent results to participant control while allowing a simpler and more repeatable method. Lansdown et al.[13] have criticized system-paced occlusion, arguing that it does not reflect actual driver behavior in which glance duration is substantially influenced by events in the driving

TABLE 9.1
Parameters of ISO Occlusion Standard

Parameter	ISO Standard	Comments
Means of occlusion (device)	Any	Literature shows variation in device type typically produces equivalent results; goggles are most frequent device
Test environment	Mock-up, vehicle buck, or real vehicle (parked)	Allows application early in design cycle through production vehicles
Interval pacing	System-paced	Equivalent results in literature, simpler experimental procedure
Vision interval (SOT)	1.5 s	Supported by literature, accepted by ISO experts
Occlusion interval (SCT)	1.5 s	Supported by literature, accepted by ISO experts
Number of participants	At least 10	Small enough to be practical and still provides statistical reliability
Participant's age	20% over 50 years of age	Decrease in abilities with increasing age is well known, sample should be representation of the wide range of driver capabilities
Participant's ability	Licensed driver	A priori requirement
Training	Two to five trials	Varies with participants' abilities
Number of trials	Five	Provides stable results
Primary task loading	None	Insufficient evidence that there is a need to load the participant during occlusion
Task	Tasks are timed from start to end without interruption, including errors and subtracting occlusion intervals	Standardized procedure
Task start	At the end of task instruction	Clear initiation of task
Task end	Participant says "done"	Clear stopping of timing
Task error	Observed by experimenter	Included in timing, provided as feedback once the trial is completed

environment. However, Lansdown et al. concede that system-paced occlusion may work as drivers rarely glance away from the roadway for more than 2 s.[14] Since the intent of the occlusion technique is to measure visual demand, the inability of occlusion to exactly mimic more intermittent natural behavior is not a hindrance to its application.

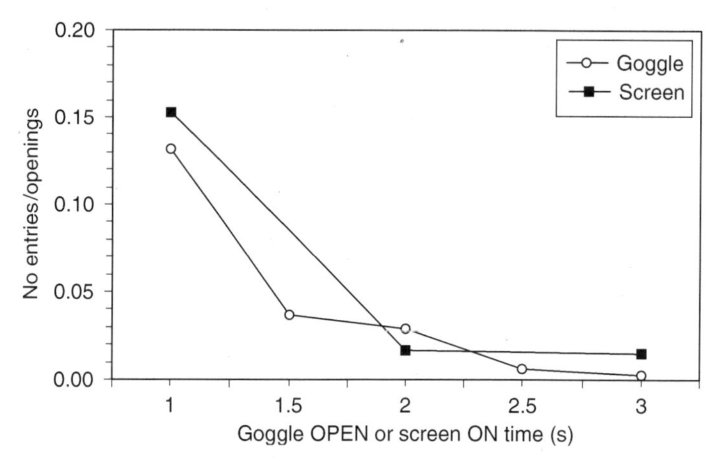

FIGURE 9.3 Increase in lack of entries for short open times.

Weir et al.[12] found that when the occluded interval is excessively long (>3 s), participants objected to the procedure and frequently refused to continue the experiment. When the interval is too short (<1 s), the procedure approaches uninterrupted viewing and loses its value (i.e., effectively there is no interruption of the participant's vision). The shutter open time (SOT) required to make an entry when interacting with a secondary task was systematically varied. Below a shutter open duration of 1.5 s, the number of shutter openings without a response rose sharply (see Figure 9.3), which indicates that the interval is too short for the participant to proceed. Secondary task entry and error rates were similar to baseline values obtained without occlusion. The 1.5 s open time is also consistent with the results of a number of on-track and on-road studies, which demonstrated that a 1.5 s glance to an in-vehicle device is sufficiently short from the standpoints of safety, functionality, and comfort.[15] Therefore, a SOT of 1.5 s is practical and acceptable for a standardized occlusion procedure.

The parameters relating to the occlusion procedure specified in ISO standard 16673 of interest are:

1. Total shutter open time (TSOT), the total time during which vision is unoccluded when using the occlusion procedure, that is, the duration of visual attention when completing a task in interrupted steps
2. TTTUnoccl, the total task time for a given task completed without occlusion (i.e., the duration when completing a task without interruption)
3. *R*, the resumability ratio, which is the ratio of the mean TSOT to the mean TTTUnoccl (see Section 9.6)

9.3 APPARATUS

The occlusion technique has an advantage in that it can be readily applied in the laboratory using only the in-vehicle device. Since there are many effective ways of providing the occlusion, even investigators with limited resources can use this technique. The most common method is the use of occlusion goggles,[16] which allows

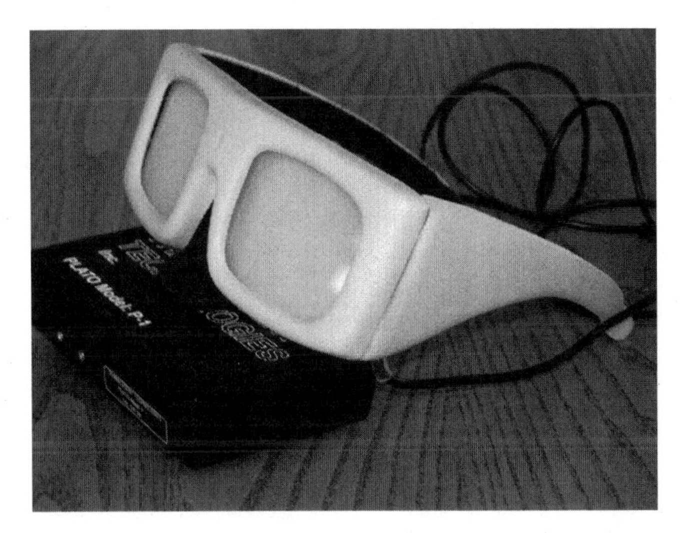

FIGURE 9.4 Visual occlusion goggles.

precise control over shutter open and shutter closed durations while causing the participant little, if any, discomfort (Figure 9.4). Senders' original method employed a helmet with a pneumatically operated visor that raised and lowered to obscure the driver's vision (see http://ppc.uiowa.edu/drivermetricsworkshop/ and access "Lunch Speaker: Professor John Senders" for a video clip of the occlusion helmet in action [accessed December 6, 2006]).

Visual occlusion can be used to measure the visual demand of an IVIS interface in the early prototyping stages. The system under investigation should be operational and fitted to a vehicle, simulator buck, or mock-up in the intended location of the interface in the vehicle (i.e., the viewing angle and control placement relationships shall be maintained). The ocular illumination levels in the vision and occlusion intervals should be comparable so that dark/light adaptation of the participants' eyes is not necessary during the procedure.[3] Occlusion can also be used in a driving simulator or on a test track with the proper safeguards.

9.4 PARTICIPANTS

As required by ISO 16673, at least 10 participants should be used, and 20% of the participants should be over the age of 50 to provide a representation of a range of capabilities.[3] Monk et al.[17] found significant differences between age groups in performance in interrupted tasks. Karlsson and Fichtenberg[18] indicated that older participants (>40 years of age) report high subjective ratings of difficulty, and Curry[19] found that participants over 56 years of age used a greater number of glances to complete IVIS tasks. Thus, it is important to include older participants to ensure that the data are representative of the overall driving population. The participants should be licensed drivers not familiar with, or technically knowledgeable about, the specific driver interface under investigation.

9.5 CRITERIA

Although the criterion value for determining excessive visual demand via occlusion is not uniformly agreed upon, several organizations have set limits. ISO standard 16673 provides consensus on the visual occlusion procedure. The specific criterion value (to aid in the decision of whether or not the task will be available to the driver while driving) is established by the user of the standard. Some criteria have been published. The Alliance guidelines state that TSOT, for a task accessible to the driver while driving,[4] should not exceed 15 s. The JAMA guidelines[5] (Section 4.2) state that "The visual information to be displayed shall be sufficiently small in volume to enable the driver to comprehend it in a short time or shall be presented in portions for the driver to scan them in two or more steps" (p. 2). The occlusion method is included in the verification techniques in Annex 3 to the JAMA guidelines, stating that the total shutter opening time (with vision intervals of 1.5 s and occluded intervals of 1.0 s) should be no more than 7.5 s. This criterion is based on the observed correlation between TSOT and total glance time (TGT), measured using direct eye glance performance.[18,20,21]

Occlusion also allows for a qualitative assessment of a task. If, for a prescribed SOT of 1.5 s, the participant cannot complete a task under occlusion or makes excessive errors, then that is a strong indication that the task should be redesigned if it is desired that the driver have access to the task while driving.

9.6 THE *R* RATIO METRIC

A second metric based on occlusion is R, the resumability ratio. If a visual or visual-manual task can be readily resumed after vision is interrupted, it is logical that performance will be better and attentional demand will be lower than if the driver cannot readily resume the task. An example of a task that would be difficult to resume is a driver scrolling through a map display to determine the distance to the next intersection. Before the driver finds the desired information, he or she would typically look back to the roadway. When the driver's attention returns to the display, it typically would take time to reorient his or her attention to the current position of the car relative to the map. This reorientation increases the length of the glance to the display. Tasks that are unsuitable to be accessible may take longer to complete when interrupted. The ease with which a task can be resumed after interruption depends on the phase of the task or subtask. Monk et al.[17] stated that a goal activation model predicts that resumption is easier near the start of the task.

The resumability ratio, R, is the ratio of TSOT to TTTUnoccl. Most tasks have a shorter TSOT than TTTUnoccl, perhaps because participants feel some time pressure with the occlusion procedure and are therefore more efficient in performing the secondary task.[18,22] Another factor is that the participant may use the shutter closed time to anticipate the next interaction.

There are still questions about the appropriate interpretation of R.[23] The value of R for a task that can readily be resumed is presumed to be ≤ 1, as the task times with and without occlusion are (nearly) equal. Lower values ($R < 1$) may suggest that a task affords some level of "blind interaction" so that parts of the task can be completed even without the participant being able to see the display—similar to a skilled typist who rarely looks at the keyboard. Or it may mean that the closed time is being used to anticipate the next steps. *R* values higher than 1.0 may indicate that the occlusion procedure interferes with task completion time. For example, for a very simple task, the shutter may remain open for a longer period of time than is needed to complete the task. However, the unoccluded procedure does not artificially extend the time period. A study that compared the interaction with four production navigation systems[24] found that the *R* value did not discriminate between task type and task difficulty. It was concluded that *R* may be useful in assessing the suitability of long duration tasks, which may be comparatively simple with a low visual demand rate, but which would not satisfy a specific TSOT criterion.

Noy et al.[25] stated that occlusion may be a necessary step in assessing the suitability of secondary tasks for time-sharing with the primary driving task, but it is not sufficient. Task resumption or interruptibility is an important aspect of distraction and is as relevant as task duration, but one measure alone does not capture the full impact of time-sharing/multitasking on primary task performance.[1]

9.7 CONCLUSION

Visual occlusion is an established technique for the measurement of the visual demand of in-vehicle secondary tasks. It can be applied in the early stages of system prototyping or used to assess a production device. It is typically easier and less costly to administer than direct eye glance measurement, and occlusion measures have been shown to be strongly correlated with TGT and other performance measures. Although the specific criterion values may vary according to the user's needs, there are values published by The Alliance,[4] SAE,[2] and JAMA[5] that the reader may find informative.

Occlusion must be applied and interpreted with care and expertise. For example, Tsimohni[22] observed that even when short SOTs (1.0 s) were used and it was predicted that it would be hard for participants to perform the task, there was almost no degradation in performance. This finding may indicate that a driver can compensate for a poor interface design, making assessment of a given interface more difficult. This is a common problem with human factors research. Participants typically want to do well even when characteristics of the interface design impede their ability to complete a task. Users of a device often assume that they are at fault rather than the design of the hardware or software. It should also be said that occlusion, by and of itself, does not provide a complete assessment of visual demand or driver distraction.[1,25] TSOT is only one of several parameters that should be examined to determine if a driver might be unduly distracted by a given task. Testing by occlusion, or by any other measure, is no substitute for a good human factors design process that considers a variety of factors and principles to create user-centered designs that do not undermine driving safety.

The use of occlusion-based measures to determine the ease of task resumption or resumability is a more involved process due to the need to collect the additional data for TTTUnoccl. The *R* ratio is not clearly supported by the literature, as there are conflicting findings reported to date. However, there is enough support for use of the *R* ratio to warrant its inclusion in ISO standard 16673, in which the occlusion procedure is described. Future research will hopefully provide a definitive position on *R*. In the meantime, occlusion will provide a useful tool to assess visual demand and, thereby, driver distraction.

9.8 SUMMARY

In summary, there are at least two applications of the occlusion method:

1. Determining the demand of the driving situation itself (as originally developed by Senders et al.[7])
2. Determining the demand that a secondary task has on the driver's ability to drive safely when engaging in that secondary task while driving as given in ISO 16673

The latter is the application described in this chapter.

The occlusion procedure is a valuable addition to the suite of human factors tools that allow for the measurement of factors that interrupt a driver's attention to the roadway. However, in many cases, a similar result could be achieved by using a simple heuristic. The heuristic is that short tasks are not a problem for a driver to engage in when driving, whereas long tasks, especially those with dynamic elements or deep menus, distract the driver. By analyzing a task it can be readily determined if it can be completed in a short period of time, for example, using a radio preselect button to tune a radio with a single keystroke. The analysis can also determine if a task has dynamic moving elements, for example, scrolling text or map, or a complicated menu structure. In the case of the simple task, the task is unlikely to consistently have a significant impact on the driver's attention. In the latter case, with dynamic elements or involved menus, it is likely to be very distracting to the driver and inadvisable to implement in a driver-vehicle interface. When using the occlusion procedure, the short task is out of scope and the task with dynamic moving elements typically cannot be completed providing an upper and a lower boundary for the application of occlusion. When a task falls between these two extremes, the occlusion procedure can provide a measure of the level of distraction associated with the task of interest.

REFERENCES

1. Crash Avoidance Metrics Partnership (CAMP), *Driver Workload Metrics Project: Task 2 Final Report*, DOT HS 810 635 USDOT, Washington, D.C., 2006.
2. Society of Automotive Engineers, *Navigation and Route Guidance Function Accessibility While Driving*, SAE Recommended Practice J2364, Society of Automotive Engineers, Warrendale, PA, 2004.
3. International Standards Organization, *ISO 16673 Road Vehicles—Ergonomic Aspects of Transport Information and Control Systems—Occlusion Method to Assess Visual Demand Due to the Use of In-Vehicle Systems*, Geneva, Switzerland, 2007.

4. Alliance of Automotive Manufacturers (The Alliance), *Statement of Principles, Criteria, and Verification Procedures on Driver Interactions with Advanced In-Vehicle Information and Communications Systems (Version 2.1)*, Author, Southfield, MI, 2003.
5. Japan Automobile Manufacturers Association (JAMA), *Guideline for In-Vehicle Display Systems—Version 3.0*, Author, Tokyo, Japan, 2003.
6. Stevens, A., Bygrave, S., Brook-Carter, N., and Luke, T., Occlusion as a technique for measuring In-Vehicle Information System (IVIS) visual distraction: A research literature review, TRL Report 609, 2004, www.trl.co.uk.
7. Senders, J.W., Kristofferson, A.B., Levison, W.H., Dietrich, C.W., and Ward, J.L., *The Attentional Demand of Automobile Driving*, Highway Research Record No. 195, 1967, 15.
8. Klauer, S.G., Dingus, T.A., Neale, V.L., Sudweeks, J., and Ramsey, D., *The Impact of Driver Inattention on Near-Crash/Crash Risk: An Analysis Using the 100-Car Naturalistic Driving Study Data*, National Highway Traffic Safety Administration, Washington, D.C., April 2006, DOT HS 810 594.
9. Smith, D.L., Chang, J., Glassco, R., Foley, J., and Cohen, D., Methodology for Capturing Driver Eye-Glance Behavior During In-Vehicle Secondary Tasks, *Transportation Research Board 84th Annual Meeting*, Washington, D.C., 2005 (CD).
10. Krems, J.F., Keinath, A., Baumann, M., Bengler, K., and Gelau, C., Evaluating visual display designs in vehicles: Advantages and disadvantages of the occlusion technique (document ISO/TC 22/SC 13/WG 8/N263), Geneva, Switzerland, 2000.
11. Senders, J.W. and Ward J.L., *Additional Studies on Driver Information Processing Final Report*, Report No. 1738, U.S. Department of Transportation, Washington, D.C., 1969.
12. Weir, D.H., Chiang, D.P., and Brooks, A.M., *A Study of the Effect of Varying Visual Occlusion and Task Duration Conditions on Driver Behaviour and Performance While Using a Secondary Task Human-Machine Interface*, Society of Automotive Engineers, SAE 2003-01-0128, 2003.
13. Lansdown, T.C., Burns, P.C., and Parkes A.M., Perspectives on occlusion and requirements for validation, *Applied Ergonomics*, 35, 2004, 225.
14. Wierwille, W.W. and Tijerina, L., Modeling the relationship between driver in-vehicle demands and accident occurrence. In *Vision in Vehicles VI*, Gale, A. et al. (Eds.), Elsevier, Amsterdam, 1998, pp. 233–243.
15. Baumann, M., Keinath, A., Krems, J.F., and Bengler, K. Evaluation of in-vehicle HMI using occlusion techniques: Experimental results and practical implications, *Applied Ergonomics*, 35(3), 2004, 197.
16. Milgram, P., A spectacle-mounted liquid-crystal tachistoscope, *Behavioural Research Methods, Instrumentation, and Computers*, 19(5), 1987, 449.
17. Monk, C.A., Boehm-Davis, D.A., and Tafton, J.G., Recovering from interruptions: Implications for driver distraction research, *Human Factors*, 46(4), 2004, 650.
18. Karlsson, R. and Fichtenberg, N., How different occlusion intervals affect total shutter open time, Reg. no. 202-0006, Volvo Car Corporation, *Exploring the Occlusion Technique: Progress in Recent Research and Applications Workshop*, Torino, Italy, available at http://www.umich.edu/~driving/occlusionworkshop2001/.
19. Curry, R., Greenberg, J., and Blanco, M., An alternative method to evaluate driver distraction, *ITS America Twelfth Annual Meeting Conference Proceedings, Intelligent Transportation Society of America*, Washington, D.C., 2002 (CD-ROM).
20. Hashimoto, K. and Atsumi, B., Study of occlusion technique for making the static evaluation method of visual distraction, *Exploring the Occlusion Technique: Progress in Recent Research and Applications Workshop*, Torino, Italy, September 2001, retrieved July 26, 2006, from http://www.umich.edu/~driving/occlusionworkshop2001/.

21. Asoh, T. and Iihoshi, A., Occlusion method to evaluate visual distraction to be caused by using car navigation systems, *Driver Metrics Workshop*, Ottawa, October 2–3, 2006, retrieved November 16, 2006, from http://ppc.uiowa.edu/drivermetricsworkshop/.
22. Tsimohni, O., *Time-Sharing of a Visual In-Vehicle Task While Driving: Findings from the Task Occlusion Method*, Technical Report UMTRI-2003013, University of Michigan Transportation Research Institute, Ann Arbor, MI, 2003.
23. Tijerina, L., CAMP driver workload metrics project: Correlation of the R metric correlation with driving performance and prior prediction, *Driver Metrics Workshop*, Ottawa, October 2–3, 2006, retrieved November 16, 2006, from http://ppc.uiowa.edu/drivermetricsworkshop/.
24. Harbluk, J.L., Burns, P.C., Go, E., and Morton, A., Evaluation of the occlusion procedure for assessing in-vehicle telematics: Tests of current vehicle systems, *Proceedings of the 50th Annual Meeting of the Human Factors and Ergonomics Society*, San Francisco, CA, 2006.
25. Noy, Y.I., Lemoine, T.L., Klachan, C., and Burns, P.C., Task interruptibility and duration as measures of visual distraction, *Applied Ergonomics*, 35, 2004, 207.

10 Distraction Assessment Methods Based on Visual Behavior and Event Detection

Trent W. Victor, Johan Engström, and Joanne L. Harbluk

CONTENTS

10.1 INTRODUCTION

A driver is following a lead vehicle at a constant headway with the intention of merging into a lane on the freeway. While he glances over his shoulder for an appropriate gap to merge into, the lead vehicle suddenly brakes. When he looks back at the lead vehicle, he cannot brake in time and a rear-end crash occurs. Alternatively, the driver could be reading a text message on his cell phone or turned around to argue with a passenger in the back seat when the lead vehicle suddenly brakes in stop-and-go traffic. In each scenario, the driver is looking away from the forward view. The first example is perhaps more interesting because it illustrates that the driving task itself, not just secondary tasks, can be distracting. Although the argument could be made that a good driver is always aware that an emergency situation could occur at any time, it is very difficult, if not impossible, to remain vigilant and pay attention to all relevant sources of information while driving. In particular, the simultaneous occurrence of an unexpected event and eye diversion has been demonstrated to play a key role in the causation of crashes and near-crashes.[1]

Drivers are regularly required to detect and respond to a variety of *predominantly visual* events, including lead vehicles braking, signs, a sharp curve over a hillcrest, an illegal turn by an oncoming vehicle, incline changes, and so on. They also engage in a variety of secondary activities involving the various senses as well as higher cognitive functions such as planning, decision making, and utilization of working memory. Vehicle control metrics (see Chapters 7 and 8 of this book) successfully capture some aspects of the impact of distraction on driver performance. However, as the body of research evidence has increased, it has become evident that vehicle control metrics are not sufficient by themselves to capture the impact of driver distraction, as illustrated in the above-mentioned examples. Because types of distraction and their manifestations vary, the measurement of distraction must also vary. Metrics based on driver visual behavior and the ability to detect relevant objects and events provide an important component in a "toolbox" of distraction assessment methods.

The aims of this chapter are to describe key conceptual factors in the measurement of distraction, to show how these can be measured by current visual behavior and event detection methodologies, and to identify and discuss outstanding issues. The methods discussed in this chapter are intended to be used in combination with other measurement metrics in various research settings, such as simulators, test tracks, and real traffic.

Three general methodological approaches are presented (1) metrics based on driver eye-movement data, (2) metrics based on natural object and event detection (OED), and (3) metrics based on artificial signal detection (the Peripheral Detection Task and related methods). The chapter concludes with a discussion of these methods, their advantages and disadvantages, and some suggested directions for future research.

Of central importance in the present context is the concept of *attention*, which broadly refers to the selection of relevant information. Thus, before going into the details of the different methods, the following section outlines a number of key factors that contribute to inattention in road crashes and presents a conceptual framework for how these factors are linked. This framework will then be used when describing and discussing the individual distraction assessment methods.

10.2 KEY FACTORS CONTRIBUTING TO INATTENTION IN CRASHES

Understanding the limitations of visual input and cognitive processes such as attention is essential to understanding the problem of distraction-related crashes. The illusion of a stable, fully detailed pictorial view of the external world is very powerful and gives us a strong subjective impression of seeing everything at all times. Thus, we often forget that our eyes are not cameras that deliver uniformly detailed, uninterrupted images (see, e.g., Chapter 5).

Attention can be defined as "the ability to selectively focus neural processing resources onto the most relevant subsets of all available sensory inputs."[2] Inattention can then be defined as improper selection of information, either a lack of selection or the selection of irrelevant information. It follows that driver inattention is the inappropriate selection of information that is relevant for safe driving, either by the lack of selection of driving information (e.g., eyes closed due to fatigue, low arousal or effort), or by the selection of irrelevant information (driving or non-driving related). Driver distraction refers to the latter case of driver inattention, i.e., the selection of irrelevant information to the extent that safety-relevant information is missed. Thus, distraction is here defined as a subset of inattention, referring to all instances when attention is misallocated, but excluding cases when attention is not allocated at all.

Inattention while driving may be associated with a number of different contributing factors. Here five key factors are described, selected because of their relevance for explaining inattention and the metrics covered in this chapter:

Stimulus saliency: The ability of a stimulus to capture the driver's attention and generate an appropriate response is highly dependent on stimulus saliency properties such as size, color, contrast, movement, and luminance.[3,4]

Shutter vision: We do not have access to a continuous stream of visual information. Blinks, saccades, and temporary occlusions result in periods of vision loss that mask visual transient responses of low-level feature detection mechanisms,[3] thus impairing bottom-up attentional capture, event detection, and response.[5]

Visual eccentricity: The visual eccentricity factor is the effect of stimuli falling on the retinal periphery instead of at the fovea (the fovea is the location at which visual processing is centered). A dramatic reduction in the performance of visual functions occurs toward the retinal periphery.[6] For example, due to this effect a decelerating lead vehicle is less likely to capture the attention of a driver who is looking away from the road at the dashboard radio, compared to looking straight ahead.

Cognitive factors: At one extreme, attention can be dominated by external events in a bottom-up, stimulus-driven manner.[7] The factors addressed so far mainly relate to this bottom-up process. On the other extreme, attention can be top-down and goal directed by cognitive factors such as knowledge, expectation, and current goals. However, attention is most often a combination of these bottom-up and top-down processes, which can be expressed as the product of competitive interactions. Recent advances in cognitive neuroscience have offered a detailed account of the neural basis for these competitive mechanisms, known as the *biased competition hypothesis*.[8] The biased competition hypothesis is based on the idea that multiple, hierarchically organized populations of neurons engage in competitive interactions that may be biased in favor of specific neurons by means of bottom-up activation or top-down signals originating from outside the perceptual system (mainly from prefrontal brain areas). Interference with the cognitive mechanisms generating this top-down bias—for example, by a cognitively demanding secondary task—will impair this ability to detect task-relevant information.

Expectancy can be seen as a cognitive factor that controls goal-directed attention.[7] To expect is to anticipate or consider probable the occurrence of an event and is related to drivers' readiness to respond. Expectations project past data or experience at all levels of the driving task. Thus, at higher levels of the driving task, expectancies are based on past experiences of occurrences (e.g., that vehicles will not respect certain traffic rules) or set by traffic signs regarding upcoming hazards on the roads or traffic rules.[9,10] At the skill-based sensorimotor level, expectation is so embedded into our senses that it alters our perceptions.[11,12] For example, we compensate for lags caused by biological feedback delays by directing actions to extrapolated *future* states of the world, such as time to collision and the future paths of objects.[13] Violations of expectancy disturb traffic throughput and increase conflicts and accidents because they cause drivers to take longer to respond to a given situation and increase the likelihood of errors.[9] Indeed, expectancy was an important factor in the 100-car study,[1] where it was found that drivers had difficulty responding appropriately when other vehicles performed unexpected or unanticipated maneuvers and when expectancies about the flow of traffic were violated, such as when there were sudden stops or lane changes. Expectancy is pointed out as a crucial factor influencing response times, errors, and traffic efficiency by numerous sources.[4,9,10,14,15] As shown in Table 10.1, traffic events can be ordered along a hypothetical continuum ranging from highly expected to highly unexpected events. These examples have been drawn from experiments (instructed) or are examples of real events (natural) that have caused crashes.[1,15–20]

Currently there is, however, no objective measure of degree of expectancy (the order of events in Table 10.1 is based on speculation only) and more research is needed to quantify its influencing factors. Thus, the purpose of Table 10.1 is not to classify each event exactly but to establish the concept of an expectancy continuum.

Many other cognitive factors are excluded from the scope of this chapter. Cognitive factors such as those involved in individual differences (e.g., skill, age, style), many aspects of decision making, and the roles of the two functions of vision (vision for action and vision for identification[21]) are relevant as contributors to inattention in crashes but are not necessarily directly measurable with the metrics covered in this chapter.

The five factors reviewed in this section could be summarized in terms of the conceptual framework outlined in Figure 10.1. The ability of a stimulus to capture the driver's attention is highly dependent on objective properties of the stimulus,

TABLE 10.1
Examples of Artificial and Natural Events along a Hypothetical Expectancy Continuum

Expectancy Continuum	Artificial and Natural Events	Response Type[a]
Expected	Signal detection task (SDT)—light/sound/vibration	I
	Signs instructing to change lanes in Lane Change Test	I
	Center hi-mounted stop light (CHMSL) onset	I
	Following vehicle turn signal	I
	Traffic lights and signs	I/N
	Low/high-beam switch at night with oncoming vehicles	N
	Lane ends due to work zone	N
	Lead vehicle (LV) braking or decelerating	N
	Sharp curve behind hillcrest (with sign)	N
	Vehicle pulls out from parallel parking	N
	LV changes mind during a lane change/turn, reentering path	N
	Sudden appearing vehicle in blind intersection	N
	Illegal oncoming or lead vehicle maneuvers	N
	Sudden debris or objects in path	N
Unexpected	Sudden animal crossing path	N

[a] I, instructed; N, natural.

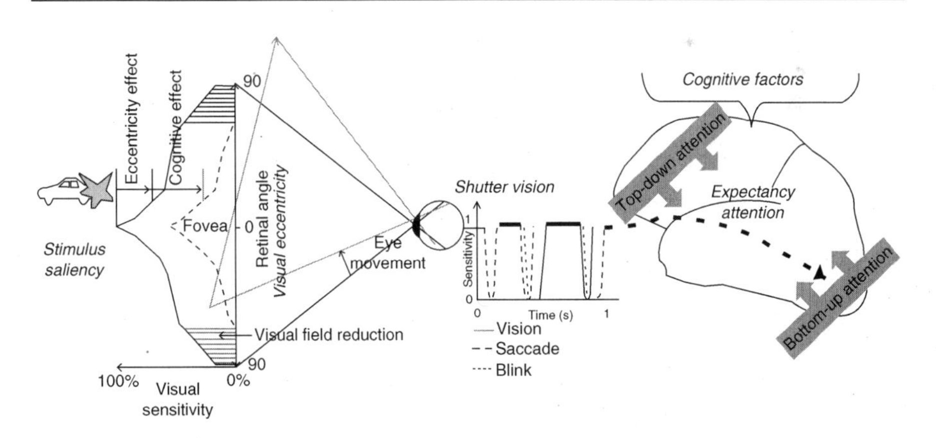

FIGURE 10.1 Conceptual framework of key factors influencing inattention in driving.

such as luminance, contrast, or motion (stimulus saliency factor). Visual sensitivity is dependent on the angle from the fovea whereby a dramatic reduction in the performance of many visual functions occurs toward the periphery (visual eccentricity factor). Transient visual inputs are also potentially masked by blinks and saccades (shutter factor). However, visual sensitivity is also effectively influenced by top-down attention selection. Impairment of these mechanisms results in a general reduction of visual sensitivity and, as a consequence, a visual field size reduction. Expectancy is closely related to top-down attention selection and a key factor influencing event detection performance. It could be expected that safety is most compromised when two or more of these factors combine. For example, when there is a simultaneous occurrence of eyes off the road, poor attention to the road scene, and an unexpected critical event. In the following sections, we will see how the three measurement approaches covered in this chapter—visual behavior measurement, OED measurement, and artificial signal detection measurement—relate to, and are affected by, different aspects of the key factors underlying inattention. Now, we turn to a discussion of these research approaches.

10.3 DISTRACTION ASSESSMENT TECHNIQUES BASED ON DRIVERS' VISUAL BEHAVIOR

10.3.1 BACKGROUND

Eye movements are motivated by the need to improve the acuity and increase the cortical processing power for (a) guiding actions and (b) identifying objects and events.[6,22] The fixation act is the most effective mechanism for attention deployment,[23] and attention and eye movements are strongly linked through shared anatomical areas in the brain.[23] Importantly, this strong attention-eye-movement link is complemented by a simultaneous *preparation to act* on the attended item.[24] Thus, eye movements are strongly indicative of where attention is allocated and of preparation to act on the fixated item.

Looking in the wrong direction at a critical moment while driving can have disastrous consequences. Yet we cannot look exclusively at the road ahead. The continuous uptake of information by foveal (central) and peripheral vision for path and headway control has to be satisfied in the presence of other tasks requiring foveal vision, such as checking moving and stationary objects in the visual periphery, reading road signs, and monitoring in-vehicle displays. When these "secondary" tasks require vision, a time-sharing behavior is exhibited with the eyes being continuously shifted back and forth between the road center and the off-path object. The main reasons to return fixations to the road center are that (a) fixating on a point on the future path about 1 s ahead is the main mechanism for trajectory aiming and gives the best coordinates for steering[25] and (b) fixating on the road center area also provides time-to-collision information regarding vehicles and objects in path. Drivers time-share not only between the road center and in-vehicle tasks, but also between the road center and other driving-related objects such as signs, bicyclists, mirrors, and scenery.[22] Visual behavior is also strongly responsive to driving demands. When drivers are faced with increased driving task demand during the performance of in-vehicle visual tasks, they adapt their glance behavior by increasing viewing time to the road[16,26,27]) or reducing vehicle speed.[26,28] The fundamental importance of eye fixations for action guidance is demonstrated by the way a person's gaze concentrates on the distant path region (i.e., the road center) during the on-road glances in visual time sharing.[27,29]

As mentioned in the previous section, there is a reduction in visual sensitivity with retinal eccentricity. Hence, looking away from the forward roadway with increasing visual eccentricity gives increasingly poor quality information on the future path and forward objects and events. It has been demonstrated that visual diversion (at eccentricities from about the speedometer level and onward) impairs path keeping[30] as well as event detection.[18*] The recent results from the 100-car naturalistic driving study[1] point to the importance of visual diversion from the forward roadway in crashes and near-crashes. In line with this, Wierwille and Tijerina[33] have shown high correlations of $r = .90$ between eyes-off-road exposure (mean glance duration × number of glances × frequency of use) and real-life crash frequency.

In addition to the basic effect of retinal eccentricity, event detection performance during visual time sharing is affected by attentional selection. As outlined in the previous section, an increased attentional focus on certain stimuli comes at the expense of a reduced sensitivity to other stimuli. This effect is also reflected in eye-movement patterns. As the visual perceptual difficulty of a secondary visual task increases, drivers look more often, for longer periods, and for more varied time durations at the in-vehicle display.[16,27] Moreover, as already mentioned, during visual time sharing the fixations made to the forward roadway are strongly spatially concentrated on the road center area,[27,29] indicating a strong attentional selection of optical features relevant for safe path control and obstacle avoidance. It is likely that this attentional selection effect is associated with impaired detection of events outside the focus of attention, but this has not (to our knowledge) been empirically investigated. Eye-movement patterns are also likely to be affected by interference with general top-down attentional control mechanisms by cognitively demanding tasks, resulting, for example, in extreme single-glance durations. However, further research is needed to better understand this effect.

Although this section so far has concentrated on visual time sharing, driven by the competition for foveal vision, eye movements are also strongly affected by purely cognitively loading tasks. In response to increases in cognitive demand, drivers increase their road center–viewing time and spatially concentrate their gaze in the road center region at the expense of peripheral glances.[27,29,34–36] Significant reductions in horizontal and vertical variability (SD) of gaze direction, longer on-road fixations, and reduced glance frequency to mirrors and speedometer are typically found. This gaze concentration effect has been demonstrated to be associated with, but not to be the direct cause of, impaired event detection.[36] Rather, empirical evidence suggests that gaze concentration and reduced event detection are largely independent effects of a general interference in cognitive mechanisms for top-down attentional control[†] (see Section 10.5 for a further discussion of this effect in the context of artificial signal detection methods).

Path control does not seem to be negatively affected by cognitive load.[37] A number of studies have even found improved lane-keeping performance during cognitive load, compared to baseline.[28,38–40] The improved lane keeping is most likely causally related to the gaze concentration effect,[27,28] although further research is needed to establish this empirically.

---

* Event detection reaction time (e.g., detection of a lead vehicle becoming dangerously close) deteriorates more quickly than lane-keeping performance as visual eccentricity (gaze angle) increases away from the road.[31] Lane-keeping performance is effected by driving experience, meaning that the effect of looking away from the road has a larger impact on unexperienced drivers' lane-keeping performance than experienced drivers.[30,32] However, event detection reaction time does not appear to be affected by driving experience.[30,32]

† The gaze concentration effect should not be confused with tunnel vision, which refers to a hypothesized perceptual narrowing of the visual field where the visual sensitivity reduction is greatest in the periphery (see Section 10.5 and Figure 10.3).

In summary, the main points for present purposes are: (1) visual time sharing occurs because path/headway control and other tasks require foveal vision; (2) the farther away from road center a driver looks (the larger the visual eccentricity) and the longer the glance, the poorer the information available for event detection and lane keeping; (3) attention and eye movements are strongly linked, and visual behavior is thus indicative of attention selection related to both the driving and secondary tasks; and (4) cognitive loading tasks result in a gaze concentration on the road center that is associated with, but not directly causally linked, impaired detection performance. We next turn to how these effects can be quantified by means of existing visual behavior metrics.

10.3.2 Using Visual Behavior Measurement Methodology

Visual behavior can be quantified by a large number of metrics: (a) detailed eye-control metrics such as within-fixation metrics (tremor, drift), saccade profiles, smooth pursuit control, pupil control, and eye closure behavior; (b) mid-level eye-movement metrics like glance behavior, area of interest, transition behavior, and semantically classified fixations (pedestrian, sign, tree); and (c) coarse visual behavior metrics such as head movement behavior (position, rotation) and facial direction (on or off road). The metrics of central interest defined here have been developed primarily to quantify the visual time-sharing behavior that is exhibited when drivers look back and forth between the road and a display during a visual task such as radio tuning. Metrics mainly quantify the amount of time spent looking on or off the road (at the object of interest), for each glance or for a period of time such as a task or time window. Of the key factors contributing to inattention in crashes (discussed previously), visual eccentricity, saccades, and blinks could also be objectively measured, but attention allocation and expectancy can only be indirectly inferred from eye movements.

Eye-movement metrics are consistently reported to be among the best performing diagnostic metrics for measuring distraction and workload (see CAMP Driver Workload Metrics project,[16] HASTE[29,41]). Eye-movement metrics are highly sensitive (i.e., discriminative, repeatable, and have predictive validity) to the demands of visual and auditory in-vehicle tasks as well as driving task demands.[16,29,41]

Glance-based metrics, such as total glance duration, glance frequency, mean single-glance duration, and total task time, have traditionally been the central metrics of interest in assessing the visual or attentional demand of in-vehicle information systems.[42,43] A *glance* describes the transition of the eyes to a given area, such as a display, and one or more consecutive fixations on the display until the eyes are moved to a new location.[42,43] The temporal characteristics of time-sharing glances between the road center and a peripheral object are remarkably constant, with glance durations typically exhibiting means between 0.6 and 1.6 s, and showing a (positively) skewed distribution toward short glances.[44,45] Many studies demonstrate that drivers generally are very unwilling to look away from the road for more than 2 s.[45]

One drawback of this methodology is that it has been notoriously difficult and time consuming to measure and analyze eye movements in real-life in-vehicle settings. The International Organization for Standardization (ISO) visual demand metrics[42,43] describe time-consuming procedures for human-rater-based manual video transcription of eye glances. For example, the recent CAMP data were reduced in this way.[16]

An alternative to the manual video transcription method is to use some form of an eye tracker, either head mounted or remote camera based. However, various technical difficulties are often associated with eye trackers, both with regard to mastering hardware, software, and calibration, and also with regard to data loss, signal quality, signal processing, and data reduction methods. For these and other reasons, eye-movement metrics are not used as frequently as other distraction metrics.

Recent efforts have resulted in the development of a more robust, more reliable, and easier to calculate computational procedure for automated analysis of eye-movement behavior.[29,46–48] This method uses a simpler procedure based on the identification of glances away from the road center area rather than glances toward a specific target. The percent road center (PRC) metric has proven to be more sensitive to the demands of visual and auditory in-vehicle tasks as well as driving task demands than the ISO-defined measures: total glance duration, glance frequency, single-glance duration, and total task time.[27,29,48] PRC was defined as the percentage of time within 1 min that the gaze falls within a road center area of 8° radius from road center (the lane ahead).[29] PRC, glance frequency, total glance duration, and total task duration are highly correlated with values ranging from $r = .90$ to $r = .97$.[27] A further advantage of PRC is that it can be directly compared with normal baseline driving and used to measure both visual and cognitive distraction, whereas glance-based metrics cannot.

Using data from manual transcription of eye-movement video, the CAMP Driver Workload Metrics project[16] found that glance frequency and total glance time performed best (were discriminative, were repeatable, and had predictive validity). The HASTE project,[29,41] which measured both the glance-based metrics and PRC, found that the latter was the most sensitive measure to the demands of visual in-vehicle tasks and was alone in being able to compare both visual and auditory-cognitive task types with baseline driving. In general, the most sensitive visual task metrics were those where glance duration and frequency are implicitly combined, namely the PRC and total glance duration metrics.

A further development of visual demand measurement was undertaken within the SafeTE project.[49] The goal of this development was to define a single visual demand metric that accounts for (1) the duration of each individual glance toward the system under study, (2) the total number of glances away from the road during the task under study, and (3) the eccentricity of the glances away from the road. It is based on a summation of individual off-road glance durations, where long glances are penalized by means of an exponential function. This sum is referred to as the weighted summed glance durations (WSGD). To account for display eccentricity, the WSGD is multiplied by a factor that penalizes eccentric display positions as a function of the radial angle of the display from the normal line of sight (i.e., toward the road center). The visual demand imposed when performing a secondary task is thus defined as:

$$\text{Visual demand}_{\text{task}} = \sum_{i}^{N} g_i^k \times E(\alpha) \tag{10.1}$$

where N is the total number of off-road glances during the task, g_i is the duration of an off-road glance i (in seconds), k is a constant, E is the eccentricity penalty function, and α is the radial gaze angle between the forward roadway and the display (or other system component requiring visual attention). The weighting of the single-glance duration is done by means of the exponent k, an idea first proposed by Wierwille and Tijerina,[33] where the exponent determines the degree to which long glances are penalized. The eccentricity penalty function $E(\alpha)$ may be derived from empirical data relating visual eccentricity to detection performance. The function used in SafeTE was estimated based on the results by Lamble et al.[18] One aspect not accounted for by this metric is the duration of glances back to the road, and further work is needed to investigate to what extent this needs to be incorporated. A reduced duration of glances back to the road may be indicative of a general impairment in attention allocation mechanisms resulting in inadequate visual sampling strategies, but the theoretical basis for this needs to be further developed.

The SafeTE visual demand metric thus takes explicitly into consideration the eccentricity factor and the effect of long glances. When put to use in evaluations of the safety of in-vehicle information systems, the SafeTE metric should promote short tasks, with few, short, and noneccentric glances toward the system. Thus, systems that enable self-paced, interruptible interaction, using centrally placed displays, should come out favorably in the assessment.

The gaze concentration effects (during cognitive load and for on-road glances during visual time sharing) previously described can be quantified by means of different metrics of spatial gaze variation. For example, Victor et al.[29] found the standard deviation of radial gaze angle to be highly sensitive to these effects, and similar metrics have been used by other authors.[36] As mentioned earlier, these effects are also captured by the PRC metric, although the sensitivity is slightly lower.[29]

It should be noted that eye-movement metrics can be calculated using a moving window and are not limited to a fixed interval such as task duration. For example, Victor[27] describes PRC, calculated with a 1 min running time window. This moving time window–based approach has been implemented in real-time distraction recognition and warning algorithms[50,51] (see Chapter 26). Implemented as moving time windows, these metrics allow task-independent analyses because there is no need to define a start or finish time for tasks.

In summary, most existing metrics of visual behavior quantify various aspects of visual time-sharing behavior mainly related to glance frequency and duration. Some time-sharing metrics also incorporate the effect of glance eccentricity. Gaze concentration can be quantified in terms of spatial variation measures, most commonly the standard deviation of gaze angle. Eye-movement metrics, despite technical difficulties associated with data collection and analysis, are consistently recommended because they are discriminative, are repeatable, and have predictive and face validity. Although metrics have been standardized by ISO, some new developments have taken place to abate current shortcomings, but further improvements are still possible, as outlined in the following section.

10.3.3 Open Issues and Future Research Needs

There is clearly much work left to do in terms of developing visual behavior metrics to become more practical and easier to use. One main issue has to do with developing measurement technology to the level that automated eye-movement measurement is robust, inexpensive, and easy to use.

The SafeTE visual demand metric represents a first step toward quantifying aspects beyond those addressed by the standard ISO metrics, incorporating penalties for eccentricity and single-glance duration. However, further efforts are needed to include more aspects of the key factors contributing to inattention in crashes, such as methods that capture the effects of attention selection due to perceptual difficulty and impairments in general attention allocation strategies. Moreover, it would be interesting to perform a closer analysis of the relationships between action outcomes and visual behavior, such as an analysis of the temporal relationships between steering wheel metrics and eye movements[52] (see, e.g., Ref. 51 for an initial attempt). Development of integrated metrics of temporal relationships between stimulus-onset and shutter vision (saccades, blinks) would be helpful in explaining missed stimuli and long response times. It is also interesting to further develop metrics that incorporate measurement of visual eccentricity (relation of gaze point and stimuli position) and saliency level (e.g., automatic, real-time, bottom-up analysis of visual field saliency) of presented stimuli. Coarse visual behavior metrics, such as head movement metrics that approximate eye movement, may also be used as a complement. Computational models of eye-movement behavior in relation to higher-level goals, top-down attention, expectation, and driving performance seem promising.[31,53]

Finally, results from naturalistic driving studies promise to bridge the gap between inattention/performance and accident risk and are very useful for identifying the visual behavior metrics that best predict crashes. Since truly unexpected real-world events (at the one extreme of the continuum in Table 10.1) preclude the deployment of anticipatory gaze allocation strategies, it seems likely that eyes-off-road would have more serious consequences in the real world than in controlled experiments, where the events are generally expected, at least to some degree. Comparisons between visual behavior in real-crash situations and the data obtained in controlled experiments could be used to investigate this further.

10.4 OBJECT AND EVENT DETECTION METHODOLOGIES

10.4.1 BACKGROUND

Drivers' responses to objects and events encountered while driving are crucial aspects of driving performance. An accurate awareness of the driving environment allows drivers to progress in a smooth fashion and, as previously discussed, interruptions by unexpected objects and events during moments of inattention are important contributors to crashes.[1] Simply stated, the use of OED methodologies for the assessment of distraction involves constructing experimental situations where drivers are presented with stimuli that require a response. This section focuses on methods employing naturally occurring objects and events, while the next section deals more specifically with methods that use artificial signals. Face validity and ecological validity of OED methods can be quite high, depending on the particular instantiation of the scenario. The results, indicating the extent to which drivers fail or are delayed in detections, have direct links to driving safety. The safety relevance of longitudinal and lateral control metrics derived from other distraction methodologies, such as drops in speed or increases in lane deviation, is less direct.

OED approaches to distraction can provide us with a sense of the driver's model of the surrounding traffic situation, beyond indicators of vehicle control. OED techniques have been used with success in a variety of distraction research studies. Response time to a lead vehicle deceleration or braking, with its clear relevance to rear-end crashes, is one of the most commonly used events in distraction studies. Perception response times can be influenced by a number of factors in addition to distraction, such as detection, identification, decision, response, driver expectancy, environmental conditions (e.g., night, poor weather), driver state (e.g., fatigue, alcohol, medications), as well as driver characteristics[54] (e.g., age, sex). It is also strongly affected by the specific kinematics of the critical situation such as initial headway, initial speed, and lead vehicle deceleration rate.[19]

Driver expectancy is particularly relevant in the present context. On the basis of his review of brake response time studies, Green[15] reported that response times increased with the level of expectation associated with the event: 0.75 s for expected events, 1.25 s for unexpected events, and 1.5 s for surprise events. As proposed in Section 10.2, events can be classified along a continuum of expectancy ranging from the fully expected to the entirely unexpected (see Table 10.1). We now turn to specific examples of OED.

Summala and colleagues[18,30,32] performed a series of experiments using the "forced peripheral driving" paradigm on a test track to examine the relationship between visual eccentricity, secondary task attentional demand, lane position maintenance, and response times to lead vehicle closing headway. Lamble et al.,[55] in a similar test track study, found delays of 0.5 s for brake response time to lead vehicle deceleration when drivers were engaged in tasks requiring *either* cognitive or visual distraction. The general finding from this series of experiments is that drivers exhibited slowed brake response times to a lead vehicle closing headway situations and reduced lane-keeping performance as visual eccentricity and secondary task attentional demand increased. Lee et al.[56] reported a 300-ms delay to the braking of a lead vehicle when drivers used a speech-based (no visual demand) email system while driving a simulator. Several studies addressing distraction arising from the use of cell phones have reported delayed braking responses for drivers conversing on either handheld or handsfree cell phones.[37,57,58]

The range of events for detection and response has been expanded in other research. In Greenberg et al.,[17] drivers responded to "front events" (severe lane violations by a vehicle two cars ahead) and "rear events" (severe lane violations in a following vehicle) by using the turn indicator. Drivers performed a range of in-vehicle tasks while driving the simulator and performing the detection tasks. Compared with baseline driving, front event detection was sensitive to aspects of both handheld and handsfree phone uses. Rear event detection also showed effects of some handheld and handsfree phone use, and climate control tasks.

Event detection methods were used in the Driver Workload Metrics project[16] carried out by the CAMP consortium. A configuration was used with a lead vehicle, subject vehicle, and following vehicle in both the track and on-road environments. Drivers were required to respond to three types of events involving the lead and following vehicles: lead vehicle deceleration, center hi-mounted stop light (CHMSL), and following vehicle turn signal. More than 20 in-vehicle tasks (auditory-vocal, visual-manual, and mixed) were investigated. When visual-manual tasks were considered, all three events discriminated multitasking from driving. To accomplish these tasks, drivers had to look away from the driving scene, resulting in increased misses over baseline ranging from 23 to 65% across the three types of events. Miss rates for the auditory-vocal tasks were "slightly elevated." It was concluded that event detection and eye glance metrics were key to evaluating the distraction impact of in-vehicle tasks (both auditory-vocal and visual-manual) on driving performance, and OED (along with visual behavior metrics) was recommended to be part of the distraction assessment toolkit.

Chisholm et al.[59] incorporated three types of events—lead vehicle braking, pedestrian crossing, and vehicle pullout requiring evasive maneuvers—in a study investigating the effects of manually dialing a cell phone on the driving behavior of novice and experienced drivers. Results generally indicated longer perception response times for the novice compared with experienced drivers for all three types of events. More collisions were observed in the lead vehicle braking events when participants were using the cell phone while driving.

Objects in the driving scene, such as traffic lights and signs, may also require a driver's response. Strayer and Johnston,[20] for example, found that participants engaged in cell phone conversations during a tracking task were more likely to miss simulated traffic signals and reacted more slowly to signals they did detect than when they were not engaged in cell phone conversations. In Smith et al.,[60] participants responded to specific road signs in a dynamic driving scene while performing various in-vehicle tasks. Response time to sign detection was slowest for the hands-free cell phone condition, and showed reliable differences to talking to a front seat passenger, conducting other in-vehicle tasks or driving with no other tasks.

This research overview demonstrates the value of OED methods for the examination of driver distraction effects resulting from interaction with in-vehicle devices. Although detection tasks require visual detection, they are sensitive to distractions presented in the auditory as well as visual modalities. In its more sophisticated implementations, this method can provide information about where drivers are attending in the driving environment (e.g., in the forward or rear direction[16,17]).

10.4.2 Using Object and Event Detection Methods: Experimental Setups and Procedures

10.4.2.1 Object and Event Detection Implementation

OED can be implemented in experimental research in many ways. Choice of the event or object varies, as does its relationship to driving safety. In the simulator, there is greater control, because the events or objects to be detected are most often incorporated in the programmed scenario. On road, the events need to be staged using a confederate in a lead vehicle or remote control of lead vehicle brake lights, for example. Timing and coordination are more difficult, as is the collection of data. Of course, the on-road situation provides increased face validity. Drivers may make their responses in any number of ways such as using a finger switch, using the turn indicator, by pressing the brakes, or undertaking evasive maneuvers. In each case, hits, misses, and response times are recorded.

10.4.2.2 Experimental Approach

These methods require a within-subject design to obtain the data required for the essential comparisons. Most commonly, the experimental design involves a comparison of detection performance while driving and performing the secondary task (e.g., operating an in-vehicle information system) with a control (nondistracted) condition. As with most research approaches, practice is important as are the instructions to drivers with respect to how they should respond in allocating their resources (usually as quickly and accurately as possible with priority to driving safely). As already mentioned, these methods are easily implemented in a simulator but are amenable to closed-track or on-road research environments.

Measures of detection and response time are the prime dependent variables collected, but differences in experimental design and equipment have resulted in variations in operational definitions. This holds in particular for brake response time.[15,16,61] The driver's brake response is composed of a series of stages starting with the detection of the object or event and ending with the depression of the brake pedal (see Figure 10.2). Total brake response time can be decomposed into three primary components: accelerator release time, movement time, and response completion time. Various combinations of these components have been used across studies. As shown by Lee et al.,[19] it is critical to consider the entire brake response process because drivers often exhibit compensatory behavior in braking situations. For instance,

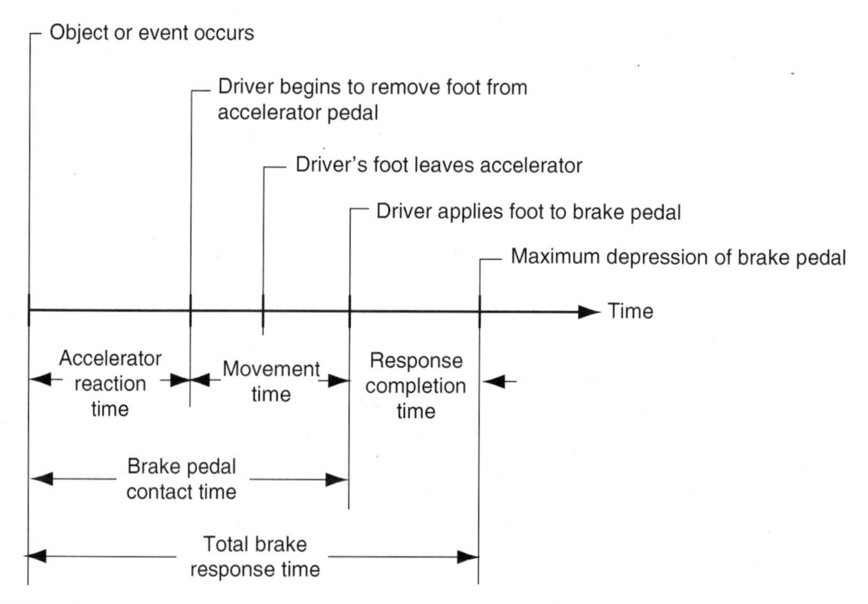

FIGURE 10.2 Components of total brake response time.[61] (From Winters, J.J., *An Investigation of Auditory Icons and Brake Response Times in a Commercial Truck-Cab Environment*, Virginia Tech, MSc Thesis, 1998. With permission.)

a late detection resulting in a late release of the accelerator pedal is often compensated for by a faster and harder braking. Lee et al.[19] found that the accelerator release time (the time from lead vehicle braking onset to accelerator release) was the most sensitive measure of braking performance, and the one most strongly related to safety impact measures such as percentage of collisions, collision velocity, or minimum time to collision.

10.4.2.3 Data Analysis

For OED methods, the usual metrics are normally hit rate and latency to detection (response time). For response time data, a response time window is usually set (e.g., a 2 s maximum window), and responses faster than 200–250 ms are considered unrealistic and usually excluded from the analyses. Response time data are usually not normally distributed (positively skewed toward shorter response times) and must be analyzed accordingly. A common approach is to perform a logarithmic transformation of the data before application of parametric statistical tests. Hit rate data are normally strongly non-normally distributed due, for example, to ceiling effects, which precludes the use of parametric statistics. Statistical comparisons are made across the conditions of interest. One concern is that there should be a sufficient number of presentations of the stimuli so that a reasonable number of responses are available for appropriate statistical analysis. In cases where only a single event or object was presented for detection, appropriate statistical methods should be used, and there should be a sufficient number of participants to make the analysis possible. More complex analyses of detection data are possible, where experiments have been designed to assess drivers' patterns of allocating attention to different locations on the road, by manipulating the locations of relevant events such as in front and behind the subject vehicle or at various locations in the forward view.[16,17]

10.4.3 Open Issues and Future Research Needs

10.4.3.1 Methodological Issues

Future research could improve OED in a number of ways. A primary concern is the lack of standardized guidelines for the use of OED, which other distraction assessment methodologies have. Such guidelines with respect to experimental design, procedure, and data analyses would benefit researchers in implementing the research and facilitate cross-experiment comparisons of results.

One important methodological concern, which has implications for several aspects of the research, has to do with the nature and number of OED stimuli presented for detection. Events for detection can range from the mundane to the rare. Unexpected or surprise events are more realistic in that they mimic the conditions under which crashes occur in the real world.[62] There is, however, a direct trade-off between the element of surprise and the number of data points that can be obtained. Clearly, catastrophic events cannot be repeated. In addition, if only one event is presented for detection and if for some reasons it is not an effective event for that driver, then the utility of that complete trial is lost. Typically, we rely on multiple data points per subject in behavioral research to obtain reliable data. Events that are presented multiple times in a scenario, however, run the risk of becoming predictable and, as a result, the drivers' expectations may turn the task into one of vigilance rather than event detection. The consequent impact on expectancy has a direct effect on the data. Response times in studies using repeated events for detection are generally much faster than for unexpected events in real-crash scenarios. For instance, Lee et al.[19] found that brake response time was reduced by 430 ms for the second exposure of an otherwise similar braking event. This has important consequences for the generalization of OED results to real-world events. Thus, one key issue for further research is the extent to which expectancy can be controlled and its influencing factors understood in a quantifiable manner. Another key issue for further research is the extent to which the effect of expectancy is simply additive or whether it interacts with other factors, such as visual or cognitive distraction. If the latter turns out to be the case, generalization of OED results should be made with great caution.

OED methodology decision-making guidelines are needed. As with most other distraction methodologies, OED does not have clear criteria for decision making. It may be possible to specify a criterion for comparison on the basis of the response time for an alert driver reacting to an expected situation. Muttart[63] has indicated that consideration of the type of event is critical. Response times must be grouped based on classes of events such as lead vehicles braking, being cut off, path intrusions and known sounds and lights.

A better understanding is required of the situations where OED methodologies are appropriate for use. Angell et al.[16] raised the possibility that OED methods might not be appropriate to assess distraction for tasks of short duration, since short tasks may not allow for the sufficient number of OED stimuli presentations to enable an appropriate assessment of distraction.

Often two or more distraction assessment methodologies are used simultaneously in an experiment. There is a need to better understand the impact that OED might have on other distraction measures, such as glance behavior, when the two methods are used concurrently. This concern was raised in the CAMP project.[16] This issue is further addressed in the discussion on artificial signal detection task (SDT) methods in section 10.5.

10.4.3.2 Theoretical Issues

The application of OED methods to investigate the effects of distraction has been a natural evolution from the general research investigating other factors that influence drivers' response times such as driving conditions and driver characteristics.[54] Vision is a prominent aspect of driving and the detection of relevant information in the research driving environment fits with the idea of detection and perception of relevant or dangerous objects and events in real-world driving. The relation to crash risk is evident. The data resulting from this methodology provide insight into to the driver's knowledge and model of the driving environment. This information feeds into the higher-order levels of driving knowledge and subsequent action.

A general framework for the classification of OED methodology is needed that can more precisely describe the relationship key factors of inattention have on performance. It would be ideal to be able to know, for each particular event, that the saliency related x_1 amount to performance, the eccentricity of the event x_2 amount, the onset of the event coinciding with a saccade or blink x_3 amount, the attentional allocation x_4 amount, and the expectancy x_5 amount.

10.5 ARTIFICIAL SIGNAL DETECTION TASKS: THE PERIPHERAL DETECTION TASK AND RELATED METHODS

10.5.1 BACKGROUND

The Peripheral Detection Task (PDT) and other methods based on artificial signal detection have recently been established as a common means for assessing driver distraction. The PDT differs from OED methods in that the stimuli to be detected are not a natural part of the driving task. Rather, when used for distraction assessment, the artificial detection task is added as a *tertiary* task to the primary driving task and the secondary task to be assessed (e.g., the operation of an in-vehicle information system). Alternatively, the detection task may be used as a secondary task in static conditions (i.e., no driving) in the laboratory (e.g., Ref. 16). The present section focuses on its application in dynamic (driving) conditions. Another key difference in relation to natural event detection is that the stimulus presentation is generally more frequent, regular, and hence, more predictable. In this sense, artificial detection tasks like the PDT could be regarded as a kind of sustained attention task on the one extreme of the expected–unexpected continuum (see Table 10.1).

The most common artificial detection task that has been used for distraction assessment is the PDT, and a number of variants of the PDT have been developed including the Visual Detection Task (VDT) and the Tactile Detection Task (TDT). The more general term *signal detection tasks* (SDTs) will henceforth be used to refer to this general class of methods.

The PDT was originally developed by van Winsum et al.[64] in the IN-ARTE EU-funded project. The method was strongly inspired by the work of Miura[65–67] who studied the effects of driving demand on peripheral vision and detection. Using artificial light stimuli reflected in the windshield with a certain spatial and temporal randomization, Miura found that response time to these stimuli increased as a function of driving demand (specifically traffic density). Moreover, the response eccentricity—that is, the distance between gaze fixation location and the target at the moment of response—decreased with increased driving demand, which was interpreted as a reduction in the functional visual field. A functional visual field reduction resulting from increased task demand has also been reported in several laboratory studies, where, in contrast to Miura's work, eye movements were not allowed to vary (usually a tachistoscope was used to keep the head and gaze direction fixed).[68–71] In addition to the field-of-view reduction, some of these studies[69,71] have demonstrated that foveal load may lead to a phenomenon known as *tunnel vision*, in which the visual sensitivity reduction (and the associated detection performance degradation) due to the foveal load is greater when it is the stimulus is presented in the periphery than when presented in the central areas. This can be compared to the *general interference* hypothesis, which states that the degradation is equal across the visual field.

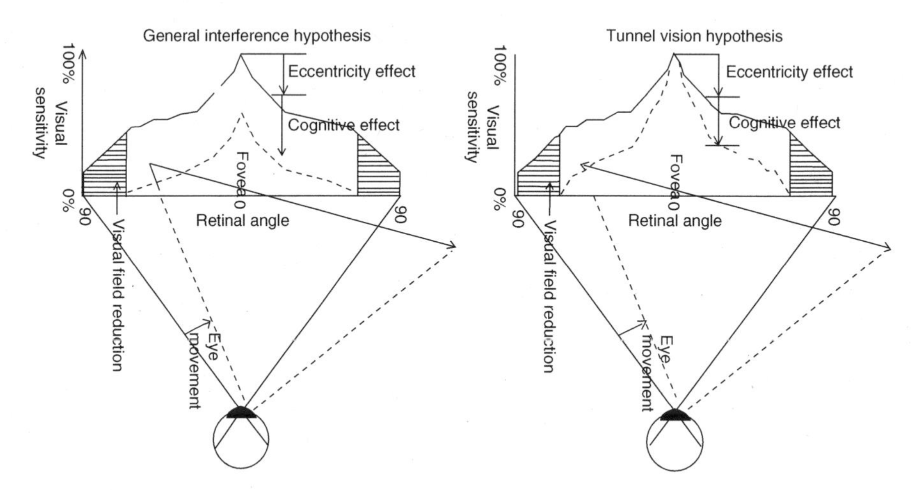

FIGURE 10.3 Illustration of the general interference (a) and visual tunneling (b) hypotheses. The former states that the degradation in visual sensitivity resulting from cognitive load is uniform across the visual field while visual tunneling implies that the degradation is greater in the periphery (i.e., that there is an interaction between load and eccentricity).

This distinction is further illustrated in Figure 10.3. Note that a visual field reduction does not imply tunnel vision, because a general reduction of visual sensitivity across the visual field will also lead to a functional visual field reduction (as illustrated in Figure 10.3a).

The key idea behind the PDT is to exploit this apparent narrowing of the visual field to obtain a method that is sensitive to workload but has a minor impact on the other tasks.* In the original PDT study, van Winsum et al.[64] tested a slightly modified version of Miura's detection task. Visual stimuli were presented in the upper-left visual field with spatial and temporal uncertainty. The area of stimulus presentation extended 11–23° horizontally and 2–4° vertically. The stimuli were presented with an interval that varied randomly between 3 and 5 s. Participants responded by pressing a button. In this simulator study, it was first verified that the method was strongly sensitive to driving demand. The authors then successfully applied it to evaluate the workload imposed by different warning messages given by driving assistance functions developed in IN-ARTE. The study also addressed the theoretical basis of the PDT, by looking at the extent to which functional visual field narrowing and visual tunneling actually occurred. To this end, they measured detection performance at different locations in the 11–23° horizontal range where demand was manipulated by classifying driving situations into high-load and low-load groups. The results did not indicate any evidence for visual tunneling. In fact, there was no difference in detection performance between the different stimulus positions; that is, there was no eccentricity effect. The authors concluded that, although the method proved highly sensitive to workload, its theoretical basis

* A persistent problem with traditional secondary task methods for workload assessment is that they often impede primary task performance.

remained rather unclear. They suggested that the effect obtained probably had more to do with general attention selection than a specific perceptual narrowing, that is, a cognitive rather than visual tunneling effect consistent with the general interference hypothesis.

The van Winsum et al. study was followed by a series of studies investigating the application of PDT in different contexts. For example, Olsson[72,73] demonstrated that the method is sensitive to both visual-cognitive and purely cognitive secondary task demand. This was also the first implementation of the PDT in the field. Harms and Patten,[74] in another field study, used PDT to compare GPS-based route guidance to memory-based navigation, and Schindhelm et al.[75] used it as one of the key methods to assess the COMUNICAR workload manager. It was also the primary method used in the major Swedish investigation into the safety effects of mobile phone use while driving.[76]

The studies cited in the preceding text all used the original PDT version, as specified by van Winsum et al.,[64] sometimes with slight modifications. However, recent work has explored presenting stimuli in other sensory modalities. This is based on the assumption, supported by the results of van Winsum et al.,[64] that the SDT detection performance degradation is due to a general, amodal interference in attention selection rather than a modality-specific visual perceptual narrowing. This has the potential advantage of yielding a "pure" measure of general attentional interference that is not "contaminated" by modality-specific factors such as visual saliency (related, e.g., to varying lighting conditions and background contrast), the direction of gaze, or the visual shutter factor (see Section 10.2). Moreover, presenting the stimuli in modalities other than the visual should not have a major effect on eye movements, which is an advantage if visual behavior is to be measured simultaneously with the detection task (as discussed later, the potential effect of the "standard" PDT on eye movements has not been systematically investigated). In a field study using visual-cognitive (phone dialing) and purely cognitive secondary tasks (answering biographic questions and counting backward with seven), Engström et al.[77] found that the TDT, where the visual LEDs were replaced by a tactile vibrator attached to the wrist (see Figure 10.4 for examples of different implementations of the tactile stimulus), was at least as sensitive as the PDT for both the visual-cognitive and the cognitive tasks. These results were replicated in a simulator study by Merat and Jamson[78] in the AIDE EU-funded project. This study also included an auditory detection task (ADT), where auditory stimuli were presented by means of loudspeakers inside the vehicle mock-up. When standardizing the effects (i.e., dividing the response time for the secondary tasks with the baseline response time), they found no differences between the three detection tasks (visual, auditory, and tactile) in terms of sensitivity to secondary task demands. The sensitivity of the TDT to both visual and cognitive secondary task load was further demonstrated, for example in the AIDE project, in a simulator study.[40] These results taken together provide strong evidence for the hypothesis that the SDT methods mainly measure a general interference in central attentional selection mechanisms that does not depend on sensory modality.

Another variant of the PDT, known as the visual detection task (VDT), was recently developed in the Swedish SafeTE project.[49] This development was

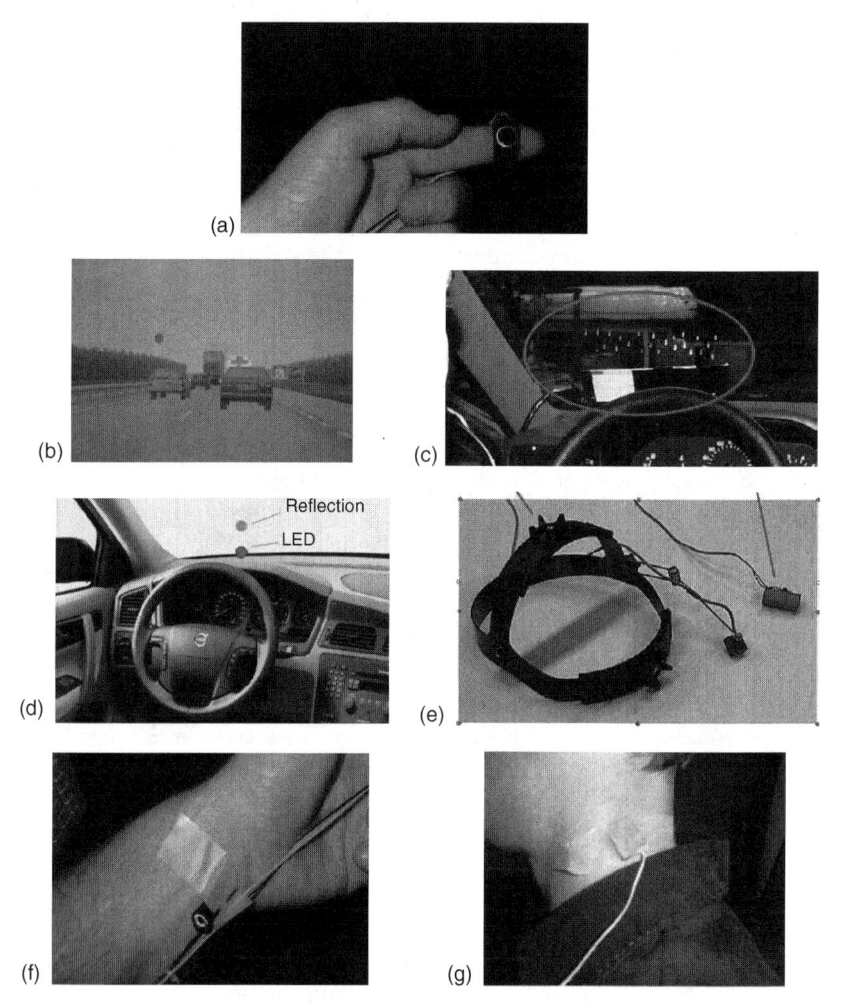

FIGURE 10.4 Response and stimulus presentation methods used in artificial detection task methods are: (a) the "standard" response button; (b) graphical stimulus presentation on the simulator screen; (c) LEDs reflected in the windshield in an instrumented vehicle (From Harms, L. and Patten, C., *Transportation Res. Part F,* 6, 23, 2003. With permission.); (d) the single, centrally positioned LED used in the VDT; (e) a head-mounted LED (From Schindhelm, R., Gelau, C., Montanari, R., Morreale, D., Deregibus, E., Hoedemaeker, M., de Ridder, S., and Piamonte, P., *Human Factor Tests on Car Demonstrator*, Deliverable no. D6.4, Final draft. COMUNICAR Programme IST KA1, Testing demonstration and pilot evaluation Work package WP6, Report No.: 2003 P 134, 2003. With permission.); (f) tactile vibrator attached to the wrist; and (g) tactile vibrator attached to the neck (From Merat, N. and Jamson, H.A., *Proceedings of the Fourth International Driving Symposium on Human Factors in Driver Assessment, Training and Vehicle Design*, Stevenson, Washington, July 9–12, 2007. With permission.).

motivated by a desire to keep the visual presentation modality (as it can be argued that this has a somewhat higher ecological validity than tactile and auditory stimulus presentation) while at the same time overcoming problems related to low stimulus saliency in real traffic (due to varying lighting conditions) and to minimize the effect on visual behavior. The VDT mainly differs from the original PDT in that it involves only a single stimulus located in the central field of view rather than in the periphery (hence the change of name). The single VDT stimulus is also substantially more intense than the LEDs typically used in the original field PDT implementation (e.g., Ref. 72; see Figure 10.4). As outlined in Section 10.2, the stimulus saliency is a key factor determining its detectability. Thus, these modifications increased the overall saliency of the stimuli that (1) reduced the impact of lighting conditions and (2) minimized the need to glance directly at the stimulus to identify it. In other words, the stimulus saliency factor was more controlled because it was adjusted to be above threshold at larger visual eccentricities. The VDT was validated on visual-manual and purely cognitive secondary tasks and proved sensitive to both.[49,79] It was also demonstrated that the VDT did not significantly affect any of the standard metrics of visual demand as long as the stimulus was positioned centrally.[80]

To summarize, SDT methods, despite their somewhat unclear theoretical basis, have repeatedly proven highly sensitive to changes in driving and secondary task demand. As demonstrated by the tactile and auditory detection task studies,[40,77,78] this sensitivity seems to be independent of stimulus sensory modality, thus indicating that the SDT primarily measures interference in general attention selection mechanisms rather than a modality-specific perceptual narrowing. In the terminology introduced in Section 10.2, SDT performance thus mainly represents effects on cognitive factors such as top-down attention selection and expectancy. This is in-line with the general interference, but not the visual tunneling, hypothesis.

The following section reviews a number of practical aspects associated with using SDT methods for distraction assessment. Section 10.5.3 then addresses some methodological and theoretical issues that remain unresolved and need to be addressed in future research.

10.5.2 Using SDT Methods: Experimental Setups and Procedures

10.5.2.1 SDT Implementation

As mentioned earlier, SDT methods have been implemented in several ways. These implementations differ primarily with respect to how stimuli are presented, whereas the response methods are generally similar (typically a micro switch attached to the index finger or the steering wheel; see Figure 10.4a). For the (visual) PDT in simulator setups,[64,78] the stimuli are normally presented as graphical objects on the screen (see Figure 10.4b). In field studies,[73,74] the stimuli are generally presented by means of small LEDs, reflected in the windshield (see Figure 10.4c). In the VDT,[49] a single LED with higher intensity is used, positioned in the central field of view (see Figure 10.4d). Another approach is to use a single head-mounted LED that ensures a fixed stimulus eccentricity (see Figure 10.4e and Ref. 75). In the TDT, a small vibrator attached to the body provides the stimulus. Different studies have used different positions of the vibrator. Engström et al.,[77] for example, attached it to the wrist (see Figure 10.4f), whereas Merat and Jamson[78] positioned it on the neck (see Figure 10.4g). Auditory stimuli are generally presented through loudspeakers in the vehicle or a simulator mock-up (e.g., Ref. 78).

10.5.2.2 Experimental Approach

The experimental approach used for SDT methods is generally similar to that used in the OED methods described in the previous section. Normally, a within-group design is used where detection performance in the three-task condition (driving + secondary task + detection task) is compared to the dual-task baseline condition (driving + detection task).

It is generally assumed that performance of the SDT has a rather steep learning curve (although this has, to our knowledge, not been systematically investigated). It is still always necessary to include a training session where the subject practices the detection task in single-, dual- (with driving), and three-task (with driving and secondary task) conditions (see Ref. 49 for a recommended training practice).

It is also important to give the subject clear instructions on how to prioritize between the different tasks, because this could strongly influence the results (like training, this issue has not been addressed experimentally). In most cases, neutral instructions are given; that is, the subjects are not explicitly told to prioritize any of the tasks, but to do their best to perform the detection task continuously without compromising safety.

10.5.2.3 Driving Scenarios

As previously mentioned, the PDT and TDT has been demonstrated to be strongly sensitive to driving demand.[64,40,77] Thus, if the objective is to assess the distraction imposed by a secondary task, the driving demand must be carefully controlled. SDT studies that have varied the driving and secondary task demands indicate that the two factors are mainly additive (i.e., there is no interaction between them), at least not for response time.[40,77] Most existing studies have used low-demand driving scenarios (rural road or motorway driving with low traffic density).

10.5.2.4 Data Analysis

As discussed, SDT performance is generally quantified in terms of response time (RT) or hit rate. Hits are normally defined as responses occurring within a certain time interval from stimulus onset $T_1 < RT < T_2$. The lower threshold (T_1) is used to prevent cheating or "lucky hits" with unrealistically low response times, while the upper threshold (T_2) defines what should be counted as a miss. In most of the studies reviewed here T_1 (if included) has been set to 200 ms and T_2 to 2000 ms.* The hit rate is then calculated as the number of hits divided by the total number of stimuli. The response time is defined as the time from stimulus onset to response.

With respect to statistical testing, the same issues apply as for the OED methods described in the previous section. Both hit rate and response time are usually non-normally distributed. The latter may be logarithmically transformed before

* T_2 was originally set to this value by van Winsum et al.[64]. However, this was questioned by Chin et al. (2004)[81] who found hits with longer response times in demanding environments. Similar results were obtained in the SafeTE project,[49] where it was decided to increase the response threshold to 3000 ms.

application of parametric statistics. For hit rate, however, nonparametric statistics should normally be used.

10.5.3 Open Issues and Future Research Needs

SDT methods have a number of advantages that make them attractive for driver distraction assessment: (1) they have proven reliable and sensitive to a wide range of secondary tasks, (2) the equipment is relatively inexpensive and easy to set up, and (3) they work well both in simulators and in the field. However, a number of methodological and theoretical issues need to be resolved before SDT can be considered a mature research methodology. These are briefly addressed in the following.

10.5.3.1 Methodological Issues

One key methodological issue concerns how to deal with two separate performance metrics (hit rate and *RT*). For example, the driver may adopt an all-or-none attention allocation between the stimuli, thus ignoring some stimuli altogether while responding quickly to the ones detected. This will result in a low hit rate and a low *RT*, a result that is very difficult to interpret. The hit rate measure itself is also problematic due to its strongly non-normal distribution, which precludes the use of parametric statistics. One way to address these issues, proposed by Engström and Mårdh[49] as part of the VDT methodology, is to use *RT* as a single performance measure, while using hit rate as a quality measure. Recall that one of the key goals of the VDT development was to boost stimulus saliency compared to the standard PDT setup. This increases the number of valid responses, which makes the data more interpretable. In this approach, data that do not show a hit rate above a certain criterion (70% was used by Engström and Mårdh[49]) are excluded from the analysis.

Another issue concerns the extent to which SDT methods affect driving performance and visual behavior. This is naturally important to consider if the SDT is intended to be used simultaneously with other distraction assessment methods. Effects on longitudinal and lateral control performance were investigated in the simulator study by Merat and Jamson[78] (these results are reported in Merat et al.[79]). However, a more critical issue concerns the potential effect on visual behavior, that is, the extent to which drivers gaze at the stimuli. Miura,[67] who performed a manual, frame-by-frame analysis of eye-movement data, reported that targets were frequently gazed at. As mentioned earlier, this is clearly problematic if SDT and visual behavior data are to be collected simultaneously (this is not a problem for TDT or ADT*). This issue was addressed in the SafeTE project,[80] where it was found that the VDT did not significantly affect the "standard" visual demand metrics (such as glance frequency, glance duration, total glance time, and PRC) as long as the stimulus was placed centrally. However, it could be expected that the PDT would have a stronger

* It could perhaps be argued that the TDT and ADT may indirectly affect visual behavior by inducing cognitive load because, as described in Section 10.3 earlier, cognitive load often leads to a gaze concentration effect. However, this seems unlikely given that these tasks require very little central cognitive resources. As further discussed later, the degrading effect of cognitive load on the SDT could be explained as a lack of top-down attention bias (rather than resulting from overload of cognitive resources).

effect on eye movements than the VDT, due to the peripheral location and lower saliency of the stimuli, which hence require more foveal vision. Thus, a similar analysis is needed for the PDT before it can be recommended for use simultaneously with visual behavior measurement.

As previously mentioned, it is generally assumed that SDTs are quickly learned and, thus, that the learning effect is not a major issue as long as a training session is included prior to the actual trials. However, this needs to be verified empirically. Further empirical work is also needed on the influence of instructions concerning task priority on PDT performance. Another important issue is the role of task allocation strategies, where subjects may exploit the fact that the stimulus occurs at relatively regular intervals. Such strategies have been anecdotally reported by Olsson[72] and need to be more systematically investigated.

10.5.3.2 Theoretical Issues

On the theoretical side, the central issue of what SDT performance actually represents remains open. The PDT method was originally developed on the basis of the idea that it could measure workload in terms of visual tunneling (which was the main reason for putting the visual stimuli in the periphery). However, this idea has been dismissed already by the inventors of the method[64] and, in general, the evidence that driving or secondary task load induces visual tunneling is weak. Miura,[65–67] who is often cited in support for visual tunneling in driving, only looked at a functional visual field reduction that is also consistent with general interference (see Figure 10.3) and does not present any results indicating an interaction between driving demand and eccentricity. Williams[69–71] found some evidence for visual tunneling in laboratory tachistoscope studies with foveal loading tasks, but only in specific conditions (e.g., instructions to strongly focus on the foveal task) and only as a local effect at small eccentricities (less than 4° from the fovea). Most of Williams' results are more consistent with a general interference interpretation. No study has (to our knowledge) yet demonstrated visual tunneling in driving as a function of either driving demand or secondary task distraction. Studies that have explicitly addressed this issue support the general interference hypothesis.[36,64] What is more, there is little evidence that eccentricity has any major effect at all on detection of PDT-type stimuli (e.g., LEDs or graphical dots).[64] Taking into account the results from studies reviewed here, demonstrating that not even sensory modality seems to play any significant role for the SDT sensitivity,[77,78] it becomes clear that visual perceptual narrowing does not seem to be the main mechanism behind of the effects found for the SDT.

Nor do limited capacity theories, such as multiple resource theory (MRT)[82] offer a satisfactory explanation. Although MRT accounts for central, amodal interference, this would not be predicted by MRT in this case because the SDTs themselves require very little attentional resources. Rather, MRT would predict that the strongest effect would occur in cases when there is a strong overlap in perceptual resources (e.g., driving + phone + PDT would yield a stronger SDT effect than driving + phone + TDT), but this prediction does not fit the data.

What, then, could be the mechanism behind the degraded SDT performance under conditions of high driving and secondary task demand? One hypothesis is that the effect occurs due to a *lack of top-down bias for attention selection*. As described in Section 10.2, recent neurophysiological studies suggest that attention selection is driven partly by bottom-up effects determined by stimulus properties (in particular saliency) and partly by top-down, task-related biases originating in frontal areas in the brain. This is known as the *biased competition hypothesis*.[8] The frontal lobe areas presumed to generate the top-down bias are also known to be involved in cognitive functions such as response selection, dealing with novel tasks, working memory as well as perception of complex visual stimuli,[83] functions that are typically needed when negotiating complex driving environments as well as in secondary tasks typically used in SDT studies such as phone dialing, phone conversation, or backward counting. Because of their regularity and high frequency, the SDT stimuli could be regarded as highly expected. Thus, detection of such stimuli would benefit strongly from sustained top-down attentional bias. In typical SDT baseline conditions (driving on a straight road with sparse traffic), there is little competition for these frontal cognitive resources, which means that detection performance can benefit strongly from the top-down bias. However, if the primary or secondary task demand increases, resulting in increased competition for the frontal cognitive resources, the top-down facilitation for the SDT will be cut off or at least reduced. In this case, the SDT stimuli will compete for bottom-up attention capture on the basis of their own properties (in particular their saliency), that is, on the same terms as unexpected stimuli (for which top-down facilitation is lacking by definition). This lack of top-down facilitation would not be expected to depend on visual stimulus eccentricity or even sensory modality, which would explain why similar effects are generally obtained for PDT, TDT, and ADT.[77,78] This hypothesis has the implication that, if the effect of secondary task distraction on the SDT is mainly due to a lack of top-down facilitation, the SDT performance would not predict the ability to detect unexpected stimuli. Rather, reduced SDT performance should then be interpreted as an impaired ability to detect and respond to *expected* events such as road signs or traffic lights, while it would say little about the ability to react when the lead car suddenly brakes. Naturally, this has strong implications for understanding the relation between SDT performance and real accidents. However, further empirical work is clearly needed to evaluate this and other possible hypotheses for explaining and interpreting the results obtained by SDT methods.

10.6 GENERAL DISCUSSION AND CONCLUSIONS

This chapter has described how characteristics of the key factors of inattention—stimulus saliency, visual eccentricity, shutter vision, attention, and expectancy—can be measured by using visual behavior, OED, and artificial signal detection methods. Recall that driver distraction was defined as inappropriate selection of information to the extent that safety-relevant information is missed. The three types of methods outlined in this chapter contribute to the measurement of "irrelevant selection" by the quantification of eye movements, and the measurement of "missed information" by the quantification of object, event, and artificial signal detection. The metrics described in this chapter constitute a key part of a toolbox of assessment methods and form an important link to naturalistic field data. Understanding the limitations of visual and cognitive processes and their relation to driving performance is essential to understanding the problem of distraction-related crashes.

Measures of visual behavior quantify various aspects of visual time-sharing behavior and degree of scanning (e.g., spatial gaze concentration). They are discriminative, are repeatable, and have predictive and face validity but need to become more practical and easier to use. They also need to include factors such as relationships between action outcomes and eye movements, relationships between stimulus-onset and shutter vision, visual eccentricity of stimuli, and the saliency-level of stimuli. Models of eye-movement behavior should include these factors in addition to attention and expectation.

Drivers' responses to objects and events are a crucial aspect of driving performance, yet the OED methodology is clearly lacking a systematic theory- and performance-based framework. A general framework for classification of OED methodologies is needed that can more precisely describe events with regard to the relationship with the key factors of inattention and performance. Also, some difficult statistical concerns about using OED remain, such as the number of presentations per driver, the learning effects, expectancy (unexpected tasks may become expected), and behavioral adaptation (drivers changing their behavior by slowing down or by becoming more cautious and looking away from the road less). Despite these concerns, OED methods have been successful regarding sensitivity and in providing unique insights into the effects of distraction.

Artificial SDTs have also proven to be highly sensitive to changes in driving and secondary task demands. The recent evidence indicating independence from sensory modality indicates that the SDT measures interference in top-down attention selection and expectancy mechanisms rather than in a modality-specific perceptual narrowing. Hopefully, future research will shed further light on these issues and eradicate the current confusion and controversy regarding the tunnel vision and general interference theories.

One of the most important tasks that distraction measurement methodology has to overcome is to bridge the gap between measurement of distraction and real-life crash causation. In the end, distraction metrics are just surrogate metrics intended to predict crash involvement. Note that the three types of methods described here can be used with each other and together with other metrics of driving performance. The greatest predictability of crashes will probably come from a combination of distraction metrics. Here, naturalistic driving data may provide invaluable, ground truth data to which this crash prediction function can be calibrated. Naturalistic data can be used to help define realistic events that are discriminative, are repeatable, and have high predictive validity—situations that have caused real crashes can be used as the basis for the design of critical events in OED studies and in driver training. Perhaps a general framework for OED would encompass a methodology that includes input from naturalistic situations and that more systematically uses different types of events. For example, expected events—such as SDT, inclines, and oncoming headlights—could be used more frequently, and unexpected events could be used in an increasingly intermittent manner depending on how difficult it is to maintain their level of unexpectancy in repeat occurrences. The increased use of cognitive neuroscience to understand the problem at a deeper level will also be helpful. Instead of using telephone operator metaphors, computer metaphors, or other engineering artifact metaphors to explain how cognitive processes function, we will hopefully find ourselves looking toward more biologically grounded models of distraction and inattention in general.

Even when we do come to a deeper understanding of distraction and a clearer understanding of the exact metrics that should be used to measure distraction, our methods will have to be cost-effective and easy to use. Distraction measurement should be seen and treated as a genuine product that should be calibrated to the needs of its users and packaged in a proper manner. A toolbox that includes formative (design) methods, guidelines, and clear decision criteria in addition to end-product evaluation methods is an essential target for future research and development in this area.

REFERENCES

1. Dingus, T.A., Klauer, S.G., Neale, V.L., Petersen, A., Lee, S.E., Sudweeks, J., Perez, M.A., Hankey, J., Ramsey, D., Gupta, S., Bucher, C., Doerzaph, Z.R., Jermeland, J., and Knipling, R.R., *The 100-Car Naturalistic Driving Study, Phase II: Results of the 100-Car Field Experiment*, DOT HS 810593, National Highway Traffic Safety Administration NHTSA, Washington, D.C., 2006.
2. Tsotsos, J.K., Itti, L., and Rees, G., A brief and selective history of attention, In *Neurobiology of Attention*, Itti, L., Rees, G., and Tsotsos, J. K. (Eds.), Elsevier, Amsterdam, 2005, pp. xxii–xxxii.
3. Stelmach, L.B., Bourassa, C.M., and di Lollo, V., Detection of stimulus change: The hypothetical roles of visual transient responses, *Perception and Psychophysics* 35(3), 245–255, 1984.
4. Rumar, K., The basic driver error: Late detection, *Ergonomics* 33(10/11), 1281–1290, 1990.
5. Rensink, R.A., Change detection, *Annual Review of Psychology* 53, 245–277, 2002.
6. Findlay, J.M. and Gilchrist, I. D., *Active Vision: The Psychology of Looking and Seeing*, Oxford University Press, Oxford, 2003.
7. Corbetta, M. and Shulman, G. L., Control of goal-directed and stimulus-driven attention in the brain, *Nature Reviews Neuroscience* 3(3), 201–215, 2002.
8. Desimone, R. and Duncan, J., Neural mechanisms of selective visual attention, *Annual Review of Neuroscience* 18, 193–222, 1995.
9. Alexander, G.J. and Lunenfeld, H., Driver expectancy in highway design and traffic operations, Report No. FHWA-TO-861, The Federal Highway Administration, Washington, D.C., 1986.
10. Dewar, R.E. and Olson, P.L., *Human Factors in Traffic Safety*, Lawyers & Judges Publishing Co., Inc., Tucson, 2002.
11. Blakemore, S.-J., Frith, C.D. and Wolpert, D.W., Spatiotemporal prediction modulates the perception of self-produced stimuli, *Journal of Cognitive Neuroscience* 11, 511–559, 1999.
12. Mack, A. and Rock, I., *Inattentional Blindness*, MIT Press, Cambridge, MA, 1998.
13. von Hofsten, C., Planning and perceiving what is going to happen next. In *The Development of Future Oriented Processes*, Haith, M., Benson, J., Roberts, R., and Pennington, B. (Eds.), University of Chicago Press, Chicago, 1995, pp. 63–86.

14. Evans, L., *Traffic Safety*, Science Serving Society, Bloomfield Hills, MI, 2004.
15. Green, M., How long does it take to stop? Methodological analysis of driver perception-brake times, *Transportation Human Factors* 2(3), 195–216, 2000.
16. Angell, L., Auflick, J, Austria, P.A., Kochlar, D., Tijerina, L., Biever, W., Diptiman, J., Hogsett, J., and Kiger, S., *Driver Workload Metrics Project: Task 2 Final Reoprt* Sponsored by National Highway Traffic Safety Administration, Washington, D.C., 2006, available at HS 810 635 http://www-nrd.nhtsa.dot.gov/pdf/nrd-12/Driver%20Workload %20Metrics%20Final%20Report.pdf.
17. Greenberg, J., Tijerina, L., Curry, R., Artz, B., Cathey, L, Kochar, D., Kozak, K., Blommer, M., and Grant, P., Driver distraction evaluation with event detection paradigm, *Transportation Research Record* 1843, 1–9, 2003.
18. Lamble, D., Laakso, M., and Summala, H., Detection thresholds in car following situations and peripheral vision: implications for positioning of visually demanding in-car displays, *Ergonomics* 42(6), 807–815,1999.
19. Lee, J.D., McGehee, D.V., Brown, T., and Reyes, M.L., Collision warning, driver distraction, and driver response to rear-end collisions in a high-fidelity driving simulator, *Human Factors* 44(2), 314–334, 2002.
20. Strayer, D.L. and Johnston, W.A., Driven to distraction: Dual-task studies of simulated driving and conversing on a cellular telephone, *Psychological Science* 12, 462–466, 2001.
21. Goodale, M.A. and Milner, A.D., *Sight Unseen: An Exploration of Conscious and Unconscious Vision*, Oxford University Press, Oxford, UK, 2004.
22. Land, M.F., Eye movements and the control of actions in everyday life, *Progress in Retinal and Eye Research* 25, 296–324, 2006.
23. Corbetta, M., Akbudak, E., Conturo, T.E., Snyder, A.Z., Ollinger, J.M., Drury, H.A., et al., A common network of functional areas for attention and eye movements, *Neuron* 21(4), 761–773, 2000.
24. Craighero, L. and Rizzolatti, G., The premotor theory of attention, In *Neurobiology of Attention*, Itti, L., Rees, G., and Tsotsos, J. K. (Eds.), Elsevier, Amsterdam, 2005, pp. 181–186.
25. Wann, J.P., and Swapp, D.K., Why you should look where you are going, *Nature Neuroscience* 3(7), 647–648, 2000.
26. Senders, J.W., Kristofferson, A.B., Levison, W.H., Dietrich, C.W., and Ward, J.L., The attentional demand of automobile driving, *Highway Research Record* 195, 15–33, 1967.
27. Victor, T., Keeping eye and mind on the road, *Digital Comprehensive Summaries of Uppsala Dissertations from the Faculty of Social Sciences 9*, Acta Universitatis Upsaliensis, Uppsala, Sweden, 2005.
28. Engström, J., Johansson, E., and Östlund, J., Effects of visual and cognitive load in real and simulated motorway driving, *Transportation Research Part F: Psychology and Behaviour* 8(2), 97–120, 2005.
29. Victor, T W., Harbluk, J.L., and Engström, J.A., Sensitivity of eye-movement measures to in-vehicle task difficulty, *Transportation Research Part F: Psychology and Behaviour* 8(2), 167–190, 2005.
30. Summala, H., Nieminen, T., and Punto, M., Maintaining lane position with peripheral vision during in-vehicle tasks, *Human Factors* 38(3), 442–451, 1996.
31. Horrey, W.J., Wickens, C.D., and Consalus, K.P., Modeling drivers' visual attention allocation while interacting with in-vehicle technologies, *Journal of Experimental Psychology-Applied* 12(2), 67–78, 2006.
32. Summala, H., Lamble, D., Laakso, M., Driving experience and perception of the lead car's braking when looking at in-car targets. *Accident Analysis and Prevention* 30, 401–407, 1998.

33. Wierwille, W.W. and Tijerina, L., Modelling the relationship between driver in-vehicle visual demands and accident occurrence. In *Vision in Vehicles VI*, Gale, A.G., Brown, I.D., Haslegrave, C.M., and Taylor, S.P. (Eds.), North-Holland, Amsterdam, 1998, pp. 233–244.
34. Harbluk, J.L., Noy, Y.I., Trbovich, P.L., and Eizenman, M., An on-road assessment of cognitive distraction: Impact on drivers' visual behaviour and braking performance, *Accident Analysis and Prevention* 39, 372–279, 2007.
35. Recarte, M.A. and Nunes, L.M., Effects of verbal and spatial-imagery tasks on eye fixations while driving, *Journal of Experimental Psychology: Applied* 6(1), 31–43, 2000.
36. Recarte, M.A. and Nunes, L.M., Mental workload while driving: Effects on visual search, discrimination and decision making, *Journal of Experimental Psychology: Applied* 9(2), 119–137, 2003.
37. Horrey, W.J. and Wickens, C.D., *The Impact of Cell Phone Conversations on Driving: A Meta-analytic Approach*, Technical Report AHFD-04-2/GM-04-1, General Motors Corporation, Warren, MI, 2004.
38. Brookhuis, K.A., de Vries, G., and de Ward, D., The effects of mobile telephoning on driving performance, *Accident Analysis and Prevention* 23(4), 309–316, 1991.
39. Östlund, J., Nilsson, L., Carsten, O., Merat, N., Jamson, H., Jamson, S., et al., Deliverable 2--HMI and safety related driver performance, Contract No. GRD1/2000/25361 S12.319626, Human Machine Interface And the Safety of Traffic in Europe (HASTE) Project, 2004.
40. Mattes, S., Föhl, U., and Schindhelm, R., Empirical comparison of methods for off-line workload measurement, AIDE Deliverable 2.2.7, EU project IST-1-507674-IP, 2007.
41. Carsten, O., Merat, N., Janssen, W., Johansson, E., Fowkes, M., and Brookhuis, K., *HASTE Final Report*, Contract No. GRD1/2000/25361 S12.319626, Human Machine Interface and the Safety of Traffic in Europe (HASTE) Project, 2005.
42. ISO, Road vehicles—Measurement of driver visual behaviour with respect to transport information and control systems—Part 1: Definitions and parameters, International Standard 15007-1, 2002.
43. ISO, Road vehicles—Measurement of driver visual behaviour with respect to transport information and control systems—Part 2: Equipment and procedures, International Standard 15007-2, 2002.
44. Wierwille, W.W., An initial model of visual sampling of in-car displays and controls. In *Vision in Vehicles IV*, Gale, A.G., Brown, I.D., Haslegrave, C.M., Kruysse, H.W., and Taylor, S.P. (Eds.), Elsevier North-Holland, Amsterdam, 1993, pp. 271–282.
45. Green, P., Where do drivers look while driving (and for how long)? In *Human Factors in Traffic Safety*, Dewar, R.E. and Olson, P.L. (Eds.), Lawyers & Judges, Tucson, AZ, 2002, pp. 77–110.
46. Victor, T., Blomberg, O., and Zelinsky, A., Automating driver visual behavior measurement, Paper presented at Vision in Vehicles 9, August 19–22, 2001.
47. Larsson, P., Automatic Visual Behavior Analysis, (ISRN LITH-ISY-EX-3259-2002), Master's thesis, Linköping University, Linköping, Sweden, 2003.
48. Kronberg, P., Victor, T.W., and Engström, J., Road-centre based measures of visual demand, Paper presented at Vision in Vehicles 11, July 27–29, 2006.
49. Engström, J. and Mårdh, S., *SafeTE Final Report*, Swedish Road Agency (SRA) Report 2007:36, 2007.
50. Zhang, H. and Smith, M., *A Final Report of SAfety VEhicles Using Adaptive Interface Technology (Phase I: Task 7): Visual Distraction Research*, 2004, available at http://www.volpe.dot.gov/opsad/saveit/docs.html.
51. Victor, T. W. and Larsson, P., Method and arrangement for interpreting a subjects head and eye activity, International application published under the patent cooperation treaty, International publication number PCT WO 2004/034905 A1, International Bureau: World Intellectual Property Organization, 2004.

52. Markkula, G. and Engström, J., A steering wheel reversal rate metric for assessing effects of visual and cognitive secondary task load, In *Proceedings of the ITS World Congress*, London, 2006.
53. Salvucci, D.D. and Gray, R., A two-point visual control model of steering, *Perception* 33(10), 1233–1248, 2004.
54. Olson, P.L., Driver perception-response time. In *Human Factors in Traffic Safety*, Dewar, R.E. and Olson, P.L. (Eds.), Lawyers & Judges Publishing Company, Inc., Tucson, Arizona, 2002, pp. 43–76.
55. Lamble, D., Kauranen, T., Laakso, L., and Summala, H., Cognitive load and detection thresholds in car following situations: Safety implications for using mobile (cellular) telephones while driving, *Accident Analysis and Prevention* 31, 617–623, 1999.
56. Lee, J.D., Caven, B., Haake, S., and Brown, T.L., Speech-based interaction with in-vehicle computers: The effect of speech-based e-mail on drivers' attention to the roadway, *Human Factors* 43, 631–640, 2001.
57. Strayer, D.L. and Drews, F.A., Profiles in driver distraction: Effects of cell phone conversations on younger and older drivers, *Human Factors* 46, 640–649, 2004.
58. Strayer, D.L., Drews, F.A., and Crouch, D.J., A comparison of the cell phone driver and the drunk driver, *Human Factors* 48, 381–391, 2006.
59. Chisholm, S.L., Caird, J.K., Lockhart, J.A., Teteris, L.E., and Smiley, A., Novice and experience driving performance with cell phones, *Proceeding of the Human Factors and Ergonomics Society 50th Annual Meeting*, October 2006.
60. Smith, R.K., Luke, T., Parkes, A.P., Burns, P.C., and Landsdown, T.C., A study of driver visual behavior while talking with passengers, and on mobile phones, In *Human Factors in Design, Safety, and Management*, de Waard, D., Brookhuis, K.A., van Egmond, R., and Boersema, T. (Eds.), Shaker Publishing, Maastricht, The Netherlands, 2005, pp. 11–22.
61. Winters, J.J., An investigation of auditory icons and brake response times in a commercial truck-cab environment, Virginia Tech, MSc thesis, 1998.
62. Wickens, C.D., Attention to safety and the psychology of surprise. In *Proceedings of the 2001 Symposium on Aviation Psychology, CD-ROM*, Ohio State University, Columbus, OH, 2001.
63. Muttart, J.W., Quantifying driver response times based upon research and real life data, In *Proceedings of the 3rd International Driving Symposium on Human Factors in Driver Assessment, Training, and Vehicle Design* 3, 8–29, 2005.
64. van Winsum, W., Martens, M., and Herland, L., *The Effect of Speech Versus Tactile Driver Support Messages on Workload, Driver Behaviour and User Acceptance*, TNO-report TM-99-C043, Soesterberg, Netherlands, 1999.
65. Miura, T., Cooping with situational demands: A study of eye movements and peripheral vision performance, In *Vision in Vehicles I*, Gale, A.G., Freeman, M., Haslegrave, C.M., Smith, P., and Taylor, S. (Eds.), Elsevier, Amsterdam, 1986.
66. Miura, T., Behaviour oriented vision: Functional field of view and processing resources. In *Eye Movements: From Physiology to Cognition*, O'Regan J.K. and Levy-Shoen A. (Eds.), Elsevier North-Holland, Amsterdam, 1987.
67. Miura, T., Active function of eye movement and useful field of view in a realistic setting. In *From Eye to Mind: Information Acquisition in Perception, Search and Reading*, Groner, R., dYdewalle, G., and Parham, R. (Eds.), Elsevier North-Holland, Amsterdam, 1990.
68. Rantanen, E.M and Goldberg, J.H., The effect of mental workload on the visual field size and shape, *Ergonomics* 42(6), 816–834, 1999.
69. Williams, L.J., Tunnel vision induced by a foveal load manipulation, *Human Factors* 27, 221–227, 1985.
70. Williams, L.J., Tunnel vision or general interference? Cognitive load and attentional bias are both important, *American Journal of Psychology* 101, 171–191, 1988.

71. Williams, L.J., Peripheral target recognition and visual field narrowing in aviators and non-aviators, *The International Journal of Aviation Psychology* 5, 215–232, 1995.
72. Olsson, S., Measuring driver visual distraction with a peripheral detection task, MSc thesis, Linköping University, 2000.
73. Olsson, S. and Burns, P., Measuring distraction with a peripheral detection task, on-line paper. NHTSA Internet Distraction Forum, 2000, available at www.nrd.nhtsa.dot.gov/departments/nrd-13/driver-distraction/welcome.htm.
74. Harms, L., and Patten, C., Peripheral detection as a measure of driver distraction: A study of memory-based versus system-based navigation in a built-up area, *Transportation Research Part F* 6, 23–36, 2003.
75. Schindhelm, R., Gelau, C., Montanari, R., Morreale, D., Deregibus, E., Hoedemaeker, M., de Ridder, S., and Piamonte, P., *Human Factor Tests on Car Demonstrator*, Deliverable no. D6.4, Final draft. COMUNICAR Programme IST KA1, Testing demonstration and pilot evaluation Work package WP6, Report No.: 2003 P 134, 2003.
76. Patten, C., Kircher, A., Östlund, J., and Nilsson, L., Using mobile telephones: cognitive workload and attention resource allocation, *Accident Analysis and Prevention* 36(3), 341–350, 2003.
77. Engström, J., Åberg, N., Johansson, E., and Hammarbäck, J., Comparison between visual and tactile signal detection tasks applied to the safety assessment of in-vehicle information systems (IVIS), In *Proceedings of the Third International Driving Symposium on Human Factors in Driver Assessment, Training and Vehicle Design,* Rockport, Maine, 2005.
78. Merat, N. and Jamson, H.A. Multisensory detection: How does driving and IVIS management affect performance, In *Proceedings of the Fourth International Driving Symposium on Human Factors in Driver Assessment, Training and Vehicle Design*, Stevenson, Washington, July 9–12, 2007.
79. Merat, N., Johansson, E., Engström, J., Chin, E., Nathan, F., and Victor, T., Specification of a secondary task to be used in safety assessment of IVIS, AIDE Deliverable 2.2.3, European Commission, IST-1-507674-IP, 2005.
80. Engström, J. and Kronberg, K., *SafeTE Deliverable 4.1: Visual Demand Measurement.* Unpublished internal project report, 2006.
81. Chin, E., Gautheret, L., and Nathan, F., *Peripheral Detection Task (PDT): RoadSense Results*, Internal document within the AIDE project, 2004.
82. Wickens, C.D., Multiple resources and performance prediction, *Theoretical Issues in Ergonomics Science* 3(2), 159–177, 2002.
83. Duncan, J., and Owen, A.M., Common regions of the human frontal lobe recruited by diverse cognitive demands, *Trends in Neurosciences* 23, 475–483, 2000.

Part 4

Effects of Distraction on Driving Performance

11 Cellular Phones and Driver Distraction

Frank A. Drews and David L. Strayer

CONTENTS

11.1 INTRODUCTION

In 1999 the World Health Organization (WHO) reported that the leading injury-related cause of death among people aged 15 to 44 years is traffic accidents.[1] Of the 5.8 million people who died of injuries in 1998, almost 1.2 million died as a direct result of injuries sustained in a motor vehicle accident. More recently, the National Highway Traffic Safety Administration (NHTSA) published data indicating that motor vehicle traffic crashes are a leading cause of death in the United States for the age group 4 through 34.[2]

According to NHTSA, driver inattention is a significant contributing factor in motor vehicle crashes (the estimated contribution is about one-third according to Stutts et al.[3]). Driver distraction is the largest contributor among the different types of driver inattention and is estimated to contribute to over half of the accidents where driver inattention is involved.[3] Driver distraction can be defined as any event or activity that negatively affects a driver's ability to process information that is necessary to safely operate a vehicle.

According to the Cellular Telecommunications Industry Association (CTIA) in the United States, there are more than 215 million cellular phone (referred to henceforth as "cell phone") subscribers and more than 100 million drivers who use cell phones while driving.[4] The NHTSA estimates that in 2005, 10% of drivers on the roadway at any given daylight moment were using their cell phones.[5] Based on data indicating that in 2000, 4% of drivers were conversing on cell phones, the number of cell phone–using drivers more than doubled between 2000 and 2005. Without legislative intervention, it is likely that the number of people using cell phones or other wireless communication and entertainment technologies while driving will increase further.

The goal of this chapter is to summarize current research regarding the impact of the use of wireless technology on driving performance in general, and cell phone use in particular (for additional reviews, see Refs 6-9). Our review of the literature is organized based on the type of distraction (e.g., cell phone conversation, text messaging) and on the methodology that was used to conduct the study (e.g., observational method, simulator-based evaluation).

11.2 HANDHELD AND HANDSFREE CELL PHONES

In this chapter we define handheld cell phones as units that house the speaker and a microphone and a numerical keypad for entering phone numbers. The keypad buttons have to be manually depressed for placing a call, and in the case of stored phone numbers, at least one or two buttons have to be used to dial a number. In recent years some handheld phones have emerged that allow the user to use voice dialing, thus minimizing the need to manually interact with the phone during dialing (often only one button has to be pressed). However, in all cases, during the conversation a handheld cell phone receiver has to be held by the operator. The use of handheld cell phones is prohibited in several counties and a number of states within the United States.[5,10]

Cell phones that do not require the user to hold the receiver to the ear are labeled as "handsfree" cell phones. This does not imply that the cell phone cannot be held when an incoming call is accepted or a number is dialed, or that people do not use their hands while conversing (the handsfree phones are often held). The use of handsfree cell phones is often promoted by cell phone providers based on the assumption that driver distraction is a result of the manual interaction with the cell phone and elimination of this interaction leads to safe use of cell phones while driving.

It is possible to convert a handheld cell phone into a handsfree cell phone by purchasing upgrade kits. These provide a separate earpiece and microphone that is placed on the person, therefore eliminating the requirement to hold the device while conversing. Another type of handsfree cell phone uses a separate microphone and speaker, which are permanently installed in the vehicle and relieve the driver of any manual interaction with the phone.

The impact of cell phones on driving performance has been evaluated using a range of tasks mimicking a conversation and using a range of more or less demanding driving tasks. Our review is organized according to the category of research method used to conduct these evaluations, starting with observational studies. Epidemiological approaches, on-road studies, simulator studies, and other approaches are then reviewed.

11.3 OBSERVATIONAL STUDIES

A number of national and regional observational studies have been conducted to assess the frequency of cell phone use while driving. For example, Glassbrenner[5] conducted a national U.S. survey to determine the proportion of drivers using cell phones while driving. Glassbrenner[5] (p. 1) estimates that in 2004 "10 percent of vehicles in a typical daylight moment" have a driver who "is using some type of phone, whether handheld or handsfree." Cell phone usage rates vary based on the region; for example, Eby and Vivoda[11] found that about 3.7% of all drivers in Minnesota were using handheld cell phones while driving during daytime. In the Washington, D.C. area, during the spring of 2004, 6% of drivers were conversing on cell phones.[7] Other studies examined the rate of drivers using different types of cell phone technologies (e.g., handheld versus handsfree). For example, McCartt et al.[7] found that in New York, despite a handheld cell phone ban, nearly 3% of the drivers were using handheld devices and only 0.4% were using handsfree devices.

Strayer and Drews[12] conducted an observational study to assess the impact of cell phone conversations on the frequency of traffic violations. The study was conducted between 5 p.m. and 6 p.m. at intersections with four-way stop signs in a residential area of Salt Lake City, Utah. Observers coded on four days the frequency of traffic violations, that is, failing to stop at the stop sign, and the prevalence of cell phone use while driving. Of the 1748 observed drivers, 6.3% were conversing on handheld cell phones. The study also found that the odds ratio for failing to stop at a stop sign was increased by a factor of 10 for the drivers who were conversing on a cell phone (the cell phone users' odds ratio was 2.93; for nonusers it was 0.27). One potential limitation of this, and of the other observational studies, is that it was not possible to judge if a person was talking on a handsfree cell phone. Thus, it is likely that the above-mentioned results are underestimating the real impact of cell phone conversations on traffic violations.

The above-mentioned observational studies indicate that the use of cell phones is widespread and that the likelihood of committing a traffic violation is higher when conversing on a cell phone while driving than when not. The studies on the frequency of cell phone use are samplings from a wide range of locations. However, more data are required to document the frequency of traffic violations committed by cell phone drivers to assess the impact of cell phone use on behavior while driving.

An alternative approach to observational studies is to investigate traffic crashes and to examine whether the driver was conversing on a cell phone or not at the time of the accident. Epidemiological studies use this approach (see also Chapters 16 and 17 of this book).

11.4 EPIDEMIOLOGICAL APPROACHES

In their seminal study, Redelmeier and Tibshirani[13] evaluated the cell phone records of 699 individuals who were involved in motor vehicle crashes over a 14-month period. The authors found that almost a quarter of these individuals were using their cell phones in the 10 min preceding the crash. The analysis of the data revealed that conversing on a cell phone while driving was associated with a fourfold increase in the likelihood of being involved in an accident compared with driving without conversing on a cell phone. The authors further investigated the difference between cell phone use of handheld and handsfree units, and found that the relative risk of a crash for handheld and handsfree devices did not differ statistically.

Other epidemiological studies draw a similar picture about the relationship between the use of cell phones while driving and accidents. In a study that analyzed the data of 100 randomly selected drivers involved in accidents within the past 2 years compared with 100 randomly selected drivers without accidents in the past 10 years, Violanti and Marshall[14] show that talking for more than 50 min on a cell phone a month while operating a vehicle is associated with a 5.6-fold increase in crash risk. A larger, follow-up study by Violanti[15] found a ninefold increase in accident risk that is associated with the use of cell phones. One explanation for the different risk estimates is that Violanti's study used a case-control study design, whereas in Redelmeier and Tibshirani's[13] study, a case-crossover analysis was utilized (see also Chapter 6 of this book for a discussion of these methods). Based on results from epidemiology, which rely on unobserved heterogeneity between cases and controls as a source of error, it appears that case-crossover studies produce more accurate results.[16]

More recently, McEvoy et al.[17] performed a case-crossover study in Australia to explore the effect on road safety of drivers' use of cell phones. More than 450 drivers participated in this 27-month study of drivers who were involved in crashes that resulted in hospitalization. The authors found a fourfold increase in the likelihood of crashing when using cell phones (odds ratio was 4.1), a result that is very similar to Redelmeier and Tibshirani's[13] findings. One finding of this study was that the type of phone used (i.e., handheld versus handsfree) did not affect the association between cell phone use and crash risk. Thus, use of both handheld and handsfree cell phones was associated with a fourfold increase in accident risk.

Laberge-Nadau et al.[18] sent out 175,000 questionnaires to drivers to explore the relationship between cell phone use and car accidents. The questionnaire addressed questions regarding exposure to risk, driving habits, opinions about safe and unsafe practices while driving, information about accident involvement in the past 2 years, and sociodemographic information. Analysis of the approximately 20% of questionnaires returned shows that the relative risk for collisions is 38% higher for users of handheld and handsfree cell phones than it is for nonusers. One important finding of the study is that it revealed the relationship between frequency of cell phone use and accident rates. In their study, the authors found that relative to infrequent users, heavy cell phone users have a relative crash risk of 2.

According to Green,[19] accident data collected by the Japanese National Police Agency Traffic Planning Department showed that activities associated with use of cell phones resulted in vehicle crashes. Interestingly, 1077 of the 2418 reported crashes were related to receiving the call, and 504 were related to dialing a phone number. As Green argues, receiving a call is challenging because often the phone is not easy to locate, and therefore, attention is directed toward finding and retrieving the phone instead of toward the driving task. The data indicate that the manipulation of the phone is a significant problem in the context of cell phone use while driving. It is important to note that even handsfree cell phones do not necessarily eliminate these problems; for example, the need to place a headset may contribute to problems.

This can be contrasted with more integrated cell phone systems that allow for voice commands and would potentially reduce the need for the cell phone manipulation (see Ref. 20).

In summary, it seems that the risk of having a car crash is increased by a factor of four or more when conversing on a cell phone while driving. It is remarkable that this applies for both handsfree and handheld cell phones, indicating that in addition to the contribution of physical interaction, the conversation in itself is a major contributor to any driver distraction caused by the use of a cell phone while driving. Finally, drivers who are engaged in cell phone conversations more often are more likely to be involved in a crash.

One problem with epidemiological studies is that it is very difficult to determine the type of conversation (e.g., superficial versus emotionally engaging conversation) and the driving conditions in which it is conducted (e.g., traffic density, weather conditions). Another issue with these studies is related to the difficulty of determining the exact timing of an accident and a cell phone call. Thus, most investigators use a rather arbitrary 10 minute window of usage of the cell phone around the time of the accident. Finally, another important limitation of these and other epidemiological studies is that often only a correlation between cell phone use and an increase in accident rate can be established based on the findings. The causal structure underlying this relationship is not clear and is therefore open to speculation. For example, it is possible that drivers who are using their cell phones are in general more risky drivers and are displaying other risky types of behavior (see Ref. 11).

To deal with the issue of identifying the causal relationship between elevated accident risk and use of cell phones, and to control for other important variables, a number of experimental and quasi-experimental studies have been conducted. What these studies have in common is that they attempt to identify the exact contributor(s) to the problems associated with conversing on a cell phone while driving. In doing so, these studies often distinguish between task components that are involved in operating a vehicle and in conversing on a cell phone.

11.5 ON-ROAD EXPERIMENTS

Brookhuis et al.[21] studied the impact of using handheld cell phones on driving under three different driving conditions in an on-road study. Twelve participants each drove in light traffic on a quiet motorway, in heavy traffic on a four-lane ring road, and in urban traffic. In their study, participants used handheld and handsfree cell phones for the conversation. The simulated conversations consisted of a 3 min paced serial addition task. This task was intended to impose a controlled mental load on the participants while operating the vehicle. Brookhuis et al. indicate that at the operational level (with the exception of steering wheel amplitude), there was little or no effect on the phone task. The exception was when participants were using handheld phones and were required to dial a number. In this case, a 10-fold increase in steering wheel amplitude was observed. A significant impact on driving performance was found at the tactical level of driving, reflecting how a driver negotiates traffic. Changes at the tactical level included slower reaction times (6.5% increase in reaction time to the onset of braking lights), a lower frequency of checking the rear-view mirror, and delayed adaptation to the speed of a car in front of the driver's car during the phone task (22.6% slower to respond to changes as compared with the control condition).

Other on-road studies have also found that dialing a cell phone while driving has a negative impact on driving,[22–24] by using subjective ratings of task difficulty. It was found that participants in an on-road study rated dialing as more difficult than a number of other tasks like conversing with passengers, adjusting a car heater or air conditioner, or tuning a car radio. However, some tasks were rated as more difficult than dialing, such as lighting and smoking a cigarette, writing down something, or reading a map.

More recently a naturalistic driving study conducted by The Virginia Tech Transportation Institute (VTTI) and the NHTSA[25] indicates that the use of wireless technology is associated with a higher number of rear-end accidents and incidents than driver interaction with any other source of distraction. In this study, the majority of participants drove their own vehicles, which were equipped with video cameras, global positioning devices, and speed and brake sensors. Data collection was continuous and focused on owner-vehicle behavior and other vehicle behaviors.

A study by Mazzae et al.[20] reported the frequency and duration of cell phone use while driving in a sample of 10 drivers observed over a 6-week period. In this study the use of cell phones while driving did not reveal significant changes in driving behavior compared with the control condition of driving only. Participants showed changes in time spent fixating on the roadway ahead when they were interacting with the cell phone and, in particular, when they were dialing. Handsfree voice-controlled dialing showed a reduction of fixations compared with the control condition, but, predictably, dialing a handheld cell phone decreased the time spent fixating on the roadway ahead even more. During conversations, cell phone use was associated with a reduction in the frequency of long-eye fixations on the roadway ahead. Interestingly, with an increase in the length of the cell phone conversation (i.e., 2 min), the time spent fixating forward on the roadway increased. However, at the same time, the percentage of time with fixations to the interior of the vehicle (i.e., dashboard, instruments), looking left and looking right, decreased. The authors interpret these findings as an indication of a reduction in situation awareness in drivers who are conversing on cell phones.

Harbluk et al.[26] selected an urban environment for their on-road study. Participants had to solve simple or difficult addition tasks simulating a conversation and use a handsfree cell phone while driving. The authors measured visual scanning patterns and recorded measures of vehicle control (e.g., braking). The results showed that increased cognitive load associated with the addition task led to changes in visual scanning with participants executing fewer saccades and fixating more centrally. Participants also reduced the time spent fixating to the right periphery, on the instruments and on the rear-view mirror. Also, changes in driving behavior were observed; for example, participants showed more hard-braking behavior when solving the mathematical problems.

Matthews et al.[27] had 13 participants drive on a 5 km rural highway and measured subjective workload associated with the use of three types of wireless devices: handheld cell phone, handsfree cell phone with single earphone and integrated microphone, and handsfree cell phone with external loudspeaker and microphone mounted in the passenger compartment. The intelligibility of the conversation was measured by the modified rhyme test. Using the NASA Task Load Index,[28] Mathews et al. found that the use of cell phones resulted in a significant increase in workload compared with the control condition and differences between the cell phone types were significant. The handsfree cell phone with single earphone and integrated microphone was associated with the lowest subjective workload followed by the handheld cell phone. Participants experienced the highest workload when using the handsfree cell phone with external loudspeaker and microphone because of problems of low intelligibility of the conversation.

Crundall et al.[29] had participants conduct passenger conversations with a passenger who was either blindfolded or able to observe traffic during the conversation. Their findings indicate that passengers respond to changes in the cognitive demand of driving. For example, they found that vehicle conversations were suppressed during demanding urban driving and that there was no impact of cognitively demanding driving situations on the conversation when conversing on a cell phone. Thus, a cell phone conversation imposes cognitive load independent of the cognitive demand resulting from the driving conditions, making it likely that the driver's cognitive limits are exceeded (see Ref. 30).

Patten et al.[31] had 40 professional drivers participate in their on-road study. Drivers had to drive an instrumented vehicle for 74 km on a motorway. The goal was to determine the impact of different types of cell phones (handheld versus handsfree) and types of conversation on the performance of a peripheral visual detection task. Driving conditions in this study were good, and driving on the motorway was of low complexity. The authors found that there was a significant impact of conversation complexity on detection performance, while the type of cell phone did not have an impact. In the complex conversation condition, the reaction time for the peripheral detection task increased by 45% from baseline, and the hit rate in this task dropped from 96% to 85% in the dual-task condition. In terms of driving performance, the authors found that participants using a handheld cell phone drove slower than in the baseline condition, whereas participants using handsfree cell phones drove faster than those in the baseline condition.

Taken together, the on-road studies clearly demonstrate a change in driving behavior when drivers are engaged in a cell phone conversation or a task that imposes cognitive demand on the driver that is intended to mimic a conversation. In addition, drivers tend to experience an increase in subjective workload when using cell phones while driving, though there are differences between handheld and handsfree cell phones; cognitive workload for handheld cell phones is usually rated as being higher.

There are some important limitations in the context of on-road studies of cell phone use while driving. One problem is that the above-mentioned studies often use simulated conversations, intended to impose a similar cognitive demand as a conversation. But, since it is not clear if this cognitive demand is comparable with a conversation, the external validity of these tasks is questionable. Another issue is related to the problem of control of the driving conditions. Because participants are usually driving in regular traffic, traffic density and driving conditions in general are out of the control of the experimenter. This creates a situation that makes it more difficult to replicate the exact conditions for each participant. Another limitation present is that ethical considerations do not allow for the creation of a cell phone conversation that is highly cognitively demanding to study the impact of cell phone conversations over a wide range of cognitive demand. Some of these limitations can be addressed by using closed-test track studies.

11.6 CLOSED-TEST TRACK STUDIES

Cooper and Zheng[32] and Cooper et al.[33] examined how handsfree conversations affected specific driving tasks on a closed-test track. They had participants weave through targets, perform a left-turn gap selection task, and respond to the onset of a red traffic light (for a similar task, see Ref. 34). During the conversation, participants received a message that contained a criterion statement, which provided the basis for an evaluation of target words. In the first two tasks, the authors found that driving performance declined in comparison with the control condition. Surprisingly, the authors found that participants conversing on a cell phone were more likely to stop at a red light, which, according to Cooper et al.,[33] was due to the fact that participants developed a strategy to anticipate events in their environment. The fact that they became more familiar with the driving task allowed them to become more proficient, even when dealing with the simulated conversation. Because the other two tasks were less predictable, participants were not able to develop a strategy, and in this condition conversing on a cell phone undermined driving performance. If predictability is central to the development of strategies and finding facilitatory effects, then in more naturalistic settings, facilitation should not be observed. This is exactly what Strayer and Drews[12] found in their observational study on cell phone–related traffic violations.

Treffner and Barrett[35] evaluated the impact of handsfree cell phone conversations in a closed-circuit driving track environment.During the conversation, participants performed either a digit reversal task, a mathematical summation task, or a categorization task. The impact of these tasks on driving performance was evaluated by analyzing critical control actions like braking or obstacle avoidance. Treffner and Barrett found that conversation negatively affected braking by delaying it and altering braking style in a nonoptimal manner. For example, cell phone drivers did not reduce their speed sufficiently when starting to decelerate, which, in turn, required them to brake more abruptly later. During the obstacle avoidance task, conversing drivers drove slower and were slower in their anticipatory response in attempting to avoid obstacles, indicating that the conversation negatively affected car control.

Similar changes in driving performance were found by Hancock et al.[36] The authors had participants perform a conversation task that involved a memory task. Hancock et al. found that participants who were conversing showed a 15% increase in failing to respond to stoplights and a slower response to those stoplights compared with the control group.

Overall, the results of test track studies show a picture similar to that of the on-road studies. Driving performance degrades in response to the increase in cognitive demand due to the conversation. Interestingly, predictability of scenarios seems to be important. If the driving conditions allow anticipation of upcoming events, a simulated conversation seems to have less of an effect on driving performance than when the driver is dealing with unpredictable conditions. Finally, based on the above-mentioned studies, it seems more and more clear that conversing on a handheld cell phone contributes to driver distraction. Noteworthy is that the above-mentioned studies focused mostly on handsfree cell phones, indicating that the cell phone conversation contributes significantly to the overall distraction of the driver.

Limitations of the above-mentioned studies relate mainly to the fact that closed-test track studies usually do not allow for highly realistic driving situations and environments. Thus, the driving task that is used is often impoverished compared with driving in real and more complex traffic. Another important limitation is that the cell phone conversations were often artificial; that is, participants answered questions rather than engaging in a naturalistic conversation. However, despite these methodological limitations, the above-mentioned studies illustrate that significant deficits in driving performance can be observed when drivers converse on cell phones.

11.7 SIMULATOR STUDIES

In the following sections, we will focus on simulator studies but distinguish between simulator studies that were conducted in high-fidelity simulators and studies that used low-fidelity simulation to evaluate the impact of phone conversations on driving performance. See Chapter 7 of this book for a discussion on simulator fidelity.

11.7.1 High-Fidelity Simulation

Burns et al.[37] studied, among other conditions, the influence of handheld cell phone conversations on driving performance. In their simulator study, they found that participants who were conversing on a handheld cell phone while driving showed significant impairments in terms of speed control and response time compared with the other groups. The authors found that participants drove slower when using handheld cell phones and their response times to traffic signs were delayed. Also, the authors found that subjects reported a significantly higher subjective workload when using cell phones as compared with not using a cell phone. One limitation of this study is that the conversations people engaged in involved verbal puzzles, sentence remembering, and monologues about some provided topic. Thus, it is possible that the conversations only partly reflect the difficulties associated with real conversations.

Impairment in situation awareness regarding the surrounding traffic when using handheld cell phones while driving was found by Ma and Kaber.[38] The authors compared the impact of using a handheld cell phone while driving with the use of an adaptive cruise control system. To measure situation awareness, the authors used the Situation Awareness Global Assessment Technique (SAGAT)[39] that queries participants about events in their environment after freezing the simulation. In their simulator study, they found that the use of a cell phone while driving led to a significant reduction in the drivers' situation awareness and a significant increase in the perceived mental workload of the driver.

Haigney et al.[40] considered the impact of handsfree and handheld cell phones on driving performance of a manual and automatic transmission car and on physiological measures like heart rate. In their study, they simulated conversation by having participants perform a cognitively more demanding version of the grammatical reasoning test.[41] In terms of physiological measures, the authors found that heart rate was elevated during the time of the conversation compared with the control condition. In terms of driving performance, Haigney et al. found that drivers using either handheld or handsfree cell phones drove slower, which may reflect some attempt at risk compensation (see also Ref. 42). It was also found that use of handheld phones resulted in an increase in off-road excursions compared with the use of handsfree cell phones. Finally, the authors found that drivers using any type of cell phone showed a reduction in accelerator pedal depression variability and a failure to change gears in the manual transmission driving condition during and after the call. The fact that drivers, during the cell phone conversation, showed a reduction in responsiveness to traffic conditions indicates that drivers may fail to respond optimally during emergency situations or when experiencing an increase in the demand of the driving task.

Alm and Nilsson[43] studied the effect of communication on a handsfree cell phone with regard to a variety of driving performance measures. In their experiment, the authors used a working memory span task[44] to simulate a conversation. They measured reaction time, lane position, speed, and workload in easy and difficult driving conditions. Interestingly, the strongest impact of the conversation was found when participants were conversing in the easy driving task. Here, reaction times were higher and the speed was slower compared with the control condition. In the more demanding driving condition, there was an impact on the lateral position of the vehicle.

A similar study was conducted by Radeborg et al.,[45] who used a working memory task that tested memory span and judgment to simulate the cognitive demand of a conversation. For this task, participants used a handsfree cell phone while driving under three conditions: a control condition, an easy driving condition, and a difficult driving condition. For this study, the authors were primarily interested in the impact of the driving task on memory performance and judgment; no measures of driving performance were collected. The results showed that driving had a significant negative impact on memory performance but not on participants' ability to judge if sentences were meaningful or not.

Horberry et al.[46] tested three age groups when either operating an entertainment system or answering questions on a handsfree cell phone while driving in simple and complex environments in a high-fidelity driving simulator. For both types of distraction, under all driving conditions and for all age groups, a degradation of driving performance in terms of speed maintenance and deviation from the speed limit was found.

Rakauskas et al.[47] used a high-fidelity driving simulator to evaluate the impact of two levels of conversation difficulty on driving performance. In their experiment, participants used handsfree cell phones and answered questions that were either easy or difficult to simulate a cell phone conversation. The authors found that both types of cell phone conversation negatively affected driving performance, as indicated by higher variation in speed (measure of speed maintenance) and reduced average driving speed. Also, participants reported higher subjective workload while conversing on a cell phone.

It is possible that specific communication training or practice has the potential to reduce the workload and, therefore, the negative consequences of talking on a cell phone. Two known studies have focused on these issues.

Hunton and Rose[48] were interested in the impact that pilot communication training has on performance in a driving task in a high-fidelity driving simulator. They tested two groups of drivers: pilots with training in effectively managing multiple-task demands during communication and a group of nonpilots without this type of training. In their study, they had participants drive only, converse on a handsfree cell phone, or converse with a passenger. The conversation was held between the participant and a facilitator who asked the participant a number of scripted questions. The analysis of crashes and incidents revealed that the cell phone conversations resulted in significantly higher rates of crashes and incidents than the control condition or the passenger conversation condition. However, the authors also report that pilot communication training has the potential to decrease the negative effect of cell phone conversations while driving (although it is not clear if this is due to the training or self-selection factors associated with aviation).

Shinar et al.[49] used a mathematical task and emotionally charged questions to simulate a conversation in their driving simulator–based study. The goal of their study was, among others, to find out how much participants were able to learn to deal with the simultaneous demands of driving and conversing. The authors reported that after approximately 2.5 h of training, participants demonstrated learning effects on most of the driving measures. Also, the authors were able to show that interference from the conversation task diminished over time. The fact that Shinar et al.[49] found training effects is interesting because training effects have not been reported before. Also, the training time in this study was rather limited and seems to be at odds with the findings of epidemiological studies demonstrating that drivers who engage in frequent cell phone conversations while driving are at higher risk of crashing. More recent work by Cooper and Strayer[50] raised these issues and failed to replicate Shinar et al.'s findings. One explanation for these findings could be related to the fact that the driving task in Shinar's study consisted of driving on a "relatively straight two-lane highway with few turns and little traffic" (p. 316).[49] It is possible that this driving task had an artificially high level of predictability and participants were able to develop skills over the short training period, resulting in improved driving performance (see also Ref. 33).

Beede and Kass[51] used a driving simulator to measure the impact on driving of a conversation task on a handsfree cell phone and a signal detection task while driving. Driving performance was measured in terms of traffic violations (running stop signs), driving maintenance (standard deviation of lane position), attention lapses (stopping in front of a green traffic light), and response time (time required to respond to an event). Driving performance measures in terms of traffic violations, driving maintenance, attention lapses, and response time were significantly impaired when participants were talking on handsfree cell phones. The authors also found that overall performance in the signal detection task was low. Finally, they found an interaction between the cell phone conversation and the signal detection task in measures of driving speed, speed variability, reaction time, and attention lapses. Conversing on a cell phone and performing the signal detection task simultaneously increased the average speed, the number of attention lapses, and reduced variability in speed maintenance. Interestingly, reaction time to a traffic light changing from red to green and to pop-up stop signs was more delayed when participants did not perform the signal detection task, indicating that participants may have traded off effort involved in the two tasks.

Strayer et al.[42] conducted four experiments in a high-fidelity driving simulator to examine the impact of naturalistic phone conversations on driving. In their first study, they found that conversations on handsfree cell phones increased drivers' reaction time to vehicles braking in front of them. In three additional experiments, the authors examined the hypothesis that the observed impairment could be attributed to a withdrawal of attention from the visual scene resulting in a form of "inattention blindness" (i.e., a fixated object is not being processed, which results in either an incomplete or no mental representation of the object). The findings indicated that cell phone conversations impaired explicit recognition memory of roadside billboards even when participants had fixated on them. The authors conclude that their findings suggest that the impairment of driving performance produced by cell phone conversations is mediated, at least in part, by reduced attention to visual inputs.

In another study, Strayer and Drews[12] were interested in a direct comparison of the impact on driving performance of cell phone conversations versus alcohol (Blood Alcohol Concentration, or BAC, of 0.08). The rationale for this study was twofold. First, earlier epidemiological studies had speculated about the risk of having a car accident while conversing on a cell phone being similar to the risk of being involved in an accident while legally drunk. Second, it was postulated that if conversing on a cell phone is associated with a similar increase in crash risk as the legally prohibited activity of drunk driving, this may have implications for the use of legislation to limit exposure to cell phone use.

The results of this comparison showed that drivers conversing on a cell phone (either handheld or handsfree) showed delayed braking reaction times and an increase in traffic accidents compared with the control group that was only driving. Interestingly, the intoxicated drivers showed a more aggressive driving style than the other drivers, as indicated by closer following distance and application of more force when braking. Drivers in this condition did not get involved in any traffic accidents, while three participants crashed in the cell phone condition. The pattern of crashes suggests that the impairment in the cell phone condition is similar to or worse than the impairment that was observed in the intoxicated driving condition. However, it is important to emphasize that the mechanisms leading to the impairment of the intoxicated driver differ from those that lead to the impairment of a driver conversing on a cell phone. The authors concluded that when driving conditions (i.e., weather and traffic density) and time on task were controlled for, "the impairments associated with using a cell phone while driving can be as profound as those associated with driving with a blood alcohol level at 0.08%" (p. 390).[12] Similar findings were reported by Burns et al.[37]

Other simulator studies have studied the impact of handheld cell phone use on driving performance. Stein et al.[52] studied lane keeping and found significant degradation of driving when placing phone calls in straight driving or on curves. Törnros and Bolling[53] found in their study of dialing and conversations with handheld and handsfree phones that dialing on a phone had a negative impact on lateral position deviation and speed with drivers driving slower.

In summary, the results of the high-fidelity simulator studies draw a clear picture of the impact of cell phone conversations on driving performance that is consistent with findings using other research methods. In general, conversing on any type of cell phone results in drivers driving at slower speed, which is interpreted as some type of risk compensation, and in slower response times to traffic events. Also, participants report higher subjective workload when engaged in a cell phone conversation while driving, which is associated with a reduction in overall situation awareness of the surrounding traffic. This finding is consistent with the idea that cell phone conversations while driving have a direct impact on the processing of visual input, which leads to inattention blindness. The impact of conversing on a cell phone while driving seems to be comparable to driving under the influence of alcohol in terms of crash risk.

Some studies report that the use of handheld cell phones results in an increase in off-road excursions. However, using a handsfree cell phone does not mitigate the above-mentioned negative consequences.

11.7.2 Low-Fidelity Simulation

McKnight and McKnight[54] used a video driving sequence that included a total of 45 highway traffic scenes (e.g., stopping or crossing vehicles, pedestrians). The 150 participants were expected to respond to the events by manipulating the controls of a vehicle. Participants were tested in five conditions: they had to place a cell phone call, engage in a conversation that was either casual or intense, tune a radio, or just respond to the traffic scenarios. The cell phone used in this study was a simulated handsfree phone that required physical interaction only when placing a call. The intense conversation was manipulated by having participants solve mathematical and short-term memory problems, whereas the casual conversation involved conversation about participants' professions or activities. The authors controlled for the age of participants by dividing them into three age groups: young, middle-aged, and older drivers. The authors reported that participants in all conditions failed to respond to traffic events. In particular, the older group of drivers were more vulnerable to multitask demands. The younger group of participants also showed a decrease in their ability to respond to traffic scenarios that was more pronounced in the intense conversation condition. Findings that provide a potential explanation of the impact of intense conversations based on perceptual processes were reported by Amado and Ulupinar.[55]

Amado and Ulupinar[55] compared the impact of a handsfree cell phone conversation, a passenger conversation, a control condition on attention, and a peripheral detection task that simulated driving. To simulate the cognitive demand of a conversation, the authors identified 56 questions of low or high complexity. These questions contained assessment of general knowledge and arithmetic skills. The authors found that both simulated conversation conditions had a negative impact on performance in the peripheral detection task compared with the control condition. The lack of a difference between the cell phone conversation condition and the passenger conversation condition is interesting because it might be related to the fact that the passenger was not responding to changes in the demand of the primary task of simulated driving. Thus, the lack of a difference is likely due to the low fidelity and lack of immersion of the passenger in the driving task (see also Ref. 29).

Consiglio et al.[56] measured reaction time of braking responses at the onset of a red light in a "laboratory station," that is, an apparatus set up to simulate foot activity in driving a vehicle with automatic transmission. Conditions examined in this experiment were conversation on a handheld cell phone and a handsfree cell phone, conversation with a passenger, and listening to a radio. The conversation was simulated by having a research assistant ask the participants questions about the nature of their school studies and other activities. Braking performance in the above-mentioned conditions was compared with performance in a single-task condition where the participants were responding to the onset of the red light only. With one notable exception, all conditions negatively impacted on braking response time; the exception being listening to the radio (see also Ref. 57).

The use of low-fidelity simulation for driving has generally resulted in findings that are similar to results from high-fidelity simulators. Participants showed impairment when conversing on cell phones, and the impairment was worse with advanced age. Again, the studies demonstrated that there was no difference between handheld and handsfree cell phones in terms of negative consequences for driving. Interestingly, the studies replicated some of the findings from high-fidelity simulator studies, showing that the impact of a cell phone conversation is worse than the impact of a passenger conversation or listening to radio, both of which had no impact on driving performance. However, it is important to emphasize that given the low fidelity, some of the findings are not generalizable. For example, it is possible that the difference in the impact of passenger conversation when comparing high-fidelity and low-fidelity simulations can be attributed to the low fidelity and the lack of immersion of the passenger. Other limitations of the above-mentioned studies are related to the nature of the conversation, which in almost all cases was a simulated conversation, which is a task that requires the participant to answer questions.

11.8 PART-TASK SIMULATIONS

An alternative to conducting the above-mentioned on-road studies, closed-track studies, or simulator studies is to identify critical components of the driving task and to use tasks that share these components to evaluate the impact of dual-task activities on these tasks. One such task that has been used in this context is a pursuit tracking task, because it reflects the basic task components of driving but does not require the use of simulation.

Briem and Hedman[58] used a pursuit tracking task that simulated driving where they varied conversation difficulty during a handsfree cell phone conversation. Also, they evaluated the impact on tracking performance under different surface conditions (i.e., by changing the control dynamics, they simulated a firm or a slippery surface) and when participants manipulated a car radio. One of their findings was that the road conditions have a negative impact on driving performance, especially when manipulating the radio. Driving while having an easy cell phone conversation had the least impact on driving performance and, in some cases, had a facilitating effect. The authors conclude that the level of difficulty of a cell phone conversation is an important factor related to impairment of driving behavior.

Strayer and Johnston[57] used a pursuit tracking task to simulate driving. In addition to following the target, participants had to respond to the onset of signals on the screen by pushing a button when a red signal flashed and ignoring it in the case of a green signal. In their study, they analyzed the impact of different types of activities on tracking performance. The activities that were studied were conversing on a cell phone, listening to a book on tape, and listening to the radio while performing the simulated driving task. Comparison of tracking performance in the single task and the dual task conditions showed that tracking and detection performance did not differ for the book on tape and radio tasks. However, when participants were engaged in naturalistic conversations on a handheld cell phone, the authors found that they were more likely to miss red signals and to respond more slowly to the onset of the light. Moreover, the authors did not find any difference between handheld and handsfree cell phones in terms of impairment.

Graham and Carter[59] were interested in the impact of cell phone conversations on peripheral target detection. They simulated the driving task by having participants track a target on a computer monitor. Conversing on a cell phone increased the target detection time by between 200 and 350 ms as a function of phone type compared with the control condition where participants had to only track. The handheld phone allowed participants to dial faster compared with a handsfree speech phone, but tracking and detection performance was worse compared with the handsfree phone.

McCarley et al.[60] investigated the scanning behavior of traffic scenes by younger and older drivers while casually conversing on a handsfree cell phone. The results of their study show that even casual conversations have a negative impact on visual scanning behavior of traffic scenes as indicated by higher error rates for change detection and higher numbers of saccades to locate a changing item on the display. Finally, the authors found that the time for fixations was reduced under dual-task conditions. The authors interpret their findings as evidence that a conversation while scanning traffic scenes has an impact on peripheral guidance of attention.

Barkana et al.[61] were interested in examining the impact of a handsfree cell phone conversation on visual field awareness. Visual field awareness was measured using the Esterman visual field examination. During the Esterman visual field examination, participants have to detect a stimulus in 100 points that are presented from 75 degrees temporal to 60 degrees nasal. Performance measures are detection rate, time for detection, rate of fixation loss, false positives, and false negatives. Participants performed the visual field awareness task in a control condition and when conversing on a cell phone with a confederate. The cell phone conversation included questions to simulate ordinary conversation and some questions that required memorizing information about a flight schedule. The cell phone conversation significantly negatively impacted all dependent variables. Overall, the findings indicate that participants conversing on a cell phone are more likely to miss points in the visual field, react slower to individual stimuli, and perform with lower levels of precision compared with a control condition (see also Ref. 62).

The above-mentioned part-task simulations cast more light on the mechanisms of the impairments. These studies indicate that the reaction time of participants is slower when conversing on a cell phone and that the scanning behavior of the environment changes significantly as indicated by higher error rates in detecting changes in the environment or the presence of signals. Some of the results indicate that problems of conversing on a cell phone while driving are a result of the drivers' reduced ability to direct attention effectively to the driving environment because of the competition for attention in the driving task and the conversation.

11.9 CELL PHONE DISTRACTIONS BEYOND CONVERSATIONS

With the increasing availability of text messaging in the newer generation of cell phones, it is no surprise that more and more people are using their cell phones to send messages. According to a survey conducted by Telstra in Australia in 2003,[63] 30% of the respondents admitted to having sent text messages while driving, and almost 20% are regularly sending text messages while driving. Given that text messaging not only requires central processing of information but also focusing on the text on the display of the phone during the process of composing a message or reading a message, it is likely that the impairments associated with the use of cell phones while driving are exceeded when people are using text messaging while driving.

Hosking et al.[64] examined the impact of text messaging on driving performance and eye movements in a high-fidelity driving simulator. Twenty young novice drivers with less than 6 months of driving experience were exposed to a number of safety-related events (e.g., a pedestrian appears from behind a car) during a driving task, which also included car following and lane changing. Both retrieving and sending text messages negatively affected driving performance. The driver's ability to control lateral vehicle position and their responses to traffic signs were significantly impaired during the messaging activity. Also, during this activity, the driver's eyes were focusing significantly less on the road compared with the control condition. The speed of the distracted drivers did not differ from the driving speed in the control condition, although the following distance increased. This increase in following distance is interpreted by the authors as the drivers attempting to compensate for the increased distractedness during driving.

Kircher[65] focused in their simulator study on receiving text messages while driving. Ten experienced drivers participated in the study where they occasionally received text messages while driving in the simulator. Participants were instructed to retrieve the messages and to respond to them verbally. Effects on driving behavior were measured based on time for braking onset. While participants were text messaging, the braking times were significantly slower compared with the times when they were only driving. Also, the drivers reported that they were driving slower during the text messaging episodes.

Text messaging is an application that can often be used in conjunction with the cell phone. An alternative to text messaging is the more common e-mail. For research concerning e-mail use while driving, see Refs 66 and 67. Given that e-mail has become an integral part of everyday life, it is no surprise that some investigators have looked at the impact on driving behavior of using it while driving. However, one important difference between text messaging and e-mail is that the e-mail systems that have been evaluated have involved the reading of the e-mail messages out loud to the driver, whereas in text messaging studies, the driver has been required to look at the cell phone display (for the impact of reading while driving, see Ref. 68). This issue is discussed further in Chapter 12 of this book.

At this point in time, few studies have addressed the issue of text messaging while driving. However, it seems that text messaging has greater potential to adversely affect driving performance than conversing on a cell phone. While conversing on a cell phone allows the driver to at least fixate on the driving environment (though this does not always lead to processing of the information in this environment), text messaging requires the driver to fixate on the screen of the cell phone, making it completely impossible to process any visual information in the driving environment at that moment.

11.10 CONCLUSIONS

Based on the literature reviewed, it is possible to answer some central questions related to the use of cell phones while driving. Three questions frequently arise: (1) whether conversing on a cell phone interferes with driving, (2) what the sources of the interference are, and (3) how significant is the interference.

In regard to the first question, the data from observational and epidemiological studies draw a clear picture about the impact of cell phone conversations on the risk of getting involved in an accident or of committing a traffic violation. Independent of the specific research design, epidemiological studies find an increase in accident risk associated with the use of cell phones. The increase in risk ranges from a fourfold increase to a ninefold increase. Data about the frequency of traffic violations while using a cell phone indicate that drivers who are engaged in a cell phone conversation are 10-fold more likely to fail to stop at a stop sign.

One of the important questions regarding cell phone use is related to the type of cell phone technology that is used (i.e., whether the crash risks are different for handheld and handsfree cell phones). One implicit assumption that is made when laws ban the use of handheld cell phones but legalize the use of handsfree cell phones is that it is the act of dialing or holding the cell phone that causes the driving impairment. At this point, there seems to be little doubt that interaction with a handheld cell phone (i.e., dialing or receiving a call) increases the risk of crash involvement. However, there is also a strong body of evidence that indicates that the difference between handheld and handsfree cell phone conversations is minimal and potentially negligible in terms of the accident risks. More recent work focuses mostly on the impact of using handsfree devices, demonstrating that the associated impairment is significant. Most researchers conclude that a significant contributor to cell phone–related driver distraction is the engagement in the conversation, which leads to a withdrawal of attention from the immediate driving environment.

When addressing the second question (i.e., what the sources of the interference are), it is interesting to look at different types of activities that are performed while operating a vehicle and to compare these with conversing on a cell phone to identify potential differences and similarities. There is some evidence that activities like listening to a book on tape or listening to a radio only minimally interfere with the driving task. One aspect these activities have in common is that they allow the driver to monitor traffic continually and to focus attention on the driving task at any given time without necessarily missing a lot of either the radio program or the text of the book. Consequently, a driver is able to allocate attention more flexibly and does this more effectively when required while driving. Comparing cell phone conversations with conversations that are held with a passenger is revealing because there is evidence that passenger conversations do not have a negative impact on driving performance. Despite the fact that in both cases the conversation imposes cognitive demand on the driver, it seems that the presence of the passenger mediates the negative influence of the conversation. One of the findings is that the conversation with a passenger changes as a function of the traffic and cognitive demand of the driving. The passenger seems to adjust the conversation with the demand of the driving. Also, in some situations the passenger actively supports the driver (e.g., in a navigation task and in directing the driver's attention to critical events). A person who is remote from the driver has no awareness of the demand of the driving environment and as a consequence is likely to unknowingly impose cognitive demand during times when the traffic requires full attention from the driver.

In terms of the cognitive mechanisms, there is more and more evidence that conversing impairs the visual processing of information. Cell phone drivers exhibit inattention blindness, and this seems to suggest that there is a bottleneck in terms of simultaneous processing of the information from the driving environment and the conversation.

Finally, to answer the third question (i.e., how significant the interference is), Redelmeier and Tibshirani[13] have suggested that "the relative risk (of being in a traffic accident while using a cell phone) is similar to the hazard associated with driving with a blood alcohol level at the legal limit" (p. 465). At this point in time, it seems that their initial assessment of the risk was correct. Studies report that the relative risk of getting into an accident while conversing on a cell phone increases by a factor of 4. Moreover, some simulator-based studies have demonstrated that the impact of a cell phone conversation on driving is comparable to driving under the influence of alcohol with a BAC of 0.08. The converging evidence indicates that using a cell phone while driving poses a significant risk to both the driver and the public at large.

REFERENCES

1. Krug, E. (Ed.), *Injury: A Leading Cause of the Global Burden of Disease*, World Health Organization, Geneva, 1999.
2. Subramanian, R., Motor vehicle traffic crashes as a leading cause of death in the United States, 2003. Research note DOT HS 810 568, Department of Transportation, National Highway Traffic Safety Administration, Washington, D.C., 2006.
3. Stutts, J.C., Reinfurt, D.W., and Rodgman, E.A., The role of driver distraction in traffic crashes: An analysis of 1995–1999 crashworthiness data system data, *Proceeding of the 45th Annual Conference of the Association for the Advancement of Automotive Medicine*, Association for the Advancement of Automotive Medicine, Barrington, IL, 2001, p. 287.

4. Cellular Telecommunications Industry Association, 2006.
5. Glassbrenner, D., Driver cell phone use in 2005—Overall results. Research note DOT HS 809 867, Department of Transportation, National Highway Traffic Safety Administration, Washington, D.C., 2005.
6. Haigney, D.E. and Westerman, S.J., Mobile (cellular) phone use and driving: A critical review of research methodology, *Ergonomics* 44, 132, 2001.
7. McCartt, A.T., Hellinga, L.A., and Geary, L.L., Effects of Washington, D.C. law on drivers' hand-held cell phone use, *Traffic Inj. Prev.* 7, 1, 2006.
8. Svenson, O. and Patten, C.J.D., Mobile phones and driving: Review of contemporary research, *Cognit. Technol. Work* 7, 182, 2005.
9. Horrey, W.J. and Wickens, C.D., Examining the impact of cell phone conversations on driving using meta-analytic techniques, *Hum. Factors* 48(1), 196–205, 2006.
10. Glassbrenner, D., Driver cell phone use in 2004—Overall results. Research note DOT HS 809 847, Department of Transportation, National Highway Traffic Safety Administration, Washington, D.C., 2005.
11. Eby, D.W. and Vivoda, J.M., Driver hand-held mobile phone use and safety belt use, *Accid. Anal. Prev.* 6, 893, 2005.
12. Strayer, D.L. and Drews, F.A. Multi-tasking in the automobile. In: Kramer, A., Wiegmann, D., and Kirlik, A. (Eds.), *Applied Attention: From Theory to Practice*, Oxford University Press: New York, 2006.
13. Redelmeier, M.D. and Tibshirani, R.J., Association between cellular-telephone calls and motor vehicle collisions, *New Engl. J. Med.* 336, 453, 1997.
14. Violanti, J.M. and Marshall, J.R., Cellular phones and traffic accidents: An epidemiological approach, *Accid. Anal. Prev.* 28, 265, 1996.
15. Violanti, J.M., Cellular phones and fatal traffic collisions, *Accid. Anal. Prev.* 30, 519, 1998.
16. Finkelstein, E.A., Chen, H., Miller, T.R., Corso, P.S., and Stevens, J.A., A comparison of the case-control and case-crossover designs for estimating medical costs of nonfatal fall-related injuries among older Americans, *Med. Care* 43, 1087, 2005.
17. McEvoy, S.P., Stevenson, M.R., McCart, A.T., Woodward, M., Haworth, C., Palamara, P., and Cercarelli, R., Role of mobile phones in motor vehicle crashes resulting in hospital attendance: A case-crossover study, *Br. Med. J.* 331, 428–430, 2005.
18. Laberge-Nadau, C., Maag, U., Bellavanc, F., Lapierre, S.D., Desjardins, D., Messier, S., and Sidi, A., Wireless telephones and the risk of road crashes, *Accid. Anal. Prev.* 35, 649, 2003.
19. Green, P., Safeguards for on-board wireless communications. Presentation at 2nd Annual Plastics in Automotive Safety Conference, Troy, MI, 2001.
20. Mazzae, E., Goodman, M., Garrott, R., and Ranney, T., NHTSA's research program on wireless phone driver, interface effects, Retrieved August 11, 2006 from http://www-nrd.nhtsa.dot.gov/pdf/nrd-01/esv/esv19/05-0375-O.pdf, 2004.
21. Brookhuis, K.A., de Vries, G., and de Waard, D., The effects of mobile telephoning on driving performance, *Accid. Anal. Prev.* 23, 309, 1991.
22. Green, P., Hoekstra, E., and Williams, M., *Further On-the-Road Tests of Drive Interfaces: Examination of a Route Guidance System and a Car Phone*, Report No. UMTRI-93-35, University of Michigan Transportation Research Institute, Ann Arbor, MI, 1993.
23. Zwahlen, H.T., Adams, C.C., and Schwartz, P.J., Safety aspects of cellular telephones in automobiles, *Proceedings of the ISATA Conference*, ISATA, Florence, Italy, 1988.
24. Kames, A.J., A study of the effects of mobile telephone use and control unit design on driving performance, *IEEE Transport. Veh. Tech.* VT-27, 282, 1978.

25. Neale, V.L., Dingus, T.A., Klauer, S.G., Sudweeks, J., and Goodman, M., An overview of the 100-car naturalistic study and findings. Paper No. 05-0400, *Proceeding 19th International Technical Conference on the Enhanced Safety of Vehicles* (CD-ROM),National Highway Traffic Safety Administration, Washington, D.C., 2005.
26. Harbluk, J.L., Noy, Y.I., and Eizenman, M., The impact of cognitive distraction on driver behaviour and vehicle control, *Road Safety Directorate and Motor Vehicle Regulation Directorate,* Transport Canada, Ottawa, ON, 2002.
27. Matthews, R., Legg, S., and Charlton S., The effects of cell phone type on drivers' subjective workload during concurrent driving and conversing, *Accid. Anal. Prev.* 35, 451, 2003.
28. Hart, S.G. and Staveland, L.E., Development of NASA-TLX (Task Load Index): Results of empirical and theoretical research, In: Hancock, P.A. and Meshkati, N. (Eds.), *Human Mental Workload*, Elsevier Science, Amsterdam, North Holland, 1988, pp. 139–183.
29. Crundall, D., Bains, M., Chapman, P., and Underwood, G., Regulating conversation during driving: A problem for mobile telephone?, *Transport. Res. Part F: Traffic Psychol. Behav.* 8, 197, 2005.
30. Drews, F.A., Pasupathi, M., and Strayer, D.L., Passenger and cell-phone conversations in simulated driving, *Proceeding of Human Factors and Ergonomic Society's 48th Annual Meeting*, Human Factors and Ergonomics Society, Santa Monica, CA, 2004, p. 2210.
31. Patten, C.J.D., Kircher, A., Ostlund, J., and Nilsson, L., Using mobile telephones: Cognitive workload and attention resource allocation, *Accid. Anal. Prev.* 36, 341, 2004.
32. Cooper, P.J. and Zheng, Y., Turning a gap acceptance decision-making: The impact of drive distraction, *J. Safety Res.* 33, 321, 2002.
33. Cooper, P.J., Zheng, Y., Richard, C., Vavrik, J., Heinrichs, B., and Siegmund, G., The impact of hands-free message reception/response on driving task performance, *Accid. Anal. Prev.* 35, 23, 2003.
34. Brown, I.D., Tickner, A.H., and Simmonds, D.C.V., Interference between concurrent tasks of driving and telephoning, *J. Appl. Psychol.* 5, 419, 1969.
35. Treffner, P.J. and Barrett, R., Hands-free mobile phone speech while driving degrades coordination and control, *Transport. Res. Part F: Traffic Psychol. Behav.* 7, 229, 2004.
36. Hancock, P.A., Lesch, M., and Simmons, L., The distraction effects of phone use during a crucial driving maneuver, *Accid. Anal. Prev.* 35, 501, 2003.
37. Burns, P.C., Parkes, A., Burton, S., and Smith, R.K., *How Dangerous is Driving with a Mobile phone? Benchmarking the Impairment to Alcohol.* TRL Report 547. Transport Research Laboratory, Berkshire, UK, 2002.
38. Ma, R. and Kaber, D.B., Situation awareness and workload in driving while using adaptive cruise control and a cell phone, *Int. J. Ind. Ergon.* 35, 939, 2005.
39. Endsley, M.R., Measurement of situation awareness in dynamic systems, *Hum. Factors* 37, 65, 1995.
40. Haigney, D.E., Taylor, R.G., and Westerman, S.J., Concurrent mobile (cellular) phone use and driving performance: Task demand characteristics and compensatory processes, *Transport. Res. Part F: Traffic Psychol. Behav.* 3, 113, 2000.
41. Baddeley, A., A three-minute reasoning test based on a grammatical transformation, *Psychoneurol. Sci.* 10, 341, 1968.
42. Strayer, D.L., Drews, F.A., and Johnston, W.A., Cell phone induced failures of visual attention during simulated driving, *J. Exp. Psychol. Appl.* 9, 23, 2003.
43. Alm, H. and Nilsson, L., Changes in driver behaviour as a function of hands-free mobile phones: A simulator study, *Accid. Anal. Prev.* 26, 441, 1994.
44. Baddeley, A., Logie, R., Nimmo-Smith, I., and Brereton, N., Components of fluent reading, *J. Mem. Lang.* 24, 119, 1985.

45. Radeborg, K., Briem, V., and Hedman, L.R., The effect of concurrent task difficulty on working memory during simulated driving, *Ergonomics*, 42, 767, 1999.
46. Horberry, T., Anderson, J., Regan, M.A., Triggs, T.J., and Brown, J., Driver distraction: The effects of concurrent in-vehicle tasks, road environment complexity and age on driving performance, *Accid. Anal. Prev.* 38, 185, 2006.
47. Rakauskas, M.E., Gugerty, L.J., and Ward, N.J., Effects of naturalistic cell phone conversations on driving performance, *J. Safety Res.* 35, 453, 2004.
48. Hunton, J. and Rose, J.M., Cellular telephones and driving performance: The effects of attentional demands on motor vehicle crash risk, *Risk Anal.* 25, 855, 2005.
49. Shinar, D., Tractinsky, N., and Compton, R., Effects of practice, age and task demands, on interference from a phone task while driving, *Accid. Anal. Prev.* 37, 315, 2004.
50. Cooper, J., and Strayer, D., Do driving impairments from concurrent cell phone use diminish with practice? *Proceedings of the 51st Annual Meeting of the Human Factors and Ergonomics Society*, Baltimore, MD, 2007.
51. Beede, K.E. and Kass, S.J., Engrossed in conversation: The impact of cell phones on simulated driving performance, *Accid. Anal. Prev.* 38, 415, 2006.
52. Stein, A.C., Parseghian, Z., and Allen, R.W., A simulator study of the safety implications of cellular mobile phone use, *Proceeding of the 31st Annual Conference of the Association for the Advancement of Automotive Medicine*, Association for the Advancement of Automotive Medicine, Des Plaines, IL, 1987.
53. Törnros, J.E.B. and Bolling, A.K., Mobile phone use—effects of hand-held and hands-free phones on driving performance, *Accd. Anal. Prev.* 37, 902, 2005.
54. McKnight, A.J. and McKnight, A.S., The effect of cellular phone use upon driver attention, *Accid. Anal. Prev.* 25, 259, 1993.
55. Amado, S. and Ulupinar, P., The effects of conversation on attention and peripheral detection: Is talking with a passenger and talking on the cell phone different?, *Transport. Res. Part F: Traffic Psychol. Behav.* 8, 383, 2005.
56. Consiglio, W., Driscoll, P., Witte, M., and Berg, W.P., Effect of cellular telephone conversations and other potential interference on reaction time in a braking response, *Accid. Anal. Prev.* 35, 495, 2003.
57. Strayer, D.L. and Johnston, W.A., Driven to distraction: Dual-task studies of simulated driving and conversing on a cellular phone, *Psychol. Sci.* 12, 462, 2001.
58. Briem, V. and Hedman, L.R., Behavioral effects of mobile telephone use during simulated driving, *Ergonomics* 38, 2536, 1995.
59. Graham, R. and Carter, C., Voice dialing can reduce the interference between concurrent tasks of driving and phoning, *Int. J. Vehicle Des.* 26, 30, 2001.
60. McCarley, J.S., Vais, M., Pringle, H., Kramer, A.F., Irwin, D.E., and Strayer, D.L., *Conversation Disrupts Visual Scanning of Traffic Scenes.* Paper presented at Vision in Vehicles, Australia, 2001.
61. Barkana, Y., Zadok, D., Morad, Y., and Avni, I., Visual field attention is reduced by concomitant hands-free conversation on a cellular telephone, *Am. J. Ophthalmol.* S138, 347, 2004.
62. Atchley, P. and Dressel, J., Conversation limits the functional field of view, *Hum. Factors,* 46, 664, 2004.
63. Telstra, Telstra, police and NRMA insurance join forces to target mobile phone use on Australian roads. Telstra News Release, www.testra.com.au/newsroom, 2003.
64. Hosking, S.G., Young, K.L., and Regan, M.A., *The Effects of Text Messaging on Young Novice Driver Performance*, Monash University Accident Research Center Report 246, Monash University, Australia, 2006.
65. Kircher, A., *Mobile Telephone Simulator Study*, Swedish National Road and Transport Research Institute, Linkloping, Sweden, 2004.

66. Jamson, A.H., Westerman, S.J., Hockey, G.R.J., and Carsten, O.M.J., Speech-based e-mail and driver behavior: Effects of an in-vehicle message system interface, *Hum. Factors* 46, 625, 2004.
67. Lee, J.D., Caven, B., Haake, S., and Brown, T.L., Speech-based interaction with in-vehicle computers: The effect of speech-based e-mail on drivers' attention to the roadway, *Hum. Factors* 43, 631, 2001.
68. Hoffman, J.D., Lee, J.D., McGehee, D.V., and Gellatly, A.W., Visual sampling of in-vehicle text messages: The effects of number of lines, page presentation and message control, *Transport. Res. Rec.* 1937, 22–31, 2005.

12 Sources of Distraction inside the Vehicle and Their Effects on Driving Performance

Megan Bayly, Kristie L. Young, and Michael A. Regan

CONTENTS

12.1 INTRODUCTION

There are many things inside the vehicle, in addition to mobile phones, that have the potential to distract the driver. This chapter will present and review information that exists on the effects of these in-vehicle sources of distraction on driving performance. In-vehicle distractions have been found to account for up to 62% of all distraction-related crashes,[1] although, as discussed in Chapters 15 and 16, the manner in which these sources of distraction are classified strongly influences the magnitude of such estimates.

Essentially, any object, event, or activity that diverts attention away from activities that are critical to safe driving toward a competing activity may be considered distracting (see Chapter 15). As discussed in Chapter 15, distraction may derive from a variety of sources inside the vehicle: objects and devices brought into the vehicle, factory-fitted fixed vehicle systems, moving objects, vehicle occupants, and internalized activities. This chapter will consider sources of distraction within the vehicle (i.e., in-vehicle distractions) as technology-based or non-technology-based, and will primarily focus on the effects of technology-based distractions on driving performance. This is not because sources of distraction deriving from technology use are any more critical or common than other in-vehicle sources of distraction. Rather, it is because the prevalence of technology use within the vehicle is increasing at a rapid rate, and there is potentially more scope to mitigate the effects of these sources of distraction—through effective design, regulation, and other measures—than everyday, non-technology-based distractions. This is likely the reason why there has been a greater research focus on the impact of technology use in the vehicle while driving. The effects of mobile phone use on driving performance are reviewed in Chapter 11 of this book and therefore will not be considered further here. Likewise, the effects on driving performance of sources of distraction located outside the vehicle are reviewed in Chapter 13.

This chapter will review the existing literature regarding the effects of in-vehicle distractions (excluding mobile phones) on driving performance. The effect of a source of distraction on driving performance will depend on many interrelated factors. These include the nature of the competing activity, the ability and experience of the user, the current complexity of the driving task, the current state of the driver, personal characteristics of the driver, and, in the case of technology, the design, location, and manner of use of the system. Exposure to a source of distraction is also important in determining to what extent any degradation in driving performance results in increased crash risk. A particular driver interaction with a source of distraction may be highly detrimental to driving performance but rarely performed, resulting in little or no increase in crash risk. Other interactions, however, may impact less on driving but be regularly performed, increasing crash risk. Epidemiological measures, such as population-attributable risk, account for these differences in providing an overall estimate of the increased risk associated with a given source of distraction for a given population.[2] These issues are further discussed in detail in Chapters 6 and 16. Although a wide range of in-vehicle sources of distraction have been identified in the literature (see Chapter 15), at present, relatively few of these have been studied experimentally to determine their effect on driving performance.

The limited available data that exists are reviewed in the remainder of this chapter. A summary of the major sources of in-vehicle distraction and their effect on driving performance measures, as identified in the literature, is provided later in Table 12.1.

12.2 TECHNOLOGY-BASED SOURCES OF IN-VEHICLE DISTRACTION

Although much of the research on distraction resulting from use of technology while driving has focused on mobile phone use, many other technologies are commonly used within the vehicle. These include "fixed" vehicle systems (i.e., those that are factory-fitted or retrofitted) and "nomadic" (portable) devices, which serve to provide a range of functions, such as entertainment, provision of information, and communication. Some technologies, such as in-vehicle navigation systems, have been designed to support the driving task. Many, however, have not been designed specifically for in-vehicle use. It is important to note that among technologies of the same type (e.g. portable auditory entertainment devices), there may be considerable differences between products. System characteristics such as the range of functions available and the design of the human-machine interface (HMI) vary greatly between devices, and this will have a significant impact on the amount of time and effort required to interact with the device and, in turn, the level of distraction that it imposes on drivers.

12.2.1 NOMADIC AUDITORY ENTERTAINMENT SYSTEMS

Drivers are bringing nomadic (portable) music players into the vehicle in increasing numbers. It has been shown that 20% of young U.S. drivers aged between 18 and 24 years select songs on their iPods while driving.[3] However, the extent to which these technologies are used by drivers in other countries has not yet been determined. These devices generally allow for a wider range of personalized music than fixed auditory entertainment systems (e.g., car radios), and provide the option of skipping and searching through extensive libraries of songs, pictures, and videos. Other audio files such as "podcasts" and "talking books" may also be stored and selected on these devices. Portable music players may be listened to through headphones, or car kits that allow the output to be played through the vehicle speakers. The portable nature of these devices means they may be placed (often unsecured) anywhere in the vehicle, including outside the driver's normal field of view. Also, they are not usually designed primarily for use in the driving environment, and as such, they often do not conform to automotive HMI design guidelines and standards. For example, the placement of these systems in the vehicle is rarely mandated by regulation, and they are not assessed in terms of whether the necessary information can be extracted from the interface in acceptably short eye glances. Some portable digital music players (e.g., Apple iPods) have been integrated into some high-end model vehicles with customized electronic interfaces. Other portable auditory entertainment devices may also be brought into the vehicle, including Walkmans and portable CD players; however, the extent to which this occurs and their effect on driving are unknown.

TABLE 12.1
Summary of the Observed Effects of In-Vehicle Sources of Distraction on Driving Performance Measures

	Lateral Control	Speed	Following Distance/Time Headway	Detection Tasks	Braking Behavior	% Time with Both Hands Off Wheel	Glance Behavior	Adverse Driving Events[a]	Subjective Mental Workload	Crashes or Errors
iPods	No effects on lateral deviation from road center for listening or watching video tasks; significant effect for selection task[5]	Decreased speed during selection tasks[4] Speed decreased while watching video but not for listening[5]	Reduced car following speed[5]	RT to hazards increased by 0.18 s (16%) during difficult operating tasks[4]			12% less time looking at roadway and 31% more time looking inside the vehicle during difficult operating task[4]			Significantly more crashes occurred during difficult operating task than easy task, and no iPod condition[4]
DVD players										
Watching	Wider mean lateral position on curves[7]	Lower mean speeds; greater speed variability[7]	Increased mean and minimum headways by 16.1 m and 0.26 s, respectively[8]	Peripheral detection task time increased by 135 ms for high-speed conditions[8]	Longer braking time for pedestrian hazards[7]		15% of total driving time spent looking at DVD during watching condition[10]		No effects[8]	No effects for time to collision[8]
	Higher average scaled lateral accelerations than no-DVD condition[10]	No effects for mean speed or speed variance[8]	Shorter headway distances of 1.99 m (combined listening/watching conditions)[9]	Perceptual RT 0.25 s slower for front seat system[9]	No effects for braking reaction time[8]					

		Decreased mean speed by almost 2 km/h (combined listening/watching conditions)[9] 1.8 km/h slower mean speeds[10]		Slower and less accurate detection when watching[10]	Sight increase in total braking time[10]
Listening only	No effects for mean or standard deviation of lateral position[7]	No effects on mean speed[7] Decreased mean speed by almost 2 km/h (combined listening/watching conditions)[9]	Shorter headway distances of 1.99 m (combined listening/watching conditions)[9]	No effect on perceptual RT for rear seat[9]	Longer braking time to one pedestrian hazard[7]
Manipulating	Significantly greater average scaled lateral accelerations[10]	"Marginally slower" mean speeds[10]		Slower and less accurate peripheral detection[10]	Slight increase in total braking time[10]

(continued)

TABLE 12.1 (Continued)

	Lateral Control	Speed	Following Distance/Time Headway	Detection Tasks	Braking Behavior	% Time with Both Hands Off Wheel	Glance Behavior	Adverse Driving Events[a]	Subjective Mental Workload	Crashes or Errors
Navigation systems										
Destination entry		Lower mean speeds for manual input[14,15]					More time with eyes off road (up to 50%); more frequent glances at device[13,15]			More mirror-checking behavior with voice activation[18]
	Fewer lateral position deviations with voice activation[18]						Less time looking at display with voice activation[18]			
Destination following	Extent of impairment depends on system configuration[16]	Lower mean speed[17]							Extent of impairment depends on system configuration[17]	Braking errors[16] Navigational errors[17]
E-mail systems	Fewer corrective steering movements; time to line crossing decreased by 2.55 s[20]	More incidences of very low speeds[18]	Longer headways of 350 ms[20]		30% increased braking RT[19] Increased braking RT by 700 ms[20]				19.10% increase in workload when system available[19]	Time to collision margin reduced by 50%[20]
Radio, CD player										
Listening	No effects[7]	No effects[7]								More crashes in simulator drive[7]
	Greater lane position deviations[22]									

Manipulating	Greater lane position deviations[26,29]	4 km/h lower mean speeds; greater speed variance from posted speed limit[27]	1.06% more time driving with both hands off steering wheel[23]	More glances over 0.5 s than mobile phone[26] More time with eyes off road than eating or phone use[29] 19.73% more time with eyes inside vehicle[23] Greater proportion of time looking at task than road; greatest total task time[28]	No effects[23]	Increased workload ratings[27]	Longest drive completion time[29]
Manipulating vehicle system controls			8.55% more time driving with both hands off steering wheel[23]	13.20% more time with eyes inside the vehicle[23]	No effects[23]		
Eating/drinking							
Preparation			3.15% more time driving with both hands off steering wheel[23]	2.91% more time with eyes inside the vehicle[23]	10.80% more adverse events per hour[23]		
Consumption (and spillage)	Greater lane position deviations[29]	Lower mean speed[29]	4.07% more time driving with both hands off steering wheel[23]	3.63% more time with eyes inside the vehicle[23]	No effects[23]		
Smoking							
Lighting or extinguishing			No effects[23]	16.55% more time with eyes inside vehicle[23]	No effects[23]		

(continued)

TABLE 12.1 (Continued)

	Lateral Control	Speed	Following Distance/Time Headway	Detection Tasks	Braking Behavior	Time with Both Hands Off Wheel	Glance Behavior	Adverse Driving Events[a]	Subjective Mental Workload	Crashes or Errors
Smoking						No effects[23]	No effects[23]	4.81% fewer adverse events per hour[23]		
Reading and writing						13.71% more time driving with both hands off steering wheel[23]	88.99% more time with eyes inside vehicle[23]	No effects[23]		
Reaching for objects						2.56% more time driving with both hands off steering wheel[23]	17.88% more time with eyes inside vehicle[23]	10.85% more adverse events per hour[23]		
Grooming						11.05% more time driving with both hands off steering wheel[23]	31.96% more time with eyes inside vehicle[23]	No effects[23]		
Passengers										
Conversation						No effects[23]	No effects[23]	No effects[23]		
Passenger behavior						No effects[23]	No effects[23]	No effects[23]		

[a] Adverse driving events, as defined by Stutts et al.,[23] are lane position deviations, other lane encroachments, and sudden braking events.

Chisholm et al.[4] investigated whether driving performance deteriorated while drivers interacted with an iPod and whether the associated deterioration was reduced by secondary task practice. Nineteen young drivers (mean age 19.4 years, whose previous iPod use experience was unstated) participated in seven driving simulator sessions and were required to respond to a number of hazardous events occurring in the roadway, such as a late amber traffic light change and a pedestrian entering the roadway while performing difficult (requiring five to seven steps) and easy (requiring one to two steps) iPod tasks. Measures of lateral and longitudinal vehicle control, eye glance behavior, and perception-response time were collected. Results revealed that performing easy iPod tasks had only minimal effects on driving performance. However, when performing the difficult iPod task, performance deteriorated: drivers' perception-response times to the hazardous events increased by 16%, or 0.18 s; they spent 12% less time looking at the road and 37% more time looking inside the vehicle; and they were involved in more crashes compared with the baseline (no iPod) condition. A decrease in perception-response times was associated with practice, although driving performance while interacting with the iPod did not return to a "safe" baseline level even with increased practice. The drivers did, however, use a number of strategies to compensate for the increased demands of using the iPod, such as decreasing their speed and stopping at traffic lights rather than driving through an intersection.

Salvucci et al.[5] also investigated the effects of interacting with an iPod on simulated driving performance. Twelve participants with previous experience of using an iPod completed selection, watching, and listening tasks on an iPod while driving in a simulator. Performance of selection tasks, such as searching for and listening to/watching music, "podcasts," or videos, was associated with increases in lane position deviation that were comparable to those observed when using a mobile phone. Significant reductions in car-following speed were also observed. Listening to songs or podcasts was not associated with changes in either lane position deviation or vehicle speed; however, watching a video was associated with significant reductions in vehicle speed but not lane position deviation.

Given the relatively recent introduction of these devices into the vehicle cockpit, their impact on driving performance and safety is not yet well understood. The effects of portable music players, even when integrated into the HMI of the vehicle, are likely to be different from those deriving from driver use of most current-generation, fixed, auditory entertainment systems (e.g., CD players and radios), as they involve scrolling and selecting through numerous levels of a hierarchical menu structure. Furthermore, as suggested by Salvucci et al.,[5] the greater level of visual attention required to operate an iPod (or similar device) renders them difficult to operate compared with other devices such as mobile phones, which have greater tactile feedback in their buttons and controls and thus may facilitate "blind" operation. Further studies are needed to investigate performance decrements and the crash risk associated with using these devices, as distinct from other auditory entertainment systems.

12.2.2 Audiovisual Entertainment Systems

Audiovisual entertainment systems, including portable TV and DVD players, have recently emerged as one of the most popular new in-vehicle devices, at least in the United States.[6] These systems may be either fixed or nomadic. Fixed systems are typically mounted to the rear of the front seats, for viewing by rear-seat passengers, whereas many nomadic devices (not necessarily designed for in-vehicle use) may be used by any vehicle occupant in any position (not necessarily legally), and are not always secured or positioned correctly. After-market systems are also available, which attach to the vehicle center console. The effects of the system on driving performance and safety will depend primarily on the position of the screen, that is, whether the driver is able to see the screen while driving, thereby creating visual distraction. Legislation in the United States and Australia prohibits any system that is not a driver's aid from having screens mounted within the driver's field of view. This legislation is designed to prevent visual distraction; however, it is reasonable to expect cognitive distraction to result from use of these entertainment systems even if the driver cannot see the screen. Hatfield and Chamberlain[7] suggest that the auditory output from an audiovisual display unit may be more distracting than that from a radio or CD, as the stimulus is incomplete without the visual display and therefore requires greater attention to follow it. It should be noted that many of these systems allow passengers to use headphones, which would eliminate this potential effect.

Given their relatively recent emergence on the commercial market, the effects of these devices have not yet been widely explored. Of the few existing studies that have explored the effects on driving performance of in-vehicle audiovisual entertainment units, only one has assessed the effects of manipulation of system controls (e.g., loading a disk, pausing, scanning) as well as the effects of the visual and auditory stimuli deriving from them. Furthermore, these studies have all investigated DVD players, not portable televisions. The first of these studies, by Kircher et al.,[8] investigated the driving performance of eight drivers in a simulator while they watched a DVD on a screen mounted on the center console. Drivers completed drives in urban and rural environments, at speeds ranging from 50 to 90 km/h. Watching the DVD while driving generally had no effect on mean speed, speed variance, braking reaction time, or perceived subjective workload, although in higher-speed conditions (90 km/h), mean headway increased by 16.10 m and average minimum headway increased by 0.26 s. Also, reaction time in a peripheral detection task increased by up to 135 ms, depending on the speed condition. The increased headway was interpreted by Kircher et al. to be a compensatory behavior.

Various configurations of in-vehicle DVD units were explored in a simulator study by White et al.,[9] where stimulus output was manipulated to represent a rear-seat or front-seat system. The front-seat system included both visual and auditory stimuli, while the rear-seat configuration only included auditory stimuli from the DVD. A total of 48 participants drove in a no-DVD condition and either the rear-seat or front-seat configuration. Presence of the DVD was associated with shorter headway distances in the order of 1.99 m (in contrast to Kircher et al.'s[8] findings) and reductions in speed of almost 2 km/h. The latter finding was interpreted as a compensatory behavior. Longer perceptual reaction times were observed for the front-seat condition (0.25 s), but this effect was not observed in the rear-seat condition.

The potential for audiovisual entertainment units to distract drivers in other ways was explored by Hatfield and Chamberlain.[7] In the context of the existing legislation, which prevents a driver from viewing a DVD player within his or her own vehicle, they explored (i) the effects on driving performance of a visual display observable through the rear window of another vehicle and (ii) the effects of a DVD playing within the driver's own vehicle (sound only) compared with a radio. In the first investigation, simulated driving performance was found to be impaired by the presence of a display visible through the rear of a car in a neighboring lane. This impairment was observed in the form of lower mean speeds, lower speed variability, greater lane position deviation on curves, slower deceleration when a pedestrian stepped onto the roadway, and a tendency to drive more toward the road edge on curves (which may result in more run-off road crashes). These effects were only observed when the driver was instructed to attend to the DVD display in the other vehicle. As with White et al.'s[9] study, the lower mean speed was interpreted as a compensatory behavior in response to the visual distraction. The second investigation explored the effects on driving performance of auditory output from a DVD player compared with that of a radio program and a no auditory stimulus condition. Few significant differences in driving performance were found between these conditions, suggesting that auditory stimuli emanating from a DVD player was no more distracting than that of a radio program, and that auditory stimuli (regardless of its source) were no more distracting than no auditory output.

A recent study assessed the effects of in-vehicle DVD players on on-road driving performance. Funkhouser and Chrysler[10] measured the driving behavior of nine participants as they completed five laps of a test track: two control laps, one lap while watching a DVD, one lap while listening to the DVD, and one lap while manipulating the system controls. All of the driving performance measures that were recorded showed degradation for at least one of the DVD conditions, although, given the small sample size, few significant differences were found. Drivers performed worse (slower and with lower hit percentages) on rear and forward detection tasks when watching or manipulating the DVD; however, these effects were not significant. Participants applied the brakes for a greater proportion of overall driving time when watching or manipulating the DVD (1.76 and 1.93%, respectively) compared with the listening (1.25%) or control laps (1.07 and 1.16%). However, these were also not significant, nor were there any significant differences in brake force between conditions. Average scaled lateral accelerations were significantly higher for the operating, listening, and watching conditions; however, only one significant effect was found—between the operating and one of the control laps. Significantly lower average speeds (1.8 km/h) were observed for the watching condition compared with the control, while the listening and operating conditions were also slower but not significant. It was also found that while driving, participants looked at the DVD for 15% of total driving time. Clearly, further research is needed to expand on the initial findings of the studies described here.

12.2.3 Navigation Systems

Navigation, or route guidance, systems have been commercially available as retrofittable, nomadic, and fixed vehicle systems for several years. These are possibly the second most well-researched in-vehicle technologies after mobile phones in terms of their effect on driving behavior and performance. Navigation systems are classified as driver aids and are therefore permissible, for example, under Australian and U.S. legislation that prevents the display screen from being visible to the driver while driving. Some navigation systems may incorporate information such as traffic congestion, delays, or high–crash risk areas or times in their suggested routes, which may negate the increased exposure to risk that has been suggested as a potential limitation to the effectiveness of these devices.[11] Interaction with navigation systems involves two primary operations: destination entry and following of directions. Destination entry may be performed manually (either by scrolling through potential destinations or typing in the destination) or through use of voice recognition technology. Directions may be presented by the system verbally, visually, or in combination. Much of the existing research has focused on determining the best modality configuration for these functions.

Destination entry can be a time-consuming process, taking in some cases up to 9 min.[12] Furthermore, this process, regardless of system type, has been found to be longer than that associated with tuning a radio and dialing a mobile phone.[13] It has been repeatedly found that compared with verbal destination entry, manual entry results in greater decrements in both simulated and on-road driving performance, such as deterioration in lateral position and lane keeping,[13,14] slower mean speeds,[14,15] more frequent glances at the device,[13] and greater periods of driving with eyes off the road.[13,15] In an on-road study, for example, Chiang et al.[15] found that drivers spent over 50% of their driving time looking at the device. Voice recognition systems have been shown to result in faster destination entry and less deterioration of vehicle control.[13,14] Tsimhoni et al.[14] found that manual entry systems resulted in 60% more deviations in lateral position than other entry input methods. It is now a feature of many systems to "lock out" this feature while the vehicle is moving,[12] regardless of input modality.

Similar to destination entry, systems that provide only visual directions are associated with greater driving performance decrements than those that provide auditory, or auditory plus visual, route guidance. Those systems that require visual attention to process the route guidance information result in more interference with the driving task.[16,17] Furthermore, turn-by-turn guidance instructions have also been deemed more acceptable and usable by drivers than complex holistic maps. For example, Dingus et al.[16] found that drivers were most impeded by a navigation system that incorporated an electronic map without voice guidance, as opposed to a turn-by-turn system with or without voice guidance or an electronic map with voice guidance. This was reflected in more braking errors and lane position deviations. The configuration that resulted in the least errors was the turn-by-turn system with voice guidance. However, the potentially adverse effects of these systems may be less detrimental to driving performance and safety than attempting to navigate while driving with a paper map or driving while lost. This effect was demonstrated by Srinivasan and Jovanis,[17] who compared the effects of driving with a head-down electronic map, head-up turn-by-turn display with vocal guidance, head-down electronic map without vocal guidance, and a paper map. It was found that the electronic map with voice guidance resulted in the least decrements in driving performance (as reflected in fastest mean speeds, fewest navigational errors, and lowest subjective workload ratings), while use of a conventional paper map was associated with the greatest decrements in driving performance on the above-mentioned measures.

Further evidence for the efficacy of voice recognition technology was provided by Gärtner et al.[18] in an on-road study exploring the effects of a driver information system incorporating a radio, CD player, telephone, and navigation system. Sixteen participants performed both simple and complex tasks on the system while using either manual or voice activation. The use of voice activation was associated with 50% fewer driving errors than manual input, including fewer lane position deviations, less time spent driving at very low speeds, less time looking at the system display, and more mirror checks.

12.2.4 Nomadic Information and Communication Systems

Nomadic information and communication devices include portable pagers, personal digital assistants (PDAs), laptop computers, and mobile phones. Functions that may be accessed when using these devices include Internet, e-mail, fax, video messaging, and mobile phone services. The capabilities of such technologies are growing rapidly, allowing the driver to create a mobile office or remote social environment within his or her vehicle. Some systems also provide navigational information, information regarding congestion and desired destinations (such as service stations, parking lots, and restaurants). However, these are not driver aids, and hence are not specifically designed to support the driving task. Many such nomadic systems have not been designed for use within the vehicle.

It is common for these systems to have speech-based capabilities; that is, they include voice recognition technology and will read aloud message content to the user. As with navigation systems, it would be expected that verbal input and output would reduce the visual attention demanded by the system. With the exception of mobile phones (see Chapter 11), however, relatively little research has been undertaken to examine the effects on driving of these interactive information and communication systems. One such study investigated the effects of simple and complex speech-based e-mail systems on driving performance.[19] Twenty-four participants completed simple and complex drives in a simulator while using voice commands to operate a simple (three-level menu) or complex (four- to seven-level menu) e-mailing system. System operation included opening an e-mail, reading an e-mail, replying to it, and exiting the system. It was found that driving performance while performing these tasks was impaired regardless of system complexity, as indicated by a 30% increase in braking reaction time. Also, a 19.10% increase in subjective workload rating was observed when the e-mail system was present, and this effect was more pronounced for the complex system.

In another simulator study, Jamson et al.[20] also found a significant increase in braking reaction time of 700 ms associated with use of a speech-based, in-vehicle e-mail system. This system vocally delivered the e-mail content to the driver, either automatically or when prompted by the driver. Other performance decrements were also observed: fewer corrective steering movements, time-to-line-crossing measures increased by 2.55 s, time-to-collision safety margins decreased by almost 50%, and drivers adopted 0.35 s longer headways with lead vehicles as a compensatory behavior when interacting with the system. It was noted, however, that less performance degradation occurred when the driver was able to control when the system was accessed. These studies have investigated the effects of speech-based e-mailing only, and as yet, the effects of interacting with an e-mail system, regardless of input and output modalities, are not well understood. It could be that interaction with any e-mail system is distracting and has the potential to degrade driving performance, regardless of modality.[21]

12.2.5 Fixed In-Vehicle Audio Entertainment Systems

Fixed in-vehicle entertainment systems include the radio, CD player, and cassette player. The in-vehicle radio has been commercially available since the 1930s, and is now a standard feature in most vehicles. Despite this, relatively little research concerning the distracting effects of these systems has occurred to date. The auditory output from radio programs (or CD players) is rarely cited, or even investigated, as a potential source of distraction,[22] despite evidence showing that audio entertainment is used for over 70% of driving time.[23] Of the studies conducted regarding the effects of audio entertainment, findings such as increases in lane position deviations[22] have been observed, whereas other studies have found no[24] or very minor[7] effects. Stutts et al.[23] found no increases in the proportion of time spent driving with eyes off the road or both hands off the steering wheel when auditory stimuli (from either the radio or CD player) were present. However, manipulation of the audio system controls was associated with significant increases in the proportion of time spent driving with both hands off the wheel (from 1.00 to 2.06%) and driving while looking inside the vehicle (from 2.85 to 22.58%).

The manipulation of radio system controls, such as tuning or adjusting volume, has been more extensively studied, usually where the action of tuning a radio is used as a reference task against which other distraction-related activities are evaluated.[21] Radio tuning is used as a benchmark activity in these studies as it is a common, socially accepted behavior. From studies such as these, it has been found, for example, that tuning a radio to find a station impacts less on reaction time than conversing on a mobile phone,[25] and operating a navigation system has been associated with less time with eyes off the road than radio tuning.[13] However, this is not always the case. Wikman et al.[26] found that tuning a radio was associated with greater proportions of driving time spent with eyes off the road for periods of more than 0.5 s at a time than when dialing a mobile phone, but less than when changing a cassette tape. However, the total time spent looking at the mobile phone while completing the task was 53%, compared with 26% for the radio task and 20% for the cassette task. Similar findings were reported by Horberry et al.,[27] who also compared the effects on driving of operating audio entertainment systems (manipulating a radio and cassette player) with a handsfree mobile phone use under various levels of simulated driving complexity. In the radio and cassette condition, participants were required to tune and adjust acoustic settings of the system, and insert and eject cassettes. In both the simple and complex driving environments, the radio and cassette conditions resulted in the greatest distraction. This was observed as reductions in mean speeds of approximately 4 km/h, increases in variance from the posted speed limit and higher reported levels of subjective mental workload. Manipulating radio system controls, such as turning on the radio and tuning and changing bandwidth, were found by Angell et al.[28] to involve greater proportions of time looking at the task than looking at the road, and the greatest total time looking at the task, compared with operating a navigation system, interacting with a CD player, reading a map, and dialing a phone, among other behaviors. From the evidence reviewed here, it seems that manipulating an in-vehicle radio can impose a substantial level of distraction.

Other studies have specifically investigated the performance decrements resulting from interaction with in-vehicle CD players. The physical act of locating, opening, and inserting a CD is one that demands high levels of physical and visual attention, and may be time-consuming. Indeed, in a simulator study, Jenness et al.[29] found that interaction with a CD player (selecting and inserting a CD, selecting a track, removing the CD) is associated with greater driving performance decrements than both eating and dialing a handheld mobile phone. This was indicated by greater lane position deviations, more frequent glances away from the road, and longer drive completion times. As with e-mail and navigation systems, it seems that voice activation technology may be useful in attenuating at least some of the distraction associated with CD players;[18] it a more challenging task, however, to automate the physical manipulation of CDs.

12.2.6 Other Vehicle-System Distractions

In addition to the systems described in the preceding sections, distraction may result from other fixed vehicle systems, either through operation of vehicle controls or instruments, or through the occurrence of vehicle mechanical problems. As yet, the effects of these on driving performance remain largely unexplored. The little research that exists, however, suggests that interactions with some fixed vehicle controls may have significant adverse effects on driving performance. Stutts et al.[23] found that manipulating vehicle system controls considered not integral to the driving task (e.g., climate controls, sun visor) was associated with an 8.55% increase in the amount of time spent driving with both hands off the steering wheel and a 13.20% increase in the amount of time driving while looking inside the vehicle.

12.3 Non-Technology-Based Sources of In-Vehicle Distraction

The primary focus of this chapter has been on sources of distraction deriving from driver interaction with technology. However, the majority of distractions within the vehicle derive from driver engagement in common, everyday activities that are considered acceptable while driving by society, such as eating, grooming, and talking to passengers[18,23,30] (see also Chapter 15). It has already been noted that the primary focus of this chapter has been on technology-based sources of distraction. However, it is also important to understand the range of other non-technology-related activities that may take place inside the vehicle and their effects on driving performance. The following discussion does not cover all potential sources of distraction deriving from within the vehicle that are unrelated to technology; a more complete list is presented in Chapter 15. Rather, the following sections describe the more commonly performed activities that have so far been explored in the distraction literature. Much of the data discussed here comes from a study by Stutts et al.,[23] which is one of the few to explore the frequency, duration, and behavioral implications of a range of distracting activities in an on-road environment. Unfortunately, no relevant performance data is currently available from the more comprehensive 100-car study,[2,31] but other findings from the 100-car study are reviewed in other chapters of the book. As for technology-related distractions, many of the distraction sources discussed here have not yet been comprehensively researched.

12.3.1 Eating and Drinking

Although eating and drinking are physically, cognitively, and visually demanding activities, they are routinely performed by the majority of drivers.[23] Examinations of crash records have revealed that eating and drinking contribute to a small but significant proportion of crashes[1,30,32] (see also Chapter 16). However, the effects of eating and drinking on driving performance remain largely unexplored. Stutts et al.[23] noted that eating and drinking involve several steps, including preparation (i.e., unwrapping), actual consumption, and sometimes spillage. Preparing to eat or drink was associated with significant increases in the proportion of time spent driving with eye gaze directed inside the vehicle (from 1.25 to 4.40%) and both hands off the wheel (from 2.61 to 5.52%) as well as an increase in the number of adverse driving events (e.g., lane position deviations and sudden braking) from 7.40 to 18.20%. Consumption and spillage were both associated with increases in time spent driving while looking inside the vehicle and driving with both hands off the wheel (4.07 and 3.63% increases, respectively), but not in adverse driving events. Jenness et al.[29] found eating was associated with greater lane position deviations and significantly lower speeds compared with driving with no secondary task.

12.3.2 Smoking

As with eating and drinking, an increased crash risk associated with smoking while driving has been demonstrated in several studies.[1,30,32] However, the driving performance effects of this activity remain largely unexplored. Like eating, smoking involves several steps: lighting or preparing, actual smoking, and extinguishing. Stutts et al.[23] found that the act of smoking itself (holding and inhaling the cigarette), while a lengthy activity (averaging 261.1 s), was not associated with changes in driving performance. Lighting and extinguishing activities (in combination), however, were associated with increased time spent driving while looking inside the vehicle (from 2.76 to 19.31%). Interestingly, actual smoking was associated with a significant decrease in adverse driving events (from 7.83 to 3.02 events per hour); however, it has been consistently found that smokers are more likely to be involved in a crash than nonsmokers (regardless of whether they were smoking when the crash occurred), even when variables such as age, gender, driving experience, and education are held constant.[33] This increased crash risk may be the result of various factors, including greater risk-taking propensity or carbon monoxide toxicity.

12.3.3 Reading and Writing

That drivers choose to perform an activity that is as obviously demanding as reading or writing while driving is alarming. These behaviors have been observed in on-road studies and have been associated with a small proportion of crashes.[1] Stutts et al.[23] found that while engaged in reading or writing, drivers had their eye gaze directed inside the vehicle for 91.50% of the time, compared with 2.51% when not reading or writing. Likewise, the proportion of time spent driving with both hands off the wheel increased from 1.39%, when not engaging in these activities, to 15.10%, when engaged. Reading and writing events tended to last around 18 s. Surprisingly, they were not associated with increases in adverse driving events. However, this may be due to a lack of data, as these behaviors were only engaged in for around 1.80% of total driving time. Furthermore, there was a tendency for drivers to engage in these activities more often during safer times; 69.50% of reading and writing occurred while the vehicle was stationary.

12.3.4 Reaching for Objects

Reaching for an object within the vehicle can be distracting. The degree to which this is so is moderated largely by the distance and concealment of the object from the driver. Glaze and Ellis[1] found that reaching activities were a contributing factor in nearly 3% of distraction-related crashes. Stutts et al.[23] found that reaching for objects was associated with significant increases in the proportion of time spent driving while looking inside the vehicle (from 1.24 to 15.10%) and with both hands off the steering wheel (from 2.22 to 20.10%), although no increase in adverse driving events was observed. A better understanding of the effects of reaching or leaning while driving is needed to better inform the positioning of in-vehicle devices.

12.3.5 Grooming

Grooming is another largely physical and visual activity. It includes actions such as shaving, brushing hair, applying makeup, or looking in a mirror. The distraction-related effects of grooming have not been widely explored, perhaps because grooming is performed less frequently than other activities. Indeed, Glaze and Ellis[1] reported that grooming accounted for only 0.40% of all distraction-related activities that resulted in a crash, and in a naturalistic driving study, Stutts et al.[23] found that such behaviors were performed for only 0.28% of total driving time. However, grooming is visually and physically demanding, and is a lengthy activity. Stutts et al.[23] found that grooming events lasted for 11.8 s on average, but ranged between 1.0 and 340.0 s. Furthermore, grooming behaviors were associated with 11.05% more time spent driving with both hands off the wheel and 31.96% more time driving while looking inside the vehicle than when not grooming.

12.3.6 Passengers

The presence of passengers has a well-known effect on crash risk, particularly for young drivers who are more at risk of a crash in the presence of peers.[34,35] However, this effect is not solely due to distraction; peer pressure, thrill seeking, and other factors also play a role.[36] Distraction may result from conversations with passengers (with or without eye contact) or distracting behavior on the part of the passenger. Adults, children, and infants create different patterns of distraction, as shown by

Stutts et al.,[23] whereby interactions with adults were almost twice as long as those with infants or children, but around eight times less frequent than interactions with infants. Likewise, varying effects are found when the passenger's age and gender relative to the driver, and relationship with the driver, are taken into account.[34,35,37] It cannot be assumed that the distracting effects of a handsfree mobile phone conversation are comparable to those deriving from a conversation with passengers. Regan[38] suggests that passengers often play a supporting role in the driving task and may limit or cease the conversation as the driving environment becomes more complex. While using a mobile phone, however, the other person cannot perform these mitigating roles as they are not aware of the complexities of the current road environment. Conversation intricacy or emotiveness may also influence the level of distraction resulting from a conversation (see Chapter 11 for further discussion of these issues).

Mixed results regarding the effects of passengers on driving performance have been reported. Stutts et al.[23] found that neither distraction resulting from passengers' behavior nor conversation with passengers (either adults, children, or infants) were associated with any impairment in driving performance. That is, there were no increases in sudden braking events, and nor were there any increases in the amount of time spent driving with both hands off the steering wheel or while looking inside the vehicle. Conversely, several studies have found negative effects on driving performance, whereby talking to a passenger has been found to result in decreased reaction times[39] or reduced speed[40] (see also Chapter 11).

12.3.7 Internal Sources of Distraction

The physiological or psychological state of the driver may also have important implications for distraction. Driver state (e.g., pain, fatigue, extreme positive or negative emotional states) may itself be distracting, and may make the driver more susceptible to other forms of distraction (see, for example, Chapter 21, for a discussion of the relationship between fatigue and distraction). Drivers may perform practically any number of secondary activities in the vehicle and may also drive while experiencing any number of physical or psychological states. Being lost in thought or daydreaming has been associated with over 4% of distraction-related crashes.[1] However, internally derived distraction may also stem from states other than being lost in thought. Rosenblatt,[41] for example, noted the lack of consideration given to emotional distraction in road safety research. Although stress and road rage have been commonly investigated as contributing factors in crashes, this research has not been undertaken in the context of distraction.

The New Zealand Crash Analysis System includes the category "emotionally upset-preoccupied" as one of its in-vehicle distraction codes. Gordon[42] (see also Chapter 16) reported that the code "emotionally upset-preoccupied" was cited in 10% of in-vehicle distraction–related crashes. Emotionality in driving was also explored by Rosenblatt,[41] who analyzed interviews with 84 participants regarding their driving behavior while grieving. It was noted that many people in bereavement may cry or experience grief while driving. Participants reported arriving at destinations unable to remember how they got there or arriving at unplanned destinations, suggesting that they had experienced extended lapses in attention due to their highly emotional state. Extremely positive or euphoric moods may be equally distracting as negative states, although there are no known data on this issue. While these emotional and cognitive states and their effects are harder to quantify and measure, this is an area that warrants further investigation.

12.4 ALTERNATIVE PERSPECTIVE OF POTENTIALLY DISTRACTING FACTORS

This chapter has focused on sources of distraction deriving from within the vehicle that have the potential to degrade driving performance. There is, however, an antithesis to this. It can at times be beneficial for drivers to willingly expose themselves to distracting activities. DVD players, for example, can placate children on long trips, and thereby reduce passenger-related distraction. Likewise, conversations, both with passengers or on mobile phones, may be beneficial in maintaining alertness and relieving boredom. Route guidance systems, when effectively designed, reduce mental workload, can reduce stress or anxiety in an unfamiliar area, and allow drivers to plan a safer route of travel. While some systems may provide such benefits, they may also distract, the extent of which, as stated earlier, depends on exposure and task and driver characteristics. Therefore, the net effect on driving performance is a balance between the benefits and costs associated with driver engagement in these secondary tasks. It would be unreasonable to suggest that conversing should be avoided altogether because it creates distraction, or that audiovisual entertainment systems should be prohibited because they have potential to distract other drivers. Rather, drivers need to be made aware of the dangers associated with activities such as these that they can willingly elect to engage in, so that they can make safer choices while driving. Drivers have always engaged, and will continue to engage, in competing activities while driving. In other chapters of this book (see Part 8), strategies are presented for preventing and mitigating the adverse effects of distraction.

12.5 CONCLUSIONS

There are many potential sources of distraction deriving from within the vehicle. Few, however, have been extensively studied to determine their effects on driving performance. This chapter has reviewed much of the research in the area (apart from research relating to mobile phones). Most of the existing research has so far focused on the effects of driver interaction with technology-based sources of distraction (largely, mobile phones) on driving performance, even though the majority of sources of distraction currently deriving from within the vehicle are unrelated to technology use (see Chapter 15). Of the studies reviewed concerned with technology-based distractions, surprisingly few have investigated the distraction potential of driving-related technologies already embedded in the vehicle cockpit. Manipulating radio system controls, for example, has been shown to adversely affect driving performance—yet this is a common, and socially acceptable, behavior. This further underscores the need for further research into the effects on driving performance of everyday activities that compete for drivers' attention.

The category of technology, other than the mobile phone, that has been most extensively researched in terms of its effect on driving performance is the in-vehicle navigation system. It seems that navigation systems and other emerging in-vehicle technologies, such as portable audio and audiovisual entertainment systems, and e-mail systems, may be problematic, depending on their design. However, there is emerging evidence to suggest that some of the adverse effects associated with drivers' use of these systems may be attenuated with additional features such as voice recognition technology. Indeed, well-designed navigation systems have been shown to be significantly less distracting than navigating using traditional paper maps. Little is known about the potentially positive effects that some sources of distraction deriving from within the vehicle may have on driving performance, such as maintaining alertness. Clearly, there is a need for further research in this area.

As noted earlier in this chapter, driving performance degradation is a function of many factors: interface design, task familiarity, task complexity, driver state, driver experience and age, and many others. Thus, the degree of performance degradation in the real world for a given source of distraction may be quite different from that observed in a controlled simulator or on-road environment. Unfortunately, there is a lack of exposure data for many of the sources of distraction described here. Hence, it is not known, for many of these, to what extent the degree of performance degradation observed actually translates into increased crash risk. As epidemiological data sets become progressively more elaborate and refined, they will shed light on this issue and provide a basis for targeting research efforts at the sources of distraction that have most potential to compromise safety.

Relatively few studies have been careful to distinguish between the specific behaviors associated with driver use of a technology-based distraction source (and, indeed, sources unrelated to technology use) that result in performance degradation; for example, whether it is the act of looking at a system, the output of the system, or physical manipulation of the system that degrades driving performance. Often, it is merely the effects of the overall presence of the system on performance that are reported. Hence, it is not yet fully understood which are the most problematic aspects of driver interaction with fixed and nomadic in-vehicle technologies. Such information is needed to inform better design strategies. Until there is standardization across studies of the methods and metrics used to quantify the impact of competing tasks on driving performance, which take into account differences in the design of the systems being tested (or the competing activities being performed), it is difficult to rank order the degree to which the various of sources of distraction that have been reviewed here impact on driving performance. This is an important challenge for future research.

REFERENCES

1. Glaze, A. and Ellis, J., *Pilot Study of Distracted Drivers*, Virginia Commonwealth University Centre for Public Policy, Virginia, 2003.
2. Klauer, S. G., Dingus, T. A., Neale, V. L., Sudweeks, J. D., and Ramsey, D. J., *The Impact of Driver Inattention on Near-Crash/Crash Risk: An Analysis Using the 100-Car Naturalistic Driving Study Data*, Report No. DOT HS 810 594, Virginia Tech Transportation Institute, Blacksburg, VA, 2006.
3. GMAC, 2006 GMAC insurance national drivers test, from www.gmacinsurance.com/SafeDriving/2006/, accessed on April 23, 2007.

4. Chisholm, S., Caird, J. K., Lockhart, J., Fern, L., and Teteris, E., Driving performance while engaged in MP-3 player interaction: Effects of practice and task difficulty on PRT and eye movements, in *4th International Driving Symposium on Human Factors in Driver Assessment, Training and Vehicle Design* Stevenson, Washington, D.C., 2007.
5. Salvucci, D. D., Markley, D., Zuber, M., and Brumby, D. P., iPod distraction: Effects of portable music-player use on driver performance, in *Human Factors in Computing Systems: CHI 2007 Conference*, ACM Press, New York, NY, 2007.
6. Technical Insights, *Human-Vehicle Interface*, Report No. 003, Technical Insights, 2001.
7. Hatfield, J. and Chamberlain, T., *The Effects of In-Vehicle Audiovisual Display Units on Simulated Driving*, NSW Injury Risk Management Research Centre, University of New South Wales, Sydney, Australia, 2005.
8. Kircher, A., Vogel, K., Tornros, J., Bolling, A., Nilsson, L., Patten, C., Malmstrom, T., and Ceci, R., *Mobile Phone Simulator Study*, Report No. 969A, Swedish National Road and Transport Research Institute, Linkoping, Sweden, 2004.
9. White, C., Fern, L., Caird, J., Kline, D., Chisholm, S., Scialfa, C., and Mayer, A., The effects of DVD modality on drivers' performance, in *37th Annual Conference of the Association of Canadian Ergonomics*, Banff, Canada, 2006.
10. Funkhouser, D. and Chrysler, S. T., *Assessing Driver Distraction Due to In-Vehicle Video Systems through Field Testing at the Pecos Research and Testing Center*, Report No. SWUTC/07/473700-00082-1, Southwest Region University Transportation Center, College Station, TX, 2007.
11. Kulmala, R., The potential of ITS to improve safety on rural roads, in *4th World Congress on Intelligent Transport Systems*, Berlin, 1997.
12. Farber, E., Foley, J., and Scott, S., Visual attention design limits for ITS in-vehicle systems: The Society of Automotive Engineers standard for limiting visual distractions while driving, in *Transportation Research Board Annual General Meeting*, Washington, D.C., 2000.
13. Tijerina, L., Parmer, E., and Goodman, M. J., Driver workload assessment of route guidance system destination entry while driving: A test track study, in *Proceedings of the 5th ITS World Congress*, Seoul, 1998.
14. Tsimhoni, O., Smith, D., and Green, P., Address entry while driving: Speech recognition versus a touch-screen keyboard, *Human Factors* 46 (4), 600–610, 2004.
15. Chiang, D. P., Brooks, A. M., and Weir, D. H., An experimental study of destination entry with an example automobile navigation system, *Society of Automotive Engineers Special Publication*, SP-1593, 2004.
16. Dingus, T., McGehee, D., Hulse, M., Jans, S., and Manakkal, N., *TravTek Evaluation Task C3 – Car Camera Study*, Report No. FHWA-RD-94-076, Office of Safety and Traffic Operations, McLean, VA, 1995.
17. Srinivasan, R. and Jovanis, P. P., Effect of in-vehicle route guidance systems on driver workload and choice of vehicle speed: Findings from a driving simulator experiment, in *Ergonomics and Safety of Intelligent Driver Interfaces*, Noy, Y. I. (Ed.), Lawrence Erlbaum Associates, Mahwah, NJ, 1997.
18. Gärtner, U., König, W., and Wittig, T., Evaluation of manual vs. speech input when using driver information systems in real traffic, in *International Driving Symposium on Human Factors in Driving Assessment, Training, and Vehicle Design*, Aspen, Colorado, 2002.
19. Lee, J. D., Caven, B., Haake, S., and Brown, T. L., Speech-based interaction with in-vehicle computers: The effects of speech-based e-mail on drivers' attention to the roadway, *Human Factors* 43 (4), 631–639, 2001.
20. Jamson, A. H., Westerman, S. J., Hockey, G. R. J., and Carsten, M. J., Speech-based email and driver behaviour: Effects of an in-vehicle message system interface, *Human Factors* 46 (4), 625–639, 2004.

21. Young, K., Regan, M., and Hammer, M., *Driver Distraction: A Review of the Literature*, Report No. 206, Monash University Accident Research Centre, Melbourne, Australia, 2003.
22. Jänke, L., Musial, F., Vogt, J., and Kalveram, K. T., Monitoring radio programs and time of day affect simulated car driving performance, *Perceptual Motor Skills* 79 (2), 484–486, 1994.
23. Stutts, J. C., Feaganes, J., Reinfurt, D., Rodgman, E., Hamlett, C., Meadows, T., Reinfurt, D., Gish, K., Mercandante, M., and Staplin, L., *Distractions in Everyday Driving*, AAA Foundation for Traffic Safety, Washington, D.C., 2003.
24. Strayer, D. L. and Johnston, W. A., Driven to distraction: Dual-task studies of simulated driving and conversing on a cellular phone, *Psychological Science* 12 (6), 462–466, 2001.
25. McKnight, A. J. and McKnight, A. S., The effect of cellular phone use upon driver attention, *Accident Analysis and Prevention* 25 (3), 259–265, 1993.
26. Wikman, A.-S., Nieminen, T., and Summala, H., Driving experience and time sharing during in-car tasks on roads of different widths, *Ergonomics* 41 (3), 358–372, 1998.
27. Horberry, T., Anderson, J., Regan, M. A., Triggs, T. J., and Brown, J., Driver distraction: The effects of concurrent in-vehicle tasks, road environment complexity and age on driving performance, *Accident Analysis and Prevention* 38 (1), 185–191, 2006.
28. Angell, L., Auflick, J., Austria, P. A., Kochhar, D., Tijerina, L., Biever, W., Diptiman, T., Hogsett, J., and Kiger, S., *Driver Workload Metrics: Task 2 Final Report*, Report No. DTFH61-01-X-00014, Crash Avoidance Metrics Partnership, Farmington Hills, MI, 2006.
29. Jenness, J. W., Lattanzi, R. J., O'Toole, M., and Taylor, N., Voice-activated dialing or eating a cheeseburger: Which is more distracting during simulated driving?, in *Human Factors and Ergonomics Society 46th Annual Meeting*, Pittsburgh, 2002, pp. 592–596.
30. Stutts, J. C., Reinfurt, D., Staplin, L., and Rodgman, E., *The Role of Driver Distraction in Traffic Crashes*, AAA Foundation for Traffic Safety, Washington, D.C., 2001.
31. Dingus, T. A., Klauer, S. G., Neale, V. L., Petersen, A., Lee, S. E., Sudweeks, J. D., Perez, M. A., Hankey, J., Ramsey, D. J., Gupta, S., Bucher, C., Doerzaph, Z. R., Jermeland, J., and Knipling, R. R., *The 100-Car Naturalistic Driving Study, Phase II: Results of the 100-Car Field Experiment*, Report No. DOT HS 810 593, Virginia Tech Transport Institute, Blacksburg, VA, 2006.
32. Neale, V. L., Dingus, T. A., Klauer, S. G., Sudweeks, J., and Goodman, M., An overview of the 100-car naturalistic driving study and findings, in *19th International Technical Conference on Enhanced Safety of Vehicles*, Washington, D.C., 2005.
33. Christie, R., *Smoking and Traffic Accident Involvement: A Review of the Literature*, Report No. GR/91-3, VicRoads, Melbourne, 1991.
34. Arnett, J. J., Offer, D., and Fine, M. A., Reckless driving in adolescence: 'State' and 'trait' factors, *Accident Analysis and Prevention* 29 (1), 57–63, 1997.
35. Doherty, S. T., Andrey, J., and MacGregor, C., The situational risks of young drivers: The influence of passengers, time of day and day of week on accident rates, *Accident Analysis and Prevention* 30 (1), 45–52, 1998.
36. Regan, M., Salmon, P. M., Mitsopoulos, E., Anderson, J., and Edquist, J., Crew resource management and young driver safety, in *Proceedings of the Human Factors and Ergonomics Society 49th Annual Meeting*, Orlando, FL, 2005, pp. 2192–2196.
37. Regan, M. A. and Mitsopoulos, E., *Understanding Passenger Influences on Driver Behaviour: Implications for Road Safety and Recommendations for Countermeasure Development*, Report No. 180, Monash University Accident Research Centre, Melbourne, Australia, 2001.

38. Regan, M. A., Driver distraction: Reflections on the past, present and future, *Journal of the Australasian College of Road Safety* 16 (2), 22–33, 2005.
39. Consiglio, W., Driscoll, P., Witte, M., and Berg, W. P., Effect of cellular telephone conversations and other potential interference on reaction time in a braking response, *Accident Analysis and Prevention* 35 (4), 495–500, 2003.
40. Fairclough, S. H., Ashby, M. C., Ross, T., and Parkes, A. M., Effects of hands-free telephone use on driving behaviour, in *ISATA Conference*, Florence, Italy, 1991.
41. Rosenblatt, P. C., Grieving while driving, *Death Studies* 28, 679–686, 2004.
42. Gordon, C., A preliminary examination of driver distraction related crashes in New Zealand, in *Driver Distraction: Proceedings of an International Conference on Distracted Driving*, Faulkes, I. J., Regan, M. A., Brown, J., Stevenson, M. R., and Porter, A. (Eds.), Australasian College of Road Safety, Sydney, 2005.

13 Distractions outside the Vehicle

Tim Horberry and Jessica Edquist

CONTENTS

13.1 INTRODUCTION: IMPORTANCE OF THE AREA

This chapter focuses on possible distractions from outside the vehicle. In particular, it examines possible driver distraction caused by billboards or other visual information that is *extraneous* to the driving task. Although it is perhaps stating the obvious, the road environment should, as much as possible, encourage safe driving. However, in most developed countries, the amount of visual information in road environments is increasing due to progressively higher traffic densities, more complex traffic management and roadway maintenance, increased commercial roadside development and roadside vending, and more commercial pressure on road authorities to permit advertising next to large roads. In addition, the increasing uptake of in-vehicle technologies and a higher number of older drivers remaining mobile lead us to believe that consideration of distraction, caused in part by the external environment, is of great importance.

Together, these factors mean that the road environment is increasingly prone to producing information that may distract a driver. This is borne out in crash rates: Stutts et al.[1] found that almost one-third of drivers who crashed due to distraction had been distracted by something outside the vehicle. However, quite surprisingly, there has been little research into how and to what extent information in the road environment can cause driver distraction.

This chapter will focus on *visual* information in the road environment. However, it is conceded that other sources of information from the external road environment (e.g., police car sirens, bumpy road surfaces, or even odors from agriculture) can also cause distraction. Likewise, although this chapter will focus purely on possible distractions from *outside* of the vehicle, it should be noted that there is likely to be an interaction between visual information seen and attended to inside and outside of the vehicle.

13.2 WHAT OBJECTS IN THE EXTERNAL ENVIRONMENT COULD DISTRACT A DRIVER?

As driving is a task that relies primarily on visual information, a sensible starting place is to define in general terms what sorts of visual information are present in many roadways. There are various means by which it can be determined which objects in the external environment could potentially distract a driver; one approach is presented in the following text. Other approaches would include examining the sources of distraction cited in police forms, using verbal protocol analysis while people are driving or using simulator data.

The possible taxonomy of visual information in the road environment suggested here classifies entities into one of four groups: *built roadway* (entities put there explicitly by road/highway engineers), *situational entities*, the *natural environment*, and the *built environment*. Table 13.1 gives a few examples within each of these groups.

TABLE 13.1
Initial Taxonomy of Visual Information in or around the Roadway

Type of Roadway Information	Examples
Built roadway	Road geometry (e.g., lane width)
	Road surface
	Traffic signs and markings
Situational entities	Moving and parked vehicles
	Pedestrians on or near the roadway
	Weather (e.g., fog, rain)
	Ambient light level (e.g., darkness at night)
Natural environment	Trees and other vegetation
	Seas, lakes, and rivers
	Hills
Built environment	Houses and other buildings
	Bus stops
	Billboards and other roadside advertising

One way to discover what things in the external environment are distracting is simply to ask drivers. Surprisingly, little research has been conducted in this area; however, recent research by Edquist et al.[2] explored the related issue of what drivers meant by "visual clutter" in the roadway environment. In this work, 54 drivers viewed a series of different road scenes and took part in a focus group discussion. The main themes that emerged can be summarized as four points of view:

1. *Everything is clutter.* Some drivers felt that scenes that are cluttered simply have too many individual objects to pay attention to.
2. *Occluding objects are clutter.* Objects that obstruct the driver's view of other objects that the driver needs to pay attention to caused clutter.
3. *Distracting objects are clutter.* Many drivers felt that objects that are conspicuous but not necessary for the driving task are a source of clutter in the road environment and should be removed or restricted.
4. *Too many driving-related objects make clutter.* Another cause of clutter is when there are too many objects to which the driver must attend to drive safely. If the driver must continually pay attention to a large number of objects (near the maximum number that the driver can attend to), it may lead to overload, with consequent negative impacts on driver safety.[2]

For the purpose of this chapter, the third point (which very roughly equates to the *built environment* in the previously mentioned taxonomy) is of most interest. Such focus group results do reinforce the view that many drivers consider aspects of the external environment to be distracting and that some of these objects (e.g., billboards) can be regulated by the road authorities to minimize distraction. To a large extent, however, using subjective opinion is limited; at best it, can merely give us an indication of the types of objects in the road environment that might cause distraction.

As such, the ongoing research being performed by Edquist et al. is now undertaking more quantitative research into how much visual clutter is contributed by each element in a scene and which attributes are most distracting, especially with reference to distraction from billboards. At the time of this writing, work is proceeding to manipulate levels of certain attributes within an advanced driving simulator and to determine the effects on drivers' ability to drive safely.[2]

Another way to assess what objects in the road environment are distracting (or, more accurately, being looked at) is by eye movement studies to assess visual behavior. There is a plethora of research in this area: recording drivers' visual behavior on real roads, in driving simulators, on test tracks, or in laboratory settings. This approach can produce percentages of total fixation time to different areas of the visual scene. An example of such work is the research by Horberry.[3] In a driving simulator and eye movement study, he found that if an advertisement or other form of visual clutter (objects not relevant to the driving task, such as graffiti) is in a road scene, then it is often looked at for quite a large proportion of the time (over 14% of the total driving time on average).[3] Other visual behavior metrics of relevance here include number and duration of fixations, scan patterns, and the distance from the driver at which an object is first attended to. However, a commonly reported limitation in such eye movement research is that fixation is not the same as attention: we can be looking at an object while thinking about something else. Likewise, we perceive information even if it has not been directly fixated on (e.g., detecting a moving object in peripheral vision). Nonetheless, eye movement studies can provide important clues about what information in the road environment can attract a driver's attention.

13.3 HOW MIGHT THE EXTERNAL ENVIRONMENT DISTRACT A DRIVER?

Although other chapters in this book will examine this topic further, for the purposes of this chapter, a general definition of distraction is "any driver involvement that takes his or her attention away from his or her intended driving task".[4] In addition, for the purpose of this chapter, three slightly different meanings of distraction can be discerned. Studies of the number of crashes due to "distraction" tend not to separate them; however, such a distinction is useful to understand what led to a particular distraction event from the external environment. "Visual distraction" or "attentional capture" refers to a situation in which the driver's attention is involuntarily attracted by a conspicuous object[5]; this depends both on the source of visual demand and on the driver's (involuntary) attention to it. "Cognitive distraction," "internal distraction," or "inattention" is when the driver is thinking about something else and not devoting full attention to the driving task. There may not be a specific environmental trigger for this, so some authors do not regard it as distraction *per se*.[1] Finally, distraction has often been used to mean a situation in which the driver voluntarily takes his or her eyes off the road to complete some in-vehicle task, such as tuning the radio.[6] This is perhaps better described as a "secondary activity" to emphasize that the driver is still controlling the direction of attention, although what the driver chooses to attend to may not be optimal for driving performance.

Objects from the external environment could potentially cause any of these three types of distraction: involuntary capture of visual attention (e.g., by flashing lights), involuntary capture of mental "attention" (e.g., thinking about a shop just passed), or voluntary direction of visual and mental attention away from the road (e.g., looking at a billboard featuring a new type of bank savings account).

Objects or lights that appear suddenly (an "abrupt onset") are well known to capture attention. Many traffic authorities around the world limit the use of flashing lights on advertising signs, in part due to the likelihood that they would interfere with drivers detecting flashing lights on emergency vehicles or important traffic signals. Crawford[7] indeed found that flashing signals against a background of steady lights are noticed more quickly than steady signals, whereas a background of flashing lights makes it very difficult to see a signal light (particularly if the signal is flashing).

Objects in the external environment high in "attention conspicuity" are also comparatively likely to capture visual attention. Cole and Hughes[8] listed the following determinants of an object's conspicuity: background complexity, contrast, eccentricity (distance from the center of fixation), color, and the boldness of the object's internal structure. Relative size is also important[9]; for example, a large traffic sign among a series of smaller signs. Where possible, this suggests that objects more salient to the driving task should be made more conspicuous, and conversely, objects less salient should be made less conspicuous. However, as visual traffic environments are not fully under the control of highway engineers, this is difficult to achieve.

13.4 HOW CAN EXTERNAL DISTRACTION BE MEASURED OR ASSESSED?

It has been asserted in this chapter that visual information outside the vehicle has the potential to significantly distract drivers. The level of external distraction in a particular environment can be assessed in a number of ways. Some techniques are as follows:

1. *Road design guidelines*, especially with reference to advertisements. For example, the Department of Main Roads Queensland (Australia) has developed a set of guidelines to minimize the possibility of roadside advertisements to distract drivers from processing important traffic signals in situations requiring particular concentration, such as merges. The main form of this is longitudinal placement controls: *clear zones* are mandated on either side of the road for a certain distance around traffic signs and areas requiring merging. On normal roads, no advertisements are permitted within a circle with a radius of $1.2 \times V$ around important traffic signs and other advertisements (where V stands for the velocity or speed environment of the road); on motorways this is extended to 2.5V. Advertisements are not permitted for a distance of 5V upstream of an on-ramp and 7.5V upstream of an exit from a motorway to prevent negative behavioral effects. In addition, further restrictions may apply in situations that require additional driver attention and decision making: those in which drivers are required to merge or weave between lanes at high speeds, for example, on large high-speed roundabouts or where a divided motorway becomes a two-way road; at complex intersections, such as where five or six roads come together; on sections of a road displaying traffic control devices that (singly or in combination) are more complex and require more time to read and interpret than would normally be expected; on pedestrian crossing facilities; and on sections of a road with a vehicle crash history higher than average.
2. *Road safety audits*. Conventional road safety audits are a systematic process for checking the safety of new roadway and traffic management schemes. The main aim is to minimize safety problems from the beginning. Safety audits are often included during the design, creation, and maintenance phases of road projects. Both the United Kingdom and Australia lead the world in undertaking such audits, although they are now becoming more commonplace worldwide. *Psychological* road audits are a less formalized area; they may take the form of assessing a driver's reactions and behavior to a new stretch of road, or applying specially constructed human factors checklists to assess likely driver behavior.
3. *Accident rates and driver behavioral responses*. Although accident rates have high face validity, it can often be difficult to directly attribute a crash

to a feature (or combination of features) in the external environment. Nonetheless, where used appropriately, accident statistics can provide compelling evidence for the safety disbenefits of general types of different environments. Likewise, driver behavioral data obtained by induction loops or similar devices in the road often produce only fairly coarse-grained data (e.g., vehicle speed or headway), which might not be of huge assistance to quantify the impact of specific external distractions. More data are gained by eye movement studies (described earlier in this chapter) or detailed, long-term observation of drivers. However, as yet no longer-term study has separated out glances to locations outside the vehicle, let alone glances to specific objects or locations within the roadside environment.

4. *Resident's and driver's opinions.* In residential areas, local people may voice strong opinions to highway authorities about the safety and acceptability of the roadside environment. Driver's opinions can be obtained in several ways: through focus groups, by stopping a driver after he or she drove past a feature and asking him or her about it, or via verbal protocols (either during the drive or while watching a recording). Although such methods can produce valuable data, they cannot be directly linked to safety effects from features of the external environment.
5. *Modeling/experimentation.* This, of course, covers a plethora of different methods ranging from computer-based modeling to driver behavioral testing. One way is to re-create the external environment in a more controlled environment (e.g., in a driving simulator) and to evaluate the impact on driving performance and behavior of external sources of distraction using a range of response measures that are otherwise too difficult or unsafe to collect on real roads (e.g., lateral lane deviation, critical incidents, eye movements, steering wheel corrections). Precisely modeling the real roadway geometry is now easier in a reasonably advanced simulator, as is obtaining a suitable measure of driver's immersion in the displayed road scene.

13.5 THE EFFECTS OF EXTERNAL DISTRACTIONS ON DIFFERENT ASPECTS OF DRIVING PERFORMANCE AND ON DIFFERENT DRIVERS

The driving task involves a range of skills and abilities; as such, from first principles it seems likely that the three types of distraction, mentioned in Section 13.3, may have different effects on driving performance. Sadly, with reference to distraction from external sources, there is a lack of research on this topic. Surprisingly, there is even a scarcity of published work about the overall effect of visual distraction by sources outside the vehicle on virtually any aspect of driving performance. Most of the work that has been done has investigated advertising billboards, which are discussed in Section 13.7. There has not been any systematic investigation of the level and effects of distraction experienced by, for example, drivers passing crash sites, although "rubbernecking" is a well-known phenomenon. What we do know is that when the driver's eyes are directed away from the forward roadway for more than 2 s, the chance of a crash rises significantly.[10] Some external distractions may actually be within the forward roadway area, which makes it particularly difficult to determine when a driver is distracted. However, when the driver attends to an external distraction at a large visual angle from the forward roadway (e.g., looking at a roadside vendor while passing it), we may expect the same result as for a distracting in-vehicle task: an increase in the crash risk especially for gazes longer than 2 s.

13.6 AGE/EXPERIENCE EFFECTS

Two groups that may be especially vulnerable to the effects of distraction are novice/inexperienced drivers and elderly drivers. Stutts et al.[1] found that young (and generally therefore inexperienced) drivers were more likely than drivers over the age of 20 to be distracted at the time of a crash (although not necessarily by external distractions). Similarly, Lam[11] found that drivers aged 16–19 years had a slightly higher risk of being involved in a crash caused by an external-to-vehicle distraction than other age groups. Owing to their inexperience, these drivers may have less attentional resources to spare from vehicle control and may not be able to divide attention effectively. There is experimental evidence to support this assertion.[12]

Regarding older drivers, Stutts et al.[1] found that for drivers who crashed as a result of distraction, the cause of the distraction was something outside the vehicle for 43% of drivers aged over 65, compared with about 30% in each of the other age groups. Older drivers may have problems in dividing and shifting attention. This is likely to lead to particular difficulties when sources of workload inside and outside the vehicle are combined, as secondary tasks impair their reaction times more than those of younger drivers.[13] However, although this negative effect for older drivers seems likely, there is only limited research evidence to support this; further research is needed.

In short, it is likely that both older and younger drivers have particular problems with external distractions (for a variety of reasons). However, it is difficult to suggest countermeasures aimed purely at these groups. To a degree, some older drivers avoid complex traffic environments, so a modicum of self-regulation does take place for this group. The effects of driver age and driving experience on vulnerability to distraction are discussed further in Chapters 19 and 20 of this book.

13.7 CASE STUDY: BILLBOARDS

In some ways, billboards are one of the easiest objects in the road environment to control. Also, most research in this general area has been performed on this type of external distraction. As such, billboards are the focus of this section.

Although their precise regulation varies across highway authorities around the world, most authorities do at least have some measure of control over them. However, advertisers and the owners of the land on which these advertisements are located often have significant financial incentives to make billboards as conspicuous as possible.

As seen in Figure 13.1, advertising billboards by their nature tend to be large, bold, brightly colored, and placed near the road. It might therefore be expected that

FIGURE 13.1 Billboards on either side of a busy freeway.

they would divert attention from the driving task. Indeed, to some extent, this is the intention of the advertiser (although, of course, this would not be labeled as "distraction" by an advertiser).

13.7.1 The Behavioral Effects of Billboards

To investigate their precise effects on driver's behavior, Johnston and Cole[14] performed a series of five laboratory experiments in which participants moved a joystick to track arrows that appeared on a screen while distracting advertisements were occasionally presented just above the arrows. They concluded that distractions from advertising billboards probably do not affect vehicle control (simulated by a tracking task) but probably do affect hazard detection (simulated by peripheral target detection, which was an additional task in three of the experiments).

More recently, eye-tracking equipment has allowed researchers to investigate how much time drivers spend looking at advertising billboards as they are driving past (as mentioned in the study by Horberry and coworkers[2]). Lee et al.[15] analyzed the visual behavior and vehicle control of participants driving a 35-mile on-road route past 30 billboard sites. They compared the data recorded in the 7 s prior to passing each site with data recorded on the approach to "comparison" sites with "logo signs, on-premise signs, etc.," and "baseline" sites with "no visual elements such as buildings or signs present." They found no significant differences in eye glances; however, there were a number of methodological flaws in the analysis. First, the "comparison" sites included other forms of outdoor advertising, which might also be expected to attract attention. The baseline sites are a better comparison; however, the study authors failed to analyze billboards on different sides of the road separately—one might expect a billboard on the left to produce more left-forward glances than one on the right, and vice versa.

Beijer et al.[16] analyzed glances at advertising signs on a Canadian expressway. Active signs (those with movable displays or components) made up 51% of signs, but received 69% of glances and 78% of glances that lasted more than 0.75 s. Unfortunately this paper did not analyze glances at any other objects, which makes it difficult to investigate whether advertising receives a disproportionate amount of visual attention. This 0.75 s figure is often considered to be the minimum perception-response time for a nonalerted driver to react to a braking vehicle, so if drivers take their eyes off the road for longer than 0.75 s, there is an increased risk that they will not detect sudden, unexpected roadway hazards (e.g., a pedestrian walking into the road in front of the vehicle).

Crundall et al.[5] recorded eye movements as participants watched video footage of driving past advertising signs, which could be situated at street level (SLA) or raised level (RLA). SLAs were fixated on more often and for longer, but this did not translate to better recognition on a subsequent memory task; in fact, memory for SLAs was worse than that for RLAs. Participants who were instructed to look for hazards fixated on SLAs more than those who were instructed to remember advertisements for the subsequent test. The authors suggest that this is because advertisements at SLA fall into the driver's search zone for potential hazards, and subsequently capture attention. Participating drivers rated the videos containing SLAs as more hazardous than those containing RLAs. As such, to some extent this would imply that raised signs are more memorable (a positive thing for an advertiser) yet less visually demanding (a positive thing for a road safety engineer). Although not all environments allow for raised advertising signage, it surely should be an avenue to be explored further.

13.7.2 Crash Rates and Billboards

In combination, the studies mentioned earlier provide evidence that some billboards may capture drivers' attention at inappropriate times and hold it for long enough that they might be unable to avoid a crash should a critical incident occur. Two recent reviews of the literature on billboards and crash rates concluded that these two factors correlate, with the usual caveat that correlation does not equal crash causation. Cairney and Gunatillake[17] reviewed studies correlating crash data with advertisement location and found that greater density of advertisements does seem to correlate with a higher crash rate, especially for changeable-message signs. Farbry et al.[18] concentrated on the effect of electronic billboards on crash rates. They concluded that crash rates are higher where electronic billboards are installed.

Wallace's[19] review of the literature noted that the presence of billboards correlated with high crash rates in some circumstances. Higher crash rates were associated with billboards in two situations: at intersections, where billboards can function as visual clutter and interfere with the driver's ability to perceive important traffic signs, and on long monotonous stretches of road, where drivers may be surprised by the sudden appearance of a billboard or fixate upon it as the brightest object in their visual field. Wallace concluded that more research should be performed into the situations in which billboards dangerously interfere with the driving task.

13.7.3 Billboard Content

One attribute of billboards that has not been studied in much detail is their content. In one sense this is the most obvious starting point (as, from first principles, we are all more likely to look at content that interests us); yet it is difficult to study experimentally. A recent study suggests that billboards containing emotive images could be more distracting than others.[20] Participants in this study watched a rapidly changing series of photographs, looking for the one that was rotated. Participants were less accurate at responding to the target when it occurred just after an emotive picture. Interestingly, in a second condition, some participants were able to fine-tune their target "template" and ignore the emotive images. This suggests that individual differences may be important in moderating the level of distraction caused by billboards. Certainly, more research is needed here; however, it will be a challenging task to make the images in billboards salient to all drivers in an experimental or field study.

13.7.4 Billboards and Traffic Signs

A traffic sign needs to be easily noticed; however, in many situations in the road traffic environment, this requirement is not met, for a variety of reasons.[21] Boersema and Zwaga[22] argued that this might be attributable to several factors: the position of the sign (e.g., too far from the normal line of sight), the sign itself (e.g., too small), objects in the vicinity of the sign (e.g., distracting buildings or advertisements), the actual layout of the road environment (e.g., a bend in the road), and the cognitive state of the user (e.g., fatigued).

As mentioned earlier in this chapter, many studies on searching for targets in cluttered visual fields have shown that distractions present in a scene generally caused a reduction in participants' visual search performance. These results lead to the conclusion that billboard advertisements can decrease the conspicuity of routing signs. To explore this, Castro and Martos (1998, cited in Ref. 21) analyzed the effect of advertisements on the conspicuity of traffic signs and the color contrast effect between the advertisement and sign. They found that increasing the number of advertisements produced an increase in sign reaction time. Color contrast between sign and advertisements also has a significant effect: the lower the color contrast between sign and advertisement, the slower the reaction time in identifying the target. Although these results are perhaps unsurprising, it must be pointed out that traffic signs and advertisements are often placed very close together, especially in cities and main roads, creating very cluttered and distracting road environments. As such, billboards and legitimate traffic control devices do often compete for drivers' visual attention (see Figure 13.2).

To assist a highway engineer to develop appropriate guidelines for the regulation of billboards, it would be useful to "benchmark" its distraction potential against other items of street furniture. However, although it would be useful to, in theory, compare the distraction potential of an advertising sign with, say, a speed camera or a roadside vendor, little firm research evidence exists. Nonetheless, one possible approach is to compare an advertising sign with an official traffic sign (e.g., a route direction sign) in terms of the distraction potential of the amount of information they display. Although such "information theory" metrics are not a current vogue in road safety research (e.g., they do not fit with contemporary information processing models in cognitive psychology), they do allow some (albeit crude) measure of quantification and comparison. Previous research by Agg[23] for direction signs found the following:

- Overload exists if a sign has more destinations than can be read in the time available, and such overload is most likely to occur when there are many

FIGURE 13.2 Illustration of the competition for visual attention in a city.

destinations on a direction sign or the time available to read a direction sign is short.

- The relationship between response time and the total number of destinations was nonlinear, but not markedly so.
- Based on earlier research, the 99th percentile time to read a sign is $T_{99} = N/3 + 2$ s (where N is the maximum number of destinations on a sign). This represents the "worse case" maximum time it would take to read a sign.
- The maximum number of destinations that are normally recommended is 6.

So, one way to compare the distraction potential of a billboard's message would be to count how many discrete "units" of information it contains and then compare that with limits for direction signs. Of course, this approach does not take account of sign position, letter height/design, information saliency, and so on. Also, it must be borne in mind that the billboard is not relevant to the driving task (whereas a direction sign generally is), so establishing that the amount of information in a billboard is no more visually demanding than a legitimate traffic control device is not necessarily a reasonable comparison. As such, it is a quantifiable, yet limited, metric.

13.8 CONCLUSIONS

Similar to distraction caused by in-vehicle systems, there is a lack of firm research evidence on which to form guidelines or standards about how much distraction from outside of the vehicle is "safe." Similarly, metrics to quantify the distraction potential of an external environment are still in their infancy.

Even more difficult is distraction caused by other factors outside the control of traffic engineers: for instance, it would be difficult to prevent attractive pedestrians from walking near roadways. Equally, because distraction may vary across drivers (due to factors such as age) and within individual drivers across time (due to temporary factors such as fatigue or motivation), findings of no significant effect from any one study must not be taken to mean that distraction caused by roadway visual information will never affect any driver's performance.

All of this to some degree paints a bleak picture. As drivers, we all "know" that aspects of the external road environment can distract us. What these aspects are vary between people and within people at different times. The available research evidence is largely either restricted to piecemeal studies investigating specific aspects (e.g., visual search) or wider meta-analyses of increased crash risks from various factors. This "gap" therefore presents a fertile ground for the researcher to explore. It is the opinion of the current authors that research in this area should take a strategic approach to investigate the situations in which distraction from visual information in the road environment is likely to be problematic for different drivers (e.g., older), for different traffic situations (e.g., junctions), and for different information types. Such work is beginning to be undertaken in various countries; however, until complete, the regulation of some types of information (e.g., billboards) in the road environment will remain more of an art than a science.

REFERENCES

1. Stutts, J.C., Reinfurt, D.W., Staplin, L., and Rodgman, E., *The Role of Driver Distraction in Traffic Crashes*, Washington, D.C.: AAA Foundation for Traffic Safety, 2001.
2. Edquist, J., Horberry, T., Regan, M.A., and Johnston, I., Visual Clutter and external-to-vehicle driver distraction, In Faulkes, I.J., Regan, M.A., Brown, J., Stevenson, M.R., and Porter, A. (Eds.). *Driver Distraction: Proceedings of an International Conference on Distracted Driving,* Sydney, Australia, 2–3 June. Canberra: Australasian College of Road Safety, 2005.
3. Horberry, T., *Bridge Strike Reduction: The Design and Evaluation of Visual Warnings,* Unpublished PhD Dissertation, University of Derby, 1998.
4. McAllister, D., Dowsett, R., and Rice, L., *Driver Inattention and Driver Distraction* (No. 15), Virginia: Virginia Commonwealth University Transportation Safety Training Center (Crash Investigation Team), 2001.
5. Crundall, D., van Loon, E., and Underwood, G., Attraction and distraction of attention with roadside advertisements, *Accid Anal Prev* 38(4), 627, 2006.
6. Horberry, T., Anderson, J., Regan, M.A., Triggs, T.J., and Brown, J., Driver distraction: The effects of concurrent in-vehicle tasks, road environment complexity and age on driving performance, *Accid Anal Prev* 38(1), 185, 2006.
7. Crawford, A., The perception of light signals: The effect of the number of irrelevant lights, *Ergonomics* 5(3), 417, 1962.
8. Cole, B.L. and Hughes, P., *Drivers Don't Search: They Just Notice.* Paper presented at the First International Conference on Visual Search, University of Durham, 1988.
9. Cole, B.L. and Jenkins, S.E., The effect of variability of background elements on the conspicuity of objects, *Vision Res* 24(3), 261, 1984.
10. Klauer, S.G., Dingus, T.A., Neale, V.L., Sudweeks, J.D., and Ramsey, D.J., *The Impact of Driver Inattention On Near-Crash/Crash Risk: An Analysis Using the 100-Car Naturalistic Driving Study Data*, Blacksburg, Virginia: Virginia Tech Transportation Institute, 2006.

11. Lam, L.T., Distractions and the risk of car crash injury: The effect of drivers' age, *J Safety Res* 33(3), 411, 2002.
12. Triggs, T.J. and Regan, M., Development of a cognitive skills training product for novice drivers, In *Proceedings of the Road Safety Research, Policing and Education Conference*, Wellington, New Zealand: Land Transport Safety Authority and New Zealand Police, 1998, pp. 46–50.
13. Lesch, M.F. and Hancock, P.A., Driving performance during concurrent cell-phone use: Are drivers aware of their performance decrements? *Accid Anal Prev* 36, 471, 2004.
14. Johnston, A.W. and Cole, B.L., Investigations of distraction by irrelevant information. *Aust Road Res* 6(3), 3, 1976.
15. Lee, S.E., Olsen, E.C.B., and DeHart, M.C., *Driving performance in the presence and absence of billboards—Executive Summary*, Blacksburg, Virginia: Virginia Tech Transportation Institute, Centre for Crash Causation and Human Factors (prepared for the Foundation for Outdoor Advertising Research and Education), 2003.
16. Beijer, D., Smiley, A., and Eizenman, M., Observed driver glance behaviour at roadside advertising signs, *Transport Res Rec* 1899, 96, 2004.
17. Cairney, P. and Gunatillake, T., Does roadside advertising really cause crashes? In *Proceedings of the 2000 Road Safety Research, Policing and Education Conference*, Brisbane, Australia, November 26–28, 2000.
18. Farbry, J., Wochinger, K., Shafer, T., Owens, N., and Nedzesky, A., *Research Review of Potential Safety Effects of Electronic Billboards on Driver Attention and Distraction*, Washington, D.C.: Federal Highway Administration, 2001.
19. Wallace, B., Driver distraction by advertising: Genuine risk or urban myth?, *Municipal Eng* 156, 185, 2003.
20. Most, S.B., Chun, M.M., and Widders, D.M, Attentional rubbernecking: Cognitive control and personality in emotion-induced blindness, *Psychon B Rev* 12(4), 654, 2005.
21. Castro, C. and Horberry, T.J., *The Human Factors of Transport Signs*, Boca Raton: CRC Press, 2004.
22. Boersema, T. and Zwaga, H.J.G., Searching for routing signs in public buildings: The distracting effect of advertisements, In Brogan, D. (Ed.), *Visual Search*, London: Taylor & Francis, 1990, p. 151.
23. Agg, H.J., *Direction sign overload*. TRL Project Report PR77. Transport Research Laboratory, Berkshire, UK, 1994.

14 Distraction and Public Transport: Case Study of Bus Driver Distraction

Paul M. Salmon, Kristie L. Young, and Michael A. Regan

CONTENTS

14.1 INTRODUCTION

Driver distraction is currently something of a buzzword within the road safety community. This appears justified given data on the frequency of road crashes in which driver distraction is a contributing factor. Early estimates of the proportion of crashes that are attributable to the driver being distracted by an object, event, or activity inside or outside the vehicle range from 10 to 12%,[1,2] although findings from the recent "100-car study" suggest that this figure may be closer to 23%.[3] Moreover, research has also concluded that, when controlling for driving conditions and time on task, the impairments associated with using a mobile phone while driving can be as profound as those associated with driving while under the influence of alcohol.[4]

14.2 DRIVER DISTRACTION

Ostensibly, driver distraction refers to those instances when a driver's attention is diverted from the primary task of driving the vehicle in a way that compromises safe driving performance. The distraction source can be either *internal* (within the vehicle, e.g., a passenger, technology) or *external* to the vehicle (outside of the vehicle, e.g., roadway advertisement, pedestrians). Within the academic literature, there have been numerous attempts at defining the construct. In Chapter 1 of this book, driver distraction is defined as a "diversion of attention away from activities critical to safe driving toward a competing activity." The American Automobile Association Foundation for Traffic Safety, however, defines driver distraction as occurring "when a driver is delayed in the recognition of information needed to safely accomplish the driving task because some event, activity, object, or person within or outside the vehicle compelled or tended to induce the driver's shifting attention away from the driving task"[5] (p. 21). Similarly, Stutts et al.[1] suggest that distraction occurs when driver inattention leads to a delay in the detection of information that is critical to safe driving performance (the reader is referred to Chapter 3 of this book for a more detailed discussion of distraction definitions).

Regan[6] differentiates between three consequences of driver distraction: visual, attentional, and physical. When distracted, drivers may take their eyes off the road to look at something, such as a billboard advertisement or ringing mobile phone. They may also divert their attention away from activities critical to safe driving and toward a competing activity, such as when conversing on a mobile phone or daydreaming. Finally, physical interference occurs when a driver's participation in a competing task interferes structurally with activities critical to safe driving, such as when using one hand to increase the radio volume while using the other to steer the vehicle. Of course, these consequences of distraction, although categorized separately, are not mutually exclusive, and one or more may be present at any one time when a driver engages in a secondary activity while driving. Dialing a mobile phone, for example, can take a driver's eyes and mind (i.e., attention) off the road and physically interfere with vehicle control.

Much of the previous research into distraction has focused on conventional automobile drivers (e.g., Refs 7 and 8). However, it is apparent that distraction could also be a major problem in the public and commercial transport sectors (i.e., bus, taxi, and train driving, aviation, heavy-goods vehicle and long-haul truck driving). This assumption is made on the basis that drivers or pilots who are *working* might be exposed to additional distracting factors as part of their work task. For example, while driving buses, drivers undertake a number of additional "bus operation" tasks, such as selling tickets, communicating with control room operators, and monitoring passengers, which often involve interaction with non-driving-related devices (e.g., ticket machines, radios and radio handsets) and also passengers.

This chapter describes the findings derived from a program of research that was undertaken by the Monash University Accident Research Centre (MUARC) for a major transport company in the Australian state of New South Wales (NSW) to investigate the potential of city bus drivers to be distracted while driving buses in and around the city. Specifically, the research aimed to identify what sources of distraction bus drivers are exposed to while operating buses, what can be done to minimize driver exposure to them, and their potential impact on performance and safety. The purpose of the chapter is to both demonstrate a novel methodology for assessing distraction within the public transport sector and communicate the findings derived from our research.

14.3 DISTRACTION IN THE PUBLIC AND COMMERCIAL TRANSPORT SECTORS

Compared with conventional automobile driver distraction research, relatively little research has focused specifically on the problem of distraction in the public and commercial transport sectors. Hanowski et al.[9] analyzed long-haul truck driver distraction by studying naturalistic data from approximately 140,000 miles of long-haul truck driving. The data were initially coded into three types of incidents: crashes, near crashes, and crash-relevant conflicts. Upon analyzing the incidents, it was concluded that the three main causes of the crashes analyzed were "judgment error," "other vehicle," and "driver distraction." Hanowski et al.[9] further analyzed the driver distraction–related incidents (178) and subsequently constructed a taxonomy of 34 distraction types, of which only four (talking on citizen's band [CB] radio, looking at CB, adjusting CB, looking at paperwork) were directly related to the driver's work tasks. The other distraction types included talking to passengers, eating, radio-related types (e.g., looking at and adjusting the radio), smoking-related types (getting cigarette, lighting cigarette, blowing smoke, etc.), adjusting the seat, cell phone–related types (e.g., answering, dialing and looking at cell phone), and personal factors such as grooming (e.g., brushing hair), using a toothpick, and rubbing one's face.

Bunn et al.[10] conducted a retrospective review of commercial motor vehicle collision data from the Kentucky Collision Report Analysis for Safer Highways (CRASH) database for incidents that took place between 1998 and 2002. Bunn et al.[10] reported that both fatigue and distraction were strongly associated with a fatal outcome in commercial motor vehicle collisions. Their findings suggested that when other contributing factors (e.g., age and restraint) were controlled for, drivers who were fatigued or fell asleep were 21 times more likely and drivers who were distracted/inattentive were 3 times more likely to be involved in a fatal commercial motor vehicle collision (CVC) than drivers involved in a collision with other human factors–related causal factors.

Barr et al.[11] describe an exploratory analysis of naturalistic driving study data that was designed to gather information related to driver distraction among truck drivers. In total, 121 h of naturalistic video data (using six truck drivers) were analyzed, resulting in the identification of 4329 "distracting events."[11] According to Barr et al., the results of the analysis indicated that drivers were engaged in potentially distracting activities for approximately 52% of the time that they spent driving (63.1 h out of a total of 121 h driving time). The most commonly occurring distracting events included drivers scratching their heads, yawning, and coughing, followed by eating, drinking, smoking, and personal grooming. In addition, the analysis revealed that long- and short-haul truck drivers spent almost one-half of their driving time engaged in either talking, eating, drinking, or smoking. Distracting activities that were found to reduce the amount of time that drivers spent looking at the road ahead included tuning the radio, conversations with a passenger, and talking on a cell phone.

Chen et al.[12] used a simulator experiment to investigate the effects of using a mobile phone and a wireless communication device on bus driver performance. They found that both devices increased the bus driver's response times to sudden events, such as a lead vehicle braking suddenly.

The small amount of literature available on driver distraction in the public and commercial transport sectors seems to indicate that in addition to the distracting factors that are relevant during conventional driving, public transport and commercial drivers also face a range of additional distractions deriving from work-related technology (e.g., CB radios) and procedures (e.g., communicating with other drivers or a control center). Clearly, further research into the sources and effects of driver distraction in the commercial and public transport sectors is required. In particular, it is our opinion that there is a requirement for additional research in two key areas. First, a comprehensive taxonomy of distraction elements is required, for both conventional and public and commercial sector driving (see Chapter 15 for further discussion of this issue). Such a taxonomy is required to inform the design of data collection and monitoring systems and the design of vehicles, road infrastructure, training programs, and procedures that minimize the potential for driver distraction. Second, a reliable and valid methodology for assessing driver distraction deriving from various sources is required. Such a methodology should be structured so that it allows analysts to identify and classify distracting elements, determine the extent to which these may distract drivers, and evaluate their effects on driver behavior.

The authors recently conducted a study that addressed both of these requirements. In response to a number of recent bus driver distraction–related incidents, an NSW-based transport company commissioned MUARC to assess the risks associated with its bus drivers engaging in potentially distracting activities while driving. Specifically, the transport company wanted to identify the different sources of distraction that currently affect their bus drivers and subsequently the level of risk associated with each source of distraction. An overview of the research, including the methodology employed and the findings derived from these investigations, is presented in the following section.

14.4 CASE STUDY ON BUS DRIVER DISTRACTION

14.4.1 Methodology

The research undertaken during the study was partitioned into three phases, as represented by Figure 14.1. The first phase involved identifying and understanding the different tasks that the bus drivers are currently required to undertake while operating the buses. This understanding of the bus operation task was then used to inform the second phase of the methodology, which was used to identify the various different sources of

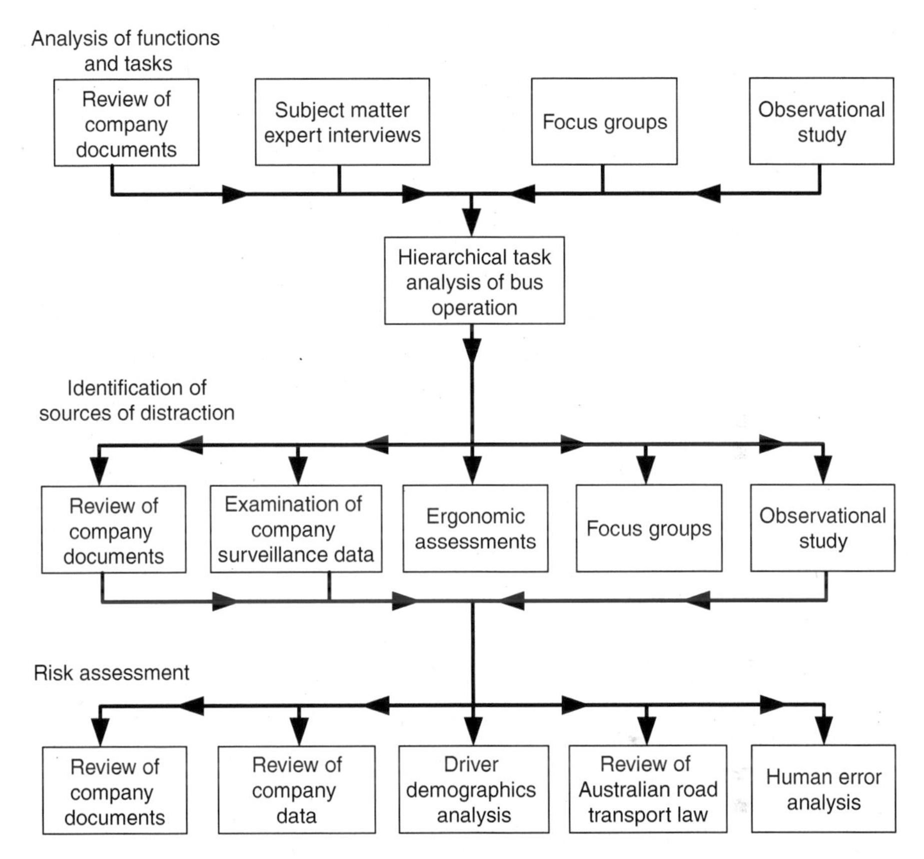

FIGURE 14.1 Bus driver distraction assessment methodology.

distraction present during the bus operation task. Finally, in line with the transport company's requirements, the third phase involved making an assessment of the level of risk associated with each of the distraction sources identified. A brief description of each of the approaches used within each phase is given in the following text.

1. *Analysis of functions and tasks undertaken by company bus drivers.* The first phase of this research involved the identification of the different tasks, both driving and nondriving, that bus drivers currently undertake while operating buses. This involved a review of relevant company documents, including training manuals and standard operating procedures, the conduct of interviews with various subject matter experts (SMEs; driver training personnel, experienced bus drivers, etc.), the conduct of a focus group discussion with a sample of current bus drivers, and the conduct of observational studies of bus driving on a range of representative routes. The information derived from this work was then used to inform the development of a hierarchical task analysis (HTA)[13] of bus operation. HTA is used to describe systems in terms of the goals, subgoals, and physical and cognitive operations required to

achieve them, including the goal-based human-machine interactions (HMI) required during task performance. It was therefore felt that describing the bus operation "system" in this manner would inform the identification of potential sources of driver distraction, since it would offer a description of the goals associated with each task, the interfaces or artifacts involved in the performance of the task, and the physical and cognitive operations required.

2. *Identification of actual and potential sources of driver distraction.* The next phase of the research involved the identification and documentation of the various sources of bus driver distraction that bus drivers are exposed to while operating buses. This phase involved a range of activities, including the conduct of a focus group discussion with a sample of company bus drivers, observational studies of bus driving on a range of representative routes, a review of the bus driving HTA, the conduct of ergonomic assessments of a range of buses that are currently used by company bus drivers, examination of relevant company documentation (e.g., policies on distraction), and a review of the range of technologies and configurations currently used by company bus drivers while operating buses. Each of the approaches used during this phase was focused on the identification of sources of distraction present during bus operation.
3. *Risk assessment.* The final phase of this research involved the assessment of risk due to company bus driver exposure to distractions. This involved the ergonomic assessment of a range of buses that are currently used by company bus drivers and a company "response vehicle" (used to respond to bus breakdowns and emergencies), a review of existing company procedures and policies on distraction, a review of current incident data held by the company on driver distractions and their contribution to crashes and incidents, a review of the demographic characteristics of company bus drivers, a review of current Australian laws relating to the use of potentially distracting devices while driving, and the conduct of a distraction-based human error identification (HEI) analysis for bus operation.

14.4.2 Results

14.4.2.1 Bus Driver Task Analysis

On the basis of the focus group, SME interview, and observational study, an HTA of bus operation was constructed. An extract of the HTA is presented in Figure 14.2. To summarize the HTA for the project stakeholders, seven categories of tasks that the bus drivers currently perform while operating buses were identified. These were preparation tasks, physical vehicle control tasks, cognitive vehicle control tasks, route/timetabling tasks, passenger-related tasks, communication tasks, and personal comfort tasks. A description of each task category is presented in the following text.

1. *Preparation tasks.* Preparation tasks include tasks that the bus driver performs to prepare the bus prior to setting off on a particular route. These include the conduct of a series of predeparture checks designed to determine whether the bus is roadworthy and is performing appropriately, such as

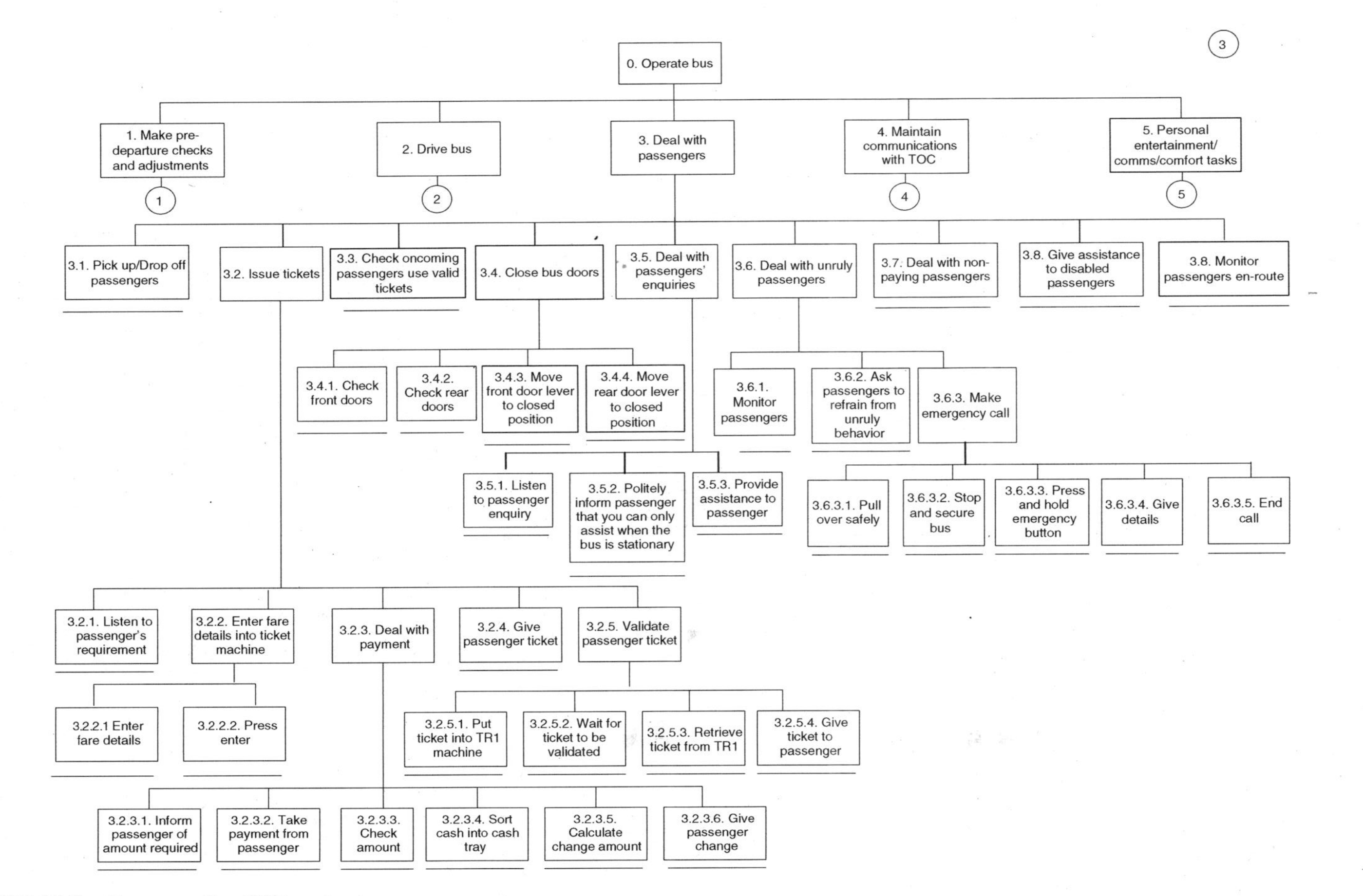

FIGURE 14.2 Bus operation HTA extract.

making engine compartment and electrical checks, checking the wheels and tires, and checking the vehicle posture, leaks, and loads. Bus drivers are also required to perform a walk-through of the bus to check instrumentation and identify any protruding objects. Once the predeparture checks are complete, the bus drivers make any predriving adjustments required. These include adjusting the seat, checking and adjusting the mirrors and visor, familiarizing themselves with the required route, and modifying the destination board.

2. *Physical vehicle control tasks.* The physical vehicle control tasks category includes the physical tasks that the bus driver has to perform while driving the bus. These include steering the bus, operating the accelerator and brake pedals, changing gears, and operating indicators and other vehicle controls.
3. *Cognitive vehicle control tasks.* The cognitive vehicle control tasks category includes the cognitive tasks that the bus driver has to perform while driving the bus. These include planning, checking the mirrors, monitoring other road users and pedestrians, forecasting and anticipating other road users' behavior, navigation, perceptual and decision-making tasks, and tasks required for situation awareness achievement and maintenance.
4. *Route/time-tabling tasks.* The route/time-tabling tasks category includes tasks that the bus driver is required to perform to keep to the desired route and timetable. These include checking the route journal and planning the route, entering the section points on the ticket machine, and also checking the current time against the time specified by the route journal.
5. *Passenger-related tasks.* The passenger-related tasks category includes the tasks that the bus driver is required to perform when dealing with passengers. These include opening and closing the bus doors, lowering and raising the bus, operating the ticket machine and issuing tickets, changing section points, checking tickets, monitoring passengers, and assisting passengers.
6. *Communication tasks.* The communication tasks category includes the tasks that the bus driver has to perform to maintain communications with the transport operations center. These include listening to general and personal broadcasts, using the radio and handset to initiate communication with the transport operations center, reporting incidents, and making emergency calls.
7. *Personal comfort tasks.* The personal comfort tasks category includes the tasks that the bus driver performs to maintain personal comfort while driving the bus. These include making adjustments to the seat, sun visor, mirrors, and driving controls, drinking and eating, and using personal entertainment equipment (e.g., portable radio).

Aside from the preparation tasks that are completed by the bus driver prior to setting off, all of the other tasks identified can be performed by the bus driver while he or she is driving the bus. According to company policy, rules, and regulations, only the physical vehicle control tasks and the cognitive vehicle control tasks should be performed while driving the bus (i.e., company policy prohibits the drivers from performing the other tasks while driving). Therefore, the other tasks are all secondary tasks that the driver should, according to company policy, rules, and regulations, perform while the bus is stationary. Data collected from the focus groups indicate, however, that a significant proportion of company bus drivers currently undertake at least some of these tasks while driving the bus. It is therefore feasible to conclude that these tasks represent a potential source of distraction to bus drivers.

It was hypothesized that the performance of passenger-related tasks, communication tasks, and personal comfort tasks while driving may result in the drivers being distracted from the primary task of driving the bus safely. We felt that the main sources of bus driver distraction are likely to arise from the passenger-related, communication, and personal comfort tasks that the bus driver performs while driving the bus and from the technologies used when performing these tasks. To investigate this further we used the data collected to identify the different sources of distraction.

14.4.2.2 Sources of Distraction for Bus Drivers

From the data collected during the focus group, SME interviews, HTA and an observational study, a taxonomy of bus driver distraction sources was constructed. The taxonomy of distraction sources contains all of the different potential sources of distraction that were identified during the study. The potential sources of distraction identified were categorized into seven main categories (see Figure 14.3). A brief summary of each category is presented in the following text.

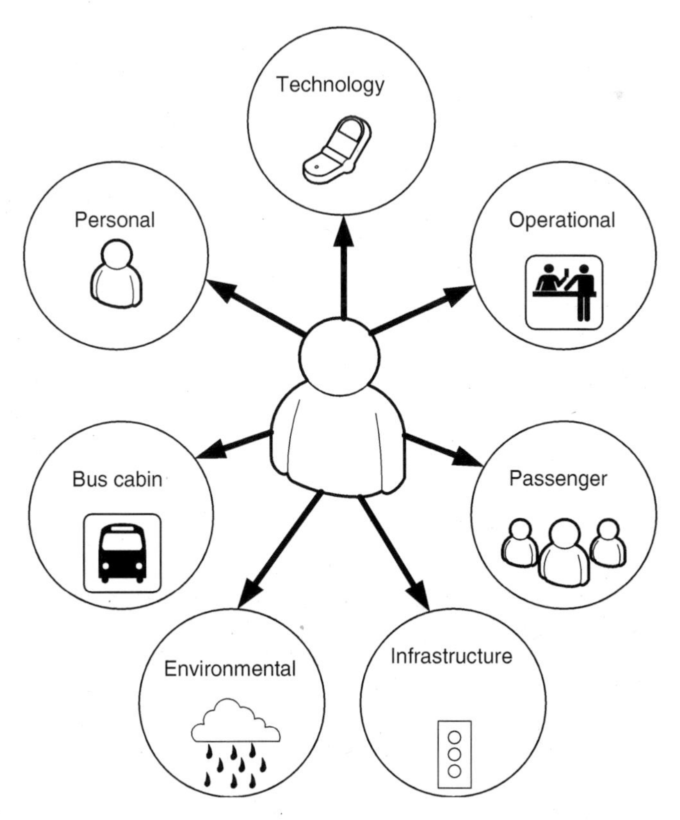

FIGURE 14.3 Sources of bus driver distraction.

1. *Technology-related distractions.* Technology-related sources of distraction include any technological devices that the driver interacts with while driving the bus, including mobile phones, CD players, the broadcast radio and handset, and the ticket machine.
2. *Operational distractions.* Operational sources of distraction include any aspects of bus operation that may be distracting, including operating the ticket machine, communicating with the transport operation center (TOC), listening to general and personal broadcasts, and reading or modifying the route journal.
3. *Passenger-related distractions.* Passenger-related sources of distraction include any aspects of managing passengers that can potentially distract the bus driver, including passenger conversations, monitoring passengers, issuing tickets, providing passenger assistance, dealing with unruly passengers, and passengers talking loudly on their mobile phones.
4. *Environmental distractions.* Environmental sources of distraction include any environmental conditions that might distract the driver, such as weather conditions (e.g., glare from the sun). Environmental conditions can become a distraction if they induce the driver to perform activities to reduce the discomfort brought about by the conditions, such as adjusting the climate controls or sun visor.
5. *Bus cabin–related distractions.* Bus cabin–related sources of distraction include any features of the particular bus and bus cabin in question that might distract the driver, including annoying rattles (e.g., cabin door, ticket machine) and adjusting the sun visor, seat, controls, and the seat belt.
6. *Infrastructure-related distractions.* Infrastructure-related sources of distraction include any features of the road infrastructure that the driver might find distracting, such as roadside advertising (e.g., on bus stops and vehicles).
7. *Personal distractions.* Personal sources of distraction include any personal factors that might distract the bus driver or make the bus driver more susceptible to distraction, such as fatigue, incapacitation, and medication, although, as pointed out in Chapter 21 of this book, the mechanisms by which these factors may mediate the effects of distraction are not yet understood.

The sources of distraction within each category are presented in Table 14.1. Within the table, those sources of distraction that are representative of violational activity (i.e., activities prohibited by company policy while the vehicle is in motion) are marked with a "V."

14.4.2.3 Ergonomic Assessment

An ergonomic assessment of the different buses that the company bus drivers currently operate was undertaken. This involved three human factors experts undertaking a walk-through-type assessment of three buses in the company of a SME. The experts

TABLE 14.1
Sources of Bus Driver Distraction

Bus Driver Distraction Sources Taxonomy						
Technology	**Operational**	**Passenger**	**Environmental**	**Bus Cabin**	**Infrastructure**	**Personal**
Radio (V)	Issuing tickets (V)	Passenger conversations	Weather conditions (e.g., glare)	Rattles (e.g., ticket machine)	Advertising	Fatigue
Radio handset (V)	General broadcasts	Passenger inquiries		Faulty sun visor	Lane width	Incapacitation
Ticket machine (V)	Personal broadcasts	Talking to passengers (V)		Adjusting seat	Road layout	Sickness
Mobile phone (V)	Recording broadcast details (V)	Unruly passengers		Adjusting seat belt	Road signage	Medication
Personal entertainment (V)	Communicating with TOC (V)	Nonpaying passengers		Adjusting steering column		Inexperience
Passenger technology	Adhering to timetable	School children		Operating climate controls		Eating
	Reading route journal (V)	Elderly passengers				Drinking
	Amending route journal (V)	Disabled passengers				
	Changing route section points	Passengers with infants				
	Bus stopping alert	Issuing tickets				
	Hand brake warning alert	Monitoring bus stops for waiting passengers				
	Raising/lowering bus (V)	Assisting passengers (V)				
	Opening and closing bus doors (V)					

Note: Violational activities refer to those activities that are prohibited by company policy to be performed while the bus is in motion. Bus drivers are instructed to only perform these tasks while the bus is stationary, thus performance of them while driving is deemed here to be a violational activity.

were taken through the tasks that the bus drivers currently perform while driving the bus and interacted with the bus controls and other devices accordingly. Instances of poor ergonomic design that could potentially lead to the drivers becoming distracted were recorded and subsequently verified by the SME. The findings from the ergonomic assessment of three bus types indicated that a range of devices on board the buses could potentially distract the driver while driving the bus. These included the broadcast radio display and handset, ticket machine, sun visor, and also audible rattles within the cabin, such as the driver cabin door and the ticket machine. A number of instances of poor ergonomic design that could potentially impair driving performance were also identified, including poor placement of vehicle controls, poor functionality and control of the bus door controls, use of nonretractable seatbelts, and little navigation assistance (which could be enhanced through the provision of route navigation systems). Finally, a number of audible alerts that could potentially distract the driver while driving were identified, including the bus-stopping alert, wheelchair ramp alert, faulty ticket machine alert, general and personal broadcast alerts (and the general broadcasts themselves), and oil pressure alert. Although these alerts are part of normal bus driving operation, it was still felt that they could potentially distract the drivers from the primary task of driving the bus safely.

A distraction checklist, developed by the authors, was used to assess the level of distraction imposed by the different devices on board the buses. Use of the checklist suggested that, of the non–driving-related devices on board the buses, the ticket machine and the broadcast radio display and handset devices could be expected to impose the greatest level of distraction when they are used by bus drivers while driving the bus. A rigorous quantitative assessment of the actual level of distraction imposed by the different devices and activities was not possible within the constraints of the study.

14.4.2.4 Human Error Identification Analysis

Human error identification (HEI) techniques allow analysts to predict potential errors that might arise during a particular human-machine interaction (HMI) and operates on the premise that an understanding of an employee's work task and the characteristics of the technology being used allows analysts to indicate potential errors that may arise from the resulting interaction.[14] A distraction-based HEI analysis was conducted using the HTA of bus operation as its input. The Systematic Human Error Reduction and Prediction Approach (SHERPA)[15] was used to predict potential errors that might arise when the bus driver is distracted visually, physically, or attentionally. Each bottom-level task step from the HTA was first classified as one of the five SHERPA behavior types (action, check, information retrieval, information communication, and selection). Each SHERPA behavior classification has a set of associated errors. The SHERPA error taxonomy is presented in Figure 14.4. The SHERPA error taxonomy and domain expertise are then used to identify, on the basis of the analyst's subjective judgment, any credible error modes for the task step in question. For each credible error identified, a description of the form that the error would take is provided, and the analyst describes any consequences associated with the error and any error recovery steps that would need to be taken

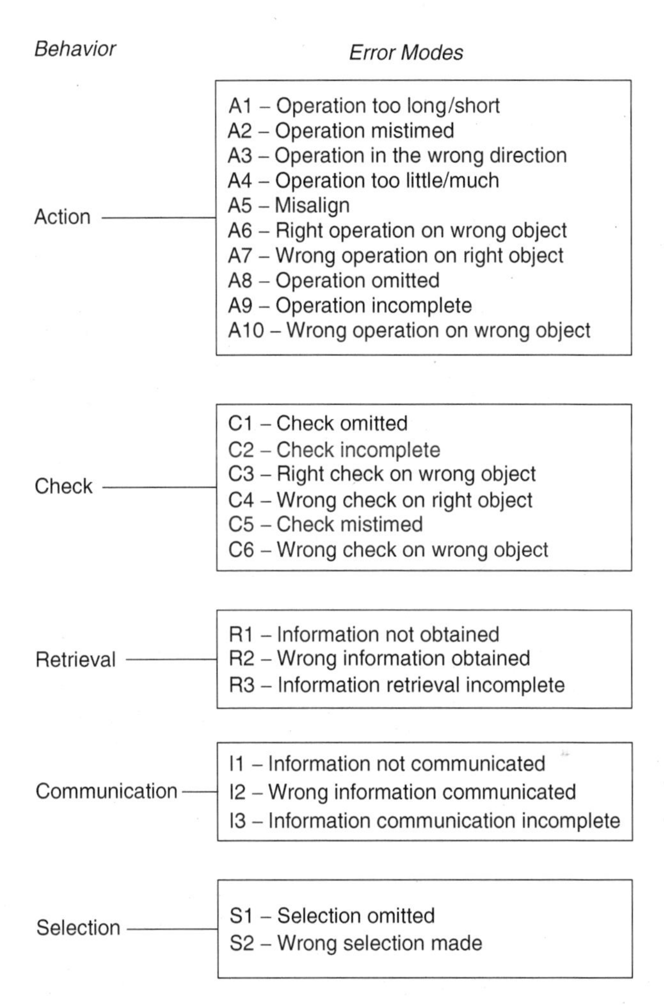

FIGURE 14.4 SHERPA error taxonomy.

in the event of the error occurring. Ratings of ordinal probability (low, medium, or high) and criticality (low, medium, or high) are then provided. The final step involves specifying any potential design remedies (i.e., how the interface or device might be modified to remove or reduce the chances of the error occurring) for each of the errors identified.

To identify the errors that could potentially arise when the driver is distracted, a modified SHERPA analysis was conducted. This involved taking each bottom-level task step from the HTA and predicting the driver errors that might arise in the event of the driver being either physically, visually, or attentionally distracted while performing the task in question (the term "physical distraction" is used here to denote physical interference). An extract of the SHERPA distraction analysis is presented in Table 14.2.

TABLE 14.2
SHERPA Distraction Analysis Extract[a]

Task Step	Distraction	Error Mode	Description	Consequence	Recovery	P	C	Remedial Measures
2.2.1.1 + 2.2.1.2	Physical							
	Visual	C1	Driver fails to check the front and back doors before closing them due to being visually distracted	There may be passengers located in or around the doors as the driver closes them		M	H	Audible prompt to check doors before opening/closing Intelligent doors that automatically open/close
	Mental	C1	Driver fails to check the front and back doors before closing them due to being visually distracted	There may be passengers located in or around the doors as the driver closes them		M	H	Audible prompt to check doors before opening/closing Intelligent doors that automatically open/close
2.2.1.3 + 2.2.1.4	Physical	A3	Driver moves door operation lever in the wrong direction	Doors remain open and do not close		L	M	Intelligent doors that automatically open/close
	Visual	A3	Driver moves door operation lever in wrong direction	Doors remain open and do not close		L	M	Intelligent doors that automatically open/close
	Mental	A8	Driver fails to close the doors before attempting to pull away	The bus will automatically stop as the doors are still open		L	M	Intelligent doors that automatically open/close
2.2.4.1 – 2.2.4.4	Physical							
	Visual	C1	Driver fails to check mirrors before pulling away	Driver may not see a pedestrian or other road user that is in close proximity to the bus		H	H	Audible prompt to check mirrors before pulling away
	Mental	C1	Driver fails to check mirrors before pulling away	Driver may not see a pedestrian or other road user that is in close proximity to the bus		M	H	Audible prompt to check mirrors before pulling away

[a] P = Probability, C = Criticality, L = Low, M = Medium, H = High

The SHERPA analysis of bus operation indicated that a number of *safety-critical* and *operational* errors could potentially be made by drivers who are physically, cognitively, or visually distracted while driving the bus. Remedial measures for the errors identified included a number of design changes to the bus cabin, the use of intelligent transport systems (ITS) technology (e.g., intelligent speed adaptation systems, following distance warning systems, lane departure warning systems, automatic lane-keeping systems, and route navigation systems) to automate some of the bus operation tasks, and the use of training on coping with the effects of distraction. Other remedial measures included the use of training and company policy to remove instances of violational activities. However, it was also concluded that a number of the errors identified would require the development of novel ITS technologies and driver support systems, such as close-proximity warning systems and situation awareness displays that present pedestrian and other road user location information to the driver.

14.5 CONCLUSIONS

This case study investigated the problem of distraction with a view to identifying the sources of distraction that have potential to degrade bus driver performance and safety. In conclusion, the findings suggest that there are a number of sources of distraction that could potentially distract bus drivers while driving buses. These include not only those that are present during conventional driving, such as eating, drinking, and roadside advertisement, but also an additional set of distraction sources that are present because of the requirements associated with bus operation, such as bus passenger–related distractions and ticketing-related distractions.

As a result of our investigations, we identified the following categories of distraction sources for bus drivers in the company in question: technology-related distractions, operational distractions, passenger-related distractions, environmental distractions, bus cabin–related distractions, infrastructure-related distractions, and personal distractions. Each of these categories was decomposed, on the basis of the data collected during our investigations, to create a taxonomy of distracting sources. The taxonomy of sources of bus driver distractions contains 51 sources of distraction that could potentially distract bus drivers from the primary task of driving the bus. Of these 51 sources, 15 were classified as representing instances where the driver is engaging in violational (i.e., prohibited by company policy, rules, and regulations) activities, such as using a mobile phone or conversing with a passenger while driving. In developing distraction mitigation strategies, it would seem appropriate to focus first on these sources. The taxonomy presented here is, however, tentative. It requires further refinement, taking into account the various issues discussed in Chapters 15 and 16 of this book.

In conclusion, the case study findings suggest that bus driver distraction is a potentially significant road safety problem within the public transport sector. The risk of distraction is particularly significant in this domain when drivers are compelled, as part of their job, to perform additional secondary tasks over and above the primary task of driving the vehicle.

This research suggests that the requirement for bus drivers to take on multiple, and at times competing, roles while driving may be exacerbating the problem of driver distraction in the public transport sector. In addition to driving the bus, bus drivers have a range of other tasks that they are required to undertake, including dealing with passengers and communicating with the TOC. A significant proportion of the distraction sources identified during this research arose from these additional tasks (over and above driving the bus) that the bus drivers are required to undertake as part of their job. In a sense, this is no different from the job of the fighter pilot, who must fly the plane while operating weapons and other tactical systems. However, there are some critical differences between the two domains. As noted by Regan,[6] additional tasks, such as operating weapons systems, are, for the pilot, regarded as an integral part of the job, and they are well prepared to deal with them. Fighter pilots are very carefully selected for their jobs—generally, only those who are superior at performing two or more tasks at the same time are chosen to fly. In addition, they are given proper training in how to perform multiple tasks at the same time. If during the design of a fighter aircraft it is determined that all of the tasks that have to be performed, even with automation, are too much for one pilot to cope with, then the aircraft is designed for two pilots—so that what those in the road safety community might regard as the "distracter tasks" can be shared or delegated to the copilot or navigator. Clearly, appropriate training and good interface design are also essential to support bus drivers if they are expected to take on roles other than that of driving the bus.

Our findings are also corroborated by other researchers' investigations into distraction in the public and commercial transport sectors. For example, Hanowski et al.,[9] Bunn et al.,[10] and Barr et al.[11] all concluded that distraction represents a problem in the public and commercial transport sectors. Hanowski et al. also concluded that operating a CB radio and completing paperwork were significant sources of distraction for long-haul truck drivers, whereas Chen et al.[12] and Barr et al. identified mobile phone use as a significant source of distraction.

Simple steps can be taken, however, by the transport company and the wider public transport sector to reduce the majority of the distractions identified. Distraction sources that are present during violational activities can be mitigated through the development and strict enforcement of company policy, rules, and regulations and the provision of training programs designed to discourage drivers from engaging in such activities and to cope with those that cannot be avoided. For example, the company in question has banned the use of *all* mobile phones (including handsfree) by drivers while driving the buses. Training and procedural design can also be used to reduce driver interactions with distracting sources while driving. Simple ergonomic bus cabin design and efficient maintenance procedures can also be used to remove sources of distraction within the bus cabin, such as faulty sun visors and annoying rattles. Finally, the provision of ITS within the cockpit could also be used to mitigate the effects of distraction. Systems such as intelligent speed adaptation, following distance warning, lane departure warning, automatic lane-keeping systems, and route navigation systems could all be used to mitigate the effects of distraction, by reducing driver workload.

This study represents the first known attempt to investigate bus driver distraction, and much further research effort is required in this area. Although it is clear, from the literature reviewed and the current study, that distraction may be a significant safety problem in the public and commercial transport sectors, our knowledge of the extent, nature, and actual impact of the problem is in its infancy. Much further research is required to identify and classify the sources of driver distraction that exist in the public and commercial transport sectors and to quantify their impact on driving performance and safety to inform and guide effective countermeasure development.

REFERENCES

1. Stutts, J.C., Reinfurt, D.W., Staplin, L., and Rodgman, E.A., *The Role of Driver Distraction in Traffic Crashes*, AAA Foundation for Traffic Safety, Washington, DC, 2001.
2. Wang, J.S., Knipling, R.R., and Goodman, M.J., The role of driver inattention in crashes: New statistics from the 1995 Crashworthiness Data System, *40th Annual Proceedings of the Association for the Advancement of Automotive Medicine*, Vancouver, Canada, 1996.
3. Klauer, S.G., Dingus, T.A., Neale, V.L., Sudweeks, J.D., and Ramsey, D.J., *The Impact of Driver Inattention on Near-Crash/Crash Risk: An Analysis Using the 100-Car Naturalistic Driving Study Data*, Virginia Tech Transportation Institute, Blacksburg, Virginia, 2006.
4. Strayer, D.L., Drews, F.A., and Crouch, D.J., A comparison of the cell phone driver and the drunk driver, *Human Factors* 48(2), 381, 2006.
5. Treat, J.R., *A Study of Precrash Factors Involved in Traffic Accidents*, Centre for Automotive Safety Research, Adelaide, 1980.
6. Regan, M.A., Driver distraction: Reflections on the past, present and future, *Journal of the Australasian College of Road Safety* 16(2), 22–33, 2005.
7. Horberry, T., Anderson, J., Regan, M.A., Triggs, T.J., and Brown, J., Driver distraction: The effects of concurrent in-vehicle tasks, road environment complexity and age on driving performance, *Accident Analysis and Prevention* 38(1), 185–191, 2006.
8. Stutts, J., Feaganes, J., Reinfurt, D., Rodgman, E., Hamlett, C., Gish, K., and Staplin, L., Drivers' exposure to distractions in their natural driving environment, *Accident Analysis and Prevention* 37, 1093–1101, 2005.
9. Hanowski, R.J., Perez, M.A., and Dingus, T.A., Driver distraction in long-haul truck drivers, *Transportation Research Part F: Traffic Psychology and Behaviour* 8(6), 441–458, 2005.
10. Bunn, T.L., Slavova, S., Struttmann, T.W., and Browning, S.R., Sleepiness/fatigue and distraction/inattention as factors for fatal versus nonfatal commercial motor vehicle driver injuries, *Accident Analysis and Prevention* 37, 862–869, 2005.
11. Barr, L.C., Yang, D.C.Y., and Ranney, T.A., Exploratory analysis of truck driver distraction using naturalistic driving data, *Proceedings of the 82nd Annual Meeting of the Transportation Research Board*, Washington, DC, 2003.
12. Chen, W.-H., Lin, T.-W., Su, J.-M., Lee, S.-W., Hwang, S.-L., Hsu, C.-C., and Lin, C.-Y., The effect of using in-vehicle communication system on bus drivers' performance in car-following tasks, *13th World Congress on Intelligent Transport Systems*, London, UK, 2006.
13. Stanton, N.A., Hierarchical task analysis: Developments, applications, and extensions, *Applied Ergonomics* 37, 55–79, 2006.
14. Stanton, N.A. and Baber, C., A systems approach to human error identification, *Safety Science* 22, 215–228, 1996.
15. Embrey, D.E., SHERPA: A systematic human error reduction and prediction approach, *The International Meeting on Advances in Nuclear Power Systems*, Knoxville, Tennessee, 1986.

Part 5

Distraction, Crashes, and Crash Risk

15 Sources of Driver Distraction

Michael A. Regan, Kristie L. Young, John D. Lee, and Craig P. Gordon

CONTENTS

15.1 INTRODUCTION

In determining to what extent driver distraction contributes to road trauma, it is essential to properly define and categorize the potential sources of distraction that exist, inside and outside the vehicle.

Given that distraction, as a concept, has been inconsistently defined in the literature (see Chapter 3), it is not surprising that there has been, between research studies and across time, considerable variation in what is, and is not, considered to be a source of distraction; and in how these sources of distraction are categorized. Consequently, there is currently no common taxonomic description of the sources and categories of driver distraction that exist at this stage in the evolution of the human-vehicle-highway system. This lack of agreement in categorization makes difficult the comparisons between specific sources of distraction and their impact on performance and safety, across studies and over time.

In this chapter we present a taxonomic description of the various sources of driver distraction that are currently known to exist. The taxonomy is derived from selected major studies that have made use of secondary crash-based data sources collected by agencies (typically the police), from selected major observational studies, and from the definition of driver distraction presented in Chapters 1 and 3 of this book. The taxonomy is intended to serve several purposes:

- To begin to resolve confusion about what are, and are not, sources of distraction inside and outside the vehicle
- To provide a common framework for categorizing sources of distraction

- To support the development of less variable protocols for collecting and categorizing crash and epidemiological data
- To provide a framework for predicting the potential impact of these sources of distraction on driving performance
- To provide a basis for the development of checklists that can be used to assess driver distraction

The chapter concludes with a final word on how the taxonomy might be expected to change as the driving task continues to evolve. Further discussion of taxonomic issues relevant to the classification of distraction-related crash data can be found in Chapter 16 of this book.

15.2 METHODS FOR IDENTIFYING SOURCES OF DISTRACTION

There are various complementary methods that can be used to identify and categorize the various potential sources of distraction that exist. The most obvious ones are discussed in the following paragraphs. Gordon, in Chapter 16 of this book, also provides some discussion of this issue.

One vitally important way is to define distraction and then use this definition *a priori* to determine what should, and should not, be regarded as a potential source of distraction. For example, in this book we define distraction as *the diversion of attention away from activities critical for safe driving toward a competing activity* (see Chapters 1 and 3). A critical assumption underlying this definition, as discussed in Chapter 3, is that some aspects of the driving task itself have potential to distract the driver. For example, according to this definition, adjusting the sun visor while driving, even though it is a driving-related activity, has the potential to draw attention away from activities critical for safe driving (e.g., detecting that the vehicle ahead has suddenly braked) and hence can be regarded as a potential source of distraction. As noted in Chapters 3 and 4, driving is a complex, multitask activity, making it likely that the demands of one element of driving will interfere with another. Considering "driving" as a single activity in defining distraction oversimplifies a complex activity and neglects important driving-related distractions drivers must manage.

Another way to create a taxonomy of distraction sources is to carefully sift through crash reports, such as police report forms and reports from in-depth crash studies, to identify sources of distraction (e.g., "using mobile phone") that have been identified as contributing factors in crashes. The problem here is that the range of potential sources of distraction is limited by what is voluntarily reported by crash victims, by the design of police report forms (by the extent to which, e.g., the report form lists as prompts different sources of distraction), and by the knowledge and expertise of police and crash investigation teams in knowing what questions to ask about the potential role of distraction as a contributing factor.

Observational studies, such as the 100-car naturalistic driving study,[1] provide another means of collecting data on potential distraction sources. Here, it is theoretically possible, through the use of video and other sensors, to observe overt driver involvement in all activities that compete with activities critical for safe driving and to record these. This technique is mainly limited by the number and types of sensors that can be used to record these interactions, identifying what outside-the-vehicle sources the driver is looking at, the difficulties associated with recording the covert diversion of attention toward competing activities, and the possible reluctance of drivers to divert their attention toward some competing activities (e.g., illegal activity) knowing that their behavior is being monitored.

15.3 CODING AND CATEGORIZATION OF DISTRACTION SOURCES

A source of distraction can be thought of as comprising two elements: a physical event or object (e.g., mobile phone) and an action of some kind that is performed on that object (e.g., dialing, talking, and listening). It is important to distinguish between these two elements, as the degree and type of distraction that derives from an object depends on the action that is performed on it. Listening to a mobile phone, for example, has been shown to interfere less with lane keeping than using it to dial a number (see Chapter 11).

In this chapter, we have reviewed the descriptions of sources of distraction reported in seven studies—five crash studies and two observational studies—that have been cited as providing evidence for the role of distraction as a contributing factor in crashes and near-crashes. From these studies, we distilled a list of actions that were identified as being associated with the various sources of distraction cited. These are shown in Table 15.1, along with the studies in which they were reported. These actions are not exhaustive, but they represent the range of actions performed by drivers of current-generation vehicles when engaging in activities that have reportedly contributed to crashes. It can be seen in Table 15.1 that there is some degree of overlap between the items in the list. The actions "manipulating," and "using," for example, share some common elements.

Noteworthy here is a generic taxonomy of task actions developed some time ago by Miller[8] (cited in Ref. 9). Miller's taxonomy, shown in Table 15.2, contains a description of the generic information transformations and control activities that are needed for system operation. These descriptions were derived from a project undertaken for the U.S. Air Force concerned with the development of a generalized taxonomy of human performance. It can be seen that there is virtually no overlap between the list of actions identified in Table 15.1 and those listed in Table 15.2. There are several reasons for this. First, the actions in Table 15.2 are largely more perceptual and cognitive in nature than those in Table 15.1. Second, many of the actions in Table 15.2 are performed covertly and hence are not likely to be captured on police report forms and even in observational studies such as the 100-car naturalistic driving study[1]. Finally, many of the terms in Table 15.2 represent actions, which are in fact subcomponents of the actions in Table 15.1. For example, the term "checking" in Table 15.1 could involve performance of several of the actions in Table 15.2: input select, filter, search, store, store in buffer, and decide/select. Reference will be made to this table again later in the chapter.

From the previously mentioned studies, it is also possible to compile and categorize the full range of reported objects and events on which one or more of the actions mentioned in Table 15.1 were performed. These are listed in Table 15.3. Again, these

TABLE 15.1
Driver Actions Associated with Sources of Distraction Cited in the Studies Reviewed

Actions	Studies in Which Cited
Adjusting	S1, S3, GE, JS, G1
Answering	S1, S3, JS, G1, K
Applying (e.g., makeup)	K
Arguing	S1, JS, G1
Attending	JS, G1
Biting (e.g., nails, cuticles)	K
Brushing (e.g., hair, teeth)	K
Changing	GE, JS
Checking (e.g., speedometer/traffic)	G1, G2, K
Cleaning	K
Combing	K
Conversing	S1, S3, G1, K
Dancing (e.g., in car seat)	K
Dialing (manually or using voice)	S1, S3, GE, G1, K
Drinking	S1, S3, GE, JS, G1, K
Dropping	S1, S3, JS, G1
Eating/dining	S1, S3, GE, JS, G1, K
Extinguishing (e.g., cigarette)	S3, K
Fixing (e.g., hair)	K
Flossing (i.e., teeth)	K
Grooming	S3, GE
Inserting (e.g., cassette, CD, contact lenses)	K
Leaning	S3
Lighting (e.g., cigarette)	S1, S3, JS, K
Listening	S3
Locating	S1, S3, GE, JS, G1, K
Looking	S1, S3, GE, JS, G1, K
Thinking (daydreaming, lost in thought)	GE, S1, JS, G1
Manipulating	S3, G1
Moving	G1
Observing (e.g., traffic)	K
Picking up	S3
Praying	GE
Preparing (e.g., to eat)	S3
Pressing	K
Putting on (e.g., jewelery)	K
Reaching	S1, S3, GE, JS, G1, G2, K
Reading	S3, K
Removing (e.g., jewelery)	K
Retrieving (e.g., cassette, CD)	K
Searching	G1, G2
Shaving	K

(continued)

TABLE 15.1 (Continued)

Actions	Studies in Which Cited
Smoking (light, smoke, extinguish)	S1, S3, GE, JS, G1, G2, K
Screening (e.g., calls)	K
Singing	K
Spilling	S3, JS
Styling (e.g., hair)	K
Swatting (e.g., a fly)	G1
Talking	S1, S3, GE, JS, K
Texting	G1
Using	S1, S3, GE, JS, G1, K
Viewing	K
Writing	S3, GE

Note: S1, Stutts et al.[2]; S3, Stutts et al.[3]; G1, Gordon[5]; G2, Gordon[5]; K, Klauer et al.[1]; GE, Glaze and Ellis[6]; JS, Joint State Commission.[7]

events and objects are not exhaustive, but are probably representative of the range of events and objects in current-generation vehicles on which one or more actions are performed by drivers when engaging in distracting activities. There was some overlap between the events and the objects identified from the studies reviewed. To overcome this, overlapping items have been incorporated as a subset of superordinate items. For example, "dental floss" appears as a subset of "personal hygiene accessory." It will be noted that under the major source labeled "Internalized activity" we have included several actions: coughing, dancing in seat, itching, praying, singing (to oneself), sneezing, and talking (to self). These are included here, rather than in Table 15.1, given that they are internally derived actions, which involve no overt interaction with any of the objects or events in Table 15.3.

There is a tendency across the studies reviewed to confound the reporting of events, objects, and actions as sources of distraction. In the Glaze and Ellis study,[6] for example, "looking at other people," "eyes not on road," "objects on the road," and "automobile mechanical problem" are all cited as sources of distraction under the category "other distractions outside the vehicle." Presumably for those sources for which no associated action is reported (e.g., objects on the road), there was insufficient data to determine what action was being performed at the time of the incident. As noted by Gordon in Chapter 16, some studies also confound the reporting of some categories of distraction in a way that mixes inside- and outside-the-vehicle sources of distraction within the same category.

From this simple exercise we were able to distil, into the following six broad categories, the various events and objects on which one or more actions were performed by the drivers in the studies reviewed: "things brought into vehicle," "vehicle systems," "vehicle occupants," "moving object or animal in vehicle," "internalized activity," "external objects, events or activities," and "other sources of distraction." The latter category is intended to accommodate sources of distraction that are yet to evolve or be discovered (see Section 15.3.5).

TABLE 15.2
Miller's (1974) Cognitive Task Transactions and the Human Information Processing Resources They Draw Upon

Cognitive Agent Task	General Category of Information Processing	Human Information Processing Resources
1. Input select. Selecting what to pay attention to next	Acquisition	Selective attention Perceptual sensitivity
2. Filter. Straining out what does not matter	Acquisition	Selective attention
3. Detect. Is something there?	Acquisition	Perceptual sensitivity Distributed attention
4. Search. Looking for something	Acquisition	Sustained attention Perceptual sensitivity
5. Identify. What is it and what is its name?	Acquisition/interpretation	Perceptual discrimination Long-term memory Working memory
6. Message. A collection of symbols sent as a meaningful statement	Handling	Response precision
7. Queue to channel. Lining up to process in the future	Handling	Working memory Processing strategies
8. Code. Translating the same thing from one form to another	Handling	Response precision Working memory Long-term memory
9. Transmit. Moving something from one place to another	Handling	Response precision
10. Store. Keeping something intact for future use	Handling	Working memory Long-term memory
11. Store in buffer. Holding something temporarily	Handling	Working memory Processing strategies
12. Compute. Figuring out a logical or mathematical answer to a defined problem	Handling	Processing strategies Working memory
13. Edit. Arranging or correcting things according to rules	Handling	Long-term memory Selective attention
14. Display. Something showing that makes sense	Handling	Response precision
15. Purge. Getting rid of irrelevant data	Handling	Selective attention
16. Reset. Getting ready for some different action	Handling	Selective attention Response precision
17. Count. Keeping track of how many	Handling/interpretation	Sustained attention Working memory
18. Control. Changing an action according to plan	Handling/interpretation	Response precision

(continued)

TABLE 15.2 (Continued)

Cognitive Agent Task	General Category of Information Processing	Human Information Processing Resources
19. Decide/select. Choosing a response to fit the situation	Interpretation	Long-term memory Processing strategy
20. Plan. Matching resources in time to expectations	Interpretation	Working memory Processing strategy
21. Test. Is it what it should be?	Interpretation	Perceptual sensitivity Working memory Long-term memory
22. Interpret. What does it mean?	Interpretation	Long-term memory Sustained attention
23. Categorize. Defining and naming a group of things	Interpretation	Long-term memory Perceptual sensitivity
24. Adapt/learn. Making and remembering new responses to a learned situation	Interpretation	Long-term memory
25. Goal image. A picture of a task well done	Interpretation	Long-term memory Processing strategies

Source: Reproduced from *Lee, J. D. and Sanquist, T. F.*, IEEE Transactions on Systems, Man, and Cybernetics, Part A, *30(3), 273–285, 2000*. Copyright 2000. With permission.

A problematic issue for those classifying distraction-related data is whether to include as sources of distraction those that pertain to events, activities, and actions associated with the driving task itself. As aforementioned, it is our view that any driving-related activity that can draw attention away from activities critical for safe driving is a potential source of distraction. The study by Klauer et al.[1] identified a number of driving-related activities that were cited as instances of "driving-related inattention to forward roadway" (p. 134). These were checking rear mirror, checking center rear-view mirror, looking through windshield (left and right of center), checking side mirrors (left and right), looking through side windows (left and right), and looking at instrument panel (to check vehicle speed, temperature, engine revolutions per minute). We believe that these activities have the potential to divert attention away from activities critical for safe driving and, in line with the definition of distraction adopted in this book, should be regarded as distractions. Hence, the objects, events, and actions associated with them have been included in Tables 15.1 and 15.3.

Of the remaining studies reviewed, the following additional driving-related activities were cited (primarily by Gordon[4]) as sources of distraction: adjusting vehicle controls/devices, driver dazzled by sun strike, driver dazzled by headlights, checking for traffic, other road users (vehicles, pedestrians, and cyclists), trying to find destination/location, scenery (persons), and police/emergency vehicles/crash scene. (The word "dazzled" in this context can be taken to mean that the driver is affected by direct glare from a bright light source within the field of vision.) With three exceptions (trying to find destination/location, driver dazzled by sun strike,

TABLE 15.3
Events and Objects on Which One or More Actions Were Performed

Major Source	Minor Source	Specific Source	Citation	Likely Actions	Notes
Things brought into vehicle	• Definition: portable objects, devices and living things brought into the vehicle				
	Animal or pet	Generic	S3, GE, G1, G2, K	• Attending to	Restrained or unrestrained
		Dog	S1, JS, G1	• Interference by	
		Insect	S3, JS, G1, K	• Reacting to	
	Document	Generic	S1, S3, GE, JS, G1	• Attending to	
		Book	GE, G1, K	• Reading	
		Newspaper	S3, GE	• Reaching for	
		Map/directions	S1, S3, GE, JS, K	• Searching for	
		Paper/s	S1, S3, JS, K	• Writing on	
		Magazine	K		
		Dockets	JS, G1		
		Mail	S1, S3, GE		
	Drink		S1, S3, GE, JS, G1, G2	• Reaching for • Searching for • Using	Open or closed container
	Eating utensils		K	• Reaching for • Searching for • Using	Pertains to food
	Food		S1, S3, GE, JS, G1, G2, K	• Reaching for • Searching for • Using	With or without a utensil

Grooming accessories	Generic	S3, GE, K	• Looking • Reaching for • Searching for • Using	
	Makeup	S1, S3, K		
	Shaver/razor	K		
	Combs/brushes	S3, K		
Lost object	Can specify the object in terms of minor and specific source	K	• Reaching for • Searching for	Objects brought into the vehicle that are "lost" and are being looked for
Personal effects	Generic	S1, S3, GE, JS, G1 G2, K	• Attending to • Reaching for • Searching for • Using • Adjusting	
	Purse	S1, S3, GE, JS, G1		
	Wallet	S1, GE, G1		
	Jewelery	K		
	Money	S3, GE, G1		
	Contact lenses	K		
	Glasses/sunglasses	S3, GE, G1		
	Clothing/shoes	G1		
	Hand bags/bags	G1		
	Pills/inhaler	S1		
Personal hygiene accessory	Generic	K	• Reaching for • Searching for • Using	
	Toothbrush	K		
	Dental floss	K		
	Cuticles	K		
Smoking device	Generic	S1, S3, GE, JS, G1 G2, K	• Attending to • Reaching for • Searching for • Using	
	Cigarette	S1, S3, JS, G1, K		
	Cigar	S1, S3, K		
	Pipe	S3, K		
	Lighter	JS, G1		

(continued)

TABLE 15.3 (Continued)

Major Source	Minor Source	Specific Source	Citation	Likely Actions	Notes
	Technology device	Cell phone	S1, S3, GE, JS, G1, K	• Attending to	Handheld or handsfree
				• Reaching for	Manual or quick dial
		Pager	S3, GE, G1	• Searching for	
		PDA	K	• Reacting to call	
		Radio telephone	G1	• Conversation	
		Data terminal	S1	• Texting	
		Technology Device	GE	• Using	
	Writing implements	Pen	S1, S3, GE	• Reaching for • Searching for • Using	
	Other device or object		S1, S3, G1, K	• Attending to • Reaching for • Searching for • Using	Caters for other things brought into the vehicle not specified in previous studies or not yet discovered
Vehicle systems	• Definition: displays, controls, objects, and devices already built into the vehicle with which the driver interacts				
	Entertainment	Generic	S1, S3, GE, JS, G1 G2, K	• Attending to • Reaching for	
		CD	S1, S3, GE, JS, G1, K	• Searching for • Using	
		Radio	S1, S3, GE, JS, G1, K		
		Cassette	S1, S3, GE, JS, G1, K		
	Mechanical problem		GE	• Attending to	E.g., engine overheating

	Vehicle controls or device	Generic	S1, S3, GE, JS, G1, G2, K	• Attending to • Reaching for • Using	Instrument panel refers to main displays on main instrument cluster (e.g., speedometer, rpm indicator, temperature gauge)
		Mirrors	S1, S3, GE, G1, K		
		Heater	S1, S3, GE, JS, G1		
		Air conditioner	S1, S3, GE, JS, G1, K		
		Sun visor	S1, S3, GE, G1, K		
		Glove box	S3, GE, K		
		Ash tray	GE		
		Vehicle lighter	K		
		Lights	S1, G1		
		Wiper/s	S1, G1		
		Seat belt	S1, G1		
		Windows	S3, G1		
		Cruise control	S3		
	Other vehicle system	Generic	S1, S3, GE, G1		Caters for vehicle systems not specified in previous studies or not yet evolved or discovered
Vehicle occupants	• Definition: vehicle occupants other than the driver with whom the driver interacts				
		Generic	S1, GE, JS, G1, G2, K	• Attending to • Reaching for • Conversation • Reacting to • Interference by • Arguing with	Front seat or back seat (S3)
		Adult	S3, G1, K		
		Young adult	G1		

(continued)

TABLE 15.3 (Continued)

Major Source	Minor Source	Specific Source	Citation	Likely Actions	Notes
		Child	S1, S3, GE, JS, G1, K		
		Baby/infant	S1, S3, G1		
Moving object or animal in vehicle	• Definition: stationary objects or animals that suddenly move in the vehicle due to hard braking, acceleration, or turning or because they have been dropped by the driver or other occupant				
	Generic		S1, S3, GE, JS, G1, K	• Reacting to • Dropped • Reaching for	Can use the things brought into the vehicle category to further breakdown the sources
	Animal		S1, S3, GE, JS		
	Insect		S1, GE, JS		
	Object normally stationary		S1, GE, JS, K		
	Object held previously by driver or occupant		S1, S3, GE, JS, K		
	Other moving object		S1, GE, JS, K		Caters for other moving objects not specified in previous studies or not yet discovered

Internalized activity	• Definition: driver behavior that has the potential to distract the driver that involves no overt interaction with an object, event, or activity			
	Coughing	G1, G2	• Reacting to	
	Daydreaming or lost in thought	S1, GE, JS, G1, G2, K		E.g., random thoughts, dwelling on emotional issues or recent events
	Dancing in seat	K		
	Itching	G1, G2		
	Medical or emotional impairment	S1, GE, G1, G2		E.g., driver goes into labor, foreign item in their eye, emotionally upset
	Praying	GE		
	Singing	K		In absence of music
	Sneezing	G1, G2		
	Talking	K		When no one else is present
	Other activity	G1, G2		Caters for other internalized activities not specified in previous studies or not yet evolved or discovered

(continued)

TABLE 15.3 (Continued)

Major Source	Minor Source	Specific Source	Citation	Likely Actions	Notes
External objects, events or activities	• Definition: Objects, events, or activities outside the vehicle that have the potential to distract the driver or the driver is paying attention to				
	Animal	Generic	S1, GE, JS, G2, K	• Looking at	On or near roadway and off-roadway
		Deer	S1, JS		
		Dog	S1		
		Elk	S1		
	Architecture		S1, G2	• Looking at	Includes buildings
	Advertising or billboards		S1, G2	• Looking at	Can be static or moving, and located on buildings, billboards, vehicles etc.
	Construction zone or equipment		S1, GE, K	• Looking at	On or near roadway and off-roadway
	Crash scene		S1, GE, G2, K		
	Incident		S1, GE, JS, G2, K		Includes near-misses, armed hold-ups, road rage, etc.
	Insect		S1, GE		
	Landmark		GE, G2		
	Road signs	Generic	S1, GE, G2		Includes all classes of road sign—destination, warning, regulatory and advertising-related

	Road users (non-vehicle)	Generic	S1, GE, G2, K	• Looking at	Includes using mirrors (rear or side) to check for road users
		Pedestrians	S1, GE, G2, K		
		Cyclists	G2		
		People	G2		
		Police	S1, G2		
	Road users (vehicle)	Generic	S1, GE, JS, G2	• Looking at	Includes using mirrors (rear or side) to check for road users
		Parked	S1, GE, JS, G2		
		Moving/in traffic	S1, JS, G2		
		Police	S1, G2		
		Emergency	S1, G2		
		Waved through	S1, JS		
	Scenery		GE, G2	• Looking at	
	Weather	Generic	GE, G2	• Looking at	
		Clouds/sky	S1		
	Other objects, events, or activities		S1, GE, JS, G2, K	• Looking at	Includes sources of distraction unrelated to the traffic environment such as sporting events, parachutists in the sky, etc
Other major sources of distraction					Caters for major sources of distraction that are yet to evolve or be discovered

Note: S1, Stutts et al.[2]; S3, Stutts et al.[3]; G1, Gordon[4]; G2, Gordon[5]; K, Klauer et al.[1]; GE, Glaze and Ellis[6]; JS, Joint State Commission.[7]

and driver dazzled by headlights), all of the actions, objects, and events associated with these driving-related activities have been included in Tables 15.1 and 15.3. In our opinion it is not, *per se*, the dazzle from the sun or from headlights that has potential to divert attention away from activities critical for safe driving. Rather, it is the actions performed by the driver in response to the dazzle (e.g., adjusting the sun visor, re-positioning the head, activating the antiglare function of the rear-view mirror, etc.). Similarly, trying to find a destination or location is not of itself a source of distraction. Rather, it is the overt and covert actions (e.g., looking, thinking) performed on the objects (e.g., road signs, maps, instructions, one's internal mental map of the route) associated with navigation that constitute the source of distraction if they have the effect of diverting attention away from activities critical for safe driving. Glare and destination represent distal sources of distraction, compared to proximal sources of distractions that are more commonly discussed.

Glaze and Ellis[6] report some remaining sources of distraction that do not conform with the definition of distraction presented here: "driver fatigue/asleep," "alcohol," "alcohol and fatigue/asleep," and "driver error (misjudgment/inexperience)." In Chapter 3, we define driver inattention as *diminished attention to activities critical for safe driving in the absence of a competing activity.* In our opinion, driver states, such as being fatigued or inebriated by alcohol, are not in themselves sources of distraction. They better fit the description of inattention as they have the effect of diminishing attention to activities critical for safe driving in the absence of a competing activity. Finally, driver error can be considered a consequence of distraction (e.g., if distraction results in a mistake), rather than a source of distraction, regardless of whether it derives from misjudgment or inexperience (the latter might also be regarded as a driver state).

It is often argued that some sources of distraction are unavoidable. If it is assumed that the classification scheme presented in Table 15.3 has some modicum of validity, then it can be used to roughly estimate the proportion of sources of distraction that might be avoidable. Such data might be used to determine which categories and sources of distraction are most amenable to countermeasure development. In Table 15.4, we have made subjective judgments about which sources of distraction we believe are, at least theoretically, avoidable—from the perspective of drivers and other stakeholders. For example, most sources of distraction under the category "things brought into vehicle" would seem to be avoidable; drivers can choose not to bring animals, drink, eating utensils, food, and other things into the vehicle that have potential to distract them. If, however, they require a document, such as a paper map, to assist them in navigating their way to an unknown destination (assuming they cannot memorize the route and do not have an in-vehicle navigation system), then they have little choice but to bring the document into the vehicle with them; the document, in this case, can be considered an unavoidable source of distraction. In developing Table 15.4, it has been assumed that "events," such as spills, mechanical problems, crash scenes, lightening strikes, and moving objects in the vehicle, will be particularly distracting, and would seem to be highly salient relative to visual attention. Hence, events have usually been classified as unavoidable sources of distraction. Of course, the judgments we have made here are, in some cases, highly context dependent.

TABLE 15.4
Potential Sources of Distraction and the Extent to Which They Can Be Avoided

Major Source	Minor Source (Events and Objects)	Avoidable Distraction?
Things brought into vehicle	Animal/pet	Yes
	Document	Yes (Although paper maps are unavoidable if the driver has to navigate through unfamiliar territory without a co-pilot or route guidance system)
	Drink	Yes
	Eating utensils	Yes
	Food	Yes
	Grooming accessories	Yes
	Lost object	No (Things get lost; people make mistakes)
	Personal effects	Yes
	Personal hygiene accessory	Yes
	Smoking device	Yes
	Technology device	Yes
	Writing implements	Yes
Vehicle systems	Entertainment	Yes
	Mechanical problem (e.g., engine overheating)	No (Some mechanical problems cannot be avoided, even with regular servicing of vehicle; and when they occur, it is instinctive to attend to them)
	Vehicle controls/devices	Yes (These sources of distraction can be avoided if the attention diverted toward them does not compete with activities critical for safe driving)
Vehicle occupants	Adult front	Yes (Via bans on passenger carriage for young drivers and by passengers moderating their conversation when activities critical for safe driving are being performed by the driver)
	Adult back	Yes (Via bans on passenger carriage for young drivers and by passengers moderating their conversation when activities critical for safe driving are being performed by the driver)
	Child front	No (Although bans on carriage of children by young and novice drivers are feasible)

(*continued*)

TABLE 15.4 (Continued)

Major Source	Minor Source (Events and Objects)	Avoidable Distraction?
	Child back	No
	Baby front	No
	Baby back	No
Moving object or animal in vehicle	Animal	Yes
	Insect	No (It is impossible to prevent all insects from entering vehicle cabin)
	Object normally stationary	No (It is not practical to restrain all loose objects)
	Object previously held by driver/occupant	Yes
Internalized activity	Coughing	No
	Daydreaming/lost in thought	No
	Dancing in seat	Yes
	Itching	No
	Medical/emotional impairment	Yes (People can elect not to drive in these states)
	Praying	Yes
	Singing	Yes
	Sneezing	No
	Talking	Yes
External objects, events or activities	Animal	Yes (Barriers preventing animals from crossing roads and signs warning of animals are feasible, at least in critical locations)
	Architecture	Yes (Architectural structures likely to distract drivers could be located out of sight of drivers, although this may be difficult to implement)
	Advertising billboards	Yes (They could be located out of sight of drivers, or only in locations in which the consequences of distracted driving are not adverse)

(*continued*)

TABLE 15.4 (Continued)

Major Source	Minor Source (Events and Objects)	Avoidable Distraction?
	Construction zone/equipment	No (Instinctive to look/attend—and it is difficult to conceal construction zones/equipment)
	Crash scene	No (Instinctive to look/attend—and it is difficult to conceal crash scenes)
	Incident	No (Instinctive to look/attend—traffic incidents are usually spontaneous events, that are difficult to foresee and avoid)
	Insect	No (Instinctive to look/attend)
	Landmark	Yes (Landmarks likely to distract drivers could be located out of sight of drivers, although this is difficult to implement)
	Road signs	No (Instinctive to look/attend—they are deliberately designed to attract attention)
	Road users	No (Instinctive to look/attend)
	Scenery	No (Instinctive to look/attend)
	Vehicle	No (Instinctive to look/attend)
	Weather (e.g., lightning)	No (Instinctive to look/attend)

For each of the main sources of distraction identified in Tables 15.3 and 15.4, the following are the proportions of sources within that category that would seem to be avoidable:

- Things brought into vehicle 92%
- Vehicle systems 67%
- Vehicle occupants 33%
- Moving object in vehicle 50%
- Internalized activity 56%
- External objects, events, activities 31%

For the category "things brought into the vehicle," for example, it was judged that 11 out of 12 (i.e., 92%) of sources of distraction in that category are avoidable.

Overall, about 55% of all specific sources of distraction identified in Tables 15.3 and 15.4 appear to be avoidable. Sixty-one percent of sources deriving from within the vehicle seem avoidable and 31% of sources deriving from outside the vehicle appear avoidable. It seems, therefore, that there is much scope for countermeasure development and that it is reasonable to concentrate effort on those sources of distraction deriving from within the vehicle, especially given the seemingly greater difficulties associated with avoiding distractions external to the vehicle (see Table 15.4). The proportions of sources of distraction that are avoidable are, of course, somewhat arbitrary. For example, if a category existed with two items and one of them was deemed to be avoidable, then that category would be 50% avoidable. The degree of "avoidability" is predicated on the assumption that each element in a category is equally frequent. However, in practice, this is not so. The numbers derived above are rough estimates only, and are presented merely to convey to the reader the simple notion that some sources of distraction may be more amenable to countermeasure development than others.

In summary, we have distilled six categories of objects and events (Table 15.3) from several published studies, which define sources of distraction that in recent years have reportedly contributed to road traffic incidents, near-misses, and crashes involving current-generation drivers driving current-generation vehicles on current-generation roads. We have also distilled from the same studies (Table 15.1) a list of actions that were identified as being associated with the sources of distraction listed in Table 15.3. From Tables 15.1 and 15.3, it is possible to derive another table, which lists all factorial combinations of actions that can be physically performed in relation to each object and event. This could serve as a basis for the development of more uniform methods for collecting, coding, and classifying distraction-related data. The take-home message for those involved in collecting, coding, and classifying sources of driver distraction is to be aware of the important distinction that can be made between events and objects on the one hand, and the actions performed on them, or in relation to them, on the other. The collection, coding, and classification of data in this way provides designers and policy makers with important specific information about those actions, for a given event or object, which are most likely to adversely affect performance and safety.

The challenge, of course, is to develop data collection tools such as police reporting forms that are parsimonious and practical enough to administer. Elaborate classification schemes such as the one presented here may be more applicable for obtaining data from in-depth crash studies and naturalistic driving studies. However, well-designed and detailed coding schemes provide flexibility in presenting or summarizing distraction information to the public or other audiences. For example, when discussing risk, the distracting activity is often described as a behavior repertoire that includes action and object, such as, dialing a mobile phone or talking on a mobile phone. If the crash-based information is well described in terms of the action and objects involved, this allows changes in summarizing the information to take place as our knowledge about distraction and risks develops.

15.4 MECHANISMS AND SOURCES OF DISTRACTION

Part II of this book focuses on theories and models of driver distraction. From that discussion it is possible to distil the main underlying dimensions that determine the extent to which a given source of distraction is likely to degrade driving performance. The present chapter has focused on *actual* sources of distraction that have been cited as contributing factors in incidents, near-misses, and crashes. Combining these two lines of thinking, it is possible to predict whether a given source of distraction will degrade driving performance when a driver interacts with it.

This line of thinking is illustrated in Table 15.5. The table shows, for a particular "object brought into vehicle" (i.e., a mobile phone), a chronological set of actions that might be performed when using the object to phone a friend and, for each of these actions, the underlying dimensions (or mechanisms) that influence the degree to which that action is likely to degrade driving performance. A whole family of such tables could be generated for the various specific sources of distraction identified in Table 15.3.

The actions in the second column from the left are derived from Table 15.1. The remaining columns represent the underlying dimensions referred to in the preceding text.

Columns 3–7 derive from the multiple resource model of attention articulated by Wickens and Horrey in Chapter 5 of this book. According to this model, the degree to which two tasks can be time-shared effectively can be predicted by their joint difficulty (labeled "demand level"), and the degree to which they overlap in demand for common resources that vary along four dimensions: processing stage (perceptual-cognitive-motor), processing code (verbal vs. spatial), perceptual modality (auditory vs. visual), and visual channel (focal vs. ambient). Wickens and Horrey assert that since driving (both hazard monitoring and lane keeping) is primarily a visual-spatial-motor task, it is predicted (and observed) to be fairly efficiently time-shared with those tasks that are auditory and language based (both in perception—hearing speech, and in action—speaking). Furthermore, they argue that because ambient and focal vision use separate resources, lane keeping and hazard monitoring can be well time-shared, given that the latter employs foveal vision.

The headings in each of columns 3–7, which derive from the model, are defined as follows.

- *Perceptual modality incompatibility* refers to the extent to which the task that competes for the driver's attention shares the same perceptual modality (auditory vs. visual) as tasks that comprise the driving task.
- *Code incompatibility* refers to the extent to which the task that competes for the driver's attention shares the same processing code (verbal vs. spatial) as tasks that comprise the driving task.
- *Visual channel incompatibility* refers to the extent to which the task that competes for the driver's attention shares the same visual channel (focal vs. ambient) as tasks that comprise the driving task. These terms are synonymous with foveal and peripheral vision, respectively.
- *Stage incompatibility* refers to the extent to which the task that competes for the driver's attention shares the same processing stages (perceptual-cognitive-motor) as the tasks that comprise the driving task.
- *Demand level* refers to the joint difficulty of the driving task and the task that competes for the driver's attention.

The greater the extent to which an action shares the same resources as driving, the higher is the degree of incompatibility between that action and driving, and the higher is the expected degree of distraction induced by performance of that action while driving.

TABLE 15.5
Mechanisms Moderating Distraction Deriving from Use of a Handheld Mobile Phone

Overall Activity	Action (Derived from Table 15.1)	Underlying Dimensions Moderating Potential for Distraction										
		Perceptual Modality Incompatibility (Visual vs. Verbal)	Code Incompatibility (Verbal vs. Spatial)	Visual Channel Incompatibility (Focal vs. Ambient)	Stage Incompatibility (Perceptual-Cognitive-Motor)	Demand Level	Nonignora-bility	Unpredicta-bility	Uninterrupta-bility	Nonadjusta-bility	Duration	Mean Score
Phone a friend (assumes driver knows number)	Looking (to locate phone in car)	Visual (High)	Spatial (High)	Focal (High)	Perceptual (Medium)	Low • Assuming that driving conditions are easy	Low • Assuming call is self-initiated	Low • Assuming there is a cradle to hold the phone	Low • Assuming call is self- initiated	Low	Low • Assuming few, short, glances	1.7
	Reaching (to pick up phone)	Visual (Medium) • Assuming minimal visual guidance is needed	Spatial (High)	Focal (Medium) • Assuming minimal visual guidance is required	Perceptual-Motor (Medium)	(Low) • Assuming that driving conditions are easy	(Low) • Assuming call is self-initiated	(Low) • Assuming there is a cradle to hold the phone	(Low) • Assuming call is self-initiated	(Low)	(Low) • Assuming minimal visual guidance required	1.5

Looking (at phone to determine whether display screen is in right mode for call) whilst holding it	Visual (High)	Verbal (Low) • Assuming low spatial demands	Focal (High)	Perceptual (Medium)	(Low) • Assuming that driving conditions are easy	(Low) • Assuming call is self-initiated	(Low) • Assuming driver is familiar with display screens	(Low) • Assuming call is self-initiated	(Low)	(Low) • Assuming one or two short glances	1.4
Dialing (friend's telephone number)	Visual (High) • Assuming manual dialing and visual guidance required	Spatial (High) • Assuming manual dialing	Focal (High)	Perceptual-cognitive-motor (High) • Assuming manual dialing, visual guidance and driver has to recall phone number	(High) • Assuming that driving conditions are easy	(Low) • Assuming call is self-initiated	(Low) • Assuming driver is familiar with device	(Low) • Assuming driver can make call when they like	(Low)	(High) • Assumes manual dialing and frequent glances	2.2
Talking (to friend) whilst holding the phone	Verbal (Low) • Assumes eyes are on the road ahead	Verbal (Low) • Assuming eyes are on the road ahead	Neither (Low) • Assuming eyes are on the road ahead	Cognitive (Medium) • Assuming eyes are on the road ahead and minimal motor resources required to speak	(Medium) • Assuming that phone conversation is casual, reception is good and driving conditions are easy	(Low) • Assuming the conversation is casual	(Low) • Assuming driver is familiar with device	(Low) • Assuming conversation is casual and they can interrupt and resume it whenever they like	(Low) • Assuming driver can slow down or terminate speech depending on changing traffic conditions	(High) • Assumes phone call spans long period (e.g., 10 min)	1.4

(continued)

TABLE 15.5 (Continued)

		Underlying Dimensions Moderating Potential for Distraction										
Overall Activity	Action (Derived from Table 15.1)	Perceptual Modality Incompatibility (Visual vs. Verbal)	Code Incompatibility (Verbal vs. Spatial)	Visual Channel Incompatibility (Focal vs. Ambient)	Stage Incompatibility (Perceptual-Cognitive-Motor)	Demand Level	Nonignora-bility	Unpredicta-bility	Uninterrupta-bility	Nonadjusta-bility	Duration	Mean Score
	Pressing ("Menu" button on phone to terminate conversation) whilst holding it	Visual (Low) • Assumes little or no visual guidance needed	Spatial (Low) • Assuming low spatial demands	Focal (Low) • Assuming little or no visual guidance needed	Perceptual (Low)	(Low)	(Medium) • Assuming that, once decision to terminate is made, task is difficult to ignore	(Low)	(High) • Assumes that, once terminated, call is not easily resumed	(Medium) • Assumes that driver may not have a total say in when conversation should be terminated	(Low) • Assumes single short-duration glance, or no glance, and simple button press	1.4
Mean Score		2.2	2.0	2.2	2.0	1.2	1.2	1.0	1.3	1.2	1.8	1.6

The remaining column headings are described in Chapter 4 and reflect a control theoretic interpretation of recent research concerning interruptions and multitask management (see, e.g., Refs. 10–14). These are defined as follows:

- A competing task is said to be *nonignorable* when it is so compelling or demanding that the driver cannot disengage from it.
- A competing task is *unpredictable* when its onset is unexpected or its consequences cannot be foreseen by the driver.
- An *uninterruptible* competing task is one that cannot be postponed or cannot be resumed after interruption.
- Where it is impossible for the driver to change the demand associated with the performance of a competing task, the task is said to be *nonadjustable.*
- *Duration* refers to the total time taken to complete a competing task. It can be viewed as a measure of total exposure to all of the above demands. In the context of distraction, a short-duration task is preferable to a long-duration task.

Each action can be rated—high, medium, or low—on each of these underlying dimensions. We have chosen this three-point rating scheme for illustrative purposes only. In combination, these ratings can be used to judge the degree to which that action is likely, through distraction, to degrade driving performance. In Table 15.5, "low" ratings imply low potential for distraction and "high" ratings imply high potential for distraction. A worked example is presented in the following paragraphs to clarify interpretation of the table.

In Table 15.5 (column 2) it is assumed that the first action performed by the driver, to phone a friend on their handheld mobile phone, is to look within the vehicle for his or her phone (i.e., looking). This is primarily a visual task, as is driving; hence, both tasks will compete for the driver's vision. Perceptual modality incompatibility (column 3) in this case is high because the two visual tasks are fundamentally incompatible with each other.

Code incompatibility (column 4), similarly, is high. Looking for the phone is primarily a spatial task, as is driving. Hence, both tasks will compete for the driver's spatial processing resources.

Visual channel incompatibility (column 5) is also high, assuming that the driving task involves frequent monitoring for potential and actual hazards. Looking for the phone and looking out for hazards both require focal, or foveal, vision. Hence the two tasks will compete for common visual processing channels. If the driver were to drive on a straight road in the middle of the Sahara Desert, with no speed limit and no potential for conflict with any actual road hazards, then the main task of driving would be lane keeping (which relies more on peripheral vision). Here, the two tasks would be more compatible.

Stage incompatibility (column 6), similarly, is medium. Looking for the phone is primarily a perceptual task, as is driving. Hence, both tasks will compete for the driver's perceptual resources.

Looking for the phone to locate it requires relatively little attention, especially if it is normally placed in the same location within the vehicle; it is not a very difficult task to perform. Hence, it is assumed that the combined demand level for both tasks (looking and driving) is low, assuming that driving conditions at the time are relatively easy.

It is assumed that looking for the phone, when the decision to phone a friend is self-initiated, is not so compelling or demanding that the driver cannot disengage from it. Hence, the rating for "nonignorability" is low.

The driver can choose when to start looking for the phone and can foresee the consequences of looking away from the road. The degree of predictability of locating the phone depends on how the driver stores the phone. If the phone is kept in a docking station then finding it would be very predictable; if the driver stores it on the passenger seat, then finding it might be less predictable. Hence, the rating for "unpredictability" is low, especially if the location of the phone is already known (e.g., if it is housed in a cradle when not in use).

Looking for the phone is a task that can be postponed and resumed after interruption by some aspect of the driving task (e.g., a braking lead vehicle), assuming that the decision to make the call is self-initiated. Hence the rating for "uninterruptability" is low.

The driver can choose to change the demand associated with the performance of the "looking" task by, for example, glancing only briefly inside the vehicle while driving. Hence, the task is low in "nonadjustability."

Finally, it is assumed that the driver knows where the phone is located and that the total time required to look for it will be minimal (less than 1 s), with only a single eye glance. Hence, the looking task is rated as a low-duration task. Of course, as discussed in Chapter 9 (in relation to the *R* metric), a looking task may, in some circumstances, be minimally distracting, even if it has a relatively long task duration. This might be so, for example, if the task is interruptible, enabling it to be broken down, or "chunked," into smaller subtasks, each requiring a short eye glance.

Similar lines of reasoning were used to derive the ratings for each of the other actions in the remaining rows of the table (i.e., rows 2–6).

It is possible to use Table 15.5 to make predictions about the likely impact on the driving performance of individual actions, or of all those actions that comprise an overall activity. For example, if the ratings high, medium, and low are converted into scores of 3, 2, and 1, respectively (where 3 denotes high distraction potential and 1 denotes low distraction potential), then the unweighted mean score for the action "looking (to locate phone in car)" in Table 15.5 is 1.7. Unweighted mean scores for this and the remaining actions are shown in the last column of the table. Across rows, the table can be used to draw attention to those actions for a given activity that have the propensity to be relatively more distracting than others. The action of dialing a friend's telephone number has the highest rating of 2.2. Across columns, the table reveals which underlying dimensions contribute most to the propensity to distract drivers. Here, the ratings suggest that, for a self-initiated call on a handheld mobile phone, it is competition for common processing resources, rather than the control characteristics of the activity, that contributes most to distraction. Averaging across rows yields a mean score of 1.6 for the six actions comprising the overall activity of phoning a friend. This score can be used to compare, for the same device (i.e., handheld phone), the relative propensity of different overall activities to distract a driver.

Of course, further work is needed to determine the relative weightings that might be given to the scores on each of the underlying dimensions. The relative weighting given to the dimension "duration" is of particular importance here, as it can be viewed as a measure of total exposure to all of the above dimensions. The ratings in Table 15.5 suggest that dialing a friend's telephone number (rating 2.2) is more distracting than talking on the phone while keeping your eyes on the road (rating 1.6), assuming that both tasks have an equally high rating for duration. This trend is consistent with the findings from the 100-car naturalistic driving study,[1] which suggest that dialing a handheld device is more risky (odds ratio 2.8) than using the device to talk and listen (odds ratio 1.3; see Chapter 16). However, data from the same study reveal that the population-attributable risk (PAR) associated with dialing a handheld mobile phone, which takes into account exposure of the population at large to this risk, is the same as that for using the device to talk and listen (3.6% for both; see Chapter 16). This implies that it might be appropriate to differentially weight scores on the "duration" dimension, to take into account not only estimated task completion times but also differential patterns of driver exposure (to the extent to which these are known).

Table 15.5 provides the basis for the development of checklists that might be used to assess the distraction potential of a device or activity. The difficulty, of course, is in using the data in Table 15.5 to make absolute judgments about acceptable levels of predicted distraction for a given action, or overall activity. For example, is an absolute score of 1.7 for "looking" to locate the phone in the car acceptable? Further work is needed to define threshold values beyond which it can be determined that an action, or overall activity, has unacceptably high potential to distract the driver. For the moment, the table is better suited to making relative rather than absolute judgments about this.

15.5 SOURCES OF DRIVER DISTRACTION: THE FUTURE

The major and specific sources of distraction identified in Table 15.3 are, of course, those that derive from the operation of current-generation vehicles, driven by current-generation drivers, on current-generation roads, in the current social, economic, and political climate. Further, they derive from selected studies in only a few countries. As the driving task, and society itself, continues to evolve, so too will the events and objects that drivers interact with and the actions they perform on them. Consequently, the sources of distraction identified in Table 15.3 will inevitably change. As various distraction prevention and mitigation strategies are implemented and begin to take effect (see Chapters 30 through 32), the extent to which the existing sources of distraction show up as contributing factors in crashes and near-crashes will also change.

How the main and specific sources of distraction listed in Table 15.3 are likely to change in future is not clear. Some will remain the same, some may disappear altogether from crash databases, and new sources will appear. Many factors will determine which ones remain, disappear, and emerge. Some of these were alluded to earlier in this chapter. These include the definitions that are used for determining what should and should not be regarded as distraction, what is voluntarily reported by crash victims, the design of data collection methods (e.g., crash studies, naturalistic driving studies), and the knowledge and expertise of police, crash investigation teams, and researchers in knowing what questions to ask about the potential role of distraction as a contributing factor in crashes. Other factors may also intervene.

Consider, for example, "things brought into the vehicle." Stricter legislation to ban specific objects from being brought into the vehicle (e.g., animals or pets) or to ban the use of specific objects brought into the vehicle while it is in motion (e.g., cigarettes) may be effective in eliminating some of these sources of distraction, perhaps in countries with strong enforcement regimes and a highly compliant driving population. Conversely, nomadic devices (with driving or nondriving-related functions) that can be brought into the vehicle will likely increase in number and functionality and, if they divert attention away from activities critical for safe driving, will surface as new specific sources. As data collection and coding schemes become more refined, it is likely that various other things brought into the vehicle will also surface as specific sources of distraction.

Sources of distraction deriving from driver interaction with "vehicle systems" are also likely to change in various ways. As data collection methods and tools become more refined, further sources of vehicle system-related distraction will be discovered for existing systems. The rapid proliferation within vehicle cockpits of factory-fitted entertainment systems (e.g., DVD players), vehicle information and communication systems (VICS; e.g., traveler information, Internet), and advanced driver-assistance systems (ADAS, e.g., adaptive cruise control, in-vehicle navigation, collision warning) may create new sources of distraction if the systems, and the functions they support, are poorly designed and located, or used inappropriately.[15,16] Conversely, some sources of distraction may disappear as system interfaces in vehicle cockpits become progressively better designed for human use, and workload managers, and other real-time distraction mitigation systems, limit driver exposure to competing activities when drivers are most vulnerable to the effects of distraction. Well-designed ADAS technologies that partially or fully automate some driving functions and tasks will make driving easier; however, in doing so, they may encourage drivers to compensate for the resultant reduction in workload by engaging in nondriving activities not yet known.

"Vehicle occupants" will always remain potential sources of distraction. Legislation can be used, however, to eliminate, at least temporarily, some vehicle occupants as sources of distraction. For example, in some countries specific legislation already exists that prohibits young novice drivers, as part of graduated licensing systems, from carrying certain passengers known to elevate their crash risk through distraction or peer pressure.

The specific sources of "moving objects or animals in the vehicle" that distract drivers are unlikely to change, although it is possible that, as more nomadic devices are brought into the vehicle, these may constitute additional items that have potential to move within the cabin if unrestrained. As noted previously, more stringent legislation may also have the effect of eliminating some of these sources of distraction.

As data collection and coding schemes become more refined, it is also likely that new sources of "internalized activity" will surface, although this may depend to some extent on the prevailing legislative regimes. If careless driving and other regulatory provisions continue to be used to blame drivers for involuntarily diverting their attention away from activities critical for safe driving toward internalized activities, drivers will continue to be reluctant to voluntarily report such episodes to enforcement authorities. As a greater amount of information is presented to the driver in the cockpit, the amount of internalized activity is likely to increase; there will be more to think about while driving.

As noted by Gordon in Chapter 16, considerably greater effort has gone into classifying sources of distraction deriving from inside the vehicle than those deriving from outside the vehicle. As data collection methods and tools become more refined, it is likely that many new "external objects, events, and activities" that have potential to distract drivers will be discovered. Advances in technology outside the vehicle have potential to influence the sources of distraction that emerge in future from crash and other studies in different ways. On the one hand, new technologies that are poorly designed and located, whether they are driving or nondriving related, may give rise to distraction and surface as additional sources of distraction. On the other hand, as more information currently displayed outside the vehicle is displayed to the driver inside the vehicle, one might expect a reduction in the number of sources of distraction that derive from outside the vehicle as shown in Table 15.3. Legislation and traffic engineering measures can play a powerful role in determining what remain as sources of distraction outside the vehicle. Legislation can be used, for example, to regulate the location, design, and operation of objects and events outside the vehicle, which have potential to distract drivers. It can be used to regulate the location and design of advertising billboards at certain road sites, to ban the presentation of advertising material on the back of buses, taxis, and other structures, and to prevent events from being staged at locations likely to distract drivers. Engineering countermeasures can be implemented to prevent animals from crossing roadways at certain locations, to divert traffic away from prominent landmarks and scenery likely to attract attraction, and to prevent or minimize interactions with other road users in high-workload areas (e.g., via installation of overhead pedestrian walkways). Tingvall, in Chapter 33, has underscored the importance of traffic engineering design in creating a distraction-tolerant road system. At least theoretically, measures of this kind have potential to eliminate several of the external objects, events, and activities listed in Table 15.3.

As the vehicle cockpit evolves, and the driving task becomes progressively more automated, the role of the driver, like that of the aircraft pilot, will inevitably change—from being less of an active controller of the vehicle to being more of a systems monitor. This will result, over time, in an increased emphasis on higher-level perceptual and cognitive activity, and an accompanying decrease in the number and types of control operations required to drive a vehicle.[17] If so, it is probable that more of the actions in the Miller[8] taxonomy will be performed on objects, events, and activities inside and outside the vehicle.

15.6 CONCLUSIONS

In this chapter, we have proposed a taxonomic description of the various main and specific sources of driver distraction that have been identified as contributing to crashes and near-misses in current-generation vehicles, driven by current-generation drivers on current-generation roads. The taxonomy is derived from several studies that have made use of secondary crash-based data sources collected by agencies (typically the police), from selected observational studies and from the definition of driver distraction presented in Chapters 1 and 3 of this book. The taxonomy is intended to serve several purposes: to begin to resolve confusion about what are, and are not, sources of distraction, inside and outside the vehicle; to provide a framework for categorizing these sources of distraction; and to support the development of less variable methods for collecting and coding crash and epidemiological data.

We have also provided a matrix for predicting the potential impact of these sources of distraction on driving performance, which provides a theoretical basis for the development of checklists that can be used to assess the level of driver distraction deriving from performance of a particular activity. The taxonomy and framework presented here will need to be updated as the driving task continues to evolve.

REFERENCES

1. Klauer, S., Dingus, T., Neale, V., Sudweeks, J., and Ramsey, D., *The Impact of Driver Inattention on Near-Crash/Crash Risk: An Analysis Using the 100-Car Naturalistic Driving Study Data*, DOT Technical Report HS 810-594, National Highway Traffic Safety Administration, Washington, D.C., 2006.
2. Stutts, J. C., Reinfurt, D. W., Staplin, L., and Rodgman, E. A., *The Role of Driver Distraction in Traffic Crashes*, AAA Foundation for Traffic Safety, Washington, D.C., 2001.
3. Stutts, J., Feaganes, J., Rodgman, E., Hamlett, C., Meadows, T., and Reinfurt, D., *Distractions in Everyday Driving*, AAA Foundation for Traffic Safety, Washington, D.C., 2003.
4. Gordon, C., A preliminary examination of driver distraction related crashes in New Zealand. In Faulkes, I. J., Regan, M. A., Brown, J., Stevenson, M. R., and Porter, A. (Eds.), *Driver Distraction: Proceedings of an International Conference on Distracted Driving*, Sydney, Australia, 2–3 June. Australasian College of Road Safety, Canberra, ACT, 2005.
5. Gordon, C., What do police reported crashes tell us about driver distraction in New Zealand? In Faulkes, I. J., Regan, M. A., Brown, J., Stevenson, M., R. and Porter, A. (Eds.), *Driver Distraction: Proceedings of an International Conference on Distracted Driving*, Sydney, Australia, 2–3 June. Australasian College of Road Safety, Canberra, ACT, 2005.
6. Glaze, A. L. and Ellis, J. M., *Pilot Study of Distracted Drivers*, Survey and Evaluation Research Laboratory, Virginia Commonwealth University, 2003.
7. Joint State Government Commission, *Driver Distractions and Traffic Safety*, Staff report, General Assembly of the Commonwealth of Pennsylvania, Joint State Government Commission, Pennsylvania, December 2001.
8. Miller, R. B., A method for determining task strategies, In *Technical Report AFHRL-TR-74-26*, American Institute for Research, Washington, D.C., 1974.
9. Lee, J. D. and Sanquist, T. F., Augmenting the operator function model with cognitive operations: Assessing the cognitive demands of technological innovation in ship navigation, *IEEE Transactions on Systems, Man, and Cybernetics—Part A: Systems and Humans*, 30(3), 273–285, 2000.
10. Altmann, E. M. and Trafton, J. G., Memory for goals: An activation-based model. *Cognitive Science: A Multidisciplinary Journal*, 26(1), 39–83, 2002.

11. Cellier, J. M. and Eyrolle, H., Interference between switched tasks, *Ergonomics*, 35(1), 25–36, 1992.
12. Jett, Q. R. and George, J. M., Work interrupted: A closer look at the role of interruptions in organizational life, *Academy of Management Review*, 28(3), 494–507, 2003.
13. McFarlane, D. C., Comparison of four primary methods for coordinating the interruption of people in human-computer interaction, *Human-Computer Interaction*, 17(1), 63–139, 2002.
14. Monk, C. A., Boehm-Davis, D. A., and Trafton, J. G., Recovering from interruptions: Implications for driver distraction research, *Human Factors*, 46(4), 650–663, 2004.
15. Regan, M. A., New technologies in cars: Human factors and safety issues, *Ergonomics Australia*, 18(3), 6–15, 2004.
16. Regan, M., Driver distraction: Reflections on the past, present and future, *Journal of the Australasian College of Road Safety*, 16(2), 22–33, 2005.
17. Fuller, R. and Santos, J. A., A note on advanced transport technology, In *Human Factors for Highway Engineers*, Fuller, R. and Santos, J. A. (Eds.), Pergamon, London, 2002, pp. 277–282.

16 Crash Studies of Driver Distraction

*Craig P. Gordon**

CONTENTS

16.1 INTRODUCTION

This chapter, using the definition of distraction provided in Chapters 1 and 3 of this book, discusses the results of several crash studies that have attempted to provide detailed categories of different types of distractions and quantify the proportion of crashes that involve driver distraction as a contributory factor.[1,2] The majority of the studies examined use traditional approaches to information capture, such as use of

* The views expressed are those of the author and do not necessarily represent the views of the Ministry of Transport.

police crash reports or the outputs of crash investigation teams. The limitations of traditional information capture methods are discussed and contrasted with the findings of one study that uses an observational method involving instrumented vehicles to collect crash and incident information. The chapter provides an estimate of the role of distraction as a contributing factor in crashes, and the relative contribution to crashes of distractions deriving from inside and outside the vehicle. The chapter concludes with a discussion of some issues regarding the role of technology as a potential source of distraction, the classification of distraction, and improved crash reporting.

16.2 DEFINING DISTRACTION

Driver distraction, as a road safety topic, is emerging as an important issue. There have been a number of attempts to define and classify the wide variety of activities, events, tasks, and objects from inside and outside the vehicle that have potential to distract the driver.[1–3] Examples that have been identified include smoking, cell phone use, adjusting in-vehicle entertainment systems or climate controls, searching for items, eating or drinking, daydreaming, attending to children, conversing with passengers, and looking at scenery or activities outside the vehicle[4] (see Chapter 15 of this book for more in-depth discussion of distraction sources).

Definitions of distraction vary in how they classify activities or objects related to driving, how they interpret distraction in relation to the concept of inattention, and how they classify driver's state such as fatigue or being drunk. Recent definitions propose that impairment from alcohol or drugs, fatigue, and other psychological states are not by themselves sources of distraction[5,6] (see also Chapters 3 and 15 of this book). However, such forms of impairment may make it easier to be distracted or may influence the effects of being distracted (see Chapter 21 of this book, regarding the link between distraction and driver impairment).

An important issue for classification, and therefore for the estimation of crash involvement, is whether to include as sources of distraction, activities, objects, or events that are related to driving. In a summary of the proceedings of a driver distraction conference held in Toronto, Canada,[6] for example, distraction is defined as "a diversion of attention from driving, because the driver is temporarily focusing on an object, person, task, or event not related to driving, which reduces the driver's awareness, decision-making, and/or performance, leading to an increased risk of corrective actions, near-crashes, or crashes" (p. 2). Distraction is thus defined in relation to events not related to driving. This suggests that tasks such as scanning for hazards, watching or scanning for other vehicles, monitoring speed, and navigation are not distractions but rather poorly performed driving task activities.

In contrast, distraction is defined in Chapters 1 and 3 of this book as "the diversion of attention away from activities critical for safe driving toward a competing activity." This definition requires the presence of a competing task, activity, or object—but an important element of it is the recognition that some driving-related activities can be potentially distracting to the driver. This chapter also discusses the impacts that the differences in definitions have on estimating the crash involvement of driver distraction.

16.3 CRASH STUDIES BASED ON POLICE REPORTS AND CRASH INVESTIGATIONS

Crash studies have provided a wide range of estimates of the contributory role of distraction in crashes.[2] Early crash studies from the 1970s, such as the Indiana trilevel study based on after-crash investigation, found that internal distraction was a definite cause in 5.7% of crashes and a probable cause in 9% of crashes.[7] Similarly, a reanalysis of the Sabey and Staughton's 1970–1974 analysis by Department for Transport, U.K., estimated that driver distraction was involved in 15.4% of driver errors.[8] More recent research has estimated driver distraction involvement in crashes to be as low as 2%[9] and as high as 25–30%.[10]

This section, using the definition of distraction proposed in Chapters 1 and 3 of this book, examines several crash studies that have used police reports or findings from crash investigation teams to provide information on a wide range of inside- and outside-the-vehicle distractions believed to have contributed to crashes. While discussing crash involvement, it is important to keep in mind that this is not the same as the crash risk of a specific distraction-related activity. Crash studies are more orientated toward providing information on the frequency and role of distraction involvement in crashes. Crash risk aims to provide information about the increase in driving risk posed by driver involvement in a distraction-related activity over and above that of the normal risk posed by driving. Estimating crash risk requires additional information on exposure, such as the amount of time spent performing different distraction-related activities while driving (see Chapter 6 of this book for detailed discussion on the issue). Crash studies typically do not collect this type of information.

16.3.1 Studies Using the National Automotive Sampling System Crashworthiness Data System

A number of crash studies have examined information from the National Automotive Sampling System Crashworthiness Data System (CDS). The CDS is a database of crash information based on an annual national probability sample of police-reported crashes in the United States.[11] The CDS samples approximately 5000 vehicles per year where the crash involved a passenger vehicle, and at least one vehicle was towed away. Trained crash investigation teams review and follow up on the police accident reports, vehicle and site information, interviews with drivers and witnesses, and available medical records. However, the CDS database still relies quite heavily on the information contained in the police reports.

In 1995, a series of codes relating to "driver's distraction/inattention" were added to the CDS. These codes included five categories of driver attention status: attentive, distracted, looked but did not see, sleepy or fell asleep, and unknown or no driver. The distraction category consisted of 13 subcategories: eating or drinking, adjusting radio/cassette or CD, other occupants in vehicle, moving object in vehicle, smoking related, talking or listening on cellular phone, using a device or object brought into vehicle, using device or controls integral to vehicle, adjusting climate controls, outside person/object or event, other distraction, and unknown distraction.[12] The CDS distraction codes are mainly related to classifying specific types of inside-the-vehicle distractions, with only one code used to capture all outside-the-vehicle distractions.

Several studies have examined the CDS for "driver distraction/inattention" and have reported on the crash involvement of the 11 driver distraction–specific codes. The estimates provided by these studies are based on nationally weighted estimates scaled up from the actual sample of crashes examined. One study, using 1995 information, identified distraction as a contributing factor in 13.3% of crashes.[13] The American Automobile Association (AAA) North Carolina study using 5 years of CDS information (1995–1999) identified distraction as a contributing factor in 10.6% of crashes.[12] A more recent analysis based on 4 years of CDS information (2000–2003) identified distraction as a contributing factor in 11.6% of crashes.[14]

Analyses based on the CDS system suggest that driver distraction is identified in around 11–12% of the crashes examined. However, these studies report high levels of "unknown" information where the driver inattention status was recorded as unknown or missing in up to 40% of the cases. Therefore, the level of distraction involvement in crashes is highly likely to be underestimated due to missing data[14] and more general issues associated with identifying distraction involvement in crashes (see Section 16.4). It has been proposed, using the CDS data, that driver distraction/inattention might be involved in up to 25–30% of crashes.[10] However, this estimate appears to be based on a combination of driver inattention factors: driver distraction, fatigue/drowsiness, and looked but did not see. Whether some or all of these factors would be considered distraction would depend on the definition being used.

A Pennsylvania state study has used the "driver distraction/inattention" CDS coding structure to examine 2 years (1999–2000) of police-reported crash data from the state of Pennsylvania.[15] This study estimated that driver's distraction was involved in 3.5% of crashes. This lower estimate of crash involvement in comparison to the CDS-based estimates is probably related to differences in the level of training of police and coders, the quality control systems in data collection involved, and region-specific crash patterns.[1]

Table 16.1 provides a comparison of the estimated level of involvement for specific distraction sources from the three studies based on the CDS information and the Pennsylvania state study. The "other distraction" and "unknown distraction" categories were combined in the table to allow comparisons across the four studies. The 1995 information was converted into a proportion for each distraction category out of all the distraction categories for comparison with the other CDS studies. The two highest distractions for all four studies are "distracted by outside person/object/event" and "other/unknown distractions." The CDS system uses only a single code to capture all outside-the-vehicle distraction sources, and the four studies suggest that outside-the-vehicle distractions constitute between 24 and 29% of all driver distractions.

The two most prominent types of inside-the-vehicle distraction are "other occupants" and "adjusting the radio/cassette/CD." There are, however, considerable differences between the 1995-1999 and 2000-2003 CDS estimates of involvement, but the reasons for the differences between analysis periods is not clear.[14] However, the more recent CDS analysis suggests that, in terms of rank order, "adjusting radio/cassette/CD" involvement has dropped significantly, and that the top four inside-the-vehicle distractions involved in crashes are "other occupant/s," "using other devices," "using/dialing cell phones," and "moving objects in the vehicle."

TABLE 16.1
Percentage Distribution of Source of Driver Distraction Based on Studies Using the National CDS Database and the Pennsylvania State Crash Information

	Proportion of Source Involvement			
Distraction Source	**CDS Data 1995[a]**	**CDS Data 1995–1999[b]**	**CDS Data 2000–2003[c]**	**PA State 1999–2000[d]**
Other occupant/s	11.5%	10.9%	20.8%	10.2%
Using other device/object brought into vehicle	1.3%	2.9%	5.2%	5.7%
Moving object in vehicle	3.8%	4.3%	3.7%	8.2%
Using/dialing cell phone	1.3%	1.5%	3.6%	5.2%
Adjusting radio/cassette/CD	15.4%	11.4%	2.9%	10.2%
Eating or drinking	1.3%	1.7%	2.8%	5.1%
Adjusting vehicle/climate controls	2.6%	2.8%	1.5%	5.2%
Smoking related	1.3%	0.9%	1.0%	4.7%
Distracted by outside person/object/event	25.6%	29.4%	23.7%	21.9%
Other distraction or unknown distraction	35.9%	34.2%	34.8%	23.6%

[a] The percentages for each distraction source from Table 1 have been created as a proportion of all distraction sources, based on 7.8% of drivers identified as distracted in crashes. Data reworked from Wang, J-S., Knipling, R.R., and Goodman, M.J. in *40th Annual Proceedings of the Association for the Advancement of Automotive Medicine*, Vancouver, British Columbia, October 7–9, 1996.

[b] Based on 8.3% of drivers identified as distracted in crashes. Data taken from Stutts, J.C. et al. The role of driver distraction in traffic crashes, Report 202/638-5944, AAA Foundation for Traffic Safety, Washington D.C., 2001. With permission.

[c] Based on 6.6% of drivers identified as distracted in crashes. Data taken from Stutts, J.C. et al. Guidance for the implementation of the AASHTO Strategic Highway Safety Plan, Volume 14: A guide for reducing crashes involving drowsy and distracted drivers, National Cooperative Highway Research Program Report 500, v14, Transportation Research Board, Washington, D.C., 2005. With permission.

[d] Based on 3.5% of crashes involving a distracted driver. Data taken from Staff report, General Assembly of the Commonwealth of Pennsylvania, Joint State Government Commission, Pennsylvania, 2001. With permission.

The AAA North Carolina study examined narrative information from the 1997/1998 CDS information, and provides an indication of the content of the grouped categories "outside-the-vehicle distraction" and "other distraction."[12] The "outside-the-vehicle" distraction category includes activities such as looking at police/crash scenes, sunlight and sunset, people playing off the roadway, drivers waving them through, vehicles moving in traffic, and animals in the roadway. Activities such as "other drivers waving them through" and "looking at other vehicles moving in traffic" could include a mixture of non-driving- and driving-related activities. By the definition coined in Chapters 1 and 3, this activity would probably be considered distraction.

The "other distraction" category appears to mix inside-the-vehicle activities, such as "looking at or reaching for objects" inside the vehicle, with outside-the-vehicle activities, such as "looking at traffic through the rear-vision mirror." By the definition provided in Chapters 1 and 3, activities such as "looking at traffic" could be considered distraction. This category also includes driver's impairment factors, such as being intoxicated/depressed, which is not considered distraction.

16.3.2 Virginia Commonwealth University Study

The Virginia Commonwealth University pilot study (VCU study) aimed to collect more specific information on crashes involving distraction in Virginia.[16] During part of 2002, the police were encouraged to report whether distraction was a contributory factor in crashes and to indicate the main source of distraction. After data quality checking, information on 2792 crashes involving 4494 drivers was received. The study did not provide an overall estimate of crash involvement for distraction, but did examine a range of specific driver distraction activities and expanded on the range of outside-the-vehicle distraction sources.

Table 16.2 provides a summary of the VCU study findings by the source of distraction. The sources identified in this study have been reorganized into four attention categories—distraction, inattention, mixed, and uncertain—using the definition of distraction coined in Chapters 1 and 3 of this book. In this study, 21% of the sources identified, such as fatigue, alcohol impairment, and driver inexperience, were placed in the "inattention" category. There are also a number of sources, such as emotional or medical impairment, indicated as "mixed," that seem likely to contain a mixture of distraction-related activities and inattention. The top four inside-the-vehicle distractions identified were "passengers/children," "adjusting the radio/changing CDs/tapes," "daydreaming or not paying attention," and "eating and drinking."

The VCU study expands the outside-the-vehicle distractions into a number of different sources. One of the sources, "driver error," is related to inattention rather than distraction. The outside-the-vehicle distraction sources also appear to include a mixture of non-driving-related activities (i.e., objects, people, and events beyond the road environment) and driving-related activities (i.e., looking at pedestrians, roadside activities, other vehicles, and road signs/traffic lights). The top two outside-the-vehicle distractions identified were "looking at a crash," "other roadside incident or traffic," and "looking at scenery/landmarks." Overall, the outside-the-vehicle distractions (if the "uncertain" and "not distraction" groups are excluded) account for 44% of the distraction sources.

16.3.3 NHTSA Large Truck Crash Causation Study

The National Highway Traffic Safety Administration large truck crash causation study (LTCC study) is an analysis of a nationally representative sample of large truck crashes from 2001 to 2003 from 24 sites across 17 states in the United States.[17] Each crash in the study involved at least one large truck where an injury was sustained. The sample consisted of 967 crashes, involving 1127 large trucks and 959 other vehicles. Information

TABLE 16.2
Distribution of Source of Distraction from the Virginia Police Data, 2002

Distraction/Inattention Source[a]	Number of Crashes (Glaze and Ellis)[a]	Inattention Type[b]	Distraction Involvement Proportion[b]
Inside-the-vehicle distraction			
Passenger/children	253	Distraction	11.5%
Adjusting radio/changing CD or tape	191	Distraction	8.7%
Daydreaming or not paying attention	125	Mixed	5.7%
Eating or drinking	123	Distraction	5.6%
Cell phone related	115	Distraction	5.2%
Adjusting vehicle controls/objects	106	Distraction	4.8%
Other personal items (i.e., wallet/purse, includes reaching for items)	86	Distraction	3.9%
Smoking related	61	Distraction	2.8%
Document, book, map, direction, newspaper	54	Distraction	2.5%
Emotional or medical impairment (i.e., driver in labor, item in eye, emotionally upset)	50	Mixed	2.3%
Grooming and other inside distractions (i.e., praying, writing, mailing)	37	Distraction	1.7%
Unrestrained pet	17	Distraction	0.8%
Technological device and pager	14	Distraction	0.6%
Fatigue and alcohol	568	Inattention	Not included
Outside-the-vehicle distraction			
Looking at a crash, other roadside incident or traffic (looking at other vehicles)	383	Distraction	17.4%
Looking at scenery/landmarks	287	Distraction	13.0%
Eyes not on road (i.e., looking at driver's blind spot, not seeing other vehicles, looking down or away from road)	101	Mixed	4.6%
Insect/animal/object entering/striking vehicle or in roadway (insect inside vehicle)	67	Distraction	3.0%
Weather conditions (heavy rain, glare, fog on windows)	56	Mixed	2.5%
Looking at other people (i.e., looking at other drivers, pedestrians, work crews)	29	Distraction	1.3%
Objects in road and other outside distractions (i.e., looking at bridge lift, vehicle dust, fireworks, yard sale)	25	Distraction	1.1%
Looking at signs (road signs, traffic lights)	25	Distraction	1.1%
Driver error (misjudgment/inexperience)	40	Inattention	Not included
Lost/unfamiliar with roads and auto mechanical problems	16	Uncertain	Not included
Unknown distraction/inattention	90	Uncertain	Not included

[a] Data taken from Glaze, A.L., and Ellis, J.M., Pilot study of distracted drivers, Report, Transportation and Safety Training Center, Centre for Public Policy, Virginia Commonwealth University, January 2003. With permission.

[b] Using the definition of distraction used in Chapters 1 and 3 of this book.

TABLE 16.3
Nationally Weighted Percentage Crash Involvement for the Critical Reasons "Driver's Recognition Errors" from the Large Truck Crash Causation Study (2001–2003)

	Estimated Proportion of Weighted Involvement in Crashes Involving Trucks	
Critical Reason	**Trucks**	**Passenger Vehicles**
Driver recognition error	35%	30%
Inattention (including daydreaming)	7%	4%
Internal distraction	3%	9%
External distraction	6%	2%
Inadequate surveillance (includes failed to look, looked but did not see)	19%	11%
Other recognition errors	0%	1%
Unknown recognition errors	1%	4%

Source: Adapted from Large Truck Crash Causation Study Summary Tables, Table 16, Federal Motor Carrier Safety Administration and National Highway Traffic Safety Administration, U.S. Department of Transport. Website accessed on October 25, 2007, http://www.ai.fmcsa.dot.gov/ltccs/default.asp?page=reports.

was collected at crash scenes and postcrash including detailed descriptions of the crash environment, vehicles involved, drivers, and other witnesses. Analysis of the contributing factors involved in crashes was divided into critical reasons (factors deemed to be critical to the crash) and associated factors (factors deemed to be present in the crash).

Table 16.3 provides a summary of the nationally weighted crash involvement for trucks and passenger vehicles involved in truck crashes for the critical reason "driver's recognition error," which includes distraction. Internal and external distractions were involved in 9% of the large truck crashes and 11% of the passenger vehicles in crashes involving large trucks. This is consistent with the estimates from the studies based on the CDS information in Section 16.3.1. Some of the other categories, such as inattention (which includes daydreaming), would appear to contain some crashes that would be considered distraction. Driver recognition error as a group of factors was involved in 30% of the passenger vehicle crashes, although this estimate includes a mixture of distraction- and inattention-related factors. More detailed information on the type of behavior that was grouped together for the inside- and outside-the-vehicle categories can be found in the LTCC study coding manual.[18]

Based on passenger vehicle drivers in the LTCC study, outside-the-vehicle distraction was involved in 18% of the internal and external distractions. This is a little lower than the CDS analysis estimates of 24–29% and much lower than the VCU estimate of 44%. Truck drivers have a higher proportion of outside-the-vehicle distraction involvement as a proportion of the driver recognition error factors than the LTCC study passenger vehicle drivers involved in crashes with large trucks. This could reflect differences in the types of distracting activities engaged in by truck drivers relative to passenger vehicle drivers, or differences between the driver groups in experience or training in managing distraction.[19]

16.3.4 New Zealand Ministry of Transport Driver Distraction Study

The New Zealand crash analysis system (CAS) is a national police-reported database.[20] All crash-related information is captured, including driver and witness interviews, scene descriptions and diagrams, attendant officer comments/notes, and other related documentation. While the report form does not provide specific reminders for distraction/inattention, the police are encouraged to provide a full narrative of the crash scene and comments on what contributed to the crash. The crash reports are processed by trained coders. Driver distraction–related information is captured under a series of codes called "attention diverted by." These codes contain a mixture of inside-the-vehicle distraction factors, possible preexisting driver's state factors such as being emotionally upset, and a range of driving- and non–driving-related outside-the-vehicle factors.

The Ministry of Transport driver distraction study (MoT study) reviewed the content and narrative of the traffic crash reports of all drivers where one of the contributing factors in the crash was "attention diverted by" over a 2-year period (2002–2003).[21,22] A total of 1964 vehicle crashes involving cars, light vehicles, motorcycles, and heavy vehicles were examined. The study categorized the information at two levels. The first level was based on the source of the distraction/inattention (i.e., the object, person, animal, or scene). The second level, within the first level, was based on the behavior involved (e.g., looking at, reaching for, conversing, using).

Table 16.4 shows the proportion of involvement for the sources identified as a proportion of all the crashes where "attention diverted by" was a contributing factor. This information is a reworking of previous analyses.[21,22] Each source has been classified based on the type of inattention and distraction (internal versus external, non-driving-related versus driving-related) using the definition of distraction coined in Chapters 1 and 3 of this book as a guide. Many of the sources of distraction identified include a mixture of non-driving- and driving-related distractions. This issue is particularly relevant when considering outside-the-vehicle distraction. Up to 45% of the sources identified involved, to some degree, driving-related distraction, and in particular two driving-related sources "driver dazzled (sun glare or headlights)" and "checking for traffic" are nearly 26% of all the sources.

The MoT study expands on the different types of outside-the-vehicle distraction sources. An estimate of the involvement of outside-the-vehicle distraction depends on whether driving-related activities are included. Under the definition of distraction provided in Chapters 1 and 3 of this book, which includes driving-related activities, outside-the-vehicle distraction is involved in 33% of all distraction-involved crashes. If driving-related outside-the-vehicle activities are excluded, then outside-the-vehicle distraction is involved in 16% of the distraction-involved crashes. This highlights the impact of the decision when defining distraction of whether or not to include driving-related activities/events as distraction.

In previous analyses of the MoT information, "driver dazzled" was included as an outside-the-vehicle distraction.[21,22] It has also been argued that "driver dazzled"

TABLE 16.4
Proportion of Crashes Involving Drivers with "Attention Diverted by" Source as a Contributing Factor to the Crash from the New Zealand CAS System (2002 and 2003)

Attention Diverted by Source	Type of Inattention or Distraction	Proportion of Involvement
Driver dazzled—sun strike	Driving-related distraction	13.1%
Passenger/s	Internal distraction	11.7%
Checking for traffic	Driving related external distraction	11.2%
Other road users—vehicles	Non–driving- and driving-related external distraction	6.5%
Cell phone or communication device	Internal distraction	5.5%
Entertainment systems	Internal distraction	5.3%
Emotionally upset, preoccupied, daydreaming	Internal distraction and driver state–related inattention	5.2%
Personal effects	Internal distraction	4.6%
Vehicle controls/devices	Non–driving- and driving-related internal distraction	4.6%
Trying to find destination/location	Driving-related external distraction	3.7%
Food—drink	Internal distraction	3.3%
Scenery—persons	Non–driving-related external distraction	2.7%
Police/emergency vehicles/crash scene	Non–driving- and driving-related external distraction	2.4%
Smoking	Internal distraction	2.3%
Animal or insect inside vehicle	Internal distraction	1.8%
Scenery—landscape/architecture	Non–driving-related external distraction	1.8%
Other road users—pedestrians/cyclists	Non–driving- and driving-related external distraction	1.7%
Driver dazzled—headlights	Driving-related distraction	1.4%
Scenery—animal outside vehicle	Non–driving and driving-related external distraction	1.1%
Sneezing/coughing/itching	Internal distraction	>1%
Other external event	Non–driving-related external distraction	>1%
Advertising/road signage	Non–driving and driving-related external distraction	>1%
Inside distraction—undefined	Internal distraction and inattention	2.9%
External distraction—undefined	External distraction and inattention	1.9%
Distraction—undefined/unknown	Distraction and inattention	6.5%

Source: Adapted from Gordon, C., What do police reported crashes tell us about driver distraction in New Zealand?, paper presented at *Australasian Road Safety Research Policing Education Conference*, Wellington, New Zealand, November 14–16, 2005. With permission.

is not distraction,[2] while Chapter 15 of this book argues that "driver dazzled" is an inside-the-vehicle distraction due to the driver actions taken in response. Because of the high level of involvement of "driver dazzled" (14.5%), classification has implications when discussing the involvement of inside- versus outside-the-vehicle distractions. If an activity related to "driver dazzled" is assumed to be an inside-the-vehicle distraction, then the top four inside-the-vehicle distraction sources are "driver dazzled," "passenger/s," "cell phones or communication devices," and "entertainment systems." The higher level of cell phone involvement reported in crashes in this study may be because New Zealand does not have a law that specifically prohibits the use of cell phones while driving.

Overall, if the driving-related distraction sources are excluded, distraction is estimated to be a contributing factor in 6% of all police-reported crashes in New Zealand. If driving-related sources, such as "driver dazzled," "checking for traffic," and "trying to find destination," are included, then driver distraction is estimated to be a contributing factor in around 10% of police-reported crashes in 2002–2003. A more recent analysis of 2006 information found that just over 11% of police-reported crashes involved diverted attention/distraction.[23] Because of the nature of the information collection method, the level of involvement in crashes is expected to be an underestimate.

16.3.5 Summary

The crash studies discussed have consistently identified distraction as a contributing factor in about 10–12% of crashes. The two studies examining the CDS information based on multiple years (1995–1999 and 2000–2003) identify distraction as a contributing factor in 11–12% of crashes. The U.S. large truck study identified distraction as a contributing factor in 9–11% of crashes. These studies, while using police reports, also involve crash investigation teams to screen and check the information. They are, however, limited to investigations of more severe crash types such as tow-away vehicle crashes or crashes involving trucks. The New Zealand MoT study, based on a national police reporting system, identified distraction as a contributing factor in 6–10% of crashes, depending on which different classification decisions are made. Under the definition cited in Chapters 1 and 3 of this book, the MoT study estimate stands at 10–11%.

The studies also provide an indication of which sources of distraction appear to be most commonly identified in crash reports or investigations, for both inside- and outside-the-vehicle distractions. Differences in the classification of distraction sources and other forms of inattention, however, mean that there is considerable variation across the studies. These issues are discussed in more depth in Chapter 15 of this book.

16.4 LIMITATIONS OF "TRADITIONAL" CRASH STUDIES

Most of the limitations associated with crash-based studies are related to the retrospective nature of the collection of crash information and the processes used to collect the information.[24] Many jurisdictions and crash-reporting systems do not routinely collect or enable information on distraction to be recorded or retrieved for analysis.[11] The crash studies discussed earlier have all made use of data collection systems and processing that has allowed driver distraction information to be analyzed. Nevertheless, the majority of these crash studies have made use of police reports of the crashes involved as one of the sources of crash information.

An important limitation of police-reported crashes is that they are a subset of all crashes, and may therefore be considered a biased sample.[11,12,24] The collection or analysis of crash information often involves the use of sampling methodologies (such as the CDS) or the use of the entire population of police-reported crashes (such as the New Zealand crash analysis system). However, these methods are limited as the police or other agencies that collect crash information do not attend all crashes. Furthermore, the reporting system may also include criteria that restrict the type of crashes recorded. For example, the CDS system only samples from tow-away police-reported crashes.

All of the crash studies report significant amounts of unknown/missing information or information that was insufficient to allow for more detailed coding of the type of distraction involved. As a result, the level of overall involvement for drivers or crashes provided in the studies is highly likely to be underestimated. It also places uncertainty on the estimates of involvement for the specific types of distraction, as the distribution of missing values could be biased in some way. Given that the studies also tend to be based on relatively low numbers of cases identified as distractions, the margin of error associated with the estimates of overall involvement and for specific types of distraction is also likely to be high.[11]

One of the difficulties faced in the collection of crash information is trying to assess the involvement of human factors–related information such as distraction/inattention in crashes after the crash has occurred.[8,24] Crash databases based on information collected after the crash, such as police report–based systems, tend to be better at providing information about the vehicle, environment (i.e., weather conditions, some driver/witness information), and physical crash scene information. The quality of information recorded and contained within crash databases on human factor involvement can be influenced by a number of factors.

Investigating crashes after they have occurred introduces a degree of subjectivity. The involvement of distraction can be inferred from evidence, such as objects in the vehicle or on the person. Usually, however, even if such evidence is present, the driver/s or witnesses involved in the crash are also relied on to provide information about what went on. A variety of potential sources of bias are possible.[14,24,25] For example, people may provide what they think are plausible reasons, or may not be honest in stating their reasons for the crash. Such reporting biases are likely to favor specific types of distraction.[14] For example, in jurisdictions where cell phones are illegal to use while driving, use of them is likely to be underreported. These biases can be minimized through careful questioning, linking to complementary sources, cross-referencing of information, and other similar techniques.

The level of training of the investigating personnel is also important to the quality of the information collected. While professional crash investigation teams are sometimes used to collect information from crash scenes, in many cases, less specialized personnel are used. In the case of the police, the level of training in crash scene investigation can vary widely between jurisdictions, and subjectivity and inferential weakness are common problems with police reports.[24,26] Where the level of training for the crash investigators is low, the processes, guidelines, and design of the reporting form, such as prompting for distraction in some way,[14] will be important to the quality of the information collected.

The quality of the information available for analysis also depends on the detail of the information provided on the source of distraction, whether the coding structures are designed to cope with the variety of potential distractions, and how much of the information is processed or retained. Retaining the raw data means that analysis is less limited by processing decisions or a lack of detailed coding structure for distraction. Check boxes on reporting forms may act as prompts and provide more consistency for considering the involvement of distraction, but they do not provide detailed information. Narrative information can provide more detail but may vary in quality of content.[26] The design of reporting forms, training of personnel, and processes around the storing and coding of raw information is important.

Finally, crash databases and crash studies typically do not collect information on exposure, such as the frequency and duration of occurrence of specific types of distraction activities. Without this additional information, crash studies are limited to estimating the frequency of involvement of distraction-related activity as a contributing factor in crashes. Statements about the relative crash risk of distracting activities require information on the frequency and duration of the activity during normal driving (see Chapter 6 of this book). Naturalistic observational crash studies may help in overcoming some of the limitations of traditional crash studies.[25]

16.5 AN EXAMPLE OF A NATURALISTIC CRASH STUDY

The 100-car naturalistic driving study is a different type of crash study in that it does not rely on secondary source data such as police reports or crash investigations. In this study,[27] one hundred drivers who commuted into or out of the Northern Virginia/Washington, D.C. metropolitan area had their vehicles instrumented or received instrumented leased vehicles. Data on drivers' behavior and incidents/crashes were collected on approximately two million vehicle miles over a 12- to 13-month period. The study collected crash and precrash information on 241 primary and secondary drivers, 69 crashes, 761 near crashes, and 8295 incidents. The findings discussed in this section are based on an analysis of the crashes and near crashes only.

The study divided inattention into the following four categories:

- Secondary task involvement (tasks that are not necessary to the primary task of driving and that are typically referred to as driver distractions)
- Driving-related inattention to the forward roadway (e.g., checking rear mirrors or blind spots, where the driver is paying attention to the driving task but not the critical aspect of the driving task)
- Nonspecific eye glance away from forward roadway (where the driver glances away from the roadway but what he or she is looking at is not discernable)
- Fatigue or drowsiness

The 100-car study estimates that 78% of crashes and 65% of near crashes involved one or more of the inattention categories as a contributing factor to the crash. Out of the crashes and near crashes where inattention was identified as a contributing factor, secondary task activity was involved in approximately 40–43%, driving-related inattention in 30–33%, nonspecific eye glancing in 29–32%, and drowsiness in 14–17%.[28]

16.5.1 Analysis of Secondary Task Distraction

Table 16.5 provides a breakdown of the number of crashes and near crashes where specific types of secondary task distraction were identified as a contributing factor. The top four secondary tasks were wireless devices (mainly cell phones), passenger-related distraction, internal distraction, and dining (eating and drinking). Internal distraction covers a variety of activities such as reading, animals in the vehicle, and reaching for or moving an object in the vehicle. Wireless device involvement

TABLE 16.5
Number of Crashes or Near Crashes by Type of Secondary Task Identified as a Contributing Factor in the 100-Car Naturalistic Study

Type of Secondary Task	Number of Crashes or Near Crashes[a]
Wireless devices	69
Passenger-related secondary task	59
Internal distraction (i.e., reading, animal or object in vehicle, reaching for or moving object)	39
Dining	21
Vehicle-related secondary task (adjusting climate controls, radio, CD, cassettes, or other integral vehicle devices)	18
External distraction (looking at object, pedestrian, crash incident, or other outside-the-vehicle object/event)	15
Talking/singing—no passenger apparent	13
Personal hygiene	13
Daydreaming	8
Others	7
Smoking	1

[a] Based on 241 drivers involved in crashes or near crashes.

Source: Adapted from Dingus, T. et al., *The 100-Car Naturalistic Driving Study, Phase II: Results of the 100-Car Field Experiment*, Report DOT HS 810-593, National Highway Traffic Safety Administration, Washington, D.C., 2006.

in crashes/near crashes is much higher in this study than the other crash studies discussed in this chapter.

The 100-car study also collected exposure information on the frequency and duration of behavior performed while driving. This meant the analysis could provide estimates of the odds ratio or crash risk associated with specific secondary task activities. From this information, the researchers also estimated the proportion of crashes and near crashes occurring in the population at large that were attributable to specific secondary task activities, which they termed "population-attributable risk." Overall, this analysis found that secondary task activities were a contributing factor in approximately 23% of crashes and near crashes that occurred in the population at large.[29]

The crash risk associated with individual secondary tasks is provided in Table 16.6. An odds ratio of 1 means that the activity is on a par with the crash risk associated with normal driving. An odds ratio above 1 means that the activity is more risky, whereas an odds ratio below 1 means that the activity is less risky. For example, "reaching for a moving object" is 8.8 times more risky than normal driving. Secondary task activities were also categorized into three levels (simple, moderate, and complex) based around the amount of complexity involved in performing them. As the level of complexity increases, so does the associated odds ratio. The odds ratios for moderate and complex secondary tasks show that drivers are at increased risk when performing these types of tasks while driving.

Population-attributable risk estimates, which take into account the odds ratio and the amount of exposure (time spent performing the activity), were also calculated for the three levels of secondary tasks and for individual secondary tasks. The population-attributable risk was calculated only for those individual secondary tasks where the odds ratio for the activity was associated with an increased level of crash risk over and above that associated with normal driving (i.e., the odds ratio was greater than 1). Moderate secondary tasks were associated with 15.2% of crashes and near crashes in the population at large, whereas simple and complex secondary tasks had similar levels of contribution at around 3–4%.

The difference between the size of the odds ratio and the population-attributable risk estimates reinforces the importance of including exposure information when discussing the role of distraction in crashes. For example, consider the difference in the odds ratio and the population-attributable risk estimate for "reaching for a moving object." This has a high odds ratio as it is a risky activity when performed, but it is not performed very often and is a short-lived activity, thus, it has a low population-attributable risk. The moderate secondary tasks have a high population-attributable risk because they tend to be more frequent or are performed for sustained periods of time.

The use of naturalistic observation techniques removes some of the important limitations of traditional crash information collection. In particular, issues related to relying on inference from evidence or on driver or witness reports of what occurred are minimized. However, there are limitations associated with the capture or interpretation of behavior related to the placement of technology. For example, it is difficult to interpret what a driver is looking at in terms of outside-the-vehicle distractions. In addition, the study was based on a low number of crashes (69), uses near crashes, and the sample was biased toward younger drivers and drivers with higher-than-average

TABLE 16.6
Population-Attributable Risk Percentage Estimates for Specific Secondary Tasks from the 100-Car Naturalistic Study

Type of Secondary Task	Odds Ratio	Population-Attributable Risk Percentage
Secondary task levels		
Simple secondary task (i.e., adjusting radio or other in-vehicle devices, singing, talking to passengers, smoking, drinking, lost in thought)	1.18	3.3%
Moderate secondary task (i.e., talking/listening to handheld device, changing CD or cassette, reaching for objects, combing hair, personal hygiene, eating, looking at external object)	2.10[a]	15.2%
Complex secondary task (i.e., dialing or answering handheld device, using a PDA, reading, animal in vehicle, reaching for moving object, applying makeup)	3.10[a]	4.3%
Individual secondary tasks		
Reaching for a moving object	8.8[a]	1.1%
Insect in vehicle	6.4	0.4%
Looking at external object	3.7[a]	0.9%
Reading	3.4[a]	2.9%
Applying makeup	3.1[a]	1.4%
Dialing handheld device	2.8[a]	3.6%
Inserting/retrieving CD	2.3	0.2%
Eating	1.6	2.2%
Reaching for nonmoving object	1.4	1.2%
Talking/listening to a handheld device	1.3	3.6%
Drinking from open container	1.0	0.0%
Other personal hygiene	0.7	na
Adjusting radio	0.6	na
Passenger in adjacent seat	0.5[a]	na
Passenger in rear seat	0.4	na
Combing hair	0.4	na
Child in rear seat	0.3	na

[a] Means that the result was statistically significant from an odds ratio of 1.

Source: Adapted from Klauer, S., et al., *The Impact of Driver Inattention on Near-Crash/Crash Risk: An Analysis Using the 100-Car Naturalistic Driving Study Data*. Report DOT HS 810-594, National Highway Traffic Safety Administration, Washington, D.C., 2006.

mileage within the metropolitan area of Washington, D.C. However, overall, this approach to information capture is likely to be less subject to a range of challenges faced by traditional data capture methods, and should provide reliable information on the involvement of distraction in crashes and near crashes.

16.6 OTHER CRASH RESEARCH FINDINGS

Crash studies also collect information such as the type of vehicle movements involved in the crashes and demographic information about the drivers. Although distraction is involved as a contributing factor in crashes across all age groups, there is evidence that young drivers are more likely to be involved in distraction-related crashes.[12,14,23,27] Specific distraction-related activity may also be more prominent as a contributing factor among different age groups. For example, using "in-vehicle entertainment systems" (i.e., adjusting the radio, changing CDs) was associated more often with drivers under 20 years of age than other age groups, "distraction by other occupants" was associated more often with drivers aged between 20 and 29, and "distraction by outside-the-vehicle objects or events" was associated more often with drivers aged 65 or more.[12,14,15] There is also some evidence that engagement in distracting activity may be associated with other contributing factors such as alcohol and fatigue.[30]

Studies examining the CDS data show that, in comparison to attentive drivers, distraction appears to be largely associated with rear-end crashes, same–travel way/same-direction crashes, single-vehicle crashes, and crashes occurring at night.[12–14] Analysis of New Zealand crash data on distraction also suggests that the type of crash involved may vary with speed zones (open road vs. urban).[23] The 100-car study found that 93% (14 out of 15) of rear-end crashes involved inattention to the forward roadway as a contributing factor.[27] Here the drivers seem to be paying attention to the driving task, but not the critical component, which would be considered distraction by the definition proposed in Chapters 1 and 3 of this book.

16.7 DISCUSSION

16.7.1 Overall Involvement in Crashes

The available crash data do not provide great certainty about the level of involvement of distraction in crashes. Overall, the traditional crash studies examined in this chapter indicate that driver distraction, as defined under the definition of distraction adopted in Chapters 1 and 3 of this book, is involved in at least 10–12% of crashes. This is known to be an underestimate due to the limitations of the information sources used, but it is unclear by how much. If other forms of inattention are also considered to be distractions, then the estimate of the involvement of distraction in crashes is considerably higher—as high as 25–30%.[7]

The 100-car study provides a reliable indication of what the level of underestimation might be for secondary task activities. Based on the population-attributable risk calculation, which takes into account exposure information, secondary task distraction is estimated to be involved in 23% of crashes and near crashes. However, if distraction is considered to include other forms of inattention, such as driving-related inattention (as proposed by the definition in Chapters 1 and 3 of this book) or non-specific eye glances (see Chapter 15 of this book), then the level of involvement in crashes will be substantially higher. In the end, how much distraction is involved in crashes depends on how broadly distraction is defined within the broader concept of inattention.

16.7.2 Classification of Distraction

The CDS and New Zealand CAS systems are examples of crash databases that have preexisting coding structures in place for distraction and inattention. However, examination of the content of these categories suggests that, in some cases, inside- and outside-the-vehicle factors are included in the same category, and that there are difficulties in separating secondary task activity or non–driving-related activity from driving-related activity that competes with the primary task of driving. The VCU study showed similar classification issues and also included factors related to impairment, such as alcohol or fatigue, which the Toronto conference and Chapters 1 and 3 (of this book) definitions of distraction would exclude.

A related issue concerns the basis on which the individual factors are grouped together. The studies discussed approach classification in different ways. In classifying distraction it is important to separate the object/event from the behavior/action involving the object/event while maintaining flexibility for discussing different aspects of distraction. For example, reaching for a cell phone in the glove compartment is part of the repertoire of behavior that the driver can perform under cell phone–related distraction. However, when investigating the role of reaching/searching for objects, it is the behavioral action of reaching/searching that is important, rather than the specific object involved.

Greater effort appears to have been placed on developing individual categorizations for inside-the-vehicle distractions than for those outside the vehicle. For example, the CDS system provides only a single-group category for all outside-the-vehicle distractions. The MoT study and the VCU study provide suggestions on what factors might be considered for an expansion of outside-the-vehicle factors for crash-based studies (also see Ref. 31 for a review of external sources of distraction). The 100-car study provides an alternative approach by specifying four types of inattention, of which two (driving-related inattention to the forward roadway and nonspecific eye glance from the forward roadway) are related to activities outside the vehicle.

Decisions on how to deal with the range of outside-the-vehicle distractions are likely to raise conceptual issues for defining and classifying distractions. Separating distractions deriving from secondary task activities from driving-related activities that compete with the primary driving task is likely to be problematic for crash studies. For example, the MoT and VCU studies show that the majority of the outside-the-vehicle distractions include a mixture of non–driving- and driving-related distractions/inattentions. As a result, decisions about whether these count as distractions or not have important implications for the estimate of their involvement in crashes.

The MoT study (Section 16.3.4) illustrated that such a difference in classification could change the estimate of crash involvement by over one-third (from 6%, excluding only driving-related activity, up to 10%, if all "attention diverted by" factors were included). Similarly, some specific activities such as driver dazzled (sun glare) may pose classification difficulties: first, whether it should be counted as distraction, and second, whether it is an inside- or outside-the-vehicle distraction. Where such activities are highly involved in crashes (such as the MoT study), this has important implications for discussions on the relative involvement of inside- versus outside-the-vehicle distractions.

The refinement of definitions such as the Toronto definition or the definition provided in Chapters 1 and 3 of this book will help to address some of the challenges involved in classifying sources of distraction. A next step would be to develop an agreed classification system using one of these definitions. This classification could then be used to review and refine existing coding structures in crash data systems and in future crash studies. The crash studies examined in this chapter highlight some issues that need to be resolved:

- Separating factors related to impairment, such as alcohol, fatigue, and psychological states, from distraction
- Separating inside- and outside-the-vehicle distractions so they are not included within the same factor
- Separating secondary task–related factors from inattention related to the primary task of driving or other forms of inattention
- Determining what individual distractions should be grouped together, and how
- Resolving how to code distractions in terms of the object/scene and the behavioral actions involved

16.7.3 Inside and Outside-the-Vehicle Distractions

The most consistent, and often the highest reported, individual inside-the-vehicle distraction in the crash studies reviewed here is "other occupants/passengers." The 100-car study, however, found that engaging with passengers was a low-risk activity in terms of the odds ratios (0.4–0.5). The authors of the study suggest that engaging with passengers may indicate that the driver is alert or that the passenger can assist the driver, and they classify talking with passengers as a simple secondary task activity.[29] It may also be because exposure is likely to be high, in the sense that, if another person is present, some form of engagement is likely.

It is known that passengers can enhance or diminish driver safety, depending on the relative age and gender of the driver and passengers, the complexity of the conversation, and the social relationship between driver and passengers.[1] For example, passengers can distract drivers in a variety of ways (i.e., converse, argue, or interfere with, require the attention of, if a baby), help keep the driver alert, or cause him or her to drive more cautiously (parents in vehicle) or recklessly (i.e., increased risk of similar-aged passengers associated with novice drivers). Different types of engagement are likely to involve different types of risk. When discussing the role of passenger-related distraction and associated crash risk, it may be important to distinguish between different types of drivers and different types of engagement.

It is difficult to provide a ranking of the involvement of other inside-the-vehicle distractions because of the inconsistencies in the way they are categorized and the differences in rank order across the studies. With this in mind, the approach of the 100-car study, which grouped different individual distractions into three levels of complexity, may be useful.

The group of distractions in the moderate group was the main contributor to crash involvement, based on an assessment of the crash risk they posed and the amount of time spent performing them while driving. Moderately complex tasks contribute to crash and near-crash involvement, not because they are the highest-risk behaviors, but because they have a substantial degree of risk (more than simple tasks) and they are frequently performed or have relatively long durations when performed. The types of secondary task included in this group are talking/listening on handheld devices, changing CDs or cassettes, reaching for objects, combing hair or personal hygiene activities, eating, and looking at external objects.

The traditional crash studies provide a range of estimates, ranging from 18 to 44% of all distractions, for the involvement of outside-the-vehicle distractions. This range is partly due to differences in how competing driving-related activity, such as mirror checking, looking at traffic or focusing on the wrong aspects of the traffic environment, are classified. The 100-car study found that external distraction was only a small proportion of the secondary task distractions (15 crashes or near crashes out of 241 drivers involved in crashes or near crashes). However, this excludes the contribution of "driving-related inattention to the forward roadway" and "nonspecific eye glances away from the forward roadway." Given this variability, it would tentatively appear that outside-the-vehicle distraction is involved in one-third of distraction-related crashes.

Many crash studies deal with only outside-the-vehicle distractions as a combined variable. The MoT and VCU studies, while providing more detailed breakdowns of outside-the-vehicle distractions, are not easily comparable. However, they suggest that an expansion of the outside-the-vehicle categories could include looking at scenery (persons, vehicles, landscape, architecture, and activities), advertising, police/emergency vehicles, crash scenes or other roadside incidents, and competing driving task activity (see Chapter 15 for further discussion of the issue).

16.7.4 The Role of Technology and Distraction

In-vehicle technology such as vehicle controls and entertainment systems, and nomadic devices brought into the vehicle such as cell phones are potential sources of distraction. Overall, technology-based distractions appear to account for at least 15–20% of all distractions identified in the traditional crash studies and around one-third of the secondary task distractions in the 100-car study. There are likely to be a number of biases that make drivers less likely to report such distractions, such as laws prohibiting their use. When combined with the difficulties associated with reporting and coding the use of technological devices in crashes, the traditional crash studies are likely to be a conservative estimate.

The rapid development and prevalence of in-vehicle technology, aftermarket additions or modifications to vehicles, and portable communication and entertainment devices (such as cell phones and iPods) that can be brought into vehicles pose problems for crash studies.[1,6] Crash analysis is not likely to be able to quickly identify the crash involvement of specific technologies. This is because the reporting systems and coding structures are not designed to cope with rapid changes in technology placed into vehicles or used in vehicles. This also poses problems for classification. For example, consider this problem: how would an iPod connected to a car radio be coded; is this a vehicle device or an entertainment system?

The crash databases, especially ones based on police reports, are also not designed for the level of detail required to identify specific technologies. Even if technological devices are reported or coded as contributory factors in crashes, sufficient information has to be reported by the crash investigator to identify the device. This information then needs to be coded or transferred into the crash information system, and there is, typically, a considerable delay involved in a report being fully coded into a crash database. Even if these problems can be addressed to some degree, it might take years for sufficient information to become available to reliably detect a potential crash issue with different types or brands of technology. However, changes in the evolution of technology are measured in terms of months.

Although crash-reporting systems should be improved to better cope with identifying technology involvement in crashes, it is unrealistic to expect them to be reliable enough for detecting changes at a detailed level of complexity or specificity. Other experimental or simulation-based methodologies are more appropriate for identifying crash risk.

16.7.5 Improved Crash Reporting

Crash studies based on retrospective crash investigation, such as police crash reports, have a number of limitations associated with them related to data availability, data quality, and reliability. Data collection, especially when it involves police reporting, can broadly be improved in three ways.

First, the way traditional crash analysis is performed can be improved. Data capture at the scene could be improved by improving the crash-reporting forms, suitably training the investigators, and making use of technology to capture some crash scene information. System processes could also be improved, such as in the capturing and storing of raw data, revising the coding structures for distraction, and using trained coders. Any changes should consider the aims, purpose of, and impact on the whole system. However, because of the nature of distraction, it is likely that the information gathered from traditional crash analysis methods will not be as accurate as for other aspects of driver's behavior that are easier to quantify and measure.[14]

The second approach is to undertake specialized crash studies to collect information on distraction, such as the 100-car study, in-depth reviews of a sample of crash reports, or the deployment of specially trained crash investigation teams. The third approach is to build into vehicles data-recording devices, "black boxes,"[2] or other technology that will record precrash and crash event information, which crash investigators can then use.

Naturalistic observational studies, using instrumented vehicles, have advantages over the traditional crash investigation approach. For example, they can collect precrash information, do not rely on self-reports of drivers or witnesses, record all the raw data, and collect exposure information. However, the technology currently used may also make it difficult to assess some distraction activities such as cognitive distraction, the role of passengers if there are privacy concerns, and the identification of external sources of distraction. While acknowledging the limitations of self-reporting of behavior, traditional approaches using trained crash investigators may be able to provide information on the role of some distractions that the use of technology cannot.

16.8 CONCLUSION

Studies based on police crash reports and, to a lesser extent, trained crash investigation teams have a number of limitations related to the retrospective approach to data collection. The results of these studies therefore need to be interpreted within the limitations of the data collection methodologies. Nevertheless, the crash studies reviewed demonstrate that distraction is involved in a substantial number of crashes and is a significant road safety issue.

Overall, the crash studies examined show that the estimate of distraction involvement in crashes is highly dependent on the method used and how broadly distraction is defined. Traditional crash studies suggest that driver's distraction is a contributing factor in at least 10–12% of crashes, and these studies are known to underestimate involvement. The 100-car naturalistic driving study estimates that secondary task distraction is a contributing factor in 23% of crashes and near crashes. Including driver-related inattention to and eye glances away from the forward roadway as distraction is likely to increase the level of crash/near-crash involvement. Approximately one-third of all distractions appear to be outside-the-vehicle distractions, and around one-fifth of all distractions appear to involve driver's interaction with technology.

The 100-car study derived exposure data, which is not available from traditional crash studies. These data showed that the greatest distractions in terms of increased crash risk are complex secondary tasks, such as reaching for moving objects, dialing/answering handheld devices, and reading and attending to animal/s in the vehicle. However, these tasks are not the ones that pose the greatest population-attributable risk. The tasks with the greatest population-attributable risk are moderate secondary tasks, such as talking/using handheld devices, reaching for objects, personal grooming, changing CDs/cassettes, and looking at external objects. These are the sources of distraction that should be a primary focus of countermeasure development.

Techniques are available to improve the quality of crash data, but the solutions are not necessarily simple. Approaches to data collection such as the 100-car study show promise and help to provide an indication of the level of underestimation involved in traditional crash studies in quantifying the role of distraction in crashes. In the end, all methods have their limitations. A combination of the different methods used to investigate crashes will need to be employed to build up a complete crash picture. Crash studies form part of the overall picture, along with laboratory research, surveys, and observational data. Together, they can build up a clearer picture of the role of distraction in road crashes. The evidence from crash studies, though variable, confirms that distraction is a contributing factor in a significant number of crashes and is therefore an important road safety issue. Distraction covers a wide range of individual activities and different behaviors. Nevertheless, a growing body of evidence suggests that as a class of behavior, in terms of crash involvement, distraction would appear to rank alongside other important road safety issues, such as drink-driving, fatigue, and speed.

REFERENCES

1. Regan, M., Young, K., and Johnston, I., MUARC submission to the parliamentary road safety committee inquiry into driver distraction, Monash University Accident Research Centre, Australia, 2005.

2. Victorian Parliamentary Inquiry, *Inquiry into Driver Distraction*, Report of the Road Safety Committee on the Victorian Parliamentary Inquiry into Driver Distraction, 2006.
3. Tasca, L., Driver distraction: Towards a working definition, Presented at the International Conference on Distracted Driving, Toronto, Canada, October 2–5, 2005. Webpage October 25, 2007, http://www.distracteddriving.ca/english/documents/LeoTasca_000.pdf.
4. Young, K., Regan, M., and Hammer, M., *Driver Distraction: A Review of the Literature*, Report No. 206, Monash University Accident Research Centre, Australia, 2003.
5. Canadian Council of Motor Transport Administrators, Strategy on distracted driving: A component of the strategy to reduce impaired driving (STRID), Report prepared for CCMTA's STRID task force and standing committee on road safety research and policies, June 16, 2006. Webpage October 25, 2007, http://www.ccmta.ca/english/committees/rsrp/strid-distraction/strid-distraction-strategy.cfm.
6. Hedlund, J., Simpson, H., and Mayhew, D., International Conference on Distracted Driving: Summary of proceedings and recommendations, Traffic Injury Research Foundation and The Canadian Automobile Association, 2006. Webpage October 25, 2007, http://www.distracteddriving.ca/english/conferenceSummary.cfm.
7. Treat, J.R., A study of the pre-crash factors involved in traffic accidents, *The HSRI Review*, 10 (1), 1–35, 1980.
8. Brown, I.D., *Review of the "Looked but Failed to See" Accident Causation Factor*, Road Safety Research Report no. 60, Department for Transport, London, UK, 2005.
9. Stevens, A., and Minton, R., In-vehicle distraction and fatal accidents in England and Wales, *Accident Analysis and Prevention*, 33, 539–545, 2001.
10. Stutts, J., How risky is distracted driving? What crash data reveals, Presented at the International Conference on Distracted Driving, Toronto, Canada, October 2–5, 2005. Webpage October 25, 2007, http://www.distracteddriving.ca/english/documents/JaneStutts_000.pdf.
11. Eby, D.W., and Kostyniuk, L.P., Driver distraction and crashes: An assessment of crash databases and review of the literature, Report UMTRI-2003-12, The University of Michigan, Transportation Research Institute, 2003.
12. Stutts, J.C. et al., The role of driver distraction in traffic crashes, Report 202/638-5944, AAA Foundation for Traffic Safety, Washington, D.C., 2001.
13. Wang, J.-S., Knipling, R.R., and Goodman, M.J., The role of driver inattention in crashes: New statistics from the 1995 crashworthiness data system, in *40th Annual Proceedings of the Association for the Advancement of Automotive Medicine*, Vancouver, BC, October 7–9, 1996.
14. Stutts, J.C. et al., Guidance for the implementation of the AASHTO Strategic Highway Safety Plan, volume 14: A guide for reducing crashes involving drowsy and distracted drivers, National Cooperative Highway Research Program Report 500, v14, Transportation Research Board, Washington, D.C., 2005.
15. *Driver Distractions and Traffic Safety*, Staff report, General Assembly of the Commonwealth of Pennsylvania, Joint State Government Commission, Pennsylvania, 2001.
16. Glaze, A.L. and Ellis, J.M., Pilot study of distracted drivers, Report, Transportation and Safety Training Center, Centre for Public Policy, Virginia Commonwealth University, January 2003.
17. Report to Congress on the large truck crash causation study. U.S. Department of Transport and Federal Motor Carrier Safety Administration, MC-R/MC-RRA, 2006.
18. *Large Truck Crash Causation Study: Analytical User's Manual*, National Highway Traffic Safety Administration, U.S. Department of Transportation, 2006.
19. Hanowski, R., Perez, M.A., and Dingus, T.A., Driver distraction in long-haul truck drivers, *Transportation Research Part F*, 8, 441–458, 2005.

20. Ministry of Transport, Motor vehicle crash data in New Zealand. Webpage October 25, 2007,http://www.transport.govt.nz/assets/NewPDFs/NewFolder/Motor-Vehicle-Crash-Data-2006.pdf.
21. Gordon, C., A preliminary examination of driver distraction related crashes in New Zealand, Presented at the International Conference on Driver Distraction, Sydney, Australia, June 2–3, 2005.
22. Gordon, C., What do police reported crashes tell us about driver distraction in New Zealand?, Paper presented at Australasian Road Safety Research Policing Education Conference, Wellington, New Zealand, November 14–16, 2005.
23. Ministry of Transport, 'Diverted attention by' crash fact sheet July 2007. Website October 25, 2007, http://www.transport.govt.nz/diverted-attention-1/.
24. Simpson, H.M., Distracted driving: How can we prove it's a problem?, Presented at the International Conference on Distracted Driving, Toronto, Ont., Canada, October 2–5, 2005. Website October 25, 2007, http://www.distracteddriving.ca/english/documents/HerbSimpsonproceedings.pdf.
25. Klauer, S. et al., How risky is it? An assessment of the relative risk of engaging in potentially unsafe driving behaviors, Report prepared for AAA Foundation for Traffic Safety, Virginia Tech Transportation Institute, December 2006.
26. Smith, M., Oppenhuis, M., and Koorey, G., Crash data collection—is it time for a rethink of process? Paper presented at the IPENZ Transportation Conference, Queenstown, New Zealand, 2006.
27. Neale, V.L. et al., An overview of the 100-car naturalistic study and findings, Paper no. 05-0400, National Highway Traffic Safety Administration, Washington, D.C., 2005.
28. Dingus, T. et al., *The 100-Car Naturalistic Driving Study, Phase II—Results of the 100-Car Field Experiment*, Report DOT HS 810-593, National Highway Traffic Safety Administration, Washington, D.C., 2006.
29. Klauer, S. et al., *The Impact of Driver Inattention on Near-Crash/Crash Risk: An Analysis Using the 100-Car Naturalistic Driving Study Data*, Report DOT HS 810-594, National Highway Traffic Safety Administration, Washington, D.C., 2006.
30. Gordon, C., Driver distraction: An initial examination of the 'attention diverted by' contributing factor codes from crash reports and focus group research on perceived risks, Presented at the IPENZ Technical Conference, Auckland, New Zealand, 2005.
31. Wallace, B., *External-to-Vehicle Driver Distraction*, Transport Research Series, Scottish Executive Social Research, 2003.

17 Epidemiological Research on Driver Distraction

Suzanne P. McEvoy and Mark R. Stevenson

CONTENTS

In this chapter, we provide an overview of the epidemiological literature on the association between driver distraction and the risk of crash. We examine the types of distracting activities that have been studied to date, summarize the results, and discuss future directions. The chapter begins with a definition of risk, an explanation of the hierarchy of epidemiological evidence, and a discussion about the types of study designs used by epidemiologists to identify whether driver distraction is a risk factor for having a crash.

17.1 EXPLAINING RISK

From an epidemiological (and road safety) perspective, it is necessary to know whether exposure to a distracting activity is associated with a significantly increased risk of having a crash. In other words, is the risk of a crash significantly higher when distracted during a driving trip compared with that when not distracted? Such information is crucial to inform strategies aimed at reducing crashes attributable to distraction. To estimate the risk, we undertake epidemiological studies to calculate the relative risk (RR) (or alternatively, the odds ratio [OR]). The former is the ratio of the crash rate in exposed drivers (probability of having a crash while distracted) to that in drivers not exposed (probability of having a crash while not distracted). In this case, the exposure of interest is a distracting activity, for example, mobile phone use.

TABLE 17.1
Hierarchy of Research Designs

Hierarchy	Research Design
Experimental	Randomized controlled trials
Quasi-experimental	Cohort study
	Case-control study
Descriptive	Case series
	Cross-sectional survey
	Ecological study

17.2 HIERARCHY OF RESEARCH EVIDENCE

There is a hierarchy of research design, from designs that are experimental and provide the best evidence for the effectiveness of an intervention to designs that are primarily descriptive and illustrate an injury event (Table 17.1).

The levels of evidence as defined by the Australian National Health and Medical Research Council[1] are as follows:

- Level I evidence is that obtained from a systematic review or metaanalysis of all relevant randomized controlled trials.
- Level II evidence is that obtained from at least one properly designed randomized controlled trial.
- Level III-1 evidence is that obtained from well-designed pseudorandomized controlled trials using alternate allocation or some other method.
- Level III-2 evidence is that obtained from comparative studies with concurrent controls where allocation is not randomized, including cohort studies, case-control studies, or interrupted time series with a control group.
- Level III-3 evidence is that obtained from comparative studies with historical controls, two or more single-arm studies, or interrupted time series without a parallel control group.
- Level IV evidence is that obtained from case series (either posttest or pretest and posttest), opinions of respected authorities, descriptive studies, reports of expert (i.e., consensus) committees, and case studies.

In the area of driver distraction, randomized controlled trials, which are the gold standard design, are not possible in the real-world setting as it is unethical and unsafe to randomly allocate drivers to engage in distracting activities while driving. However, it may be possible to use an experimental design in which drivers could be randomly allocated to be exposed to distracting activities (or not) in a simulator setting. In this chapter, we focus on epidemiological research designs (Level III-2), in particular case-control studies and case-crossover studies, a variant of the case-control design.

17.3 IDENTIFYING RISK FACTORS

The most common epidemiological research designs that have been used to assess the role of driver distraction in crashes have been the case-control study and the case-crossover study. In a case-control study, groups of individuals, who are defined on the basis of whether they have sustained a given injury outcome, are compared with respect to their exposure to a proposed risk factor. For example, if we wished to investigate whether passenger carriage is associated with an increased risk of crash, then cases would be drivers involved in crashes and controls would be drivers who had not crashed and were selected from the same population that gave rise to the cases. We would then compare the exposure (in this case, passenger carriage) between the two groups to calculate the odds of crashing while carrying passengers. The case-control study represents an efficient design: it can be completed in a relatively timely fashion; it can be used to test hypotheses about rare events, such as crashes; and it is less costly than other longitudinal designs such as cohort studies.

Case-crossover studies differ from case-control studies in that cases act as their own controls. Thus, the design controls for individual-level factors that may affect risk but do not change over a short period of time, for example, driver age or driving experience. This means that adjusting for these potential confounders either in the recruitment phase (e.g., by matching cases and controls on the basis of certain characteristics) or in the analysis phase (e.g., by using multiple logistic regression techniques), as is warranted in a case-control study, is unnecessary. A case-crossover study is appropriate to use when a brief exposure causes a transient rise in the risk of a rare outcome. Accordingly, case-crossover studies have been used to examine the association between mobile phone use (exposure) and the risk of crash (outcome).

The potential limitations of case-control and case-crossover studies include selection and information biases. Selection bias refers to systematic differences in characteristics between those who are selected for the study and those who are not, and information bias refers to systematic differences arising from inaccurate measurement.[2] Both can lead to results that differ from the true value, which is a threat to the internal validity of the study. Therefore, adequate assessment of the internal validity of these studies is paramount when interpreting the literature.

Both of these study designs can identify associations between an exposure and an outcome of interest. To confirm a causal relationship, a number of additional factors need to be considered, as described by the Bradford-Hill criteria.[2] These factors include the strength of the association, the temporal relationship (i.e., the exposure must precede the outcome), biological plausibility, dose-response, the consistency of the association (i.e., the study results must have been replicated in different settings and using different methods), and the coherence of the association (i.e., the association is compatible with current knowledge and theories).

Ultimately, risk factor identification has several uses. If the risk factors are causal and potentially modifiable, then the information can be used to plan, and be used as evidence for, the implementation of countermeasures. However, even the identification of nonmodifiable risk factors can be useful, as these can identify high-risk groups, who may benefit from targeted strategies. Young drivers are one such group.

The remainder of this chapter describes the key risk factors for driver distraction that have been identified from the epidemiological literature.

17.4 PASSENGER CARRIAGE AS A RISK FACTOR

Whether passengers are a risk factor for crashes has been an area of interest in recent years, particularly in relation to the role of passengers in crashes involving novice drivers. Not only is the presence of a passenger suggested to be important but also the number, gender, and ages of those passengers. In some jurisdictions, including many states in the United States, passenger restrictions have been introduced as part of graduated driver licensing programs to reduce the overrepresentation of novice drivers in crash statistics.

17.5 EVIDENCE IN RELATION TO YOUNG DRIVERS

Chen et al.[3] assessed whether carrying passengers was a risk factor for fatal crashes among 16- and 17-year-old drivers in the United States by performing an incidence study that compared data from three sources: the Fatality Analysis Reporting System and General Estimates System (1992–1997) and the Nationwide Personal Transportation Survey (1995). The investigators found that the risk of fatal driver injury increased with the number of passengers carried. Among 17-year-olds, compared with drivers of the same age without passengers, the RR of driver death per 10 million trips was 1.5 (95% confidence interval [95% CI] 1.4–1.6) for drivers with one passenger, 2.6 (95% CI 2.2–3.0) for drivers with two passengers, and 3.1 (95% CI 2.5–3.8) for drivers with three or more passengers. The increased risk was not altered by the time of day of the crash. Crashes were more likely to be fatal if the passengers were male, teenagers, or aged between 20 and 29 years.

In a Canadian study, Doherty et al.[4] examined the role of passenger carriage and passenger numbers in crashes by comparing police-reported crash data for over 306,000 licensed drivers aged between 16 and 59 years in 1988 with exposure data from a provincial travel survey conducted in the same year. Both sources were made available by the Ontario Ministry of Transport. The crash rates were presented by driver age group, gender, and type of crash (property damage only, injury crash, and fatal crash); some of the comparisons were statistically significant, whereas others were not. The investigators found that the overall crash involvement rate, among 16- to 19-year-old drivers, was twice as high with passengers as without. However, this difference was not statistically significant ($.05 < p < .10$). Passenger carriage was protective for female drivers aged 25–59 years across all three types of crashes examined; among male drivers aged 25–59 years, it was protective for fatal crashes only. In drivers aged 16–19 years, the crash rates rose with increasing numbers of passengers (none versus one versus two or more). The effect of passenger age and gender was not assessed in the study.

A total of 273,054 fatal crash–involved drivers of passenger vehicles for the years 1990–1995 were identified in the Fatality Analysis Reporting System to assess the role of passenger presence in at-fault fatal crashes.[5] All drivers involved in single-vehicle crashes or in multiple-vehicle crashes who had committed one or more driver errors were coded as at-fault. Not-at-fault drivers were involved in multiple-vehicle crashes in which the driver was deemed not to be at-fault. In the induced exposure method, the assumption is that the distribution of not-at-fault drivers is a surrogate measure of the travel exposure of all drivers in the study population. Among drivers aged 16–24 years, passenger presence was associated with proportionately more at-fault fatal crashes. For example, drivers aged 16 years were almost five times more likely to have an at-fault fatal crash when traveling with passengers than drivers aged 30–59 years traveling with passengers (RR 4.7, 95% CI 4.3–5.2). By comparison, 16-year-old drivers were twice as likely to have an at-fault fatal crash while traveling alone as drivers aged 30–59 years traveling alone (RR 2.3, 95% CI 2.1–2.5). The RR was particularly high among teenage drivers traveling with two or more teenage passengers (RR 7.9, for 16-year-olds), irrespective of time of day. The effect of passenger gender was not examined.

Over 77,000 drivers aged 16–20 years involved in police-reported crashes in Kentucky between 1994 and 1996 were studied to examine the effect of passenger carriage on the crash propensity of young drivers.[6] The induced exposure method (described earlier) was used to calculate a relative accident involvement ratio (RAIR). Young drivers had an increased risk of a single-vehicle crash when traveling with peers (RAIR 1.3). Crash propensity increased with increasing numbers of peer passengers. In this study, the gender of the passenger had little effect on crash risk. Young drivers were least likely to be involved in a single-vehicle or at-fault multiple-vehicle crash when carrying adults, children, or both (RAIR 0.7).

Using the California Highway Patrol database between January 1993 and June 1998, a case-control study was conducted to examine the association between passenger carriage and the rate of injury crashes to young drivers.[7] Driver injury crashes were analyzed, and an induced exposure method was used to estimate driving exposure. Cases were 10,795 visibly injured young drivers (16 and 17 years) in single-vehicle crashes and at-fault multiple-vehicle crashes as documented by the reporting police officer. Controls were 12,906 young drivers in multiple-vehicle crashes who were deemed not culpable. Relative to driving alone, carrying three or more male teenage passengers was associated with an increased risk of an injury crash among young drivers (adjusted odds ratio [adj. OR] 1.27, 95% CI 1.03–1.57) in an analysis that adjusted for driver age, gender, alcohol use, and time of crash. Young drivers carrying children, adults, or female teenagers were not at an increased risk of an injury crash. Indeed, the findings showed that the presence of adults was associated with a protective effect, suggesting a supervisory role for these passengers.

A case-control study of car crash injury was conducted between 1998 and 1999 in Auckland, New Zealand that investigated the effects of passenger carriage on the risk of car crash injury and involved 571 cases and 588 controls.[8] In that study, carrying two or more passengers, irrespective of the ages of the passengers, significantly increased the risk of car crash injury among drivers aged less than 25 years. The crude OR for two or more same age passengers was 5.8 (95% CI 2.3–14.3). The crude OR for two or more other age passengers was 5.2 (95% CI 1.7–16.2). After adjusting for driver gender, nighttime driving, self-reported alcohol consumption, average number of kilometers driven per week, and driver sleepiness, the adjusted ORs were 15.6 (95% CI 5.7–42.0) and 10.2 (95% CI 2.8–36.7) for two or more same, or other, age passengers, respectively.

The studies described earlier had limitations. First, the studies by Chen et al.[3] and Doherty et al.[4] used a number of different data sources to establish exposure and crash outcome, none of which was specifically designed to collect information for studies on passenger carriage. Second, exposure data came from the results of travel surveys, which are self-reported and not validated,[3,4] or induced exposure based on whether a driver was deemed to be at-fault in a crash or not.[5–7] In the study by Preusser et al.,[5] the comparison group was drivers aged 30–59 years in whom passenger carriage was relatively protective. Third, the studies provide limited information about what it is about carrying passengers, that increases the risk of a crash to young drivers (aside from the number, and in some studies, the age and gender of the passengers). Importantly, confounding by other human factors, for example, speeding, was not analyzed, although police-reported alcohol use was assessed in the study by Rice et al.[7] In relation to the study by Lam et al.,[8] there are two notable caveats. First, the CIs were wide because of small numbers within certain groups. Second, the unit of interest was car crash injury, irrespective of whether the driver or passenger had sustained the injury. Having additional occupants in a vehicle is likely to increase the probability of someone being injured in the event of a crash. Thus, using car crash injury as the unit of interest may have inflated the risk associated with passenger carriage.

An alternative way to examine the role of passenger carriage among novice drivers is to assess the effects of the introduction of passenger restrictions for this group. One example is the evaluation of California's graduated driver licensing program conducted by Masten and Hagge.[9] In July 1998, a 6-month restriction from driving with passengers under the age of 20 years was introduced, unless supervised. This was one of the several enhancements to the existing program in California. The others were a minimum 6-month instruction permit period, guardian-certified completion of at least 50 h behind the wheel (10 h at night), and a 12-month restriction on driving between 12 a.m. and 5 a.m. unless supervised by an adult. Monthly per capita fatal and injury crash rates for novice drivers involving a passenger aged less than 20 years were analyzed using intervention time-series analysis. The proportion of such crashes fell significantly and these restrictions were estimated to result in an annual saving of 816 fatal or injury crashes. However, by restricting passenger carriage, the relative proportion of crashes involving an unaccompanied driver would be expected to increase and this needs to be taken into account in any calculations. In the study by Masten and Hagge,[9] the savings were calculated based on the assumption that the graduated driver licensing passenger restriction did not cause a change in nonpassenger crashes. Additionally, teasing out the effectiveness of a single component within a program is difficult.

Overall, the research to date indicates that the presence of passengers increases the risk of a crash among young drivers and that this risk increases with increasing passenger numbers and, in some studies, in the presence of peer-age and male passengers. Importantly, teenage passengers may increase the risk of crash for young drivers by multiple pathways, which include not only driver distraction but also peer influence; the latter of which may be direct or indirect. Two recent studies support this view.[10,11] In a study of fatal crashes among 16-year-old drivers in the United States in 2003, there was an increasing trend for the crashes to be single-vehicle events and to involve speeding and driver error when accompanied by increasing numbers of teenage passengers, compared with driving alone.[10] In an observational study that examined the driving behavior of teenage drivers,[11] these drivers, on average, drove faster than the general traffic and allowed shorter headways, particularly in the presence of a male teenage passenger.

17.6 OTHER EVIDENCE IN RELATION TO THE NUMBER OF PASSENGERS CARRIED

In addition to the epidemiological evidence on the role of passengers in crashes involving young drivers, driving with two or more passengers has been associated with an increased likelihood of fatal driver injury on Friday and Saturday nights (adj. OR 2.3, 95% CI 1.1–4.7) when compared to driving with a single passenger, irrespective of driver age.[12] Drivers with two or more passengers, irrespective of driver age group, have also been shown to be twice as likely to have a crash requiring hospital attendance by a driver (adj. OR 2.2, 95% CI 1.3–3.8) compared with unaccompanied drivers in a metropolitan setting during the weekday hours of 8 a.m. to 9 p.m.[13] Another study has reported an increased risk of an at-fault police-reported crash among drivers aged 75 years and above carrying two or more passengers.[14]

In summary, although further, well-designed epidemiological studies are warranted to investigate the effects of passenger carriage more fully, the available evidence has important implications for road safety. Graduated driver licensing systems are in place in several countries, however, not all include restrictions on the carriage of peer passengers by young drivers. Studies, such as the ones reviewed here, can highlight the importance of such restrictions, independent of other measures, including nighttime driving restrictions. Advocates can use this evidence to influence policy makers to adopt restrictions with the consequent benefit of reducing crashes and injury in this vulnerable road user group.

17.7 MOBILE (CELLULAR) PHONE USE AS A RISK FACTOR

With increasing mobile phone ownership, there has been concern about whether mobile phone use while driving is a significant factor that contributes to crashes. Experimental studies have provided evidence that mobile phone use results in driving impairment (Chapters 6 and 11). Case-control studies, case-crossover studies, and most recently, a naturalistic driving study have examined the risk associated with mobile phone use in crashes or near-crashes in the real-world setting. A recent review of the evidence has been published by McCartt et al.[15]

Violanti[16] conducted a case-control study to explore the association between the use or presence of a mobile phone and fatal road crashes. Data from 223,137 police-reported crashes during 1992–1995 in Oklahoma, USA, were used for the analysis. There were 1548 driver fatalities, of whom five drivers were reportedly using a phone at the time of the crash. After adjusting for age, gender, alcohol use, speed, inattention, and driving left of center, police-reported mobile phone use (type of phone not specified) was associated with a ninefold increase in the odds of a fatal (case) versus nonfatal (control) crash (adj. OR 9.3, 95% CI 3.7–23.1). A study of 206,639 police accident reports in Oklahoma in the same period that compared drivers based on the noted presence or absence of a mobile phone in the vehicle found that drivers with phones had a significantly higher crude rate ratio of crashes involving unsafe speeds, inattention, and driving on the wrong side of the road.[17] No information on the role of potential confounders was available. A small case-control study undertaken by Violanti and Marshall[18] in New York State, USA, involving 60 cases who had crashed in the past 2 years and 77 controls who had not crashed in the past 10 years, examined the role of mobile phones in reportable road crashes and found a five and a half-fold increase in the odds of having a crash among drivers using their mobile phone more than 50 min/month (adj. OR 5.6, 95% CI 1.2–37.3).

The three papers authored by Violanti had considerable limitations including no phone billing information to demonstrate that drivers were using their phones at the time of the crash;[16–18] reliance on police accident reports that may have involved more thorough investigations into fatal crashes than nonfatal ones, resulting in an inflated risk estimate;[16] no adjustment for some variables that may affect the risk of crash, including distance traveled each year;[16–18] and small sample size with only 14 mobile phone users (seven cases and seven controls) in one study.[18]

Laberge-Nadeau et al.[19] assessed the role of mobile phones and crashes by comparing crash data for mobile phone users and nonusers in Quebec, Canada. Of a mail out of 175,000 license holders, 36,078 (20.6%) responded. The completed questionnaires were merged with phone activity records from wireless phone companies (for a period of 12–25 months between August 1998 and August 2000) and with 4 years of drivers' records and police reports (January 1996 to August 2000). In a logistic regression model, being a mobile phone user increased the risk of crash modestly (males: adj. RR 1.1, 95% CI 1.0–1.2; females: adj. RR 1.2, 95% CI 1.0–1.4). Heavy phone users had a higher risk of crash than those who used their phone infrequently, after adjusting for factors such as age and kilometers driven annually (adj. RR between 2.2 and 2.7, depending on the gender of the driver and the total number of calls per month). The study did not confirm that drivers were using their phones at the time of a crash; rather, the authors assumed that the overall frequency of phone use was correlated with the frequency of use while driving. The low response rate is likely to be relevant as crash rates were higher for nonrespondents than respondents, suggesting volunteer bias.

Wilson et al.[20] studied a sample of 3869 drivers in Vancouver, Canada, following a snapshot roadside observation of whether these drivers were seen using a mobile phone (users) or not (nonusers). Observations were collected during daylight hours between August and November 1999. Vehicle license plate, gender, and estimated age group details were used to obtain records of insurance claims and police-reported crashes and violations (available for 54% of the original sample of 7169). The nonuser sample was matched to the user sample for date, time, and location of observation. Drivers observed using a phone while driving had a higher risk of an at-fault crash than nonusers (adj. OR 1.2, $p = .04$) after controlling for estimated age, gender, alcohol offences, aggressive driving offenses, and not-at-fault crash claims, a surrogate for driving exposure. However, the following caveats should be noted. First, the study did not establish whether drivers were using their phones at the time of an at-fault crash. Second, the definition of nonusers as drivers not observed to be using a phone on a single occasion on a city street is flawed. Third, not-at-fault claims may not be a valid surrogate for the amount of driving undertaken.

Redelmeier and Tibshirani[21] used a case-crossover design to study 699 drivers in Toronto, Canada, who had mobile phones and were involved in a crash that resulted in property damage but no personal injury. Phone use in the 10 min before a crash increased the risk of having a crash by fourfold (RR 4.3, 95% CI 3.0–6.5). The risk was increased for both handheld (RR 3.9) and hands-free (RR 5.9) phones, although the proportion of drivers with a hands-free device was low. Similarly, the risk was raised irrespective of age group and gender. As described earlier, this design eliminates the need to adjust for factors that may affect risk but do not change over a short period of time, for example, driving experience, age, and risk-taking propensity. A potential limitation of the study is misclassification bias (in this case, misclassification of calls occurring after the crash as having occurred beforehand). This is possible because the exact time of a crash may not be known, mobile phone use in the aftermath of a crash is common and a call to the emergency services may not be the first call made after the event. Such misclassification would overestimate the risk associated with phone use. However, the authors estimated the time of collision using information from up to three sources (self-report, police records, and time of call to emergency services) and, where discrepancies were noted (67% of cases), chose the earliest recorded time to reduce the likelihood of misclassifying calls. As the hazard interval for the primary analysis was a period of 10 min, drivers were not necessarily on the phone at the time of their crash and the results demonstrate a statistical association only.

McEvoy et al.[22] used the same type of study design to examine the role of mobile phone use in serious crashes resulting in hospitalization of the driver. This study involved 456 drivers who used or owned mobile phones and found that drivers' use of a mobile phone up to 10 min before a crash was associated with a fourfold increased likelihood of having a serious crash (OR 4.1, 95% CI 2.2–7.7, $p < .001$). The risk was raised irrespective of whether or not a hands-free device was available for use in the vehicle (hands-free: OR 3.8, 95% CI 1.8–8.0, $p < .001$; handheld: OR 4.9, 95% CI 1.6–15.5, $p = .003$). As in the study by Redelmeier and Tibshirani,[21] the risk was similar in men and women and across age groups. The sample size of the study was not large enough to assess whether certain types of hands-free devices might be safer than others.

Both of these studies[21,22] obtained phone activity records from the participants' telecommunication companies to ascertain mobile phone use while driving during the hazard and control intervals. As discussed earlier, misclassification bias may be a factor because the precise time of a crash may not be known with certainty. In the study by McEvoy et al.,[22] in addition to using several sources to ascertain the time of collision and choosing the earliest reported time, the level of agreement between self-reported phone use and phone activity records was assessed and inconsistencies reconciled. Although the response rates were high in both studies (70% in Ref. 21 and 91% in Ref. 22), it is possible that drivers who refused to take part differed from the participants. For example, an individual may have refused to participate because of concern about the potential liability of admitting to phone use before a crash. The overall effect of such a bias would be to underestimate the risk of crash associated with phone use.

The major limitation with all of the studies presented in this section is that it is inherently difficult to determine conclusively whether a driver has actually been on the phone at the time of a crash and, as such, the results indicate a statistical association only. However, the finding that phone use increases the likelihood of a crash is consistent across a number of epidemiological studies and is supported by the experimental literature that has demonstrated driving impairment resulting from phone use in the simulator setting. Moreover, the naturalistic driving design, which films whether a driver is using a mobile phone at the time of a crash, near-crash, or other critical incident, has been developed recently and can overcome this difficulty.[23]

Naturalistic driving studies use video cameras to monitor the behavior of drivers in their everyday driving. In the recently published 100-car naturalistic driving study, the use of handheld wireless devices (primarily mobile phones) was associated with the highest frequency of secondary task distraction-related events.[24] A case-control data set, derived from the collected data, included an event database containing information on crashes and near-crashes (case data) and a baseline database containing information on multiple, randomly selected epochs during which no crashes, near-crashes, or critical incidents occurred (control data). The authors found that dialing a handheld device was associated with an increased likelihood of an at-fault crash or near-crash (OR 2.8, 95% CI 1.6–4.9). Talking or listening to a handheld device was not associated with a significantly increased risk of an at-fault crash or near-crash (OR 1.3, 95% CI 0.9–1.8). However, the population-attributable risk percentages were identical (each factor contributing to 3.6% of crashes and near-crashes) because drivers spent a much greater percentage of the time talking or listening to handheld devices than dialing.[23]

Just as the risk of crash pertaining to passenger carriage may be influenced by multiple pathways, it is also possible that factors aside from distraction play a role in the increased risk of crash observed for mobile phone users. There is some evidence to suggest that drivers who use a phone while driving have more permissive attitudes to a range of risk-taking behaviors on the road, including speeding,[25] driving with a blood alcohol concentration of 0.05 g/dL (the legal limit in Australia),[25] and seat belt use.[26] Of course, the case-crossover design takes into account individual-level factors such as these because cases act as their own controls.

Another consideration is whether young drivers are particularly vulnerable to the effects of technology-related distractions, including mobile phones. Although the epidemiological studies conducted to date have not found a difference in the level of risk from phone use by age group, no observational study has focused specifically on novice drivers. However, although the risk is raised irrespective of age group, young drivers are significantly more likely to use a phone while driving,[25,27,28] to send text messages while driving,[25] and to report more frequent use of a mobile phone while driving.[25,29] Because young drivers' exposure to this risk factor is higher, crashes resulting from phone use would be expected to occur at greater frequency among this group.

In conclusion, notwithstanding the limitations, mobile phone use while driving has been shown to be associated with an increased risk of crashing, including property-damage only and injury crashes, as well as at-fault near-crashes. Risk estimates in studies that have used phone activity records or video evidence of phone use at the time of crash are generally between three- and fourfold. The risk of crashing appears to be increased whether or not a hands-free device is available for use in the vehicle. Further research to examine whether certain types of hands-free devices are safer than others is warranted. However, given that hands-free devices do not eliminate certain distracting effects, namely, those relating to the act of conversing, it is doubtful that any device will be free of risk. If one type of device is found to be safer (but not free of risk), its increased use may paradoxically increase the number of crashes associated with phone use, as the impact of a risk factor on road safety is a function not only of the risk estimate but also the prevalence of use.

Information on the risks associated with mobile phone use while driving has resulted in restrictions in many jurisdictions. These restrictions are generally, but not always, limited to handheld phone use. In some jurisdictions, the level of restriction while driving will vary according to the driver's license status. For example, in the states of New South Wales and Queensland in Australia, a ban on all phone use for learner and first year provisional drivers was introduced in July 2007. Provided adequate education and enforcement (including both police and, where applicable, employer enforcement) are in place to reduce exposure to phone use while driving, road safety benefits in the form of lower numbers of crashes and associated injuries are likely to accrue.

17.8 OTHER DISTRACTING ACTIVITIES AS RISK FACTORS

In the 100-car naturalistic driving study, secondary task distraction contributed to over 22% of all crashes and near-crashes.[23] In relation to crashes, looking at or reaching for an object in the vehicle (12%), passengers (9%), and wireless devices (9%) were the most common distraction-related secondary tasks. In that study, to estimate the risks attributable to certain types of distractions, distracting activities were divided into three groupings: complex, moderate, and simple. Complex secondary tasks were defined as tasks requiring multiple steps, multiple eye glances away from the forward roadway, and multiple button presses. These tasks included dialing a handheld device, locating/reaching/answering a handheld device, operating a personal digital assistant (PDA), viewing a PDA, reading, dealing with an animal or object in the vehicle, reaching for a moving object, dealing with an insect in the vehicle, and applying makeup. Moderate secondary tasks were defined as those requiring up to two glances away from the roadway or up to two button presses. These tasks included talking or listening to a handheld device, inserting or retrieving tapes or compact disks (CDs), reaching for an object except a handheld device, combing or fixing hair, other personal hygiene-related activities, eating, and looking at external objects. Simple secondary tasks were defined as those requiring no or one button press or one glance away from the forward roadway. These tasks included adjusting the radio and other vehicle-related devices, talking, singing, drinking, smoking, and lack of concentration.

Engaging in complex secondary tasks increased the likelihood of an at-fault crash or near-crash by threefold (OR 3.1, 95% CI 1.7–5.5) and engaging in moderate secondary tasks increased the likelihood of an at-fault crash or near-crash by twofold (OR 2.1, 95% CI 1.6–2.7). Simple secondary tasks were not associated with an increased risk of having an at-fault crash or near-crash (OR 1.2, 95% CI 0.9–1.6). ORs for some specific types of distracting activities were also calculated. Reaching for a moving object was associated with the highest risk of an at-fault crash or near-crash (OR 8.8, 95% CI 2.5–31.2). However, the risk attributable to many specific distracting activities remains unknown.

The naturalistic driving study described earlier has several limitations: the relatively small, nonrepresentative, volunteer sample; the difficulty in reliably capturing some types of secondary distracting tasks, such as drivers' level of cognitive attention, the role of passengers (for privacy reasons), and some outside distractions; issues with inter-rater reliability in coding distracting activities and assigning fault for crashes and near-crashes; and a lack of data on the role of driver distractions in more serious crashes resulting in driver injury.

The case-control study by Brison[30] has examined the association between smoking status and the likelihood of an at-fault crash. The study found that drivers who reported a history of smoking while driving were at an increased risk of an at-fault crash (occasionally smoke and drive: RR 1.4; frequently smoke and drive: RR 1.6). In a logistic regression model that adjusted for miles driven annually and age, smokers had an increased risk of an at-fault crash (RR 1.5, $p = .01$). However, the study did not establish the reason for the increased risk. The author proposed three possible mechanisms: driver distraction related to the act of smoking, behavioral differences between smokers and nonsmokers, and carbon monoxide toxicity. Accordingly, further research is needed to determine whether the act of smoking truly increases the risk of a crash or not. The response rate for the case-control study was about 54%.

17.9 FUTURE DIRECTIONS

There is still much work to be done to evaluate the role of distracting activities on the risk of a crash. For example, what is it about carrying passengers that increases the risk of having a crash among young drivers? Possibilities might include the actual interaction with passengers (talking to or looking at passengers) or an increased risk-taking propensity in the presence of certain passenger groups (such as young, male passengers). Moreover, research to clarify the circumstances during which distracting activities present the greatest risk is also warranted. Large, well-designed, naturalistic driving studies, using representative driving populations and sound epidemiological methods, may be useful in this regard and may provide estimates of the risk of crash associated with a range of distracting activities.

Despite the need for further research, a strategy to minimize distracting activities while driving is indicated, given the current state of knowledge, particularly in relation to the carriage of peer passengers for novice drivers and the use of mobile (cellular) phones. Components of the strategy could include driver education about the known risks, legislation to reduce the carriage of peer passengers by young drivers and limit phone use while driving, more effective enforcement of existing laws including those that require drivers to maintain proper control of their vehicles, occupational health and safety measures by employers to restrict phone use while driving during work hours, systematic recording of the presence and types of distracting activities contributing to police-reported crashes for surveillance purposes, continued efforts by motor vehicle manufacturers to develop early warning systems to prevent collisions that may result from driver distraction, and thorough assessment of the safety aspects of novel in-vehicle technologies.[31]

REFERENCES

1. National Health and Medical Research Council, A Guide to the Development, Implementation and Evaluation of Clinical Practice Guidelines, Canberra, 1998.
2. Last J.M., *A Dictionary of Epidemiology*, 3rd Edition, Oxford University Press, New York, 1995.
3. Chen L.H., Baker S.P., Braver E.R., and Li G., Carrying passengers as a risk factor for crashes fatal to 16- and 17-year-old drivers, *JAMA*, 283, 1578–1582, 2000.
4. Doherty S.T., Andrey J.C., and MacGregor C., The situational risks of young drivers: the influence of passengers, time of day and day of week on accident rates, *Accid Anal Prev*, 30, 45–52, 1998.
5. Preusser D.F., Ferguson S.A., and Williams A.F., The effect of teenage passengers on the fatal crash risk of teenage drivers, *Accid Anal Prev*, 30, 217–222, 1998.
6. Aldridge B., Himmler M., Aultman-Hall, and Stamatiadis N., Impact of passengers on young driver safety, *Transport Res Rec*, 1693, 25–30, 1999.
7. Rice T.M., Peek-Asa C., and Kraus J.F., Nighttime driving, passenger transport, and injury crash rates of young drivers, *Inj Prev*, 9, 245–250, 2003.
8. Lam L.T., Norton R., Woodward M., Connor J., and Ameratunga S. Passenger carriage and car crash injury: a comparison between younger and older drivers, *Accid Anal Prev*, 35, 861–867, 2003.
9. Masten S.V. and Hagge R.A., Evaluation of California's graduated driver licensing program, *J Safety Res*, 35, 523–535, 2004.
10. Williams A.F., Ferguson S.A., and Wells J.K., Sixteen-year-old drivers in fatal crashes, United States, 2003, *Traffic Inj Prev*, 6, 202–206, 2005.
11. Simons-Morton B., Lerner N., and Singer J., The observed effects of teenage passengers on the risky driving behavior of teenage drivers, *Accid Anal Prev*, 37, 973–982, 2005.
12. Keall M.D., Frith W.J., and Patterson T.L., The influence of alcohol, age and number of passengers on the night-time risk of driver fatal injury in New Zealand, *Accid Anal Prev*, 36, 49–61, 2004.
13. McEvoy S.P., Stevenson M.R., and Woodward M., The contribution of passengers versus mobile phone use to motor vehicle crashes resulting in hospital attendance by the driver, *Accid Anal Prev*, doi:10.1016/j.aap.2007.03.004, 2007.
14. Hing J.Y.C., Stamatiadis N., and Aultman-Hall L. Evaluating the impact of passengers on the safety of older drivers, *J Safety Res*, 34, 343–351, 2003.
15. McCartt A.T., Hellinga L.A., and Bratiman K.A., Cell phones and driving: Review of research, *Traffic Inj Prev*, 7, 89–106, 2006.
16. Violanti J.M., Cellular phones and fatal traffic collisions, *Accid Anal Prev*, 30, 519–524, 1998.
17. Violanti J.M., Cellular phones and traffic accidents, *Public Health*, 111, 423–428, 1997.
18. Violanti J.M. and Marshall J.R., Cellular phones and traffic accidents: an epidemiological approach, *Accid Anal Prev*, 28, 265–270, 1996.
19. Laberge-Nadeau C., Maag U., Bellavance F., Lapierre S.D., Desjardins D., Messier S., and Saïdi A., Wireless telephones and the risk of road crashes, *Accid Anal Prev*, 35, 649–660, 2003.
20. Wilson J., Fang M., and Wiggins S., Collision and violation involvement of drivers who use cellular telephones, *Traffic Inj Prev*, 4, 45–52, 2003.
21. Redelmeier D.A. and Tibshirani R.J., Association between cellular telephone calls and motor vehicle collisions, *N Engl J Med*, 336, 453–458, 1997.

22. McEvoy S.P., Stevenson M.R., McCartt A.T., Woodward M., Haworth C., Palamara P., and Cercarelli R., Role of mobile phones in motor vehicle crashes resulting in hospital attendance: a case-crossover study, *Br Med J*, doi:10.1136/bmj.38537.397512.55, 331, 428–430, 2005.
23. Klauer S.G., Dingus T.A., Neale V.L., Sudweeks J.D., and Ramsey D.J., *The Impact of Driver Inattention on Near-Crash/Crash Risk: An Analysis Using the 100-Car Naturalistic Driving Study Data* (Technical Report No. DOT HS 810 594), National Highway Traffic Safety Administration, Washington, D.C., April 2006.
24. Dingus T.A., Klauer S.G., Neale V.L., Petersen A., Lee S.E., Sudweeks J., Perez M.A., Hankey J., Ramsey D., Gupta S., Bucher C., Doerzaph Z.R., Jermeland J., and Knipling R.R., *The 100-Car Naturalistic Driving Study, Phase II—Results of the 100-Car Field Experiment* (Technical Report No. DOT HS 810 593), National Highway Traffic Safety Administration, Washington, D.C., April 2006.
25. McEvoy S.P., Stevenson M.R., and Woodward M., Phone use and crashes while driving: a representative survey of drivers in two Australian states, *Med J Aust*, 185, 630–634, 2006.
26. Eby D.W. and Vivoda J.M., Driver hand-held mobile phone use and safety belt use, *Accid Anal Prev*, 35, 893–895, 2003.
27. Sullman M.J.M. and Baas P.H., Mobile phone use amongst New Zealand drivers, *Transport Res F*, 7, 95–105, 2004.
28. Beirness D.J., Simpson H.M., and Desmond K., The Road Safety Monitor 2002. Risky Driving. The Traffic Injury Research Foundation, 2002.
29. Lamble D., Rajalin S., and Summala H., Mobile phone use while driving: public opinions on restrictions, *Transportation*, 29, 223–236, 2002.
30. Brison R.J., Risk of automobile accidents in cigarette smokers, *Can J Public Health*, 81, 102–106, 1990.
31. McEvoy S.P., Stevenson M.R., and Woodward M., The impact of driver distraction on road safety: results from a representative survey in two Australian states, *Inj Prev*, 12, 242–247, 2006.

18 Driver Distraction Exposure Research: A Summary of Findings

Kristie L. Young and Michael A. Regan

CONTENTS

18.1 INTRODUCTION

The degree to which a competing activity undermines driving safety is a function of the degree of distraction associated with the activity and the extent to which the driver is exposed to that activity. With regard to driver distraction, exposure can be defined as the amount of time spent engaging in an activity. It is measured by determining the duration of each single activity and the frequency of engagement in that activity during the period of interest. Consequently, exposure data has two important elements: duration and frequency. It is not possible to obtain an accurate picture of a driver's exposure to distracting activities without measuring both of these elements (see Chapter 5 for more details on elements of exposure). An accurate estimate of exposure is critical to extrapolate performance decrements in controlled settings to assess their effects on overall driving safety.

As noted in Chapter 5, an important element to consider when measuring drivers' exposure to distracting activities is the magnitude of threat or the level of distraction associated with an activity. An activity may be extremely risky on a task basis (i.e., relative risk high) but may be engaged in infrequently or for only short durations (i.e., total exposure is low), and as such, it may have a low overall crash risk. Alternatively, an activity that imposes a low level of distraction may be engaged in frequently or for long durations (i.e., total exposure is high) and, thus, its overall crash risk would be higher. Mobile phone use is a good example to illustrate this point. Dialing a mobile phone imposes a high level of distraction on drivers, but exposure to this activity is typically very low and hence its crash risk is also likely to be low. Conversing on a phone, although less demanding than dialing, generates high levels of exposure and, therefore, may have a greater contribution to crash rates. When determining the overall risk of certain activities, it is important to consider both the level of demand imposed by the task and drivers' exposure to the task.

The collection of exposure data is important for a number of reasons. It allows us to identify activities that contribute to distraction-related crashes and incidents, to calculate the risk of being involved in a crash or near-crash when engaging in an activity (odds ratios), and to quantify the impact of drivers' engagement in these activities on the population as a whole (population-attributable risk, see Chapter 5). Such data can be used to evaluate the effectiveness of countermeasures designed to prevent and mitigate distraction. Exposure data also provide insights into which driving populations are engaging in distracting activities and under what circumstances this occurs, which can be used by policy developers and road authorities to better design and implement countermeasures that target "at-risk" driver groups.

A number of studies have attempted to quantify the amount of time drivers spend engaging in distracting activities. This research is still in its infancy and the majority of the studies have focused on mobile (cell) phone use only, although some data does exist for other distraction sources (e.g., Refs 1 and 2). Studies examining driver exposure to distracting activities and patterns of interaction have used a range of methods to collect exposure data: telephone and mail surveys, roadside observation, and naturalistic driving studies. Chapter 5 discusses the various methods by which exposure data can be collected and the advantages and disadvantages of each method. This chapter focuses on the findings from the various exposure studies that have been conducted and should be read in conjunction with Chapter 17, which provides additional data on this topic.

18.2 TELEPHONE AND MAIL SURVEYS AND INTERVIEWS

Surveys are a relatively quick and inexpensive means of establishing general patterns and trends in drivers' exposure to distracting activities and their opinions regarding how these distractions affect their driving performance and safety. Telephone, mail, and face-to-face surveys have been undertaken in a number of countries, including the United States,[1,3] Australia,[4–7] Canada,[8,9] New Zealand,[10] Sweden,[11] Finland,[12,13] and Spain,[14] to estimate the prevalence of, and patterns of exposure to, distracting activities while driving. Many of these studies have focused on drivers' use of mobile phones and have found that the use of these devices while driving is widespread.

In the United States, the 2003 Motor Vehicle Occupant Safety Survey of 6000 drivers found that, during 2003, over two-thirds (68%) of the drivers surveyed usually have a mobile phone in their vehicle while driving (up from 17% in 1994).[15] Of these drivers, almost three-quarters said they talk on the phone while driving at least "occasionally," whereas 23% said they never talk on their mobile phone while driving, despite carrying it in their vehicle. Of the drivers who indicated that they occasionally use a phone while driving, 60% said that they use a handheld phone, whereas 39% reported that they typically use a handsfree phone. A statewide telephone survey of drivers in the U.S. state of North Carolina found similar phone usage rates and patterns among drivers.[3] Almost 60% of drivers in this survey reported that they had used a mobile phone while driving, and one in four of these drivers reported that they often, not always, used a handsfree phone while driving.

Surveys conducted in other countries reveal that phone use is common among drivers the world over. The findings from a representative survey of drivers in the Australian states of New South Wales and Western Australia revealed that 57.3% of Australian drivers use mobile phones while driving and almost 40% of these drivers do not have handsfree devices. The survey also revealed that 12.4% of drivers have written and sent text messages while driving, and 1.6% do so on at least half of their trips.[7]

Telephone and mail surveys conducted in Canada from 1998 to 2001 have revealed that between 20 and 40% of drivers use mobile phones while driving, with phone use higher among drivers who are male and are young to middle-aged.[8,9] In Finland, phone use among drivers increased from 68% in 1999[12] to 81% in 2002.[13] Again, young male drivers had higher phone usage rates than female or older drivers. A mail survey conducted in Sweden found that 30% of drivers use mobile phones while driving.[11] A New Zealand survey[10] found high phone usage among drivers (57%), and the usage rates among Spanish drivers were similar (60%).[14] In all of these countries, phone usage rates were highest among young male drivers, and the vast majority of phone users use handheld rather than handsfree phones to make and receive calls and read/send SMS messages. Finally, a survey of heavy vehicle drivers in Denmark found that almost all of the heavy vehicle drivers surveyed (99%) use a mobile phone while driving and, among these drivers, 31% use a handheld phone despite legislation banning their use while driving.[16]

Although a high proportion of drivers admit to using mobile phones while driving, many studies have found that drivers who use phones tend to do so infrequently and for short durations. Thulin and Gustafsson,[11] for example, found that drivers in Sweden reported using a mobile phone an average of 1.1 times a day while driving, with an average conversation length of 10 min. McEvoy et al.[7] found that drivers who use their phone on more than half of their trips spend around 8%, or 10 min, of their driving time per day on the phone. Beirness et al.[8] found that 58% of Canadian drivers spend less than 10 min a week using a mobile phone when driving, whereas only about 5% of Finnish drivers spend 16 min or more per day using a phone.[12]

Survey data on drivers' exposure to potentially distracting activities other than mobile phone use is sparse, but some data does exist. McEvoy et al.,[5] in an Australian study of the impact of driver distraction on safety, reported the top five most commonly reported distracting activities engaged in by drivers during their most recent driving trip: lack of concentration (thinking about other things and daydreaming) (71.8% of drivers); adjusting in-vehicle equipment (68.7%); viewing outside objects, people, or events (57.8%); talking to passengers (39.8%); and reaching for objects in the vehicle (23.1%). Relatively few drivers reported using a mobile phone (9.0%) during their last trip. Laberge-Nadeau et al.[9] also found that 63% of drivers in Canada "very often" or "often" drive with passengers in the vehicle, that 95% listen to the radio, CD, or cassette tapes very often or often, and that 65% will manipulate the radio, CD, or tapes "very often" or "often" while driving. Moreover, the majority of these drivers reported that talking to passengers or listening to the radio, CD, or tapes could be "somewhat harmful" to driving. Finally, in a U.S.-based survey, Royal[17] found that, although around one quarter of drivers use a mobile phone when driving, relatively few use other types of technology, such as personal digital assistants (PDAs) or email (2% of drivers) or pagers (3%). The majority of drivers (66%) did, however, report adjusting the radio or looking for CDs or tapes while driving. The proportion of drivers engaging in nontechnology-based activities was higher, with 81% of the drivers talking to passengers, 50% eating or drinking, and 12% reading a map or directions when driving.

Although they are quick, easy, and inexpensive to use, exposure surveys have a number of limitations, including self-reporting bias, low-response rate, sample bias, and a limited shelf life given increasing ownership of electronic devices (for further information on the limitations of surveys the reader is referred to Chapter 5 of this book).

18.3 OBSERVATIONAL ROADSIDE SURVEYS

Roadside observational surveys can be a reliable means of collecting data on driver engagement in potentially distracting activities at a particular point in time. In these surveys, an observer stands on the side of the road and records the number of drivers who are or are not engaged in a particular activity. As with the mail and telephone surveys, observational studies have tended to focus on drivers' use of mobile phones and also suggest that mobile phone use is widespread among drivers.

In 2005, the observational probability-based National Occupant Protection Use Survey conducted in the United States found that, at a given daylight moment, 6% of passenger vehicle drivers were using a handheld phone, compared with 3% of drivers in 2000.[18,19] In addition, 0.8% of drivers were observed talking with headsets on, and 0.2% were seen manipulating a handheld device.

An Australian study by Taylor et al. observed drivers' use of handheld mobile phones on major roads in the city of Melbourne, where it is illegal to use a handheld phone while driving.[20] They found that 2% of the drivers observed were using a handheld mobile phone, and that these drivers were predominantly younger males. An earlier Western Australian study found that 1.5% of the drivers observed were using a handheld mobile phone while driving.[21] This study also found that the drivers observed using a mobile phone were predominately male drivers aged less than 40 years.

Roadside observations collected in other countries or states where it is illegal to use handheld phones while driving have found similar rates of handheld phone use to those found in Australia.[22–25] In the United Kingdom, roadside observations revealed that, prior to the introduction in December 2003 of legislation banning handheld phone use, 1.85% of drivers were observed using a mobile phone during the evening rush hour. This figure dropped to 0.95% 10 weeks after the introduction of the phone ban.[22]

In the United States, prior to the introduction of handheld phone bans, 6.1% of drivers in Washington, D.C.,[24] and 2.3% of drivers in New York[26] were observed using handheld mobile phones while driving. These usage rates fell to 3.5% in Washington and 1.1% in New York several months after the introduction of the bans in 2004 and 2001, respectively. The observed higher phone usage rate in Washington compared to New York is likely to be due to the different time periods during which the data was collected; the Washington data were collected during mid- to late-2004, more than 2 years after the New York observations were made, in late 2001 to early 2002. Indeed, other data show that phone usage rates among drivers increased rapidly over this period.[23] A 4-year analysis of drivers' handheld mobile phone use in the U.S. state of Michigan (where no handheld phone bans were in force) revealed that, between the years 2001 and 2005, the percentage of drivers observed using handheld phones was more than doubled from 2.7 to 5.8%.[23] Extrapolating from this data (and assuming that no phone bans are introduced), the authors predicted that, by the year 2010, 8.6% of drivers in Michigan would be using handheld phones during any given daytime hour.

Direct observation of drivers' handheld phone use in New Zealand, where there are also no restrictions on handheld phone use while driving, revealed that, in March 2002 and again in March 2003, 3.9% of drivers were using handheld phones.[25] These usage rates are almost double the rates observed during 2002 in the Australian city of Melbourne, where handheld phone use is banned while driving.[20]

A study by Johnson et al.[27] examined drivers' use of mobile phones and their engagement in other distracting activities by examining high-quality photos taken of vehicles and drivers on the New Jersey Turnpike between March and July 2001. They found that just less than 5% of the photos contained drivers who were engaging in non-driving-related activities. Mobile phones were the most common source of potential distraction observed, accounting for one-third of potential distractions (1.5% of drivers), followed by interacting with a passenger (0.7%), adjusting controls (0.3%), and finally "other" distractions (0.3%).

As is the case with the telephone and mail surveys, the observational surveys have typically found that mobile phone usage rates are highest among younger drivers and very low among drivers estimated to be aged 60 years or more.[18,20,26,27] The results relating to gender differences are mixed, with many studies not finding any differences in phone use across male and female drivers,[18,25,27,28] one finding that rates were higher among female drivers,[26] and others finding that males had higher usage rates than females.[7,20,21] It is not clear why these differences exist across studies.

There are a number of limitations associated with using roadside observation methods to estimate drivers' exposure to distracting activities. First, data is typically collected only during daylight hours and on sections of the road where speeds are lower, such as at intersections or freeway exit ramps. This restricted sampling range can lead to under- or overestimates in overall usage rates depending on the time at which the observations are made. Second, observational studies can only examine use or nonuse of a device or activity at one time point; they cannot provide information regarding the duration of the distracting activity or the frequency with which it is engaged in. Third, observation surveys are typically restricted to estimating the rates of those distracting activities that are directly observable from outside of the vehicle. Most of the observational studies published to date have examined drivers' use of handheld phones only. It is, of course, far more difficult to accurately determine, through roadside observation, the proportion of drivers at any one time who are using handsfree mobile phones, as drivers who are observed to be talking could be talking to themselves, to an unseen passenger, to an animal, or to themselves (e.g., when singing). Finally, like other survey methods, observational studies have a short shelf-life, as the ownership rate of mobile phones and other electronic devices is increasing rapidly as is their use in vehicles.[18,23]

18.4 NATURALISTIC DRIVING STUDIES

The naturalistic driving study is a relatively recent research method that has been used to collect data on drivers' exposure to a range of distracting activities. By using motor vehicles instrumented with a wide range of sensors (e.g., video and vehicle state), researchers can obtain accurate data on drivers' exposure to a wide range of distracting activities in everyday driving conditions, including the frequency and duration of these activities, the conditions in which drivers engage in them, and the effect these activities have on driving behavior, performance, and safety. To date, three naturalistic driving studies known to the authors have been conducted to examine driver involvement in distracting activities.

In the United States, Stutts et al.[2,29] examined drivers' exposure to potentially distracting activities in their natural driving environment. Video camera units were installed in the vehicles of 70 volunteer drivers in North Carolina and Philadelphia (where use of handheld phones was legal) and recorded the frequency and duration with which the drivers engaged in distracting activities over the course of a week. Participants included equal numbers of males and females from five age groups ranging from 18 to more than 60 years. Three unobtrusive cameras continuously recorded the driver's face, the interior of the vehicle, and the forward roadway. Three hours of randomly selected video data was analyzed for each participant.

The results revealed that, excluding the time spent conversing with passengers, drivers were engaged in some form of potentially distracting activity for 14.5% of the time when their vehicle was in motion. All participants engaged in distracting activities deriving from within the vehicle, including manipulating vehicle controls (other than radio/audio controls) and reaching for objects, and spent almost 4% of their total driving time engaged in these activities. Over 90% of drivers manipulated music/audio controls while driving, although they only spent 1.5% of their total driving time performing this activity. Around three-quarters of the drivers ate or drank and conversed with passengers while driving and almost half engaged in reading or writing and grooming activities. Approximately 34% of drivers were observed using their mobile phone (to talk, dial, and answer) while their vehicle was in motion, and these drivers spent almost 4% of their total driving time using the phone. The study did not specify whether the mobile phones used by drivers were handheld or handsfree. Interestingly, while only 7% of drivers in the study engaged in smoking activities, these drivers spent over 21% of their driving time lighting, extinguishing, or smoking cigarettes.[2]

Only 5.7% of drivers in the study were judged as being distracted by external events or objects, and this occurred for about 2.0% of their driving time. Although a wide age range was examined, no age-related differences were found in drivers' willingness to engage in the various potentially distracting activities; however, it was revealed that drivers were more likely to engage in certain activities, such as reading and writing, manipulating vehicle controls, and reaching for objects, when the vehicle was stopped. The performance of potentially distracting activities generally led to higher rates of adverse driving events, such as sudden braking and lane encroachments, and a higher percentage of time spent with no hands on the steering wheel and eyes directed inside the vehicle, although these increases rarely reached significance due to the small sample sizes in some distraction categories. The results did show, however, that drivers spent significantly more time with their hands off the steering wheel and their eyes directed inside the vehicle when using a mobile phone, eating or drinking, manipulating audio and vehicle controls, reading, writing, grooming, or reaching for objects.

The 100-Car naturalistic driving study conducted by the Virginia Tech Transportation Institute is the largest naturalistic driving study undertaken to date.[1,30] The instrumented vehicles collected approximately 2 million vehicle miles, or 43,000 h, of driving data from 241 drivers in the Northern Virginia/Washington, D.C. metropolitan area over a 12- to 13-month period. Drivers involved in the study ranged in age from 18 to 55 years and older. Five video cameras and a range of vehicle state sensors continuously recorded driver, vehicle, roadway, and other road user behavior throughout the study. The study revealed a wealth of data on drivers' exposure to potentially distracting activities and the role of driver distraction in crashes, near-crashes, and incidents that required an evasive maneuver. However, at the time this chapter was being written, detailed results regarding drivers' exposure to distracting activities had not been published. Detailed data is available, however, on the relationship between various distracting activities and adverse driving outcomes.

In total, 69 crashes, 761 near-crashes, and 8295 incidents were recorded during the study. Results revealed that 78% of the crashes and 65% of the near-crashes involved driver inattention as a contributing factor. Secondary task distraction was found to be the largest of four subcategories of inattention identified from analysis of the data (the other three categories were "fatigue," "inattention to the forward roadway," and "nonspecific eye glance"). The sources of secondary task distraction that contributed to the greatest proportion of near-crashes were wireless devices (mobile phones and PDAs), followed by passenger-related tasks (primarily conversing) and internal distractions (manipulating or locating miscellaneous objects not related to wireless devices, in-vehicle systems, passengers, eating/drinking, or smoking). These three sources of distraction also contributed to the greatest proportion of crashes; however, internal distractions accounted for a slightly higher proportion of crashes than wireless devices and passenger-related tasks.[30] More specific analysis of the 100-car study secondary task data was conducted by Klauer et al.[1] and revealed that the three secondary tasks with the highest crash risk were "reaching for a moving object" (odds ratio [OR] = 8.82); "looking at external object" (OR = 3.70); and "reading" (OR = 3.38). This means, for example, that when an adverse event occurred, drivers were approximately nine times as likely to be reaching for a moving object as when an adverse event did not occur.

Finally, Mazzae et al.[31] conducted a naturalistic driving study to examine drivers' mobile phone use patterns and the effects of phone interface type (handheld, handsfree headset and handsfree with voice dialing) on driving performance and behavior under real-world conditions. Ten participants drove instrumented vehicles equipped with video cameras for 6 weeks. Drivers used each of the three phone types for a period of 2 weeks. Results revealed that, on average, drivers engaged in 2.25 calls per hour, and these calls lasted for an average of 2.4 min. Handheld phones were associated with more frequent and longer calls than the handsfree interfaces. Phone use also differed across driving conditions, with drivers making fewer phone calls in heavy traffic conditions compared to light traffic, particularly when using the handsfree phones. Driving performance did not differ significantly across phone type or the phone versus no phone conditions; however, drivers did spend less time looking at the roadway when dialing and more time looking straight ahead when conversing compared to baseline (no phone) driving, suggesting that there was a decrease in driver situational awareness when using the phone.

Naturalistic driving studies provide a means to collect detailed and accurate real-world data on drivers' patterns of engagement in distracting activities and the contribution of these activities to adverse safety outcomes, such as crashes. This method does, however, yield an enormous amount of data, which can be very expensive and time consuming to code and analyze. Furthermore, the highly varied and uncontrolled driving conditions experienced by drivers during naturalistic studies can make analysis of the data complicated and reduce the sensitivity of the study. Drivers' everyday driving behavior can also be altered in response to being monitored, regardless of how unobtrusive is the data collection equipment. Another limitation of the naturalistic studies conducted to date is their relatively small sample sizes, which reduce the degree to which the sample used is representative of the general driving population. Finally, although conventional video sensors used in naturalistic studies are capable of recording driver engagement in a wide range of potentially distracting activities, they cannot accurately record instances of cognitive distraction, which occur when drivers daydream or become absorbed in internal thought. Developments in driver state recording technologies, discussed in Chapters 26 and 28, are likely to resolve this problem in future naturalistic driving studies.

18.5 SUMMARY OF FINDINGS ACROSS THE EXPOSURE METHODS

On the basis of findings discussed, it is possible to obtain best estimates of drivers' exposure to various sources of distraction across the different exposure methods. These data are summarized in Table 18.1. It is important to highlight the different exposure methods all yield different estimates of exposure. This occurs because each method measures slightly different aspects of exposure. For example, observational roadside surveys are measuring driver exposure to an activity at a single point in time, and as such, yield lower exposure estimates than the other two methods that collect more complete exposure information (duration and frequency) over a longer time frame. In essence, roadside observations are measuring the proportion of drivers engaged in a distracting activity at any one point in time (point prevalence), whereas the other methods seek to establish not only the presence of an activity but also its frequency and duration. From Table 18.1, we can see that while many drivers engage in distracting activities, the amount of driving time spent engaged in them is typically quite low. It can also be seen that the exposure estimates for handheld and

TABLE 18.1
Exposure Estimates (Percentages of Drivers and Driving Time) to Various Activities across Exposure Methods

	Exposure Method				
	Phone/Mail Survey		Roadside Observations	Naturalistic Study	
Distraction Source	Drivers	Driving Time	Drivers	Drivers	Driving Time
Phone (undefined)[a]	50.0–80.0	8.0–14.0		34.0	4.0–9.0
Handheld phone (talking, dialing, answering)	20.0–60.0		2.0–6.0		
Handsfree phone (talking, dialing, answering)	40.0–80.0		0.8		
Text messaging	12–14[b]				
Passengers	40.0–60.0		0.7	75.0	20.0
Adjusting vehicle controls	65.0		0.3	100	4.0
Radio/CD	95.0			90.0	1.5
Reaching for objects	23.0				
Eating/drinking	50.0			70.0	2.0–5.0
Reading/writing	12.0			40.0	1.8
Grooming				45.0	0.6
Smoking				7.0	21.0
External distraction	60.0			6.0	2.0

[a] Study did not specify phone type.
[b] Younger drivers aged below 25 years.

handsfree phone use are very broad. It is very difficult to obtain best estimates of drivers' exposure to handheld and handsfree phones based on the phone and mail survey data, because the percentage of drivers who report using these devices differs greatly across regions and jurisdictions with and without handheld phone legislation.

18.6 EFFECTS OF HANDHELD PHONE LEGISLATION ON EXPOSURE ACROSS TIME

The use of handheld mobile phones while driving has been prohibited in all Australian states, in most countries in the European Union, the Canadian provinces of Newfoundland and Labrador, and a number of states in the United States. But has this legislation had any effect on handheld phone exposure rates among drivers? Research in a number of countries has been conducted to examine the short- and long-term effects of mobile phone laws on handheld phone use.[22,28,32,33]

In the United States, the long-term effect of handheld phone laws on phone use has been examined using observational studies in the District of Columbia (DC)[32] and in the state of New York.[28] Both studies found that, before the legislation was introduced, approximately 6.0% of drivers in District of Columbia and 2.0% in New York State were observed using a mobile phone. Shortly after the law was introduced (about 3 months after), these usage rates decreased to 3.5%[24] and 1.1%[26] of drivers, respectively. However, these short-term decreases were not sustained 1 year after the introduction of the ban, with the proportion of drivers observed using handheld phones increasing over this period to 4.0% in District of Columbia[32] and 2.1% in New York State.[28] Possible reasons for these increased usage rates are an increase in the number of phone users over the testing period or that the enforcement of the bans may not have been aggressive enough to deter drivers from using handheld phones.

Similar trends were found in Finland, where handheld phone use while driving was banned in early 2003.[33] According to self-reported data, prior to the introduction of the ban, 55.6% of the drivers surveyed reported that they occasionally used handheld phones while driving. This figure decreased to 15.2% immediately after the law took effect, but 16 months later had increased significantly to 20.0% of drivers, an increase of 32.0%. The results of the roadside observational data showed even stronger upward trend, with the proportion of drivers observed using handheld phones increasing from 3.1% immediately after the law was introduced to 5.8% 16 months later. Again, these increases could be due to a greater number of mobile phone users or inadequate enforcement of the bans.

Several distraction exposure studies have examined whether and how drivers' exposure patterns to potentially distracting activities (mainly mobile phones) varies according to different driver characteristics. As discussed earlier, while findings regarding the effect of gender on distraction exposure are mixed, a consistent trend across the survey, observational and naturalistic driving data is that exposure to distracting activities is highest among younger drivers and very low among drivers aged 60 years or older.[6,18,20,26,27] Research also suggests that exposure to distracting activities (primarily mobile phones) is highest among drivers traveling in urban areas,[5,7,8,10,14,18] who have a high annual mileage,[8,10,13,14] commercial drivers and drivers of sport utility vehicles (SUVs),[18,32] and drivers with fewer years of driving experience.[6,14]

18.7 CONCLUSIONS

Driver engagement in potentially distracting activities, and especially mobile phone use, is common and widespread, particularly among drivers who are young, are inexperienced, travel in urban areas, and have high annual mileage rates. Indeed, it has been estimated that in the United States, 60% of the time is spent using mobile phones in the vehicle.[34] What's more, the rapid proliferation of technology into vehicles and increasing mobile phone and MP3 player (e.g., iPod) ownership rates are likely to increase drivers' exposure to potentially distracting activities even further.

The collection of accurate distraction exposure data is important for a number of reasons. It is vital to understand what distracting activities the drivers are engaged in and the patterns and circumstances of this engagement. It is also essential to estimating the level of overall risk associated with driver engagement in various potentially distracting activities. Further, data regarding drivers' exposure to distracting activities can be invaluable for guiding and targeting driver distraction countermeasure development. For instance, findings that suggest that mobile phone usage rates are particularly high among young, inexperienced drivers can be used to target distraction policy and education toward this high-use, high-risk group. Indeed, research findings that show that young, inexperienced drivers are more likely to engage in, and be negatively affected by, mobile phone use, is already influencing distraction policy in a number of jurisdictions. In the United States, a number of states and the District of Columbia have restricted mobile phone use among young, inexperienced drivers as part of their graduated licensing systems.[35] Similarly, the Australian states of Queensland, New South Wales, and Victoria (from July 2008), as part of their new graduated licensing schemes, prohibit the use of *all* phones by drivers during the learner period and for the first year of their probationary license (P1 phase).[36]

Changes in drivers' exposure to distracting activities over time and in response to legislation can also provide insights into the effectiveness of bans or restrictions and other distraction countermeasures. Exposure studies have found substantial reductions in cell phone use immediately after the introduction of bans, but it seems that these reductions are not sustained and return to near preban levels after 1 year. Such findings suggest that, for handheld bans to be effective, they must be strictly and vigorously enforced over an extended period of time, which is a costly and time-consuming activity, especially as the number of devices brought into the vehicle continues to rise. They also suggest that training and education or the use of signal-blocking technology may be more effective approaches than legislation in reducing mobile phone use among drivers. Ultimately, real-time distraction mitigation technologies, such as those described in Chapters 26 and 28 of this book, might provide the only practical means by which the potential risk associated with exposure to mobile phone use can be reliably minimized.

The use of telephone, mail and roadside observational surveys, and naturalistic driving studies has generated a wealth of data regarding drivers' usage rates and patterns of use of mobile phones. However, distraction exposure research is in its infancy, and data on drivers' exposure to other potentially distracting activities are scarce. It is important that future research focuses on the collection of exposure data for a wide range of potentially distracting activities, not just mobile phone use, to determine the level of risk associated with driver engagement in these activities and to guide future countermeasure development.

ACKNOWLEDGMENT

We are grateful to John Lee, Suzanne McEvoy, and Karen Stephan for their helpful and insightful comments on earlier versions of this chapter.

REFERENCES

1. Klauer, S. G., Dingus, T. A., Neale, V. L., Sudweeks, J. D., and Ramsey, D. J., The impact of driver inattention on near-crash/crash risk: an analysis using the 100-Car Naturalistic Driving Study data, Virginia Tech Transportation Institute, Blacksburg, Virginia, 2006.
2. Stutts, J., Feaganes, J., Reinfurt, D., Rodgman, E., Hamlett, C., Gish, K., and Staplin, L., Drivers' exposure to distractions in their natural driving environment, *Accident Analysis & Prevention* 37, 1093–1101, 2005.
3. Stutts, J., Hunter, W. W., and Huang, H. F., *Cell Phone Use While Driving: Results of a Statewide Survey*, in Transportation Research Board, Annual Meeting CD-ROM, 2003.
4. Baker, S. and Spina, K., Drivers' attitudes, awareness and knowledge about driver distraction: research from central Sydney communities, In *Driver Distraction: Proceedings of an International Conference on Distracted Driving*, Faulkes, I. J., Regan, M. A., Brown, J., Stevenson, M. R., and Porter, A. (Eds.), Australasian College of Road Safety, Sydney, Australia, 2005.
5. McEvoy, S. P., Stevenson, M. R., and Woodward, M., The impact of driver distraction on road safety: results from a representative survey in two Australian states, *Injury Prevention* 12(4), 242, 2006.
6. McEvoy, S. P., Stevenson, M. R., and Woodward, M., The prevalence of, and factors associated with, serious crashes involving a distracting activity, *Accident Analysis & Prevention* 39(3), 475–482, 2007.
7. McEvoy, S., Stevenson, M. R., and Woodward, M., Phone use and crashes while driving: a representative survey of drivers in two Australian states, *Medical Journal of Australia* 185, 630–634, 2006.
8. Beirness, D., Simpson, H. M., Pak, A., *The Road Safety Monitor: Driver Distraction*, The Traffic Injury Research Foundation, Ottawa, Ontario, 2002.
9. Laberge-Nadeau, C., Maag, U., Bellavance, F., Lapierre, S., Desjardins, D., Messier, S., and Saidi, A., Wireless telephones and the risk of road crashes, *Accident Analysis & Prevention* 35, 649–660, 2003.
10. Sullman, M. J. M. and Baas, P. H., Mobile phone use amongst New Zealand drivers, *Transportation Research Part F* 7, 95–105, 2004.
11. Thulin, H. and Gustafsson, S., *Mobile Phone Use While Driving. Conclusions from Four Investigations*, Report No. VTI Report 490A, Swedish National Road and Transport Research Institute, Linkoping, Sweden, 2004.
12. Lamble, D., Rajalin, S., and Summala, H., Mobile phone use while driving: public opinions on restrictions, *Transportation* 29, 233–236, 2002.
13. Poysti, L., Rajalin, S., and Summala, H., Factors influencing the use of cellular (mobile) phone during driving and hazards while using it, *Accident Analysis & Prevention* 37, 47–51, 2005.
14. Gras, M. E., Cunill, M., Sullman, M. J. M., Planes, M., Aymerich, M., and Font-Mayolas, S., Mobile phone use while driving in a sample of Spanish university workers, *Accident Analysis & Prevention* 39(2), 347–355, 2007.
15. Boyle, J. M. and Vanderwolf, P., *2003 Motor Vehicle Occupant Safety Survey. Volume 4: Crash Injury and Emergency Medical Report*, Report No. DOT HS 809–857, National Highway Traffic Safety Administration, Washington, D.C., 2005.
16. Troglauer, T., Hels, T., and Falck Christens, P., Extent and variations in mobile phone use among drivers of heavy vehicles in Denmark, *Accident Analysis & Prevention* 38, 105–111, 2006.
17. Royal, D., *National Survey of Distracted and Drowsy Driving Attitudes and Behavior: 2002*, Report No. DOT HS 809 566, National Highway Traffic Safety Administration, Washington, D.C., 2003.

18. Glassbrenner, D., *Driver Cell Phone Use in 2005, Overall Results*, Report No. DOT HS-809–967, U.S. National Highway Traffic Safety Administration, Washington, D.C., 2005.
19. Utter, D., *Passenger Vehicle Driver Cell Phone Use*. Results from the Fall 2000 National Occupant Protection Use Survey, Report No. DOT HS 809 293, National Highway Traffic Safety Administration, Washington, D.C., 2001.
20. Taylor, D., Bennett, D.M., Carter, M., and Garewal, D., Mobile telephone use among Melbourne drivers: a preventable exposure to injury risk, *The Medical Directory of Australia* 179(3), 140–142, 2003.
21. Horberry, T., Bubnich, C., Hartley, L., Lamble, D., Drivers' use of hand-held mobile phones in Western Australia, *Transportation Research Part F* 4, 213–218, 2001.
22. Johal, S., Napier, F., Britt-Compton, J., and Marshall, T., Mobile phones and driving, *Journal of Public Health* 27, 112–113, 2005.
23. Eby, D. W., Vivoda, J. M., and St. Louis, R. M., Driver hand-held cellular phone use: a four-year analysis, *Journal of Safety Research* 37(3), 261–265, 2006.
24. McCartt, A. T., Hellinga, L.A., and Geary, L.L., Effects of Washington, D.C. law on drivers' hand-held cell phone use, *Traffic Injury Prevention* 7, 1–5, 2006.
25. Townsend, M., Motorists' use of hand held cell phones in New Zealand: an observational study, *Accident Analysis & Prevention* 38(4), 748–750, 2006.
26. McCartt, A. T., Braver, E. R., and Geary, L. L., Drivers' use of handheld cell phones before and after New York State's cell phone law, *Preventive Medicine* 36(5), 629–635, 2003.
27. Johnson, M. B., Voas, R. B., Lacey, J. H., McKnight, A. S., and Lange, J. E., Living dangerously: driver distraction at high speed, *Traffic Injury Prevention* 5(1), 1–7, 2004.
28. McCartt, A. T. and Geary, L. L., Longer term effects on New York State's law on drivers' handheld cell phone use, *Injury Prevention* 10, 11–15, 2006.
29. Stutts, J., Feaganes, J., Rodgman, E., Hamlett, C., Meadows, T., and Reinfurt, D., *Distractions in Everyday Driving*, AAA Foundation for Traffic Safety, Washington, D.C., 2003.
30. Dingus, T. A., Klauer, S. G., Neale, V. L., Petersen, A., Lee, S. E., Sudweeks, J., Perez, M. A., Hankey, J., Ramsey, D., Gupta, S., Bucher, C., Doerzaph, Z. R., Jermeland, J., and Knipling, R. R., *The 100-Car Naturalistic Driving Study, Phase II: Results of the 100-Car Field Experiment*, Virginia Tech Transportation Institute, Blacksburg, Virginia, 2006.
31. Mazzae, E. N., Goodman, M., Garrott, W. R., and Ranney, T. A., NHTSA's research program on wireless phone driver interface effects, In *Proceedings of the 19th International Technical Conference on the Enhanced Safety of Vehicles*, National Highway Traffic Safety Administration, Washington, D.C., 2005.
32. McCartt, A. T. and Hellinga, L. A., Longer-term effects of Washington, D.C., law on drivers' hand-held cell phone use, *Traffic Injury Prevention* 8(2), 199–204, 2007.
33. Rajalin, S., Summala, H., Poysti, L., Anteroinen, P., and Porter, B.E., In-car cell phone use and hazards following hands free legislation, *Traffic Injury Prevention* 6, 225–229, 2005.
34. Hahn, R. W., Tetlock, P. C., and Burnett, J. K., Should you be allowed to use your cellular phone while driving?, *Regulation* 23(3), 46–55, 2000.
35. Insurance Institute for Highway Safety, U.S. Licensing systems for young drivers, available at http://www.iihs.org/laws/state_laws/pdf/us_licensing_systems.pdf, accessed on June 27, 2007.
36. Victorian Government, Graduated Licencing System: Information for P1 Probationary drivers, available at http://www.arrivealive.vic.gov.au/c_youngGLS_4.html, accessed on June 27, 2007.

Part 6

Factors Mediating the Effects of Distraction

19 Factors Moderating the Impact of Distraction on Driving Performance and Safety

Kristie L. Young, Michael A. Regan, and John D. Lee

CONTENTS

19.1 INTRODUCTION

The fact that a driver's attention is diverted away from activities critical to safe driving and toward a competing activity does not by itself guarantee that performance and safety will be compromised. This point was underscored in Chapter 5. There are many factors that moderate the outcome of such an interaction: the complexity of the competing activity and current driving demands; how often, and for how long, the driver is exposed to the competing activity; and driver characteristics such as age, gender, driving experience, driver state, and willingness to engage. All of these factors affect the driver's ability to prevent and mitigate the impact of the competing activity. As noted in Chapter 7, the risk of crash is influenced by extrinsic and intrinsic factors. Extrinsic factors may include the driving task, traffic density, speed, and weather conditions, and intrinsic factors may include an individual's risk-taking propensity, driving experience, age, and state (e.g., fatigued, drowsy, inebriated). Many of these factors may change between episodes of exposure.

Understanding the factors that make drivers more or less vulnerable to the distracting effects of competing activities is important when designing countermeasures to prevent and mitigate the effects of distraction. The potential for a competing activity to distract the driver and degrade safety is determined by the complex interaction of a number of factors. This chapter will examine a number of these moderating factors, both intrinsic and extrinsic, for which there is some accumulated knowledge. These include drivers' willingness to engage in distracting activities, their ability to compensate for the increased demands imposed by a competing activity (self-regulation), driving task demands, driver characteristics (e.g., age, gender, driving experience), task familiarity, and driver state. Other moderating factors, such as exposure to, and the complexity of, distracting activity, are discussed in other chapters of this book (see Chapters 3 and 7 for exposure and Chapters 12 and 13 for secondary task complexity) and hence will not be reviewed here.

19.2 RELATIONSHIP BETWEEN MODERATING FACTORS AND DISTRACTION

As alluded to earlier, the potential for a secondary activity to distract drivers and degrade driving performance is moderated by a number of complex and interacting factors. Figure 19.1 displays a number of key moderating factors and the mechanisms by which they might influence drivers' willingness to engage in a secondary activity, their self-regulatory behavior, and the effect of the activity on driving performance and safety. It is important to note that the figure is not intended to contain an exhaustive list of all possible moderating factors. Rather, it is provided to illustrate the relationship between some key moderating factors discussed in the chapter and the mechanisms underlying these relationships. It also provides a means to draw together and summarize, in a simplified manner, the various concepts discussed throughout the chapter.

19.3 SELF-REGULATION

A fundamental question regarding the effect of competing activities on driving performance is whether and how drivers compensate for any decrease in attention to the driving task (i.e., self-regulate) to maintain adequate safety margins. Surprisingly, little research has directly addressed this issue. Indeed, as noted in Chapter 5, much of the distraction research has tended to view drivers as passive receivers and processors of distracting information. Drivers, however, can, and do, actively adjust their driving behavior in response to changing or competing task demands to maintain an adequate level of safe driving.[1]

Self-regulatory behavior can occur at a number of levels ranging from the strategic (e.g., choosing not to use a mobile phone while driving) to the operational level (e.g., reducing speed)[2] (see also Chapter 4). At the highest level, drivers can moderate their exposure to risk by choosing not to engage in potentially distracting activities while driving. Research has shown, for example, that the driving performance of older drivers is impaired to a greater degree than that of younger drivers when using a mobile phone, and this appears to encourage them to engage in compensatory

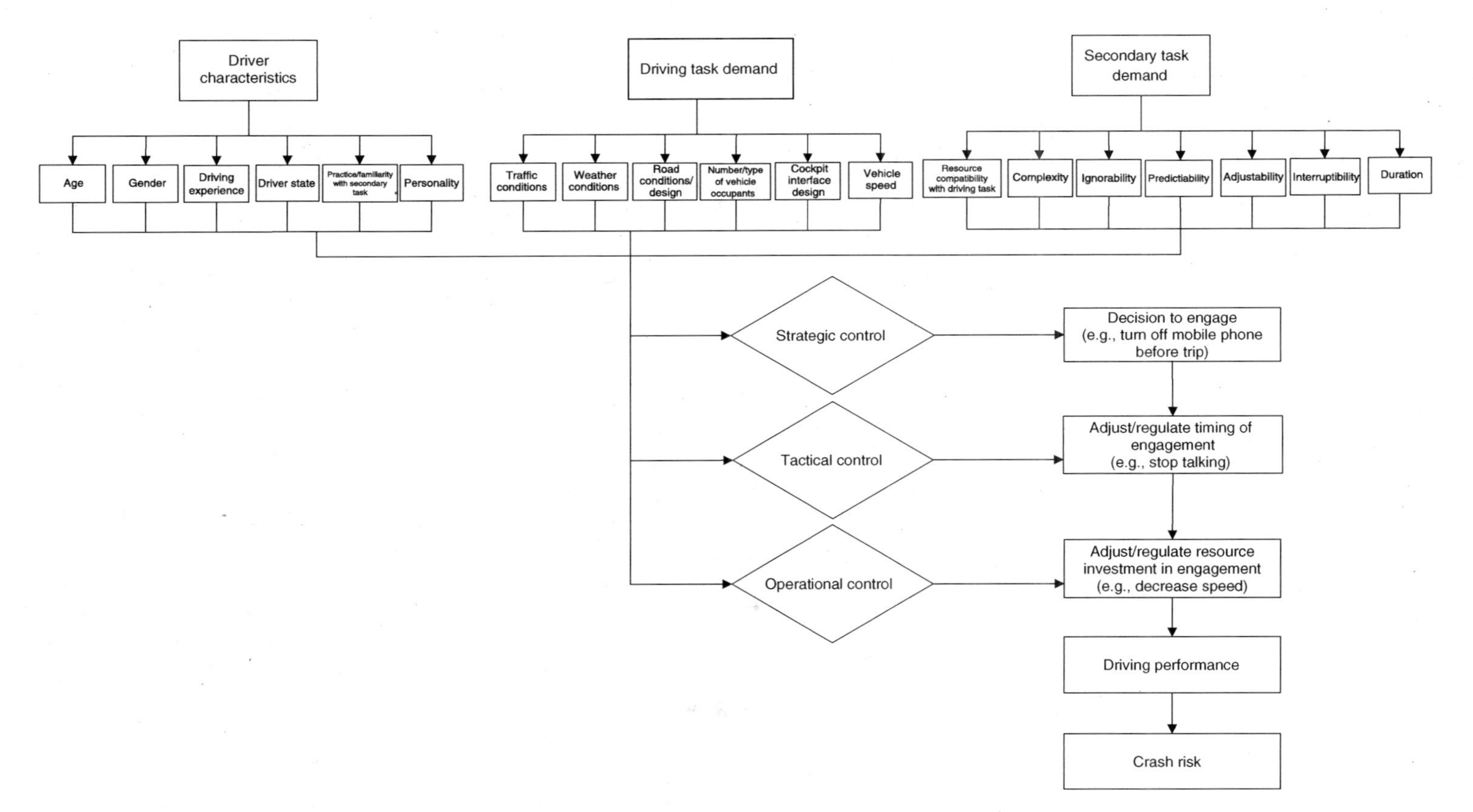

FIGURE 19.1 Factors that moderate the effects of distraction on driving performance and crash risk.

behavior at the strategic level; many older drivers choose to not use a mobile phone while driving.[3,4] Research that requires older drivers to use in-vehicle devices while driving may, therefore, overestimate the risks of these devices for this driving population because they tend not to use them in real driving situations (see Chapter 19 for further discussion of older drivers and distraction).

At the tactical and operational levels, research has shown that drivers attempt to reduce workload and moderate their exposure to risk while engaging in secondary activities, through a number of means: by decreasing speed,[1,5–7] by increasing intervehicular distance,[8–10] and by reducing or ceasing to engage in certain driving tasks, such as checking mirrors and instruments less frequently.[11–13] These self-regulatory behaviors can be viewed as examples of performance trade-offs, because by performing these behaviors, drivers are changing the relative level of priority that they assign to the driving task to accommodate performance of the competing activity.

A number of on-road and simulator studies have found that drivers typically decrease their driving speed when engaging in a secondary task. Haigney et al.[1] examined the effects of handheld and hands-free mobile phone tasks on simulated driving performance. Thirty participants completed four simulated drives while completing a grammatical reasoning task designed to simulate a mobile phone conversation. The results revealed that the mean speed and standard deviation of accelerator travel decreased while participants were conversing on the mobile phone. More recent research carried out in a driving simulator by Rakauskas et al.[7] also found that drivers' mean speed decreased while carrying out a naturalistic conversation on a mobile phone. Drivers also tend to reduce speed when using other in-vehicle devices. Chiang et al.,[14] for example, found that drivers decreased their speed when entering destination details into a route navigation system, whereas Horberry et al.[15] found that drivers' mean speed decreased when interacting with an in-car entertainment (radio and CD player) system.

An increase in following distance is another compensatory behavior that has been displayed by drivers while they are interacting with in-vehicle devices. Using a driving simulator, Strayer et al.[10] found that conversing on a hands-free mobile phone while driving led to an increase in following distance from a lead vehicle, and this increase was particularly pronounced under high traffic density conditions. Strayer and Drews[9] also found that drivers' following distance to a lead vehicle increased by 12% when drivers were conversing on a hands-free mobile phone under simulated driving conditions. Finally, in a driving simulator study, Jamson et al.[8] revealed that drivers adopted longer headways from a lead vehicle while processing e-mails using a speech-based e-mail system. Interestingly, although the drivers in all three studies attempted to compensate for their reduced attention to the roadway by adopting longer following distances, in many cases this increased headway was often inadequate to avoid collisions with other road users.

Another compensatory behavior that drivers engage in when interacting with in-vehicle devices is to reduce or cease their engagement in certain, and perhaps less critical, driving tasks, such as monitoring mirrors and instrument panels. Brookhuis et al.,[11] for example, found that drivers paid less attention to other traffic (as measured by the frequency of checking the rearview and side mirrors) on a quiet motorway when engaging in a mobile phone conversation. In another on-road assessment, Harbluk et al.[13] also found that when drivers were performing demanding cognitive tasks (adding double-digit numbers), they reduced their monitoring of vehicle mirrors and instrument panel and some drivers abandoned these tasks altogether.

There are a number of possible mechanisms underlying drivers' self-regulatory behavior when they engage in competing activities. The observed changes in speed and headway while engaging in secondary activities could be the result of drivers increasing their safety margins to compensate for their diminished ability to respond to roadway demands. They could also simply be exhibiting diminished driving performance resulting from them allocating too much attention to the secondary task and insufficient attention to the primary driving task. Or it may be the case that drivers are willingly and consciously accepting a diminished driving performance criterion. All of these explanations can have road safety implications. For example, if a driver does not allocate sufficient resources to the driving task, he or she may fail to detect hazards in the road environment, or it may be that the driving performance standard that drivers are willing to accept may fall below societal norms for safe driving.

There are a number of conditions under which it is difficult or impossible for drivers to self-regulate their driving in response to a competing activity. These conditions occur when tasks, either the driving or the secondary task, are unpredictable, nonignorable, uninterruptible, and nonadjustable (see also Chapter 2). A task is unpredictable when its onset is unexpected or its consequences cannot be foreseen by the driver. A task is nonignorable when it is so compelling or demanding that the driver cannot disengage from it. An uninterruptible task is one that cannot be postponed or cannot be resumed after interruption. And finally, a task is nonadjustable when it cannot be altered to lower the demand it places on the driver. When a task, whether it is a driving or secondary task, has these four characteristics, it is difficult or impossible for drivers to regulate or adapt their pattern of interaction or behavior in relation to the task. When a driving situation and a secondary task, each possessing these four characteristics, overlap with each other temporally, the consequences for safety may be adverse, as the driver is not able to trade off performance on either task. Further discussion of these four task characteristics and their affect on distraction is contained in Chapter 2.

Overall, a number of factors can influence a driver's self-regulation strategies in response to a competing task and, thus, their vulnerability to being distracted by this task. It is to these factors that we now turn.

19.4 DRIVING TASK DEMANDS

The demands of the driving task itself, such as increases in traffic density and the complexity of the traffic environment, can influence the distracting effects of secondary activities. For instance, the use of mobile phones on a quiet country road may have a considerably different effect on driving performance than mobile phone use in a busy urban environment. The complexity of the driving environment can moderate the effects of distraction on driving performance and safety in two ways: (1) by increasing or decreasing the driver's mental workload and, hence, reducing or increasing the amount of cognitive resources available for performing competing activities, and (2) by modifying the probability that the driver will have to react rapidly to an unexpected critical event that can give rise to a collision.

A number of studies have examined the interaction between performance of a secondary activity (primarily mobile phones) and the complexity of the driving task.[10,11,15–18] Strayer and Johnston,[18] for example, examined the effects of increasing the complexity of the driving environment on simulated driving (pursuit tracking) performance while using a mobile phone. Participants were required to converse on handheld and hands-free mobile phones while performing easy and difficult pursuit tracking tasks, which involved participants using a joystick to keep a cursor as closely aligned as possible to a moving target on a computer screen. The results revealed that when using a mobile phone, participants missed almost twice as many tracking targets as when they were not using a mobile phone and that this effect was more pronounced when the participants performed the difficult tracking task. Consistent with Lee et al.,[19] Strayer et al.[10] also found that conversing on a hands-free mobile phone while driving led to an increase in reaction times to a lead braking vehicle, and this impairment in reaction times became more pronounced as the density of the traffic increased. One interesting aspect of this finding is that neither the test car nor the lead vehicle interacted with other vehicles on the road, suggesting that simply increasing the perceptual complexity of the road environment can intensify the distracting effects of engaging in a phone conversation while driving.

Adverse weather conditions have also been shown to influence distracted drivers' ability to make safe cross-traffic turning decisions.[16] When drivers were engaged in a mobile phone task, they did not take into account the road surface condition (whether it was wet or dry) when deciding whether to accept or reject a gap. Indeed, on the wet road surface, participants were estimated to have initiated twice as many potential collisions when distracted by verbal messages.

Research by Horberry et al.,[15] however, failed to reveal any interaction between the complexity of the driving environment and conversing on a hands-free mobile phone. They manipulated the complexity of the driving environment by increasing the number of billboards and advertisements placed on the roadside and the number of buildings and oncoming traffic. Participants drove along the simple and complex driving environments while interacting with the mobile phone and while not performing any secondary task. Interacting with the mobile phone affected driving performance, by decreasing mean speed, increasing speed variability, and decreasing responses to a pedestrian hazard. However, no interaction between the distracter task and environmental complexity was revealed, suggesting that driving performance while interacting with a mobile phone was not further degraded by increased complexity of the traffic environment.

The type of objects used to increase the complexity of the driving environment may explain why the Horberry et al.[15] study failed to find that more complex driving environments further degrades driving performance when distracted. Horberry et al. used objects that were not central to the driving task to increase the complexity of the drives, such as billboards and buildings, whereas other research has tended to increase the complexity of the driving environment by manipulating objects central to driving such as other traffic and the difficulty of the driving terrain. It is possible that increasing the number of objects that are not central to the driving task has little effect on increasing the demands of the driving task because drivers simply ignore environmental features that are not essential to the driving task when already under increased load (e.g., when performing a secondary activity).

The moderating effect of driving task demand on driver distraction raises a number of questions regarding the evaluation of in-vehicle systems. In particular, it raises questions regarding the driving conditions under which in-vehicle systems should be evaluated and whether we can validly set specific pass/fail criteria for in-vehicle systems, given that the environment in which the system is evaluated will modify its effect on driving performance. This is an important issue that warrants further investigation and discussion in the literature and in design and evaluation guidelines. The moderating effect of the roadway environment also has important implications for developing adaptive vehicle-based systems to mitigate distraction, an issue considered in Chapter 27 of this book.

19.5 DRIVER CHARACTERISTICS

There is a large body of evidence that driver characteristics can influence the distracting effects of secondary activities.[20–25] Characteristics such as driver age, driving experience, and gender can affect drivers' willingness to engage in distracting activities, their ability to divide attention appropriately between multiple tasks, and their ability to self-regulate their driving to maintain suitable safety margins when distracted.

19.5.1 Age and Driving Experience

It is difficult to separate the effects of age and driving experience on driving performance, because they are highly correlated (e.g., young drivers are inexperienced drivers, and older drivers typically have many years' driving experience). Moreover, most of the driver distraction research has not attempted to separate the moderating effects of age and driving experience on distraction. As such, these two factors will be discussed together in this section. Chapter 20 discusses issues related to older drivers and distraction in more detail; however, older drivers will also be discussed briefly here.

Older drivers often have a decreased ability to share their attention between two concurrent tasks due to decreases in their visual and information processing capacities,[26] and hence, they may be more susceptible to the distracting effects of using a device while driving than their younger counterparts. Similarly, it is widely recognized that young or inexperienced drivers often lack the driving skills necessary to operate and maneuver a vehicle using only minimal attentional resources and, therefore, do not have sufficient spare attentional capacity to devote to secondary activities.[27] Young drivers also have a greater propensity for risk taking, are relatively poor at judging risk, and are vulnerable to the effects of peer pressure, all of which can encourage them to engage in risky behaviors, including distracting activities.[28]

Research has examined the moderating effects of age at the strategic level, that is, on drivers' decisions to engage in distracting activity while driving. Survey research by McEvoy et al.[29] found that young drivers, aged between 18 and 30 years, are more likely than 50- to 65-year-old drivers to engage in distracting activities while driving, including using a mobile phone (including text messaging), handling in-vehicle equipment, and attending to events, objects, or people outside the vehicle. Moreover, the young drivers rated many of the distracting activities as being less risky than did the older drivers, and they were significantly more likely to report that they had been involved in a distraction-related crash in the past 3 years.[29] Other research by Lamble et al. also found that young drivers (aged 15–24 years) reported a much higher level of mobile phone use while driving than did older drivers (55 to 65+ years).

In addition to the effects of age on willingness to engage in distracting activities, numerous studies have examined the interactive effects of driver age and distraction on driving at the tactical and operational levels.[3,21,22,25,30] Using a simulated driving task, McKnight and McKnight[21] found that drivers aged 46–80 years demonstrated a greater deficit in being able to respond to traffic signals while conversing on a mobile phone than did younger (17–25 years) and middle-aged (26–45 years) drivers. Drivers in the youngest age group demonstrated a similar level of decline in responsiveness to traffic signals as middle-aged drivers when engaged in a simple or intense phone conversation, but they responded to significantly less signals than both the older and middle-aged drivers when tuning the radio. Alm and Nilsson also found that during simulated driving, phone use increased drivers' reaction times to a braking lead vehicle and that this effect was more marked for older drivers (aged 60 years or older) than for the younger drivers (below 60 years).

More recently, Greenberg et al.[31] reported that when compared with drivers aged 25–66 years, teenage drivers (16–18 years) detected fewer events occurring in a simulated roadway when dialing a handheld phone and had a higher lane violation rate when accessing voice mails. In addition, Schreiner et al.[24] also found, in a closed-course study, that older drivers' (mean age 57 years) ability to detect forward and peripheral events while concurrently driving and using a voice recognition system to dial phone numbers was impaired compared with their baseline performance. The younger to middle-aged drivers (mean age 23 years), however, did not demonstrate a performance decrement when interacting with the voice recognition system. Similarly, McPhee et al.[22] found that compared with the younger to middle-aged drivers (aged 17–33 years), older drivers (56–71 years) were less accurate and slower at identifying target signs in a digitized image of a traffic scene when engaging in a simulated conversation (e.g., listening to and answering questions about a short paragraph). Finally, driving simulator research by Shinar et al.[25] demonstrated that older drivers' (60–71 years) driving performance (e.g., speed control and lane keeping) was more adversely affected by phone conversations than that of middle-aged (30–33 years) and young, inexperienced (18–22 years) drivers. The driving performance of the young and middle-age groups when distracted was similar.

Naturalistic driving studies conducted in the United States have revealed that younger drivers have a higher rate of involvement in inattention- and distraction-related crashes and incidents than older drivers.[32,33] The recent 100-car naturalistic driving study, for example, found that the rate of inattention-related* crash and

* In the 100-car study, inattention included secondary task distraction, inattention to the forward roadway (e.g., driving behavior that directs driver's attention from the forward field of view), drowsiness, and nonspecific eye glance away from the forward roadway.

near-crash events decreases dramatically with age. Furthermore, the rate of being involved in an inattention-related crash or near-crash was as much as four times higher for the 18- to 20-year-old age group compared with the 45–54 and 55+ driver groups.

An Australian study has also examined, for drivers of different ages, the association between distraction inside and outside the vehicle and the risk of being involved in a crash.[20] Fatal and injury crash data collected by New South Wales police during the years 1996 and 2000 were examined, and crashes were categorized as resulting from no distraction or distraction inside or outside the vehicle. In-vehicle distractions included using a handheld phone, attending to passengers, tuning the radio, and adjusting the CD player and smoking. Results revealed that, with the exception of mobile phones, the risk of being involved in a fatal or injury crash resulting from in-vehicle distractions increased with increasing age. In relation to mobile phones, drivers in the 25–29 year age group had the highest risk of being involved in a fatal or injury crash while using a handheld phone. However, Lam suggested that this finding is likely to have resulted from differential exposure to mobile phone use across age groups, rather than to differences in attention sharing ability; that is, young drivers may be more likely to use their mobile phones while driving than older drivers, and this increased exposure heightens their crash risk.

It is difficult to draw firm conclusions regarding the moderating effects of age on driver distraction, given that the classification of younger, middle-aged, and older drivers varies considerably across studies. For example, younger drivers have been defined as ranging from 17 to 25 years,[21] 16 to 18 years,[31] and 18 to 22 years,[25] whereas older drivers have been defined as ranging from 46 to 80 years,[21] 56 to 71 years,[22] and 60 to 71 years.[25] It is particularly difficult to draw conclusions regarding whether and how younger drivers' driving performance decrements when distracted differ from that of middle-aged drivers, as a number of studies have grouped younger and middle-aged drivers together. In the studies that have compared these two age groups, their driving performance decrements when distracted appear to be largely similar; however, the younger drivers, particularly teenage drivers, do exhibit greater performance decrements on some measures of driving performance (e.g., hazard detection rates and steering variability) than middle-aged drivers.[25,31] The wide age range that defines "young" drivers across studies is particularly problematic given the dramatic change in their crash rates in the first months and years of driving.[34]

It is interesting to note that although many driver distraction studies collect information on both driver age and driving experience, they typically only report the effects of age on distraction and do not attempt to separate the moderating effects of age versus driving experience on distracted driving performance. This is likely to be due to the difficulties inherent in discriminating between the effects of age and inexperience, particularly for younger drivers. Age and driving experience are highly correlated; young drivers are typically inexperienced drivers, and experienced drivers are typically older. You can always find exceptions, however. In some U.S. states, for example, drivers can receive a driver's license as young as 14 years. By the time these drivers reach the age of 16, they have more driving experience than their newly licensed counterparts in other states, making distinguishing the effects of age versus experience easier. Then there is also the problem that factors other than age and driving experience may confound the relationship between these two factors and crash risk; personality characteristics, for example, may lead some drivers to obtain their license early or late, and that may also contribute to increased crash risk.

A number of studies in the general young driver literature have investigated the relative contribution of age versus driving experience to increased crash involvement,[35,36] although it has not been examined widely in distraction research. These studies have found conflicting results, with some finding that age is a greater contributor to young driver crashes,[37,38] whereas others have found that driving experience is more important than age in determining crash involvement.[35] Many of these studies suffer from methodological limitations such as imprecise measurement of driving experience and the assumption that differences in crash involvement between drivers of different ages and driving experience levels are due solely to age and driving experience and not other factors that may lead drivers to obtain a license early or late. Despite the difficulties inherent in discriminating the relative effects of age versus driving experience on driving performance and crash involvement, research should be conducted to establish the relative influence of age versus driving experience in moderating the effects of distraction on driving performance. Such knowledge may assist with countermeasure development; for example, in targeting training methods and education campaigns for reducing young drivers' involvement in risky driving behaviors rather than targeting time-sharing skill development if age was found to be a more important moderating factor.

Overall, it appears that older drivers demonstrate greater decrements in driving performance when engaged in competing tasks than do middle-aged and young drivers. It is important to note that although younger drivers may perform better than older drivers when engaging in secondary activities (i.e., because they are better at multitasking), they are also more likely to engage in nondriving tasks, and this greater level of exposure can increase their crash risk. However, drawing firm conclusions regarding the moderating effects of age on driver distraction is difficult, given that the classification of younger, middle-aged, and older drivers has varied considerably across studies. It is also not possible to draw firm conclusions regarding the relative influence of age versus driving experience in moderating the effects of distraction on driving performance, as most distraction studies have either ignored driving experience as a factor or have not separately examined the effects of these two variables.

19.5.2 Task Familiarity and the Effects of Practice

In addition to driving experience, a driver's familiarity, or experience, with a competing activity can influence how this activity affects driving performance. Practice improves task performance, and this is particularly true for dual-task performance.[39] Practice can improve dual-task performance in a number of ways. First, practice on a task may permit it to be performed almost automatically or in a more economical manner, reducing the amount of resources required to perform it and thereby reducing its ability to interfere with, or be affected by, the concurrent performance of another task.[40,41] Second, practice allows a person to develop strategies for executing each task and prioritizing the performance of multiple tasks in an optimal manner.[42] This research suggests that with increasing practice on both the driving and competing activities, drivers can learn to better prioritize the driving and competing tasks and perform the competing activity with less interference to the driving task than would be expected from the performance of an unfamiliar or unpracticed competing activity.

In many distraction studies, the effect of competing activities on driving performance is typically examined only over a limited number of trials. Participants are not usually given the opportunity to interact with the in-vehicle device or perform the secondary activity over multiple trials, and therefore, any learning effects, whereby drivers learn to effectively time-share the competing and driving tasks, are not assessed. Although the results of this research might provide important insights into how the performance of secondary tasks can affect driving for novice users, it very likely overestimates the effects of these tasks on driving for more experienced users.

Research by Dingus et al.,[43] as part of the TravTek study, examined the effects of repeated experience with the TravTek route guidance system on driving performance. Participants were tested once before they had any previous experience with the TravTek system and once after they had used the system everyday for 6 weeks. The results revealed that, with practice, the drivers developed strategies for using the system in a more economical and safer manner. Specifically, after experience, drivers glanced at the navigation display fewer times and for shorter durations and made a smaller number of large steering reversals than they did before experience with the navigation system. Performance on lateral and longitudinal driving measures and subjective workload did not change with increased experience.

A more recent study by Shinar et al.[25] examined whether repeated practice of conversing on a mobile phone led to a learning effect, whereby drivers became better able to share the phone and driving tasks, thus reducing the effects of the secondary task on driving performance. Thirty participants carried out two mobile phone tasks (a mathematical operation task and emotionally engaging conversation) over five driving sessions. As expected, the use of the mobile phone had a negative impact on driving performance, with drivers displaying lower mean speeds and greater speed and steering variability. However, over the course of the five sessions, the negative effects of the phone tasks on driving performance diminished so that, on several of the driving measures, there was no difference observed between performance in the distraction and no-distraction conditions on later trials.

In short, it appears that the studies that examine the effects of a competing activity over a limited number of trials may be overestimating the detrimental effects of particular competing activities on driving performance, particularly for experienced users. Of course, certain properties of the competing task, such as how redundant it is (e.g., Ref. 44), will greatly influence the degree to which it becomes automated with practice and, hence, the degree to which it will interfere with driving, even after further dual-task practice. Clearly, further research is needed to determine the extent to which practice is capable of diminishing the adverse effects of distraction. Nevertheless, the epidemiological and crash data reviewed in other chapters of this book suggest that even experienced drivers, who engage in highly overlearned competing activities (e.g., talking, smoking, daydreaming), are vulnerable to the effects of distraction.

19.5.3 Gender

Gender differences in drivers' exposure to, and ability to cope with, distraction have been relatively underexamined in the literature compared with other driver characteristics, such as age. The results regarding the effects of gender on distraction exposure and distractibility are mixed.

Sullman and Baas[45] and Poysti et al.[2] found that males reported that they use mobile phones more often when driving than females, whereas Wogalter and Mayhorn[46] found that a greater number of females reported using mobile phones while driving. Differences between these studies may result from age-related differences in the samples surveyed. Sullman and Bass' participant sample was older, by an average of 10 years, than Wogalter and Mayhorn's sample, and these age differences may influence the use of phones while driving across genders. In terms of gender differences on driving performance, some studies have found that distraction has a greater impact on the driving performance of female than male drivers,[16,47,48] whereas other studies have found no gender differences in the effects of distraction.[18,21,23,49] Again, age differences between the study samples may be driving the discrepancies in the results. The studies that found gender effects tended to have an older participant sample than studies that found no effects. Indeed, Hancock et al.[48] found an interaction between age and gender, whereby no differences in the effect of distraction were found between younger male and female drivers, but that distraction had a greater effect on older female drivers than on older males.

The limited, and more general, body of literature on dual-task performance and gender does not appear to indicate that one gender is any better than the other in maintaining primary task performance in the presence of a competing task. It is probable that if there are gender differences in drivers' ability to cope with distraction, these are less likely to derive from biological differences and more likely to derive from differences between genders in such things as the amount of practice they have in driving while distracted, differences in their propensity to engage in particular activities that may distract them, and differences in the extent to which their attention is unwillingly diverted away from the driving task (e.g., by the content of advertising billboards).

19.6 INTERACTION BETWEEN DRIVER STATE AND DISTRACTION

There has been very little research examining the interaction between various driver states (e.g., fatigue, drowsiness, intoxication by drugs or alcohol, emotional state, mood) and distraction. A number of driver states, including distraction, increase crash risk, and this crash risk could be increased further if drivers are affected by a combination of driver states. Driver state has the potential to interact with distraction, by increasing or decreasing a driver's willingness to engage in distracting activities, by interfering with the self-regulation strategies normally adopted by drivers to compensate for being distracted, or by influencing the degree to which a competing activity will affect driving performance and safety. The relationship between driver fatigue and distraction is discussed in Chapter 20. This section will focus on what is known about the interactive effects of alcohol and distraction.

A number of studies have been conducted to compare the effects on driving performance of engaging in secondary activities with that of driving with a blood alcohol concentration (BAC) at or above a particular level. Very few studies, however, have examined the combined effects of alcohol and engagement in a secondary activity (e.g., mobile phone use) on driving. It is likely that alcohol interacts with distraction in a number of ways: (1) by making drivers more or less likely to engage in distracting activities, (2) by affecting how drivers self-regulate their driving when distracted, and (3) by increasing the degree to which engagement in a distracting activity will affect driving performance (e.g., both alcohol and distraction reduce drivers' reaction times; thus, in combination, these two factors might have a cumulative effect on reaction time).

A number of studies have found, under dual-task conditions, that alcohol intoxication affects the capacity to divide attention.[50] However, these studies have typically examined the effects of alcohol on dual-task performance using laboratory-based tasks (i.e., shadowing messages heard through headphones). Studies examining the effects of alcohol on real or simulated driving performance under dual-task conditions are scarce. A study by Brewer et al.,[51] however, examined a South Australian in-depth crash database to determine the extent to which crashes involving alcohol-intoxicated drivers were characterized by the driver's attention being diverted to a nondriving activity before the crash. The study found that a greater proportion of drivers who had a BAC above 0.05 (Australian legal limit) were involved in a secondary activity shortly before the crash than drivers who were not intoxicated (50% of intoxicated drivers versus 38% of nonintoxicated drivers). The authors concluded that the data suggest that intoxicated drivers may be more likely than sober drivers to engage in distracting activities while driving. However, given the lack of data on secondary activity involvement among crash-free drivers, it is not possible to establish a causal relationship between intoxication and precrash secondary activity. Furthermore, it is not clear whether the effect of alcohol was to encourage a diversion of attention away from driving toward a competing activity, to exacerbate the adverse effects of distraction on dual-task performance, or both.

A more recent study by Rakauskas and Ward[52] examined the combined effects of alcohol impairment and engagement in a distracting task on simulated driving performance. Drivers performed a mobile phone task and other in-vehicle tasks (e.g., interacting with the radio or heating, ventilation, and air-conditioning system) while sober or with a BAC of 0.08. The results indicated that although the performance of the secondary tasks degraded driving performance, these degradations were similar, in direction and magnitude, across the sober and intoxicated drivers, indicating that the presence of alcohol did not further affect driving performance when distracted. However, a number of factors may have attenuated the interactive effects of alcohol and distraction on driving in this study. First, only 9 of the 24 drivers in the alcohol condition reached a BAC over 0.08. It is possible that alcohol may only exacerbate the effects of distraction on driving at higher levels of intoxication. Second, only males aged over 21 years participated in the study. It is possible that male drivers have a higher tolerance to alcohol than female drivers and that younger drivers are more adversely affected by the combined effects of alcohol and distraction than older drivers. From an exposure perspective, it is also possible that males have more experience in dealing with distraction when inebriated than females, given their generally greater tendency to drink and drive. These are all issues that merit further investigation.

Overall, despite a lack of empirical evidence, it is likely that driver state is a factor that moderates the effects of distraction on driving performance and safety. More research should be devoted to examining the interaction between distraction and various other driver states and their combined effects on driving performance.

19.7 CONCLUSIONS

A range of factors influence whether drivers will engage in distracting activities while driving and how they will cope with these activities once engaged. These moderating factors interact in complex ways to determine whether and how driving performance will be affected by distracting activities. Factors such as age, driving demands, and driver state, for example, all influence how drivers self-regulate their driving when engaged in a competing activity; and self-regulation can, in turn, influence the degree to which distraction affects driving performance.

Understanding the factors that moderate distraction and the mechanisms underlying these relationships can inform the development of distraction countermeasures and mitigation strategies. For example, it has been found that both older and young, experienced drivers demonstrate greater decrements in driving performance when engaged in secondary tasks than do middle-aged drivers. Thus, it appears that age is an important factor moderating distraction. However, the mechanisms underlying the relationship between age and distraction are different for the two age groups, suggesting that different countermeasures are needed for these two driving groups. The larger observed decrements in older drivers' driving performance are believed to result from diminished visual and cognitive capacity and physical limitations associated with aging, which decrease their ability to perform multiple tasks concurrently. Distraction countermeasures for older drivers should be aimed at addressing and supporting the visual, physical, and cognitive limitations, such as the use of larger fonts and buttons on in-vehicle displays and limiting the amount of information presented to drivers at any one time.

For younger drivers, driving inexperience and their greater propensity to engage in risky behaviors are believed to underlie their increased vulnerability to distraction. Inexperienced drivers will often lack the skills necessary to operate a vehicle using only minimal attentional resources, leaving them with limited spare attentional capacity to devote to secondary activities. Graduated licensing systems, which ban drivers from engaging in certain distracting activities (e.g., mobile phone use and passenger carriage) for the first year or two of licensure, are one category of countermeasures that can mitigate the effects of distraction for young drivers.[53] Training and educational campaigns that make young drivers aware of the dangers of distraction and inform them of ways in which they can regulate their behavior to better cope with distractions are another measure that may be effective in reducing their willingness to engage in distracting activities and the negative impact of distraction on them as a driving population.

Much of the distraction research to date has focused on the negative effects of distraction on driving performance. Few studies have directly examined the factors that moderate this relationship. It is important that researchers and practitioners understand not only the factors that moderate distraction but the mechanisms through which this moderation occurs. Such knowledge can play an enormous role in guiding the development of effective countermeasures that are targeted at addressing the capabilities and limitations of different driver populations.

REFERENCES

1. Haigney, D., Taylor, R. G., and Westerman, S. J., Concurrent mobile (cellular) phone use and driving performance: task demand characteristics and compensatory processes, *Transportation Research Part F* 3, 113–121, 2000.
2. Poysti, L., Rajalin, S., and Summala, H., Factors influencing the use of cellular (mobile) phone during driving and hazards while using it, *Accident Analysis & Prevention* 37, 47–51, 2005.
3. Alm, H. and Nilsson, L., The effects of a mobile telephone on driver behaviour in a car following situation, *Accident Analysis & Prevention* 27(5), 707–715, 1995.
4. Lamble, D., Rajalin, S., and Summala, H., Mobile phone use while driving: public opinions on restrictions, *Transportation* 29, 233–236, 2002.
5. Alm, H. and Nilsson, L., Changes in driver behaviour as a function of hands-free mobile telephones: a simulator study, *Accident, Analysis and Prevention* 26, 441–451, 1990.
6. Burns, P. C., Parkes, A., Burton, S., Smith, R. K., and Burch, D., How dangerous is driving with a mobile phone? Benchmarking the impairment to alcohol, TRL Limited, 2002.
7. Rakauskas, M. E., Gugerty, L. J., and Ward, N. J., Effects of naturalistic cell phone conversations on driving performance, *Journal of Safety Research* 35(4), 453–464, 2004.
8. Jamson, A. H., Westerman, S.J., Hockey, G.R.J., and Carsten, O.M.J., Speech-based e-mail and driver behaviour: effects of an in-vehicle message system interface, *Human Factors* 46(4), 625–639, 2004.
9. Strayer, D. L. and Drews, F. A., Profiles in driver distraction: effects of cell phone conversations on younger and older drivers, *Human Factors* 46(4), 640, 2004.
10. Strayer, D. L., Drews, F. A., and Johnston, W. A., Cell phone-induced failures of visual attention during simulated driving, *Journal of Applied Psychology* 9(1), 23–32, 2003.
11. Brookhuis, K. A., de Vries, G., and de Waard, D., The effects of mobile telephoning on driving performance, *Accident Analysis & Prevention* 23(4), 309–316, 1991.
12. Harbluk, J. L., Noy, Y. I., and Eizenman, M., The impact of cognitive distraction on driver visual behaviour and vehicle control, Report No. TP No. 13889 E, Road Safety Directorate and Motor Vehicle Regulation Directorate, Ottawa, Canada, 2002.
13. Harbluk, J. L., Noy, Y. I., Trbovich, P. L., and Eizenman, M., An on-road assessment of cognitive distraction: impacts on drivers' visual behavior and braking performance, *Accident Analysis & Prevention* 39(2), 372–379, 2007.
14. Chiang, D. P., Brooks, A. M., and Weir, D. H. D. H., On the highway measures of driver glance behavior with an example automobile navigation system, *Applied Ergonomics* 35(3), 215–223, 2004.
15. Horberry, T., Anderson, J., Regan, M. A., Triggs, T. J., and Brown, J., Driver distraction: the effects of concurrent in-vehicle tasks, road environment complexity and age on driving performance, *Accident Analysis & Prevention* 38(1), 185–191, 2006.
16. Cooper, P. J. and Zheng, Y., Turning gap acceptance decision-making: the impact of driver distraction, *Journal of Safety Research* 33(3), 321–335, 2002.
17. Liu, B.-S. and Lee, Y.-H., In-vehicle workload assessment: effects of traffic situations and cellular telephone use, *Journal of Safety Research* 37(1), 99–105, 2006.

18. Strayer, D. L. and Johnston, W. A., Driven to distraction: dual-task studies of simulated driving and conversing on a cellular telephone, *Psychological Science* 12(6), 462–466, 2001.
19. Lee, J. D., Caven, B., Haake, S., and Brown, T.L., Speech-based interaction with in-vehicle computers: the effect of speech-based e-mail on drivers' attention to the roadway, University of Iowa, Iowa City, 2001.
20. Lam, L. T., Distractions and the risk of car crash injury: the effect of drivers' age, *Journal of Safety Research* 33(3), 411–419, 2002.
21. McKnight, A. J. and McKnight, A. S., The effect of cellular phone use upon driver attention, *Accident Analysis & Prevention* 25(3), 259–265, 1993.
22. McPhee, L. C., Scialfa, C. T., Dennis, W. M., Ho, G., and Caird, J. K., Age differences in visual search for traffic signs during a simulated conversation, *Human Factors* 46(4), 674, 2004.
23. Reed, M. P. and Green, P. A., Comparison of driving performance on-road and in a low-cost simulator using a concurrent telephone dialling task, *Ergonomics* 42(8), 1015–1037, 1999.
24. Schreiner, C., Blanco, M., and Hankey, J. M., Investigating the effect of performing voice recognition tasks on the detection of forward and peripheral events, *Human Factors and Ergonomics Society 48th Annual Meeting*, New Orleans, Louisiana, 2004, pp. 2354–2358.
25. Shinar, D., Tractinsky, N., and Compton, R., Effects of practice, age, and task demands, on interference from a phone task while driving, *Accident Analysis & Prevention* 37(2), 315–326, 2005.
26. Eby, D. W., Trombley, D. A., Molnar, L. J., and Shope, J. T., The assessment of older driver's capabilities: A review of the literature, Report No. UMTRI-98-24, The University of Michigan Transportation Research Institute, Ann Arbor, MI, 1998.
27. Regan, M. A., Deery, H., and Triggs, T. J., Training for attentional control in novice car drivers, *Proceedings of the 42nd Annual Meeting of the Human Factors and Ergonomics Society (Volume 2)*, Chicago, IL, 1998, pp. 1452–1456.
28. Williamson, A., Young drivers and crashes: why are young drivers over-represented in crashes? Summary of the issues, Paper prepared for the Motor Accidents Authority of NSW, Sydney, Australia, 1999.
29. McEvoy, S. P., Stevenson, M. R., and Woodward, M., The impact of driver distraction on road safety: results from a representative survey in two Australian states, *Injury Prevention* 12(4), 242, 2006.
30. Angell, L. S., Auflick, J. L., Austria, P. A., Kochhar, D. S., Tijerina, L., Biever, W., Diptiman, D., Hogsett, J., and Kiger, S., *Driver Workload Metrics: Task 2 Final Report*, National Highway Traffic Safety Administration, Washington, D.C., 2006.
31. Greenberg, J., Tijerina, L., Curry, R., Artz, B., Cathey, L., Grant, P., Kochhar, D., Kozak, K., and Blommer, M., Evaluation of driver distraction using an event detection paradigm, *Journal of the Transportation Research* Board No. 1843, 1–9, 2003.
32. Dingus, T. A., Klauer, S. G., Neale, V. L., Petersen, A., Lee, S. E., Sudweeks, J., Perez, M. A., Hankey, J., Ramsey, D., Gupta, S., Bucher, C., Doerzaph, Z. R., Jermeland, J., and Knipling, R. R., *The 100-Car Naturalistic Driving Study, Phase II—Results of the 100-Car Field Experiment*, Virginia Tech Transportation Institute, Blacksburg, VA, 2006.
33. Stutts, J., Feaganes, J., Rodgman, E., Hamlett, C., Meadows, T., and Reinfurt, D., Distractions in everyday driving, AAA Foundation for Traffic Safety, Washington, D.C., 2003.
34. Mayhew, D. R., Simpson, H. M., and Pak, A., Changes in collision rates among novice drivers during the first months of driving, *Accident Analysis & Prevention* 35(5), 683–691, 2003.

35. Catchpole, J. E., MacDonald, W. A., and Bowland, L., Young driver research program: The influence of age-related and experience-related factors on reported driving behaviour and crashes, Report No. CR 143, Monash University Accident Research Centre, Clayton, Vic., 1994.
36. Cooper, P. J., Pinili, M., and Chen, W., An examination of the crash involvement rates of novice drivers aged 16 to 55, *Accident Analysis & Prevention* 27(1), 89–104, 1995.
37. Levy, D. T., Youth and traffic safety: the effect of driver age, experience and education, *Accident Analysis & Prevention* 22, 327–334, 1990.
38. Mayhew, D. R. and Simpson, H. M., New to the road: Young drivers and novice drivers: Similar problems and solutions?, Traffic Injury Research Foundation of Canada, 1990.
39. Ruthruff, E., Johnston, J. C., and Van Selst, M., Why practice reduces dual-task interference, *Journal of Experimental Psychology: Human Perception and Performance* 27, 3–21, 2001.
40. Allport, D. A., Antonis, B., and Reynolds, P., On the division of attention: a disproof of the single channel hypothesis, *The Quarterly Journal of Experimental Psychology* 24, 225–235, 1972.
41. Shaffer, L. H., Multiple attention in continuous verbal tasks, in *Attention and Performance V*, Rabbitt, P. M. A. and Dornic, S. (eds.), Academic Press, London, UK, 1975, pp. 157–167.
42. Eysenck, M. and Keane, M., *Cognitive Psychology: A Student's Handbook*, Psychology Press, 2005.
43. Dingus, T. A., Hulse, M. C., Mollenhauer, M. A., Fleischman, R. N., McGehee, D., and Manakkal, N., Effects of age, system experience, and navigation technique on driving with an advanced traveler information system, *Human Factors* 39, 177–199, 1997.
44. Gladstones, W. H., Regan, M. A., and Lee, R. B., Division of attention: the single-channel hypothesis revisited, *Quarterly Journal of Experimental Psychology* 41A, 1–17, 1989.
45. Sullman, M. J. M. and Baas, P. H., Mobile phone use amongst New Zealand drivers, *Transportation Research Part F* 7, 95–105, 2004.
46. Wogalter, M. S. and Mayhorn, C. B., Perceptions of driver distraction by cellular phone users and nonusers, *Human Factors* 47(2), 455, 2005.
47. Briem, V. and Hedman, L. R., Behavioural effects of mobile telephone use during simulated driving, *Ergonomics* 38, 2536–2562, 1995.
48. Hancock, P. A., Lesch, M., and Simmons, L., The distraction effects of phone use during a crucial driving maneuver, *Accident Analysis & Prevention* 35(4), 501–514, 2003.
49. Woo, T. H. and Lin, J., Influence of mobile phone use while driving: the experience in Taiwan, *International Association of Traffic and Safety Sciences* 24, 5–19, 2001.
50. Moskowitz, H. and Depry, D., Differential effect of alcohol on auditory vigilance and divided attention tasks, *Quarterly Journal of Studies on Alcohol* 29, 54–63, 1968.
51. Brewer, N. and Sandow, B., Alcohol effects on driver performance under conditions of divided attention, *Ergonomics* 23, 185–190, 1980.
52. Rakauskas, M., & Ward, N., Behavioural effects of driver distraction and alcohol impairment, *Human Factors and Ergonomics Society 49th Annual Meeting*, Orlando, FL, 2005.
53. Lee, J. D., Technology and teen drivers, *Journal of Safety Research* 38(2), 203–213, 2007.

20 Distraction and the Older Driver

Sjaanie Koppel, Judith L. Charlton, and Brian Fildes

CONTENTS

As stated in Chapter 1, driver distraction is the diversion of attention away from activities critical for safe driving toward a competing activity. There is converging evidence, as noted in other chapters, that distraction is a significant contributing factor in vehicle crashes. However, the degree to which distraction is problematic for older drivers is less known. In this chapter, we address this issue within the context of the known crash risk and driving patterns of older drivers. Consideration is also given to the effects of aging on functional abilities and other factors that might predispose older drivers to the effects of distraction and their propensity to engage in distracting activities. In addition, we consider strategies for reducing the potentially negative effects of distraction on older driver performance and safety.

20.1 OLDER DRIVERS

The target group of interest in this chapter is older drivers. However, there is no consensus regarding a lower threshold age for defining this group; indeed, there is a level of controversy on this issue. Some studies have identified the onset of aging processes related to driving performance at 55 years; in other studies criteria such as frailty have been of more interest and thus a higher defining age of 70 or 75 years has been adopted.[1] Interestingly, although age 75 is where crash rates begin to increase significantly,[2] the most commonly accepted age for defining the older driver is 65 years, the traditional age of retirement. Although this serves as a useful reference for categorizing the older driver, issues relating to older drivers and distraction shall not be limited to this group.

20.2 THE OLDER DRIVER CRASH PROBLEM

Over the next four or five decades, there will be a substantial increase in both the number and proportion of older people in most industrialized countries.[3] With the aging of the population, it is also anticipated that there will be an increase in older drivers' licensing rates.[4,5] Further, the private car is likely to remain the principal mode of transportation for the emerging cohorts of older drivers who, it is predicted, will be undertaking longer and more frequent journeys.[3] Demographic growth, increased licensing rates, and increased car use will combine to produce a marked increase in the number of older drivers on the road. Increased older driver numbers and increased driving exposure have led to expectations of a commensurate increase in future crash levels. It has been predicted that, by 2025, older driver fatalities in the United States[6] and Australia[7] will triple, relative to 1995 levels.

Although older drivers are currently involved in few crashes in terms of absolute numbers, they represent one of the highest risk categories for crashes involving serious injury and death per number of drivers and per distance traveled,[3] largely due to their greater frailty and reduced tolerance to injury.[8–13] Li et al.[2] have estimated that at least 60% of the increase in death rate per distance traveled for those aged 60 and over can be accounted for by increases in fragility.[2]

A more detailed analysis of crash types can be instructive in understanding the factors that predispose older drivers to crash and injury risk and may provide some insight into the role of distraction in older driver crashes. Older drivers have noticeably different crash patterns from those of their younger counterparts. For example, older drivers are more likely to crash during daylight hours, at low speeds, with low blood alcohol concentration (BAC) levels, at intersections, with other vehicles (multivehicle crashes), and with a severe injury outcome.[14,15] In addition to the frailty factor, much of this crash profile has been attributed to their diminishing cognitive, perceptual, attention, and motor processes brought about by the aging process, their overexposure in specific places and times, and their interface with vehicle or road systems, which are largely designed for the "average" driver.[16,17] In addition, it has been argued that the behaviors that lead to older driver crashes seem to be more related to inattention or slowed perception and responses than to the deliberate, unsafe actions that are more common to younger drivers, such as speeding and drunk driving.[18] However, the extent to which older drivers are willing to engage in potentially distracting behavior while driving is largely unexplored. This issue is considered in more detail later in the chapter.

20.2.1 Older Driver Licensing and Driving Rates

Older people continue to have travel needs after retirement, although the nature, frequency, and duration of their trips may change. Overall, as people age they make fewer journeys, mainly due to reductions in the number of work journeys, and the average length of all journeys consistently decreases. The number of journeys made for nonwork activities remains almost constant to the age of 75 and decreases thereafter, with the length of these journeys also reducing with increasing age. For example, the Insurance Institute for Highway Safety in the United States reported that drivers aged 70 and older drove 55% fewer annual miles, on average, than drivers aged 35–69.[19] Similarly, a survey of self-reported distances driven each week by a sample of younger (17–40 years) and older (60 years and older) drivers in Australia reported that older drivers drove shorter distances each week than younger drivers,[14] a point possibly to their detriment in terms of sustaining a higher crash risk according to Hakamies-Blomqvist.[20] For example, Janke[21] notes that high mileage drivers are more likely to use freeways and multilane divided roadways with limited access. By implication, low mileage drivers do more of their driving on local roads and streets, which have a greater number of potential conflict points and hence higher crash rates per unit of road distance. Janke noted that there were 2.75 times more crashes per mile on nonfreeways than freeways. For older drivers, urban travel is more likely to result in crashes because of the greater numbers of possible traffic conflict points, especially intersections.[22]

20.2.2 Future Cohorts of Older Drivers

Although licensing rates and driving distances are relatively low among current cohorts of older drivers compared with younger drivers, researchers have predicted that these patterns are likely to change substantially over the next few decades, particularly for women.[3,6,7] For example, recent figures show that drivers aged 65 years and older account for 15% of all licensed drivers in the United States, and the numbers in this group of drivers have increased by 17% compared with the previous decade.[23] In addition, as the current middle-aged baby boomers reach old age, the private car is likely to remain the dominant form of transportation. Hence, they will be more mobile, travel more frequently, and travel greater distances, and will have higher expectations with regard to maintaining personal mobility[3] than earlier cohorts. Moreover, future cohorts of older drivers might also differ in the way they engage with vehicles, technologies, and other aspects of the road environment, which in turn may influence their driving patterns and willingness to engage in potentially distracting activities. This issue is discussed further in the following sections of the chapter.

Given the predicted increase in numbers of older drivers on the roads, their higher crash rates, and their increased likelihood of injury, safe and efficient mobility for older drivers has become a challenging social problem for most Western societies.[3] There is now a convincing body of research demonstrating that older road users have distinct risk factors relative to young and middle-aged drivers. Foremost of these is their frailty and hence vulnerability to injury in the event of a crash and, for many, a general decline in physical, sensory, and cognitive functioning. The next section of this chapter outlines the typical changes in sensory, cognitive, and physical abilities that occur as part of the aging process, how these are known to affect safe driving, and whether the aging process makes older drivers particularly vulnerable to distraction.

20.3 WHAT MAKES OLDER DRIVERS DIFFERENT?

Discussing the issue of older driver distraction assumes that this group is different from other groups of drivers.[24] Age is often described in terms of either chronological or functional status. Chronological age is uncomplicated—it is simply the number of years since birth. However, chronological age is an inaccurate indicator of performance. There is now well-documented evidence of wide individual differences in performance among older adults of the same chronological age.[25] Functional status (commonly identified in terms of level of functional impairment) is more relevant, and its use in the context of driver distraction recognizes that humans of the same age differ in their ability to perform tasks across a range of areas, involving cognition, physical skill, and perception. Although there are many individual differences in the aging process, even relatively healthy older adults are likely to experience some level of functional decline in sensory, cognitive, and physical abilities: a decline in visual acuity and contrast sensitivity; visual field loss; reduced dark adaptation and glare recovery; loss of auditory capacity; reduced perceptual performance; reductions in motion perception; a decline in attentional and cognitive processing ability; reduced memory functions; neuromuscular and strength loss; postural control and gait changes; and slowed reaction time.[25,26] Of relevance to older drivers is how the degradation of these skills relates to safe driving and whether these skill degradations put them at an increased risk of driver distraction.

Current evidence for causal relationships between specific medical conditions and increased crash risk is limited.[25,27–30] Clearly, not all medical conditions affect injury risk on the road to the same extent and not all individuals with the same condition will be affected in the same way.[27] The severity of the condition and other characteristics of the disorder are likely to be important determinants of crash risk. Indeed, it is not necessarily the medical condition or medical complications *per se* that affect driving, but rather the *functional impairments* that may be associated with these conditions. In discussing the merits of focussing on impairments in assessing risk, Marottoli[31] noted that functional impairments are "the common pathway through which ... medical conditions affect driving capability and ... can be relatively easy to test" (p. 11). Moreover, the extent to which individuals may be able to adapt or compensate for their impairment while driving will undoubtedly have some bearing on their likelihood of crash involvement.[27]

However, there is mounting evidence that a number of age-related functional impairments may be of sizeable concern to road safety. For example, these functional impairments may predispose older drivers to the effects of distraction and, without appropriate compensatory strategies, may lead to an increased risk of distraction-related crashes. A summary of research on the association between driving performance and vision, cognitive, and physical impairments follows.

20.3.1 Vision

Vision is critical to driving, with some researchers suggesting that vision makes up to 95% of the sensory input needed to drive.[32] Owsley et al.[33] describe the driving task as "a visually cluttered array, both primary and secondary visual tasks, and simultaneous use of central and peripheral vision. In addition, the driver is usually uncertain as to when and where an important visual event may occur" (p. 404). Vision is one area where there is compelling evidence for age-related changes.[24]

Difficulties of various kinds resulting from age-related vision changes relevant to driving are known: focusing on distant objects (e.g., oncoming vehicles or traffic lights in the distance) and on near objects (e.g., speedometer) is more problematic; distortion of focus on near and far objects; and difficulty adjusting focal length between objects in the near and far field of view.

Because the deterioration of visual function that occurs with advancing age is recognized, considerable research attention has been focused on the relationship between visual capability and the driving performance of older adults. Age-related declines in dynamic and static visual acuity, visual field, resistance to glare, contrast sensitivity, visual processing speed, visual search, low-light sensitivity, perception of angular movement, and movement in depth have all been associated to varying degrees with increased crash risk.[32,34,35] However, attempts to relate specific visual functions (such as visual acuity) to driving ability in general have not been particularly effective. Influential early research by Burg and colleagues examining drivers' visual functions, including static visual acuity, shows a weak correlation with crashes, but only for older drivers.[36,37] This is not surprising, since visual acuity does not adequately reflect the full range of visual capacities required in dynamic and complex traffic. Burg and colleagues argue that more explanatory power might be attributable to age-related declines in higher-level processing abilities.

Indeed, the *processing* of visual information has been shown to be a problem in older adults.[38] For example, performance deficits on assessments such as the useful field of view (UFOV), which examines visual processing and attentional control functions that may be symptomatic of numerous neurological and visual disorders, are consistently and significantly associated with crash risk even after other factors have been adjusted for.[39] In their study of risk factors of injury crashes among older drivers, Owsley et al.[39] found that drivers with a reduction of UFOV of more than 40% were at least 20 times more likely to be involved in a crash involving injury than were those with minor visual limitations. In addition, compared with young participants, older adults tend to be worse at tasks which require detection of target objects amongst distracters, find it more difficult to identify fragmented or incomplete objects[40] and are poor at recognizing objects that are imbedded in other objects.[41] Without appropriate compensatory strategies, these age-related vision and visual-attention impairments are likely to have serious consequences for older driver distraction.

20.3.2 Cognitive Ability

The safe operation of a motor vehicle in traffic also requires the effective execution of a combination of perceptual and cognitive skills.[28] For example, the requirements for safe driving include monitoring information, identifying hazards, making accurate and timely decisions, understanding and remembering traffic rules and signs, following directions, problem solving, maintaining concentration and attention, and minimizing distraction. Moreover, driving involves higher mental abilities, or "executive functions," that supervise all movements and decisions taken by the driver. Executive functions are a variety of loosely related, higher-order cognitive processes like planning, initiation, and regulation of actions, cognitive flexibility, judgment, feedback utilization, and self-perception. These processes are critical for carrying out effective and contextually appropriate behaviors.[42]

One of the most well-established research findings in the developmental psychological literature is the slowing of performance with age. Strong evidence suggests that aging results in slower and inadequate detection and registration of sensory information, slower cognitive processing of that information, slowed integration of the relevant information, reduced memory and attentional capability, and slowed initiation of movement and execution of responses.[35,43,44] In addition, there is a growing body of evidence suggesting that executive functions tend to decline with age.[45–51]

Given the complexity of the driving task, it is not surprising that diminished capability in any of these facets of cognition has the potential to compromise driving performance and lead to an increase in crash risk. For example, Ball et al.[35] reported that visual attention and mental status are the two best predictors of crash frequency, in a model incorporating measurement of eye health, visual function, visual attention, and cognitive functions. Daigneault and colleagues have demonstrated a strong relationship between executive function disorder and motor vehicle crashes for older drivers (65 years and older).[42] For example, older drivers with a crash history showed more deficits that reflected mental rigidity and poor planning ability compared with older drivers without a crash history.

According to the *complexity hypothesis*, older adults are proportionally slower than younger adults when the complexity of the tasks being performed increases.[43,52–54] This is regardless of whether the tasks are performed in single or dual-task paradigms. However, a substantial body of literature also illustrates large and robust age-related deficiencies in dual-task environments.[54–56] A related finding is that older adults have more difficulty than younger adults in the management or coordination of multiple tasks.[56] Although there is consensus about the strength of dual-task and divided attention effects and the results of complex tasks, there is disagreement about the mechanisms that underlie age differences in the performance of these tasks.

What does seem clear is that, through some component of cognitive decline, when cognitive load increases, processing speed decreases.[57] This has serious implications for driver distraction because the additional requirements of having to process different sources of information in a serial mode may protract the decision process to dangerous levels, resulting in higher involvement in certain types of crashes.[4,58,59] Indeed, these deficits have been implicated in older driver crashes.[4,26,60–62]

20.3.3 Physical Limitations

In addition to vision and cognitive impairments, there are physical effects of aging, which may compromise an older individual's ability to drive safely. These include reduced flexibility, strength and coordination, increased difficulty in moving limbs and extremities, and increased discomfort, pain, and fatigue. These changes may affect the ability to initiate and regulate movement accurately, resulting in difficulties with important vehicle operations such as the ability to use vehicle controls, braking, accelerating, and steering control. Turning the head to scan for vehicles when at intersections, in merging traffic, or when changing lanes is another problem experienced by many older drivers. Additionally, older drivers may experience difficulty manipulating small control buttons and knobs on the vehicle control panel resulting in longer time to complete secondary tasks such as tuning the radio or adjusting the heating. Physiological conditions that can result in these effects include general physical weakening, loss of agility, cardiovascular degeneration, and musculoskeletal decline.[63,64] However, the extent to which these limitations impact safety is unclear, and there is no evidence linking these impairments with either distraction or crashes among older drivers.[27]

20.3.4 Summary

Although age *per se* is not a good predictor of driving performance, most older adults are likely to experience some level of functional decline in sensory, physical, and cognitive areas.[25,26] The age-related functional declines outlined in this section have several implications for driver distraction: (1) the increased demands of the driving task, brought about by age-related declines, will leave relatively little spare capacity to deal with competing activities; (2) competing activities are likely to be more distracting for older drivers, as extra time, visual and attentional capacity, and physical effort will be required to perform them; and (3) it will be more difficult for them to trade-off between driving and a competing task to maintain safe driving because of impoverished time-sharing skills. However, it should be noted that current evidence for causal relationships between declines in specific abilities and reduced driving performance or increased crash risk is limited. Meyer[24] has proposed that one possible explanation for this is that drivers can be highly adaptive and can compensate for deficiencies in certain areas by changing their behavior (i.e., using different driving techniques, changing the conditions in which they drive, or using technologies to assist with some of their deficiencies). Importantly, such adaptive behaviors may also have a positive influence on crash risk by reducing exposure to potentially distracting situations. Self-regulation strategies adopted by older drivers, and the way in which they may reduce the possibility of driver distraction, are discussed in the next section.

20.4 SELF-REGULATION STRATEGIES AMONG OLDER DRIVERS

Many older drivers are aware of their decline in functional capacities and adapt their driving patterns to match these changes by self-regulating when, where, and how they drive.[18,65–71] For example, older adults typically choose to reduce their exposure by driving fewer annual kilometers, making shorter trips, and making fewer trips by linking multiple trips together.[72–74] Older drivers have also been found to avoid complex traffic maneuvers that are cognitively demanding,[75,76] to limit their peak-hour and night driving, to restrict long-distance travel, to take more frequent breaks, and to drive only on familiar and well-lit roads.[66,77] In addition, several studies have shown that most older drivers recognize that good vision is one of the most important elements for safe driving and often cite poor vision as a major determinant for reducing driving at night or in poor weather.[78–80] This evidence suggests that at least some older adults are able to compensate well for limitations in their abilities in such a way that is likely to minimize exposure to difficult driving situations to reduce their crash risk.

A process of self-regulation related to the issue of distraction is older drivers' willingness to engage in tasks or risky activities while driving. Lerner[81] investigated drivers' willingness and perceived risk of engaging in various technology-related (e.g., performing different functions with a mobile phone, a navigation system, and a personal digital assistant or PDA) and nontechnology related (e.g., eating, drinking, and conversing with a passenger) tasks. Participants drove their own vehicles over a specified route and, at specified points, rated their willingness to engage in specific tasks at that time and place, but were not required to actually engage in the tasks. In general, teenaged drivers (16–17 years) and younger drivers (18–24) expressed more willingness than middle-aged (25–59) or older (60 and older) drivers to use in-vehicle technologies. Younger drivers also perceived this use as less risky than middle-aged and older drivers. Across all participants, drivers were more willing to use mobile phones than PDAs or navigational systems while driving and were more willing to have conversations on mobile phones than to answer or dial phones. Lerner concluded that older drivers' reluctance to engage in distracting tasks while driving may be a process of self-regulation, in that the older drivers may be aware of some functional decline and opt not to engage in activities that may increase the complexity of the driving task.[5]

Although self-regulation seems to be an effective way to alleviate some of the problems that arise with age, it is by no means perfect.[24] For example, Rothman et al.[82] argue that people of all ages are poor at recognizing the relationship between their own actions and potential risks. This may lead to optimism about one's invulnerability, underestimation of risk and overestimation of one's driving ability.[83] In support of this argument, several studies have demonstrated that some older drivers do not adequately compensate for age-related changes in vision and cognitive abilities when driving and therefore place themselves at an increased risk of crash involvement.[84,85] Interestingly, once drivers are made aware of their declining abilities, many make appropriate adjustments to their driving.[85] This suggests an important role for education/training. In addition, at least some self-regulation changes are counterproductive from the viewpoint of crash reduction. For example, older drivers' apprehension about high-speed travel has led to greater travel on low-speed and urban roads. However, as noted earlier, urban travel is more likely to result in crashes because of the greater numbers of possible traffic conflict points, especially intersections,[22] which may increase older drivers' crash risk.[20,86]

Although the process of self-regulation is likely to minimize older drivers' exposure to difficult driving situations, thereby reducing their crash risk, it is also likely to lead to a reduction in mobility. For many older adults, the loss in one's ability to drive is associated with a loss of independence, an increase in depression and a decline in overall quality of life.[87,88] In addition, for many, driving cessation precipitates a decline in health status over and above any decline associated with normal aging. In addition, the same health conditions and functional impairments that caused a change in driving patterns may also limit access to other transport options,[3,89] thereby further contributing to restricted community mobility and diminished social connections. With the negative impact associated with driving cessation, there is a growing interest in exploring measures that may enable older adults to continue to drive safely for as long as possible. In particular, modifications to the driving environment, including recent advancements in vehicular technology, have potential to compensate for some of the age-related vision, cognition, and physical changes that can affect the safe operation of a motor vehicle[90,91] while still maintaining mobility.

20.5 OLDER DRIVERS AND DISTRACTION

Despite the fact that older adults represent the fastest growing segment of the driving population and that because of age-related degradations they may be particularly vulnerable to the effects of distraction, an extensive literature search revealed only a limited number of studies that had specifically investigated older drivers and their propensity for distraction. In addition, although driver distraction can be caused by driver interaction with a number of sources (see Chapter 15), most of the available literature on older driver distraction has focused on the use of mobile phones and navigation systems, with a few studies examining the impact of collision-warning devices and vision-enhancement systems on their driving performance and crash risk. The following sections explore the literature on this topic with a particular emphasis on the current use and appreciation of mobile phones and in-vehicle technologies and the role of distraction on driving performance and crash risk among older drivers.

20.5.1 OLDER DRIVERS AND TECHNOLOGY

Despite the fact that increasing technology use by older adults is a worldwide trend in developed countries,[92] older adults still tend to lag behind the general population in adopting new technologies.[93] Although the stereotypical view may be that older adults are less able to learn or are not interested in new technologies, research suggests that they are willing to use new technology if they are provided with adequate training[94] and if the benefits of the technology are clear to them.[95]

In-vehicle technology holds considerable promise in the area of older driver safety, particularly since it lends itself to be adapted to individual drivers' needs and areas of vulnerability. Research is also starting to address whether or not older drivers will appreciate the benefits of in-vehicle technologies and find the systems user-friendly.[91] Consistent with Lerner's findings,[81] Wochinger and Boehm-Davis[96] report that older drivers express more reluctance to use navigation-related technology than do younger drivers; the authors attribute this reluctance to a "fear of using technology." Similarly, Caird et al.[97] report that older drivers are more skeptical of the utility of vision-enhancement systems (VES) than younger drivers. However, other research suggests that older drivers are significantly more favorable toward using crash avoidance technologies than younger drivers'[98] with the authors attributing this finding to the perceived benefits of the concepts as well as to the tendency of older drivers to buy vehicles that are fully equipped with comfort and safety features. In addition, other research has shown that that initial reluctance by older drivers to use navigation systems diminished as they gained more experience with the systems.[99] Moreover, Barham et al.[100] and Oxley[101] found that navigation and location systems improve older driver confidence and extend their willingness to drive to unfamiliar locations or to drive more frequently. Oxley[101] also reported that older drivers found VES easy to use and more than 60% of drivers said they would use the systems to drive in the future.

20.5.2 Role of Distraction on Driving Performance and Crash Risk among Older Drivers

Studies that have investigated the impact of distraction on the driving performance of drivers in general, including older drivers, are reviewed in Chapters 11 through 13. As outlined earlier, there is a paucity of research into the impact distraction has on driving performance and crash risk among older drivers. Most of the research in the area of older drivers and distraction has focused on the use of mobile phones and navigation systems, with a few studies examining the impact of collision-warning devices and vision-enhancement systems on their driving performance and crash risk. The limited literature that exists is outlined here.

20.5.2.1 Mobile Phones

There are many potential in-vehicle sources of distraction. One of the most frequently reported is the use of a mobile, or cellular, phone. Several studies have demonstrated that the distracting effect of concurrent mobile phone use on driving performance measures is greater for older drivers compared with other age groups[102–109] (see Chapter 11 for a broader discussion of mobile phone use and distraction). More specific areas of degradation that have been shown to be exaggerated among older drivers include detection time,[105] visual scanning,[110] lane keeping and driving speed,[103,111] visual fixation and recognition memory,[104,111] and time to dial and answer the phone.[112] In contrast, in some studies (e.g., Ref. 113) no such age differences have been found. These authors reported that the effects of handsfree phone conversation tasks on reaction time, following distance, and speed recovery after braking did not differ between drivers aged 18–25 years and those aged 65–74. A possible reason for this finding may be that older drivers were compared with young drivers, who demonstrate similar degradations in driving when distracted. In addition, Shinar et al.[114] (see also Chapter 11) demonstrated that the initial deleterious effects of handsfree phone conversations on many simulated driving tasks were reduced or eliminated with continued practice, albeit at a faster rate for younger drivers (aged 18–33) than for older drivers (aged 60–71). More specifically, in the course of five sessions, participants were given two kinds of distracting phone tasks while driving: (1) an arithmetic operations task (identical to the one used by McKnight and McKnight[105]) and (2) a conversation in which participants were asked a series of questions about information they had provided before the task (i.e., personal information such as social habits, hobbies, and interests) to generate conversations that would be emotionally challenging. The authors report that the effects of the distracting task on driving were greatest when the distracting task was difficult (e.g., mathematical operations), the driver had no experience in performing the dual tasks (day 1), and the driver was older (60–71 years). All participants owned mobile phones and all reported having used their phone while driving, with reported usage rates varying from rarely to frequently.

Although many studies have documented the performance decrements associated with driver distractions, few have examined drivers' awareness of these distraction effects. This is an important issue because the perception or awareness of distraction effects may influence drivers' willingness to engage in distracting activities. For example, drivers who do not calibrate their driving with respect to the magnitude of distraction effects may engage in activities because they do not realize their performance is compromised.[115] Lesch and Hancock[116] conducted a study examining the extent to which different driver groups are aware of the distracting effects of mobile phone use while driving. The authors focused on older (55–65 years) and younger (25–36 years) drivers' *a priori* ratings of confidence in their ability with regard to mobile phone use and the relationship between their confidence level and the observed actual decrement in their driving performance. Most participants (67%) reported feeling comfortable dealing with distractions while driving, with younger drivers reporting the greatest confidence. The association between actual performance and drivers' perceptions was weak. Many drivers were relatively unaware of the decrements in their actual driving performance resulting from concurrent mobile phone use. This was especially true for older female drivers where, as confidence increased, performance decreased. Additionally, when drivers were matched in terms of confidence level, brake responses of older females were slowed to a much greater extent (0.38 s) than were the brake responses of any other group (0.10 s for younger males and females and 0.07 s for older males). Although the results of this study should be considered as suggestive because they are based on a relatively small number of participants and relatively crude measures of confidence, the findings highlight the consequences of a mismatch between older (especially female) drivers' awareness of the possible effects of distraction and their capacity to overcome them.

Expanding on the research by Lesch and Hancock,[116] Horrey et al.[115] investigated whether subjective estimates of distraction and actual distraction varied as a function of phone type (handheld, handsfree). Younger and older drivers drove an instrumented van on a closed test track while performing a continuous mental arithmetic task administered over a handheld or handsfree mobile phone. Subjective measures of distraction effects were recorded and compared with actual performance on multiple measures of driving performance. Although their driving performance suffered in dual-task conditions, drivers were generally not well-calibrated to the magnitude of the distraction effects. In some cases, estimates of distraction were the opposite of the observed effects (i.e., smaller estimates of distraction corresponded to larger performance deficits). In general, a disconnection between performance and awareness was observed across driving measures and phone type. Analysis of the driver groups, though limited by a small sample size, showed some differences. For female drivers, there were no significant relationships across all measures and phone types. The results for males, however, were less clear. On some measures young and older males revealed a nearly equal but opposite relationship between estimated and actual distraction effects. Younger males, in particular, were poorly calibrated, suggesting that this may be an important group for targeted remediation.

Although mobile phone use has attracted the greatest media interest, other advanced electronic devices also have the potential to increase perceptual and cognitive demand when driving. Relatively simple tasks, such as tuning the radio, also have safety implications. Indeed, Horberry et al.[117] examined the effects of distraction on driving performance for drivers in three age groups: younger (under 25); middle-aged (30–45 years); and older (60–75 years). Participants were required to perform two secondary tasks while driving: a handsfree mobile phone task, in which they answered a series of general-knowledge questions, and an entertainment system task, in which they were required to tune the radio, change the radio's bass/treble and speaker balance, and insert and eject cassettes while driving in both simple (no billboards and few buildings and traffic) and complex (many billboards, buildings, and oncoming vehicles) simulated driving environments. Measures of mean speed, speed deviation from the posted speed limit, perceived workload, and responses to hazards were recorded. The authors noted that both in-vehicle tasks impaired several aspects of driving performance, with use of the entertainment system distracter having the greatest negative impact on performance. They further noted that these findings were relatively stable across drivers' age groups and environmental complexities. However, the authors also noted that, although participants were instructed to maintain the posted speed limit, older drivers traveled at lower mean speeds in the complex highway environment, achieved a lower minimum speed with respect to hazards, and had higher speed standard deviations compared with younger drivers. The authors suggested that older drivers had more difficulty performing the driving task while distracted and compensated by slowing their speed. In addition, older drivers performed as well as younger drivers on the mobile phone task, indicating that they did not trade off mobile phone performance to enable them to drive safely. Rather they slowed down to give themselves an increased margin for error, possibly because they knew they could respond to hazards as quickly. Although the authors reported participants' level of experience with different technologies, it should be noted that the sample size for each group was quite small (n = 10–11).

20.5.2.2 Other In-Vehicle Technology-Based Distraction

As outlined in other parts of this book, in-vehicle information systems (IVIS) and advanced driver assistance systems (ADAS) have the potential to assist drivers with the complex demands associated with the driving task.[91] For older drivers, in particular, these systems have the potential to help drivers avoid hazardous driving situations by accommodating for age-related vision, and cognitive and physical changes that can affect the safe operation of a motor vehicle.[90] However, although certain in-vehicle technologies have much to offer in terms of improving safe transportation for older drivers, there is some evidence that age can magnify the relative distracting effects of in-vehicle devices, leading to impaired driving performance. This is likely to result in increased crash risk relative to younger and middle-aged drivers. The evidence for this is discussed in more detail in the next section.

20.5.2.2.1 Navigation or Route Guidance Systems

Navigation or route guidance systems offer considerable promise for reducing cognitive load and therefore potentially reducing older drivers' crash risk in complex traffic situations.[118] However, research evidence for this is equivocal. Several studies have reported that older drivers demonstrate difficulty with the dual-task of following a route guidance system while driving.[100,119,120] For example, Dingus et al.[119] reported that older drivers (65 years and older) drove more slowly and cautiously, while making more safety-related errors (e.g., increased lane departures) compared with younger drivers (16–18 years) when using an advanced traveler information system (ATIS). Similarly, Oxley[101] reported that navigation systems that required significant attention caused older drivers to reduce speed and steer off course, and that, as the complexity of route guidance increased, basic driving task performance declined. Others have reported that, compared with younger drivers, older drivers looked more frequently and for longer periods at in-vehicle displays[121] and took 40–70% longer to complete some ATIS map-reading tasks.[45,120]

Although older drivers demonstrated an increase in safety-related errors, Dingus et al.[119] concluded that the ATIS may benefit older drivers by allowing them to spend more time visually scanning the roadway, although the operating demands of these systems must be congruent with the abilities of the older driving population.[91]

Exploring the design compatibility issue further, Liu[122] used a driving simulator to examine age-related differences in performance when the driver was using various ATIS features. In this study, the ATIS features were presented to drivers in three modes: visually only, aurally only, and by a multimodality display (i.e., visual and auditory). Participants were instructed to perform a push-button task and a navigation task while using the displayed traffic information to drive through each scenario. For all participants, both the auditory and multimodality displays resulted in significantly fewer errors, and produced better performance in terms of responding to hazard warnings and retaining better control of their simulated vehicles. However, the authors noted that older drivers demonstrated slower response times, drove more slowly, and made more navigation-related errors.

Verwey[123] investigated the potential for advanced transport telematics (ATT) to reduce the mental workload of drivers navigating road segments of varying difficulty. An on-road evaluation measured driver, based on the driver's ability to perform two secondary, ATT-related tasks: visual detection of a target stimulus on a dashboard-mounted display and an auditory response-based task. Although there were no significant age differences for overall visual detection and auditory addition performance, older drivers (60–79 years) demonstrated more errors on the auditory performance task relative to younger drivers (27–45 years) during the most difficult driving task conditions. Verwey proposed that older drivers have less processing resources available relative to younger drivers and recommended that telematic devices should be designed so that messages are postponed or canceled if the road situation warrants such action. Systems capable of doing so are reviewed in Chapter 26. Such systems may assist older drivers to remain focused on the primary task of driving, rather than on secondary tasks, such as those associated with ATT technology.

20.5.2.2.2 Heads-Up Display and Vision-Enhancement Systems

Heads-up display (HUD) technology may be a component of various in-vehicle systems for displaying information such as route directions, collision warnings, traffic signs, enhanced forward vision, and vehicle status. By displaying information superimposed on the forward road scene, the HUD may help reduce the driver's eyes-off-road time—a potential benefit to older drivers with a limited field of view and difficulty attending to spatially separated sources of information. However, researchers are concerned that HUD systems may have characteristics detrimental to older drivers, particularly if the display contains too much information, which could be distracting. In support of this contention, Yoo et al.[124] reported that it took older drivers 40% longer than younger drivers to respond to warning signals from HUDs. This is consistent with the findings of Wolffsohn et al.,[125] who reported that increased age (across 19–74 years) was associated with an overall increase in mean response times and detection of changes in HUD images or the scene outside. Neither study specifically reported the older drivers' baseline performance, so it is not possible to know if the HUD warning enhanced or degraded the response of older drivers compared with driving without HUD.

One of the common concerns of older drivers is the difficulty of driving at night, because of headlight glare and the difficulty of seeing very far with conventional head lighting systems. In a recent study of the age-related effects of VES, Caird et al.[97] conducted a comparative study of the effects of two types of VES on driving performance: conformal (the direct overlay of HUD images to highlight obstacles) and nonconformal (display of driving information, such as vehicle speed), where images only highlight the presence of obstacles, not where they are located (e.g., cars or pedestrians in the roadway). In their study, Caird et al. found that, although conformal displays had an advantage over nonconformal displays for both younger (18–32 years) and older (67–86 years) drivers, older drivers were more likely than the younger drivers to run the stoplight in the intersection scenario. Caird et al. suggest that these findings are consistent with the findings that older drivers have difficulties with intersections,[14,15] which may be the result of compromised performance because of increased complexity or slower response capabilities. Their findings emphasize the need for further research to determine the adverse consequences of visual enhancement systems on driver behavior, particularly among the growing number of older drivers.[91]

20.5.2.3 External-to-Vehicle Distraction

Driver distraction is not just related to technologies or events *inside* the vehicle. It can also be related to technologies or events *outside* the vehicle, such as the complexity of the driving environment.[117] This is an important issue because, as Horberry et al.[117] note, the amount of visual information presented to drivers is increasing in most industrialized countries and the potential for distraction caused by increased visual clutter in the road environment may be exacerbated in older drivers with decreased visual and cognitive capacities and slower reaction times. For example, Ho et al. [126] used latency and eye movement measures to examine the effects of aging and clutter on visual search for traffic signs embedded in digitized images of driving scenes for younger (20–27 years) and older (56–71 years) drivers. The findings of this study showed that older drivers were slower and less accurate and required more fixations to acquire a traffic sign. The authors suggested that in time-limited situations involving visually complex scenes (e.g., a busy intersection), older drivers are more likely than younger drivers to misidentify a sign or miss a sign altogether. Interestingly, older drivers were not more adversely affected by increased clutter than were young drivers. In contrast, Schieber and Goodspeed[127] reported age deficits in sign acquisition in more cluttered driving scenes. However, as Horberry et al.[117] note, there were important differences between the two studies in terms of the operational definition of clutter.

The findings of Ho et al.[126] are consistent with that of Horberry et al.,[117] who reported that older drivers traveled at lower mean speeds in the complex highway environment compared with younger drivers. Horberry et al.[117] proposed that the results of these studies suggest that although visual clutter in the driving environment may have a distracting effect on older drivers, more research is needed to quantify the effect and possible safety implications of increased driving environment complexity for older drivers and how this might be exacerbated by additional in-vehicle tasks and technologies.

20.5.3 Distraction and Older Driver Crashes

As discussed earlier, older and younger driver crash types differ, and the behaviors that lead to older driver crashes may be more related to inattention or slowed perception and responses than to deliberate unsafe actions that are more common in younger drivers.[18] The role of distraction in crashes generally is discussed in Chapter 16. Here, the focus is on the few studies that have investigated the role of distraction in older driver crashes.

Stutts et al.[128] conducted an analysis of the 1995–1999 Crashworthiness Data System (CDS) data to determine the role of driver distraction in U.S. police reported crashes (where at least one vehicle was towed away) and the specific sources of this distraction. The authors reported that 8.3% of the drivers were distracted at the time of their crash; after adjustment for the large percentage of drivers with unknown distraction status, the figure rose to 12.9%. The most frequently cited sources of driver distraction were persons, objects, or events outside the vehicle (29.4% of distracted drivers), adjusting the radio, tape, or CD player (11.4%), and other occupants in the vehicle (10.9%). Younger drivers (under 20) were more likely than older drivers to be identified as distracted at the time of their crash: 11.7% of younger drivers were found to be distracted, compared with 7.9% of older drivers (65 years and older). In contrast, drivers aged 65 and older were more than three times more likely to have "looked but did not see" (16.5%) listed as a contributing factor in crashes compared with younger drivers (5.4%). However, Stutts et al. reported that these differences were not statistically significant. The authors also reported detailed information on the specific types of distractions at the time of the crash for the various driver age groups. For example, drivers aged 65 and older were more likely to have been distracted by objects and events outside the vehicle (other vehicles, people, signs, animals, etc.; 43%) compared with all other age groups (27–33%). This finding may suggest that visual clutter or a reduced capacity to attend to relevant driving information while filtering out irrelevant stimuli has a role in older driver crashes. The findings also showed that older drivers were less likely to have been distracted by adjusting the radio/cassette/CD (0.2%) compared with all the other age groups (0.6–29%). Similarly, older drivers were less likely to be distracted by other occupants (2.6%) in the vehicle compared with the other age groups (9–11%), except the 50–64 year group (1.5%). This finding may reflect a deliberate strategy of older drivers to refrain from adjusting their radio and other controls while driving.

More recently, an Australian study examined the association between distraction, both inside and outside the vehicle, and the risk of being in a crash for drivers of different ages.[129] Fatal and casualty crash data collected by New South Wales police during 1996–2000 were examined, and crashes were categorized as resulting from no distraction, from distraction inside the vehicle, from distraction outside the vehicle, and from using a handheld phone. However, it should be noted that it is not clear whether *using* was defined as *talking on the phone* or *using the phone for another purpose* (e.g., text messaging). The in-vehicle distractions included attending to passengers, tuning the radio, adjusting the CD player, and smoking. Distraction outside the vehicle was defined as any circumstance on the road that constituted a distraction from the driving task. Lam[129] reported that there was no significant increase in the risk of being killed or injured in a crash for drivers using a handheld phone in most age groups, when compared with those crashed drivers without any distraction (RR range = 0.05–1.69), except for 25- to 29-year-old drivers (RR = 2.37, CI 1.31–4.27). In contrast to the findings for the handheld phone, other in-vehicle distractions increased the risk of crash injury for drivers of nearly all age groups, except 40- to 49-year-old drivers, compared with those crashed drivers without any distraction (see Figure 20.1). However, outside vehicle distractions had no significant effect on increasing the risk of crash injury for drivers of all ages (RR range = 0.74–0.94).

Lam[129] noted that, although drivers in the 25 to 29 age group had the greatest risk of being involved in a fatal or injury crash when using a handheld phone, they may be more likely to use handheld phones while driving than older drivers and therefore their heightened exposure may have been responsible for their increased crash risk. Lam concluded that the finding of other inside distractions *increasing* with increasing age is the result of a decreased ability of older drivers to share attention between two concurrent tasks.

In contrast to the findings of Stutts et al.,[128] Lam reported that drivers of all ages tended to be more distracted from internal than external events. However Lam notes that a possible explanation for the failure to find a significant effect for external

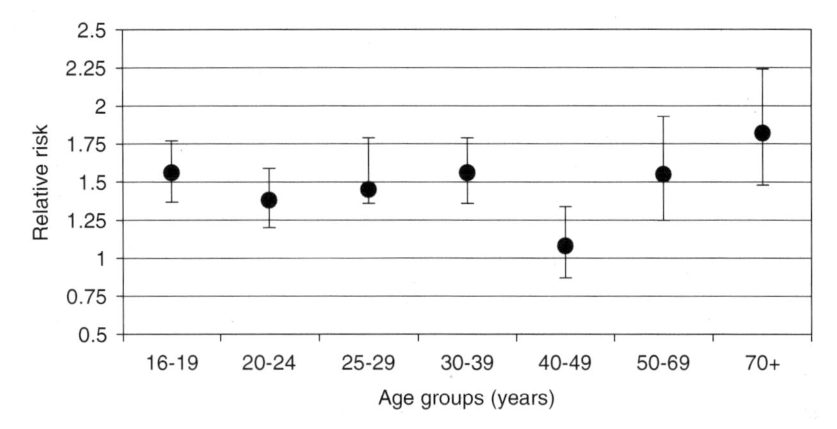

FIGURE 20.1 Relative risk of being killed or injured in an in-vehicle distraction-related crash compared with a nondistraction crash by age group.

vehicle distractions may be related to reporting bias. That is, "self-generated internal distractions are easily identified as inattention-due-to-distraction, whereas responding to stimuli outside the vehicle can be considered as part of the normal driving tasks and would less likely be identified as inattention" (p. 417). It is also important to note that the findings of Lam are based on fatal and casualty crash data, whereas the findings from Stutts et al. were based on crashes where at least one vehicle was towed away, and it is not possible to determine if these crashes resulted in fatalities or casualties.

Most recently, McEvoy et al.[130] examined the prevalence and type of distracting activities involved in serious injury crashes. Interviews were conducted with hospitalized drivers within hours of their crash. Types of distracting activities before the crash included passengers (11.3%), lack of concentration (10.8%), an outside object/person/event (8.9%), adjusting in-vehicle equipment (2.3%), mobile phone (2.0%), other object/animal/insect in vehicle (1.9%), smoking (1.2%), eating or drinking (1.1%), or "other" (0.8%). Crashes involving a distracting activity were more likely to be reported by younger drivers (17–29 years) compared with drivers aged 50 years and older (39.1% vs. 21.9%). Driving experience was also a strong predictor of distraction-related crashes (38.0% for drivers with less than 10 years' experience vs. 21% for drivers with more than 30 years'). As expected, age was highly correlated with driving experience. For each additional year of driving experience, a driver was 2% less likely to have a crash involving a distracting activity. The authors also report similar findings from a survey study that showed younger and novice drivers were more likely to report a distraction-related crash than older drivers.[131] Interestingly, the number of estimated distraction-involved crashes reported by McEvoy et al.[130] is higher than that reported by Lam.[129] One potential reason for this difference may be differences in crash severity across the two studies, with McEvoy et al. reporting only serious crashes, where the driver was hospitalized.

The findings of McEvoy et al.[130] are consistent with the findings of Hakamies-Blomqvist,[5] who found that drivers 65 years and older were significantly less likely to be distracted by a nondriving activity (such as eating, drinking, smoking, listening to the radio, conversing) immediately preceding a crash (42%) than younger drivers (26–40 years; 57%).

Similarly, the National Highway Traffic Safety Administration 100-Car Study[132] demonstrated that the rate of inattention-related crashes and near-crashes (where inattention activities included drivers eating, writing, conversing with a passenger, or looking away from the forward roadway at rear-view mirrors, or at objects inside or outside the vehicle) decreased dramatically with age. Inattention rates were found to be as much as four times higher for younger drivers (18–20 years) relative to older drivers (35 years and older). Unfortunately, due to the limited number of older drivers studied, the specific rate of inattention-related crashes and near-crashes was not reported. It would also be informative to examine separately the rates of driving-related inattention such as looking at rear-view mirrors or following a navigator's direction from nondriving activities. Notwithstanding these limitations, naturalistic driving study methods offer considerable promise for examining drivers' attentional status during real-world driving and for understanding the role of distraction in older-driver near-misses and crashes.

As outlined in earlier chapters, many researchers have shown that younger drivers are adversely affected by the presence of passengers. However, recent research suggests that older drivers may be *positively* affected by the presence of passengers. For example, Vollrath et al.[133] reported a 28% reduction in the risk of a driver being responsible for a crash in the presence of passengers. A possible explanation for this finding is that older drivers use passengers as copilots to help them navigate to unfamiliar destinations or to alert them to potential hazards.[134] However, although Vollrath et al. showed that this benefit was strongest in some situations (e.g., keeping a safe distance from other cars), it was weaker in other situations (e.g., at crossroads, in situations involving right of way, while overtaking, while turning). Similarly, Bedard and Meyers[135] reported that, although the presence of passengers exerted an overall protective effect for older drivers (65 years and older), the beneficial effect was not present for all types of actions. For drivers aged 65 to 79 years, the presence of passengers was associated with a reduced risk for some unsafe actions (e.g., driving the wrong way) but a higher risk of other actions (e.g., ignoring signs, warnings, or right of way). Bedard and Meyers noted that this is a crucial finding because many researchers have shown that intersections are problematic for older drivers,[4,68,72] and given the growing body of evidence suggesting that older individuals may have difficulties in dividing attention,[54–56] it raises the possibility that, in some critical situations, the presence of passengers may act as a source of distraction for the older drivers, thereby increasing their crash risk.[133] Bedard and Meyers concluded that, to maximize safe driving for older drivers, it would be desirable to enhance the role of passengers in situations where they may play a beneficial role and minimize the potential negative effect they have in others. This area warrants further research, although research on the topic has been undertaken in relation to the role of young passengers in influencing the safety of young drivers.[136,137]

20.5.4 Summary

Although older adults represent the fastest growing segment of the driving population, limited research has been conducted to evaluate the impact of distraction on driving performance and crash risk among older drivers. Most research in the area of older drivers and distraction has focused on the use of mobile phones and navigation systems, with a few studies examining the impact of collision-warning devices and vision-enhancement systems on their driving performance. As noted by Vrkljan and Miller-Polgar,[91] many of these studies utilized small sample sizes and complex design methods that would be difficult to replicate, and many used driving simulators to evaluate the effect of in-vehicle technologies on driving performance. Consequently, we should exercise caution when extrapolating these findings to on-road driving performance. Despite their methodological issues, these studies provide important knowledge regarding in-vehicle technologies and older drivers.[91]

20.6 CONCLUSIONS AND RECOMMENDATIONS

This chapter has highlighted several emerging themes relevant to older driver distraction, including the growing older driver population, the scope and nature of their crash involvement, their driving patterns, and their propensity to adopt self-regulatory driving behaviors. Although physical frailty accounts for most variance in older drivers' over-involvement in serious injury crashes, age-related changes in functional abilities in various cognitive, visual-attention, and physical domains are also implicated. As a result of these declines, it is likely that at least some older drivers operate their vehicles under considerably higher cognitive loads than their younger counterparts. This is particularly likely to be true in complex traffic conditions where there are numerous simultaneous demands on the drivers' attention.

The age-related functional declines outlined in this chapter have several consequences for driver distraction: (1) increased demands of the driving task leave relatively little spare capacity to deal with competing activities; (2) competing activities are likely to be more distracting for older drivers, as extra time, visual and attentional capacity, and physical effort are required to perform them; and (3) it is more difficult for older drivers to make the trade-off between driving and a competing task to maintain safe driving because of impoverished time-sharing skills.

Despite the fact that older adults represent the fastest growing segment of the driving population and that, due to age-related functional declines, they may be particularly vulnerable to the effects of distraction, an extensive literature review revealed only a limited number of studies that had specifically investigated older drivers and their propensity for distraction. From the limited research available, there is evidence demonstrating greater performance decrements for older drivers compared with younger age groups while performing a concurrent task such as using a mobile phone.[102–115,117] There is also some indication that older drivers have more difficulty extracting relevant information from road signs, particularly when driving in complex traffic and in time-limited situations[126] and where high levels of visual clutter exist in the driving environment.[127]

In terms of the role that distraction plays in older driver crashes, the findings from analysis of crash data has shown that older drivers are significantly *less* likely to engage in distracting activities at the time of the crash compared with younger age groups.[5,128,130] These findings are consistent with the assertion that the behaviors that lead to older driver crashes may be more related to inattention or slowed perception and responses than to deliberate unsafe actions, which are more common in younger drivers.[18] Interestingly, distractions due to objects and events outside the vehicle have been implicated in older driver crashes[128] although evidence for this is equivocal.[129]

It is important to note that many older drivers offset their functional limitations by adapting their driving behaviors in a variety of ways.[18,65–71] One common strategy used to achieve reduced exposure and risk is to simply reduce the frequency and duration of trips. Another self-regulatory strategy is to avoid known difficult driving situations such as nighttime driving and driving in complex traffic in an attempt to reduce the potential risks associated with failing vision and slowed responses. Of particular interest to this discussion is the willingness of older drivers to engage in distracting activities. Evidence for this, albeit limited, suggests that older drivers are less willing than younger drivers to engage in distracting activities such as use of mobile phones and navigation systems while driving, and they perceive this activity as being more risky than do younger drivers.[81]

Although older drivers are not currently overrepresented in distraction-related crashes, it is important to note that researchers have predicted older driver cohorts will have substantially different driving patterns in the future.[3,6,7] As baby boomers get older, they will be more mobile than preceding generations at that age, will travel more frequently, will travel greater distances, will have high expectations with regard to maintaining personal mobility,[3] and may be more comfortable with new technologies and therefore more willing to engage in potentially distracting activities, such as using a mobile phone while driving. In addition, over the next 10 years at least 25% of vehicle purchasers will be over 50 years, and they will purchase a large share of the premium luxury vehicles,[138] the majority of which are fitted with new in-vehicle technologies. To address these impending trends, it is important to identify effective strategies to prevent older drivers from becoming more involved in distraction-related crashes.

Several strategies for reducing the potentially negative effects of distraction on older drivers' performance and safety are outlined in the following section.

20.6.1 Recommendations for Managing Older Driver Distraction

20.6.1.1 Legislation, Enforcement, and Licensing

Although some evidence suggests that older drivers are less willing than younger drivers to engage in distracting activities while driving,[81] future older driver cohorts may be more technologically savvy and therefore more likely to engage in such activities, despite the fact that they demonstrate greater performance decrements while performing concurrent tasks such as using a mobile phone[102–115,117] compared with younger age groups. Current legislation should be reviewed with regards to the aging population and potential sources of distraction. The effectiveness of police enforcement of current legislation should be monitored given that future cohorts of older drivers may be less compliant. In addition, licensing authorities have the opportunity to bring to the attention of motorists, including older drivers, the potential risks of being distracted while driving through licensing handbooks and by testing motorists' knowledge of these risks. They can design practical driving tests to identify, and prevent from being licensed, drivers who are incapable of compensating for the effects of driving while distracted.[139]

20.6.1.2 Education and Training

Educational and training programs designed to promote safe driving strategies for older drivers have become a popular approach to addressing the older driver safety problem. One reason for their growing popularity may be the evidence that driving errors related to lack of knowledge or driving experience can be overcome through training and education.[140] This may be especially true for the current cohort of older drivers, many of whom had little formal driving education.[141] For example, there is some evidence suggesting that training can assist older drivers by identifying and promoting adaptive or self-regulatory strategies such as minimizing the amount of driving conducted under conditions that impose a heavy perceptual and cognitive load (e.g., avoiding extensive driving or driving in unfamiliar areas;[78,142] enlisting the cooperation of others to help share the driving load [e.g., having a passenger navigate or read the road signs[78,142]]; and exercising alternatives to reduce perceptual and cognitive load [e.g., using less-traveled roads[140]]). Given the fact that many older drivers self-regulate when, where, and how they drive, based on their changing functional abilities, there is considerable scope for training older drivers to adjust their driving styles and use of technologies to minimize distraction. Indeed, Shinar et al.[114] demonstrated that older drivers were able to minimize or eliminate initial deleterious effects of handsfree phone conversations on many driving tasks with continued practice.

Vehicle manufacturers may also play a significant role in educating older drivers about minimizing distraction.[24] For example, manufacturers could develop a delivery process that familiarizes older drivers with the technologies of the vehicle at the time of the vehicle purchase and train older drivers to maximize the benefits of these systems and minimize the potential for distraction.[143] This is particularly relevant as older drivers are willing to use new technology if they are provided with adequate training[94] and if the benefits of the technology are clear to them.[95] In addition, Meyer suggests that older drivers will need to have the technologies customized to their particular needs based on an objective evaluation of the driver's needs by a specialist.

20.6.1.3 Vehicle Design

If used appropriately, in-vehicle technologies (IVIS and ADAS) could assist all drivers with the complex demands associated with the driving task.[91] For older drivers in particular, these systems could help them reduce their exposure to hazardous driving situations by compensating for age-related vision, cognitive, and physical decline.[90] However, these technologies can only benefit older drivers if their design is congruent with the complex needs and diverse abilities of this driving cohort.[91] For example, as noted by Meyer,[24] because lowered acuity and contrast sensitivity are common among older drivers, displays should be at a brightness level that differs as much as possible from the background. In addition, intelligent on-board "workload manager" technologies could determine whether a driver is overloaded or distracted, and if so, alter the availability of telematics and the operation. For example, the system may temporarily suppress calls and prevent access to phone functions and controls when distraction potential is estimated to be high.[139] Several authors[45,91,144] have proposed that if the underlying principles of universal design and human factors are integrated from the beginning of the design process for devices, such as IVIS, it may increase the likelihood that drivers of varying abilities, including older drivers, will be able to use the product. However, as outlined earlier, more research is needed to define how vehicle design and technologies affect older drivers' cognitive load and their attention to the driving task, and the way in which these might enhance or comprise older drivers' driving, as well as their ability to make safe decisions in different driving environments.[144]

20.6.1.4 Road Design

Road design also contributes to the level of risk that older drivers face on the road, particularly because of the combination of complex road environments and diminished functional abilities.[145] Although current design features are unlikely to be the primary cause of many older driver crashes, design changes to create a more forgiving road environment will have substantial road safety benefits. Moreover, as the population ages, it will become increasingly important to design roads that will accommodate the needs and capabilities of older road users. A number of engineering solutions could address many of these problems and should be given high priority as the population ages. For example, traffic control devices need to be more visible and conspicuous to compensate for the inherent limitations of the aging driving population. It should be noted that a safer road environment for older road users generally means a safer road environment for all road users.[145] Road and traffic engineers, therefore, have a significant role to play in improving the comfort and safety of older drivers. In particular, Oxley et al.[145] note that there is a need for

- Greater awareness of the declining physical and mental abilities of the older driver population
- Appreciation of how these changing abilities affect the driving task
- Appropriate modification of design and maintenance standards to enable older drivers to cope more easily with the increasingly complex task of driving

In terms of driver distraction, reductions in the complexity of the road environment could reduce the cognitive load of older drivers and thereby reduce the potential for distraction.

20.6.1.5 Data Collection and Future Research

As outlined earlier, research investigating older drivers and their propensity for distraction and the role distraction plays in vehicle crashes has been limited. Much of the current knowledge on older driver distraction comes from four key sources: survey data on *previous driving behaviors and distractions*; drivers' *willingness to engage in distracting activities* in hypothetical driving situations; *simulator-based studies* of driving while undertaking concurrent tasks; and *retrospective information on distractions* derived from reported crash circumstances from real-world crash research. Although these study methods have provided important insights into the scope of the problem among older drivers, there remain a number of gaps in our knowledge on this topic.

Importantly, we need to know more about the profile of the self-regulating older driver—how and why some drivers refrain from engaging in potentially distracting activities while others habitually engage in risky activities that distract them from the driving task. With the proliferation of new vehicle and infrastructure technologies it will be critical to evaluate the impact of these systems before their universal adoption, focusing on the effect of these systems on older drivers' safety. To understand these processes better, it is necessary to conduct further research to confirm and extend current findings and, particularly, to undertake observational studies of real-world driving to validate survey and experimental work from laboratory-based and driving simulation studies.

Three key research priorities are proposed to promote older drivers' safe mobility:

- Identify key factors or problem areas (driver-, vehicle-, road infrastructure-related) that lead to distraction for older drivers.
- Identify vehicle technology solutions to offset those areas.
- Evaluate ways of enhancing the safety benefits of vehicle-based technologies and minimizing any disbenefits for older drivers.

Simulator and instrumented vehicle-based research and naturalistic driving study methods offer considerable promise for examining drivers' attentional status and the role of other driver, vehicle, and road infrastructure attributes in relation to distraction. Longitudinal studies will also be highly valuable to study the effects of distraction under a range of conditions of traffic complexity and cognitive load in drivers across the age span.

Research outcomes pertaining to older driver distraction should be made available to technology designers within the motor vehicle industry and to road and traffic engineers and licensing authorities, with the expectation that the consideration of aging and functional abilities will be a significant component of all technology design and other policy consideration for the future.

REFERENCES

1. Charlton, J., Fildes, B., Koppel, S., Andrea, D., Newstead, S., Oxley, P., and Pronk, N., Evaluation of a referral assessment tool for assessing functionally impaired drivers, VicRoads, Victoria, Australia, 2002.
2. Li, G., Braver, E. R., and Chen, L., Fragility versus excessive crash involvement as determinants of high death rates per vehicle-mile of travel among older drivers, *Accident Analysis and Prevention* 35 (2), 227–235, 2003.
3. OECD, *Aging and Transport. Mobility Needs and Safety Issues*, OECD Publications, Paris, France, 2001.
4. Hakamies-Blomqvist, L. E., Fatal accidents of older drivers, *Accident Analysis and Prevention* 25, 19–27, 1993.

5. Hakamies-Blomqvist, L. E., Compensation in older drivers as reflected in their fatal accidents, *Accident Analysis and Prevention* 26, 107–112, 1994.
6. Hu, P., Jones, D., Reuscher, T., Schmoyer, R., and Truett, L., *Projecting Fatalities in Crashes Involving Older Drivers, 2000–2025*, Oak Ridge National Laboratory, Tennessee, 2000.
7. Fildes, B., Fitzharris, M., Charlton, J., and Pronk, N., Older driver safety—A challenge for Sweden's 'Vision Zero', *Australian Transport Research Forum*, Hobart, Australia, 2001.
8. Padmanaban, J., Crash injury experience of elderly drivers, *Aging and Driving Symposium, Association for the Advancement of Automotive Medicine*, Des Plaines, IL, 2001.
9. Evans, L., *Traffic Safety and the Driver.* Van Nostrand Reinhold, New York, NY, 1991.
10. Mackay, M., Occupant protection and vehicle design, *Proceedings from the Association for the Advancement of Automotive Medicine Course on the Biomechanics of Impact Trauma*, Los Angeles, 1998.
11. Viano, D. C., Culver, C. C., Evans, L., Frick, M., and Scott, R., Involvement of older drivers in multi-vehicle side impact crashes, *Accident Analysis and Prevention* 22, 177–188, 1990.
12. Dejeammes, M. and Ramet, M., Aging process and safety enhancement of car occupants, *15th International Technical Conference on the Enhanced Safety of Vehicles*, Melbourne, Australia, 1996, pp. 1189–1196.
13. Augenstein, J., Differences in clinical response between the young and the elderly, *Aging and Driving Symposium, Association for the Advancement of Automotive Medicine*, Des Plaines, IL, 2001.
14. Fildes, B., Corben, B., Kent, S., Oxley, J., Le, T., and Ryan, P., *Older Road User Crashes*, Monash University Accident Research Centre, 1994.
15. Eberhard, J., Do older drivers have a heightened crash risk?, *Licensing Authorities' Options for Managing Older Driver Safety: Practical Advice from the Researchers, Transportation Research Board 86th Annual Meeting*, Washington, D.C., 2007.
16. Charlton, J. L., Fildes, B., and Andrea, D., Vehicle safety and older occupants, *International Journal of Gerentechnology* 1 (4), 274–286, 2002.
17. Morris, A. P., Welsh, R., Frampton, R., Charlton, J. L., and Fildes, B., Vehicle crashworthiness and the older motorist, *Aging and Society* 23, 395–409, 2003.
18. Eberhard, J., Safe mobility of senior citizens, *Journal of International Association of Traffic and Safety Sciences* 20 (1), 29–37, 1996.
19. Federal Highway Administration, 2001 National Household Travel Survey, Washington, D.C., 2001.
20. Hakamies-Blomqvist, L. E., Ageing Europe: The challenges and opportunities for transport safety, *5th European Transport Safety Lecture, European Transport Safety Council*, Brussels, 2003.
21. Janke, M., Accidents, mileage and the exaggeration of risk, *Accident Analysis and Prevention* 32 (2), 183–188, 1991.
22. Keall, M. and Frith, W., Older driver crash rates in relation to type and quantity of travel, *Traffic Injury Prevention* 5, 26–36, 2004.
23. NHTSA, *Traffic Safety Facts—Older Population*, NHTSA, 2005.
24. Meyer, J., Personal vehicle transportation, in *Technology for Adaptive Aging*, Pew, R. L. and van Hemel, S. B. (eds.), The National Academies Press, Washington, D.C., 2004.
25. Janke, M., *Age-Related Disabilities That May Impair Driving and Their Assessment: Literature Review*, California Department of Motor Vehicles, Sacramento, 1994.
26. Stelmach, G. and Nahom, A., Cognitive-motor abilities of the elderly driver, *Human Factors* 34, 53–65, 1992.

27. Charlton, J., Koppel, S., O'Hare, M., Andrea, D., Smith, G., Khodr, B., Langford, J., Odell, M., and Fildes, B., *Influence of Chronic Illness on Crash Involvement of Motor Vehicle Drivers*, Monash University Accident Research Centre, Melbourne, 2003.
28. Dobbs, B., *Medical Conditions and Driving: Current Knowledge*, National Highway Transportation Safety Administration and The Association for the Advancement of Automotive Medicine Project, Washington, D.C., 2001.
29. Marottoli, R., Richardson, E., Stowe, M., Miller, E., Brass, L., Cooney, L., and Tinetti, M., Development of a test battery to identify older drivers at risk for self-reported adverse driving events, *Journal of the American Geriatrics Society* 46 (5), 562–568, 1998.
30. Staplin, L., Lococco, K., Stewart, J., and Decina, L., *Safe Mobility for Older People: Notebook*, National Highway Traffic Safety Administration, Washington, D.C., 1999.
31. Marottoli, R., Health issues for older road users, *Mobility and Safety of Older People Conference*, Melbourne, 2001.
32. Shinar, D. and Scheiber, F., Visual requirements for safety and mobility of older drivers, *Human Factors* 33, 507–519, 1991.
33. Owsley, C., Ball, K., Sloane, M. E., Roenker, D. L., and Bruni, J. R., Visual/cognitive correlates of vehicle accidents in older drivers, *Psychology & Aging* 6 (3), 403–415, 1991.
34. Kline, D., Kline, T., Fozard, J., Kosnik, W., Scheiber, R., and Sekuler, R., Vision, aging and driving: The problems of older drivers, *Journal of Gerontology: Psychological Sciences* 47 (1), 27–34, 1992.
35. Ball, K., Owsley, C., Sloane, M., Roeneker, D., and Bruni, J., Visual attention problems as a predictor of vehicle crashes in older drivers, *Investigative Ophthalmology and Visual Science* 34, 3110–3123, 1993.
36. Burg, A., The relationship between vision test scores and driving record: General findings, Department of Engineering, UCLA, Los Angeles, 1967.
37. Hills, B. L. and Burg, A., A reanalysis of California driver vision data: General findings, Transport and Road Research Laboratories, Crowthorne, Berkshire, 1977.
38. Merat, N., Anttila, V., and Luoma, J., Comparing the driving performance of average and older drivers: The effect of surrogate in-vehicle information systems, *Transportation Research Part F: Traffic Psychology and Behaviour* 8, 147–166, 2005.
39. Owsley, C., McGwin, G., and Ball, K., Vision impairment, eye disease, and injurious motor vehicle crashes in the elderly, *Opthalmic Epidemiology* 5 (2), 101–113, 1998.
40. Salthouse, T. A. and Prill, K. A., Effects of aging on perceptual closure, *American Journal of Psychology* 101, 217–238, 1988.
41. Capitani, E., Della Sala, S., Lucchelli, F., Soave, P., and Spinnler, H., Perceptual attention in aging and dementia measured by Gottschaldt's hidden figure test, *Journal of Gerontology Series B – Psychological Sciences & Socials Sciences* 43, 157–163, 1988.
42. Daigneault, G., Joly, P., and Frigon, J.-F., Executive functions in the evaluation of accident risk of older drivers, *Journal of Clinical and Experimental Neuropsychology* 24 (2), 221–238, 2002.
43. Welford, A., Motor performance, in *Handbook of the Psychology of Aging*, Birren, J. and Schaie, K. (eds.), Van Nostrand Reinhold, New York, NY, 1977, pp. 450–496.
44. Wickens, C., Attention and skilled performance, in *Human Skills*, Holding, D. (ed.), Wiley, Chichester, 1989, pp. 71–105.
45. Brennan, M., Welsh, M. C., and Fisher, C. B., Aging and executive function skills: An examination of a community-dealing older adult population, *Perceptual and Motor Skills* 84, 1187–1197, 1997.
46. Daigneault, S. and Brown, C. M., Working memory and the self-ordered pointing task: Further evidence of early pre-frontal decline in normal aging, *Journal of Clinical and Experimental Neuropsychology* 15, 881–895, 1993.

47. Daigneault, S., Brown, C. M., and Whitaker, H. A., Early effects of normal aging on perseverative and non-perseverative pre-frontal measure, *Developmental Neuropsychology* 8, 99–114, 1992.
48. Graf, P., Uttl, B., and Tuokko, H., Color- and picture-word Stroop tests: Performance changes in old age, *Journal of Clinical and Experimental Neuropsychology* 17, 390–415, 1995.
49. Mejia, S., Pineda, D., Alvarez, L. M., and Ardila, A., Individual differences in memory and executive function abilities during normal aging, *International Journal of Neurosciences* 95, 271–284, 1998.
50. Parkin, A. J. and Java, R., Deterioration of frontal lobe function in normal aging: Influences of fluid intelligence versus perceptual speed, *Neuropsychology* 13 (4), 539–545, 1999.
51. Raz, N., Gunning-Dixon, F. M., Head, D., Dupuis, J., and Acker, J. D., Neuroanatomical correlation of cognitive aging: Evidence from structural magnetic resonance imaging, *Neuropsychology* 12, 95–114, 1998.
52. McDowd, J. M. and Craik, F. I. M., Effects of aging and task difficulty on divided attention performance, *Journal of Experimental Psychology: Human Perception and Performance* 14 (2), 267–280, 1988.
53. Inui, N., Simple reaction times and timing of serial reactions of middle-aged and old men, *Perceptual and Motor Skills* 84, 219–225, 1997.
54. Sit, R. A. and Fisk, A. D., Age-related performance in a multiple-task environment, *Human Factors* 41 (1), 26–34, 1999.
55. Salthouse, T., *A Theory of Cognitive Aging*, Elsevier, Amsterdam, 1991.
56. Korteling, J., Effects of ageing, skill modification and demand alternation on multiple-task performance, *Human Factors* 36 (1), 27–43, 1994.
57. Lamble, D., Kauranen, T., Laakso, M., and Summala, H., Cognitive load and detection thresholds in car following situations: Safety implications for using mobile (cellular) telephones while driving, *Accident Analysis and Prevention* 31 (6), 617–623, 1999.
58. Fildes, B., Pronk, N., Langford, J., Hull, M., Frith, B., and Anderson, R., *Model Licence Re-assessment Procedure for Older and Disabled Drivers*, Austroads Canberra, 2000.
59. Stamatiadis, N., Taylor, W., and McKelvey, F., Elderly drivers and intersection accidents, *Transportation Quarterly* 45 (3), 559–571, 1991.
60. Cooper, P., Tallman, K., Tuokko, H., and Beattie, B., Vehicle crash involvement and cognitive deficit in older drivers, *Journal of Safety Research* 24, 9–17, 1993.
61. Planek, T. and Fowler, R., Traffic accident problems and exposure characteristics of the aging driver, *Journal of Gerontology: Psychological Sciences* 26 (2), 224–230, 1991.
62. Transportation Research Board, *Transportation in an Aging Society: Improving Safety and Mobility of Older Drivers*, National Research Council, Washington, D.C., 1988.
63. Brummel-Smith, K., Falls and instability in the older person, in *Geriatric Rehabilitation*, Kemp, B., Brummel-Smith, K., and Ramsdell, J. (eds.), College-Hill Publishers, Boston, MA, 1990, pp. 193–208.
64. Bishu, R., Foster, B., and McCoy, P., Driving habits of the elderly—A survey, *Human Factors Society, 35th Annual Meeting*, 1991, pp. 1134–1138.
65. Evans, L., Older driver involvement in fatal and severe traffic crashes, *Journal of Gerontology: Psychological Sciences* 43, 186–193, 1988.
66. Smiley, A., Adaptive strategies of older drivers, *Transportation in an Aging Society: A Decade of Experience,* Transportation Research Board, Maryland, 1999.
67. Preusser, D., Williams, A., Ferguson, S., Ulmer, R., and Weinstein, H., Fatal crash risk for older drivers at intersections, *Accident Analysis and Prevention* 30 (2), 151–159, 1998.
68. McGwin, G. and Brown, D. B., Characteristics of traffic crashes among young, middle-aged and older drivers, *Accident Analysis and Prevention* 31, 181–198, 1999.

69. Charlton, J. L., Oxley, J., Scully, J., Koppel, S., Congiu, M., Muir, C., and Fildes, B., *Self-Regulatory Driving Practices amongst Older Drivers in the ACT and NSW*, Monash University Accident Research Centre, Melbourne, 2006.
70. Charlton, J. L., Oxley, J., Fildes, B., Oxley, P., Newstead, S., O'Hare, M., and Koppel, S., *An Investigation of Self-Regulatory Behaviours of Older Drivers*, Monash University Accident Research Centre, Melbourne, 2003.
71. Baldock, M. R. J., Mathias, J., McLean, J., and Berndt, A., Self-regulation of driving and older drivers' functional abilities, *Clinical Gerontologist* 30 (1), 53–70, 2006.
72. Benekohal, R., Michaels, R., Shim, E., and Resende, P., Effects of aging on older drivers' travel characteristics, *Transportation Research Record* 1438, 91–98, 1994.
73. Rosenbloom, S., Travel by the elderly, *Nationwide Personal Transportation Survey, Demographic Special Reports*, Management, O. f. H. I. U.S. Department of Transportation, Federal Highway Administration, Washington, D.C., 1995, pp. 3–49.
74. Rosenbloom, S., The mobility of the elderly: There's good news and bad news, *Transportation in An Aging Society; A Decade of Experience Conference*, 1999.
75. Hakamies-Blomqvist, L. and Wahlström, B., Why do older drivers given up driving?, *Accident Analysis and Prevention* 30 (3), 305–312, 1998.
76. Ball, K., Owsley, C., Stalvey, R., D., Sloane, M., and Graves, M., Driving avoidance and functional impairment in older drivers, *Accident Analysis and Prevention* 30 (3), 313–322, 1998.
77. Ernst, R. and O'Connor, P., *Report on Accident Countermeasures Focusing on Elderly Drivers*, South Australian Department of Road Transport, Adelaide, 1988.
78. Persson, D., The elderly driver: Deciding when to stop, *Gerontologist* 33, 88–91, 1993.
79. Marottoli, R., Ostfield, A., Merril, S., Perlman, G., Foley, D., and Cooney, L., Driving cessation and changes in mileage driven among elderly individuals, *Journal of Gerontology, Social Sciences* 48, 8255–8260, 1993.
80. Kostyniuk, L. and Shope, J., *Reduction and Cessation of Driving among Older Drivers: Focus Groups*, The University of Michigan Transportation Research Institute, 1998.
81. Lerner, N. D., *Driver Strategies for Engaging in Distracting Tasks Using In-Vehicle Technologies: Final Report*, National Highway Traffic Safety Administration, Washington, D.C., 2005.
82. Rothman, A., Klein, W., and Weinstein, N., Absolute and relative biases in estimations of personal risk, *Journal of Applied Social Psychology* 26, 1213–1236, 1996.
83. Matthews, M., Aging and the perception of driving risk and ability, *Human Factors Society, 30th Annual Meeting*, 1986, pp. 1159–1163.
84. Stutts, J., Do older drivers with visual and cognitive impairments drive less?, *Journal of the American Geriatrics Society* 46 (7), 854–861, 1998.
85. Holland, C. and Rabbitt, P., People's awareness of their age-related sensory and cognitive deficits and the implications for road safety, *Applied Cognitive Psychology* 6, 217–231, 1992.
86. Langford, J., Methorst, R., and Hakamies-Blomqvist, L., Older drivers do not have a high crash risk—A replication of low mileage bias, *Accident Analysis and Prevention* 38 (3), 574–578, 2006.
87. Marottloi, R., Mendes de Leon, C., Glass, T., Williams, C., Cooney, L., and Berkman, L., Consequences of driving cessation: Decreased out-of-home activity levels, *Journal of Gerontology, Social Sciences* 55B (6), S334–S340, 2000.
88. Fonda, S. J., Wallace, R. B., and Herzog, A. R., Changes in driving patterns and worsening depressive symptoms among older adults, *Journal of Gerontology Series B—Psychological Sciences & Socials Sciences* 56B (6), S343–351, 2001.
89. Peel, N., Steinberg, M., and Westmoreland, J., *Older Drivers and Cognitive Functioning: Experiences, Perceptions and Management Needs*, Centre for Accident Research and Road Safety, Queensland, 2000.

90. Caird, J. K., Chugh, J. S., Wilcox, S., and Dewar, R. E., A design guideline and evaluation framework to determine the relative safety of in-vehicle intelligent transportation systems for older drivers, University of Calgary, Calgary, 1998.
91. Vrkljan, B. H. and Miller-Polgar, J., Advancements in vehicular technology: Potential implications for the older driver, *International Journal of Vehicle Information and Communication System* 1 (2), 88–105, 2005.
92. Pew Global, *Truly a World Wide Web*, Available at: http://www.pewglobal.org, accessed on 07/06/2007.
93. Pew Internet & American Life, Older Americans and the Internet, available at http://www.pewInternet.org/PPF/r/117/report_display.asp, accessed on 07/06/2007.
94. Rogers, W. A., Fisk, A. D., Mead, S. E., Walker, N., and Cabrera, E. F., Training older adults to use automatic teller machines, *Human Factors* 38, 425–433, 1996.
95. Melenhorst, A., Rogers, W. A., and Caylor, E. C., The use of communication technologies by older adults: Exploring the benefits from the users perspective, *Proceeding of the Human Factors and Ergonomic Society 46th Annual Meeting*, 2001, pp. 221–225.
96. Wochinger, K. and Boehm-Davis, D., Navigational preference and acceptance of advanced traveler information systems, in *Ergonomics and Safety of Intelligent Driver Interfaces*, Noy, I. Y. (ed.), Lawrence Erlbaum Associates, Mahwah, NJ, 1997.
97. Caird, J. K., Horrey, W. J., and Edwards, C. J., Effects of conformal and nonconformal vision enhancement systems on older-driver performance, *Transportation Research Record* 1759, 38–45, 2001.
98. Charles River Associates Incorporated, *Consumer Acceptance of Automotive Crash Avoidance Devices—A Report of Qualitative Research*, U.S. Department of Transportation, Boston, MA, 1998.
99. Campbell, J., Kinghorn, R., and Kantowitz, B., *Driver Acceptance of System Features in an Advanced Transportation Information System (AITS)*, ITS America, Washington, D.C., 1995, pp. 967–973.
100. Barham, P., Alexander, A., and Oxley, P., What are the benefits and safety implications of route guidance systems for elderly drivers, *Seventh International Conference on Road Traffic Monitoring and Control*, London, 1994, pp. 137–140.
101. Oxley, P., Elderly drivers and safety when using IT systems, *IATSS Research* 20 (1), 102–110, 1996.
102. Schreiner, C., Blanco, M., and Hankey, J. M., Investigating the effect of performing voice recognition tasks on the detection of forward and peripheral events, *Human Factors and Ergonomics Society's 48th Annual Meeting*, Santa Monica, CA, 2004.
103. Nilsson, L. and Alm, H., *Effects of Mobile Telephone Use on Elderly Drivers' Behaviour Including Comparisons to Young Drivers' Behaviour*, Swedish National Road and Transport Research Institute (VTI), Linkoping, Sweden., 1991.
104. McPhee, L. C., Scialfa, C. T., Dennis, W. M., Ho, G., and Caird, J. K., Age differences in visual search for traffic signs during a simulated conversation., *Human Factors* 46, 674–685, 2004.
105. McKnight, A. J. and McKnight, A. S., The effect of cellular phone use upon driver attention, *Accident Analysis and Prevention* 25, 259–265, 1993.
106. Hancock, P. A., Lesch, M., and Simmons, L., The distraction effects of phone use during a crucial driving maneuver, *Accident Analysis and Prevention* 35 (4), 501–514, 2003.
107. Cooper, P. J. and Zheng, Y., Turning gap acceptance decision-making: The impact of driver distraction, *Journal of Safety Research* 33, 321–335, 2002.
108. Cooper, P. J., Zheng, Y., Richard, C., Vavrik, J., Heinrichs, B., and Siegmund, G., The impact of hands-free message reception/response on driving task performance, *Accident Analysis and Prevention* 35, 23–35, 2003.
109. Alm, H. and Nilsson, L., The effects of a mobile telephone task on driver behaviour in a car following situation, *Accident Analysis and Prevention* 27, 707–715, 1995.

110. McCarley, J. S., Vais, M. J., Pringle, H., Kramer, A. F., Irwin, D. E., and Strayer, D. L., Conversation disrupts change detection in complex traffic scenes, *Human Factors* 46, 424–436, 2004.
111. Reed, M. P. and Green, P. A., Comparison of driving performance onroad and in a low-cost simulator using a concurrent telephone dialing task, *Ergonomics* 42, 1015–1037, 1999.
112. Ranney, T., Watson, G. S., Mazzae, E., Papelis, Y. E., Ahmad, O., and Wightman, J. R., Examination of the *Distraction Effects of Wireless Phone Interfaces Using the National Advanced Driving Simulator—Final Report on a Freeway Study*, National Highway Traffic Safety Administration, Washington, D.C., 2005.
113. Strayer, D. L. and Drews, F. A., Profiles in driver distraction: Effects of cell phone conversations on younger and older drivers, *Human Factors* 46, 640–649, 2004.
114. Shinar, D., Tractinsky, N., and Compton, R., Effects of practice, age, and task demands on interference from a phone task while driving, *Accident Analysis and Prevention* 37, 315–326, 2005.
115. Horrey, W. J., Lesch, M. F., and Garabet, A., Awareness of performance decrements due to distraction in younger and older drivers, *Fourth International Driving Symposium on Human Factors in Driver Assessment, Training and Vehicle Design*, 2007.
116. Lesch, M. F. and Hancock, P. A., Driving performance during concurrent cell-phone use: Are drivers aware of their performance decrements?, *Accident Analysis and Prevention* 36, 471–480, 2004.
117. Horberry, T., Anderson, J., Regan, M. A., Triggs, T. J., and Brown, J., Driver distraction: the effects of concurrent in-vehicle tasks, road environment complexity and age on driving performance, *Accident Analysis Prevention* 38 (1), 185–191, 2006.
118. Suen, S. L. and Mitchell, C. G. B., The value of intelligent transport systems to elderly and disabled travellers, *8th International Conference on Mobility and Transport for Elderly and Disabled People, TRANSED*, Perth, Australia, 1998.
119. Dingus, T. A., Hulse, M. C., Mollenhaur, M. A., Fleischman, R. N., McGehee, D. V., and Natarajan, M., Effects of age, systems experience and navigation technique on driving with an advanced traveler information system, *Human Factors* 39 (2), 177–199, 1997.
120. Green, P., Variations in task performance between younger and older drivers: UMTRI research on Telematics, *Association for the Advancement of Automotive Medicine Conference on Aging and Driving*, Southfield, MI, 2001.
121. Scialfa, C., Ho, G., Caird, J., and Graw, T., Traffic sign conspicuity: The effects of clutter, luminance, and age, *Annual Meeting of the Human Factors and Ergonomics Society*, Houston, TX, 1999.
122. Liu, Y., Effect of advanced traveler information system displays on younger and older drivers' performance, *Displays* 21, 161–168, 2000.
123. Verwey, W. B., Online workload estimation: effects of road situation and age on secondary task measures, *Ergonomics* 43 (2), 187–209, 2000.
124. Yoo, H., Tsimhoni, O., Watanabe, H., Green, P., and Shah, R., Display of HUD warnings to drivers: Determining an optimal location, Report No. UMTRI-99-9, University of Michigan Transportation Research Institute Ann Arbor, 1999.
125. Wolffsohn, J. S., McBrien, N. A., Edgar, G. K., and Stout, T., The influence of cognition and age on accommodation, detection rate and response times when using a car head-up display, *Ophthalmic and Physiological Optics* 18 (3), 243–253, 1997.
126. Ho, G., Scialfa, C. T., Caird, J. K., and Graw, T., Visual search for traffic signs: The effects of clutter, luminance, and aging, *Human Factors* 43, 194–207, 2001.
127. Schieber, F. and Goodspeed, C. H., Nighttime conspicuity of highway signs as a function of sign brightness, background complexity and age of observer, *Proceedings of the Human Factors and Ergonomics Society*, Human Factors and Ergonomics Society, Santa Monica, CA, 1997, pp. 1362–1366.

128. Stutts, J. C., Reinfurt, D. W., and Rodgman, E. A., The role of driver distraction in crashes: An analysis of 1995–1999 Crashworthiness Data System Data, *Annual Proceedings Association for the Advancement of Automotive Medicine* 45, 287–301, 2001.
129. Lam, L. T., Distractions and the risk of car crash injury: the effect of drivers' age, *Journal of Safety Research* 33 (3), 411–419, 2002.
130. McEvoy, S. P., Stevenson, M. R., and Woodward, M., The prevalence of, and factors associated with, serious crashes involving a distracting activity, *Accident Analysis and Prevention* 39 (3), 475–482, 2007.
131. McEvoy, S. P., Stevenson, M. R., and Woodward, M., The impact of driver distraction on road safety: results from a representative survey in two Australian states, *Injury Prevention* 12, 242–247, 2006.
132. Dingus, T. A., Klauer, S. G., Neale, V. L., Petersen, A., Lee, S. E., Sudweeks, J., Perez, M. A., Hankey, J., Ramsey, D., Gupta, S., Bucher, C., Doerzaph, Z. R., Jermeland, J., and Knipling, R. R., *The Impact of Driver Inattention on Near-Crash/Crash Risk: An Analysis Using the 100-Car Naturalistic Driving Study Data*, National Highway Traffic Safety Administration, Washington, D.C., 2006.
133. Vollrath, M., Meilinger, T., and Krüger, H.-P., How the presence of passengers influences the risk of a collision with another vehicle, *Accident Analysis and Prevention* 34, 649–654, 2002.
134. Wallace, R. B. and Retchin, S. M., The evaluation of the older driver, *Human Factors* 34, 16–24, 1992.
135. Bedard, M. and Meyers, J. R., The influence of passengers on older drivers involved in fatal crashes, *Experimental Aging Research* 30 (2), 205–215, 2004.
136. Regan, M. A. and Mitsopoulos, E., *Understanding Passenger Influences on Driver Behaviour: Implications for Road Safety and Recommendations for Countermeasure Development*, Monash University Accident Research Centre, 2001.
137. Regan, M., Salmon, P. M., Mitsopoulous, E., Anderson, J., and Edquist, J., Crew resource management and young driver safety, *Human Factors and Ergonomics Society 49th Annual Meeting*, Orlando, FL, 2005, pp. 2192–2196.
138. Coughlin, J., Not your father's auto industry? Aging, the automobile and the drive for product innovation, *Generations* 28 (4), 38–44, 2005.
139. Regan, M., Preventing traffic accidents by mobile phone users, *Medical Journal of Australia* 185 (11/12), 628–629, 2006.
140. McKnight, A. J., Driver and pedestrian training, in *Transportation in an Ageing Society: Improving Mobility and Safety for Older Persons (Volume 2)*, Transportation Research Board, National Research Council, Washington, D.C., 1988.
141. Goggin, L. and Keller, M. J., Older drivers: A closer look, *Educational Gerontology* 22, 245–256, 1996.
142. Kostyniuk, L., Streff, F. M., and Eby, D. W., The older driver and navigation assistance systems, The University of Michigan Transportation Research Institute, Ann Arbor, MI, 1997.
143. Coughlin, J. F. and Reimer, B., *New Demands from an Older Population: An Integrated Approach to Defining the Future of Older Driver Safety*, SAE Convergence, 2006.
144. Regan, M., Oxley, J., Godley, S., and Tingvall, C., *Intelligent Transport Systems: Safety and Human Factors Issues*, Royal Automobile Club of Victoria, Melbourne, Australia, 2001.
145. Oxley, J., Corben, B., Fildes, B., O'Hare, M., and Rothengatter, T., *Older Vulnerable Road Users—Measures to Reduce Crash and Injury Risk*, MUARC, Melbourne, Australia, 2004.

21 The Relationship between Driver Fatigue and Driver Distraction

Ann Williamson

CONTENTS

Fatigue is a major road safety problem of similar dimensions to drinking while driving. In Australia, for example, current estimates suggest that fatigue plays a role in up to 30% of fatal crashes,[1] which is a similar level of involvement to that of alcohol. A number of studies have demonstrated the role of fatigue in crashes. For example, the Auckland car crash case control study showed that the risk of injury-related crashes increased significantly for drivers who reported that they were sleepy while driving, for drivers with 5 h sleep or less, and for drivers who were on the road between 2:00 a.m. and 5:00 a.m.[2] Similarly, a case control study in the United States[3] showed a clear relationship between long-distance driving, increasing fatigue, and increased crash risk, with drivers doing more than 600 mile journeys showing a more than 10 times increased risk of crashing and drivers who reported falling asleep at the wheel showing a 14-fold increase in crash risk.

Fatigue presents greater problems for road safety, however, than other driver behavior-related problems like alcohol and speeding, as we understand less about its causes and, consequently, the countermeasures for mitigating its effects. Fatigue is a hypothetical process, which cannot be measured directly. Fatigue measurement relies on indirect measures of its effects, such as on self-rated feelings, driver performance, and changes in physiological state, rather than on direct measures such as those that exist for speeding and for the measurement of the amount of alcohol in the body. In contrast, definitions of fatigue emphasize factors like tiredness, adverse effects on performance in response to repeated stimulation by the same stimulus, problems of sustained attention, and a range of effort-related experiences such as unwillingness to continue with the task or the inability to continue putting effort into the task. These characteristics make management of driver fatigue one of the significant challenges for road safety.

Driver distraction is also increasingly being recognized as a significant road safety problem. It is generally accepted that the potential for distraction is a common experience for most drivers, even though there is considerable debate about the effects of specific distractors. Both driver distraction and driver fatigue may affect drivers, and both may have adverse effects on their capacity to drive safely.

The effects of both fatigue and driver distraction relate to the role of attention while driving. Driving is a task that requires constant vigilance as a primary component of the task. The problem of fatigue lies in decreasing overall attention to the task, whereas the problem of driver distraction is in the diverting of attention away from the driving task. The major differences between the two phenomena lie in how and why the attentional changes occur, whether it is an internal process or caused by external stimuli, and in the time frame over which the attention is likely to occur. Driver distraction relates to shorter-term, even momentary, attentional diversion, whereas driver fatigue can involve long periods of suboptimal attention to the task of driving. The two phenomena can also be distinguished in terms of the volitional role of the loss of attention from the task of driving. Although driver distraction is often characterized as a dual-task problem, where drivers are actually attempting to do two things at once and so intend that their attention be diverted to another activity (like using a mobile phone or reading a navigational aid), loss of attention due to fatigue is entirely unintended. Although driver distraction can be unintended, the inattention to the driving task due to fatigue will always be unintentional.

There has been very little consideration in the literature of the relationship between driver distraction and driver fatigue. We know comparatively little about the effects of each phenomenon in relation to each other and whether or not they interact. For example, do tired drivers experience more or less distraction? Does distraction make you more or less tired? Are the ways of managing driver distraction and driver fatigue compatible with one another? This chapter attempts to draw together the available evidence on driver fatigue and discusses its relevance and likely implications for driver distraction. Specifically, the discussion proceeds on two levels. The first examines the nature of the relationship between the two phenomena and looks at the argument for fatigue being viewed as a type of internal distraction from driving. The second looks at the influence of fatigue on the vulnerability of drivers to distractors and looks at the compatibility of the countermeasures being advocated for managing driver distraction and fatigue.

21.1 FATIGUE AS A DISTRACTOR

Many definitions of driver distraction specify that it is a form of inattention that shifts attention away from the task at hand. On the basis of this definition, fatigue could be included as a driver distraction. For example, the U.S. National Highways and Transport Safety Administration categorized four distinct types of driver distraction: visual, auditory, physical, and cognitive distraction.[4] The first three categories could be classified as external distractors, as they result from stimuli occurring outside or external to the person. Cognitive distraction, however, is due to internal stimuli as it is defined as "any thoughts that absorb the driver's attention to the point where they are unable to navigate through the road network safely and their reaction time is reduced."[5] This last category is particularly relevant to the current discussion of fatigue and distraction.

Definitions of fatigue similarly include reductions in attention, especially under conditions requiring sustained attention and in tasks with little variety. For example, Brown[6] defined driver fatigue as a "disinclination to continue performing the task at hand and a progressive withdrawal of attention from road and traffic demands." Such definitions are consistent with the idea that fatigue is an internal distractor from the driving task as they include attentional withdrawal from the driving task.

Theories of the causes of fatigue while driving relate mainly to the requirements for sustained attention and the often monotonous nature of the driving task. A number of theories have been advanced to attempt to account for changes in performance with sustained attention. These include arousal theory, resource theory, and the effort-compensation theory. These theories are distinguished by their conceptualization of the role of attention. Arousal theory[7] holds that decreases in arousal occur due to high monotony of stimulus presentation, which results in a decrement in performance. Resource theory,[8] however, considers that information processing resources, especially relating to attentional capacity, are finite and the requirement to maintain or sustain attention and to perform a task over time results in depletion of the pool of attentional resources which has adverse effects on performance. The related effort-compensation theory[9,10] argues that the vigilance or fatigue effect is related to stress effects and occurs because the prolonged effort required to sustain attention is taxing and stressful and over time this depletes capacity to perform at high levels.

The research on the causes of fatigue in general (as distinct from the causes of fatigue while driving) includes a broader range of factors, but it tells us relatively little about the relationship between fatigue and distraction while driving. The causes of fatigue have been conceptualized mainly as relating to internal or endogenous factors that affect the individual's level of alertness. In fact Thiffault and Bergeron[11] distinguish endogenous factors that they argue stem from within the organism and exogenous factors that stem from the person's interaction with his or her task and environment. These authors argue that most of the research on the causes of fatigue has highlighted the role of endogenous factors. These include three main causes: time on task, as seen in the discussion of fatigue while driving; time-of-day or circadian influences; and the length of time awake or amount of sleep obtained recently. The effects of fatigue differ somewhat depending on the cause, although slowing of responses, missing of signals, and the tendency to apply less effort seem to be outcomes of fatigue no matter what the cause. The differences between the endogenous causes of fatigue lie in when they exert their effects. Although fatigue due to increasing time on a task takes time to be generated, circadian and sleep loss influences can have very early onset. The worst effects occur when these causes combine.

The emphasis on the endogenous characteristics of fatigue is a major reason for the lack of consideration of the interaction between fatigue states and other task factors such as external distraction (i.e., distractions occurring outside or external to the person). The evidence on the effects of fatigue on performance, however, reveals a number of similarities with the effects of external distractors since both result in withdrawal of attention from the driving task. Fatigued drivers show slowing of reaction times and missing of relevant information, especially visual signals, compared with drivers who are not fatigued.[12] These performance effects tend to increase markedly with increasing time on task, an effect called the vigilance decrement.[13,14] This effect is accentuated when the task is monotonous, such as is often the case when driving. There appear to be two separate aspects to the vigilance effect:[15] an overall decline in the person's sensitivity to perceptual stimuli or his or her ability to discriminate target stimuli from nontarget stimuli and a change in response criteria such that they tend to become more cautious about reporting a target. Both aspects increase with longer time on a task, but changes in perceptual sensitivity seem to occur only when the task involves the discrimination of a change in some aspect of a repetitive standard stimulus in combination with fast event rates. The task of driving, especially on monotonous roads, would fall into this category. In addition, the continuous requirement to sustain attention at a sufficiently high level to maintain good driving performance produces a high workload for the driver.[16] The level of workload increases with increasing time at the wheel, also making fatigue effects increasingly more likely.

Other effects of fatigue include changes in mood states,[17] attentional narrowing,[18] less analytical processing of information, especially poorer planning and a tendency to perseverate on particular strategies,[19] and reduced effort in the task.[20] All of these effects are likely to have adverse consequences for driver performance and safety and contribute to the evidence that fatigue and the effects of fatigue will distract the driver away from the primary task of driving.

Overall, therefore, it seems that fatigue, like driver distraction, has adverse effects on driving because of withdrawal of attention from the driving task, although the causes are due to the driver's internal state, not to an external distraction. The question then is whether fatigue really is a form of internal distraction.

As discussed earlier, changes in internal or cognitive state, such as loss of attention due to daydreaming, are included with external factors in the categorization of driver distraction,[4] as they are linked by their effects of diverting attention away from the driving task. There are problems with doing so, however, as the origins and effects of the changes in internal state like daydreaming or being lost in thought are different to those of external distractors. Skilled drivers are able to think of other things and daydream while driving due to the spare attentional capacity they have because they are skilled at doing the task. Often the diversion of thoughts away from the driving task occurs without intention, but the driving task can be unaffected. In addition, during cognitive distraction or daydreaming, the extent and nature of the diversion of attention to other thoughts is internally controlled and more likely to be calibrated according to the demands of the primary task of driving. Externally generated distractors, however, may result in less controlled allocation of attention as they can *capture* attention at any time regardless of the demands of the primary driving task. This is what makes them dangerous for driving.

Although at one level fatigue can be thought of as a distractor because it diverts attention from the driving task, there are some important differences between fatigue and other internal or cognitive distractors like daydreaming or being lost in thought. Most importantly, fatigue results in an overall decrease in attentional capacities that is increasingly less possible to resist regardless of events in the driving task, whereas cognitive distractions like daydreaming affect the selection of the object of attention, are short-term or even transient, and may be countered by events in the driving task. Fatigue affects overall alertness and attention, and cognitive distraction such as daydreaming affects what you are thinking about. Although fatigue produces changes in internal state that certainly divert attention from the driving task, it does not share a number of other characteristics of distractors. Clearly, the definition of distraction in the context of driving could do with some review.

21.2 FATIGUE AND VULNERABILITY TO DISTRACTIONS

A second consideration in looking at the relationship between fatigue and distraction while driving is the nature of the relationship. Of particular interest to road safety is the extent to which tired drivers are more or less vulnerable to the effects of distractors. The available research on this question suggests that fatigued drivers may be less vulnerable to distractions while driving. This section looks at this evidence.

Findings from research on the nature of the effects of fatigue suggest that these effects will also moderate the driver's response to external distractors. For example, a number of studies have looked at the effect of fatigue on the usable visual field. These have demonstrated significant loss of the ability to detect signals in the periphery of the visual field with increasing fatigue and with increasing time on a monotonous task. Studies have shown that one night without sleep results in significant narrowing of the visual field,[21,22] with increasing proportion of signals being missed the further they are in the peripheral visual field. Roge et al.[21] showed that extended periods of driving, especially under monotonous conditions, also reduced the useful visual field. Although reduction in the size of the visual field has obvious implications for safe driving, it also influences vulnerability to distractions while driving. If drivers are not picking up visual information in the periphery very well anyway, they are also likely to be less vulnerable to the effects of distractors that occur in these parts of the visual field.

Furthermore, there is evidence that mental fatigue reduces executive control processes. Work by van der Linden et al.[19] showed that, when fatigued, individuals were poorer at planning, poorer at working toward a target in their task performance, and more likely to persevere with behavior even when they had feedback demonstrating that different strategies were needed. These authors concluded that the reduced executive control meant that behavior was less goal directed and more automatic, and less likely to be affected by changes in the context in which it occurred when the person was fatigued. These results also suggest that the tendency to focus only on certain well-learned aspects of the task when fatigued will increase the chance that other, irrelevant, aspects of the task environment will be ignored.

Similar conclusions can be drawn from studies of *highway hypnosis*, or driving without attention,[23] which has been reported to occur after long periods of driving, especially in highly predictable environments. When drivers experience this effect, they are unaware of even the essential requirements of the driving task, much less any extra stimuli within the driving environment that could be distractors.[23] Related to this is the finding in work by Hancock and Warm[16] of fatigue-related changes in effort being applied to the task. This effect has been shown to result in concentration on aspects of the task that are simpler and require less effort. This may also result in drivers focusing only on the main task of driving, so again making them less vulnerable to external influences that are potential distractors.

Different characteristics of individual drivers are likely to influence the vulnerability of fatigued drivers to distractors. There is considerable evidence that older people[22] and less experienced drivers[24] are more prone to attentional tunneling or narrowing of attention, even in the absence of fatigue. Other evidence demonstrates that older drivers are also more likely than younger drivers to show a vigilance effect of lowered alertness with increasing time on task.[25] These findings suggest that these fatigue-related changes occurring for older drivers and less experienced drivers may make them less vulnerable to the effects of distractors.

Further, there may be personality differences in the effect of distractors during fatigue states. The introversion-extroversion theory,[26] for example, suggests that extroverts are more prone to the vigilance effect in performance in prolonged and monotonous situations because they are inherently under-aroused, but would be more vulnerable to distractors in monotonous driving because they seek to increase external stimulation. Although the evidence for this theory is somewhat limited, a study by Thiffault and Bergeron[27] showed that extraversion and sensation-seeking characteristics predicted falling asleep at the wheel in a simulation study of driving under monotonous conditions. These studies suggest that where fatigue may reduce the tendency to be distracted while driving, the effect may be greater for some individuals than others.

Rather than being distracted by external stimuli, there is some evidence that tired drivers use external factors to overcome the effects of fatigue by increasing their arousal and alertness levels. A number of studies have shown that people attempt to control their increasing fatigue levels, especially those relating to monotony and the requirements for long periods at the same task, by increasing the amount of stimulation available in the task environment. For example, Davenport[28] showed that performance on a visual vigilance task, which showed the usual vigilance decrement, was improved when irrelevant music was presented at random intervals during the task. Notably, the randomness of the music was an important determinant of its effectiveness. This suggests that, rather than being distracting, the additional stimuli had beneficial effects by reducing the effects of fatigue on performance. A similar result was obtained in a study where changes in self-rated fatigue were tracked during a physical performance task.[29] The results showed that study participants who were presented distracting sounds while doing the physical task experienced less fatigue. In addition, research on long-distance truck drivers shows that they employ a range of different strategies to help them overcome the effects of fatigue.[30] These include listening to the radio, talking on the mobile telephone or CB radio, eating, drinking, or smoking cigarettes. Most of these strategies have been implicated as potential external sources of distraction. It is possible that the strategies that tired drivers use to moderate the effects of fatigue increase the amount of distraction, in turn further increasing their level of inattention to the main task of driving. On the other hand, we know that external stimuli are important for maintaining alertness and, consequently, for enhancing safe driving performance in tired drivers. This means that attempting to limit access to strategies that are not central to the driving task may not be of benefit for tired drivers.

The effect of fatigue on internal or cognitive distraction is likely to be similar to external distraction. The decreased vulnerability of drivers to external distractors when tired is also likely to apply to internal distractors like daydreaming. For example, in the early stages of doing a monotonous task, people use internal *distraction* or thinking about something else to help maintain alertness, but when fatigue begins due to a long period of time on the task, lack of sleep, or time-of-day influences, attentional capacity decreases, thus decreasing the vulnerability to cognitive distractors like thinking of other things.

In their discussion of the factors influencing driver distraction effects, Ranney et al.[31] identified *willingness to engage* as an important factor in determining whether drivers will consciously or unconsciously make a decision to submit to a distraction and thereby undertake a secondary task while driving. The present analysis of the relationship between fatigue and distractions while driving suggests that the situation may not be so simple when the driver is experiencing fatigue. Certainly, the arousal theory explanation of the effects of fatigue due to time on task is consistent with the concept of willingness to engage because it predicts that tired drivers will engage with any external stimuli due to the benefits of increasing arousal and overcoming the fatiguing effects of long, monotonous tasks like driving. Resources theory only partially supports the willingness to engage concept, however, since it suggests that external stimuli or distractions are only of benefit in reducing fatigue effects if they occur in a different modality to that being used in the main task. This prediction is based on research suggesting that attentional resources are sensory modality specific. Resources in one modality might be depleted, but those of another can be high. In the case of driving, long periods at the wheel would deplete visual attentional resources due to the visual nature of the task, whereas a secondary auditory task would be beneficial in increasing alertness. The effort-compensation account of the effects of fatigue on performance is probably most directly relevant as it takes into account the motivational aspects of the role of willingness to engage. This theory predicts, however, that distraction would never have benefits for driver fatigue, since coping with distraction would require even more effort, which the driver would be unwilling or unable to expend. Clearly, further research is needed to determine whether external stimuli or other distractors can or do overcome the effects of fatigue and whether this can be of benefit for driving performance.

21.3 DISTRACTION AS A CAUSE OF DRIVER FATIGUE

The final question of interest is whether driver distractions and the need or perceived need to deal with them can actually make drivers more tired. The most obvious answer lies in the fact that fatigue onset is most likely when doing monotonous tasks, so distractions while driving probably prevent fatigue or at least stave off its onset for a longer period. The role of external stimuli in assisting tired drivers to reduce their fatigue, which was discussed in the previous section, also suggests that distractions while driving are likely to be of benefit in minimizing fatigue. In situations where a driver has been at the wheel for very long periods and is already fatigued, or where they are already tired due to lack of sleep or time-of-day factors, distractions are unlikely to have positive effects as under these conditions short-term strategies that increase arousal for short periods are unlikely to be of great benefit. Very tired drivers need to stop, sleep, and recover. Under these conditions, it is most likely that distractors will have little effect at all as drivers will minimize effort and only pay attention to the basic essentials of the driving task.

21.4 CONCLUSIONS

Fatigue may be related to driver distraction due to its similar effects of withdrawing attention from the main task of driving. In this sense, fatigue could be considered to be an internal distractor due to its effect on the current state of arousal and alertness of the driver. Fatigue, however, is not like other cognitive activities that are classified as distractors because it affects general attentional and alertness capacity rather than selective attention and cannot be overcome while continuing to drive, whereas cognitive distraction can be overcome by events within the driving context. The concept of cognitive distractors and the role of fatigue needs further research and definition.[32] In addition, some of the characteristics of the effects of fatigue may actually reduce the vulnerability of a fatigued driver to attentional capture by external features in the driving environment. These include effects of fatigue, like tunneling of attention and the tendency to move to simpler and less effortful approaches to the driving task, which may reduce the inclination for drivers to be distracted by external stimuli. Some of the strategies that drivers use to manage fatigue while driving fall into the category of external distractors (including using mobile phones and conversing on the CB radio) and so may increase the withdrawal of attention from the driving task in drivers who are beginning to experience fatigue. In fact there may be additive effects on performance when tired drivers choose to engage in a potentially distracting task, as both distraction and fatigue can produce attentional tunneling. Yet there is good evidence that external stimulation is of significant benefit for drivers who are becoming tired, as it assists in maintaining alertness. This means that strategies aimed at reducing driver distraction may inadvertently be limiting the strategies that tired drivers can use to help maintain alertness and safe driving performance, especially if they are resisting the most beneficial strategy of taking a break from driving for rest and sleep. Further research is needed to establish whether fatigue effects do moderate the effects of external distractors while driving and the role of distractions in managing the onset of driver fatigue.

REFERENCES

1. Australian Transport Safety Bureau, *Road Safety in Australia: A Publication Commemorating World Health Day*, 2004.
2. Connor, J., Norton, R., Ameratunga, S., Robinson, E., Civil, I., Dunn, R., Bailey, J. and Jackson, R., Driver sleepiness and risk of serious injury to car occupants: population based case control study. *Br. Med. J.*, 324, 1125, 2001.
3. Cummings, P., Koepsell, T.D., Moffat, J.M. and Rivara, F.D., Drowsiness, countermeasures to drowsiness and the risk of a motor vehicle crash. *Inj. Prev.*, 7, 194–199, 2001.

4. Ranney, T.A., Garrott, R. and Goodman, M.J., NHTSA driver distraction research: Past, present, and future. *Proceedings of the 17th International Technical Conference on the Enhanced Safety of Vehicles* (Report No. 233, CD-ROM). US Department of Transportation: Washington, D.C., 2001.
5. Young, K., Regan, M. and Hammer, M., *Driver Distraction: A Review of the Literature.* Monash Accident Research Centre Report No. 206, November, 2003.
6. Brown, I.D., Driver fatigue, *Hum. Factors*, 36, 298, 1994.
7. Mackworth, J.F., *Vigilance and Habituation: A NeuroPsychological Approach*, Harmondsworth: Penguin Books, 1969.
8. Wickens, C.D. and Kessel, C., Processing resource demands of failure detection in dynamic systems, *J. Exp. Psychol: Hum. Percept. Perf.*, 6, 564, 1980.
9. Hockey, G.R.J., Compensatory control in the regulation of human performance under stress and high workload: A cognitive-energetical framework, *Biol. Psychol.*, 45, 73, 1997.
10. Matthews, G., Levels of transaction: A cognitive science framework for operator stress, In Hancock, P.A. and Desmond, P.A. (Eds.), *Stress, Workload and Fatigue*, Mahwah, NJ: Erlbaum, 2001.
11. Thiffault, P. and Bergeron, J., Monotony of road environment and driver fatigue: A simulator study, *Accid. Anal. Prev.*, 35, 381, 2003.
12. Dinges, D.F., Pack, F., Williams, K., Gillen, K.A., Powell, J.W., Ott, G.E., Aptowicz, C. and Pack, A.I., Cumulative sleepiness, mood disturbance and psychomotor vigilance performance decrements during a week of sleep restricted to 4–5 hours per night, *Sleep*, 20, 267, 1997.
13. Davies, D.R. and Parasuraman, R., *The Psychology of Vigilance*, London: Academic Press, 1982.
14. Warm, J.S., *Sustained Attention in Human Performance*, London: Wiley, 1984.
15. Parasumaman, R., Memory load and event rate control sensitivity decrements in sustained attention, *Science*, 205, 924, 1979.
16. Hancock, P.A. and Warm, J.S., A dynamic model of stress and sustained attention. *Hum. Factors*, 31, 519, 1989.
17. Broadbent, D.E., Is a fatigue test now possible?, *Ergon*, 22, 1277, 1979.
18. Easterbrook, J.A., The effect of emotion on cue utilization and organization of behaviour, *Psychol. Rev.*, 66, 183, 1959.
19. van der Linden, D., Frese, M. and Meijman, T.F., Mental fatigue and the control of cognitive processes: Effects on perseveration and planning. *Acta Psychol.*, 113, 45, 2003.
20. Smit, A.S., Eling, P.A.T.M. and Coenen, A.M.L., Mental effort causes vigilance decrease due to resource depletion, *Acta Psychol.*, 115; 35, 2004.
21. Roge, J., Pebayle, T., El Hannachi, S. and Muzet, A., Effect of sleep deprivation and driving duration on the useful visual field in younger and older subjects during simulator driving, *Vis. Rec.*, 43, 1465, 2003.
22. Mills, K.C., Spruill, S.E., Kanne, R.W., Parkman, K.M. and Zhang, Y., The influence of stimulants, sedatives and fatigue on tunnel vision: Risk factors for driving and piloting, *Hum. Factors*, 43, 310, 2001.
23. Wertheim, A.H., Highway hypnosis: A theoretical analysis, In Gale, A.G. et al. (Eds.), *Vision in Vehicles-III*, Elsevier North-Holland, Oxford, 467, 1991.
24. Crundall, D., Underwood, G. and Chapman, P., Driving experience and the functional field of view, *Perception*, 28, 1075, 1999.
25. Campagne, A., Pebayle, T. and Muzet, A., Correlation between driving errors and vigilance level: Influence of the driver's age, *Physiol. Behav.*, 80, 515, 2004.
26. Eysenck, H.J., *The Biological Basis of Personality*, Springfield, IL: Thomas, 1967.
27. Thiffault, P. and Bergeron, J., Fatigue and individual differences in monotonous simulated driving, *Personal. Indiv. Diffs.*, 34, 159, 2003.

28. Davenport, W.G., Arousal theory and vigilance: Schedules for background stimulation, *J. Gen. Psychol.*, 91, 51, 1974.
29. Pennebaker, J.W. and Lightner, J.M., Competition of internal and external information in an exercise setting, *J. Person. Soc. Psychol.*, 39, 165, 1980.
30. Williamson, A.M., Feyer, A.-M., Friswell, R. and Finlay-Brown, S., *Driver Fatigue: A Survey of Long Distance Heavy Vehicle Drivers in Australia.* Australian Transportation Safety Bureau, CR 198, 2001.
31. Ranney, T.A., Mazzae, E., Garrott, R. and Goodman, M.J., NHTSA Driver distraction research: Past, present and future, www.nhtsa.dot.gov, 2000.
32. Sheridan, T.B., Driver distraction from a control theoretic perspective. *Hum. Factors*, 46(4), 587, 2004.

Part 7

Design and Standardization

22 European Approaches to Principles, Codes, Guidelines, and Checklists for In-Vehicle HMI

Alan Stevens

CONTENTS

22.1 INTRODUCTION

With driver distraction an increasing concern in the context of the proliferation of in-vehicle displays and information systems, this chapter focuses on the design and assessment of in-vehicle human-machine interaction (HMI)* within Europe with specific examples drawn from the United Kingdom. The European eSafety program is shaping the approach to the design and assessment of both driver information and assistance systems, which are expected to provide major contributions to accident reduction targets. Even though these systems are expected to have a positive effect on safety, a key issue is to ensure that any negative consequences from driver distraction are minimized or eliminated. This chapter describes the development and content of the European Statements of Principle (ESoP) on HMI, which provide high-level design guidelines for information and communication systems and aim to promote systems that are usable and safe, taking full account of the potential for driver distraction. The challenge of quantitative measurement of distraction is noted and a U.K. approach to assessment of HMI, based on an expert checklist, is described. Finally, the chapter discusses the design and assessment of driver assistance systems and the emergence of a code of practice being developed within a manufacturer-led European project.

22.2 OVERALL EUROPEAN APPROACH

22.2.1 European Market Situation

There has been substantial development of in-vehicle information systems (IVIS) in Europe in the last 10 years although the market size is still small compared with that in Japan, but much larger than in the United States. The latest navigation systems, for example, incorporate dynamic, real-time traffic and travel information and rerouting. There is a growing demand for lower-cost *nomadic* (portable) systems, such as those that provide navigation guidance on personal digital assistants (PDAs), as well as off-board navigation obtained via telecommunications service providers. Figure 22.1 shows examples of the TomTom, a navigation system implemented on a PDA platform.

These systems are undergoing rapid development in response to customer demand and developments in technology; for example, systems are increasingly able to respond to voice commands. Against this background, any form of detailed guidance or regulation has the potential to constrain innovation because it would be framed within a context of current technology and with current understanding of likely future development paths. Nevertheless, responding to social pressure and

* Throughout this chapter, HMI is used to include both the hardware interface and the dialogue that supports communication between the driver and the system.

FIGURE 22.1 Examples of nomadic information systems.

safety concerns, specific regulations have been enacted by some member state governments in Europe. Probably the most high-profile debate has focused on in-vehicle telephones, with the subsequent banning of the use of handheld mobile phones in many European countries.

Advanced driver assistance systems (ADAS), such as adaptive cruise control, have yet to achieve the market position of IVIS systems. In part this is due to the relative immaturity of their technological capability, but price and lack of customer demand are also factors. Such systems, however, are generally recognized as having a much greater potential than pure information systems for reducing accidents. The distinction between information and assistance systems is somewhat blurred. Hence, there are some common HMI issues that can be considered together.

22.2.2 European Approach to Human-Machine Interaction and eSafety

In September 2001 the European Commission (EC) presented the White Paper European Transport Policy for 2010,[1] setting a very ambitious target for road safety: a 50% reduction of road fatalities by 2010, which would bring the number of deaths per year down to 20,000.

The application of information and assistance systems is seen as playing an important role in achieving this goal, and in 2002 the EC established an eSafety working group, bringing together all the key stakeholders. On the basis of the group's report and other consultations, the EC adopted, in 2003, a communication on eSafety.[2] This presents recommendations to accelerate the development and deployment of advanced information and communication technologies (ICTs) for road safety and the wider use of active safety systems. The eSafety Forum[3] was established by the EC in close collaboration with industry, industrial associations, and public sector stakeholders to address both safety and market issues in the implementation of driver information and assistance systems as a contribution to European road safety improvement targets. The eSafety Forum created several working groups to focus on priority actions as set out in the eSafety communication; one of these was HMI.

The European approach to in-vehicle HMI issues focuses on long-term safety goals with a balanced approach to risks and benefits. The importance of a safe human-machine interface for IVIS and for safe HMI with ADAS is a common understanding of all stakeholders. Governments, industry, and consumer groups share the vision of the availability of a wide range of systems providing services to the driver and contributing to the goal of reducing European fatalities. The European philosophy is that functionality not directly related to driving should not introduce additional risks. For functionalities related to driving, potential benefits should not be masked by poorly designed HMI that introduces distraction or inappropriate driver behavior. ADAS and IVIS will vary in their safety potential and even a specific system that is expected to provide a safety benefit may also introduce a certain risk—the balance of risks and benefits needs to be acceptable in order to contribute to the ultimate goal of reducing the number of fatalities. ADAS, in particular, are likely to change, to some extent, the existing dynamic relationship between drivers, their vehicles, and the road environment, and, at a detailed and statistical level, changes may be expected in the nature of accidents. It should be noted that safety-enhancing features such as airbags and antilock brakes are not without their problems in specific situations. So, although an overall improvement in safety can be expected from the introduction of ADAS and IVIS, research may reveal subtle changes in the type of accident, its contributory factors, and the road users involved.

Appropriate testing is considered vital for both ADAS and IVIS before market introduction to optimize the HMI and eliminate aspects that could compromise safety, such as driver distraction. However, it may be impossible in terms of time and costs to undertake sufficiently large field trials to identify the full impact on driver performance and safety of every system feature/function. Thus, it can be regarded as virtually impossible to *prove* enhanced safety before market introduction. Consequently, system and policy development cannot, in practice, be entirely evidence based and must draw on expert knowledge and consensus.

Developments in technology, design, and understanding of drivers' needs and behaviors also contribute to more mature products. For example, route guidance systems are today much easier to use than earlier versions and, on balance, may offer safety advantages.[4] (Of course, this is difficult to prove as accidents avoided—e.g., by being aware of an upcoming traffic queue—are not recorded.) If initial implementations had been excluded from the market, the current advantages offered by the systems may not have materialized. Similarly, the assumption that more information and functionality is automatically *worse* than less cannot be accepted: An assessment needs to be made of the impact of the information and functionality on both distraction and other driver behavior.

In assessing safety, consideration of drivers' behavior in a broader context is important. If a driver wants to obtain a route, for example, the driver will decide how to achieve this goal. Restricting functionality within a specific system in an attempt to reduce distraction may result in an increased use of less-restricted (and less well-designed) devices or greater use of conventional paper maps.

22.3 EUROPEAN STATEMENT OF PRINCIPLES FOR IN-VEHICLE INFORMATION SYSTEMS

22.3.1 Development and Comparisons

The European approach to the design and assessment of HMI has been to develop relatively high-level guidelines, called principles. The expectation is that the broad approach afforded by these principles will be technologically neutral or, at least, more resilient to developments in technology. As such, they complement international standards and United Nations Economic Commission for Europe (UNECE) regulations (discussed in Chapter 24).

In December 1999, the EC adopted a recommendation incorporating the ESoP,[5] and in 2001 it published an expansion of the principles by its expert group.[6] The documents contain 3 overall design goals relating to HMI and 32 principles covering the topics of system installation, information presentation, interaction with displays and controls, system behavior, and information about the system. In 2001 the European Automobile Manufacturers' Association (ACEA) made a voluntary agreement to fully respect the ESoP.

Under the eSafety initiative, one of the key areas of focus for the HMI working group was a review of the ESoP. Its application by car manufacturers and suppliers of original equipment was judged to be positive, but it was concluded that the impact of the ESoP could be improved for other stakeholders, specifically nomadic device manufacturers and service providers. Information from the EC member states concerning impact of the ESoP was also studied and, based on these responses and further reflection, the working group made a number of recommendations concerning ESoP development.[7]

The HMI working group also compared European, Japanese, and U.S. approaches to the development and use of HMI guidelines. In the United States, an initiative from the Alliance of Automobile Manufacturers (AAM) responded to a government challenge to develop guidelines for the HMI of IVIS to minimize distraction and promote safety. The AAM guidelines build on the previous European work of the ESoP. In Japan, the Japanese Automobile Manufacturers Association (JAMA) guidelines apply to in-vehicle display systems. Another Japanese document, developed by the Ministry of Land, Infrastructure, and Transport, replaces and includes the previous JAMA guidelines and provides more advice concerning distraction. Using a measurement technique for driver distraction called occlusion, with specific setup parameters, it states that the occlusion measure of total shutter opening time should not exceed 7.5 s. The Japanese and U.S. approaches are discussed in greater detail in the next two chapters.

When the European, Japanese, and developing U.S. guidelines are compared, some small and some more substantial differences emerge. All three approaches seek to minimize driver distraction, but the European approach is less prescriptive. Introducing quantitative pass/fail criteria, as both the U.S. and the Japanese guidelines do with respect to the driver's visual behavior, was regarded by the European working group as problematic; criteria that are too simplistic may be spurious—that is, they may exclude good HMI solutions and may not exclude bad HMI solutions.

As an example of an overly simplistic guideline, a discussion within the eSafety working group on HMI is provided. Limiting the number of button presses to achieve a specific goal may initially seem attractive if it avoids a driver reaching to a location on the dashboard and making a number of movements with an index finger; however, simple repetitive use of fingertip paddles on the steering wheel to control auditory volume may actually be a good solution. So, providing simple guidelines may cause a system provider to exclude some functionality completely. This then eliminates the challenge for researchers and industry to strive for an HMI solution that permits these tasks to be performed while the vehicle is in motion. Simplistic criteria for HMI may also imply that it is safe to undertake tasks while driving, whereas, no matter how restrictive, there will always be traffic situations in which the operation of these tasks may cause critical situations.

The draft AAM guidelines available to the European eSafety Group contained some helpful additions and clarifications compared with the (then current) European documents, so the AAM text influenced subsequent development of the ESoP. For example, the revised 2006 ESoP (discussed later) benefits from clear cross-referencing from the individual principles to relevant standards, rules, and directives; it is clearly helpful to link available, scientifically well-founded, and accepted assessment methods to specific principles.

In responding to the working group report, the commission made some funding available through existing HMI-related projects (HUMANIST[8] and AIDE[9]) for a small group of specialists to implement the HMI working group recommendations concerning a revised ESoP and hold Europe-wide workshops. The following describes what was published as part of a commission recommendation and available on the eSafety website.[3]

22.3.2 Stakeholders Involved in System Design and Construction

The principles are written for those who design and manufacture information and communication systems or provide functionality intended to be used while driving. The principles also presume that those applying them have technical knowledge of the products as well as access to the resources necessary to apply the principles in designing these systems. Introductory text reminds users that the principles are not a substitute for regulations and standards.

The principles are intended to apply to systems and functionalities in original equipment manufacturer, aftermarket, and nomadic (portable) systems. In general, a number of organizations are involved in designing, producing, and providing elements of such systems and devices. These include the manufacturers of parts enabling the use of nomadic devices by the driver while driving (e.g., cradles, interface connectors) and providers of broadcast information meant to be used by the driver while driving (e.g., RDS information and radio program information presented as running text).

Where systems are provided by a vehicle manufacturer, it is clear that the manufacturer is responsible for the overall design. In other cases, the product-responsible organization will include the organization introducing a product or functionality into the market, part or all of which may have been designed and produced by different parties. Consequently, the responsibility may often be shared between organizations. For ease of reference, the term *manufacturer* is used in the following text, although this may include several product-responsible organizations.

22.3.3 European Statement of Principles Contents

The ESoP opens with a number of overall design goals, including the statement that *the system does not distract or visually entertain the driver.* The main part of the ESoP provides system design and construction guidelines that aim to contribute to the overall design goals and includes

- Installation principles
- Information presentation principles
- Principles on interaction with displays and controls
- System behavior principles
- Principles on information about the system

The principles are short statements summarizing specific and distinct HMI issues. Each principle is followed by an elaboration with the following sections:

Explanation includes some rationale and further explanation for the principle.
Examples (*Good* and *Bad*) further explain implementation.
Application describes which specific systems or HMI functionality is being addressed.
Verification requires that, where possible, a suitable method is outlined and interpretation of the resulting metric is given:

- Where the result can be expressed as *Yes/No*, this indicates the availability of a clear identification of compliance with a principle.
- In other cases the approach/methods identified do not lead to simple pass/fail criteria but offer the opportunity of increased optimization of the HMI.
- If regulations are addressed, the appropriate EC directive is mentioned (and the product-responsible organization is required to comply with the current version of the directive).

References are included, for example, to international standards including, sometimes, draft standards.

The ESoP development group did not believe that the current state of scientific development was sufficient to robustly link compliance criteria with safety for all the principles; this reflects, generally, the view of most European stakeholders. In terms of assessment of specific systems for their overall contribution to safety, most stakeholders are extremely cautious; there does not appear to be a sufficiently well-established base of scientific knowledge to provide unequivocal assessment in this dynamic area.

22.3.4 Scope and Limitations within the European Statements of Principle

The ESoP provides a relatively high-level approach to HMI design and assessment. This produces relative technology insensitivity—user-centered principles are likely to remain valid across technology developments. However, the high-level approach does have weaknesses: Principles can have exceptions; technology can influence principles; the application of principles relies on human judgment, and there can be few specific measurable criteria.

To illustrate the difficulty, consider attempting to write a guideline for the order of presentation of congestion information. Immediately, there are multiple possibilities. Present the most important information first, present information concerning the closest congestion problem to the driver first, and so on.

Thus, in terms of a guideline, the agreement that *information should be presented in a logical order* may be at too high a level to be of real use (to whom should the presentation order be logical? does it depend on the situation? how could it be implemented? how could it be tested? and so on). Such was the challenge facing the ESoP development group throughout its work.

There is no ideal and universally agreed way to divide up the field of HMI and there are, inevitably, relationships between the principles. It is likely that any group with similar background and expertise would come up with broadly similar principles but expressed in a different way. Within the ESoP itself, there is an uneven depth of treatment between principles. In part, this reflects the degree of scientific knowledge in each area and the scope of the principles and their relationship to each other. For example, there is a lot of advice concerning provision of information to the driver about operation of a system, but in the area of distraction and screen complexity, there is only general guidance.

In terms of scope, there are some declared and implied limitations:

- The principles apply specifically and only to vehicles of class M and N, including passenger cars, trucks, and buses—but not motorcycles or other vehicles.[10]
- The principles apply to information and communication systems and not specifically to ADAS, although the ESoP may be applicable to some aspects of ADAS.
- Head-up displays and night-vision systems are excluded, due to lack of scientific knowledge.
- Voice-controlled devices are similarly excluded. (Experience with these has been mixed; well-designed systems appear to offer advantages to some drivers in terms of operating some functions, but system performance may be speaker-dependent and there is insufficient experience to inform guidelines at this time.) Clearly this is an area for future study.

Integration of functions and systems appears very likely in the future. This may necessitate some of the principles being revised. The issue of integration is discussed in a later section.

The ESoP includes few quantitative measurement and assessment criteria. This is an obvious limitation, but it is unclear to what extent this could be achieved without seriously impeding innovative design. Overall, therefore, the ESoP can be considered as useful in providing design advice but not sufficient, in itself, for undertaking a safety impact assessment of any specific function or system.

22.3.5 Role of Drivers and Their Employers

The driver can be supported in safe operation of in-vehicle systems by making other aspects of the context of use as benign as possible. Such aspects can be directly influenced by fleet managers or the driver's employer, and additional recommendations have been formulated to inform and influence organizations. Consequently, the ESoP design principles are supplemented in the EC recommendation by a document on system use that includes recommendations on context/definition, driver training, use by drivers, and assessment of use.

Driver education is considered very important. European and national laws assert that the driver should always be in control of the vehicle, so drivers need to be informed and educated about new technology and warned against misuse. Where the manufacturer's intention has been clearly stated (such that the driver can reasonably be expected to be aware of it) and the driver subsequently uses the system in a way that is not intended by the manufacturer, this can be considered as misuse.

22.4 HUMAN-MACHINE INTERACTION GUIDELINES AND CHECKLISTS FOR IN-VEHICLE INFORMATION SYSTEMS IN THE UNITED KINGDOM

22.4.1 Development of U.K. Guidelines

Since the mid 1990s, the U.K. Department for Transport, recognizing the potential problem of driver distraction, has commissioned the development of information to assist providers of IVIS to design safer products. Initial work began with the development of a code of practice and design guidelines for IVIS. Shortly afterward, the European Committee of Ministers of Transport (ECMT) discussed adopting the U.K. code in lieu of international standards and, following a consensus forming process involving representatives from European governments and vehicle manufacturers, the ECMT Statements of Principle was published.[11] The Department for Transport then asked the British Standards Institute to publish the code (using the ECMT wording) and IVIS design guidelines,[12] which reviews key human factors design considerations (e.g., control and display location, system interaction) and provides normative data and references to supporting text.

The U.K. IVIS design guidelines were updated in 2002[13] to provide designers, manufacturers, and others in the design chain with a summary of legal and ergonomic issues relevant to safety and ease of use. These guidelines cover several issues:

- Description of the design process and how to include system assessment
- Documentation and user instructions to be provided with the system
- System installation, including compatibility with other systems and integration issues

- Controls, visual displays, and auditory information
- Information presentation and dialogue issues, including setting to individual preferences
- General safety aspects of IVIS (e.g., driver distraction, risk compensation by behavioral adaptation, system faults, and accurate and timely information provision)

The U.K. design guidelines are compatible with but more detailed than the European guidelines (the ESoP) and also contain information about the design and assessment process.

22.4.2 Development of the Transport Research Laboratory Human-Machine Interaction Checklist

In 1999 the Transport Research Laboratory (TRL), working for the U.K. Department for Transport, developed a safety checklist,[14] which provides a tool for assessing IVIS with regard to HMI and road safety. The checklist allows experts to make a rapid and structured assessment of the key safety-related features of an IVIS in a standard way and provides an initial assessment of whether an IVIS complies with the ESoP.[5] This checklist was also designed to help experts assess:

- A manufacturer's documentary evidence of good design
- Clarity of information concerning a driver's responsibility for road safety
- Driver workload to a limited degree: since a complete assessment of this important aspect of driver performance would require substantial effort, the checklist, instead, assesses various components that are likely to contribute to workload when interacting with controls and visual displays

The checklist was developed iteratively through international consultation and application. It was widely distributed for comment to the European human factors community. At a workshop held in Stockholm in 1999 under the auspices of the International Harmonized Research Activities working group on Intelligent Transport Systems (IHRA-ITS),[15] the checklist was used in small groups on a fleet of identical vehicles equipped with identical IVIS. Although there was no formal and independent evaluation, there was good intergroup reliability in the safety assessments, and the workshop participants found the checklist comprehensive and useful. A final version was completed in January 2000.

22.4.3 Checklist Description

The TRL checklist is split into six sections: documentation, installation, controls, auditory, visual, and dialogue and safety. The issues are addressed through a series of specific questions, some of which have a short list of statements that can be used to identify relevant features. In the assessment *pro forma*, each question has the response box options shown in Figure 22.2.

None ☐	Minor ☐	Serious ☐	NA ☐

FIGURE 22.2 TRL checklist response options.

None is the appropriate response when interaction with the IVIS is not anticipated to compromise the driver's ability to control the vehicle, for *most* drivers under *most* conditions. *Minor safety concerns* means that interaction might, sometimes, compromise vehicle control for *some* drivers under *some* conditions. A second-level assessment may be made following completion of the checklist, to assess the number, type, and relationship between the minor concerns. The cumulative effect of several minor safety concerns might lead the assessor(s) to an overall major concern. *Serious safety concerns* indicates that the system is likely to prevent a significant number of drivers under normal driving conditions from maintaining full control of the vehicle.

22.4.4 Recommendations for Use

The checklist assessment *pro forma* is accompanied by supportive information, which provides recommendations for use and additional rationale for the questions. It includes some examples of good and bad design, and provides technical references and a glossary of terms and abbreviations.

Although the checklist can be used by a single assessor, it is recommended that several be involved. Pairs of assessors were found to be practical because this allowed one to use and rate the system while the other recorded the results. Also, small questions or areas of difference could be discussed and having two assessors is more likely to ensure that all the *pro forma* is completed. A good scheme is to have two pairs of assessors using the checklist during daylight and another pair undertaking the assessment at night. Performing a day and night assessment is required to explore display visibility issues under different lighting conditions. The six assessors then agree on a consensus position. Following completion of the individual sections, an assessment summary report should be completed. Here, both good and bad features of the system and design are noted, and comments can be made about any safety concerns identified (and possibly recommendations made about how they might be overcome).

22.4.5 Properties and Limitations of Checklists

The TRL HMI safety checklist is designed for use by human factors experts to provide a rapid initial assessment of a system and to highlight areas where more detailed studies and measurements would be required. It has been applied to a number of commercial systems (either in prototype, prelaunch or deployed form), as summarized in Figure 22.3, on behalf of the U.K. government.[16]

Overall, the checklist has been found to provide a structured approach to safety assessment. It can be completed in a few hours and is useful in identifying areas (specific IVIS functions or design features) that need to be studied further using in-depth and more quantitative assessment techniques.

System type	
Navigation	7
Mobile data terminal	5
Handheld data terminals	3
Traffic information	2
Mobile telephone	2
Congestion warning	1
Reversing aid	1
Taxi-meter	1
Radar speed camera detector	1
	Total = 23

FIGURE 22.3 Systems to which the TRL checklist was applied.

For assessing an in-vehicle system, a checklist, if used alone, has a number of limitations:

- It cannot assess implications following a malfunction of part or the whole system. This requires specific interventions and experimental tests.
- Experts using the checklist cannot fully assess visual or cognitive distraction. Again, this is an area where field tests and agreed variables and acceptable limits are required.
- It cannot fully assess users' understanding of instructions or of system operation. Although experts can pass an opinion, field trials with different subjects would be required for a more complete analysis.
- It assesses use as intended by the system producer. The extent of misuse of a system would be a matter of speculation without very careful studies.

In summary, a checklist cannot be regarded as an entirely sufficient tool for assessing the safety or risk of using a specific IVIS, although it has to be pointed out that no single tool or tools in combination currently have sufficient consensus to perform that function. What a checklist can do is help human factors experts form an initial opinion and help identify where more detailed measurements may be required. Perhaps its greatest benefit is providing a common language and focus between system producers and human factors professionals to highlight issues that are the most safety critical.

22.5 HUMAN-MACHINE INTERACTION GUIDELINES FOR ADVANCED DRIVER ASSISTANCE SYSTEMS

The eSafety HMI working group and the U.K. checklist principally concentrate on IVIS. ADAS are fundamentally different from IVIS and therefore require a different approach. IVIS can be addressed in terms of installation, information presentation, interaction, and use, always aiming at minimizing their demands on driver attention and hence distraction. ADAS address the primary task of driving. Consequently, the HMI for these systems is designed to attract sufficient driver attention (e.g., when providing warnings). Other additional issues involve the cooperation with the driver in execution of the driving task. In contrast with IVIS, ADAS are so closely integrated with vehicle controls that there is very limited scope for aftermarket systems; ADAS are principally within the domain of mainstream vehicle manufacturers.

Compared with IVIS, ADAS technologies are still relatively novel, at least in Europe. There is still a need to clearly define what constitutes an ADAS. However, the classification of *systems* will always be problematic, particularly with increasing integration. It is therefore better to consider the safety of a specific system in terms of its individual functions. Drivers buy and use systems, such as a navigation system, an adaptive cruise control system, or a collision warning system. All systems perform functions, such as obtaining an address from the driver, warning of a collision, or adjusting headway. Additionally, some systems may include information functions, warning functions, and assistance functions and also inbuilt stability functions such as antilock brakes.

Figure 22.4 introduces a fourfold classification of in-vehicle functions covering IVIS and ADAS. These are

- *In-built*—where the function is automatically initiated by driver or vehicle actions
- *Informing*—where the driver is presented with information and a key issue is distraction
- *Warning*—where the function is designed to attract driver attention
- *Assistance*—where the driver initiates and supervises an automated aspect of driving

Figure 22.4 also compares a number of key issues associated with each of the four functions. The issue of drivers' locus of control refers to the interaction of the driver

Issue \ Function	In-built	Informing	Warning	Assisting
Example	ABS ASC Collision mitigation	Route guidance Mobile phone	LDWS ISA advisory	ACC
Focus	Vehicle stability	Information to the driver	Warning the driver	Aspects of longitudinal and lateral control
Driver's locus of control	None	Full	Depends	Overrideable
System supplier	OEM	OEM after market Nomadic	OEM after market	OEM
Safety issue	Technical	Distraction	Understandability	Controllability
Typical human interface	None (or via existing controls)	Screen + audio	Buzzer Symbol	Button Small display Existing controls

FIGURE 22.4 Comparison of key HMI issues for different IVIS and ADAS functionalities.

with the function: with inbuilt functions there is no interaction; for an informing function, the driver is fully in control; the driver has little control over when a warning function is activated but may choose whether or not to respond; and, finally, the assistance function allows for driver override.

It is possible to define relatively high-level human factors principles for ADAS.[17,18] Descriptions at such a general level are, however, of limited use to practicing designers. Design principles for ADAS are currently being addressed in a number of European research projects; for example, AIDE[9] is considering interfaces that are adaptive to the driver's dynamic state.

Involving users in the process of design is even more important with ADAS than with IVIS. This is because ADAS are relatively unfamiliar and drivers will not, initially, be aware of the technical possibilities. However, once informed, they can help shape the functions and systems that will be of most benefit to them. Involving users at an early stage can also clarify drivers' needs for information about how a function works and how drivers will interact with the function, both in the learning phase and when they are more experienced with it. Similarly, having a full, structured program of testing is very important for ADAS. In the EC-funded RESPONSE project, for example, a code of practice for the design, development, and testing of ADAS has been developed.[18] Figure 22.5 presents the process-oriented approach.

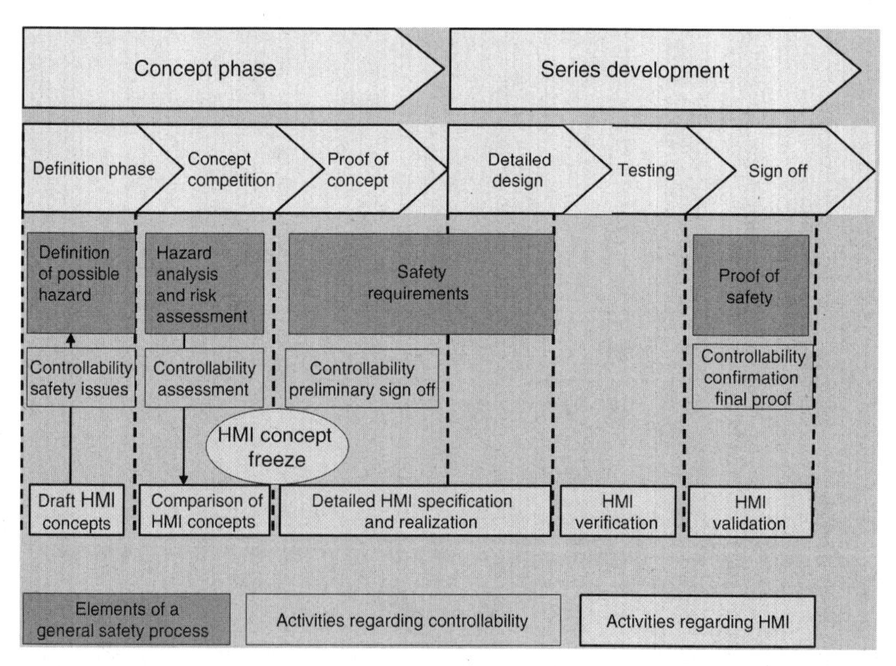

FIGURE 22.5 Process-oriented approach to HMI assessment of ADAS from RESPONSE project. (From Schwartz, J., Code of practice for the design and evaluation of ADAS. Response 3 Consortium v.1.6, 2006-08-11. RESPONSE: http://www.prevent-ip.org/en/prevent_subprojects/horizontal_activities/response_3/. With permission.)

22.6 CONCLUSIONS AND FUTURE CHALLENGES

The European approach to developing high-level design principles is a pragmatic approach to the consideration of HMI issues for IVIS and may well be the best approach for ADAS (although experience is limited at present). Clearly, some IVIS design principles, as distilled in the ESoP, also apply to ADAS, but such principles would need to be reexamined and extended for assistance systems because ADAS actively assist the driver in performing the primary driving task by providing information (e.g., *vehicle ahead*), warnings (e.g., *high closing velocity*), and assistance (e.g., applying the brakes). For example, the design principles developed for IVIS do not fully address the issues of controllability and the driver's role in supervising semiautomatic ADAS functionality.

As noted in the discussion of the limitations of ESoP, there appears to be an increasing integration of different functions (information sources, communication options, driver assistance) within in-vehicle systems. Integration can involve sharing of resources, such as a display screen (by sharing parts of the display or time sharing its use) but deeper integration can involve sharing information between functions and managing how and when information is presented to a driver. As our understanding develops, functional and systematic integration may assist in reducing driver distraction and may have an impact on the driver's awareness of the surrounding situation. This may necessitate revision or extension of existing design guidelines as our experience develops.

In this dynamic situation, where scientific knowledge and consensus concerning HMI are developing, the European approach is one of caution concerning interpretation of quantitative measurements in terms of safety. Many approaches to assessment can be taken, including modeling studies, laboratory, simulators, track and field trials (e.g., to quantify visual glance behavior and distraction), and analysis of vehicle and accident data. There remains a research gap concerning how to combine individual assessment methods into an overall integrated methodology to make predictions about safety in use. An integrated methodology would need to address all of the safety issues with IVIS and ADAS functionality (as shown in Figure 22.4). This would need to include technical safety performance (particularly for inbuilt functions); the impact of driver distraction; an assessment of the safety implications of drivers' understanding (particularly for warning functions); and the impact of partial automation of the driving task under driver supervision (for assistance functions). Beyond these measures, a full evaluation of societal impact would need to take into account penetration rates (possibly by user groups), the impact on other road users, and any changes to the nature of accidents. This is an extremely complex and challenging scenario!

The checklists described here could provide a framework within which different measurements are summarized, but a clear consensus would still be required concerning which measurements were important and how they could be interpreted as a whole. As a final conclusion, a pragmatic approach for both IVIS and ADAS is to make use of existing guidelines during development and then to undertake structured and extensive evaluations culminating in user trials.

REFERENCES

1. Commission of the European Communities, *European Transport Policy for 2010: Time to Decide*, white paper, COM (2001) 0370, December 9, 2001.
2. Commission of the European Communities, Information and communication technologies for safe and intelligent vehicles, COM (2003) 542 Final, September 15, 2003.
3. eSafety: http://ec.europa.eu/information_society/activities/esafety/index_en.htm.
4. Luke, T. and Reed, N., Visual distraction of paper map and electronic system based navigation. Human Factors and Ergonomic Society (HFES) Europe Chapter—Annual Meeting, 2004.
5. Commission of the European Communities, Recommendation on safe and efficient in-vehicle information and communication systems: a European statement of principles on human machine interface. OJ L19, January 25, 2000, p. 64.
6. Report of an independent expert group on the expansion of the principles laid down in the Commission Recommendation of 21 December 1999 on "Safe and efficient in-vehicle information and communication systems" (2000/53/EC), July 2001.
7. eSafety Forum for Directorate-General Information Society of the European Commission. *Recommendations from the eSafety-HMI Working Group – Final Report*, February 2005, available at [3].
8. Humanist: http://www.noehumanist.org/.
9. AIDE: http://www.aide-eu.org/.
10. United Nations Economic and Social Council, Classification and definition of power-driven vehicles and trailers. TRANS/WP.29/Rev.1/Amend.2/Annex 7, April 16, 1999.
11. European Conference of Ministers of Transport, Statement of principles of good practice concerning the ergonomics and safety of in-vehicle information systems, in *New Information Technologies in the Road Transport Sector: Policy Issues, Ergonomics and Safety*, Paris: OECD Publications, 1995, pp. 35–42.
12. British Standards Institute, Guide to in-vehicle information systems. DD235:1996.
13. Stevens, A., Quimby, A., Board, A., Kersloot, T., and Burns, P., Design guidelines for safety of in-vehicle information systems, TRL Report PA3721/01, Berkshire, UK: Transport Research Laboratory, 2002. http://www.trl.co.uk/content/download.asp?pid=76.
14. Stevens, A., Board, A., Allen, P., and Quimby, A., A safety checklist for the assessment of in-vehicle information systems: A user's manual, TRL Project Report PA 3536/99, Berkshire, UK: Transport Research Laboratory, 2000.
15. Noy, Y.I., International Harmonized Research Activities (IHRA) Working Group on ITS: Status Report, *16th International Technical Conference on the Enhanced Safety of Vehicles*, Windsor, Canada, 1998.
16. Stevens, A., Quimby, A., Board, A., and Rai, G., Development of human factors guidelines for in-vehicle information systems. In *Proceedings 9th World Congress on ITS*, Chicago, October 14–17, 2002. (CD-Rom) ITS America www.itsa.org.
17. Lansdown, T.C. and Stevens, A., Driver assistance systems: guidance for designers regarding human capabilities. In *Proceedings International Ergonomics Association 15th Triennial Congress'*, Seoul, Korea, August 24–29, 2003.
18. Schwartz, J., *Code of Practice for the Design and Evaluation of ADAS*. Response 3 Consortium V. 1.6, 2006-08-11. RESPONSE: http://www.prevent-ip.org/en/prevent_subprojects/horizontal_activities/response_3/.

23 North American Approaches to Principles, Codes, Guidelines, and Checklists for In-Vehicle HMI

Peter C. Burns

CONTENTS

23.1 INTRODUCTION

The road-traffic environment has little tolerance for error. Unexpected situations routinely occur, and these become more dangerous if drivers are less aware. It is essential that people concentrate fully on the driving task, and activities that distract attention from driving must be kept at a minimum. There are two broad approaches to minimizing driver distraction focusing on (1) the driver or (2) the equipment. The first type of countermeasure targets the distracted driver's behavior. Possible techniques for changing driver behavior include social marketing, education, and enforcement. The problem with this approach is that it places the responsibility mainly on the drivers. Norman has recently described this as the *blame and train* philosophy; whereby we blame drivers for the problem, punish them appropriately, and insist they get more training.[1] This approach is difficult to justify because distraction does not occur independently; it requires a distracter. Education and enforcement may not be able to extinguish a driver's reflexive diversion of attention to an in-vehicle display that is flashing and beeping. Drivers reasonably assume that any equipment that comes with their vehicle can be used while driving, even if it has potential to be distracting. An emphasis on the driver's responsibility for avoiding distraction does not encourage safe design. Using a poorly designed device that was not optimized for use while driving, can increase crash risk, and the victims of poor design should not be penalized. To minimize distraction, the distracters must be eliminated or redesigned so that they are compatible with driving. Targeting the source is the second broad tactic for minimizing driver distraction, and will be the main focus of this chapter (see also Chapters 22 and 24).

Safety principles, codes of practice, and guidelines are needed to establish desirable design practices for limiting distraction risk. This chapter describes the current status of guideline-based distraction countermeasures in Canada and the United States. The focus is primarily on guidelines for new in-vehicle information and communications systems, although these have many design requirements in common with conventional vehicle displays and controls. The chapter provides a history of the development of these countermeasures, highlights common features, and draws comparisons with efforts elsewhere.

In-vehicle information and communication systems (IVIS) can be separate, dedicated aftermarket systems or integrated, in design and function, and installed as factory-fitted devices. Technical advances are enabling in-vehicle systems to offer more, and increasingly complex, functions. It is no longer possible for product designers to rely on common sense engineering. For example, there could be endless permutations for the design of a menu structure for a navigation system. Decisions on the appearance, categorization, organization, depth, and breadth of the menu items are not obvious. Nor is it obvious which type of controller and control layout should be used. With a lack of human factors practitioners in the automotive industry, there is a strong demand for better guidance to reduce the uncertainty of design decisions.

Designers and engineers need accessible and usable guidelines that are applicable during the early stages of design and expressed in usable terms understood by product engineers (p. 71).[2] A guideline is a practice that is desirable to follow, but not one that must be followed in all cases.[3] Guidelines can be general or very specific. They are not a substitute for regulations and standards (see Chapter 25), which would prevail over guidelines when they are conflicting. There are currently no federal motor vehicle safety regulations in North America directly addressing driver distraction from in-vehicle information and communication systems. There is some provincial, territorial and state legislation restricting the use of certain types of visual display devices in vehicles, most commonly television screens.[4] In Canada, for example, there are nine jurisdictions that prohibit or restrict the use of television screens in vehicles.

Human factors guidance has been available for automotive engineers for some time, but this has not directly targeted the design of advanced in-vehicle information and communication systems. They can refer to specific International Standards Organization (ISO) or Society of Automotive Engineers (SAE) standards, and some companies have internal guideline documents. There are also some relevant military standards (MIL-STD-1472F)[5] and advice is available in human factors reference books such as Sanders and McCormick[6] or Salvendy.[7]

23.2 UMTRI DESIGN GUIDELINES FOR DRIVER INFORMATION SYSTEMS

Green examined the available international interface design guidelines, and two sets of detailed guidelines for in-vehicle devices existed in North America before 2001.[3] The first set of comprehensive North American guidelines was published by the University of Michigan Transportation Research Institute (UMTRI) in 1993 and was titled the *Suggested Human Factors Design Guidelines for Driver Information Systems*.[8] The document includes principles, general guidelines, and specific design criteria with an emphasis on navigation interfaces. The guidelines were organized according to the following topics:

- Primary and secondary design principles (e.g., operations that occur most often or that have the greatest impact on driving safety should be the easiest to perform)
- Guidelines for manual controls (e.g., eliminate the need for manual user input while driving)
- Spoken input, information needs, and dialogue guidelines (e.g., design dialogues that enable users to speak brief commands)
- Guidelines for visual displays (legibility, understandability, organization, content)
- Guidelines for auditory displays (loudness, discriminability of warnings, speech output)
- Navigation guidelines (visual and auditory displays, presentation modality, input)
- Traffic information (e.g., text or graphics used to present traffic information)
- Guidelines for in-vehicle warning systems (e.g., red used for highly critical warnings demanding immediate action by the driver)
- Car phone guidelines (e.g., handsfree dialing preferred over manual dialing)
- Guidelines for system integration (e.g., direct access to features provided where possible)

The authors of the UMTRI guidelines studied the properties of good guidelines and found there is a trade-off between comprehensiveness and ease of finding critical information. Their aim was to provide accessible answers to questions that emerged when in-vehicle interfaces were designed. This established a good model on which to frame subsequent guidelines. No specific human factors expertise was required to apply these guidelines.

23.3 BATTELLE HUMAN FACTORS DESIGN GUIDELINES FOR ADVANCED TRAVELER INFORMATION SYSTEMS

The most substantial guidelines for IVIS in North America is the U.S. Federal Highway Administration's (FHWA) *Human Factors Design Guidelines for Advanced Traveler Information Systems (ATIS) and Commercial Vehicle Operations (CVO)*, published in 1998.[9] This handbook was intended to address the information gap between new automotive technologies and the availability of human factors guidelines. The handbook is available in paper, in interactive electronic versions online, and on compact disk. It is based on the traditional human factors principles that the effectiveness of new in-vehicle technology depends on driver acceptance, its compatibility with the driving task, and its conformity to the physical and cognitive capabilities of the driver. The handbook summarizes human engineering data, guidelines, and principles and the broad target audience is anyone responsible for the conceptualization, development, design, testing, or evaluation of IVIS systems.

The handbook emerged from a 6-year research program, conducted for the FHWA by Battelle, that included extensive literature reviews, empirical analyses, and user requirements analyses. The guidelines are presented in a two-page format that includes an introduction, the guideline itself, a figure demonstrating the guideline or a table of parameters, the supporting rationale, a rating of the level of empirical support for the guideline, cross references, and key references.[10] It provides design guidelines for 75 distinct design parameters, which are concise, unambiguous, and traceable to specific references. They also highlight implications for driver performance where appropriate, and it applies to both private and commercial vehicle drivers. The end result is design guidelines that successfully generalize new and existing research data into design parameters that suit the needs of designers.[10] The guidelines are organized according to the following topics:

1. General guidelines for displays
    - Symbols (contrast, height, uniformity, font, stroke and width-to-height ratios, spacing, color)
    - Visual displays (color coding and contrast, sensory modality)
    - Auditory message length (e.g., messages that require an urgent action should be a single word or a short sentence with the fewest number of syllables possible)
    - Complexity of information (e.g., text messages presented when the vehicle is in motion should be no longer than 4 information units, to minimize the eyes-off-road time)
    - Symbols, text, and message styles (e.g., notification rather than command style messages should be used for presenting low-criticality information)
    - Head-up displays (e.g., provide a luminance adjustment control for the image)
    - Individual preferences, special needs (e.g., drivers should be able to have the system repeat messages)
    - User interface design (e.g., objects, actions, and options should be visible to the user)

    - Alerts (e.g., high-priority visual information should be preceded by an auditory alerting tone)
2. General guidelines for controls
    - Control type (e.g., continuous rotary knob or thumbwheel where precise adjustment is needed)
    - Control coding (e.g., proper coding of controls such as location, shape, and size to help driver locate them more quickly and accurately)
    - Movement and compatibility (e.g., control movements that correspond to the expectations of the user—ON is up, right forward, or push)
    - Keyboards (e.g., variable-function keyboards when several subsets of functions are frequently used)
    - Speech input (e.g., give immediate feedback of recognized speech)
3. Routing and navigation guidelines (e.g., routes should be clearly distinguishable from other map features)
4. Motorist service guidelines (e.g., services/attractions information should only be presented if requested by the driver)
5. Route safety warning guidelines (e.g., distance to congestion should be presented with an alerting tone, then speech)
6. Augmented signage information guidelines (e.g., messages should allow sufficient time to interpret and make an appropriate response)
7. Guidelines for commercial vehicle related systems (e.g., critical tasks should not require more than three control actions, such as button presses, while the vehicle is in motion)
8. Tools designed for sensory allocation

23.4 GUIDELINES FOR IN-VEHICLE DISPLAY ICONS

Another more recent document was created to provide designers of in-vehicle technologies with a set of human factors guidelines specifically for designing in-vehicle icons.[11] Icons are visual representations or images that symbolize an object, action, or concept. This report addresses issues such as the legibility, recognition, interpretation, and evaluation of graphical and text-based icons and symbols. For example, text labels of two to three words should accompany icons that have no conventional or broadly understood meaning. It follows a two-page structure similar to the ATIS guidelines and aims to present them in a clear, concise, and user-centered format. This includes guidance and procedures for evaluating icons and a design tool to help determine the most appropriate display modality for presenting in-vehicle information. Although icon legibility, recognition, and interpretation are relevant features for driver distraction, they are more critical concerns for designing effective warning systems.

23.5 ALLIANCE GUIDELINES

In 2000, the National Highway Traffic Safety Administration (NHTSA) sponsored several events focusing on driver distraction. There was a Driver Distraction Internet Forum, which was held over a 5-week period during the summer, two expert workshops and a public meeting. The Internet forum was set up as a virtual conference with technical papers and discussion topics for experts and the public.[12] Although guidelines for the design and evaluation of technology were not discussed in detail on the Internet forum, there were several papers on the site reviewing existing practices and guidance.

At the workshops, separate groups of experts met to discuss five key topics. One of these meetings concerned human factors guidelines to aid in equipment design. The meeting report stated, "Designers and engineers need accessible and usable guidelines for creating and evaluating interfaces that are compatible with safe driving. Guidance should be applicable during the early stages of design to prevent costly reengineering once a product is brought to market, and should be expressed in terms useful for product design engineers. The aim is to produce systems that are usable and safe by establishing rigorous design protocols to ensure that in-vehicle systems do not pose safety risks to drivers" (p. 40).[2]

The expert group discussed existing guidelines and their application. The experts said that the few relevant existing guidelines tended to be vague and lacked formal evaluation procedures. There was a need to ensure that current guidelines are effectively applied to interface design. The experts were concerned that guidelines failed to account for the specifics of every environment and context of interface use. They also wanted guidelines to be more accessible to their users and sensitive to product differentiation needs. The experts identified research priorities to support the development of guidelines for benchmarking, measuring distraction, evaluation methods, interruptible tasks, speech recognition, system integration, adaptive interfaces, and improvements to infrastructure and data collection.

NHTSA also invited representatives of the public, industry, government, and safety groups to a public meeting on July 18, 2000 to share viewpoints, information, and recommendations regarding strategies to limit driver distraction. NHTSA challenged stakeholders to find a solution to the rising concern in this area. At this meeting, the Alliance of Automobile Manufacturers (Alliance) offered to develop a code of practice for telematics devices. The Alliance, at the time of the meeting, was a trade association of 13 automobile manufacturers that sell vehicles in the United States. They indicated that vehicle manufacturers already used internal guidelines to improve their vehicles, and that it may be possible to develop some best practices based on these guidelines.[13] They proposed that voluntary industry design guidelines would assure that the best requirements would be put in place in the shortest possible time. Preliminary guidelines were released later that year, but this first draft was based mainly on the 1999 European Statement of Principles (ESoP) (see Chapter 22) rather than best practices among the Alliance members. These were intended as a starting point for more complete design and performance requirements.

In early 2001, the Alliance established the Driver Focus-Telematics Working Group, which was comprised of representatives from the major domestic and foreign automobile manufacturers. Experts from NHTSA, Transport Canada (TC), and other stakeholder groups were only invited to attend some of this group's early meetings as observers. The group produced a draft document in April 2002 called the *Statement of Principles, Criteria and Verification Procedures on Driver Interactions with Advanced In-Vehicle Information and Communication Systems*.[14]

This version of the Alliance guidelines contained 24 principles, 11 of which were provided with measurement and performance criteria. The stated objective was for systems to be designed "to minimize adverse effects on driving safety; to enable the driver to maintain sufficient attention to the driving situation while using the system; and to minimize driver distraction and not to visually entertain the driver while driving" (p. 8).

The document applies to advanced information and communication systems (e.g., navigation systems) and the visual–manual interaction while driving. It does not apply to speech-user interfaces, advanced driver assistance systems (ADAS; e.g., lane departure warning), or conventional systems such as radio or climate controls, unless they share the same controls and displays as advanced functions. Technical knowledge of the products under evaluation is required to apply these guidelines. Alliance members voluntarily committed to design production vehicles in accordance with these principles. The guidelines were organized according to the following topics:

1. Installation principles
   - Regulations, standards, and manuals (e.g., systems should be located and fitted in accordance with relevant regulations)
   - Field-of-view and access to vehicle controls (e.g., no part of the system should obstruct the driver's field-of-view)
   - Location and glare (e.g., reduce or minimize glare and reflections)
2. Information presentation principles
   - Amount of interaction (e.g., visual displays should be designed such that the driver can complete the desired task with sequential glances that are brief enough not to adversely affect driving)
   - International display standards (e.g., standards for legibility, icons, symbols, words, acronyms, or abbreviations should be used)
   - Accuracy, timing and controllability (e.g., information relevant to the driving task should be timely and accurate)
3. Principles for interactions with displays and controls
   - One-handed or handsfree interaction (e.g., should allow the driver to leave at least one hand on the steering control)
   - Task interruption and pacing (e.g., drivers should be able to control the pace of interaction with the system)
   - Feedback (e.g., system's response following driver input should be timely and clearly perceptible)
4. System behavior principles
   - Lock out of dynamic information (e.g., distracting visual information not related to driving should be disabled while the vehicle is in motion)
   - Malfunctions (e.g., safety-relevant status information should be presented to the driver)
5. Principles for system instructions and manuals (e.g., adequate instructions for safe use should be provided)

NHTSA and TC raised some concerns with the Alliance document.[15] They were concerned that the draft guidelines allowed unduly demanding tasks to be carried out by drivers while driving. There was particular concern with the principle that set limits on visual distraction. These limits do not restrict some in-vehicle tasks that require exceptionally long glances to complete (i.e., a certain number of glances lasting longer than 2 s are acceptable). Another significant issue was the radio tuning reference task proposed by the Alliance. A reference task is a secondary task that results in a level of driving performance, which is used for comparison purposes. The chosen reference task was considered to be exceedingly difficult and unlike real radio tuning. At that stage, TC did not endorse the draft Alliance document because there were some concerns that it might provide false comfort to the public that the hazards of driver distraction were being adequately addressed by industry.

The Alliance guidelines were viewed as a work in progress that would be refined and updated as more scientific support became available. The assumption that manufacturers will follow a recognized product development process was later added to the Alliance guidelines. The last update of the guidelines was Version 2.1, released in June 2006. A final version has not yet been released.

23.6 GUIDELINES IN COMPARISON

Four guideline documents have been described in this chapter. The UMTRI document was the first comprehensive collection of guidelines in North America and it places an emphasis on navigation interfaces.[3] The ATIS guidelines from Battelle are larger and provide physical ergonomics guidelines; and commercial vehicle systems are included in the scope. The UMTRI and two Battelle documents have a broad scope and do not strictly focus on driver distraction. They tend to deal with the traditional human factors priorities, which are, nonetheless, closely related to driver distraction. These guidelines support the design of human-machine interfaces (HMI) that are compatible with the driving task and meet the physical and cognitive capabilities of the driver. Task and interface compatibility and suitability are crucial for limiting driver distraction. Unlike the more general UMTRI and Battelle guidelines, the Alliance principles were developed for the specific purpose of limiting driver distraction. Both the UMTRI and the Battelle documents place an emphasis on providing clear, accessible, and objective information with traceable references. This may have influenced how the Alliance Driver Focus-Telematics Working Group evolved their guidelines from the European principles on which they were based.

Even with the many available guidelines in a clear and accessible format, evidence suggests that they are not getting to the right people, being understood, or being applied effectively.[2] It is clearly evident from the current fleet of vehicles in the field that many textbook human factors principles are being ignored. A human factors expert will find HMI problems with most vehicles, particularly those with advanced information and communication systems. A mechanism is needed to give more impetus to the application of these guidelines. These are discussed in the next section.

23.7 CANADIAN ACTIVITIES

In 2003, TC consulted industry stakeholders and the Canadian public regarding the issue of driver distraction from in-vehicle telematics devices. The main goal of the consultations was to solicit feedback regarding various potential initiatives that would limit driver distraction from in-vehicle telematics devices. TC invited industry, the provinces and territories, road safety interest groups, and the public to comment on potential initiatives and to provide feedback on alternative approaches for reducing driver distraction. Consultations began with the publication of a discussion document on driver distraction from in-vehicle telematics.[15] The discussion document defined the problem, reviewed research, and outlined possible regulatory and nonregulatory countermeasures. It reported that limited scientific understanding exists for the objective and accurate evaluation of driver distraction, that few of the available standards and guidelines attempt to set out performance-based requirements, and that compliance with them is voluntary. The available guidelines and recommendations are not satisfactory because they are underspecified, incomplete, and unverifiable. However, they do offer some guidance to designers, or evaluators of telematics devices, and give direction for some initiatives to limit driver distraction.

Nonregulatory options in the discussion document include public awareness initiatives and a Memorandum of Understanding (MOU) or advisory between government and industry concerning appropriate design guidelines and design processes to be implemented by manufacturers. Safety standards can be design-based, performance-based, or process-oriented in their approach.[15] Design standards provide precise specifications for a vehicle or vehicle system in terms of, for example, physical attributes or geometry. Because they are design restrictive, their use is limited to instances when compatibility or consistency is crucial, for example, dimensional standards to ensure the proper fit of replacement tires and rims. Performance-based standards, as they apply to motor vehicles, set out the minimum level of performance that a vehicle or its components and equipment must meet when tested in accordance with the prescribed test method. The advantage of a performance-based standard is that it provides an objective basis for evaluating the safety of a product. Because this type of standard does not specify precise physical attributes, it allows design flexibility and, therefore, does not hinder innovation. However, performance-based standards rely on the existence of reliable and valid test procedures and criteria. Efforts to develop performance-based requirements to limit the potential for driver distraction are ongoing, for example limits on the amount of visual attention needed to perform an in-vehicle task. In contrast to design and performance-based standards, a process-oriented safety standard does not set out requirements that apply to the end product, but rather it outlines the general principles and procedures that should frame the product's design, development, evaluation, manufacture, and installation. This type of standard is concerned with the systems and procedures that a manufacturer should establish and follow during its development and implementation cycle to ensure that its products reflect best practice and minimize potential risk and likely misuse. Like their performance-based counterparts, process-oriented standards allow flexibility in product design and do not impede innovation. With respect to limiting distraction, a systematic and comprehensive process would be established to prioritize safety and human factors considerations during device design and development.

The Canadian discussion document on driver distraction from in-vehicle telematics presented several regulatory countermeasures. These included standards that limit the access of drivers to certain device functions, impose limits on the amount of visual distraction, or prohibit certain features of telematics devices (e.g., open architectures) that would allow the use of untested, aftermarket applications.

The responses to these consultations indicated that a government-industry MOU, including guidelines and human factors process requirements, was the preferred option to limit driver distraction from in-vehicle telematics devices.[16] Both industry and public groups also expressed strong support for public awareness and education initiatives related to distracted driving in general, including that caused by in-vehicle telematics. Finally, more objective, carefully designed scientific research into the issue was recommended, especially that which assesses the impact of telematics device use on collision frequency.

A government-industry working group was established in Canada in 2004 to negotiate the terms of an agreement on the safety of in-vehicle telematics devices. The group consisted of representatives from TC, the Canadian Vehicle Manufacturer's Association (CVMA), and the Association of International Automobile Manufacturers of Canada (AIAMC). Most members of these Canadian industry associations have connections to the Alliance. The agreement would include commitments by industry to recognize a safety design and development process, and adhere to industry-developed performance guidelines in telematics device design and development (i.e., the Alliance statement of principles). No agreement could be reached between the associations and Transport Canada on the terms of an MOU. The main reasons were that Transport Canada was not prepared to endorse an incomplete and unseen statement of principles document, and the manufacturers would not commit to following a safety management systems process when designing telematics devices.

TC contracted an evaluation of the draft Alliance guidelines to assess their validity and reliability.[17] Two approaches were taken to this evaluation conducted by Humansystems Inc. In the first approach, experts critically reviewed the Alliance guidelines, and commented on and gave ratings of their utility. In the second approach, four 2005 model vehicles equipped with OEM navigation devices were examined for their compliance with the guidelines by a group of inspectors. The four vehicles complied with most of the guidelines, but none was fully compliant. The guidelines were found to be valid but insufficiently detailed in places. This led to a lack of combatability between the inspectors' ratings. Suggestions were given for further development, and there was a particular need for more detail among the elaborations supporting the principles was identified. It was believed that, with the appropriate revisions and test alternatives, the Alliance guidelines would help to produce safer telematics devices.

The results of this work were presented to the Alliance Driver Focus-Telematics Working Group, but it is uncertain whether the feedback will influence the development of their principles. The group is critical of the Humansystems Inc. evaluation because Alliance automotive engineers, the target user for these principles, did not perform the tests. This is an irrational critique because the Alliance document is largely based on the ESoP, which was not limited to that specific experience or association. There is hope of extending these same requirements to the manufacturers of aftermarket and portable consumer electronics devices. This would not be possible for the manufacturers of aftermarket and portable consumer electronics devices to do if only Alliance engineers can apply the principles. Furthermore, the requirements would be impossible to validate or enforce because impartial third parties could not perform compliance testing.

An International Conference on Distracted Driving was held in Toronto in 2005.[18] A number of supportive recommendations emerged from this event. It was recommended that industry guidelines should continue to be developed and refined, and that manufacturers should follow this code of practice in their product development. Later, the Canadian Council of Motor Transport Administrators (CCMTA), an organization with representatives from the provincial, territorial, and federal governments, published a strategy document for reducing distracted driving.[19] The document also recommends that, as one of the strategies, best practice guidelines for manufacturers be developed and that manufacturers should be encouraged to follow these guidelines.

23.8 COMPARISON WITH EUROPEAN STATEMENT OF PRINCIPLES

The emphasis that the previous North American guidelines[8,9] placed on concise, unambiguous, and traceable references may have influenced the Alliance Driver Focus Task Force's effort to develop their guidelines. The Alliance requirements resulted in a better standard of information than is given in the ESoP.[20]

Distraction mitigation was a central concern of the original ESoP on the Design of HMI.[21] At one point, European experts working on the further specification of the principles considered a visual task demand criterion of four glances off the road for not longer than 2 s.[21] The ESoP draft that emerged in 2005[22] was far less demanding, mainly descriptive, and lacked suitable test procedures and safety criteria. The result is that the ESoP offers much looser guidance than the Alliance document and will likely prove less effective in limiting driver distraction.

Without a clear test for compliance, designers must rely on their own interpretation of the principles. The problem is that these interpretations will vary considerably, and they may not stand up to challenges from competing vehicle design priorities. These priorities include safety, usability, comfort, cost, and styling. For example, visual displays compete with the many other features in a vehicle for the limited space in front of the driver. The dashboard must accommodate an essential set of controls (e.g., steering wheel, instrument stalks), safety equipment (e.g., window defrost vents, airbags, and padding), and comfort and convenience systems (e.g., audio and climate systems). ESoP Principle 2.4 states: *Visual displays should be positioned as close as practicable to the driver's normal line of sight* (p. 14). This alone is not very persuasive against the intense competition for dashboard real estate and the safety hazards of poor display location. The equivalent principle from the Alliance (Principle 1.4) has a precise criterion for the downward viewing angle of the display, and it provides procedures for determining this value. Although it could be more stringent, the result is an objective and quantifiable requirement for a more prominent display location that should prevail over less specific priorities such as interior styling or the placement of vents.

The Alliance guidelines should also lead to safer devices because, unlike the ESoP, they try to account for the total amount of interaction with an information and communication system. Clearly the goal of minimizing overall interaction time is a critical safety issue in the design of HMI for in-vehicle use. Shorter tasks with fewer interruptions while driving are inherently less distracting than longer tasks and many interruptions. Limiting exposure to risk is central to any effective safety strategy and failing to account for system interaction times is a serious oversight with inevitably fatal consequences. The four vehicles that TC evaluated had tasks that failed to comply with the Alliance guidelines.[17] They also would have failed the requirements established by the Japan Automobile Manufacturers Association *Guideline for In-vehicle Display Systems* (see Chapter 24). Yet even the most distracting tasks would have passed the latest ESoP requirements.[22]

On the positive side, Europe appears to have had more success in extending their principles to aftermarket and portable devices. Aftermarket devices have greater constraints on size and their stand-alone design necessitates an HMI that is not thoroughly integrated with the vehicle. Unlike some original equipment and aftermarket devices, portable consumer electronics devices are not specifically designed to be suitable for use while driving. With portable devices, the priorities are to be small, lightweight, and energy-efficient, and these features tend to make them more distracting to use while driving. Consumer electronics manufacturers seem to place greater emphasis on interface usability than the original equipment manufacturers. The interface is a more prominent feature on consumer electronics devices, so ease-of-use and acceptability are more likely to influence purchase decisions in a rapidly changing and highly competitive market with shorter product life cycles. Telematics devices are a secondary feature of vehicles and consumers may not even notice them during their test drive. It may also be easier to justify the need for good usability because the relative cost of usability testing is reduced across the higher unit sales volume associated with consumer electronics devices.

23.9 SUMMARY AND CONCLUSIONS

This chapter described the current status of guideline-based distraction countermeasures in Canada and the United States. It reviewed the development of these countermeasures, highlighted common features, and made comparisons with efforts elsewhere. Numerous standards and guidelines are available; however, designers still lack the necessary information to support the design of safe and usable devices.[3] Telematics have improved significantly in the past 10 years, but an astonishing gap still exists between good and poor designs. It is clearly evident from the current fleet of vehicles in the field that many textbook human factors principles are not being applied. This gap emphasizes the need to have standard and stringent requirements for the safety and usability of in-vehicle systems; system designers are either unaware of the requirements or they are failing to apply them appropriately.

There is strong pressure to differentiate products from others on the market and to creatively partition them by trim levels and equipment packages. This emphasis on features and styling appears to have trumped any requirements for usability or concerns about distraction. It is surprising how some manufacturers fail to recognize the importance of usability in the design of automotive products. There are cases of groupthink in which a manufacturer fixated on an interface design and largely ignored feedback on their vehicle interface, even after uniformly negative reviews. The economics of usability engineering is established, yet some believe that user-friendly products are too costly to produce. Pride may make designers unreceptive to restrictive guidelines and the harsh realities of user feedback. Human factors guidelines, safety testing, and customer input are crucial to improving the usability and safety of prototypes and the evolution of products. Safety and usability are priorities for consumers and they must also be a priority for manufacturers.

It is essential to have human factors design guidelines for in-vehicle information and communication systems, but the guidelines by themselves are insufficient. A mechanism is needed to ensure that the designers are aware of these guidelines and procedures, and have the resources and skills to apply them effectively. A mechanism is also needed within the product development process so that the risks of driver distraction are routinely and systematically considered during product design, development, and testing. Lastly, a mechanism is needed to prioritize and implement more effective usability testing and engineering in the automotive industry.

The most appropriate mechanism is a formal requirement for manufacturers to follow a safety management system, a systematic process that prioritizes safety and human factors throughout the design cycle. Vehicle manufacturers have recognized that a process-based code of practice is an essential part of developing safe driver assistance systems and avoiding liability concerns.[23] Manufacturers also need to recognize that a similar process is needed to address the problem of driver distraction from in-vehicle information and communication systems. This process should involve the analyses of user needs and risk, setting safety and usability objectives, and conducting evaluations. These activities should be assigned to qualified personnel with clearly defined roles and responsibilities, including process oversight. Finally, device performance should be monitored in the field and this information should be used to set future design targets.

REFERENCES

1. Norman, D., *The Design of Future Things*, Basic Books, New York, NY, 2007, p.12.
2. Driver Distraction Expert Working Group Meetings: Summary & Proceedings, Washington, D.C., NHTSA, November 10, 2000.
3. Green, P., Synopsis of Driver Interface Standards and Guidelines for Telematics as of Mid-2001 (Technical Report UMTRI-2001-23), Ann Arbor, MI, The University of Michigan Transportation Research Institute, 2001.
4. Wilson, J., Legislation and regulation in Canada with respect to driver distraction. Paper presented at the International Conference on Distracted Driving, Toronto, Canada, October 2–5, 2005.
5. United States Department of Defense, Department of Defense Design Criteria Standard Human Engineering (Military Standard MIL-STD 1472F), Washington, D.C.: U.S. Department of Defense, 1999.
6. Sanders, M.S. and McCormick, E.J., *Human Factors in Engineering and Design* (7th edition), McGraw-Hill, New York, NY, 1993.
7. Salvendy, G., *Handbook of Human Factors*, Wiley, New York, NY, 1987.
8. Green, P., Levison, W., Paelke, G., and Serafin, C., Preliminary Human Factors Guidelines for Driver Information Systems (Technical Report UMTRI-93-21 and FHWA-RD-94-087). The University of Michigan Transportation Research Institute, Ann Arbor, MI, 1993.

9. Campbell, J.L., Carney, C., and Kantowitz, B.H., Human Factors Design Guidelines for Advanced Traveler Information Systems (ATIS) and Commercial Vehicle Operations (CVO). (USDOT FHWA report no. FHWA-RD-98-057), www.fhwa.dot.gov/tfhrc/safety/pubs/atis, 1998.
10. Monk, C.A. and Moyer, J., The Customer-Driven Development of Human Factors Design Guidelines. Public Roads, United States Department of Transportation – Federal Highway Administration, January/February 2000, Vol. 63, No. 4, 2000.
11. Campbell, J.L., Richman, J.B., Carney, C., and Lee, J.D., In-Vehicle Display Icons and Other Information Elements, Volume I: Guidelines, September 2004, FHWA-RD-03-065, www.tfhrc.gov/safety/pubs/03065/index.htm, 2004.
12. Llaneras, Robert E., NHTSA Driver Distraction Internet Forum: Summary & Proceedings, Draft Report Submitted by Westat, July 5–August 11, 2000.
13. Public Meeting on the Safety Implications of Driver Distraction when Using In-Vehicle Technologies, Transcript from the Public Meeting. US DOT, NHTSA. Washington, D.C., July 18, 2000.
14. The Alliance of Automobile Manufacturers, Statement of Principles on Human Machine Interface (HMI) for In-Vehicle Information and Communication Systems, 2002, www.autoalliance.org.
15. Burns, P.C., Strategies for Reducing Driver Distraction from In-Vehicle Telematics Devices: A Discussion Document, Transport Canada (TP 14133 E), June 2003, www.tc.gc.ca/roadsafety/tp/tp14133/menu.htm.
16. Rudin-Brown, C., Strategies for Reducing Driver Distraction from In-Vehicle Telematics Devices: Report on Industry and Public Consultations (TP14409 E), June 2005, www.tc.gc.ca/roadsafety/tp/tp14409/menu.htm.
17. Go, E., Morton, A., Famewo, J., and Angel, H., Final Report: Evaluation of Industry Safety Principles for In-Vehicle Information and Communication Systems, Report prepared for Transport Canada, Contract No. T8080-040240, Humansystems Inc., 2006.
18. Hedlund, J., Simpson, H., and Mayhew, D., International Conference on Distracted Driving: Summary of Proceedings and Recommendations (April 2006), 2006.
19. Strategy on Distracted Driving: A Component of the Strategy to Reduce Impaired Driving (STRID). The Canadian Council of Motor Transport Administrators, June 16, 2006, www.ccmta.ca/english/pdf/strid_distraction_strategy.pdf.
20. EEC, Commission Recommendation of 21 December 1999 on safe and efficient in-vehicle information and communication systems: A European statement of principles on human machine interface, Official Journal of the European Communities L 19/64, 25.1.2000, European Commission DGXIII, Brussels, 1999.
21. Janssen, W., Driver Distraction in the European Statement of Principles on In-Vehicle HMI: A Comment, NHTSA Driver Distraction Internet Forum, 2000.
22. ESoP, European Statement of Principles on the Design of Human Machine Interaction (ESoP, 2005): Draft, European Commission Information Society and Media Directorate-General – G4 ICT for Transport, 2005.
23. RESPONSE 3, Code of Practice for the Design and Evaluation of ADAS, V 2.0. European Integrated Project, Preventive and Active Safety Applications, Contract number FP6-507075, 2006.

24 Japanese Approaches to Principles, Codes, Guidelines, and Checklists for In-Vehicle HMI

Motoyuki Akamatsu

CONTENTS

24.1 INTRODUCTION: A BRIEF HISTORY OF THE DEVELOPMENT OF CAR NAVIGATION AND OTHER IN-VEHICLE SYSTEMS

Driving a car involves several parallel or time-shared tasks: stabilization, maneuvering, and route following. *Stabilization* is the basic task of controlling a vehicle, accelerating, decelerating, and so on, to make it run smoothly. *Maneuvering* involves keeping the vehicle in the road zone and avoiding collisions with other traffic. While operating a vehicle, drivers also perform a *route-following* task to get to their destination. This strategic task involves identifying intersections and selecting the appropriate branch for reaching the destination.

The road map has always been an important tool for the route-following task. The road map was already in widespread use for bicycle journeys when the automobile was invented at the end of the nineteenth century. To read a map, the driver must identify the vehicle's current position on the map. The landscape and the direction of the vehicle are cues for this location identification. Usually a driver stops the car to get these cues and read the map (Figure 24.1). Reading a map is an important skill for a driver. A typical tool for assisting with map reading is a magnetic compass used to estimate direction. An in-vehicle magnetic compass was built into the Ford Edsel (1958), as seen in Figure 24.2.

At present the most common in-vehicle device that is not relevant to the driving task is an entertainment device such as a car stereo system. The first in-car entertainment device was the car radio. Radio broadcasting started in 1922 in the United States. Radios were introduced in vehicles in the 1920s, and by the 1930s some models came with car radios installed (Figure 24.3). The driver had to turn a knob for channel tuning in early car radios. In the 1940s, button selectors for tuning were introduced to make tuning easier (Figure 24.4). A car stereo using four-track or an eight-track tape was invented in the early 1960s and became popular in the

FIGURE 24.1 *La Route*: Painting in watercolor by Louis Sabattier (ca. 1905). (The original painting is collection of Mr. Michel Legrand. With permission.)

FIGURE 24.2 Instrument panel of Ford Edsel (1958).

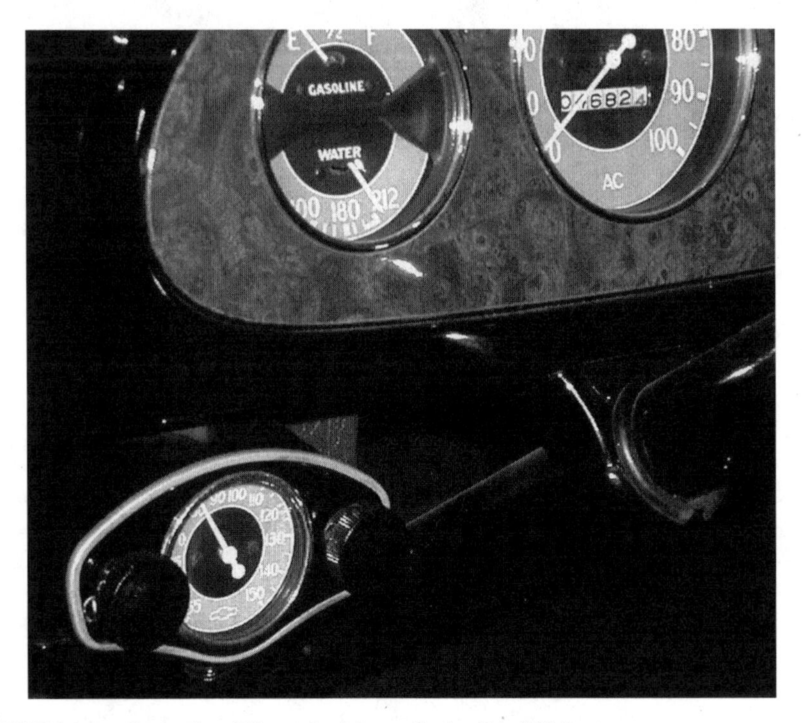

FIGURE 24.3 Car radio of Chevrolet Master Series DA (1934).

FIGURE 24.4 Car radio of Chrysler Newport (1941).

late 1960s in the United States and Japan. The music source transitioned to compact cassette tape in the 1970s and then to compact disks in the 1980s. Car entertainment devices have since become high-tech devices with multiple functions, and there are many small buttons on the control panel.

The development of modern in-vehicle navigation devices began around 1970.[1] One of the earliest navigation systems is the Automatic Route Control System (ARCS), developed by Robert L. French (Figure 24.5).[2] The current position of the vehicle was calculated from a speed sensor and a direction sensor. The system was developed for field tests in a newspaper delivery truck in 1971, but it was not made commercially available. Since the 1970s, the technology of electrical devices has been rapidly developing in the automotive engineering field. Honda developed a navigation system called Gyrocator in 1981. The movement of the vehicle was sensed by a gas rate sensor. The current position was indicated through a map printed on transparent film (Figure 24.6). In the same year, the in-vehicle navigation system for passenger cars was delivered by Toyota. The system, called NAVICOM, used a geomagnetic sensor to detect the vehicle's position.[3] NAVICOM indicated the direction and the remaining distance after the initial direction and distance to the destination were put into the system (Figure 24.7). Since then, there has been further development of sensing devices and digital maps. One practical navigation system was the Electro Multi

FIGURE 24.5 Automatic Route Control System installed in a newspaper delivery vehicle. (Photograph provided courtesy of Robert L. French.)

FIGURE 24.6 Honda Gyrocator. (Photograph is provided by Honda R&D Co. Ltd. Automobile R&D Center.)

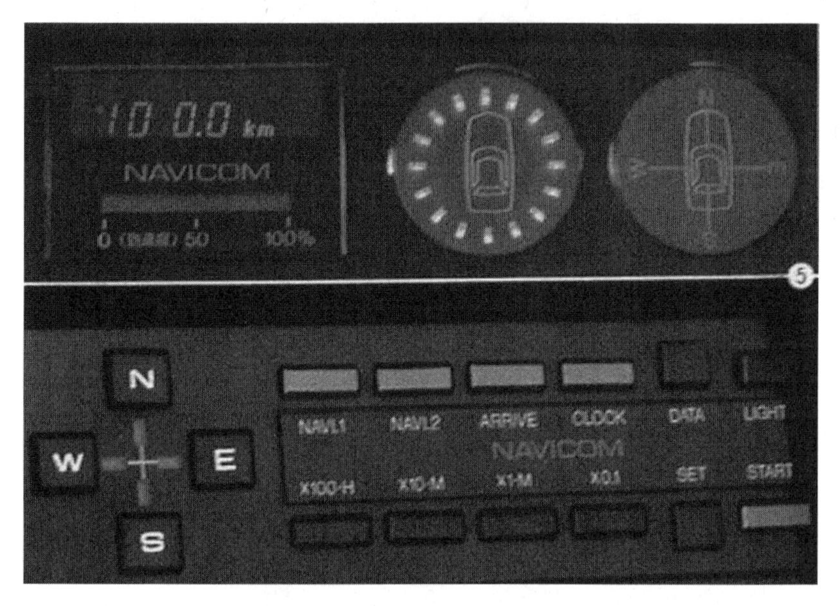

FIGURE 24.7 Control panel of Toyota NAVICOM. (From catalog of Toyota Celica XX 1972.)

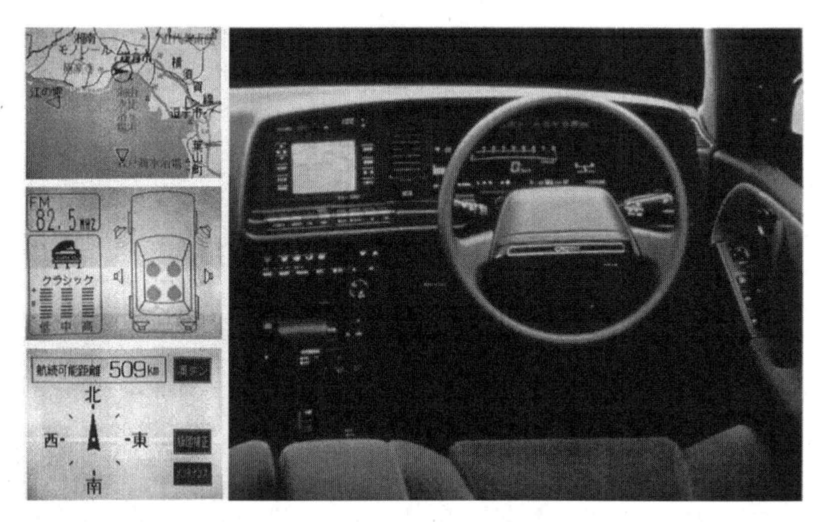

FIGURE 24.8 Toyota Electro multivision system. (From Arai, H., Ito, H., and Azuma, S., *The Journal of The Institute of Television Engineering of Japan*, 44, 507, 1990. With permission.)

Vision system developed by Toyota (1982) (Figure 24.8). It used a digital map stored on a CD-ROM. The Electro Multi Vision System could also provide various other information, including vehicle conditions, a maintenance guide, and a TV image. A navigation system using GPS and map-matching technology to identify the position of the vehicle was introduced into the market in 1990. After GPS became officially available for free civilian use in 1993, navigation systems quickly became common in the market. Annual shipments of navigation systems into Japan exceeded 1 million units in 1998 and reached about 4.5 million units in 2007 as per statistics report from Japan Electronics and Information Technology Industries Association (JEITA).

24.2 CHRONOLOGICAL REVIEW OF CODES AND GUIDELINES IN JAPAN

24.2.1 Background

Before the widespread use of car navigation systems, the authors conducted a series of experiments to examine how drivers would use a navigation system. It was found that drivers looked at the navigation display often, especially before or after the intersection to be turned at, to identify whether or not they had turned at the target intersection.[4] An in-vehicle display has a high potential for displaying various information to the driver, such as navigation information, traffic information, vehicle condition information, and entertainment. If all available information is provided to the driver, the driver has to look at the display all of the time to get all the information. It was not difficult to imagine that presenting all the information to the driver might lead to driver distraction. However, we did not have any data regarding how much information could be given to the driver before it affected driving performance. Therefore, the Japanese car industry created its own, voluntary guidelines for designing the presentation of information on an in-vehicle display and for the operation of the system, the Japan Automobile Manufacturer's Association (JAMA) Guideline.[7] The first version was produced in 1990, before the widespread use of navigation systems (Table 24.1). In fact, the annual shipment of the navigation system in 1989 was about 50,000; in 1990 it was about 100,000 units (statistics report from JEITA). It was believed that the JAMA Guidelines would promote the development of technology and the market introduction of navigation systems.

Almost in parallel with the development of car navigation systems, the cellular phone has become common in Japanese life. The automobile telephone was introduced in Japan in 1979, but at that time it was for backseat passengers, not for the driver. The first mobile phone for personal use, called the "shoulder phone," because it was carried like a shoulder bag due to its size, was made available in 1985. Since the digital cellular phone system became available in 1993, it has grown rapidly and is widely used, even by drivers. The cellular phone ownership rate per household was 10% in 1995, and it exceeded 60% in 1999 (Information and Communication in Japan 2005 [white paper]). The number of injury-causing traffic crashes in which the use of a cellular phone was a factor was 2648 in 1998 (1999 National Police Agency [NPA] report). Drivers' use of cellular phones and TVs while driving became a major concern for the NPA, which is responsible for traffic safety. The NPA revised the law relating to cellular phone use in 1999 to deal with this problem (Table 24.1).

24.2.2 JAMA Guideline Version 1.0 (1990)

The first version of the JAMA guideline was delivered in 1990. In this version, detailed information and moving images were prohibited from being presented on the display while the vehicle was in motion because it took time to read and might

TABLE 24.1
Timeline of Navigation System Development in Japan

Year	Navigation Technology in Market	Road Traffic Law	Guideline
1981	NAVICOM (Toyota) Gyrocator (Honda)		
1986	Navigation System Researchers' Association		
1987	Electro multivision (Toyota) Digital map in CD-ROM		
1988	Japan Digital Road Map Association		
1990	GPS navigation system (Mitsubishi Electric/Mazda, Pioneer Corporation)		JAMA guideline version 1.0: prohibition of detailed information and display monitor operation
1991	TFT-LCD display, route guidance, touch panel		
1992	Voice guidance		
1995			JAMA guideline version 1.1: addressing VICS information
1996	VICS		
1998	Telematics: MONET (Toyota), InterNavi (Honda), COMPASSLINK (Nissan)		
1999	Digital map in DVD	Road Traffic Law, Article 71: prohibiting use of cellular phones and gaze at video monitor	JAMA guideline version 2.0: allowing presenting minor roads when on the road
2002		Road Traffic Law, Article 109: principles of presentation, operation, and presenting information	JAMA guideline version 2.1: viewing angle of monitor is less than 30°
2003	Digital map in HD		
2004		Road Traffic Law, Article 71 revised: punitive clause for use of cellular phone becomes strict	JAMA guideline version 3.0: TGT is to be less than 8 s. TSOT shall not exceed 7.5 s.

capture the driver's attention. Navigation system operational tasks that require long or multiple operations were also prohibited. The following are specific items that were prohibited in this initial guideline.[5]

Display information prohibited while the vehicle is in motion:

- Minor roads in cities shall not be displayed on the navigation map screen.
- The navigation system shall not scroll maps according to the vehicle speed.
- TV pictures and other video images shall not be displayed on the screen.
- Addresses and telephone numbers shall not be displayed as guidance information.
- Descriptive information for hotels and restaurants shall not be displayed.

Display monitoring operations prohibited while the vehicle is in motion:

- Setting or revising destination using cursor switch
- Manual map scrolling
- Map search by area name or point of interest
- Cellular phone keypad operation
- Data input such as memoranda and addresses

(Above list is quoted with permission from Ref. 5.)

24.2.3 JAMA Guideline Version 1.1 (1995)

Version 1.1 was published in 1995, just before the operational launch of The Vehicle Information and Communication System (VICS), which provided road traffic information such as congested areas via an FM multiplex-broadcasting system and roadside beacons.[6] Because VICS information is time-variable, it is referred to as *dynamic information*. To cover the functions of VICS, the 1995 version of the JAMA Guidelines added the following prohibitions.[5]

Display Information prohibited while the vehicle is in motion:

- Dynamic information on the display exceeding 30 characters
- Automatic scrolling of dynamic information

Display monitoring operations prohibited while the vehicle is in motion:

- Manual scrolling of dynamic information
- Display area selection for dynamic information

(Above list is quoted with permission from Ref. 5.)

24.2.4 Revised JAMA Guideline Version 2.0 (1999)

The presentation of minor roads had been prohibited in version 1.0, but version 2.0 allowed it if the vehicle was on that road to reduce the stress of the drivers while using the navigation system.

Display information allowed while the vehicle is in motion:

- If the scale of the navigation map is 1:20,000 or more detailed, minor roads may be displayed when the vehicle is running on that road.

- If the scale of the navigation map is 1:5000 or more detailed, minor roads may be displayed.
- Relevant and easily recognizable static images for driving shall be displayed.

Display information prohibited while the vehicle is in motion:

- FM multiplex-broadcast information not relevant to driving

(Above list is quoted with permission from Ref. 5.)

24.2.5 Article 71 of Road Traffic Law (1999)

The same year that JAMA guideline version 2.0 was released, the Road Traffic Law was revised. In Article 71 of this law, using a handheld cellular phone when the vehicle was in motion was prohibited. In addition to the use of a cellular phone, gazing at a video monitor, including a navigation display, was also prohibited. If a traffic crash happened while the driver was using a cellular phone or gazing at the video monitor, the driver could be fined an amount not exceeding 50,000 Japanese yen.

24.2.6 Revised JAMA Guideline Version 2.1 (2002)

In 2002, the JAMA guideline was slightly revised to version 2.1, to provide the following numerical criterion for the optimal installation position for display monitors to make them easily visible to the driver. The guideline said "The display shall be mounted in a position at which the downward viewing angle is less than 30 degrees."

24.2.7 Article 109 of Road Traffic Law (2002)

In the same year, the Road Traffic Law was also revised. In this revision, Article 109, "Provide traffic information," was added. This article was for the industry and information providers, not for drivers, to improve safety and to smooth the flow of traffic using VICS. It covered various aspects of VICS information and aimed at ensuring the accuracy of information, defining traffic congestion, and outlining the limitations of a guided road. It also included principles of information presentation to a driver, the operation of in-vehicle devices, and means of presenting information.

The principles of information presentation were as follows:

- Use abbreviations or shortening of words for long and complex names of places or roads.
- Use signs and symbols when text presentation is confusing due to a complex background or when it takes time to read.
- Use simplified map when the real map is too complex to be comprehended.
- Use color when traffic information is displayed on the road map. Orange and red were recommended to be used for traffic congestion.

The principles for operating in-vehicle devices were as follows:

- The position of the operation device of the in-vehicle system is to be selected so as not to interfere with the driver's steering and pedal operation when the driver uses the device.

- Functions that require complex operation are prohibited while the vehicle is in motion. A complex operation here refers to multilayer menu input and continuous input of numerals and characters.

The principles of presenting information to avoid driver distraction were as follows:

- The position of the display is to be selected so as not to interfere with the driver's forward view and to minimize the driver's eye movement to look at the display.
- The information content should not be voluminous or too complex, so as to avoid extended gazing by the driver (e.g., looking at a display for more than 2 s).
- TV images and moving images from CD-ROM (e.g., movies) cannot be displayed while the vehicle is in motion. A change of icon or characters for comprehension, a scroll of the map, and the blinking of an item and images of the surrounding vehicles for safety assistance are beyond the scope of this guideline.
- Advertisements and other information not related to car driving cannot be displayed while the vehicle is in motion.
- Additional auditory information presentation is recommended to allow the driver to obtain information without looking at the display.

24.2.8 Revision of JAMA Guideline to Version 3.0 (2004)

Three years later, the JAMA Guideline underwent a major revision to version 3.0. Quantitative criteria for the use of navigation systems were introduced in this version. This version is the current version of JAMA Guideline. Details of the guideline are presented in Section 24.3.

24.2.9 Revision of Article 71 of Road Traffic Law (2004)

The 2004 revision of the Road Traffic Law was regarding Article 71, which focused on the use of cellular phones. The content is basically the same as the previous one, but the punitive clause became stricter. In the previous version, the punishment was applied only if it caused an accident but under the current law it is applied regardless of whether an accident occurs or not. The use of a cellular phone with a hands-free device is not prohibited. The section involving the use of in-vehicle navigation system displays is unchanged in this revision. The number of injury-causing traffic crashes in which the use of a cellular phone was a factor was 2597 in 2003. This decreased to 1868 in 2004, and 946 in 2005, perhaps as a consequence of applying the more strict punishment regime.

24.3 JAMA GUIDELINE VERSION 3.0

The JAMA Guidelines up to version 2.1 consisted of principles and prohibitions with respect to the information to be displayed and the operations to be conducted. However, when a new system is developed, the system developers must examine new information content and new operational procedures to see whether any are prohibited. Sometimes, it is not easy to decide whether a particular item could be applied to the new functions. In addition, when the new functions and items are not covered by the prohibited items, the guideline must be revised if it is in the form of a list of prohibited items. On August 18, 2004, JAMA published version 3.0, which features the world's first quantitative criteria for displaying monitor operations while a vehicle is in motion. This was done to make the guideline more applicable to recently introduced functions of new navigation systems. The English version of the JAMA Guideline is available at the referenced website.[7]

The principles and scope of JAMA Guideline version 3.0 are the same as previous versions.

Principles:

1. Preferably, the display system is designed in such a way that its adverse effect on safe driving will be minimized.
2. The display system should be installed in such a position that driving operation and the visibility of the forward field will not be obstructed.
3. The types of information to be provided by a display system should be such that the driver's attention will not be distracted from driving; information that is essentially entertainment is to be avoided.
4. The display system should be operated by the driver without adversely affecting his or her driving performance.

Scope:

1. This Guideline applies to display systems, whether factory-installed or installed by a vehicle manufacturer-designated dealer, that are installed in vehicles (excluding motorcycles) and are located at a position visible to the driver.
2. A "display system" in this Guideline refers to a system capable of displaying diagrams, letters, numbers and images that have been either memory-stored in advance or are received via telecommunications.

Version 3.0 is composed of three parts: (1) "Display Monitor Location," (2) "Content and Display of Visual Information While Vehicle Is In Motion," and (3) "Operation of Display Monitors While Vehicle Is In Motion." The quantitative criteria for the Display Monitor Location section were already introduced in version 2.1. The qualitative criteria for the Content and Display of Visual Information section are still in force, although numerous attempts have been made to introduce quantitative criteria for display information.

Quantitative criteria were introduced in this version in the Operation of Display Monitors While Vehicle Is in Motion section. The operation of an in-vehicle system is prohibited if the driver's operational task fails to comply with the following criteria:

1. The total time of the driver's glance at the monitor between the start and completion of the operational task shall not exceed 8 s; and

2. When the above-mentioned total time is measured by a bench test using the occlusion method, the total of shutter open time shall not exceed 7.5 s.

JAMA selected these quantitative criteria based on experimental data. The series of experiments used to obtain evidence for selecting the criteria are described in the next section.

24.4 EXPERIMENTAL EVIDENCE FOR THE CRITERIA IN JAMA GUIDELINE VERSION 3.0

JAMA has conducted a series of experiments for determining the quantitative criteria for the use of in-vehicle systems in the guideline. One series examined the total glance time (TGT) on actual roads, and the other, the total shutter open time by use of the occlusion method.[5]

24.4.1 Upper Limit of Total Glance Time Required for Use of Vehicle Navigation System While Driving

During the TGT experiment, glance time was measured during operation of the navigation systems while driving on real roads. Four types of navigation systems with different operating methods (e.g., touch panel, joy stick, remote control, and rotary knob) were tested in four different vehicle models (e.g., Toyota Crown, Nissan Cedric, Toyota Mark II, and BMW 520). The data were collected under four difference road conditions: (1) Urban arterial road (total four lanes/two lanes each side, speed limit 60 km/h), (2) Urban road (total two lanes/one lane each side, speed limit 60 km/h), (3) the Joban Highway (total six lanes/three lanes each side, speed limit 100 km/h), and (4) the Urban Expressway (total four lanes/two lanes each side, limiting speed 80 km/h). The subjects were asked to operate the navigation system to perform the following tasks:

- Map scale change (two-step operation)
- Display of nearby facilities
- Destination entry (input of three to seven characters for the name of destination)
- Destination entry (input of 10-digit telephone number of the destination)
- Continuous map scrolling
- Selection of text information from VICS (eight-step operation)
- Selection of text information from VICS (12-step operation)
- Page skip in text information (eight-step operation)
- Page skip in text information (12-step operation)

Measures of visual behavior, vehicle control performance, and subjective evaluation were obtained from this experiment. Glance time was obtained using video images of the driver's face, defined as the elapsed time between the shifting of the driver's gaze from the forward field of view to the navigation system and then back to the forward field. Since the driver alternately glances at the navigation system and back at the forward field a number of times while completing a single operational task (e.g., destination entry), individual glance times were accumulated to obtain a TGT for each operation. The vehicle position is measured in relation to lane markings using video images of the road surface taken by a rooftop video camera. The vehicle's lateral displacement was obtained as a quantitative measure of driving performance while completing the navigation system tasks. The control data were obtained from a 40 s span of regular driving during which the subject driver was instructed not to perform any navigation system task. The control range of the lateral displacement was about 300 mm in rural road conditions and about 500 mm in highway or expressway conditions. The lateral displacement falling within the control data range in each condition was assumed to be a permissible amount of lateral displacement. In addition to lateral displacement, indices of vehicle control such as vehicle speed, steering angle, accelerator operation, headway distance, and lateral acceleration were examined. The driver stress was also measured based on the subjective evaluation by the driver. After completion of the test routine, the driver was asked to rate his stress level while operating the navigation system, on the basis of a 7-point scale (7, very stressed; 4, neutral; 1, very relaxed). Scores of 4 (i.e., neutral=acceptable) or lower were assumed to indicate permissible stress levels. A total of 10 adult male drivers, ranging in age from 26 to 52 with an average age of 34, participated in the experiments.

Figure 24.9 presents the relationship between driver stress levels during a prescribed navigation system operation and TGT to complete the operation. A high

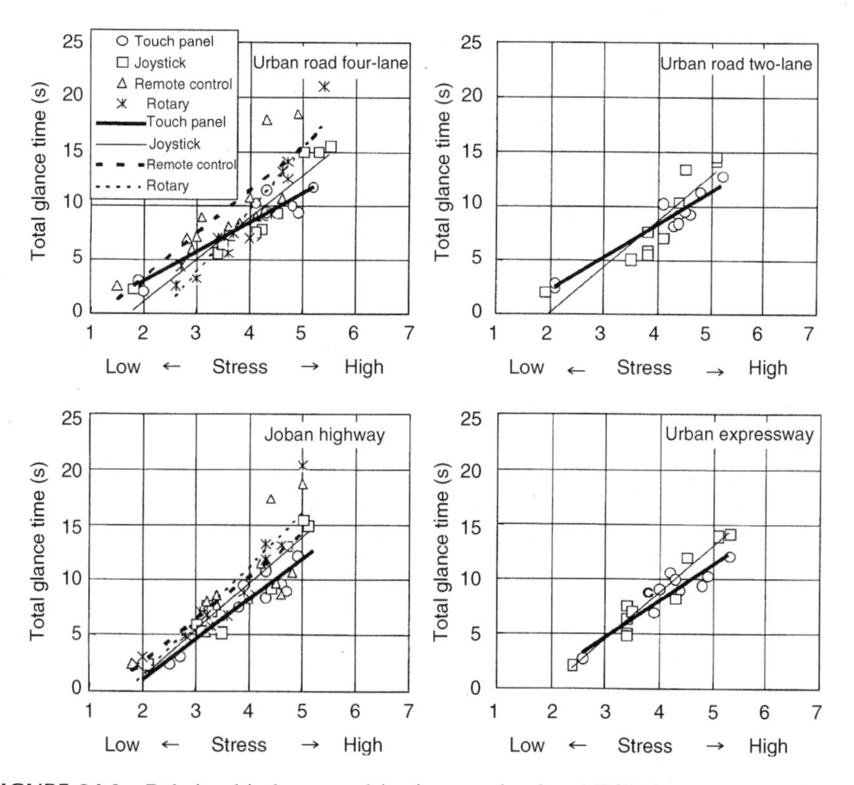

FIGURE 24.9 Relationship between driver's stress level and TGT for each type of road. (From Kakihara, M. and Asoh, T., in *Proceedings of the 12th World Congress on ITS*, 3231, 2005. With permission.)

correlation between the two variables was revealed for all four road types, indicating that the drivers' stress levels increased as the TGT lengthened. As can be seem from the regression lines in Figure 24.9, a neutral stress level of 4 on the 7-point evaluation scale corresponds to a TGT of between 7.9 and 11.4 s for all road types and all the navigation system types tested.

Among the indices of vehicle control performance, the vehicle's lateral displacement showed the highest correlation with the TGT and with the driver's stress level. A high positive correlation between the vehicle's lateral displacement and TGT was observed for all test-road types (Figure 24.10). The vehicle's lateral displacement was within the control data range when the TGT was between 8.2 and 14.3 s for all road types and navigation systems tested. This means that navigation system operation does not affect vehicle behavior if the TGT is within 8.2 and 14.3 s.

These two variables—driver's stress level and the vehicle's lateral displacement—were used as measures of the effects of glancing behavior on the driver and his or her driving. To determine the maximum permissible TGT that neither induces the driver's stress nor affects the vehicle's behavior while driving on a public road, the two measures were combined. Table 24.2 lists the maximum permissible TGTs that satisfy these two conditions, that is, the stress level did not exceed the neutral level and the vehicle's

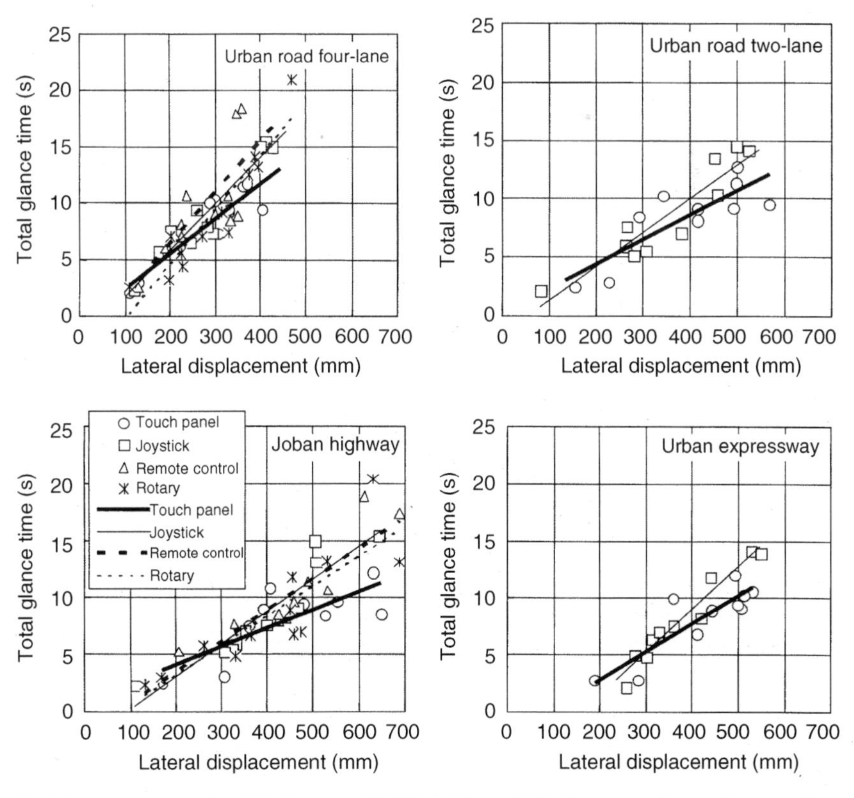

FIGURE 24.10 Relationship between TGT and lateral displacement for each type of road. (From Kakihara, M. and Asoh, T., in *Proceedings of the 12th World Congress on ITS*, 3231, 2005. With permission.)

TABLE 24.2
TGTs That Do Not Exceed the Neutral Stress Level or the Control Level of Lateral Displacement of the Vehicle

	Urban Road with Four Lanes	Urban Road with Two Lanes	Joban Highway	Shuto Expressway	Maximum Permissible TGT Value
Touch panel	8.4	8.2	8.2	7.9	7.9
Joystick	8.9	8.6	9.7	8.3	8.3
Remote control	10.2	N/A	10.2	N/A	10.2
Rotary	8.2	N/A	10.6	N/A	8.2
					s

Source: From Kakihara, M. and Asoh, T., in *Proceedings of 12th World Congress on ITS*, 3231, 2005. With permission.

lateral displacement was within the control range. Since there were differences among the road types, the lowest TGT was selected for each navigation system, and all were between 7.9 and 10.2 s. The strictest case (i.e., the lowest value) among the navigation systems was 7.9 s, close to 8.0 s. It was therefore concluded that in-vehicle system operation does not cause stress and does not affect vehicle control regardless of the navigation system type and road type as long as the TGT does not exceed 8 s. A TGT of 8 s can be considered a criterion for navigation system operation while driving.

In addition to the TGT, they examined the single glance time. However, they could not observe a clear correlation between the mean single glance time and the driver's stress or the vehicle behavior. It can be concluded that the TGT at the navigation display is a good measure of the visual demand for the navigation system.

24.4.2 Occlusion Method: Bench Test on Total Glance Time

The measurement of the TGT described in the previous section used a direct method to assess the effect of navigation system operation on the driver and driving performance. However, because it takes time to conduct this experiment, the TGT criterion is not easy to apply in the design stage of an in-vehicle system. It would be helpful if the guideline were to provide a criterion that could be examined using a bench test, so that the system could be tested before being installed. This is why the visual occlusion method is useful.

Drivers usually operate an in-vehicle device while alternately glancing at the device and looking back at the forward field. The visual occlusion method (see also Chapter 9) simulates this back-and-forth process using a liquid crystal shutter goggle. The shutter open time corresponds to the time it takes to glance at the navigation system, and the shutter close time corresponds to forward field observation time. The total shutter open time (TSOT) is the sum of the shutter open time periods from the start to the completion of a navigation system task. As the occlusion method requires preset shutter open and closed patterns, several combinations of the shutter open time (1.0 and 1.5 s) and the shutter close time (0.5 s, 1.0 s, 3.0 s) were examined when measuring TSOT.

In the experiment, TSOT was measured for the navigation tasks using different navigation systems; the same systems used for the TGT experiment. The TSOT for each task obtained from the occlusion experiment were compared with the TGT. The TSOT showed a high correlation with the TGT (Table 24.3). The highest correlation was obtained with a shutter pattern of 1.5 s open/1.0 s closed. The TSOT value corresponding to an 8 s TGT was calculated using the regression line between the TGT and TSOT. The corresponding TSOT ranged from 7.1 to 8.0 s for the navigation systems tested, as listed in Table 24.4. Therefore, the shortest TSOT of 7.1 s was selected as the TSOT upper limit. This means that any navigation operation that can be completed in five steps, each 1.5 s long—that is a TSOT of 7.4 s—is considered permissible.

A shutter open time of 1.5 s and shutter closed time of 1.0 s under the occlusion method showed the highest correlation with the TGT. This result was close to the driver's glancing behavior during actual driving, where the mean plus one standard deviation for the single glance time at navigation system was around 1.5 s, and about 1.0 s for the forward field.[8] From this, it was reasonable to select a shutter open time of 1.5 s and shutter closing time of 1.0 s for the occlusion method.

After the first series of experiments, there was an argument over the necessity to have a wider range of subjects, especially elderly drivers. When the same experiment was conducted using elderly drivers, the elderly drivers exhibited similar tendencies and confirmed that the criteria could be applied to them as well.[9]

TABLE 24.3
Correlation Coefficient Values between TGT and TSOT

	Open 0.5 s Closed 1.0 s	Open 1.0 s Closed 1.0 s	Open 1.5 s Closed 0.5 s	Open 1.5 s Closed 1.0 s	Open 1.5 s Closed 3.0 s
Touch panel	0.858	0.845	0.847	0.857	not tested
Joystick	0.867	0.891	0.885	0.888	not tested
Remote control	0.887	0.868	0.900	0.889	0.862
Rotary	0.890	0.932	0.926	0.939	0.924
Average	0.876	0.884	0.890	0.893	0.893

TABLE 24.4
TSOT (Open 1.5 s/Closed 1.0 s) Equivalent to a TGT of 8 s

	Shutter: Open 1.5 s/ Closed 1.0 s				
	Urban Road with Four Lanes	Urban Road with Two Lanes	Joban Highway	Shuto Expressway	Average
Touch panel	7.7	7.7	8.0	7.9	7.8
Joy stick	6.7	7.2	7.2	7.7	7.2
Remote control	8.0	N/A	7.9	N/A	8.0
Rotary	7.0	N/A	7.2	N/A	7.1

24.5 CHECKLIST APPROACH

The Japanese automotive industry has much experience in developing navigation systems. Each company has a wide range of models for their system. The checklist of human machine interface (HMI) design can be a good tool to promote efficiency in the design process. The items of the JAMA Guideline (Sections 24.2.2 through 24.2.4 and Section 24.3) can be applied to create checklist items. However, the JAMA Guideline covers only the principal part of the design. Therefore, each company may have its own checklist for in-vehicle HMI design. However, each company's checklist is confidential and no checklist is available to the public.

24.6 CONCLUSION

The first government intelligent transport system (ITS) project, Comprehensive Automobile Traffic Control System, began in 1973. Since then, government and industry have worked together on many projects to develop ITS for practical use. Japan has been the world's leader in making navigation system technology pervasive in the automotive industry. People who were engaged in this field were convinced that this technology had a high potential for supporting drivers by reducing driver workload and making driving and navigation less stressful. Thus, they realized that they must grow this technology carefully, with a good balance of competition and collaboration. This was the motivation for issuing the JAMA Guideline at an early stage of the market introduction of navigation systems.

As different manufacturers worked together to prepare the guidelines, they became aware of their responsibility for ensuring safety and of the importance of human factors in ITS development. Also, the guideline established in this industry segment has worked as a good communication tool between government and industry. The published guideline demonstrates consideration by the automotive industry for the safety of drivers using in-vehicle devices. The government is able to understand the methods and procedures used to determine the criteria in the guideline by referring to the appendix that explains the basis for its development. The public announcement of this basis can make communication more effective. In addition, it has had effects on the users. As described previously, in early versions of the guideline the driver could not scroll the map and could not see minor roads while the vehicle was in motion. This was sometimes inconvenient, but this self-imposed restraint sent a message to users about the safety concerns of the automotive manufacturers. The users might learn that the prohibited operations have potential to threaten driving safety. The guideline has played an important role in the growth of navigation systems and other in-vehicle devices in Japan, among government users, and in industry.

The latest version of the JAMA Guideline provides quantitative criteria for use of an in-vehicle device. The criteria were determined based on field and laboratory experiments. The basis for the criteria was findings from the field experiments. These experiments were conducted under different road conditions and navigation or vehicle system conditions (four types of navigation systems in different vehicles). The time it took to collect data limited the number of subjects. Although additional experiments were conducted using elderly drivers, experimental data from other types of drivers, such as female drivers, novice drivers and novice navigation system users, will be necessary to confirm the criteria. Results from a large number of subjects will enable the derivation of criteria based on population percentiles, not on the subject mean. As an objective measure, the experimenters used the lateral displacement of the vehicle. This is just one of the indices for the quality of the stabilization task of car driving. It will be useful to conduct further studies to establish other measures from the maneuvering task, such as the delay of braking onset or lane change.

The most important thing for the criteria in the guidelines is that the quantitative data that are used to derive them be made public. Once data have been published, people (from government, industry, and users) can discuss the validity of the criteria. If they agree that there is a need for new criteria or more experimental data, the criteria in the guideline can be revised based on any new findings.

REFERENCES

1. Rosen, D. A., Mammano, F. J., and Favout, R., An electric route guidance system for highway vehicles, *IEEE Trans. Veh. Technol.*, VT-19, 142, 1970.
2. French, R. L. and Lang, G. M., Automatic route control system, *IEEE Trans. Veh. Technol.*, VT-22, 36, 1973.
3. Itoh, H., Magnetic field sensor and its application to automobiles, *SAE800123*, 1980.
4. Akamatsu, M., Yoshioka M., Imacho, N., Daimon, T., and Kawashima, H., Analysis of driving a car with a navigation system in an urban area, in *Ergonomics and Safety of Intelligent Driver Interfaces*, Noy, Y. I. (Ed.), Lawrence Erlbaum Associates, Mahwah, NJ, 85–96, 1997.
5. Kakihara, M. and Asoh, T., JAMA Guideline for in-vehicle display systems, in *Proceedings of the 12th World Congress on ITS*, 3231, 2005.
6. http://www.vics.or.jp/english/index.html
7. http://www.jama-english.jp/release/release/2005/In-vehicle_Display_GuidelineVer3.pdf.
8. Asoh, T., Kamiya, H., and Ito, H., Cognitive performance of visual messages on in-vehicle displays and driving behavior, in *Proceedings of the 6th World Congress on ITS*, Paper No. 00055, 1999.
9. Asoh, T. and Iiboshi, A., Study of the car navigation tasks for elder drivers while driving, in *Proceedings of the 2003 JSAE Annual Congress (Spring),* No. 19-03, Paper No. 20035195, 2003 (in Japanese).

25 Driver Interface Safety and Usability Standards: An Overview

Paul Green

CONTENTS

25.1 WHAT IS THE GOAL OF THIS CHAPTER?

Guidelines and standards provide a collection of the accepted wisdom of an industry, but knowing which guidelines and standards to apply to a particular product is often difficult. The goal of this chapter is to help engineers and others determine which guidelines and standards to consider when designing and evaluating a driver interface for motor vehicle communication and information systems to minimize distraction. Some of the systems of interest are navigation, traffic information, voice communication, text messaging, and entertainment. How much these systems distract drivers from driving depends on the task and its demands, and not on the device, who made it, or how it got into the vehicle. Therefore, there should be no distinction between original equipment manufacturers (OEM) and aftermarket products, or between installed and carry-in devices, even though some standards that were reviewed limit the types of devices to which they apply.

Acquiring guidelines and standards is expensive and time consuming, but the cost of not using them may be greater. Noncomplying interfaces can be dangerous to use, a magnet for lawsuits, and undesired by customers. Customers, especially in the United States, expect motor vehicles to be safe and easy to use. They do not expect to read manuals or take a class to learn how to operate motor vehicle features. In the United States, the accepted liability benchmark is that products must be designed for reasonable and expected use and misuse. Driver interfaces that are not safe or easy to use may be so because they are distracting, which is the focus of this book.

25.2 WHAT KINDS OF GUIDELINES AND STANDARDS EXIST?

In common parlance, guidelines describe how something *should* be done, whereas standards describe how something *must* be done. As will become apparent later in the chapter, because some standards are not enforced, they are viewed as guidelines, and, therefore, exist as standards in name only.

Guidelines and standards usually pertain to either product design or performance. Design guidelines and design standards concern physical characteristics, such as tolerances, maximum loads, operating temperatures, and so on. Design guidelines and standards are appropriate where a topic has been reasonably well investigated and for which accumulated knowledge is stable, such as requirements for minimum character size to assure legibility or where physical interfaces occur, such as bumper heights (so that bumpers from different vehicles contact one another).

Performance guidelines and performance standards concern how things should be measured and how well they should function, such as time and errors when entering a destination into an in-vehicle navigation system. Performance guidelines and standards are appropriate where there are many methods to achieve a goal or where the underlying technology is rapidly evolving.

Less commonly, guidelines and standards may concern not what is produced, but the process of how it is produced. Such guidelines and standards can cover requirements for the credentials of personnel involved in design and testing, when and how design reviews and tests are done, and how the entire process must be documented.

25.3 WHO DEVELOPS MOTOR VEHICLE SAFETY GUIDELINES AND STANDARDS, AND HOW?

Guidelines and standards are developed by professional and trade associations, national and international standards organizations, governments, and others (Schindhelm et al.[1]). This section describes the most prominent of these organizations, with Table 25.1 describing the process each organization uses to develop guidelines and standards.

The Society of Automotive Engineers (SAE) is the leading worldwide professional organization of automotive engineers. The development of SAE driver interface standards is the purview of the Intelligent Transportation Systems (ITS) Safety and Human Factors Committee (SAE).[3] SAE develops information reports (literature compilations), recommended practices, and standards, all of which are considered to be "standards" by SAE and follow the same development process.[4]

TABLE 25.1
Guidelines and Standards Development Processes

Who	Process
SAE	Drafts of SAE standards are developed by working groups who report to committees. Ballots to approve occur at the working group, committee, and the Vehicle Systems Group. At each stage, editing can occur to resolve negative comments, and if the changes are substantial, reballoting occurs. The approved document receives editorial review (by SAE headquarters) before publication.
ISO	Proposals for ISO standards can originate in national standards organizations of the major vehicle-producing countries such as DIN (Deutsches Institut für Normung) in Germany or from trade and professional associations such as SAE. ISO documents follow a well-defined, six-stage, process involving a series of documents — Preliminary Work Item (PWI), New Work Item (NWI), Working Draft (WD), Committee Draft (CD), Draft International Standard (DIS), Final Draft International Standard (FDIS), and International Standard (IS)—that can take up to three years to develop as a proposal moves from the working group to the subcommittee to the technical committee to the secretariat for review and balloting (ISO[2]). The most important effort comes from the roughly five people on the working group task force that write the first draft. As with the SAE process, there are opportunities for editorial changes after ballots. More informational documents become technical reports instead of standards.
U.S. DOT	These standards are developed following the process described in the Administrative Procedures Act (http://usgovinfo.about.com/library/bills/blapa.htm, retrieved July 15, 2007). The act requires proposed rules (initiated by the U.S. DOT) to be published in the Federal Register, along with listings and responses to public comments, hearings, and the final rule. The "public" includes manufacturers (who provide the most detailed comments), suppliers, consumer organizations, and government agencies (including foreign governments), not just individuals.
AAM	Their standards development process is open only to members, and as best the author can tell, AAM produces only guidelines.
JAMA	There is little information on the process in the open literature, though some insights appear in Chapter 24. Some aspects of these standards have been developed as a result of "suggestions" from the National Police Agency.

Standards are for procedures or specifications that have broad engineering acceptance. Though compliance with SAE recommended practices and standards is voluntary, deviations from them are very difficult to defend in U.S. courts, so suppliers and manufacturers comply with them.

The International Organization for Standardization (ISO) has developed about 12,000 global consensus standards for a wide range of applications as well as a significant number of information reports. ISO driver interface standards are developed primarily by Technical Committee 22/Subcommittee 13 (Road Vehicles/ Ergonomics Applicable to Road Vehicles) and, in particular, Working Group 8 (Transport Information and Control Systems on Board—Man-Machine Interface).

Similar to compliance with SAE standards, ISO standards are voluntary. However, some countries require "type certification" for vehicles to be sold (which includes compliance with ISO standards). Global manufacturers find that producing ISO-compliant, globally marketable vehicles is less costly than producing noncompliant, country-specific vehicles. Hence, manufacturers comply with them.

The National Highway Traffic Safety Administration (NHTSA), an agency of the U.S. Department of Transportation (U.S. DOT), is responsible for the Federal Motor Vehicle Safety Standards (NHTSA[5]). Because the United States is currently the largest market for vehicles, these safety standards are particularly important. Furthermore, Canadian standards (Transport Canada[6]) tend to mirror those of the United States'. Failure to comply with U.S. DOT safety standards can result in fines and mandatory recalls. U.S. DOT activity on safety standards depends on which political party holds the presidency and controls Congress. The Republicans favor less government activity (fewer standards or the use of industry/commercial standards). The Democrats favor consumer protection (more federal standards and more rigorous requirements). (For an elaboration, see Joan Claybrook's interview on *Frontline* [Public Broadcasting System[7]].) The U.S. DOT standards do not address issues of distraction in any substantial way.

A number of highly detailed guidelines were written under contract to the U.S. DOT (e.g., Green et al.[8]; Campbell et al.[9]; see also Chapter 23), but there was no open process for them to be evaluated. The guidelines are interesting and well documented. However, because they are informational and not cited as good practice to be followed by standards development organizations, there is not much incentive to comply with them.

The Alliance of Automobile Manufacturers (AAM) is the trade association of the U.S. automobile manufacturers. AAM guidelines are voluntary and publicly available, and AAM does not enforce them. The major incentive for manufacturers to comply with the AAM guidelines (in some sense, their own guidelines) is the potential negative outcome of a product liability action.

The Japan Automobile Manufacturers Association (JAMA) also has developed guidelines (see Chapter 24). The guidelines were developed specifically to minimize distraction and, although they are theoretically voluntary, as a matter of accepted practice, they are followed by all Japanese OEMs and sometimes by aftermarket suppliers. Some of these guidelines are particular to Japan (e.g., concerning very narrow roads).

25.4 COMMENTS ON THE GUIDELINES AND STANDARDS DEVELOPMENT PROCESS

There are differences in how standards development organizations function. Some, such as SAE, ISO, and the U.S. DOT, follow an open process, and for SAE and ISO, consensus is important. Other organizations, such as AAM, are closed to those who are not employees of AAM member companies.

Full participation in standards development requires travel to meetings, which for ISO involves international travel. Only major manufacturers and government agencies have the resources to fund professional time and travel, so such committees have few members from academic organizations or citizen groups. The result is that those with a vested interest in product development have a disproportionate influence on standards development.

Further, although organizational charters require that standards be developed by technical experts, in a few instances participant expertise is weak (e.g., they are not full members of the Human Factors and Ergonomics Society or certified by the Board of Certification in Professional Ergonomics).

Finally, assembling the relevant information and developing a consensus takes time. Guidelines and standards, particularly those of SAE, JAMA, AAM, and ISO, take at least three years to develop. Employers are often not willing to provide much support for the background work to aid standards development (such as literature reviews or studies with subjects), especially on short notice, and most committees only meet a few times a year. Thus, design and performance standards are often produced years after the products they affect.

25.5 WHAT DESIGN AND EVALUATION STANDARDS EXIST FOR DRIVER INTERFACES?

25.5.1 U.S. DOT, EU, AAM, JAMA, and SAE Guidelines and Standards

Table 25.2 summarizes the design and performance guidelines and standards of many organizations, except those of the ISO, which are described later. Copies of these documents can be retrieved from the University of Michigan Transportation Research Institute (UMTRI) Driver Interface website (www.umich.edu/~driving), although in some cases final copies are not provided due to copyright protection.

The U.S. DOT effort that funded the previously mentioned UMTRI guidelines (Green et al.[8]) led to a follow-on project, the Battelle guidelines (Campbell et al.[9]). The UMTRI guidelines were intended to supplement standard human factors references such as Military Standard 1472 (U.S. Department of Defense[20]), which is regarded as the human engineering bible, whereas the Battelle guidelines included information, such as required character sizes (see Chapter 23 for an extensive description of the UMTRI and Battelle guidelines). As a follow-on to those efforts, the Virginia Tech Transportation Institute developed a model to predict driver performance (Hankey et al.[21]). In parallel, the HARDIE guidelines (Ross et al.[13]) were developed in Europe. There are no data on how much these documents specifically influenced design practice though they did help organize the literature, identify gaps, and indirectly led to efforts to develop SAE and ISO standards.

At about the same time, the European Union (EU) human-machine interface guidelines were developed and later enhanced (Commission of the European Communities[11,12]). The original EU guidelines were extremely brief, but were expanded and renamed as the European Statement of Principles (ESoP) in 2005 (see Chapter 22). The 1999 EU guidelines also served as the basis for the AAM guidelines (which AAM called principles[10]) and may have influenced the development of a checklist and guidelines by the Transport Research Laboratory (TRL) (Stevens et al.[18]; Stevens et al.[19]). The AAM work is described in greater detail later

TABLE 25.2
Major Non-ISO Telematics Guidelines and Recommended Practices

Common Name	Authors and Year	Pages	Comments
Alliance Guidelines, Version 3	Alliance of Automobile Manufacturers (AAM) (2003)[10]	67	This elaboration of original EU principles contains a rationale, criteria, verification procedures, and examples for each guideline. These guidelines are likely to be used by U.S. OEMs. Key sections are principles 2.1 and 2.2, which concern visual demand and are still in need of refinement.
Battelle Guidelines	Campbell et al. (1997)[9]	261	These guidelines emphasize heavy vehicles with chapters concerning navigation (3), warning systems, in-vehicle signs, trucks, and other topics. The guidelines include physical ergonomics (e.g., control sizes). Each guideline is accompanied by the rationale, application notes, references, and a four-star rating of the supporting evidence. Example: Chapter 5 (routing and navigation guidelines): *Road segments should be color coded (green, yellow, red) to indicate the mean speed of the traffic flow.*
EU Guidelines	Commission of the European Communities (1999)[11]	2	These 24 very brief guidelines are mostly "motherhood" statements concerning overall system design, installation, information presentation, interaction with controls and displays, system behavior, and other topics. Example: Section 4 (overall design) *The system should be designed to support the driver and should not give rise to potentially hazardous behavior by the driver or other road users.*
European Statement of Principles (ESoP)	Commission of the European Communities (2005)[12]	59	This expansion of original EU guidelines (renamed as a statement of principles) contains the now 35 reworded guidelines with added explanations. Example: 1.3 Design Goal III: *The system does not distract or visually entertain the driver … The goal … is to ensure that the driver is not distracted … such that their ability to be in full control … is not compromised … Visual entertainment may occur by visually displaying images which are attractive …*
HARDIE Guidelines	Ross et al. (1996)[13]	480	These European guidelines contain chapters on road and traffic information, navigation, collision avoidance, adaptive cruise control, and variable message signs. They contain less data than UMTRI or Battelle guidelines but are broader as they consider driver assistance systems. Obtaining a copy of this handbook is very difficult.

JAMA Guidelines	Japan Automobile Manufacturers Association (2004)[14]	15	Although these guidelines are brief, they contain several very specific requirements concerning display location (not permitted 30° or more below the driver's viewing plane) and constraints on what can be shown in a moving vehicle (e.g., no broadcast TV, no descriptions for hotels and restaurants, 30 characters or less of traffic information).
SAE J2364 (15-s Rule)	Society of Automotive Engineers (2004)[15,16]	13	This recommended practice specifies the maximum allowable task time and test procedures for navigation system tasks performed while driving for systems with visual displays and manual controls. It also describes an interrupted vision (visual occlusion) method. Designers will find SAE J2678, the rationale for SAE J2364, helpful in understanding when and how to apply SAE J2364.
SAE J2365 (SAE Calculations)	Society of Automotive Engineers (2002)[17]	23	This document describes a method to compute total task time for visual-manual tasks not involving voice and is used early in design to estimate compliance with SAE J2364 and other purposes. The method utilizes time estimates for mental operations, keying, searching, and so forth.
TRL Check List	Stevens et al. (1999)[18]	18	Simple checklist Example: Controls, item C1 *Are the IVIS controls easily reached by the user when driving?* *All controls needed when driving can be reached from the normal driving position …*
TRL Guidelines	Stevens et al. (2002)[19]	70	These design guidelines focus on how to design and assess for safety. There are sections concerning compliance with regulations, packaging and instructions, installation, driver input, visual displays and information presentation, auditory information presentation, information comprehension, menus, and temporal information.
UMTRI Guidelines	Green et al. (1993)[8]	111	This first set of U.S. guidelines includes principles, general guidelines, and specific design criteria with chapters on manual controls, speech, visual displays, auditory displays (4), navigation interfaces (3), traffic information, phones, vehicle monitoring, and warning systems. For every guideline the supporting literature is described. Example: Guideline 7.14 *For expressway ramps, give both the route name and direction, and a city locator* *For example, the display should show "I-275 North" and "Flint." The names should be on separate lines, so … makes in-vehicle displays compatible with signs."*

in this chapter and in Chapter 23. Finally, in parallel with all of those efforts, was the development of the JAMA guidelines[14] (covered in detail in Chapter 24). In a somewhat different direction was the development of two SAE Recommended Practices (Green[22–24]; Nowakowski and Green[25]; Nowakowski et al.[26]; Society of Automotive Engineers[15–17]), although the SAE efforts (described in detail later in this chapter) were aware of the prior guidelines and ISO work.

Thus, over a five-year period, a large number of design guidelines were produced, some of which were connected with the original EU guidelines. However, these guidelines and standards differ quite widely in their scope and specificity, with the amount of detail provided increasing over time. Clearly, a guideline that states how far from an intersection a turn signal should be presented is readily applied, but in situations without specific guidelines, one must rely on general principles (e.g., "present information consistently"). Experimentally investigating every possible safety and usability design issue is not cost effective, so general principles are often used to make design decisions. Unfortunately, some industry practitioners do not know these guidelines even exist. Furthermore, smaller organizations may not have staff with professional training in human factors, so how those companies would apply the general guidelines is unknown. The lasting contribution of these guidelines could be to spur the development of standards that designers are more likely to follow. Compliance with these guidelines should make driver interfaces safer and easier to use and, in doing so, reduce the extent to which distraction is a problem.

Potentially the AAM principles could have a major impact on driver interface design. Version 2 is the current set, and Version 3 is reportedly forthcoming (AAM[10,27]). Although their scope states that the AAM principles apply only to advanced information and communication systems, these principles should also apply to traditional information or communications systems (e.g., entertainment systems) because the underlying goals (reading displays, pressing buttons, etc.) and method of execution are the same, as is the manner in which tasks distract people from driving.

Of the AAM principles, Principle 2.1 most directly addresses distraction. That principle states, "Systems with visual displays should be designed such that the driver can complete the desired task with sequential glances that are brief enough not to adversely affect driving." There are two alternative verification methods. For alternative A, the criterion is that the 85th percentile of single glance durations should not exceed 2 s, and the total glance time for a task should not exceed 20 s. There is considerable debate as to what the percentile criterion for a single glance and the maximum task time should be (Go et al.[28]), in particular, that the AAM guidelines are too permissive. Compliance can be verified by a visual occlusion procedure (see Chapter 9) or from eye-fixation data collected in either a divided attention condition or on-road test (by a camera aimed at the face or an eye-fixation monitoring system). (In the visual occlusion procedure, subjects wear LCD glasses or in some other manner have their vision to the device periodically interrupted, simulating looking back and forth between the road and the device. In the AAM guidelines the viewing time is 1.5 s and the occlusion period 1.0 s.)

For alternative B, the criterion is that the number of lane departures does not exceed the number for a reference task, such as manual radio tuning (and at the same time, car following headway should not degrade). The compliance test involves driving on a real or simulated divided road at 45 mph or less in daylight, on dry pavement, with low to moderate traffic. The AAM principles document provides details concerning the location of the radio, the stations among which to choose, what constitutes a trial, and subject selection. Even though manual tuning seems like a potential benchmark, the range of task times and interface usability varies quite widely among products, making selection of the standard radio-tuning task difficult. In part, this is because radios have evolved from two knobs, five buttons, and a dial to complex multifunction systems with touch screens and menus in which the radio function is embedded. If forced to choose an alternative, the author would choose A because the data are more revealing in terms of when drivers are distracted (which can serve to improve the design) and because the test method and criteria are better defined.

SAE Recommended Practice J2364[16] presents two test methods and criteria to determine if visual manual tasks should not be performed while driving. Though the practice scope statement constrains application to navigation-system–related data entry tasks, there is no reason why J2364 should not apply to other visual manual tasks and other systems. The practice does not, and should not, apply to voice interfaces because the task methods are fundamentally different. The idea behind this practice is that the longer an in-vehicle task takes to complete, the greater is the time drivers are distracted from keeping their eyes on the road, and the greater is the probability of a crash.

The static test procedure in J2364 requires 10 test subjects between the ages of 45 and 65, which means that the reasonable worst case of elderly drivers is not considered. Each subject completes five practice trials and three test trials of the task in question (e.g., entering a street address) in a parked vehicle, simulator, or laboratory mockup. In the static method, the task is acceptable if the mean task time (actually the antilog of the mean of the logs to reduce the influence of outliers) is less than 15 s ("the 15 s rule," a duration that has been the subject of considerable discussion). To minimize distraction, the author would recommend 10 s. Compliance with the 15 s limit can be estimated using a calculation procedure described in SAE Recommended Practice J2365[17] based on task element times (keying, mental operations, etc.) popular in the human-computer interaction literature and adjusted for automotive applications (see also Pettitt et al.[29]).

In an alternative procedure involving occlusion, the device is visible for 1.5 s and occluded for 1–2 s, with 1.5 s being recommended. The criterion (again determined using logs) is 20 s. Recently, Pettitt et al.[30] have proposed a calculation for estimating compliance.

25.5.2 ISO Standards

Table 25.3 shows the documents developed or being developed by ISO Working Group 8. Most of the documents can be quite general, sometimes not containing the detail found in the Battelle, HARDIE, or UMTRI guidelines. Because reaching agreement is difficult, measurement methods are often presented without acceptance criteria for safety. To promote international harmonization and to comply with trade

TABLE 25.3
ISO Standards

Document, Reference	Short Title	Pages	Comment
Std 15005: 2002(E) (ISO[31])	Dialogue Management Principles and Compliance Procedures	15	This standard provides high-level ergonomic principles (compatibility with driving, consistency, simplicity, etc.) to help design driver and in-vehicle system interactions in moving vehicles. Compliance is determined mostly by observation. Example: "*5.2.4.2.4 Individual TICS dialogues shall be designed to guide the driver in giving a priority to the information displayed. Example 1: A collision-avoidance system will rapidly attract the driver's attention (but without startling the driver) when a collision is imminent …*"
Std 15006:2004 (ISO[32])	Specifications and Compliance Procedures for Auditory Presentation	18	This standard provides requirements for in-vehicle auditory messages including signal levels, appropriateness, coding, and so on, along with compliance test procedures. In a suggested procedure for tone discernment, the standard suggests 90% correct (out of 10 trials for 10 subjects) as an acceptance criterion.
Std 15007-1:2002(E) (ISO[33])	Visual Behavior Measurement 1: Definitions and Parameters	8	This standard defines common terms related to driver eye glances and shows areas of interest in the vehicle. Examples: "*direction of gaze—target to which the eyes are directed*" "*dwell time—sum of consecutive individual fixation and saccade times to a target in a single glance.*"
TS 15007-2:2001 (ISO[34])	Visual Behavior Measurement 2: Equipment and Procedures	14	This trial standard describes video-based equipment (cameras, recording procedures, etc.) and procedures (subject descriptions, experiment design parameters, tasks, performance measures, etc.) used to measure driver visual behavior.
Std 15008:2003 (ISO[35])	Legibility (Visual Presentation of Information)	24	Example specifications include character size (20 arc minutes recommended), contrast (5:1 for night, 3:1 for twilight and diffuse daylight, 2:1 for direct sunlight), width-to-height ratio (0.08 to 0.16), ISO 2575 symbol resolution (32×32 if 1 bit, 24×24 if gray scale), as well as how character dimensions are measured.

TR 16352:2005 (ISO[36])	Warnings Literature Review	137	In 17 chapters this ca. 2004 review of warning systems ergonomics covers topics such as alarm theories, the design of visual, auditory, tactile warnings, and redundancy. The report contains 188 references.
TS 16951:2004 (ISO[37])	Message Priority	35	This trial standard describes how to determine a priority index for in-vehicle messages (e.g., navigation turn instruction, collision warning) presented to drivers. It includes evaluator and scenario selection, message content, and analysis. Priority is based on two four-point scales, criticality (the likelihood of injury or vehicle damage if the event occurs), and urgency (required response time).
Std 17287:2002 (ISO[38])	Suitability of Interfaces While Driving	38	This standard describes a process to assess if a driver interface is suitable for use while driving using hierarchical task analysis, and encourages the use of driver performance tests. It considers the user's task, context of use, assessment, and documentation. The approach is extremely general and contains no performance criteria.
Std 16673:2007 (ISO[39])	Occlusion Method to Assess Distraction	19	This standard describes a test to assess the visual demand of a display by periodically blocking (occluding) the driver's view of the display (1.5 s visible, 1.5 s occluded). It includes requirements for 10 subjects and their training, their age (at least 2 over 50), test hardware, and two performance measures (total shutter open time, resumability).
CD 26022 (ISO[40])	Lane Change Test to Assess Distraction	34	This committee draft standard describes a PC-based driving simulator test of the demand of a driver interface. Subjects drive and, when signaled by signs, change lanes. Some of those changes occur while using an in-vehicle device. The desired performance metrics are still being discussed, though several are proposed. The procedure contains considerable detail, such as instructions to subjects.

agreements, national standards organizations, technical societies (e.g., SAE), and government organizations (e.g., U.S. DOT) often permit ISO standards to supersede their own standards. Hence, ISO standards are very important.

ISO standards concern both design (15005—dialogue management; 15006—auditory information; 15008—legibility; 16352—warnings literature review; 16951—message priority) and design process issues (17287), as well as performance assessment (15008—visual behavior; 16673—occlusion method; 26022—lane change test). A review of the design standards shows that few have assessment criteria, and most content is quite general. That does not mean, however, they are not useful for addressing distraction.

The lack of performance acceptance criteria in either 16673 or 26022 is particularly noteworthy. Standard 16673 states, "If TSOT is … a criterion then the age of the participants should be taken into account. If $R > 1$, … participants may be having difficulty in resuming the task …. It is for the users … to determine the exact value to be used as a criterion …." (TSOT is the total shutter open time, the time a driver interface is visible in an occlusion test. For more extensive descriptions of TSOT and R, see Chapter 9.)

For the lane change test (ISO 26022, see Table 25.3 and also Chapter 8), there is no agreement as to which performance measures to use—the mean delay in missed lane change initiation, number of missed lane changes, mean task completion time, total number of errors, and numerous measures of lane position and path error. In part, this reflects the fact that the test is new and research is ongoing. However, it also reflects the lack of a useful model of how people drive that can predict the effects of performing secondary tasks while driving.

Quite different from all the other standards is ISO 17287, which describes an ISO 9000-like process to promote safety and usability (and reduce driver distraction) (see Table 25.3). Though some OEMs and suppliers naturally do some of what is suggested in the standard as good practice, no one seems to pay much attention to complying with it. In part, this is because the administrative burdens are perceived to outweigh the benefits.

25.6 CLOSING THOUGHTS

Even after decades of research, the human factors community is not able to predict crash risk from the many test methods proposed (not the occlusion or the lane change test described here, or the peripheral detection task, eyes off the road measures, or other methods; see Part 3 of this book for a discussion of these methods), though there have been proposals for connecting task time to crashes. Somehow, those connections must be made, and research on such is of the highest priority. In particular, equations are needed that relate performance in various tests to fatalities and injuries likely at that level of performance (although see Young and Angell[41] for some interesting ideas about the relationships between measures). Further, the notion that a single, low-cost, test can assess the interference with driving of a secondary task—regardless of its visual, auditory, cognitive, and psychomotor demands— may need to be abandoned, and that is reflected in ISO's adoption of multiple standards for driver interface assessment.

As described in this chapter, and in Chapters 22, 23, and 24, there are numerous guidelines and standards to assist in designing driver interfaces to enhance safety and ease of use, and reduce distraction, but only a few have performance criteria. Following the design guidelines for ease of use can reduce opportunities for distraction, although performance standards more directly address the distraction problem. Both types of documents, however, are needed.

When selecting a safety or usability guideline or standard, consider the process by which it was developed: the openness of the process (open leads to a greater driver protection), technical expertise of the developers, criterion for safety thresholds (do no harm, be generally safe, is safe unless shown otherwise, etc.), the quantity and quality of the supporting data, the difficulty of applying the data or process, and the relevant application context. These all affect the reliability of the results and the extent to which the standard is biased toward assessing something as safe or unsafe (or distracting or not distracting). In the context of this chapter, should only systems that are flagrantly distracting not be allowed, or those likely to be distracting, or those that are potentially distracting, or should the criterion be something else? What then should a designer do to design a minimally distracting driver interface?

During the initial phases of design, designers should read and apply all of the ISO guidelines that specify the physical characteristics of the driver interface (e.g., 15006, 16951, and especially 15008). They also should read and apply the relevant design guidelines for their intended market: UMTRI, Battelle, and AAM for the United States, HARDIE for Europe, and JAMA for Japan. For Europe, the ESoP document should also be considered, but because it is so general (as are the AAM principles), those without human factors training will find these documents difficult to apply. However, because all of these guidelines contain good human factors practice, there is no reason why they should not be applied globally.

However, to minimize distraction, the focus should not be on the physical aspects of the driver interface, but on the tasks the driver performs, especially how often the driver needs to look away from the road and how long the task takes. Accordingly, there should be ongoing calculations of task times and visual demands using the SAE J2365 and Pettitt et al.[30] procedures, and modifications of driver tasks to reduce distraction based on those calculations. Those calculations need to occur early in design when the interface is still a paper concept that can be easily changed, and continue to be updated as the design evolves. Also needed are measurements of cognitive demand when they are independent of visual demand.

Finally, when a prototype interface is available, its safety, usability, and potential for distraction need to be assessed using the procedures in SAE J2364, ISO 16673 (occlusion), ISO CD 26022 (lane change test), and, for U.S. markets, the AAM principles, most likely 2.1. Of these procedures, the author would suggest checking compliance with SAE J2364 first because it is the easiest of the procedures to complete, especially the static task time procedure. Admittedly, static task time does not capture all the key characteristics of task interference (Young et al.[42]), but it is the only measure for which an accepted standard exists with specific criteria.

It must be kept in mind that compliance with all these guidelines and standards is voluntary, although concerns of negative outcomes in product liability actions result in OEMs and major suppliers complying with the SAE Recommended Practice in the United States, and pressure to conform leads to compliance with JAMA guidelines in Japan. ISO design guidelines are used worldwide but, except for the legibility standard (15008), there are no concrete performance criteria, only personal judgments.

With engineering decisions, hard specifications (e.g., numeric values for operating temperatures or maximum weight) always take precedence over soft specifications (making it "safe," "easy to use," or "not distracting"). The situation is even worse when no specifications or no performance criteria are provided.

If ISO and other organizations are not willing to set numeric performance criteria for safety, safety will be shortchanged, and systems that are distracting to drivers will continue to be produced.

REFERENCES

1. Schindhelm, R., Gelau, C., Keinath, A., Bengler, K., Kussmann, H., Kompfner, P., Cacciabue, P.C., and Martinetto, M., Report on the review of available guidelines and standards (AIDE deliverable 4.3.1), European Union, Brussels, Belgium, 2004. Retrieved on July 15, 2007, available at:www.aide-eu.org/pdf/aide_d4-3-1_draft.pdf.
2. International Organization for Standardization, Stages of the development of international standards, 2007. Retrieved July 15, 2007, available at: www.iso.org/iso/en/stds-development/whowhenhow/proc/proc.html.
3. Society of Automotive Engineers, ITS safety and human factors committee public forum, 2007. Retrieved on July 15, 2007, available at: forums.sae.org/access/dispatch.cgi/TEITSSHFpf/saveWS/automotive.
4. Society of Automotive Engineers, Technical standards board governance policy (revision 8), Society of Automotive Engineers. Warrendale, PA, 2007. Retrieved on July 15, 2007, from www.sae.org/standardsdev/tsb/tsbpolicy.pdf.
5. National Highway Traffic Safety Administration, Federal motor vehicle safety standards, 2007. Retrieved July 15, 2007, available at: www.access.gpo.gov/nara/cfr/waisidx_01/49cfr571_01.html.
6. Transport Canada, Motor vehicle safety regulations, 2007. Retrieved on July 15, 2007, available at: www.tc.gc.ca/acts-regulations/GENERAL/m/mvsa/regulations/mvsrg/ toc_mvsrg.htm.
7. Public Broadcasting System, *Frontline* program—Rollover: The hidden history of the SUV (aired February 21, 2002), 2002. Retrieved August 26, 2007 available at: www.pbs.org/wgbh/pages/frontline/shows/rollover/.
8. Green, P., Levison, W., Paelke, G., and Serafin, C., Preliminary human factors guidelines for driver information systems, Technical Report FHWA-RD-94-087, U.S. Department of Transportation, Federal Highway Administration, McLean, VA, 1995.
9. Campbell, J.L., Carney, C., and Kantowitz, B.H., Human factors design guidelines for advanced traveler information systems (ATIS) and commercial vehicle operations (CVO), Technical Report FHWA-RD-98-057, U.S. Department of Transportation, Federal Highway Administration, Washington, D.C., 1997.
10. Alliance of Automobile Manufacturers, Statement of principles on human-machine interfaces (HMI) for in-vehicle information and communication systems, Version 3.0. Washington, D.C.: Alliance of Automobile Manufacturers, 2003. Retrieved July 15, 2007, available at: www.umich.edu/~driving/guidelines/AAM_Statement2003.pdf.
11. Commission of the European Communities, Statement of principles on human machine interface (HMI) for in-vehicle information and communication systems (Annex 1 to

Commission Recommendation of 21 December 1999 on safe and efficient in-vehicle information and communication systems: A European statement of principles on human machine interface), European Union, Brussels, Belgium, 1999.

12. Commission of the European Communities, European statement of principles on the design of human machine interaction (ESoP 2005) draft, 2005, (Retrieved August 15, 2007, available at: ec.europa.eu/information_society/activities/esafety/doc/esafety_library/esop_hmi_statement.pdf), Commission of the European Communities, Brussels, Belgium.
13. Ross, T., Midtland, K., Fuchs, M., Pauzie, A., Engert, A., Duncan, B., Vaughan, G., Vernet, M., Peters, H., Burnett, G., and May, A., HARDIE design guidelines handbook: Human factors guidelines for information presentation by ATT systems, Commission of the European Communities, Luxembourg, 1996.
14. Japan Automobile Manufacturers Association, JAMA guideline for in-vehicle display systems, version 3.0. Japan Automobile Manufacturers Association, Tokyo, Japan, 2004 Retrieved July 15, 2007, available at: www.jama.or.jp/safe/guideline/pdf/jama_guideline_v30_en.pdf.
15. Society of Automotive Engineers, Navigation and route guidance function accessibility while driving (SAE recommended practice 2364), Society of Automotive Engineers, Warrendale, PA, 2004.
16. Society of Automotive Engineers, Rationale document for SAE J2364 (SAE information report J2678), Society of Automotive Engineers, Warrendale, PA, 2004.
17. Society of Automotive Engineers, Calculation of the time to complete in-vehicle navigation and route guidance tasks (SAE recommended practice J2365), Society of Automotive Engineers, Warrendale, PA, 2002.
18. Stevens, A., Board, A., Allen, P., and Quimby, A., A safety checklist for the assessment of in-vehicle information systems: A user's manual, Report No. PA 3536/99, Transport Research Laboratory, Crownthorne, Berkshire, UK, 1999.
19. Stevens, A., Quimby, A., Board, A., Kersloot, T., and Bur, P., Design guidelines for safety and in-vehicle information systems, Technical Report PA3721/01, Transport Research Laboratory, Crowthorne, UK, 2002.
20. U.S. Department of Defense, Department of Defense design criteria standard—Human engineering (Military Standard 1472F), U.S. Department of Defense, Washington, D.C., 1999. Retrieved on July 15, 2007, available at: http://hfetag.dtic.mil/docs-hfs/mil-std-1472f.pdf.
21. Hankey, J.M., Dingus, T.A., Hanowski, R., Wierwille, W.W., and Andrews, C., In-vehicle information systems behavioral model and design support: Final report, Technical Report FHWA RD-00-135, U.S. Department of Transportation, Federal Highway Administration, McLean, VA, 2000.
22. Green, P., Estimating compliance with the 15-second rule for driver-interface usability and safety, Proceedings of the Human Factors and Ergonomics Society 43rd Annual Meeting (CD-ROM). Human Factors and Ergonomics Society, Santa Monica, CA, 1999.
23. Green, P., Navigation system data entry: Estimation of task times, Technical Report UMTRI-99-17, University of Michigan Transportation Research Institute, Ann Arbor, MI, 1999.
24. Green, P., The 15-second rule for driver information systems, ITS America Ninth Annual Meeting Conference Proceedings (CD-ROM), Intelligent Transportation Society of America, Washington, D.C., 1999.
25. Nowakowski, C. and Green, P., Prediction of menu selection times parked and while driving using the SAE J2365 method, Technical Report 2000–49, University of Michigan Transportation Research Institute, Ann Arbor, MI, 2000.
26. Nowakowski, C., Utsui, Y., and Green, P., Navigation system evaluation: The effects of driver workload and input devices on destination entry time and driving performance

and their implications to the SAE recommended practice, Technical Report UMTRI-2000–20, University of Michigan Transportation Research Institute, Ann Arbor, MI, 2000.
27. Alliance of Automobile Manufacturers, Statement of principles on human-machine interfaces (HMI) for in-vehicle information and communication systems, Version 2.0, Alliance of Automobile Manufacturers, Washington, D.C., 2002. Retrieved July 15, 2007, available at: www.autoalliance.org/archives/driver_guidelines.pdf.
28. Go, E., Morton, A. Famewo, J. and Angel, H., Final report: Evaluation of industry safety principles for in-vehicle information and communication systems, Transport Canada, Ottawa, Canada, 2006.
29. Pettitt, M., Burnett, G. and Karbassioun, D., Applying the keystroke level model in a driving context, Proceedings of the Ergonomics Society Annual Meeting, 2006.
30. Pettitt, M., Burnett, G. and Stevens, A., An extended keystroke level model (KLM) for predicting the visual demand of in-vehicle information system, CHI 2007, 2007.
31. International Organization for Standardization, Road vehicles—Ergonomic aspects of transport information and control systems—Dialogue management principles and compliance procedures, ISO Standard 15005:2002(E), Geneva, Switzerland, 2002.
32. International Organization for Standardization, Road vehicles—Ergonomic aspects of transport information and control systems—Specifications and compliance procedures for in-vehicle auditory presentation, ISO Standard 15006: 2004, Geneva, Switzerland, 2004.
33. International Organization for Standardization, Road vehicles—Measurement of driver visual behavior with respect to transport information and control systems—Part 1: Definitions and parameters (ISO Standard 15007-1: 2002), Geneva, Switzerland, 2002.
34. International Organization for Standardization, Road vehicles—Measurement of driver visual behavior with respect to transport information and control systems—Part 2: Equipment and procedures (ISO Technical Specification 15007-2: 2001), Geneva, Switzerland, 2001.
35. International Organization for Standardization, Road vehicles—Ergonomic aspects of transport information and control systems—Specifications and compliance procedures for in-vehicle visual presentation (ISO Standard 15008: 2003), Geneva, Switzerland, 2003.
36. International Organization for Standardization, Road vehicles—Ergonomic aspects of in-vehicle presentation for transport information and controls systems—Warning systems (ISO Technical Report 16352: 2005), Geneva, Switzerland, 2005.
37. International Organization for Standardization, Road vehicles—Ergonomic aspects of transport information and control systems (TICS)—Procedures for determining priority of on-board messages presented to drivers (ISO Trial Standard 16951: 2004), Geneva, Switzerland, 2004.
38. International Organization for Standardization, Road vehicles—Ergonomic aspects of transport information and control systems—Procedure for assessing suitability for use while driving (ISO Standard 17287:2002), Geneva, Switzerland, 2002.
39. International Organization for Standardization, Road vehicles—Ergonomic aspects of transport information and control systems—Occlusion method to assess visual distraction due to the use of in-vehicle information and communication systems (ISO Standard 16673: 2007), Geneva, Switzerland, 2007.
40. International Organization for Standardization, Road vehicles—Ergonomic aspects of transport information and control systems—Simulated lane change test to assess in-vehicle secondary task demand (ISO Committee Draft Standard 26022: 2007), Geneva, Switzerland, 2007.

41. Young, R.A. and Angell, L., The dimensions of driver performance during secondary manual tasks, Driving Assessment 2005, University of Iowa, Iowa City, Iowa, 2003. Retrieved on August 25, 2007, available at: ppc.uiowa.edu/driving-assessment/ 2003/ Summaries/Downloads/Final_Papers/PDF/25_Youngformat.pdf.
42. Young, R., Aryal, B., Muresan, M., Ding, X., Oja, S. and Simpson, N., Road-to-lab: Validation of the static load test for predicting on-road driving performance while using advanced in-vehicle information and communication devices, Driving Assessment 2005, University of Iowa, Iowa City, Iowa, 2005. Retrieved on July 15, 2007, available at: ppc.uiowa.edu/driving-assessment/2005/final/index.htm.

Part 8

Prevention and Mitigation Strategies

26 Real-Time Distraction Countermeasures

Johan Engström and Trent W. Victor

CONTENTS

26.1 INTRODUCTION

In general terms, distraction can be defined as misallocated attention.[1] In the context of driving, distraction could be induced by a range of activities such as looking after children, looking for road signs, applying makeup, and using in-vehicle information systems. Distraction may also be purely "internally" triggered, for example when daydreaming. A large body of empirical research links driver distraction to degraded driving performance, for example, reduced lateral control, reduced event and object detection performance, and impaired decision making (see, e.g., Young et al.[2] for a review). Distraction has also repeatedly been identified as a major contributing factor in crashes,[3–5] although direct causal links between distraction-induced performance degradation and actual crash risk have been difficult to establish empirically. This is due mainly to the methodological difficulties associated with collecting sufficiently detailed precrash data. However, recent results from naturalistic field studies, such as the 100-car study,[6,7] have contributed to bridging this gap (see Chapters 16 and 17), by demonstrating significant increases in (relative) risk resulting from driver engagement in a variety of distracting activities.

The driver distraction problem can be tackled in many different ways, for example by raising public awareness of the risks and optimizing the design of in-vehicle systems to reduce the potential for distraction. However, the present chapter focuses on the further possibility of distraction countermeasures that operate in *real time* while driving. In recent years, significant progress has been made in this area, and some first-generation systems have even entered the market.

The main objective of this chapter is to provide a general introduction to the emerging field of real-time distraction countermeasure (RDC) research by reviewing the key developments and providing some detailed examples of how RDC systems might work in practice. The next section defines the RDC concept. Following that, the two main classes of existing RDC functions—real-time distraction *prevention* (with the focus on *workload management)* and real-time distraction *mitigation* (with the focus on *distraction warning/feedback)*—are discussed, including a review of existing developments in each area. In addition, the use of distraction monitoring technologies for enhancing warning functions is addressed in Section 26.5. The chapter ends with a general discussion and some suggestions for future research.

26.2 THE CONCEPT OF REAL-TIME DISTRACTION COUNTERMEASURES

RDCs can be categorized in different ways. Donmez et al.[8,9] adopt a broad perspective where *distraction mitigation strategies* are classified along three dimensions: (1) degree of automation (low-high), (2) type of initiation (system initiated versus driver initiated), and (3) type of task (driving versus non–driving related). This taxonomy includes a broad range of driving support functions, from collision warning and mitigation (moderate to high level of automation, system initiated, driving related) to, for example, prioritization and filtering of information (moderate level of automation, system initiated, non–driving related). It also includes driver-initiated strategies, which rely on the driver to modulate his or her own level of distraction on the basis of the perceived level of distraction.

In this chapter, a slightly different perspective is adopted. First, the more general term "countermeasures" is used rather than "mitigation," since the chapter deals with both functions that *prevent* and functions that *mitigate* distraction. Second, for the present purposes a somewhat simpler taxonomy is adopted, which is based on when the countermeasures take effect in the causal chain leading to an accident. RDCs are viewed here as a subset of the more general category of *driving support functions*; that is, in-vehicle functions with the primary purpose of supporting one or more aspects of the driving task. RDC functions are distinguished from other driving support functions, such as collision warning and mitigation functions (e.g., lane-keeping support, forward collision warning, blind spot warning, and collision mitigation) in that they intervene earlier in the causal chain. Whereas collision warning and mitigation functions address critical decrements of driving performance (such as lane departures or close encounters with lead vehicles), or attempt to mitigate the

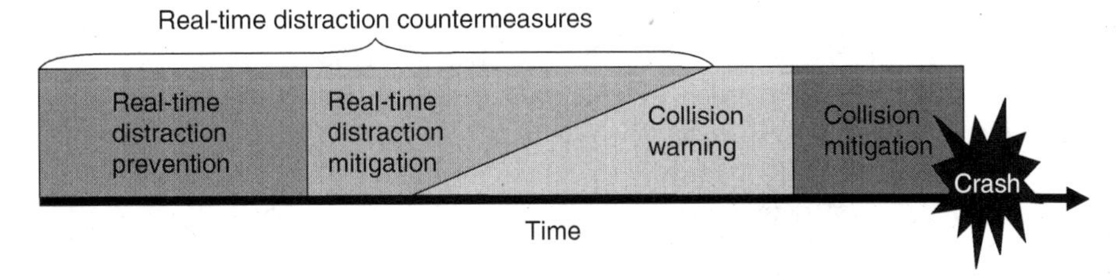

FIGURE 26.1 Proposed relation between the two proposed types of RDC functions and collision warning/mitigation functions. As indicated, there is a certain overlap between distraction mitigation and collision warning functions since it is possible to envision distraction mitigation functions that also take into account performance decrements in near-crash situations. However, most existing functions reviewed in this chapter are focusing specifically on distraction mitigation.

consequences of the crash (e.g., by automatic application of the brakes when the crash is unavoidable), the goal of RDC functions is to prevent or mitigate distraction that might lead to performance decrements and, ultimately, crashes. A further distinction is made between two general types of RDC functions:

- *Real-time distraction prevention*: This includes functions that serve the main purpose of *preventing* mental overload or distraction from occurring in the first place; for example, by prioritizing and scheduling system-initiated information according to the current driving situation or driver state. These types of functions are also commonly known as *workload managers* (e.g., Ref. 10).
- *Real-time distraction mitigation*: This refers to functions that serve the main purpose of *mitigating* distraction once it occurs; for example, by redirecting the driver's attention to the relevant aspects of the driving task, for example by means of visual or sound alerts.

Figure 26.1 describes the relationship between the two proposed general types of RDCs.

Another type of functionality, which is sometimes considered part of workload management, is *driver-vehicle-environment* (*DVE*)*-adaptive collision warning*. This refers to the adaptation, for example, of the timing and intensity of warnings from a forward collision warning function based on real-time monitoring of the DVE state (e.g., distraction). According to the present taxonomy, this should be viewed as an enhanced collision warning function, rather than as an RDC function. DVE-adaptive warnings functions are addressed separately in Section 26.5.

26.3 REAL-TIME DISTRACTION PREVENTION BY MEANS OF WORKLOAD MANAGEMENT

Whereas distraction is defined on the basis of attention allocation, mental workload refers to the amount of resources required to perform one or more tasks, relative to the resources available.[11] Thus, mental workload is causally linked to distraction by the fact that exhausted mental resources increase the risk for misallocated attention.

However, high workload is not a necessary precondition for distraction, since misallocated attention may be caused by low-workload activities as well, for example, daydreaming or looking at roadside billboard signs.

The workload imposed on the driver while driving, and the associated potential for distraction, changes dynamically with the driving situation. This leads to short "workload peaks" in demanding driving situations (e.g., when negotiating intersections or when overtaking), which may coincide with other resource-demanding tasks (e.g., talking to a passenger or answering an incoming phone call). Normally, the driver manages these workload peaks by adapting to the driving situation (e.g., by slowing down, interrupting the conversation, or refraining from answering the phone). However, drivers may not always be able to anticipate a demanding or risky situation and adapt accordingly. Moreover, the psychological attraction of many secondary tasks is very strong; for instance, many people find it hard to resist answering an incoming phone call, even in very demanding driving situations.[10] The basic idea behind workload management (WM) functions is to prevent excessive workload and distraction by dynamically supporting the driver to manage the different driving- and nondriving-related tasks, in particular by controlling the information initiated by in-vehicle systems and by limiting the system functionality available to the driver in demanding, or potentially demanding, situations.

The basic ideas behind workload management can be traced back to the aviation domain, in particular the Pilot's Associate system developed by the U.S. Army.[12] The first comprehensive effort in this area in the automotive domain was the Generic Intelligent Driver Support (GIDS), an EU-funded project, conducted between 1989 and 1992 as part of the PROMETHEUS program.[13] This was followed by a number of European projects, including CEMVOCAS,[14] COMUNICAR,[15] and the German SANTOS project.[16] The most recent effort in this line of European research is the Adaptive Integrated Driver-Vehicle Interface (AIDE) project.[17] In the United States, similar research is being conducted in the SAfety VEhicle(s) using adaptive Interface Technology (SAVE-IT) project[8,9,18] funded by the U.S. National Highway Traffic Safety Administration (NHTSA). In addition, several organizations have conducted extensive in-house development of WM systems, including CoDrive by the Dutch institute TNO[19] and Motorola's Driver Advocate.[20] However, to date, only two WM systems have entered the market: the Saab Dialogue Manager and the Volvo Car Intelligent Driver Information System (IDIS).[21] Both systems focus on the rescheduling of information in demanding driving situations on the basis of real-time workload estimation from vehicle sensor signals such as speed, acceleration, steering wheel angle, foot pedal position, gear position, and the use of input controls. The next generation of IDIS will feature more advanced workload estimation as well as centralized management of information from all types of onboard applications (see Ref. 21). The reader is referred to Green[10] and Hoedemaeker et al.[22] for general reviews of existing workload manager developments.

26.3.1 Workload Management Functions

There are several different kinds of WM functions. The most common is *information scheduling*, with the general purpose to ensure that the driver receives information only when it is needed and when he or she is available to receive it. This may involve prioritization between concurrent system messages as well as the interruption or rescheduling of system-initiated information in certain driving situations. An example of the latter is the delay of an incoming phone call or an e-mail received in the middle of a busy intersection until the intersection has been passed. A further possibility is *function lockout*, which involves the entire disabling of a function or subfunction in certain conditions.

Another type of WM function is *adaptation of information format*, which involves altering the way the information is presented (not just the timing) to the current context (i.e., the driving situation or the presence of concurrent messages). For example, if the driver is highly visually loaded, it may be desirable to present an important text message in auditory format (e.g., as a text-to-speech message). Moreover, the graphical presentation of information could be simplified (e.g., through decluttering) in demanding driving environments (e.g., in city driving).

26.3.2 Basic Principles of Workload Management: An Example from the AIDE Project

A key component of most existing WM systems is a centralized manager that implements the global prioritization and scheduling logic. To illustrate how this might work, we will focus on an example from the AIDE project, mentioned earlier. AIDE is an EU-funded project, finished in 2008, dealing with technological development, evaluation methods, as well as behavioral effects related to adaptive and integrated automotive driver-vehicle interfaces (see Ref. 17 for an overview). In the solution developed in AIDE, the main part of the WM functionality is controlled by the *Interaction and Communication Assistant (ICA)* module, which works as a central scheduler of information initiated by individual applications.[23] A central concept in AIDE is the *action*. An action is defined as any event that is initiated by the driver or an application, for example, issuing a warning (application initiated), displaying a text message (application initiated), or entering a destination on the navigation system (driver initiated). Every time an application wants to initiate an (HMI-related) action, it has to ask the ICA for permission, using a predefined protocol known as the *Application Request Vector*. The ICA then checks the current driver–vehicle–environment (DVE) state, represented in the *DVE vector*, as well as the presence of other ongoing or incoming requests and sends back a reply to the application (using a prespecified *Reply Vector*). The application then acts accordingly (e.g., by executing the action as planned or by modifying, postponing, or terminating the action).

The *Application Request Vector (ARV)* contains an abstract description of the intended action; for example, in terms of its Initiator {User, System}, Duration {Transient, Sustained}, Safety Criticality {None, Low, High}, or Time Criticality {None, Low, High}. The ARV also contains information about the requested output channels (display, part of a display, sound system, etc.). The *Reply Vector* (RV) contains the answer {YES, NO} and (if "YES") information on which of the requested output channels are available.

The general role of the *DVE modules* is to provide a real-time estimate of the relevant aspects of the current, and predicted, DVE state. Whereas many existing WM systems (e.g., GIDS,[13] CEMVOCAS,[14] and COMUNICAR[15]) use a single, generic, "workload," "situation," or "distraction" estimate, the AIDE approach is based on the assumption that different WM functions may require different, often quite specific,

DVE parameters. Thus, the DVE parameters used in AIDE have been derived on the basis of a detailed analysis of the use cases and scenarios to be handled by the system. In total, the AIDE system contains five DVE modules. The outputs from these modules are concatenated into the *DVE vector*, which is continuously provided both to the ICA and directly to the applications. The DVE modules take input from a shared array of sensors and other sources of information such as a digital map database and stored information on driver's profiles. The roles of the individual modules are summarized in Table 26.1.

TABLE 26.1
Overview of the AIDE DVE Modules

Module	Output	Supported WM Functions
Driver availability assessment module (DAA)	Computes an estimate of the driver's availability to receive information, as determined by the demand of the primary driving task. This is based on vehicle sensors, as well as digital map data	Information scheduling. May also be used to decide on function lockout and adaptation of information format
Traffic and environment risk assessment (TERA) module	Computes estimates of traffic and environment risk mainly based on environment sensing, but also digital map data. This also includes estimation of drivers' intent to perform driving maneuvers, partly based on input from the CAA module	The traffic risk estimate is mainly used for information scheduling. The main purpose of the intent estimate is to support DVE-adaptive warning strategies
Driver state degradation (DSD) module	Computes an estimation of the current drowsiness state, based on a combination of eye closure monitoring, lane tracking and vehicle sensor data	DVE-adaptive warning strategies
Cockpit activity assessment (CAA) module	Computes online estimates of the driver's secondary task activity and (visual and cognitive) distraction level. The module also provides input to the intent computation in the TERA. The main input is eye/head tracking data, but other information sources such as lane tracking data and in-vehicle button presses are used as well	All the CAA outputs can be used to support DVE-adaptive warning strategies. However, the visual time-sharing estimate may also be used for certain information scheduling purposes (e.g., postpone presentation of noncritical information until the driver has finished his/her task)
Driver characteristics (DC) module	Maintains a static record of driver characteristics and preference, based on driver identification (e.g., by means of a smart card) but also computes a dynamic profile of driver characteristics (e.g., reaction time, preferred following distance, etc.) computed from sensor data	This information is used to personalize the ICA strategies as well as the functionality of the individual applications (e.g., warning timing and intensity)

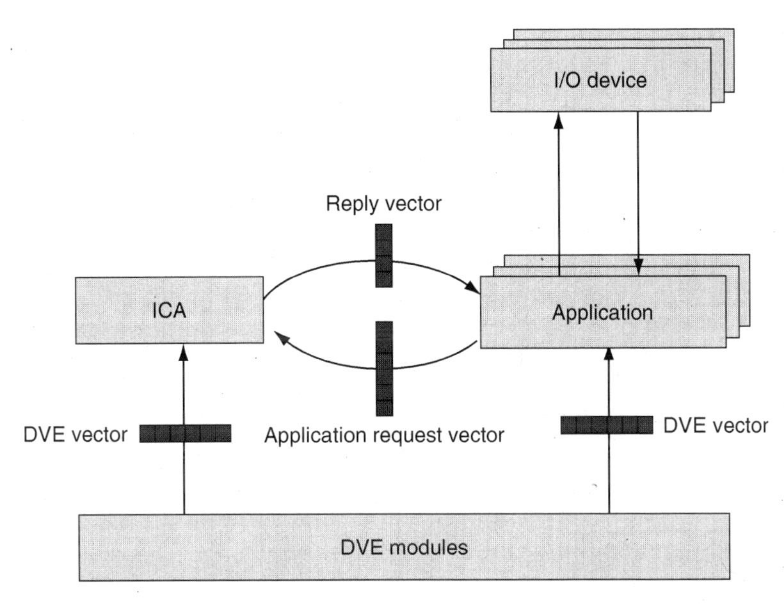

FIGURE 26.2 Illustration of the basic principles behind the AIDE workload management system.

The ICA information management logic works in four basic steps. First, the requested action (defined by the ARV) is assigned a priority class (this solution gives the vehicle manufacturer, rather than the system supplier, control over the message prioritization). Second, the information is filtered on the basis of the current DVE conditions (defined by the DVE vector), and it is decided whether the action could be executed as planned or delayed/canceled. Third, the format of the action is adapted according to the DVE conditions (if needed). In the final step, it is decided which input or output devices could be allocated to the requesting system (see Ref. 23 for further details regarding the ICA logic).

The general principles behind the AIDE architecture are illustrated in Figure 26.2. The reader is referred to Kussmann et al.[24] for a more detailed description, and to Broström et al.[21] for an industrially developed solution based on similar principles.

26.3.3 Driver-Vehicle-Environment Monitoring for Workload Management

As described in the previous section, most WM functions depend on the monitoring of relevant aspects of the DVE state, for example, driving demand, secondary task demand, driver impairment, traffic risk, and individual driver characteristics (see Table 26.1). The relevance of the different parameters naturally depends on the needs of the specific WM functions to be implemented. In the following example, we will focus specifically on real-time DVE monitoring parameters relevant for realizing information-scheduling functions.

It should be noted that very accurate estimation of driver workload is neither a necessary nor sufficient condition for realizing WM functions. Rather, the key issue is to identify the *situations* where a certain type of information presentation should be altered (e.g., when an incoming phone call should be put on hold). These situations *may* involve high driver workload, but this does not have to be the case, as further illustrated.

The first step in defining a DVE monitoring algorithm for information scheduling is thus to define the target situations to be detected; that is, situations where certain types of information should be rescheduled. One way of doing this is to let experts annotate video recordings of driving situations (recorded from the driver's view) and indicate the situations with respect to where the driver is "available to receive a message of type X."[14,25,26] The target situations can also be identified analytically, for example, by using some formal taxonomy of driving situations such as the Fastenmeier scheme.[27]*

The second step is to define an algorithm for detecting the target situations in real time on the basis of available sensor data, for example, vehicle signals such as speed, acceleration, steering wheel movements, and pedal use. Another type of data that could be very useful is information derived from digital maps.[16,21,25] Several types of situation-detection algorithms have been proposed in the literature, including statistical models such as neural networks trained with the expert annotations as "ground truth,"[14,26] or rule-based approaches.[25,26] Another method is to use *lookup tables*, which map specific driving situations (defined, e.g., in terms of the Fastenmeier scheme[27]) to demand or workload values that have been previously determined in empirical studies. The lookup table approach was adopted, for example, in the GIDS and SANTOS projects.[13,16]

Figure 26.3 illustrates some real data from neural-network and rule-based situation detectors, described in Lööf.[26] The graph also includes expert annotations of the target situations (representing situations when the experts judged that an incoming phone call should be put on hold) and the momentary driver workload measured by the Tactile Detection Task (TDT) methodology (see Ref. 28 for further details on the TDT). It could be noted in the figure that there are several target situations (annotated by the experts) in which the actual workload (as measured by the TDT) stays low. One example is the second situation encountered, where the driver passes through a green light at an intersection. Even though the workload is low, it could be reasonable to block incoming phone calls in such situations because the risk of encountering unexpected events (e.g., a pedestrian entering the road) is relatively high. Since the situation detection algorithms in this example are based on vehicle data only, they are able to detect only those target situations where vehicle handling is affected (i.e., the first, third, and fourth situation). Situations like the second one encountered in Figure 26.3 are possible to detect by means of digital maps. In this example, it could also be noted that the rule-based model has the important advantage of reacting slightly faster to the situation, compared with the neural network.

26.3.4 Evaluation of Workload Management Functions

Several studies have demonstrated a high level of user acceptance for WM functions. For example, Bellet et al.[14] found that 43% of the nonmanaged messages in the CEMVOCAS system were considered "disturbing" by the driver compared with

* The Fastenmeier scheme is a general taxonomy for describing driving situations, mainly in terms of their demands for information processing and vehicle handling.

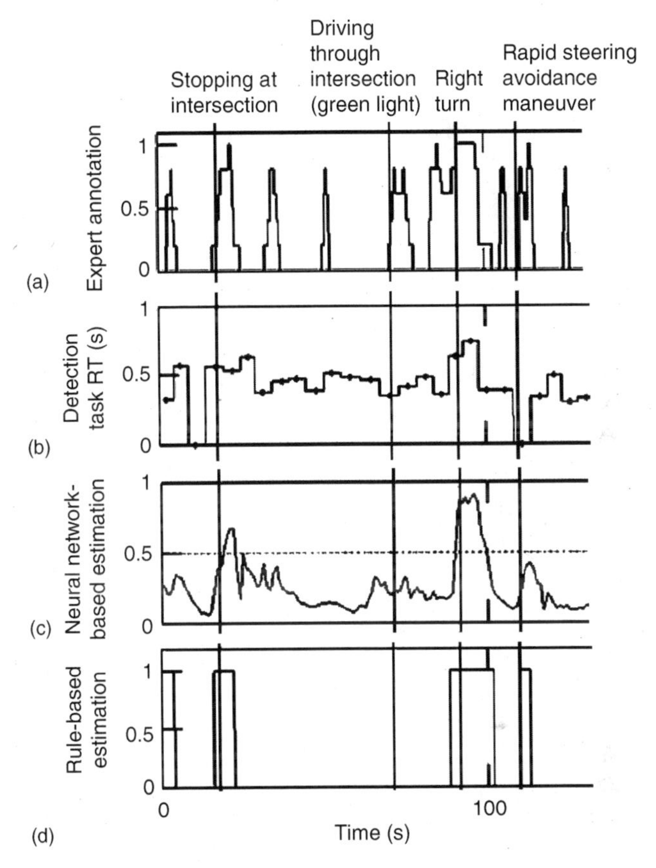

FIGURE 26.3 Examples of real-time driving situation detection for workload management in real traffic, adapted from Lööf.[26] The two upper panels show two different types of "ground truth": (a) average values of experts' manual annotations of situations when an incoming phone call should be blocked, on the basis of video recordings (1 = blocking situation) and (b) momentary workload measured in terms of performance on a tactile detection task (where zero values mean missed detection). The two lower panels show the output of two different real-time situation detectors: (c) A neural network model trained to approximate the expert annotations and (d) a rule-based model where parameters have been determined by hand. The estimation in these models is based on vehicle data only. The vertical lines indicate different driving situations encountered. (Adapted from Lööf, J., *Statistical Models for Real-time Driver Workload Estimation*, MSc thesis, Chalmers University, 2003, with permission.)

only 5% of the messages scheduled by the system. Moreover, the majority of the participants in the study considered the CEMVOCAS system to contribute "significantly" or "very significantly" to road safety. However, results demonstrating actual workload reduction and driving performance improvement effects of WM systems are less common. Hoedemaeker et al.[29] performed an evaluation of information scheduling, as implemented by the COMUNICAR system. The study compared the subjective workload (as measured by the Rating Scale Mental Effort method) in two conditions: (1) multiple messages presented in demanding driving situations (such as when overtaking or driving through intersections) and (2) the same situations with the messages rescheduled by a workload manager until just after the situations. However, no significant differences between the conditions were found.

Two recent studies, performed in the AIDE and the SAVE-IT projects, respectively, have demonstrated that workload management can indeed improve driving performance. In the AIDE study, it was found that different WM functions, in particular information scheduling, had a beneficial effect on driving performance.[30] In this driving simulator study, driving performance was assessed by the INVENT Traffic Safety Assessment (I-TSA) methodology, where a large number of performance metrics have been combined into a set of traffic safety indicators. The I-TSA safety indicators used in this study were (1) longitudinal control (time headway, time to collision); (2) lateral control, where the left- and right-hand sides were treated separately (standard deviation of lane position, time-to-line-crossing, lane departure, lateral distance to vehicle while passing); (3) objective mental workload (steering entropy, steering wheel reversal rate, speed); and (4) subjective assessment of mental workload. It was found that the AIDE workload management yielded significantly higher safety scores for all but one indicator (lateral control left) compared with non-managed interaction with different in-vehicle information systems. However, these effects were only obtained in more difficult driving scenarios (where difficulty was varied in terms of traffic density and curvature).

The SAVE-IT study[8] investigated the impact of function locking and advising strategies on driving performance. We will focus here on the results for function locking. In the study, distraction was imposed by means of visual and auditory secondary tasks. In the locking condition, the secondary tasks were interrupted in demanding situations (lead vehicle suddenly braking and curve entry). Performance improvements were found for both visual and auditory locking (compared with performing the secondary task without interruption), although the effects differed between the two task types. For the auditory task it was found that locking resulted in drivers adopting greater safety margins as indicated by longer minimum time to collision (TTC) and less deceleration in the braking situations. By contrast, locking out the visual task led to improved lateral control in curves (in terms of less erratic steering). The results from the study also pointed to the importance of taking driver adaptation into account when evaluating these types of functions. For instance, because of risk compensation when performing the visual secondary task, the speed when entering the curve was higher (riskier behavior) with the locking strategy compared with secondary task with no locking.

Two important conclusions from these studies are that (1) the benefits of information scheduling are mainly found in highly demanding driving situations and (2) that the effects depend strongly on the type of information that is scheduled. Moreover, driver adaptation plays a key role in determining the net safety effect. More research is needed to better understand the mechanisms underlying drivers' responses to WM functions and how the potential safety benefits of these types of systems are best quantified.

26.4 REAL-TIME DISTRACTION MITIGATION

How do you recognize when a driver is excessively distracted? What would you do as a passenger if your driver was distracted? This section describes work aimed at giving vehicles the ability to recognize driver distraction when it occurs and the means with which to do something about it. The basic idea behind real-time distraction mitigation is to provide feedback to help the driver shift attention back to driving when he or she is judged as being "too distracted" according to predetermined criteria set by the system, the driver, or the owner. It is challenging to define in real time what is "too distracting," but, as we shall see, several successful attempts to do so have been made. The distraction feedback, signifying that more attention should be allocated to driving right now, is provided to help the driver realize that he or she is being "tricked" into a distractive behavior. In addition to the immediate effect of redirecting attention toward the critical aspects of the driving situation, distraction feedback may also result in positive long-term behavioral changes (e.g., safer visual allocation strategies).

The first efforts in this area can be traced to a national Swedish project conducted by Volvo called VISREC,[31] which was initiated to investigate the use of promising developments in dash-mounted eye-tracking technology to create distraction countermeasure functions.[32,33] This line of research was followed by SAAB-initiated project work,[34] and current collaborative efforts, in particular SAVE-IT,[8,9,18,39] AIDE,[17,35] and Fletcher.[36]

26.4.1 Distraction Mitigation Functions

The most commonly investigated distraction mitigation function is the *visual distraction alert*. The basic idea here is to help the driver to realize that he or she is glancing away from the road for too long or too often, and to "train" the driver to recognize a limit (see algorithm descriptions in the following sections). As such, it alerts the driver to inappropriate visual behavior and does not necessarily have a direct coupling to driving performance deterioration. A distraction alert could be issued in many different ways. The VISREC project tested several different possibilities, such as flashing light emitting diodes (LEDs), icons, tones, seat vibration, and voice messages[32,33,37] in desktop, driving simulator, and on-road environments. In one implementation (Figure 26.4a), five strings of LEDs were attached along the upper interior of the doors and the dashboard to create waves of light toward the road center. A later version used only three brighter LEDs (see Figure 26.4b). A "tickle" warning, provided by means of sparkling LEDs reflected in the windshield, was activated as a gentle prewarning during less serious distraction levels.

In the Volvo implementation,[38] visual distraction was estimated in real time on the basis of a combination of the following parameters: (1) the percentage of time that the gaze falls within a road center area (percent road center, PRC) over a 1 min running window, (2) a single-glance duration, and (3) detection of visual time-sharing behavior, calculated as a PRC running average using a 10 s time window.

A similar approach has been adopted in the SAVE-IT project,[39] where distraction was estimated on the basis of the current off-road glance duration (β_1) and the total off-road glance duration during the past 3 s (β_2), resulting in the function $\gamma = \alpha\beta_1 + (1 - \alpha)\beta_2$, where α determines the weight of the current glance duration.[39] Alerts were then given at two levels, defined by thresholds for γ (2 and 2.5 s were used as threshold values in this study).

The visual distraction alert prototype system developed by Saab Automobile is described by Karlsson.[34] The system featured two types of distraction feedback: a row

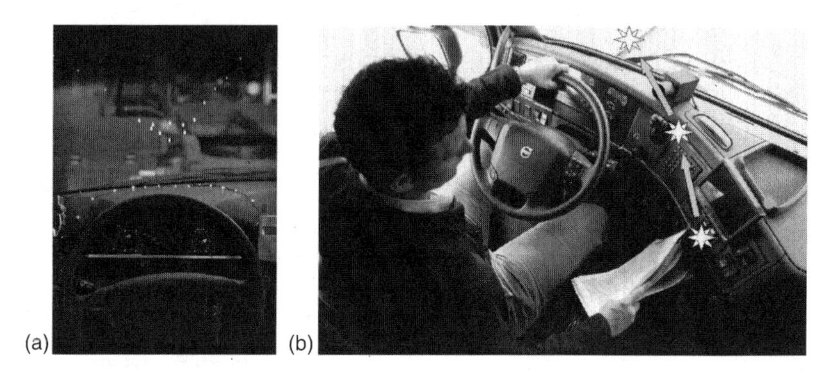

FIGURE 26.4 Examples of a visual distraction alert that redirects gaze using flashing LEDs. The first implementation (a) used strings of smaller LEDs running along the doors and center console toward the road center. The LEDs at the road center were reflected on the windshield (Photo: Lars Ardarve). A later version used three strong LEDs, as illustrated in (b) with the last LED reflecting in the windshield (Photo: Volvo).

of blue flashing LEDs reflected in the windshield at road center position, and a single kinesthetic brake pulse. The Saab distraction detection algorithm used the concept of an *attentional budget* that runs out if the driver looks away from the road too much, and receives "funding" when the driver looks back at the road.[34] This model has three parameters: (1) initial budget limit, (2) the rate at which the "budget" runs out when the driver is looking away, and (3) the rate at which the driver receives "funding" upon looking back to the road.

Similarly, Fletcher[36] used the Percent Road Center metric to reset a counter. Once the driver is observed to have a stable gaze at the road ahead, the counter and the warning is reset until the next diversion. As the driver's gaze diverges, the counter begins. The time period of permitted distraction is an inverse square function of speed. When the gaze has been diverted for more than a specific time period, an audible warning is given.

A further possibility is to provide a *cognitive distraction alert* in situations where the driver is cognitively distracted, that is, when excessive attention is directed to internal thoughts or auditory content (e.g., talking on a mobile phone, conversation, thinking), without a need for vision to achieve task goals. There are several behavioral indicators that may be used for real-time detection of cognitive distraction. In particular, it has been repeatedly demonstrated that cognitive distraction leads to a strong gaze concentration toward the road center (see Victor[33] for an overview of gaze concentration effects). Cognitive distraction has also been shown to induce a *reduction* in lane-keeping variance[40] (the opposite effect to that induced by visual distraction) as well as an increase in small steering wheel reversals, mainly below 2°.[41] Ongoing work in the AIDE project is focusing on combining such indicators with measures of gaze concentration for real-time cognitive distraction detection, on the basis of statistical pattern classification implemented by means of support vector machines (see Ref. 35 for preliminary results). The VISREC algorithm for cognitive distraction detection was based on gaze only and employed the same PRC metric that was used for the visual distraction alert described above, wherein a cognitive distraction alert was issued when PRC reached a certain threshold. The alert used LEDs reflected in the windshield in the center and toward the visual periphery. The LEDs flashed three times at road left and road right, thereby encouraging the driver to increase scanning of the road environment. Whereas this type of cognitive distraction alert is technically feasible, there are some confounding factors that might make it difficult to single out cognitive distraction from other effects. For example, like cognitive distraction, both increased driving demand and decreased road environment clutter lead to a gaze concentration to road center.[33] Also, the benefits of a cognitive distraction alert seem less obvious than those gained from the visual distraction alert. Other, probably more important, applications of real-time cognitive distraction monitoring are workload management (see Section 26.3) and warning adaptation (see Section 26.5 and Ref. 35).

26.4.2 Evaluation of Distraction Mitigation Functions

The Volvo visual distraction alert function, illustrated in Figure 26.4a, was evaluated by 30 truck drivers who tested a three-level warning strategy. All drivers stated that they noticed the alerts and responded to them by looking up at the road center. They ranked the distraction alert highly in comparison with other driving support functions (second to drowsiness alert) and indicated that they might like to purchase it if it was not too expensive, although they did not judge the function as essential.[37] By contrast, Karlsson[34] conducted an assessment of the potential driving performance enhancements of the Saab visual distraction alert system. However, no significant such effects were demonstrated. Further evaluation of the Saab distraction alert function is currently being performed within the Intelligent Vehicle Safety Systems (IVSS) research program in Sweden, but no public results are yet available.

A study with a similar scope was also conducted in the SAVE-IT project.[8] The distraction feedback led to a significant reduction of glance frequency to the display as well as longer glances to the road. However, as in the Saab study, no significant effects were found with respect to driving performance (which was measured in terms of accelerator release time, minimum TTC, minimum acceleration during lead vehicle braking events, and steering performance in curves). The study also investigated driver acceptance of the visual distraction alert function by means of the van der Laan usefulness and satisfaction scale[42] and found positive ratings for both dimensions, but only significantly for usefulness. Finally, drivers were asked if the visual distraction alert enhanced their driving performance on a Likert scale ranging from "strongly disagree" to "strongly agree" (where assessment was done for several versions of the system). Overall, the drivers answered "agree" or "strongly agree" for 57% of the responses, "neutral" for 34%, and "disagree" for 9%.

Another SAVE-IT study[39] evaluated visual distraction feedback that was given as a strip of LEDs on top of the dashboard/steering wheel and as yellow or orange strips on the top portion of an LCD display mounted on the center console. A significant change in drivers' interaction behavior with IVIS was shown, as drivers looked at the in-vehicle display less frequently. No significant driving performance benefits (in terms of lateral or longitudinal control) were observed. It was concluded that the distraction alert positively altered drivers' engagement in distracting activities, helping them attend to the roadway.

The lack of driving performance improvements in these studies may partly be due to methodological issues, such as difficulties in eliciting truly distracting situations *without driver expectation* of the driving performance-probing events. In general, more research is needed to determine the effectiveness of distraction mitigation functions in terms of performance and safety improvements. Studies are also needed that address the potential of distraction alert functions with respect to long-term improvements of driving performance. Possible routes for improvement of the alerting functions include further testing of alternative countermeasures, such as graded intensities of brake pulses, and creating more safety-critical distraction-detection algorithms that assess the *simultaneous* occurrence of distractive eye and head movement behavior and driving performance deterioration.

26.5 DRIVER-VEHICLE-ENVIRONMENT-ADAPTIVE COLLISION WARNING FUNCTIONS

Imagine that your spouse, who typically leaves you to handle critical driving situations, suddenly shouts and points at a hard-braking car because he or she sees that you are not paying attention to it. This situation illustrates the basic idea behind DVE-adaptive advanced collision warning functions, for which *DVE adaptation* potentially improves warning functionality with regard to effectiveness and acceptance. DVE states that warrant adaptation of warnings include eyes-off-road, visual time-sharing, cognitive distraction, high driving demand, driver impairment, driver intent of maneuvering, high traffic risk, and driver characteristics. Possible types of adaptation include altering the timing, intensity, duration, complexity, or modality of warnings. According to the present taxonomy, DVE adaptation is better considered as an enhancement of collision warning functionality than as an RDC function. This is because the primary *role* of the warning adaptivity is not to counteract distraction *per se* but rather to optimize collision warning functionality (see Figure 26.1).

The two main motivations for DVE adaptation are response time variability and driver acceptance issues. Response time (RT) varies significantly according to three main factors: (a) the degree of expectation, as unexpected and surprise events add an average of at least 0.4 s to RT[43]; (b) driver state, such as distraction, which also adds an average of 0.4 s to RT[44–46]; and (c) other driver-related factors, such as individual differences. It is particularly the simultaneous occurrence of loss of forward roadway vision and an *unexpected* event that has been shown to be the key causal factor in crashes and near-crashes.[6] A static (i.e., nonadaptive) warning will likely warn too late for an unexpected event that occurs when the driver is distracted because of the increased RT associated with unexpected events. Conversely, just as we become irritated if our passenger repeatedly warns us of a danger we are already aware of, we may become irritated if a collision warning system only confirms an already identified danger. Although the warning may be issued correctly, it may be viewed as a nuisance warning simply because it is not experienced as being relevant. Timing and relevance mismatches may lead to reduced effectiveness, increased annoyance, and poor acceptance.

Distraction-adaptive forward collision warning (FCW) is the most investigated DVE-adaptive warning function.[32,44–50] Such FCW systems generally adjust the warning *timing* based on where attention is allocated, although other types of adaptation (e.g., of warning intensity) are possible as well. Toyota has introduced a distraction-adaptive forward collision warning system on the market that uses horizontal facial orientation to estimate eyes-off-road.[49–51] An early warning in an FCW system is shown to significantly reduce the number of collisions and collision severity, compared with a late warning.[44,45] Opperud et al.[48] found that although a majority of the test respondents were highly appreciative of the idea of distraction-adaptive driving support systems, there was no clear indication of increased preference of distraction-adaptive FCW over static FCW. This was likely due to difficulties in identifying when an FCW was distraction adaptive. Roland et al.[47] found that the earlier warning is significantly more appreciated and useful, despite the number of warnings issued being significantly higher (almost doubled in the distraction-adaptive condition). Adaptive forward collision warnings were also investigated in a series of AIDE studies reported in Brouwer and Hoedemaeker.[46] Whereas no effects were found for distraction-adaptive warnings in this study, clear benefits in terms of safety and acceptance were found for *road friction-adaptive FCW* and *driving style-adaptive FCW* (aggressive/nonaggressive).

Distraction-adapted lane departure warnings either cancel a warning when the driver is attentive[32,33,48] or provide an earlier lane departure warning if distraction is detected.[47] Results are in favor of adaptation despite the facts that earlier warnings generally cause more warnings and that canceled warnings may create some confusion (Why didn't it warn now, is it because it malfunctioned or because I was not distracted?). A potential solution, which will be tested within the AIDE project, is to increase warning intensity when distraction is detected. For example, a subtle visual feedback or steering wheel vibration can be given when the driver is attentive and a stronger warning can be given when distraction is detected.

Further, *distraction-adapted curve speed warning (CSW)* involves issuing an earlier warning when distraction is detected.[47] Speed-keeping performance was shown to be significantly improved by the adaptation. However, this adaptation was considered less useful than distraction-adapted FCW and LDW possibly because of the higher frequency of alerts associated with the earlier distraction-adaptive warnings.[47] *Distraction-adaptive adaptive cruise control* involves automatically and gradually increasing the headway to the vehicle in front or changing the set speed if distraction is detected[32,33] to create more "reaction space." This adaptation has generally been considered less useful than others[48] but has not been fully investigated.

In conclusion, there is evidence that DVE-adaptive collision warnings improve safety. However, the results also indicate the importance of reducing the irritation that could be caused by higher warning frequency as a result of earlier warnings. Another conclusion is that system transparency, that is, the extent to which the driver is able to predict and understand the current adaptive state, is a key issue. Nontransparent, unpredictable system adaptation may seriously reduce the usability of the system, and thus, care must be taken when designing DVE-adaptive warnings to ensure that they have the intended effects, i.e., that they increase rather than reduce warning efficiency.

26.6 DISCUSSION AND CONCLUSIONS

This chapter has provided a review of existing developments in the area of real-time distraction countermeasures (RDCs), focusing on two general types of RDC functions: (1) distraction *prevention* by means of workload management and (2) distraction *mitigation* by means of distraction alert functions. The further possibility of using distraction monitoring for optimizing collision warnings was also briefly addressed. RDC is currently a very active research field, with some first-generation products already available on the market. However, the field is still rather immature, as evidenced in particular by a lack of common terms and concepts. The RDC field is also still to a large extent technology driven, and there is yet little consensus on which RDC functions are the most efficient and useful. For instance, there are many technological possibilities to enable real-time HMI adaptivity to the DVE state, but the full benefits of adaptive functions are not yet clear.

As reviewed earlier, several existing studies have demonstrated a high level of user acceptance of RDC functions. To date, driving performance improvements have mainly been demonstrated for WM functions, although studies of distraction alert functions have at least demonstrated safety-enhancing effects on visual time-sharing.[8,39] The lack of results showing actual driving performance improvement is probably at least partly due to the methodological difficulties associated with the evaluation of these types of systems in controlled experiments. A particular difficulty here concerns the construction of realistic critical driving scenarios. In the studies cited earlier, the scenarios (e.g., a lead vehicle braking) were always somewhat expected and it is likely that a distraction alert would have its strongest effect in unexpected critical situations. Moreover, the current studies have not addressed the potentially positive long-term effects of distraction feedback. Ultimately, the best way to investigate the real safety potential of RDC functions is by means of large-scale field operational tests (FOTs). Moreover, closer analysis of data from existing naturalistic driving (ND) studies (such as the 100-car study[6]) could also help to identify the RDC functions with the greatest safety potential. In particular, data from naturalistic driving studies could help identifying the key distraction-induced critical driving scenarios that lead to crashes. In addition, such data could be very useful for the iterative development and testing of workload management and distraction alert algorithms (as well as for the development of other types of driving support functions).

To conclude, RDC functions are potentially very valuable complements to non-technical distraction countermeasures (such as public awareness campaigns), although further work is needed to further demonstrate their benefits in terms of enhanced driving performance and safety. Further work is also needed to identify the RDC functions that have the largest impact on driving safety, efficiency, and comfort and to develop suitable methods for their evaluation. If the recent conclusions from ND studies are correct regarding the impact of inattention to the forward roadway on traffic safety,[6] then the RDCs outlined here certainly offer great potential to save lives.

REFERENCES

1. Smiley, A., What is distraction? *Proceedings of the International Conference on Distracted Driving*, Toronto, Ontario, Canada, October 2–5, 2005.
2. Young, K., Regan, M., and Hammer, M., *Driver Distraction: A Review of the Literature*, Monash University Accident Research Centre, Report No. 206, 2003.
3. Treat, J. R., Tumbas, N. S., McDonald, S. T., Shinar, D., Hume, R. D., Mayer, R. E., Stanisfer, R. L., and Castellan, N. J., Tri-level study of the causes of traffic accidents: Final report—Executive summary (Technical Report DOT HS 805 099), Department of Transportation, National Highway Traffic Safety Administration, Washington, D.C., 1977.
4. Wang, J., Knipling, R., and Goodman, M., The role of driver inattention in crashes; new statistics from the 1995 crashworthiness data system, *40th Annual Proceedings of the Association for the Advancement of Automotive Medicine*, Vancouver, Canada, October 7–9, 1996.
5. Stutts, J. C., Reinfurt, D. W., Staplin, L., and Rodgman, E. A., The role of driver distraction in traffic crashes. AAA Foundation for Traffic Safety, Washington, D.C., 2001.
6. Dingus, T. A., Klauer, S. G., Neale, V. L., Petersen, A., Lee, S. E., Sudweeks, J., Perez, M. A., Hankey, J., Ramsey, D., Gupta, S., Bucher, C., Doerzaph, Z. R., Jermeland, J., and Knipling, R. R., *The 100-Car Naturalistic Driving Study, Phase II—Results of the 100-Car Field Experiment*, DOT HS 810 593, National Highway Traffic Safety Administration, Washington, D.C., 2006.
7. Klauer, S. G., Dingus, T. A., Neale, V. L., Sudweeks, J. D., and Ramsey, D.J., *The Impact of Driver Inattention on Near-Crash/Crash Risk: An Analysis Using the 100-Car Naturalistic Driving Study Data*, Report No. DOT HS 810 59, U.S. Department of Transportation, Washington, D.C., 2006.
8. Donmez, B., Boyle, L. N. and Lee, J. D., The impact of driver distraction mitigation strategies on driving performance, *Human Factors*, 48(4), 785–804, 2007.
9. Donmez, B., Boyle, L. N., Lee J. D., and McGehee, D., Drivers' attitude toward imperfect distraction mitigation strategies. *Transportation Research Part F* 9, 287–398, 2006.
10. Green P., Driver distraction, telematics design, and workload managers: Safety issues and solutions, SAE paper No. 2004-21-0022, 2004.
11. de Waard D., The measurement of drivers' mental workload, Traffic Research Centre. University of Groningen, ISBN 90-6807-308-7, 1996.
12. Miller, C., Hannen, M., and Guerlain, S., The rotorcraft pilot's associate cockpit information manager: acceptable behavior from a new crew member? *Proceedings of the American Helicopter Society 55th Annual Forum*, Montreal, 1999.
13. Michon, J. A. (Ed.). *Generic Intelligent Driver Support: A Comprehensive Report on GIDS*. Taylor & Francis, London, 1993.
14. Bellet, T., Bruyas, M. P., Tattegrain-Veste, H., Forzy, J. F., Simoes, A., Carvalhais, J., Lockwood, P., Boudy, J., Baligand, B., Damiani, S., Opitz, M., "Real-time" analysis of the driving situation in order to manage on-board information, *e-Safety Conference Proceedings*, Lyon, 2002.
15. Montanari, R., COMUNICAR final project report, public version. European Commission: Contract number IST-1999-11595, 2003.
16. Piechulla, W., Mayser, C., Gehrke, H., and König, W., Reducing drivers' mental workload by means of an adaptive man-machine interface. *Transportation Research Part F, 6*, 233, 2003.
17. Engström, J., Arfwidsson, J., Amditis, A., Andreone, L., Bengler, K., Cacciabue, P. C., Janssen, W., Kussman, H., and Nathan, F., Towards the automotive HMI of the future: Mid-term results of the AIDE project, *Advanced Microsystems for Automotive Applications*, Valldorf, J. and Gessner, W. (Eds.), Springer, Berlin, 2006.

18. SAVE-IT, SAfety VEhicle(s) using adaptive Interface Technology (SAVE-IT) program. DTRS57-02-20003. US DOT, RSPS/Volpe National Transportation Systems Center (Public Release of Project Proposal), 2002. Available at http://www.volpe.dot.gov/opsad/saveit/index.html.
19. Zoutendijk, A., Hoedemaeker, M., Vonk., T., Schuring, O., Willemsen, D., Nelisse, M., and van Katwijk, R., Implementing multiple intelligent services in an intelligent vehicle with a workload aware HMI, *Proceedings of the ITS World Congress*, Madrid, Spain, 2003.
20. Remboski, D., Gardner, J., Wheatley, D., Hurwitz, J., MacTavish, T., and Gardner, R. M., Driver performance improvement through the driver advocate: A research initiative toward automotive safety, *Proceedings of the 2000 International Congress on Transportation Electronics, SAE P-360*, 509, 2000.
21. Broström, R., Engström, J., Agnvall, A., and Markkula, G., Towards the next generation intelligent driver information system (IDIS): The Volvo cars interaction manager concept, *Proceedings of the ITS World Congress*, London, 2006.
22. Hoedemaeker, M., de Ridder, S. N., and Janssen, W. H., *Review of European Human Factors Research on Adaptive Interface Technologies for Automobiles*, TNO Report TM-02-C031, 2002.
23. Deregibus, E., Andreone, L., Bianco, E., Amditis, A., Polychronopoulos, A., and Kussman, H., The AIDE adaptive and integrated HMI design: The concept of the interaction communication assistant, *Proceedings of the ITS World Congress*, London, 2006.
24. Kussmann, H., Engström, J., Amditis, A., and Andreone, L., Software architecture for an adaptive integrated automotive HMI, *Proceedings of the ITS World Congress*, London, 2006.
25. Tattegrain Veste, H., Bellet, T., Boverie, S., Kutila, M., Bekiaris, E., Panou, M., and Engström, J., Development of a driver situation assessment module in the AIDE project, *Proceedings of the XVI IFAC World Congress*, Prague, Czech Republic, July 4–8, 2005.
26. Lööf, J., Statistical models for real-time driver workload estimation, MSc thesis, Chalmers University, 2003.
27. Fastenmeier, W., Die Verkehrssituation als Analyseeinheit im Verkehrssystem, *Autofahrer und Verkehrssituation*, W. Fastenmeier (Ed.), 27, Verlag TÜV, Köln, 1995.
28. Engström, J., Åberg, N., Johansson, E., and Hammarbäck, J., Comparison between visual and tactile signal detection tasks applied to the safety assessment of in-vehicle information systems (IVIS), *Proceedings of the Third International Driving Symposium on Human Factors in Driver Assessment, Training and Vehicle Design*. Rockport, Maine, 2005.
29. Hoedemaeker, M., Schindhelm, R., Gelau, C., Belotti, F., Amditis, A., Montanari, R., and Mattes, S., COMUNICAR: Subjective mental effort when driving with an information management system, *Human Computer Interaction: Theory and Practice (part 2): Proceedings of the HCI International 2003*, Stephanidis, C., and Jacko, J. (Eds.), 2003, pp. 88.
30. Rimini-Döring, M., ITSA—Traffic safety assessment in a simulator experiment with integrated driver information and assistance systems. Abstract submitted to the 2007 Driving Assessment Conference, Stevenson, Washington, USA.
31. Victor, T. W., Driver support by recognition of visual behavior, VISREC project description, Program Board for Swedish Automotive Research, VINNOVA Dnr 2001-02551 P12881-1, 1999.
32. Victor, T., United States Patent No. US 6,974,414 B2, System and method for monitoring and managing driver attention loads, 2005.
33. Victor, T., Keeping eye and mind on the road, *Digital Comprehensive Summaries of Uppsala Dissertations from the Faculty of Social Sciences 9*, Acta Universitatis Upsaliensis, Uppsala, 2005.

34. Karlsson, R., Evaluating driver distraction countermeasures, MSc thesis, Linköping University, Sweden, 2004.
35. Markkula, G., Markkula, G., Kutila, M., Engström, J., Victor T. W., and Larsson, P., Online detection of driver distraction—Preliminary results from the AIDE project, *Proceedings of the International Truck & Bus Safety & Security Symposium*, Alexandria, Virginia, USA, November 14–16, 2005.
36. Fletcher, L., An automated co-driver for advanced driver assistance systems: the next step in road safety, PhD thesis, Australian National University, 2007.
37. Victor, T, Åberg, N, and Tevell, M., Can inattention-induced accidents be reduced by real-time attention support: Roadgaze reminders, attention sensitive driving support, workload management, drowsiness warnings, and attention histories, Manuscript in preparation.
38. Victor, T. W., and Larsson, P., Method and arrangement for interpreting a subject's head and eye activity. International application published under the patent cooperation treaty. International publication number PCT WO 2004/034905 A1. International Bureau: World Intellectual Property Organization, 2004.
39. Donmez, B., Boyle, L. N., and Lee, J. D., Safety implications of providing real-time feedback to distracted drivers, *Accident Analysis and Prevention*, 39(3), 581–590, 2007.
40. Engström, J., Johansson, E., and Östlund, J., Effects of visual and cognitive load in real and simulated motorway driving. *Transportation Research Part F*, 8, 97, 2005.
41. Markkula, G. and Engström, J., A steering wheel reversal rate metric for assessing effects of visual and cognitive secondary task load, *Proceedings of the ITS World Congress*, London, 2006.
42. van der Laan, J., Heino, A., and de Ward, D., A simple procedure for the assessment of acceptance of advanced transport telematics, *Transportation Research, Part C*, 5(1), 1–10, 1997.
43. Green, M., How long does it take to stop? Methodological analysis of driver perception-brake times, *Transportation Human Factors*, 2(3), 195–216, 2000.
44. Lee, J. D., Ries, M. L., McGehee, D. V., Brown, T. L., and Perel, M., Can collision warning systems mitigate distraction due to in-vehicle devices? *Proceedings of the Internet Forum on The Safety Impact of Driver Distraction When Using In-Vehicle Technologies*, NHTSA, DOT, Washington, D.C., available at www-nrd.nhtsa.dot.gov/departments/nrd-13/driver-distraction/Welcome.htm, 2000.
45. Lee, J. D., McGehee, D. V., Brown, T. L., and Reyes, M. L., *Driver Distraction, Warning Algorithm Parameters, and Driver Response to Imminent Rear-End Collisions in a High-Fidelity Driving Simulator*, National Highway Traffic Safety Administration, Washington, D.C., 2002.
46. Brouwer, R. F. T. and Hoedemaeker, D. M. (Eds.), *Driver Support and Information Systems: Experiments on Learning, Appropriation, and Effects of Adaptiveness*. AIDE IST-1-507674-IP Deliverable D1.2.3, 2006.
47. Roland, J., Horst, D., and Paul, A., User evaluations of adaptive warning strategies. AIDE IST-1-507674-IP, Manuscript in preparation.
48. Opperud, A., Jeftic, Z., and Jarlengrip, J., Real-time adaptation of advanced driver assistance systems to driver state, *Proceedings of the 2005 International Truck and Bus Safety and Security Symposium*, November 14–16, 2005, Washington, Alexandria, Virginia, 2005.
49. Toyota, Toyota enhances precrash safety system with driver-monitoring function, Toyota Press Release, September 6, 2005.
50. Hansen, P., Driver distraction offers huge safety opportunity: Volvo and Toyota lead the way. *The Hansen Report on Automotive Electronics*, No. 7, Vol. 19, 2006.
51. Shiraki, N., A rapid facial-direction detection system based on symmetrical measurement, *Proceedings of the ITS World Congress*, San Francisco, 2005.

27 Driving Task Demand–Based Distraction Mitigation

Harry Zhang, Matthew R.H. Smith, and Gerald J. Witt

CONTENTS

27.1 INTRODUCTION

Driver distraction is a major contributing factor to automobile crashes. In the United States, the National Highway Traffic Safety Administration (NHTSA) has estimated that approximately 25% of crashes are attributable to driver distraction and inattention.[1] Similar estimates have been obtained from other studies.[2,3] A recent U.S. field study of 100 cars has found that 78% of crashes and 65% of near crashes are attributable to driver inattention, which includes factors such as driver engagement in secondary tasks, driver drowsiness, and eye glances away from the forward road.[4,5] The 100-car study indicates that visually and manually complex distraction tasks, especially those that require long off-road eye glances, pose a significant risk to highway safety. The problem of driver distraction may become worse in the next few years because more electronic devices (e.g., navigation systems, cell phones, wireless Internet, and e-mail devices) may be used while driving.

Two general approaches have been taken to mitigate the driver distraction problem.[6] One approach is the nonadaptive approach. With this approach, the distraction potential of nondriving tasks (e.g., manually dialing a 10-digit phone number) is assessed with performance metrics, models, and procedures that are based on task analysis, laboratory experiments, test track evaluation, and field studies.[7,8] These metrics can guide designers to create a less distracting task (e.g., speed dialing with a single button press rather than manually dialing 10 digits). In addition, a threshold line can be drawn on the performance scale to discriminate between highly distracting and less distracting tasks. Highly distracting tasks may be locked out when the vehicle is in motion (e.g., lock out manually entering an address in the navigation system), whereas less distracting tasks are allowed (e.g., radio tuning). Several design guidelines have been put forward to help minimize the distraction potential and improve the usability of in-vehicle devices. These include general design principles by the Alliance of Automobile Manufacturers[9] and the concept of limiting the total completion time of a nondriving task to 15 s[8] (see also Part 7 of this book).

Because the level of attention required for a driving task may vary with the driving environment, and the amount of driver distraction stemming from a nondriving task may differ among individual drivers at different times, the nonadaptive approach is insufficient. Instead, an adaptive interface technology approach should be adopted in which the amount of distraction is allowed to vary with the dynamic driving environment. Commonly known as the workload management approach,[6,10–13] this adaptive approach uses the sensors that are installed in vehicles to assess the attention demand imposed by the driving task and the amount of distraction that stems from nondriving tasks. It attempts to mitigate driver distraction either by intervention or by prevention. With the intervention strategy, when the amount of distraction that is exhibited by the driver exceeds a threshold for a particular driving situation, alerts are delivered to the distracted driver to reorient the driver's attention back to the driving task. With the prevention strategy, if a nondriving task has the potential to generate a distraction level that exceeds a threshold for a particular driving situation, the nondriving task may be locked out, or an advisory may be issued to discourage the driver from engaging in this nondriving task. It is worth noting that, with the adaptive approach, the distraction threshold is contingent on the driving situation, and the user interface is modifiable.

This chapter will focus on the adaptive interface technology approach of mitigating driver distraction. First we will review the major research programs that have been undertaken to develop and evaluate adaptive interface technologies. We will describe an experiment that has been carried out to determine driving task demand using variables that are collected from in-vehicle sensors. We will discuss research on the two distraction mitigation strategies: the intervention strategy that provides demand-based distraction feedback to the drivers; and the prevention strategy that locks out certain nondriving tasks and delivers demand-based advisories, which are designed to discourage drivers from engaging in certain nondriving tasks.

27.2 ADAPTIVE INTERFACE TECHNOLOGIES

An early program adopting the adaptive interface technology approach was the Generic Intelligent Driver Support (GIDS) program,[13] which applied Wickens' multiple resource theory of attention[14] to develop a lookup table to estimate the workload for driving tasks such as turning and overtaking and scheduled nondriving task messages based on the estimated workload and the message priority. A follow-up program, the Communication Multimedia Unit Inside Car (COMMUNICAR) program, used vehicle data such as brake and accelerator positions and neural network models to estimate the workload for the driving task.[11]

Recently, two large-scale research programs have been initiated. One is the United States–based SAfety VEhicle(s) using adaptive Interface Technology (SAVE-IT) program, which is led by Delphi Electronics & Safety under the sponsorship of the NHTSA. The goal of the SAVE-IT program is to demonstrate and evaluate the potential safety benefits of adaptive interface technologies that manage the information from various in-vehicle systems using real-time monitoring of the roadway conditions and the driver's capabilities.[6,12,15] Specifically, roadway and traffic conditions are taken into account to determine the driving task demand, certain distracting tasks are locked out, and advisories are issued against the use of distracting tasks over a certain demand level. The other is the European Adaptive Integrated Driver–vehicle Interface (AIDE) program that is sponsored by the European Commission.[16] The AIDE program aims to reduce the level of driver distraction and enable the potential benefits of new in-vehicle technologies and nomadic devices to be realized (see Chapter 26 for a more detailed discussion of the AIDE project). In addition to minimizing driver distraction, both the SAVE-IT and AIDE programs take into account the driver distraction status (e.g., whether a driver is distracted) to adjust crash avoidance systems (e.g., forward collision warning systems) to enhance the system effectiveness and driver acceptance. For example, forward collision warnings may be delivered earlier to a distracted driver to compensate for the long reaction time that results from driver distraction.

Figure 27.1 illustrates the general concept of the adaptive interface technology approach.[6,12] The three vertical lines indicate the low, moderate, and high level of driving task demand. The dark solid bars on the left side of the figure indicate the

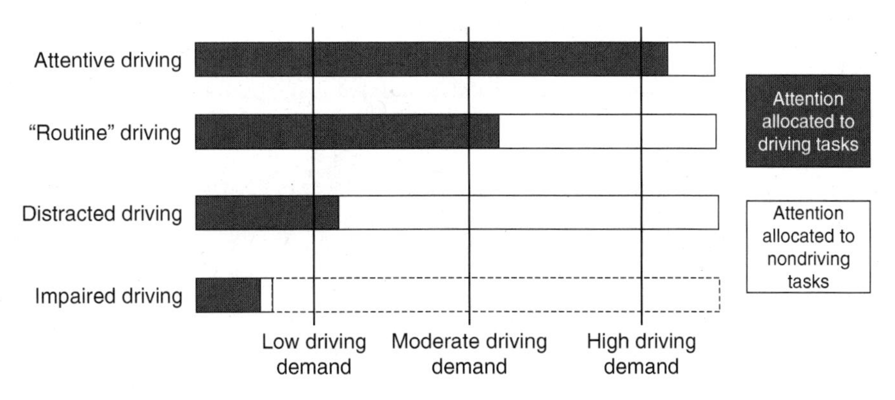

FIGURE 27.1 Driving task demand and attention allocation to driving and nondriving tasks.

attention that is allocated to driving tasks (e.g., turning and overtaking). The unfilled solid bars on the right side of the figure indicate the attention that is allocated to the nondriving task (e.g., manually dialing a 10-digit phone number). The dashed bar, at the bottom of the figure, indicates the attention deficit that is caused by driver impairment (e.g., driver fatigue). To maximize driver safety, the attention allocated to the driving task must be commensurate with the demand imposed by the driving task. By the same token, driving task demand should be commensurate with the driver's limited attentional capability. Crash risk increases to the extent that the attention allocated to driving is insufficient for a particular driving task demand.

As indicated in Figure 27.1, both the driving task demand and the attention that is allocated to the driving task must be assessed to ensure that the attention allocation matches the driving task demand. This assessment may be accomplished manually or automatically. The manual assessment approach assumes that drivers are aware of their attentional limits. It relies on drivers' knowledge about their own driving and allows drivers to manually set the policy for permitting or prohibiting certain nondriving tasks. For example, one driver may choose to prohibit incoming phone calls and reroute all phone calls to a voice message system, and another driver may choose to permit all incoming phone calls to go through. One implementation of this method is to notify the remote caller at the other end of the phone conversation that the caller is driving or the caller needs to reduce the phone conversation intensity to concentrate on the driving task. Several studies have revealed that the remote caller frequently reduces the conversation intensity with this type of notification.[17–19]

Alternatively, workload management can be accomplished automatically by an adaptive interface technology system. Both intervention strategies and prevention strategies may be adopted. With the intervention strategy, *post hoc* distraction feedback will be delivered to a driver when the level of distraction exhibited by the driver exceeds a certain threshold (see Chapter 29). For example, when the total eyes-off-the-road time over a time window (e.g., 5 s) exceeds a threshold (e.g., 50%), the driver may receive a feedback message indicating that they should reduce the level of distraction. With the prevention strategy, particular in-vehicle device features may be locked out, or drivers may be discouraged from using them, so that these distracting tasks will not be initiated in the first place. In order for the potential of adaptive interface technologies to be realized, research must be conducted to determine (1) the amount of attention that is dynamically allocated to the driving and the nondriving tasks, (2) the distraction potential of nondriving tasks, and (3) the task demand that is imposed by the driving task. The first area of research includes the identification of visual and cognitive distraction in real time using diagnostic metrics such as eyes-off-the-road time and saccadic eye movements.[6,20] The second area of research attempts to determine the demand of nondriving tasks such as manually dialing phone numbers and reading SMS text messages,[21] which overlaps with the nonadaptive approach.[7] The third area of research has been performed in driving simulators.[22] Although simulator studies have examined the effects of factors such as road curvature on the driving task demand, the effect of traffic has not been adequately examined. In addition, driving task demand may be more adequately determined from realistic roadway and traffic conditions. In the following section, we describe an experiment in which driving task demand is estimated from roadway and traffic conditions.

27.3 ASSESSMENT OF DRIVING TASK DEMAND

To develop and evaluate an adaptive distraction mitigation system for the SAVE-IT program, an experiment was performed to estimate the driving task demand directly from data from in-vehicle sensors. An instrumented vehicle was driven on public roads to acquire data that are related to the vehicle and the driving environment (e.g., speed, brake use, and forward objects as measured by forward-looking radar) and video data of forward scenes. Short video segments were presented in a driving simulator environment to obtain an estimate of driving task demand as rated by participants. An algorithm was then developed to predict the rated demand from the data from in-vehicle sensors. Once implemented in an instrumented vehicle, the algorithm can be deployed to determine the driving task demand in real time as the vehicle is driven on public roads. The predicted driving task demand can support distraction mitigation strategies, such as the intervention and prevention strategies. In the following sections, we describe the method and the results from this experiment.

27.3.1 METHOD

To assess driving task demand using data from in-vehicle sensors, a 1992 Buick LeSabre was equipped with a long-range, forward-looking radar, a forward-looking camera, a yaw sensor, a steering wheel angle sensor, a photo cell, and sensors that were used for determining the brake pedal depression, turn signal activation, wiper status, and ambient temperature. The forward-looking radar had a 15° field of view horizontally. The radar detected up to 20 objects that were located within 150 m in the forward scene. For each of these objects, the range (distance between the object and the host vehicle), range rate (relative speed), and angle were measured. The radar system also identified the lead vehicle that was located in the same path as the host vehicle. The range rate can be used to determine whether a forward object moved at the same speed as the host vehicle. If the range rate was zero, the object ahead moved at the same speed as the host vehicle. If the range rate was positive, the speed of the object ahead was greater than that of the host vehicle. If the range rate was negative, the speed of the object ahead was slower than that of the host vehicle, or the object ahead was in the oncoming lane. The angle was positive if the object was located to the right side of the longitudinal path of the host vehicle, and negative if the object was located on the left side of the longitudinal path of the host vehicle.

An experimenter drove the instrumented host vehicle in the central Indiana areas (Indianapolis and Kokomo, Indiana) for approximately 8 h. The roads driven included urban, suburban, and rural surface streets and highways. The drives covered both peak and off-peak times. Vehicle data such as speed, object range, and range rate were recorded in text format for subsequent processing. The forward-looking camera was routed to a video recording device to record forward scenes in black-and-white format.

After reviewing the forward scene videos, the experimenter selected 103 videos that spanned a wide range of driving conditions, including from no or little traffic to high traffic density; urban, suburban, rural road types; two, four, or six lanes; and various driving maneuvers (e.g., lane change, merge, and highway entering and exiting). Each video was 8 s long. 8 s videos were chosen because each was sufficiently long for reliable measurement of vehicle sensor data and subjective ratings and sufficiently short to avoid changes in driving task demand.

27.3.2 Participants

To acquire subjective ratings of driving task demand, ten 35- to 53-year-old participants (five males and five females) were recruited from the salaried employee pool at Delphi Electronics & Safety in Kokomo, Indiana. Several participants were electrical engineers who had extensive knowledge about in-vehicle devices such as the navigation system. They possessed a valid driver's license and had a minimum visual acuity of 20/40 (vision correction with eyeglasses permitted) as tested with the Snellen eye chart. They were given a $50 gift card for their participation in the 2 h experiment. Participants sat in the driver's seat of the driving simulator and were presented with 103 videos that were displayed on the forward screen one at a time. The first three videos were practice and the remaining hundred videos were used to estimate the driving task demand. Participants were asked to rate the driving task demand of each, using the following scale: "On a scale of 1–7 (1, low; 4, medium; 7, high), rate how attentive you need to be to detect unexpected events and hazards, control vehicle speed and lane position, and avoid crashes and road departures." Note that three anchors (low, medium, and high) were used in the demand scale.

27.3.3 Results: Demand Ratings

Because there were individual differences among the demand ratings provided by the 10 participants (the standard deviation averaged across the videos was approximately 1.2), the median rating was chosen to represent the rated demand for a particular video. The median demand ratings ranged from 1.5 to 6 and were classified into the following three categories:

1. Thirteen videos with a low demand level (median rating ≤ 3)
2. Sixty-one videos with a medium demand level (median rating between 3.5 and 5)
3. Twenty-six videos with a high demand level (median rating ≥ 5.5)

27.3.4 Results: Predicting Demand Ratings from Vehicle Sensor Data

The vehicle sensor data were analyzed for the middle 5 s of the video to predict the median demand ratings. The rated demand was highly correlated with several vehicle sensor variables. It had a Pearson correlation coefficient of $r = -.6$ with the mean range value. It had a moderate correlation with variables including mean time headway ($r = -.34$), standard deviation of time headway ($r = .33$), and speed variability for the host vehicle ($r = .31$). It had a low correlation with variables, including speed variability among targets ($r = -.11$) and speed variability for lead vehicles ($r = .15$).

The vehicle sensor variables that were correlated with the rated demand were combined to derive an equation that can predict driving task demand. The equation that produced the best prediction was as follows:

$$\text{Weight} = \sum_{i=1}^{\text{Objects}} \left(\frac{K}{\text{normalized angle}_i \times \min(\text{normalized range}_i, \text{time headway}_i)} \right)$$

where k was a constant, which was 1 for objects that moved in the same direction as the host vehicle, 5 for objects that moved in the opposite direction (e.g., in oncoming lanes), and 0 for stationary objects. The normalized angle_i was defined as follows:

$$\text{Normalized angle}_i = 1 + \frac{\text{abs}(\text{angle}_i - \text{lead angle})}{4}$$

where angle_i was the angle for forward objects that were detected by the forward-ing-looking radar, and lead angle was the angle for the moving lead vehicle. The normalized range was defined as follows:

$$\text{Normalized range}_i = \frac{\text{range}_i}{25}$$

where range_i was the range for forward objects that were detected by the radar system. The time headway was defined as follows:

$$\text{Time headway}_i = \frac{\text{range}_i}{\text{speed}}$$

where speed was the host vehicle speed.

This weight equation was easy to understand. The minimum between the normalized range and the time headway represented the range variable that was most demanding to participants. In general, driving task demand increased with a reduction in either the normalized range or the time headway. The normalized angle variable indicated that the driving task demand was lower for objects that were laterally farther away from the longitudinal axis of the host vehicle. The summation across objects indicated that the driving task demand increased with an increasing number of objects in the forward scene.

The preceding weight equation was adjusted on the basis of the following factors:

- *Yaw rate*. Weight was adjusted up when the yaw rate exceeded a threshold that indicated turning and driving on curves (e.g., yaw rate >2 or 4°). A larger weight adjustment was made when the yaw rate was larger, indicating sharp curves or turns (e.g., yaw rate >7 or 10°).
- *Highway entrance or exits*. Weight was adjusted up at highway entrance, exits, or interchanges that were detected from the GPS information.
- *Lane width*. Weight was adjusted up when the lane width was narrower than a threshold (e.g., 3 m).
- *Speed*. Weight was adjusted down when the host speed was lower than a threshold (e.g., 10 mph).
- *Brake pedal depression*. Weight was adjusted up when the host vehicle braked.

The adjusted weights were classified into three demand categories as follows:

- Low demand (when the adjusted weight was <0.5)
- Medium demand (when the adjusted weight was ≥ 0.5 and <2.18)
- High demand (when the adjusted weight was ≥ 2.18)

Figure 27.2 presents the median demand ratings against the predicted weights. The Pearson correlation coefficient between the predicted weights and median demand ratings was .79. The high level of correlation indicated that the demand algorithm was able to predict the rated demand. A close examination of Figure 27.2 indicates that most of the mismatches were borderline cases. Among the 15 videos that were predicted to have a high demand level but were rated as a medium demand level, 11 videos had a rated demand of 5, which was very close to a high demand rating (with a medium rating ≥ 5.5). Among the videos that were predicted to have a low demand level but were rated as a medium demand level, several of them had a rated demand of 3.5, which was very close to a low demand rating (with a medium rating ≥ 3).

The predicted demand categories and median demand ratings are tabulated in Table 27.1. Table 27.1 indicates that the predicted demand matched the rated demand for 74% of the videos. There was a mismatch for the remaining 26%. When the mismatch occurred, the predicted demand and the rated demand were different by one category (e.g., the predicted demand was high but the rated demand was medium). There are no gross mismatches in Table 27.1 (e.g., the predicted demand was high, but the rated demand was low).

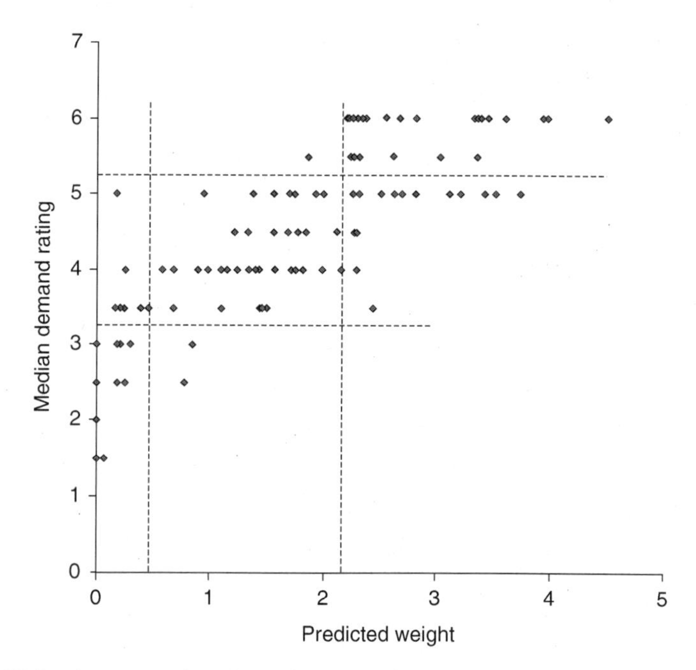

FIGURE 27.2 A scatter plot of median demand ratings against the predicted demand weight.

TABLE 27.1
Tabulation of Predicted and Rated Demand Categories

	Rated Demand		
Predicted Demand	**High**	**Medium**	**Low**
High	25	15	0
Medium	1	38	2
Low	0	8	11

It is worthwhile to note that, in Figure 27.2 and Table 27.1, the demand ratings are the median values taken from a sample of 10 participants. Therefore, the correspondence between the predicted demand and subjectively rated demand represents the average case. Because of individual differences, this level of correspondence may not be achieved for every participant.

27.4 THE INTERVENTION STRATEGY: DEMAND-BASED DISTRACTION FEEDBACK

As described earlier, the real-time assessment of driving task demand is a major enabler for demand-based distraction mitigation. In this section, we will discuss the intervention strategy. Another enabler for the intervention strategy is the real-time assessment of driver distraction. Significant progress has been made in this area. Zhang et al.[6] have identified a diagnostic measure of visual distraction using the proportion of off-road glance time that is averaged across a short time window (e.g., 60 s). This glance measure may be based on eye gaze or head pose. When a driver's eye gaze or head pose is within a 24° by 24° rectangular area that is centered at the focus of expansion, the driver's glance is classified as forward. Otherwise, it is classified as off-road. The proportion of off-road glance time over a time window is found to be highly correlated with standard deviation of lane position, number of lane departures, and reaction time to a lead vehicle braking event.

With the intervention strategy, feedback may be provided to indicate to the driver that the proportion of total off-road glance time has exceeded a particular threshold.[6,12] Because driving performance degrades as the proportion of total off-road glance time increases, the threshold can be reconfigurable to optimize both driving performance and driver acceptance. For example, a low threshold (e.g., 25%) could be used as an aggressive value to discourage off-road glances, and a high threshold (e.g., 50%) could be used as an alternative value to enhance driver acceptance and minimize false alarms.

As illustrated in Figure 27.1, the amount of permissible distraction should vary with driving task demand. This concept has been tested in the experiment that was described earlier. For each video, the same participants were asked to determine whether visual distraction should be curtailed when driving under the condition that was portrayed by the video. Figure 27.3 presents the number of "yes" responses to curtail off-road glances or the amount of visual distraction at various demand levels. When the median demand was low, few participants desired to curtail the amount of visual distraction. When the median demand was high, a majority of the participants

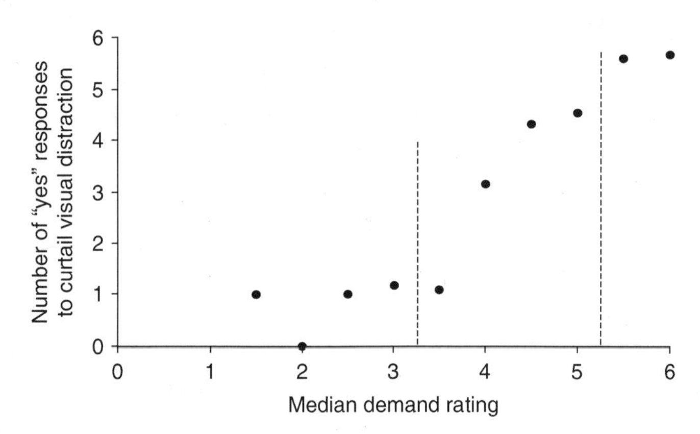

FIGURE 27.3 Number of "yes" responses to curtail visual distraction based on driving task demand.

wanted to curtail the amount of visual distraction. The correlation coefficient between the number of "yes" responses and the median demand rating was 0.76, indicating a strong correlation. As shown in Figure 27.3, as the driving task demand increased, more drivers believed that visual distraction should be curtailed. In other words, the allowable visual distraction should be reduced with an increasing level of driving task demand. These results strongly supported the distraction mitigation concept.

27.5 THE PREVENTION STRATEGY

27.5.1 Demand-Based Advisories and Lockouts against Using Certain Nondriving Tasks

As described earlier, an algorithm has been developed that predicts the demand of the driving task based on real-time traffic and road conditions. With this algorithm, driving task demand will be classified into one of three groups: low, medium, and high level. Using the prevention strategy, certain nondriving tasks are discouraged or locked out when the driving task demand exceeds a given threshold. Several experiments have provided support for this prevention strategy.[23,24]

To assess driver acceptance of driving task demand-based advisories and lockouts, additional responses were acquired from the demand assessment experiment that was described earlier. To determine if certain in-vehicle information system features should be advised against or locked out, we asked the same participants who originally rated the video clips for driving task demand to decide which of the following features should be allowed (assigned a value of 1), advised not to be used (assigned a value of 2), or locked out (assigned a value of 3) when driving conditions are as portrayed by the video clip. The system features discussed were tuning the radio (with seek or preset buttons), tuning a satellite radio (selecting by genre or favorites), playing a CD (via track selection, random selection), selecting an MP3 item (searching by artist, album, genre, and selecting from a playlist), reading text messages (SMS), manually dialing a 7- or 10-digit phone number, answering a phone call, talking on a cell phone, entering the nearest points of interest on a navigation

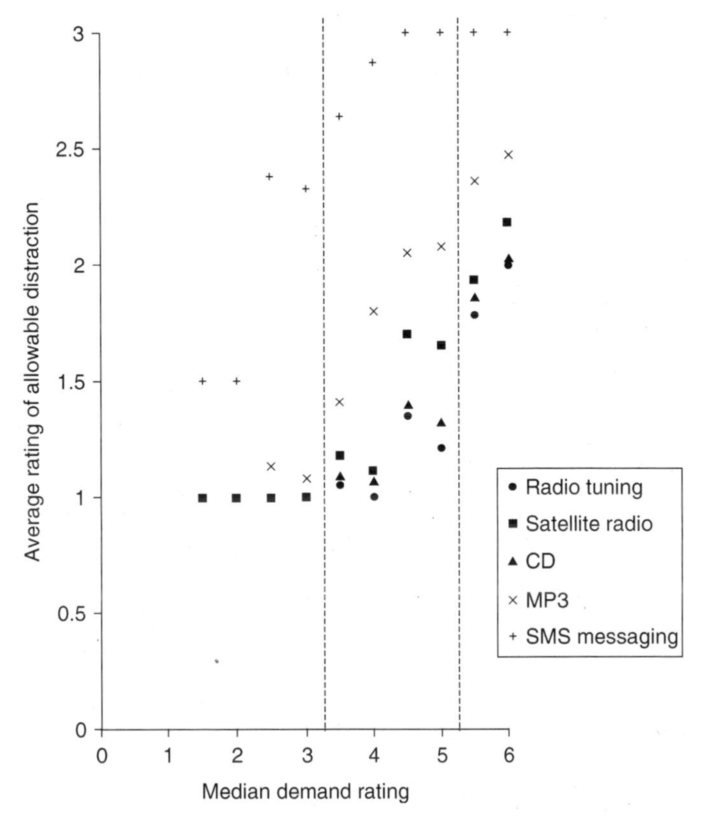

FIGURE 27.4 Average rating of allowable distraction as a function of driving task demand for radio, CD, MP3, and text messaging tasks.

system, reading a map on a navigation system, and using turn-by-turn directions on a navigation system.

Figure 27.4 presents the average ratings of allowable distraction for traditional radio tuning, satellite radio, CD, MP3, and SMS text messaging. The results were similar for traditional radio tuning and CD: the average ratings were 1 (allowed) or close to 1 at the low and medium demand levels and close to 2 (advised not to be used) at the highest level of driving task demand. Similar results were obtained for satellite radio at the low and medium demand levels, but the average ratings exceeded 2 (advised not to be used) at the highest demand level. This difference indicates that satellite radio is perceived to be more demanding than traditional radio tuning, probably because satellite radio is a recent technology and a higher level of visual attention is required in navigating the satellite radio menu structure. For MP3, the average ratings were close to 1 (allowed) at the low demand levels, approached 2 (advised not to be used) at the medium demand levels, and exceeded 2 at the high demand levels. For SMS messaging, the average rating was 1.5 initially and exceeded 2 (advised not to be used) when the driving task demand was at the low level. It approached 3 (locked out) when the driving task demand was at the medium or high level.

Figure 27.5 presents the average ratings of allowable distraction for phone and navigation tasks. The results were similar for manual phone dialing and manually entering an address for a point of interest on a navigation system. For both of them, the average rating was 1.5 initially, dipped to 1 (allowed), but quickly approached 2 (advised not to be used). The dip may not be reliable because only one video clip was used. It exceeded 2 (advised not to be used) when the driving task demand was at the medium levels and approached 3 (locked out) when the driving task demand was at the high levels. For answering a phone call, the average rating was 1 at the low demand levels, increased to below 2 (advised not to be used) at the medium demand levels, and exceeded 2 at the high demand levels. For phone conversation, the average rating was 1 or close to 1 when the driving task demand was at the low or medium level, and approached 2 (advised not to be used) when the driving task demand was at the high level. For map reading on a navigation system, the average rating was approximately 2 (advised not to be used) when the driving task demand was at the medium level and 3 (locked out) when the driving task demand was at the high level. For the turn-by-turn function on a navigation system, the average

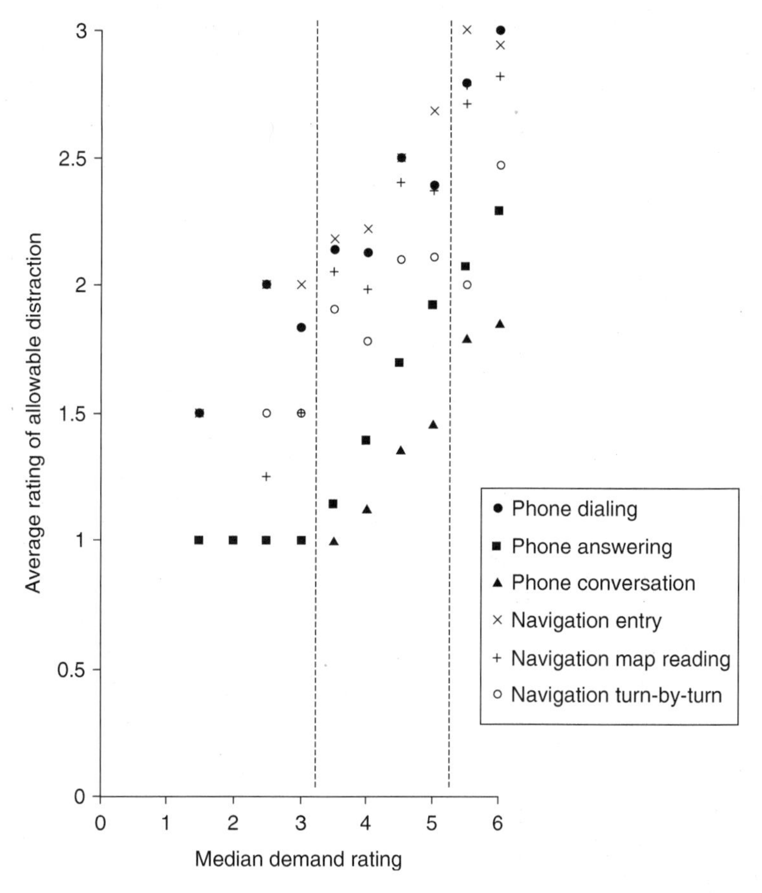

FIGURE 27.5 Average rating of allowable distraction as a function of driving task demand for telephony and navigation tasks.

rating was approximately 2 when the driving task demand was at the high or medium level. Because reading maps is perceived to be more distracting than the turn-by-turn function, drivers are encouraged to use the turn-by-turn function rather than navigation maps to minimize the amount of driver distraction.

To summarize the results from Figures 27.4 and 27.5, Table 27.2 illustrates whether lockouts and advisories against using certain nondriving functions should be provided at a particular level of driving task demand. When driving (even with a low driving task demand), advisories should always be issued to discourage manually dialing 7- or 10-digit phone numbers, manually entering destination addresses, and reading and sending a text message. When the driving task demand is at the medium level, functions such as manually entering destination addresses into a route navigation system and reading and sending a text message should be locked out. In addition, advisories should be issued to discourage the use of MP3 music players, manually dialing 7- or 10-digit phone numbers, reading navigation maps, and using turn-by-turn navigation systems. When the driving task demand is at the high level, functions such as manually dialing 7- or 10-digit phone numbers, manually entering destination addresses and reading and sending a text message should be locked out. In addition, advisories should be issued for all other device functions except tuning traditional radios or playing CDs. It should be pointed out that Table 27.2 represents recommendations rather than design requirements because the perceived rather than actual risk was investigated in the present study. Furthermore, the distraction potential for a particular task (e.g., manual phone dialing) may also vary according to the specific interface design implementations.

TABLE 27.2
Driving Task Demand–Based Prevention Strategies for the Use of Certain Device Functions

	Level of Driving Task Demand		
Nondriving Task	**Low**	**Medium**	**High**
Radio tuning	No advisories	No advisories	No advisories
Satellite radio	No advisories	No advisories	Advisories
CD	No advisories	No advisories	No advisories
MP3	No advisories	Advisories	Advisories
Manual dialing (7 or 10 digits)	Advisories	Advisories	Lockout
Answering phone	No advisories	No advisories	Advisories
Phone conversation	No advisories	No advisories	Advisories
Navigation POI address entering	Advisories	Lockout	Lockout
Navigation map reading	No advisories	Advisories	Advisories
Navigation turn-by-turn	No advisories	Advisories	Advisories
Text messaging	Advisories	Lockout	Lockout

Note: Advisories, advisories against using a certain device function under a particular driving task demand; lockout, locking out a certain device function under a particular driving task demand; POI, points of interest.

27.6 CONCLUSION

In this chapter, a workload management system adopting an adaptive interface technology approach to mitigate driver distraction has been described. For safe driving, drivers must allocate sufficient attention to the driving task. A major enabler for the adaptive interface technology approach is the real-time determination of driving task demand. We have developed an algorithm for estimating the driving task demand using the traffic and roadway variables that can be obtained from in-vehicle sensors. These variables include distance and time headway between the host vehicle and the vehicles in the forward path, vehicle speed, brake status, yaw rate, and roadway lane width. The estimated demand automatically calculated by this algorithm may be categorized into three groups (e.g., low, medium, and high demand). This algorithm is validated by the correspondence between the estimated demand that is produced by the algorithm and the subjective demand that is rated by the participants.

The amount of attention that is allocated to the driving task must be commensurate with the demand for attention that is imposed by the driving task under a particular driving environment (e.g., roadway and traffic conditions). If the driving task demand is high (e.g., driving on a multilane highway with heavy traffic during rush hour), drivers must allocate a large amount of attention to the driving task. If the driving task demand is low (e.g., driving on a straight, level road with little traffic in good weather conditions), however, drivers may not need to allocate all of their attentional resources to driving and a certain amount of attention may be reallocated to nondriving tasks. This adaptive allocation strategy is borne out by subjective rating data from the present study. When the driving task demand is low, very few participants desire to curtail the amount of driver distraction and most participants believe that most in-vehicle device features should be allowed. As the driving task demand increases, many participants desire to curtail the amount of driver distraction and limit the use of certain in-vehicle device features. This mitigation strategy is applicable to all types of distraction, although it is arguably more difficult to determine cognitive distraction in real time.

The present study has several limitations. First, we have examined the perceived subjective risk that can be tolerated rather than the actual risk associated with driving and nondriving tasks. Because a driver assistance system should ultimately be accepted by the driver, it is reasonable to estimate the attentional demand from the perceived risk. However, it should be pointed out that the estimated risk might be different from the actual risk for a driving task.

Second, the driving demand algorithm is applied to predict the median demand ratings rather than the individual's ratings. Third, the study was implemented using videos that were presented in a driving simulator rather than on real roads. This design was used to maintain the same driving conditions across participants. It is conceivable that different results may have been obtained from real-road driving studies. Fourth, because no prototype was given to the participants when they were asked to determine the allowable amount of distraction for various nondriving tasks (e.g., reading SMS messages), participants did not make the determination based on a specific system. Because device manufacturers may have different interface design implementations for the same system (e.g., SMS messaging), some caution must be exercised when generalizing the present results. To overcome some of these limitations, research is under way to evaluate the effectiveness of this mitigation strategy in reducing crashes and crash risk. This mitigation strategy has been implemented in driving simulators and in instrumented vehicles in the SAVE-IT and AIDE programs. These programs will shed new light on the system effectiveness and driver acceptance of such a mitigation strategy.

ACKNOWLEDGMENTS

This chapter is based on research performed in the SAVE-IT program that was awarded to the Delphi Corporation by the NHTSA (project manager: Michael Perel) and administered by the John A. Volpe National Transportation Systems Center (project manager: Mary D. Stearns). The writing of this chapter benefited from the suggestions and comments from all three editors of this book.

REFERENCES

1. Wang, J.-S., Knipling, R. R., and Goodman, M. J., The role of driver inattention in crashes; New statistics from the 1995 crashworthiness data system, *40th Annual Proceedings of the Association for the Advancement of Automotive Medicine*, Vancouver, BC, 1996.
2. Eby, D. W. and Kostyniuk, L. P., SAfety VEhicles using adaptive Interface Technology (Task 1). Distracted-driving scenarios: A synthesis of literature, 2001 crashworthiness data system (CDS) data, and expert feedback, 2004. http://www.volpe.dot.gov/hf/roadway/saveit/docs.html.
3. Stutts, J. C., Reinfurt, D. W., Staplin, L., and Rodgman, E. A., The role of driver in traffic crashes, Report for AAA Foundation for Traffic Safety, 2001, www.aaafoundation.org.
4. Klauer, S. G., Dingus, T. A., Neale, V. L., Sudweeks, J. D., and Ramsey, D. J., *The Impact of Driver Inattention on Near-Crash/Crash Risk: An Analysis Using the 100-Car Naturalistic Driving Study Data,* DOT HS 810 594, 2006.
5. Neale, V. L., Dingus, T. A., Klauer, S. G., Sudweeks, J., and Goodman, M., An overview of the 100-car naturalistic study and findings, *Proceedings of the 19th International Technical Conference on the Enhanced Safety of Vehicles*, Washington, D.C., DOT HS 809 825, 2005.
6. Zhang, H., Smith, M. R. H., and Witt, G. J., Identification of real-time diagnostic measures of visual distraction with an automatic eye tracking system, *Human Factors*, 48, 805–821, 2006a.
7. Deering, R. K., Annual report of the Crash Avoidance Metrics Partnership: April 2001–March 2002, US DOT HS 809 531, U.S. Department of Transportation, Washington, D.C., 2002.
8. Green, P., The 15-second rule for driver information systems, *Intelligent Transportation Society of America Conference Proceedings [CD]*, Intelligent Transportation Society of America, Washington, D.C., 1999b.
9. Driver Focus-Telematics Working Group, Statement of principles, criteria and verification procedures on driver interactions with advanced in-vehicle information and communication systems, Version 2.0, April 12, 2002, Alliance of Automobile Manufacturers, Washington, D.C., 2002.
10. Green, P., Driver distraction, telematics design, and workload managers: Safety issues and solutions (SAE 2004-21-0022), Society of Automotive Engineers, Warrendale, PA, 2004.

11. Hoedemaeker, M., de Ridder, S. N., and Janssen, W. H., Review of European human factors research on adaptive interface technologies for automobiles, TNO Rep. TM-02-C031, 2002. Retrieved September 8, 2006. Available at http://www.volpe.dot.gov/opsad/saveit/docs/hoedemaeker.pdf.
12. Zhang, H., Smith, M. R. H., and Witt, G. J., Driver state assessment and driver support systems, *SAE Convergence 2006 Proceedings (CD)*, Detroit, MI, October 16–18, 2006b.
13. Michon, J. A. (Ed.), *Generic Intelligent Driver Support*, Taylor & Francis, London, 1993.
14. Wickens, C. D., Processing resources in attention, In Parasuraman R. and Davies D. R. (Eds.), *Varieties of Attention*, pp. 63–102, Academic Press, New York, 1984.
15. Zhang, H. and Smith, M., SAfety VEhicles using adaptive Interface Technology (Task 7): A literature review of visual distraction research, 2004. Available at http://www.volpe.dot.gov/hf/roadway/saveit/docs.html.
16. Engström, J., Adaptive integrated driver-vehicle interfaces: The AIDE integrated project, *ITS in Europe Conference*, Budapest, May 24–26, 2004. http://www.volpe.dot.gov/hf/roadway/saveit/research.html.
17. Esbjörnsson, M. and Juhlin, O., Combining mobile phone conversations and driving—Studying a mundane activity in its naturalistic setting, *Proceedings of ITS'2003 – 7th World Congress on Intelligent Transport Systems*, 2003.
18. Esbjörnsson, M., Juhlin, O., and Weilenmann, A., Drivers using mobile phones in traffic: An ethnographic study of interactional adaptation, *International Journal of Human Computer Interaction*, 22, 39–60, 2007.
19. Manalavan, P., Samar, A., Schneider, M., Kiesler, S., and Siewiorek, D., In-car cell phone use: Mitigating risk by signaling remote callers, *CHI Extended Abstracts on Human Factors in Computing Systems*, ACM Press, New York, 790–791, 2002.
20. Liang, Y., Reyes, M. L., and Lee, J. D., Real-time detection of driver distraction using support vector machines, *IEEE Transactions on Intelligent Transportation Systems,* 8(2), 340–350, 2007.
21. Green, P., *Visual and Task Demands of Driver Information Systems*, UMTRI Report 98-16, 1999a.
22. Tsimhoni, O. and Green, P., Visual demand of driving curves determined by visual occlusion, *Vision in Vehicles 8 Conference*, Boston, MA, 1999.
23. Donmez, B., Boyle, L., Lee, J. D., and McGehee, D. V., SAfety VEhicles using adaptive Interface Technology (Task 4): A literature review of distraction mitigation strategies, 2004a. Available at http://www.volpe.dot.gov/hf/roadway/saveit/docs.html.
24. Donmez, B., Boyle, L., Lee, J. D., and McGehee, D. V., SAfety VEhicles using adaptive Interface Technology (Task 4): Experiments for distraction mitigation strategies, 2004b. Available at http://www.volpe.dot.gov/hf/roadway/saveit/docs.html.

28 Adapting Collision Warnings to Real-Time Estimates of Driver Distraction

Matthew R.H. Smith, Gerald J. Witt, Debbie L. Bakowski, Dave Leblanc, and John D. Lee

CONTENTS

To control a vehicle in the dynamic roadway environment, a varying portion of the driver's attention must be allocated to the driving task. Under many circumstances, the flow of traffic may stabilize or disperse, making the following few seconds seem quite predictable and freeing the driver's attention to intermittently engage in nondriving activities for a few moments. Many tasks that are unrelated to driving may compete for the driver's attentional resources, such as talking with passengers, conversing on a cellular phone, or interacting with cellular phones or other nomadic devices. Even different driving-related tasks may compete with each other such that, while checking the blind spot, the driver is unable to simultaneously survey the forward road scene for potential threats. Drivers' expectations typically guide attention to potential threats in an efficient manner. Although many miles may pass without event, inevitably a situation will suddenly emerge that violates the driver's expectations. When such an unexpected and sudden situation develops, a driver may fail to devote sufficient attention to the roadway to support a timely and appropriate response. Collision warning systems can support the driver in these situations by directing the driver's attention to the unexpected or unnoticed situation on the roadway.

Data from actual driving suggest that unexpected situations can suddenly emerge and jeopardize driving safety if they coincide with a lapse in attention to the roadway. The results of the 100-car study imply that many crashes occur when the driver makes an inadequate response to an unexpected event just after the driver has been glancing away from the forward roadway.[1] Whereas the majority of noncollision lead-vehicle incidents* did not appear to be directly related to driver inattention, the linkage between "inattention to the forward roadway" and lead-vehicle crashes was compelling. In 11 of the 15 lead-vehicle crashes, the drivers' eyes were away from the forward scene just before or during the onset of the precipitating factors of the collision. Dingus et al.[1] suggests that inattention converted incidents into collisions by interfering with drivers' avoidance responses. This result indicates that inattention to the forward roadway is an important contributing factor in lead-vehicle and single-vehicle crashes, perhaps even to a greater extent than conventional collision statistics had implied (e.g., Ref. 2).

Although emerging technologies, such as nomadic devices and increasingly elaborate cellular phones, may interfere with the driving task, other innovations are being developed to enhance automotive safety. Many innovations have focused on improving the design of driver-vehicle interfaces to minimize driver head-down and hands-off-wheel time. Human factors principles have been applied to vehicle interface design to match the interface more closely to user expectations and abilities (e.g., Ref. 3). Such innovations are likely to reduce the demands of a given task substantially, such as dialing a phone number or reading a text message, but they do not directly address the timing of glances toward the in-vehicle technology in relation to the events on the roadway.[4] Another approach is needed.

One such approach is the development of safety-enhancing systems, such as forward collision warning (FCW) and lane departure warning (LDW), which are entering the automotive market. These systems support safety by warning the driver about immediate, unexpected conflicts. A major reason for the relatively slow introduction of collision warning systems into the passenger vehicle market is the potential rejection by drivers due to nuisance alerts. A system that drivers do not accept will provide no safety benefit. Both extended exposure to these systems on the road and relatively short exposures in the simulator suggest that nuisance alerts undermine driver acceptance (e.g., Refs. 5 and 6). Nuisance alerts often stem from the difficulty these systems have in detecting the driver's current state of awareness. For example, even if a system performs exactly as the designers intended and correctly identifies a potential threat associated with a slowing lead vehicle, attentive drivers may still view the situation as a nuisance alert. Whether a driver views an alert as useful or annoying depends on the driver's state of mind as well as the traffic situation. As a consequence, the next step in the adaptive-vehicle cockpit will be to measure not only the traffic situation, but also the driver's situation. To avoid annoying the driver, collision warning systems may need to adapt their warnings

* Lead-vehicle incidents were defined by Dingus et al.[1] as conflicts that required a crash-avoidance response that was smaller in magnitude than a rapid evasive maneuver but beyond the 99% confidence limit for a control input by the particular driver.

according to whether the driver is attending to the road or not. Such adaptive systems may greatly improve driver acceptance and safety.

Government-sponsored projects, that will evaluate the potential for adaptation to increase the effectiveness of conventional collision warning systems, are under way in both United States and Europe.[7] Whereas the European program Adaptive Integrated Driver-vehicle InterfacE (AIDE) has evaluated the concepts of adaptation to a host of different variables (e.g., driver preference, surface friction, and driver state; Brouwer and Hoedemaeker[8]; see also Chapter 26), the U.S. program (Safety Vehicles Using Adaptive Interface Technology, or SAVE-IT) has focused specifically on adaptation of collision warnings to the driver's state of distraction.[9–11] Although the results of the AIDE program appear to be relatively mixed for most types of adaptation, perhaps due to the extremely sensitive baseline collision warning algorithms,* the early results from the SAVE-IT program suggest that adapting collision warning systems to the driver's state may enhance the safety benefit of warning systems while simultaneously improving driver acceptance. This chapter explores how FCW and LDW systems have been adapted to take into account the driver's visual orientation.

28.1 VISUAL DISTRACTION AS A CATALYST FOR COLLISION

On global or national scales, automotive crashes occur with regularity, incurring significant costs and excessive human suffering. Severe crashes can produce devastating consequences, ending or changing forever the lives of the people involved. Automotive crashes are the most common cause of death for Americans between the ages of 4 and 34.[12] Yet for a single individual during any given mile of travel, the chances of a collision are extremely low. Although near-misses may be quite frequent, according to the National Highway Traffic Safety Administration (NHTSA) police-reported crashes in the United States occur on average only 2.1 times every million miles or once every 32 years of driving.[13] It appears that for many of these crashes to occur, a confluence of unfortunate circumstances must work together simultaneously.

As an extension of Heinrich's triangle,[14] Dingus et al.[1] argued that crashes are most often the result of the driver failing to adequately respond to a precipitating event due to various contributing factors. Figure 28.1 displays a simple model of lead-vehicle crash causation that is based on the observations of the 100-car study.[1] The central idea of this model is that crashes are typically caused when contributing factors, such as inattention or weather, interfere with an avoidance response to a precipitating event. When precipitating events (such as a lead vehicle unexpectedly braking) occur in the absence of the contributing factors, they are usually resolved by the flexible and adaptive response of the driver, resulting in what may be a near-miss, but usually not a collision. However, when contributing factors such as visual distraction are added, they act as a catalyst for a crash by interfering with the driver's response, converting a mere incident, or near-miss, into a collision. Other examples

* When the AIDE FCW and LDW systems were combined, drivers received alerts at a rate of approximately 70/h.

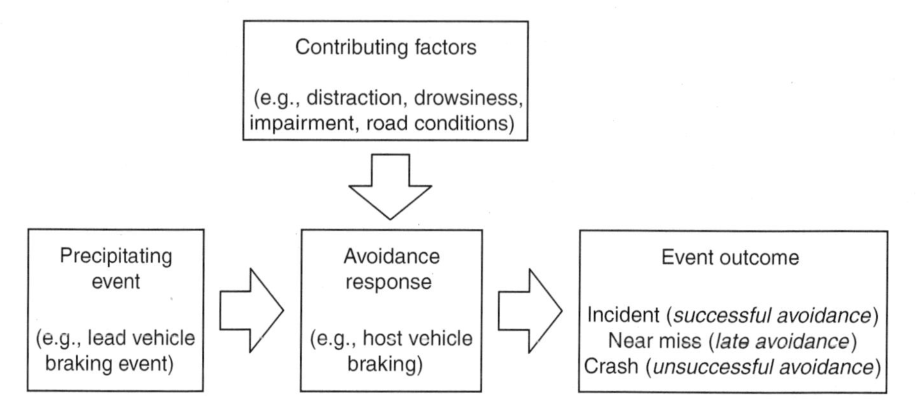

FIGURE 28.1 A simple model of distraction as a catalyst for collision.

of catalysts that can degrade driver responses might be poor roadway conditions, mechanical failure, and driver impairment due to fatigue, alcohol, or other factors.

This model is clearly an oversimplification and only considers the feedback of the driver in response to events rather than the ability of drivers to proactively drive in a manner to reduce risk. However, to the extent that this simple model approximates the reality of lead-vehicle or single-vehicle crashes, it would predict that, in the absence of a catalyst such as visual distraction, drivers are unlikely to benefit significantly from warnings. If the driver is attending to the forward roadway at the moment a precipitating event occurs, the driver usually detects the event and responds appropriately. In such a circumstance, there is usually little opportunity to improve the process. A collision warning would only present information to which an attentive driver is already aware. Even if the driver is visually attentive but is likely to react slowly due to age or intoxication, the warning system may still be unable to hasten the process, because ultimately drivers must confirm the threat for themselves before applying the brake.[6] The data presented in Figure 28.2 support this conclusion, showing that in this driving simulator study, drivers who were attentive to the forward scene released the accelerator at approximately the same time regardless of whether or how they were warned.[11] Although this situation was inherently threatening, with a collision rate between 10 and 15%, the collision warning system was unable to improve the response of visually attentive drivers.

It should be noted, however, that Lee et al.[15] found a different result in a similar experiment, in which a collision warning system eliminated the collisions which otherwise occurred at a rate of 14% for attentive drivers. This discrepancy may just represent differences in testing sensitivity or may stem from the relative ease with which drivers could see the lead vehicle in the Lee et al. study. The Lee et al. study used projectors of relatively low resolution and contrast, which may have made the deceleration of the lead vehicle relatively hard to detect and the benefit of the warning relatively great. Other factors that might contribute to the difference include the fact that the events in Lee et al. occurred at a lower speed than the SAVE-IT results showed in Figure 28.2 and the warnings were provided earlier. The discrepancy in these results demonstrates the importance of considering how drivers' perceptual capacity and attentional state interact with the onset of potential hazards.

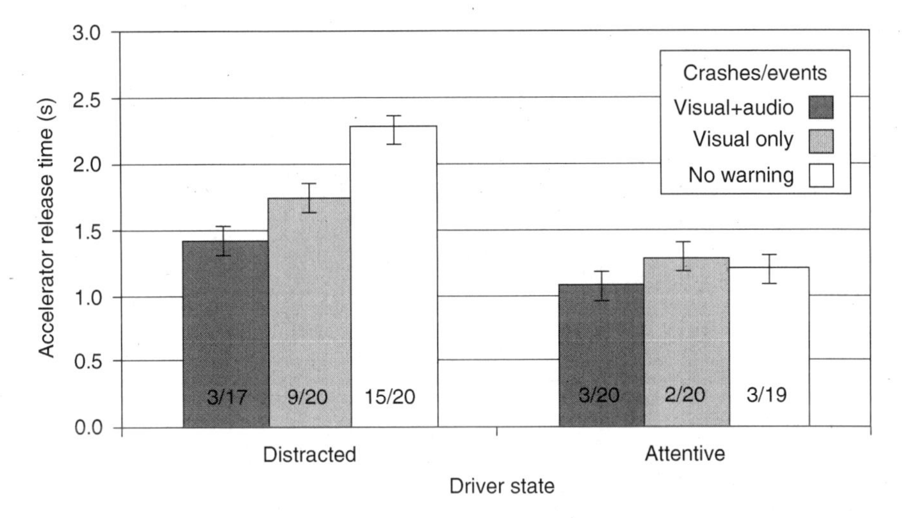

FIGURE 28.2 Accelerator release times and driving simulator crash rates of distracted drivers versus attentive drivers in the Safety Vehicles using adaptive Interface Technology program. Drivers were provided with either a visual-and-auditory, visual-only, or no forward collision warning alert. The bold numbers show the number of crashes/the number of events.

To the extent that FCW or LDW alerts are only useful for drivers who are not visually attending to the forward roadway, suppressing warnings when the driver's head pose is forward is likely to be an effective strategy for reducing nuisance alerts. In the context of the simple model shown in Figure 28.1, such an adaptive warning system monitors the essential ingredients for a collision, the precipitating event in the environment,* and the most common contributing factor that interferes with a successful avoidance response. By monitoring both environmental and driver facets of a potential collision, adaptive warning systems may provide alerts when drivers need them most while simultaneously reducing the overall rate of alerts.

If visually attentive drivers have little need for warnings, even when they are cognitively distracted, the strategy of suppressing alerts while the driver's visual attention is oriented to the forward scene is unlikely to compromise safety. The Collision Avoidance Metrics Partnership (CAMP) workload metrics project[16] data suggested that auditory-vocal tasks can actually lead to a reduction in the standard deviation of lane position (SDLP), supporting the conclusion that cognitive distraction does not degrade lane keeping.[11] The case for LDW may therefore be quite clear-cut: when an awake driver's visual attention is oriented toward the forward scene, if the vehicle is departing the lane, the driver is likely to be already aware of it. The optic flow specifying lane departure is extremely salient, and only deeply ingrained stimulus-response cognition is usually necessary for the driver to produce a counteracting response.

* For example, a lead vehicle braking or the host vehicle drifting out of the lane.

The extent to which FCW alerts may assist drivers who are cognitively distracted may be more uncertain. Many studies have demonstrated that cognitive distraction can have a detectable, yet relatively small, effect on driver reaction times to a lead-vehicle conflict.[17] Reyes and Lee[18] examined whether an auditory-vocal task increased driver reaction times to a lead-vehicle conflict. Although the impact of the auditory-vocal task was small for conflicts that could not be anticipated, when the conflict could be anticipated, the effect was far greater, increasing accelerator release times by as much as 600 ms. Reyes and Lee proposed that although auditory-vocal tasks had relatively little effect on the control level of driving, these tasks had a greater detrimental effect at the tactical level (see Chapter 5). Whereas cognitively attentive drivers were receptive to the clues that the conflict was about to occur, responding earlier when the clues were present, cognitively distracted drivers responded as if the clues were absent. Therefore, although cognitive distraction clearly has a smaller effect than visual distraction on a driver's ability to respond to rear-end conflicts, it is still uncertain whether FCW alerts are completely redundant for situations involving cognitive distraction alone (e.g., talking on a cellular phone). In the 100-car study data, two out of the 15 rear-end crashes occurred just after the driver appeared to be "lost in thought" or "daydreaming." It is difficult to predict whether an FCW system would have been successful in preventing these collisions, but this uncertainty may suggest that even if cognitive distraction cannot easily be measured, perhaps it should still be accounted for. For example, rather than suppressing an FCW alert completely in the absence of visual distraction, perhaps the alert should be softened or delayed.

28.2 FCW AND LDW COUNTERMEASURES

The U.S. Department of Transportation (DOT) has funded two large field operational tests (FOTs) to investigate the driver acceptance and potential safety benefits of FCW and LDW systems. The first of these was the Advanced Collision Avoidance Systems Field Operational Test (ACAS FOT) program that investigated adaptive cruise control (ACC) in combination with FCW.[6] In this program, 96 drivers used ACAS-equipped vehicles for a period of 4 weeks and a total of 137,000 miles. The FCW system used a forward-looking radar (FLR) to detect the range, range rate, and azimuth angle of several vehicles in front of the driver. When it appeared that the driver's vehicle needed to brake hard to avoid colliding with the lead vehicle, the ACAS system provided the driver with an icon on a head-up display (HUD) in conjunction with a series of rapid auditory tones.

During periods of manual (non-ACC) driving, FCW alerts were experienced at a rate of 14 alerts per 1000 miles. Analyses revealed that the alerts roughly broke down into thirds, including 36% of the alerts resulting from out-of-path events,* 32% of the alerts resulting from transitioning-path events,† and 27% of the alerts

* Out-of-path events were defined as situations wherein the target was never in the host vehicle's lane, and included mostly stationary objects, such as bridges and other objects either above the roadway or on side of the road.

† Transitioning-path events were defined as situations wherein a moving vehicle was is in the same lane as the host vehicle for some period of the conflict and out of the host vehicle's lane for another period.

resulting from in-path* vehicles. Determining which of these alerts was useful is an inherently subjective task that is likely to vary among drivers. Common types of nuisance alerts included situations where the host vehicle approached a lead vehicle that was vacating the lane (either turning or changing lanes) or when the host vehicle was approaching the lead vehicle with an intention to pass. By asking the drivers in the study to review alerts that were issued during their own driving experiences, the ACAS program revealed that usefulness ratings differed significantly between the three scenario types described earlier.[19] Whereas in-path events yielded alerts that were judged to be useful 53% of the time, only 33% of alerts that occurred in response to transitioning-path events were judged to be useful. Out-of-path events produced alerts that were judged to be useful only 14% of the time. Drivers' open-ended responses suggested that the level of perceived risk in these situations was often qualitatively different, even between in-path and transitioning-path events.

Using Lees and Lee's[5] terminology, this analysis suggests that a large percentage of the alerts were either unnecessary (alerts corresponding to situations judged as hazardous by the algorithm but not by the driver) or false (alerts corresponding to random activation of the system that does not correspond to a threat). Acceptance of the FCW system varied widely, with many drivers commenting that the rate of nuisance alerts contributed to their negative perceptions of the system. The two most frequent suggestions drivers made for improving the system were to reduce the rate of nuisance alerts and to allow the system to be turned off under certain circumstances.

A second FOT, the Road Departure Crash Warning System Field Operational Test or RDCW FOT was conducted using a similar design (78 drivers for 4 weeks) to investigate a LDW and curve speed warning system (CSW).[21,22] The CSW alerted drivers of a detected need to reduce vehicle speed in advance of a sharp curve. Unlike most LDW systems currently on the market, this system was more elaborate and featured not only lane-boundary sensing but also radar sensors to detect whether there was an obstacle beyond the lane boundary. The driver-vehicle interface for LDW used auditory cues for lateral movements of the vehicle over either solid painted lane markers or movements over dashed markers with an obstacle near that lane boundary. Haptic (seat vibration) cues were designed to give an impression similar to that of traveling over rumble strips. This alert occurred when the driver traveled over a dashed boundary with no detected obstacle. Common alerts, which might be considered to be unnecessary, included alerts when the driver knowingly strayed over the lane marker without consequence, or when the driver intentionally changed lanes without using the turn signal. Although the LDW alert rate per distance traveled was approximately six times greater for the RDCW FOT than it was for the ACAS FOT, these drivers provided more favorable acceptance ratings than the ACAS FOT drivers had for the FCW system.[22] Drivers of the LDW system also rated false alerts as having less utility than alerts in which they had drifted from their lane and later corrected their position.

* In-path events were defined as those wherein the host and lead vehicles occupied the same lane throughout the conflict. Almost all of the objects in this category were moveable targets (objects that the radar had previously seen to move).

One hypothesis for why the RDCW FOT yielded higher acceptance ratings for LDW than the ACAS FOT had yielded for FCW, in spite of the much higher alert rate, was that LDW is based on a criterion that drivers were able to understand and accept (crossing lane makers), compared with the FCW system, which could potentially leave drivers confused.* Whereas lane markers provide an unambiguous threshold that is visible to the driver for LDW, there is no immediately visible or universally agreed-upon threshold for FCW. Opinions about when an alert is warranted are likely to vary far less across drivers for LDW compared with FCW, and drivers can accurately predict when LDW alerts are likely to occur.

Perhaps as a result of the large amount of variability in these naturalistic studies, the two FOTs were unable to find a statistical link between nuisance alerts and driver acceptance. Lees and Lee (in preparation) examined the effects of nuisance alerts in a more controlled setting, manipulating the number of unnecessary alerts and the number of false alerts for an FCW system (see preceding definitions). In the context of this study, whereas systems prone to unnecessary alerts had relatively little effect on the driver's trust compared with a perfect FCW system, systems prone to false alerts reduced the drivers' trust compliance. Lees and Lee suggest that the predictability of the alerts is an important factor in determining both the trust in the system and the effectiveness of alerts. This result may imply that the ability to suppress some types of nuisance alerts may assist not only in improving system acceptance, but that this acceptance may, to some extent, determine the safety benefit of the warning system.

28.3 ADAPTIVE COUNTERMEASURES

When an FCW system does not take into account the driver's state, it cannot assume that the driver is completely attentive or completely distracted all the time. Driving simulator studies demonstrate that when a driver is distracted, single-exposure brake reaction times for unexpected lead-vehicle events often surpass 3 s.[9,15] Although far smaller mean reaction times (<1 s) have been recorded in test track events that were designed to surprise the driver (e.g., Ref. 23), these results likely reflect a heightened level of awareness due to the presence of an experimenter in the vehicle, the novelty of the test track, and the use of a surrogate target.† For this reason, it may be safer to assume that the upper bound for brake reaction times, which reflects a distracted driver who does not expect an event to occur in the next few moments, is closer to 3 s than it is to 1.5 or 2 s. Yet an FCW algorithm cannot invariably assume a brake reaction time of 3 s, because it would produce alerts at a rate that would both annoy drivers and degrade their confidence that the system provides anything other than false positives. But neither can an FCW system invariably assume that the driver will be able to react in less than 1 s. Such an assumption would likely result in a system that produces few nuisance alerts; however, when a necessary warning does occur, a distracted driver may require significantly more than 1 s to respond to avoid the collision.

* For example, alerts caused by overhead out-of-path objects such as signs or peripheral objects such as vehicles in other lanes or cones on the roadside.

† A research tool that appears as the rear end of a lead vehicle but is only a towed rear façade that is attached to a collapsible beam, capable of absorbing low-velocity collision impacts.

If the system cannot assume that the driver is either completely attentive or completely distracted and does not have access to data regarding driver state, it must adopt warning parameters that reflect the middle ground (e.g., a brake reaction time of 1.5 s). This middle ground represents a compromise between providing sufficient time for distracted drivers to respond while preventing an excessive rate of nuisance alerts. Because the nuisance alert rate is still likely to be greater than what many drivers are willing to accept, the selection of warning stimuli must similarly reflect a compromise between safety benefit and driver acceptance. Because of the high positive correlation between stimuli that are able to capture the driver's attention and those stimuli that annoy the driver,[20] rather than selecting warnings that are best able to quickly acquire the driver's attention, designers of the conventional warning system must select a set of less urgent stimuli, such as pleasant warning tones or subtle haptic seat pulses. The end result is a system that at best provides moderate safety benefit and produces moderate driver acceptance.[24]

Recent advances have allowed real-time driver-state monitoring technologies to evolve from a research tool into a system that is able to function within the constraints of the automotive environment. For several decades, driver-state monitoring technologies have existed as research tools, detecting driver drowsiness, head pose, and eye gaze. These technologies have typically been quite expensive, often costing as much as an entire vehicle, or have made requirements that are unreasonable for automotive applications, such as requiring physical contact with the driver. Reductions in the price of cameras and computing power are bringing the prices of these technologies down to a level that automotive consumers are more likely to accept and, in Japan, head-pose monitoring technology is now entering the automotive consumer market.[25]

The most common head-pose monitoring systems utilize computer vision algorithms that discriminate between a head pose that is forward or not forward. Further engineering development will soon provide head-pose monitoring technologies that will have relatively little impact on the cost of the safety warning countermeasures and may even allow for less expensive systems by reducing the requirements of external sensing. Because the SAVE-IT system utilizes head-pose monitoring technology to assess the driver's state of attention, the adaptive countermeasures discussed in this section adapt the state of the warning systems to the drivers' visual rather than cognitive distraction. Early experiments in the SAVE-IT program suggested that the techniques that are described in this section were not suitable for cognitive distraction and were thus applied to visual distraction only.[9]

The SAVE-IT program developed an initial list of adaptation strategies for the FCW and LDW systems (see Table 28.1) and selected the most promising candidates for more in-depth evaluation. The four adaptation strategies were Differential Display Location, Differential Display Modalities, Differential Alert Timing, and Alert Suppression. Whereas the Differential Display Location and Differential Display Modalities adaptations modify the nature of the driver-vehicle interface, the Differential Alert Timing and Alert Suppression adaptations modify the algorithms that generate the alerts. The Differential Display Location adaptation positions the visual stimulus of the alert in the location of the visual distraction to which the driver is currently attending, and the Differential Modalities adaptation provides a more

TABLE 28.1
Safety Vehicles Using Adaptive Interface Technology Program Positive and Negative Adaptation Strategies for Forward Collision Warning and Lane Departure Warning

Attention-Based Adaptation Strategy		Negative Adaptation—Attention Forward Goal: Improved Acceptance	Positive Adaptation—Attention Not-Forward Goal: Improved Safety
Nonadaptive		Nominal alert	Nominal alert
Modify human machine interface	Differential display location	Nominal alert	Visual alert in location of driver's attention
	Differential alert stimuli	Less intrusive or urgent stimuli	More intrusive or urgent stimuli
Modify algorithm	Differential alert timing	Later alert (less likely)	Earlier alert (more likely)
	Alert suppression	No alert	Nominal alert

urgent or attention-capturing stimulus when the driver is distracted. Whereas the Alert Suppression adaptation simply prevents alerts from being generated when the driver is attentive, the Differential Alert Timing adaptation modifies the likelihood that alerts will be generated, by providing earlier alerts when the driver is distracted and later alerts when the driver is not.

Most types of adaptation can be either negative or positive. Whereas negative adaptations diminish the warnings when the driver's attention is on the road, positive adaptations accentuate warnings when the driver's attention is away from the road. Specifically, adaptations in the "Attention Forward" column are negative, in that they feature methods for suppressing or softening alerts, and the modifications under the "Attention Not-Forward" column are positive adaptations, in that they include methods for accentuating or promoting the alerts. The primary goal of negative adaptation is to improve driver acceptance by reducing the potential nuisance of unnecessary and false alerts. Although unnecessary alerts are targeted more directly by the negative adaptations, the reduction in the overall number of alerts during the attentive periods of the drive will also significantly decrease the rate of false alerts. The primary goal of positive adaptation is to improve the safety benefit of the warning systems. Although the primary goals are separate for negative and positive adaptation, the dimensions of driver acceptance and safety benefit are not independent. For a system to be successful in achieving a safety benefit, the driver, to some extent, must accept the alerts. Lees and Lee (in preparation) demonstrated that drivers complied less with their FCW system when the system was prone to false alerts. This "cry wolf" effect has been consistently shown to undermine response to warning systems.[26–28] Furthermore, it is likely that if the system apparently fails to achieve a safety benefit, drivers may be less likely to accept the system because they do not perceive the system as being useful.

28.3.1 Differential Display Location

The Differential Display Location adaptation strategy modifies the placement of the alerting visual stimulus as a function of the driver's focus of attention. In the

SAVE-IT example of this adaptation, when the drivers were distracted by their interactions with center-console applications and either the FCW or LDW systems produced an alert, the safety warning systems presented a visual warning stimulus (an FCW or LDW icon) on the center console, temporarily replacing the material that was currently being displayed. This icon was presented redundantly with the other visual-and-auditory stimuli only when the driver was already interacting with the center console, so that if the driver was attending to the forward roadway, only the conventional visual-and-auditory warning stimuli appeared. The rationale for this adaptation is that if the driver glances at the center console, an alert icon placed in the center console is more likely to acquire the driver's attention, thus increasing the utility of the visual stimulus. The reason that warning icons should not be indiscriminately positioned in the center console is that positioning an icon away from the frontal location is likely to draw the driver's attention away from the external threat. This adaptation attempted to circumvent this problem by only presenting an icon in the center console when the driver's attention was already away from the forward scene. As shown in Table 28.1, drivers who were attending to the forward scene at the time of the alerting event received a nominal (nonadaptive) warning. Although this type of adaptation appears to be reasonable, SAVE-IT testing suggests that it is not beneficial to the driver. Perhaps the center-console icon delayed the drivers from returning their gaze back to the forward scene, and thus the subsequent decision to brake.

28.3.2 Differential Display Modalities

Like the Differential Display Location adaptation strategy, the Differential Display Modalities strategy operates by modifying the driver interface of the FCW and LDW systems. In the SAVE-IT program, this adaptation modified the interface by providing a visual-only alert when the driver was attending to the forward scene and a visual-plus-auditory alert when the driver was not. The reasoning behind this strategy is that drivers who are already looking in the forward direction are likely to be able to detect a visual alert located near the forward scene and thus may not need the auditory stimulus. Whereas, drivers with a forward head pose may be adequately alerted from an orientation-dependent stimulus (such as a visual alert), distracted drivers may require an orientation-independent stimulus (such as an auditory alert) to reacquire their attention. An auditory alert is more likely to produce annoyance than a visual stimulus because it is usually more intrusive than a visual stimulus and cannot be localized to the driver, thus potentially interrupting conversations or undermining the passenger's confidence in the driver. The Differential Display Modalities adaptation seeks to reduce annoyance by suppressing the most intrusive component of the alert, which is usually the auditory component, when the driver is attentive to the forward roadway.

In the initial phase of the SAVE-IT program, this strategy not only used a negative adaptation of suppressing the auditory component of the alert, but also provided a positive adaptation by providing a voice stimulus (saying "lead vehicle braking" or "drifting left" or "drifting right") at an early threshold when the driver was attending to the forward scene. This particular type of positive adaptation (using voice stimuli) was quickly dismissed after the experiment demonstrated that it led to excessive driver annoyance; however, other, less annoying, alternatives might be conceived. The data in Figure 28.2 imply that when drivers attend to the forward roadway, neither the visual nor auditory stimuli are likely to provide a significant benefit. Although this result may represent a lack of sensitivity rather than an actual lack of a difference, it suggests that the opportunity to benefit attentive drivers is relatively small. Thus, rather than supporting a Differential Display Modalities strategy, it seemed to suggest that the strategy of suppressing the alert entirely (Alert Suppression) may be viable.

One potential benefit of Differential Display Modalities adaptations is that they may be able to help reveal the underlying functionality of the warning system. For example, rather than completely suppressing an LDW alert when the driver is unlikely to require it, an LDW system might only suppress the more intrusive auditory component, still providing a more private haptic (e.g., seat vibration) stimulus. That way, even though the auditory stimulus is suppressed and drivers are saved from the potential annoyance, the core LDW system functionality is revealed, and drivers will still be able to observe that the system detected the lane crossing. It also allows for the possibility of providing some benefit for drivers who are severely cognitively distracted or drowsy if such drivers are able to benefit from the warnings. Whereas the total suppression of an alert when the driver is visually oriented to the forward scene would not provide assistance for drivers' states other than visual distraction, Differential Display Modalities adaptations still provide an alert to drivers who are oriented to the forward scene. These alerts may potentially benefit a driver who is visually attentive but whose response may be degraded in some other way (e.g., cognitive distraction), even though the other source of degradation is not identified. In this way, the Differential Display Modalities adaptation may accommodate other types of degradation that cannot be measured.

28.3.3 Differential Alert Timing

The Differential Alert Timing strategy provides earlier alerts for distracted drivers and later alerts for drivers who appear to be attentive to the forward scene. In the first phase of the SAVE-IT program, the Differential Alert Timing strategy was applied to both the FCW and LDW systems. Whereas the FCW algorithm was adapted by altering the assumed driver brake reaction time in response to the alert, the LDW system was adapted by narrowing the thresholds for lane crossings. This strategy is based on data demonstrating that distracted drivers require more time to respond to an alert than attentive drivers (e.g., Refs. 9 and 15). By providing differential predictions for how quickly a driver will respond to the warning, an FCW algorithm can provide distracted drivers with sufficient time to respond and prevent a large percentage of unnecessary alerts from being given to attentive drivers.* SAVE-IT research demonstrated that the rate of nuisance alerts could effectively be reduced by later timing when the driver was attentive (negative adaptation), and that early alerts for distracted drivers could reverse the negative effects of distraction (positive

* Attentive drivers are likely to respond to the threatening conditions, or the fleeting pseudo-threat is more likely to dissipate before the alert threshold is reached.

adaptation). Like Lee et al.'s[15] results, the SAVE-IT program demonstrated that earlier warnings translated into significantly earlier responses, with distracted drivers who experienced the earlier alert* braking 2.3 s after the lead vehicle braking event compared with the nominally alerted distracted drivers who began braking 3.1 s after the event.[11]

28.3.4 Alert Suppression

Alert Suppression is perhaps the most simple and obvious negative adaptation strategy, and it directly governs whether the warning is issued rather than the manner in which it is issued. When the driver attends to the forward roadway, the alert is suppressed, but otherwise the alert functions like a nominal warning system. Figure 28.2 implies that the safety benefit of an FCW system is compromised little by suppression of alerts when the driver is attentive to the forward roadway. One potential shortfall with the Alert Suppression strategy for FCW is that it does not allow FCW alerts to provide benefit in situations other than when the driver is visually distracted. When the driver's visual attention is forward, regardless of the driver's cognitive state, the alert will be suppressed. Alert Suppression may be a better candidate for LDW systems. The data collected during the SAVE-IT program suggest that drivers who are attentive to the forward roadway receive no benefit from an LDW system.[11] Even without an LDW system, attentive drivers reacted so quickly to a simulated wind gust in a driving simulator experiment, that there appears to be little room for improvement. A wind gust that was quite threatening to distracted drivers simply did not pose a threat to the drivers who were not engaged in a secondary task. Many drivers corrected with the lateral disturbance so rapidly that they did not even notice the wind gust at all.

The SAVE-IT program evaluated the Alert Suppression strategy in a small on-road study. In this study, 14 Delphi employees† drove a vehicle for 160 miles each, experiencing an adaptive LDW system for half of the time and a nonadaptive LDW system for the other half.‡ The adaptation suppressed 95% of the alerts, with the 14 drivers experiencing a total of 81 alerts (78 alerts per 1000 miles) while the system was in nonadaptive mode and only 4 alerts (4 alerts per 1000 miles) while the system was in the adaptive mode. Participants indicated that the adaptive system produced alerts at a rate that was more likely to be acceptable, and 12 of the 14 participants preferred the adaptive system to the nonadaptive system. The remaining two drivers, who preferred the nonadaptive system, indicated that they preferred the consistency of receiving alerts every time that they crossed the lane. Further subjective measures suggested that these drivers might have preferred a Differential Display Modalities adaptation that suppressed only the auditory component when they were attentive. Although the system only employed a negative adaptation technique, participants

---

* When the driver was distracted the adaptive FCW algorithm assumed a 3-s reaction time, compared to a 1-s reaction time for a nonadaptive algorithm. The earlier reaction time assumption translated to an alert that was provided 2 s earlier for the adaptive system.

† Employees were screened so as to exclude those who worked on products related to LDW or driver monitoring.

‡ The order of the trials was counterbalanced.

indicated an average rating* of 3.9 for the adaptive system in response to the statement "the LDW system enhances on-road safety" compared with an average rating of 3.1 for the nonadaptive system.

For the FCW system, the SAVE-IT program selected a version of the Differential Timing strategy that, on the negative side of adaptation (alert reduction), was quite similar to an Alert Suppression strategy. Rather than suppressing an alert outright, the alert timing was implemented in such a way that only in rare circumstances would drivers with a forward head pose receive an alert. By selecting a short brake reaction time (0.5 s), the FCW system could prevent the vast majority of nuisance alerts from occurring while the driver was attentive to the forward scene. Yet in rare cases, where an attentive driver could potentially benefit from an alert, a late alert was still able to occur. The SAVE-IT program also evaluated the adaptive FCW system in comparison with a nonadaptive FCW system in a small on-road study. In this study, 14 drivers experienced the two versions of the FCW system over a total of 1698 miles. Unlike the adaptive LDW system, the adaptive FCW system could provide alerts in some circumstances that might not have occurred in the nonadaptive baseline condition (due to the earlier timing during nonforward head poses). Whereas the nonadaptive system produced a total of 64 alerts (75 alerts per 1000 miles), the adaptive system produced 19 alerts (22 alerts per 1000 miles), representing a 70% reduction in alerts. Out of the 13 of 14 participants who indicated a clear preference, 10 participants preferred the adaptive mode to the nonadaptive mode. Whereas eight of 14 participants indicated that more than 50% of the nonadaptive alerts were a nuisance, only two participants indicated that the adaptive nuisance alert rate was greater than 50%. Participants also indicated more favorable ratings for alert timing and greater likelihood to recommend the system to others. Surprisingly, even though the FCW adaptive system featured both positive and negative adaptations, and the LDW system only featured negative adaptation, unlike the LDW system, participants rated similarly the extent to which the adaptive and nonadaptive systems enhanced safety.

28.4 CONCLUSIONS AND FUTURE CONSIDERATIONS

The future of adaptive collision warnings is likely to be primarily influenced by two factors: the driver state–sensing technology and the marketing of adaptive systems to consumers. As sensing technologies provide increasingly accurate and sensitive information about the state of the driver, the methods for adaptation will likely evolve to make use of the new information. For example, whereas head pose provides a coarse indication of visual distraction only, future systems will provide information about where the driver's eyes are focused. This more fine-grained information will likely support the detection of both cognitive distraction (Reyes and Lee[28]) and driver intention.[30] Although the price of automotive-grade technology with sufficient resolving power to support the detection of cognitive distraction may currently be excessive, such technology is likely to become increasingly affordable,

* The questionnaire used a four-point scale (1-strongly disagree, 2-somewhat disagree, 3-somewhat agree, 4-strongly agree).

soon providing adaptive systems with information regarding the driver's cognitive state (see Refs. 29–32). Increased resolving power is also likely to support the detection of driver impairment due to alcohol or other drugs in the near future. The adaptation techniques that were reviewed in this chapter did not appear to be suitable for the presentation of collision warnings when drivers are cognitively distracted[9]; however, more subtle and sophisticated techniques may be developed to make use of this information regarding the driver's cognitive attention. One of the challenges of adapting collision warning systems to the driver's cognitive state may be that, whereas the criteria for a driver's visual orientation might be easily observed and understood, the criteria for more complex phenomena such as the degree to which the driver is mentally engaged in the driving task may be more subjective and less easily understood by the driver.

The extent to which drivers accept adaptive collision warnings may also be influenced by how these systems are marketed to the public. Drivers who have an inadequate understanding of the system might perceive the warning behavior as inconsistent when an alert is provided in one instance (when the driver's head pose is *not* forward) but not in another (when the driver's head pose *is* forward). Such drivers might view this system as failing to provide a safety benefit. If unnecessary alerts do not degrade trust in the system,[5] drivers may prefer to witness alerts, even when the system may be able to predict that the alert will be unnecessary. Drivers who understand the concept of adaptation may prefer a system that provides a subtle alert (e.g., haptic or visual-only) when they are attentive rather than one that suppresses the alert completely, because it may continue to reinforce that the system is providing the protection and is accurately detecting the lane change or the lead vehicle braking. The recent SAVE-IT results suggest that drivers who have adequate knowledge of how the system is intended to operate may appreciate that their own behavior is taken into account when the system decides when to issue an alert. The driver's mental model is thus a crucial factor in determining the acceptance of different adaptation techniques. How an adaptive system is perceived or which type of adaptation is preferred may therefore be highly dependent on how these systems are marketed or how they are sold on the showroom floor.

Adaptive collision warning systems face a challenging trade-off. Without adaptation, the systems are likely to warn drivers unnecessarily or fail to warn drivers in a timely manner. As a consequence, trust and acceptance may decline. With some types of adaptation, drivers may feel that the system operates in a capricious and arbitrary manner, a feeling that also leads to a decline in trust and acceptance. The theoretical basis of trust may offer a way to manage the trade-off.[31] Trust depends on the driver's assessment of the performance, process, and purpose of a system. For collision warnings, performance depends on the number of warning failures. Adaptation could enhance the performance basis of trust. For collision warnings, the process basis of trust depends on the driver's ability to understand the algorithms and mapping between environmental conditions and the warning occurrence. Adaptive systems involve more complex algorithms and so may undermine the process basis of trust if they are not implemented carefully. For collision warnings, the purpose basis of trust depends on the driver's understanding of why the system was developed and that the complexity of adaptive warnings could impact this basis in an unpredictable manner. To achieve the greatest possible benefit of adaptation, the next generation of warning systems must carefully consider all three bases of trust, perhaps using a combination of the different positive (accentuating) and negative (diminishing) adaptation alternatives.

ACKNOWLEDGMENTS

This research was conducted as a part of the SAVE-IT program by Delphi Electronics and Safety, in collaboration with the University of Michigan Transportation Research Institute (UMTRI) and the University of Iowa, sponsored by the U.S. DOT, NHTSA, Office of Vehicle Safety Research, and administrated by the Volpe Center. The authors gratefully acknowledge Mike Perel (NHTSA), Mary Stearns, and Tom Sheridan (Volpe) for their assistance and guidance in this program.

REFERENCES

1. Dingus, T. A., Klauer, S. G., Neale, V. L., Petersen, A., Lee, S. E., Sudweeks, J., Perez, M. A., Hankey, J., Ramsey, D., Gupta, S., Bucher, C., Doerzaph, Z. R., Jermeland, J., and Knipling, R. R., *The 100-Car Naturalistic Driving Study, Phase II: Results of the 100-Car Field Experiment*, NHTSA DTNH22-00-C-07007, 2006.
2. Campbell, B. N., Smith, J. D., and Najm, W. G., *Examination of Crash Contributing Factors Using National Crash Databases*, DOT HS 809 664, National Highway Transportation Safety Administration Report, Washington, D.C., 2003.
3. Campbell, J. L., Carney, C., and Kantowitz, B. H., *Human Factors Design Guidelines for Advanced Traveler Information Systems (ATIS) and Commercial Vehicle Operations* (CVO), No. FHWA-RD-98-057, Federal Highway Administration, Washington, D.C., 1998.
4. Wiese, E. E. and Lee, J. D., Attention grounding: A new approach to IVIS implementation. *Theoretical Issues in Ergonomics Science*, 8(3), 255–276, 2007.
5. Lees, M. N. and Lee, J. D., The influence of distraction and driving context on driver response to imperfect collision warning systems, *Ergonomics*, 50(8), 1264–1286, 2007.
6. National Highway Transportation Safety Administration, *Automotive Collision Avoidance System Field Operational Test: Final Program Report*, No. DOT HS 809 886, National Highway Traffic Safety Administration, Washington, D.C., 2005.
7. Witt, G. J, Zhang, H., and Smith, M. R. H., Safety Vehicle(s) using adaptive Interface Technology (SAVE-IT): Phase I Progress Report, *International Workshop on Progress and Future Directions of Adaptive Driver Assistance Research*, Washington, D.C., 2004. http://www.volpe.dot.gov/hf/roadway/saveit/workshop.html.
8. Brouwer, D. M. and Hoedemaeker, D. M. (Eds.), *Driver Support and Information Systems: Experiments on Learning, Appropriation and Effects of Adaptiveness*, 2006. http://www.aide-eu.org/pdf/aide_d1-2-3.pdf.
9. Smith, M. R. H. and Zhang, H., *Safety Vehicle(s) Using Adaptive Interface Technology (SAVE-IT) Task 9 Final Report: Safety Warning Countermeasures*, 2004b, http://www.volpe.dot.gov/hf/roadway/saveit/docs/dec04/finalrep_9b.pdf.
10. Zhang, H., Smith, M. R. H., and Witt, G. J., Identification of real-time diagnostic measures of visual distraction with an automatic eye tracking system, *Human Factors*, 48(4), 805–822, 2006.
11. Smith, M. R. H., Bakowski, D. L., and Witt, G. J., *Safety Vehicle(s) Using Adaptive Interface Technology (SAVE-IT) Task 9 Final Report: Safety Warning Countermeasures*, in preparation.

12. Subramanian, R., Motor vehicle crashes as a leading cause of death in the United States, 2004, http://www-nrd.nhtsa.dot.gov/pdf/nrd-30/NCSA/RNotes/2006/810568.pdf, 2006.
13. National Highway Transportation Safety Administration, Traffic safety facts, 2006.
14. Heinrich, H. W., Petersen, D., and Roos, N., *Industrial Accident Prevention*, McGraw-Hill, New York, 1980.
15. Lee, J. D., McGehee, D. V., Brown, T. L., and Reyes, M. L., Collision warning timing, driver distraction, and driver response to imminent rear-end collisions in a high fidelity driving simulator, *Human Factors*, 44, 314–334, 2002.
16. Angell, L., Auflick, J., Austria, P. A., Kochhar, D., Tijerina, L., Biever, W., Diptiman, T., Hogsett, J., and Kiger, S., *Driver Workload Metrics Project: Final Report*, Sponsored by National Highway Traffic Safety Administration, DOT HS 810 635, Washington, D.C., November 2006.
17. Horrey, J. H. and Wickens, C. D., Examining the impact of cell phone conversations on driving using meta-analytic techniques, *Human Factors*, 48(1), 196–205, 2006.
18. Reyes, M. L. and Lee, J. D., The influence of IVIS distractions on tactical and control levels of driving performance, *Proceedings of the 48th Annual Meeting of the Human Factors and Ergonomics Society*, Vol. 2, pp. 2369–2373, Human Factors and Ergonomics Society, Santa Monica, CA, 2004.
19. Ervin, R., Sayer, J., LeBlanc, D., Bogard, S., Mefford, M., Hagan, M., Bareket, Z., and Winkler, C., *Automotive Collision Avoidance System (ACAS) Field Operational Test Methodology and Results*, US DOT HS 809 901, National Highway Traffic Safety Administration, Washington, D.C., 2005.
20. Lerner, N., Dekker, D., Steinberg, G., and Huey, R., *Inappropriate Alarm Rates and Driver Annoyance*, DOT HS 808 532, National Highway Traffic Safety Administration, Washington, D.C., 1996.
21. Emery, L., Srinivasan, G., Bezzina, D., LeBlanc, D., Sayer, J., Bogard, S., and Pomerleau, D., Status report on USDOT project "An intelligent vehicle initiative road departure crash warning field operational test," US DOT HS 809 825, *Proceedings of the 19th International Technical Conference on the Enhanced Safety of Vehicles*, Washington, D.C., 2005.
22. LeBlanc, D., Sayer, J., Winkler, C., Bogard, S., Devonshire, J., Mefford, M., Hagan, M., Bareket, Z., Goodsell, R., and Gordon, T., *Road Departure Crash Warning System (RDCW) Field Operational Test Final Report*, US DOT report, Washington, D.C., 2006.
23. Kiefer, R., LeBlanc, D., Palmer, M., Salinger, J., Deering, R., and Shulman, M., *Development and Validation of Functional Definitions and Evaluation Procedures for Collision Warning/Avoidance Systems*, DOT-HS-808-964, U.S. Department of Transportation, Washington, D.C., 1999.
24. Lee, J. D., Hoffman, J. D., and Hayes, E., Collision warning design to mitigate driver distraction, *Proceedings of CHI 2004*, pp. 65–72, ACM, New York, 2004.
25. Bliss, J., Dunn, M., and Fuller, B. S., Reversals of the cry-wolf effect: An investigation of two methods to increase alarm response rates, *Perceptual and Motor Skills*, 80, 1231–1242, 1995.
26. Bliss, J. and Acton, S. A., Alarm mistrust in automobiles: How collision alarm reliability affects driving, *Applied Ergonomics*, 34, 499–509, 2003.
27. Young, K. L., Regan, M. A., Triggs, T. J., Tomasevic, N., Stephan, K., and Mitsopoulos, E., Impact on car driving performance of a following distance warning system: Findings from the Australian TAC SafeCar Project, *Journal of Intelligent Transportation Systems*, 11, 121–131, 2007.
28. Reyes, M. L. and Lee, J. D., Effects of cognitive load presence and duration on driver eye movements and event detection performance, *Transportation Research Part F*, in preparation.

29. Smith, M. R. H. and Zhang, H., Safety Vehicle(s) using adaptive Interface Technology (SAVE-IT) Task 8 Phase I Report: Intent, http://www.volpe.dot.gov/hf/roadway/saveit/docs/dec04/finalrep_8b.pdf, 2004a.
30. Lee, J. D. and See, K. A., Trust in technology: Designing for appropriate reliance, *Human Factors*, 46(1), 50–80, 2004.
31. Liang, Y., Reyes, M. L., and Lee, J. D., Real-time detection of driver cognitive distraction using support vector machines, *IEEE Intelligent Transportation Systems*, 8(2), 340–350, 2007.
32. Victor, T. W., Harbluk, J. L., and Engstrom, J. A., Sensitivity of eye-movement measures to in-vehicle task difficulty, *Transportation Research Part F*, 8, 167–190, 2005.

29 Designing Feedback to Mitigate Distraction

Birsen Donmez, Linda Boyle, and John D. Lee

CONTENTS

29.1 INTRODUCTION

The rapid development of sensor, wireless communication, and computing technology has given rise to a range of devices that are capable of entertaining, informing, and supporting the driver (e.g., MP3 players, cellular phones, and navigation systems). However, these devices may also undermine safety due to conflicts between the demands of the in-vehicle system and the demands of driving. Part 7 of this book describes design approaches that reduce the demands associated with using in-vehicle information system (IVIS) functions while driving. Technology that can assist in mitigating distraction in real time include warning the driver about dangerously high levels of distraction, locking out functions (see Chapter 26), or having a system adapt appropriately to the degree of distraction experienced by the driver (see Chapter 28). Such technology mainly focuses on enhancing immediate driving performance (i.e., real-time performance when the technology takes action to mitigate distraction). Another approach to mitigate distraction is to provide feedback to the driver to enhance immediate performance as well as to induce a positive behavioral change, such as diminishing the willingness,to engage in future distracting activities to enhance long-term driving performance.

Feedback within the context of this chapter can be defined as the information provided to the driver regarding the state of the driver-vehicle system. Immediate driving performance (e.g., lane position) is feedback inherent in the driving task that can be enhanced with additional feedback provided via an in-vehicle system. For example, feedback can be provided as alerts to warn the driver of critical roadway situations (e.g., lane drift signaled by virtual rumble strips) or high levels of distraction. Changes can occur very rapidly in the driving environment, and the driver may fail to track these changes, particularly if the driver's attention is directed toward a nondriving-related activity or if the driver is cognitively loaded.[1–4] In such situations, feedback can help the driver respond to these changes (e.g., lead vehicle braking) more appropriately. Feedback can provide a warning to the driver based on a hazardous situation (e.g., lane deviation warning), can help the driver learn what is unsafe (e.g., failure to reduce speed during bad weather conditions), and can ultimately alter driver behavior (e.g., inhibiting knowingly risky behavior such as speeding).

Drivers currently receive feedback that shapes their immediate response to the driving situation and their long-term behavior. For example, there are educational messages provided on billboards or in radio or television media (e.g., "don't drink and drive," "click it or ticket"), variable message signs, radar speed display signs, and by law enforcement and driving instructors. Each medium or person provides some feedback on what drivers should be doing or have already done in the hope that they will correct their behavior on future drives. However, such feedback is dependent on the environment, is not tailored to the behavior of the individual, and may be absent in some situations. For example, there is no means by which to consistently provide feedback when a driver looks away from the road for a dangerously long period of time. Furthermore, existing feedback is not tailored to the driver, and drivers do not receive moment-to-moment feedback regarding their performance. Therefore, such feedback may have little influence on drivers' behavior. Emerging technology can circumvent the limits of current feedback and may provide an effective means by which to mitigate distraction as well as alert the driver to other inappropriate behavior. Different feedback characteristics (e.g., positive or negative feedback) can facilitate these outcomes differently. This chapter describes characteristics of different feedback types and the benefits they can provide for enhancing performance and modifying driver behavior such as reducing drivers' willingness to engage in distractions.

A three-dimensional taxonomy has previously been proposed to define different distraction mitigation strategies that focus on enhancing immediate driving performance.[5–8] These dimensions include the degree of automation of the mitigation strategy, the type of initiation, and the type of task that is being modulated by the strategy. The degree of automation can range from a simple driver alert signal to complete system control. These automation levels can be initiated either by the driver or the automation and can modulate either the driving task or the in-vehicle task. This chapter proposes another dimension in designing distraction mitigation strategies—the temporal dimension—which considers the immediate effect of the system on driving performance as well as the long-term effect on drivers' willingness to engage in a distracting activity. Changing behavior, such as the willingness to engage in a distracting activity, may have a particularly powerful influence on safety.[9] Providing feedback is promising because drivers may not always realize the potential hazards created from decisions to engage in a distracting activity, may not always make the safest choice in doing so, and often experience no negative consequences for a poor choice. Feedback may enhance immediate performance, but this effect may not always be sustained once feedback is removed, unless it can provide information that updates the driver's internal model of safe driving, thereby resulting in a behavioral change. There is a need to develop design strategies that can mitigate the effects of driver distraction on immediate driving performance and to encourage safer long-term driving habits. Current driving literature is very limited with respect to different feedback timescales. Therefore, much of this chapter is based on theoretical assertions that are grounded in research conducted in other domains. More research is needed to investigate the effects of different feedback timescales on driving performance and behavior.

29.2 TIMESCALES OF FEEDBACK

There are four major timescales that can be designed in a distraction mitigation system: concurrent (milliseconds), delayed (seconds), retrospective (minutes, hours), and cumulative (days, weeks, months). Figure 29.1 suggests that as the timescale of feedback extends, the goal of feedback will change from improving immediate driving performance to inducing safer driver behavior. For example, concurrent (i.e., real-time) feedback can improve performance of the driver who has just departed his or her lane and

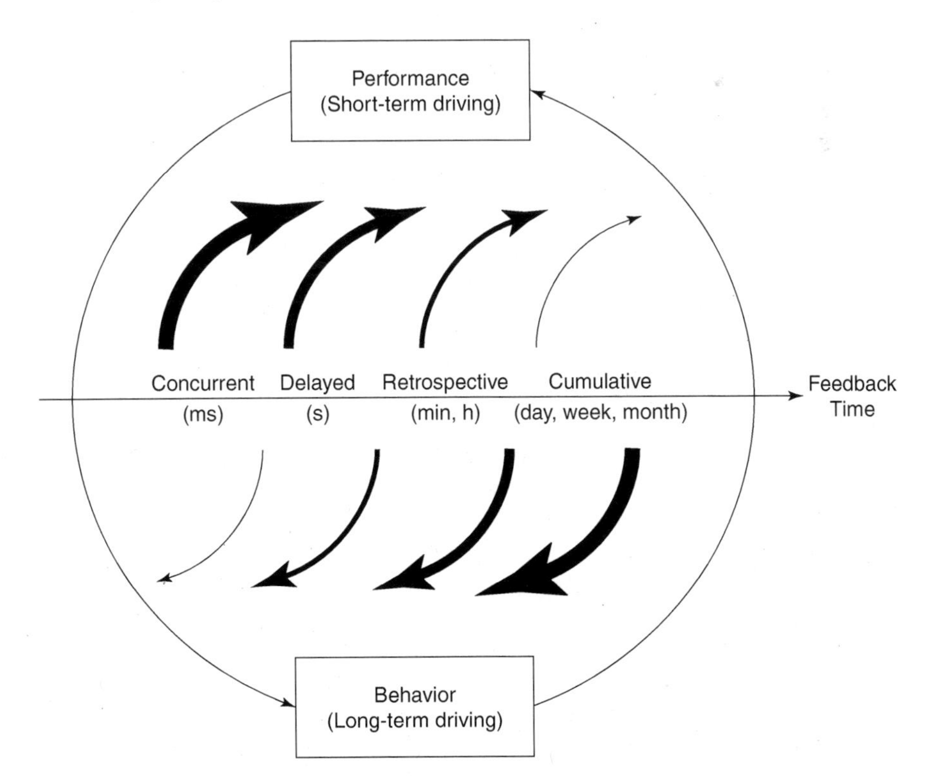

FIGURE 29.1 Levels of feedback timing and the magnitude of targeted influence (indicated by arrow thickness).

should discontinue a cell phone conversation. This concurrent feedback may improve the driver's lane-keeping performance, but might not influence the driver's willingness to use a cell phone during the next trip. However, accumulating and reporting to the driver the number of lane departures over months can make the driver more aware of how cell phone conversations cause these departures, and this may diminish the driver's willingness to engage in this activity while driving. The following sections discuss the pros and cons of each feedback timescale, which are summarized in Table 29.1. Each of these feedback timescales is also incorporated into a driver information–processing model that provides insights for designing distraction mitigation systems.

29.2.1 Driver Information–Processing Model with Temporal Feedback

The driver information–processing model with temporal feedback encompasses five stages (Figure 29.2). Four of these stages (i.e., intention, perception, cognition, and action) and the disturbances to these stages from the physical or cognitive distracters (e.g., cell phones) are based on a model presented by Sheridan.[10] Sheridan's information-processing model is modified here in two ways: (1) the revised model alters the interaction between distracters and the intention stage, and (2) it also adds a new stage—"internal model of safe driving."

Sheridan[10] defines the intention stage as creating a "priority-ordered sequence of near-term driving goals." This definition focuses on driving only and excludes driver intentions to engage in distracting activities. However, the driver's intentions will set goals for both the driving and the nondriving tasks. To capture the intent to engage in distracting activities (both driving and nondriving related), this model includes a link from the intention stage to the distracters, suggesting that distractions can be initiated by the driver. Distracters can also create disturbances that influence the driver's intentions, and thus, the link is bidirectional. Providing feedback to inhibit the driver's intentions to engage in distractions can enhance safety.

Sheridan's[10] model assumes that the basic intention of the driver is to drive safely regardless of any additional tasks undertaken. However, the definition of safe driving can vary among individuals and change over time. To capture this effect, a fifth stage is included in the model to represent the internal model of safe driving (i.e., a driver's belief of acceptable behavior while driving). Interactions also exist among these five stages. For example, the driver's perception of the environment can update intentions, and cognition can guide perception by directing attention to different aspects of the environment. Feedback to the internal model of safe driving has potential to alter driving behavior. Feedback that directly helps with intention and perception has potential to enhance immediate driving performance. Each feedback timescale has a different degree of influence on these stages. The following sections define each feedback timescale in detail and describe how these timescales affect behavior and immediate performance.

29.2.2 Concurrent Feedback

Concurrent feedback can be presented to the driver in real time when there is a resource conflict between driving and distracters. For example, if the driver is distracted, or if the driver fails to respond appropriately to a roadway demand,

TABLE 29.1
Potential Pros and Cons for Different Feedback Timescales

Timescale	Pros	Cons
Concurrent feedback	• Immediate implications for enhancing driving performance when feedback is present • Can help the driver learn safe maneuvers (e.g., safe following distance)	• Driver may adapt to feedback inappropriately • May elevate the level of cognitive distraction if feedback is not intuitive to the driver • Low acceptance can lead to disuse of feedback • Overreliance on feedback can result in dangerous situations if feedback fails • Can interfere with immediate task performance • Unexpected lags can undermine the effect of feedback • Deterioration of productivity (i.e., in-vehicle information system [IVIS] task performance)
Delayed feedback	• Informs the driver about correct and incorrect driving behavior while avoiding cognitive overload • Can enhance driving performance during a trip for upcoming events • Can help the driver learn safe maneuvers (e.g., safe following distance)	• Feedback is not provided at the time of the incident and can therefore not enhance immediate driving performance • Unexpected lags can undermine the effect of feedback • Deterioration of productivity (i.e., IVIS task performance) • Low acceptance may lead to disuse
Retrospective feedback	• Intentions leading to unsafe driving behavior can be explained to the driver without cognitive overload • Can enhance driving performance for future trips • Can refresh drivers' memory on performance for the completed trip • Can calibrate driver's subjective performance by presenting a connection between intentions and events that occurred during a trip	• Feedback is not provided at the time of the incident and can therefore not enhance immediate driving performance • Requires the driver to be an active recipient of information • Driver may fail to link feedback with incident • Low acceptance may lead to disuse
Cumulative feedback	• Intentions leading to unsafe driving behavior can be explained to the driver without cognitive overload • Can enhance driving performance for future trips • Can refresh driver's memory on performance for past trips • Can calibrate driver's subjective performance by highlighting persistent behavior that leads to errors	• Feedback is not provided at the time of the incident and can therefore not enhance immediate driving performance • Requires the driver to be an active recipient of information, however people may not take the time to review this type of feedback • Driver may fail to link feedback with incident • Low acceptance may lead to disuse

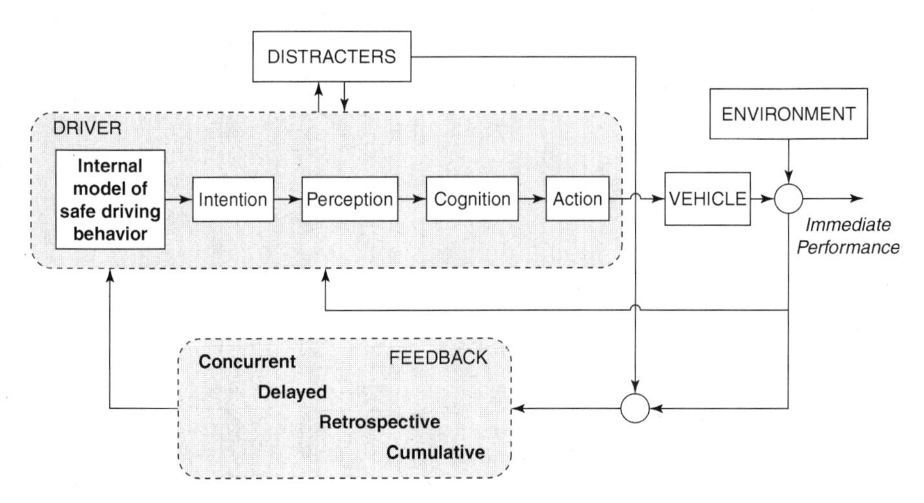

FIGURE 29.2 The process by which driver responds to feedback at different timescales.

concurrent feedback would remind the driver to discontinue the in-vehicle task and direct attention to the roadway. Therefore, concurrent feedback can directly enhance driving performance. Warnings are forms of concurrent feedback, and the literature related to warnings in the driving domain is vast.[11–14] A comprehensive discussion of concurrent feedback in the driving domain is also provided by Donmez.[15]

The effects of concurrent feedback based on the driver's momentary distraction level has been previously investigated.[8] The results suggest that concurrent feedback can help drivers modulate their distracting activity. Specifically, the study showed that concurrent feedback (a visual alert on the in-vehicle display) based on drivers' distraction level decreased the glance frequency to the in-vehicle display while increasing the glance duration to the road between in-vehicle glances. However, concurrent feedback may not be completely effective in mitigating distraction and may even exacerbate distraction. Drivers may become dependent on feedback to identify hazardous situations and may not respond appropriately if the feedback mechanism fails. For example, drivers may become more comfortable with engaging in a cell phone conversation while they are driving if they depend on the collision avoidance system to warn them when their distraction places them in a dangerous situation. This dependence can compromise driver safety if the warning system is unreliable. It is also possible that, with increased dependence, drivers may eventually filter out warnings when they are distracted.

In addition to drivers' inappropriate dependence on feedback, another concern with unreliable feedback is its potential to undermine driver acceptance,[5] which may lead drivers to ignore (or disuse) concurrent feedback.[16] Unreliable feedback can include false positives and false negatives. False-positive feedback (false or nuisance alarms) is information provided when there is no need for it. False-negative feedback is information not provided when there is a need for it. High false alarm rates can lead to driver frustration, which can also undermine traffic safety.[17] However, not all false-positive alarms are harmful. Such alarms can be used to train novice drivers, and help drivers become familiar with the system. False-positive alarms may also lead to more cautious driving and thereby result in reduced false alarm rates.[13,18] Thus, for a warning system to be effective, an acceptable false alarm rate should be established. The reliability of feedback is a major issue regardless of feedback timing. The concerns about driver acceptance, trust, and reliance also hold for feedback in longer timescales.[19]

Another reason why concurrent feedback may not be completely effective in mitigating distraction is that it may interfere with immediate task performance. Research confirms this in radar monitoring[20] and in driving.[21] Because of the limited processing time and resources available during driving, concurrent feedback may impose additional task load on the driver. One way to avoid possible information overload but still inform the driver about inappropriate behavior is to delay feedback for several seconds, until the demand decreases. Feedback at this timescale is described further in the following section.

29.2.3 Delayed Feedback

Delaying feedback by even a few seconds might avoid overloading the driver, but this delay may also diminish possible improvements to immediate driving performance. Therefore, when compared with concurrent feedback, delayed feedback may center more on altering driver behavior and less on enhancing immediate driving performance. Altering behavior, such as decreasing the driver's willingness to engage in distractions, would in turn enhance immediate driving performance.

In the driving domain, delayed feedback has been investigated by only a few researchers.[2,21] In one such system, Car Cognitive Adaptive Computer Help (CarCoach), feedback provided is canceled or delayed when the driver is cognitively overloaded.[21] The algorithms used in CarCoach define cognitive overload as occurring in two situations: (1) the driver has been making many mistakes in spite of receiving much feedback, or (2) the driver appears unusually busy with a particular driving task not generally performed while driving, such as backing up. Considering only these two situations may not define cognitive overload adequately. Drivers may be cognitively overloaded even if they are not making a lot of mistakes. Thus, a better way to evaluate cognitive overload may be to assess convergent data from physiological and performance measures. Using CarCoach, Sharon et al.[2] demonstrated that better performance was possible if undistraced drivers were guided to a more gradual acceleration by slightly delaying instructional messages on their acceleration behavior (until the acceleration maneuver was over) when compared to concurrently presenting the messages. These results suggest that notifying drivers about abrupt braking or steering maneuvers might be an effective way to enhance driver behavior. This study had limitations, including the lack of a baseline condition (i.e., no feedback), which makes it difficult to recommend any of the feedback timings (i.e., concurrent, delayed) as capable of enhancing driving performance when compared with no feedback. The existence of only a few studies (e.g., Refs 2 and 21) on delayed feedback and the limitations of these studies suggest that further research is needed on this feedback timescale.

There are some additional concerns that need to be investigated to ensure effectiveness of delayed feedback. Because driving and in-vehicle tasks are carried out in an interlaced fashion, these tasks can be viewed as mutually interrupting.[22] A potential concern of these interruptions is the initial decrease in performance as the interrupted task is resumed.[23] In addition to the safety considerations associated with the interruption of the primary task of driving by an in-vehicle task, productivity issues may also arise as in-vehicle tasks are interrupted by the need to shift attention back to the road. Therefore, even if the main objective of concurrent and delayed feedback is to enhance safety, a successful design should also aim to enhance driver productivity in interacting with the IVIS, or at least protect this productivity from deteriorating. Concurrent feedback and delayed feedback are likely to undermine productivity because they occur during the course of a trip and would therefore have the potential to interfere with IVIS interactions.

Delayed or concurrent feedback can help drivers understand whether a maneuver they are performing is unsafe. For example, providing concurrent feedback, such as a warning when the driver gets too close to a lead vehicle, can help the driver learn safe minimum stopping zones. Other unsafe driving behavior, such as talking on a cell phone while changing lanes, can also be corrected with concurrent or delayed feedback. However, due to the limited processing capacity of a driver, it may be difficult to fully explain the reasons for feedback while he or she is driving.

29.2.4 Retrospective Feedback

Retrospective feedback is defined on the temporal scale as information provided immediately after a trip is complete and not while driving. It provides information to the driver about appropriate and inappropriate behavior for the most recently completed trip. Measures can include the duration of eyes off the road, and number of distracting tasks performed during dangerous situations. Providing retrospective feedback can influence future driver behavior. The driver can learn what constitutes safe driving, when not to engage in distractions, what speed to maintain in different driving conditions, and how to diminish risk for future trips.

Drivers can only attend to concurrent and delayed feedback for very short periods, making it impossible to provide detailed information regarding an event that triggers feedback. As a consequence, concurrent or delayed feedback may not be able to convey the information necessary to understand the reason for feedback. For example, if there is a relationship between two driving incidents that occur at different times, then presenting this link via concurrent or delayed feedback may be too complex for the driver to interpret while driving. However, this information can be useful in helping drivers assess their overall driving performance by highlighting the persistent behavior that leads to errors. In the absence of feedback, drivers tend to forget their roadway incidents very quickly. Chapman and Underwood[24] found that an estimated 80% of near-accidents are forgotten after 2 weeks. This suggests that driver behavior may be changed by refreshing drivers' memory of their driving performance as well as calibrating their subjective performance (i.e., how safe they think they drive). This information or feedback may be better presented retrospectively.

29.2.5 Cumulative Feedback

Cumulative feedback is a comprehensive summary of past driving performance and driver behavior and is not provided during driving. Cumulative feedback integrates driving data over many trips that may span several weeks or months. Similar to retrospective feedback, cumulative feedback has the potential to change driver behavior. Both retrospective and cumulative feedback can present information on several incidents over time. This can help the drivers assess their overall driving performance by highlighting those persistent behaviors that lead to errors.

McGehee et al.[25] examined the effects of feedback on training teenage drivers in a naturalistic driving study. An event-triggered video device was installed in each vehicle that recorded abrupt driving maneuvers and these events were reviewed weekly with parents for approximately 6 months. The data from 26 teenage drivers showed an 89% decrease in the number of incidents for the more at-risk teen drivers. In addition, the effect of this feedback persisted even after the device was removed. This suggests that cumulative feedback can lead to lasting behavioral changes. However, because there was no baseline group that was also monitored for 6 months (i.e., drivers with no feedback), more research is still needed to assess the exact benefits of such feedback. Neither retrospective feedback nor cumulative feedback has been systematically studied in the driving domain, and both require further research (see Refs 25 and 26 for some preliminary examples). For example, the effectiveness of cumulative feedback should be investigated to determine the most appropriate time lag between the event and feedback.

29.3 COMBINATIONS OF DIFFERENT FEEDBACK TIMESCALES

Feedback mechanisms need not be mutually exclusive. The little research that has considered feedback timing at different timescales has compared one level to another, but has not assessed the potential benefits of providing both levels together. Presenting feedback at multiple timescales can provide redundancy and refresh a driver's memory of an incident. This redundancy is useful since, as already noted, there is research that suggests that drivers forget the majority of near-accidents very rapidly.[24] Combined feedback timescales can also complement each other in enhancing performance and changing behavior. For example, providing concurrent feedback in the form of a short alert cannot explain the specific problems associated with changing lanes while simultaneously talking on a cell phone in a congested area. This can be better conveyed in more detail as retrospective feedback. However, concurrent feedback can indicate that some response is needed immediately.

Another advantage of combining feedback timescales is that receiving feedback in shorter timescales can help the driver understand feedback over longer timescales. However, this support may diminish as the time between the feedback increases. For example, concurrent feedback provided for an incident and cumulative feedback provided weeks or months later regarding the same incident may not be easily connected by the driver. If concurrent feedback has been strengthened in memory by the help of retrospective feedback, the driver can better relate cumulative feedback to a particular incident. For the driver to easily relate to different feedback timescales, the representation of feedback should promote a consistent mental model.[27] For example, if a high level of distraction is presented with an orange warning light for concurrent feedback, then this color should also be used in retrospective feedback for the same information.

29.4 FEEDBACK TIMING AND FEEDBACK TYPE

The type of feedback can have different influences on the effectiveness of each feedback timescale. Feedback provided to the driver can be positive or negative.[28,29] Positive feedback is provided for correct actions.[30] Negative feedback is provided for errors and includes error flagging, directive feedback, and explanatory feedback[28,31]: (1) error flagging includes acknowledging and identifying the error occurrence; (2) directive feedback includes providing instructions on how to correct the error; and (3) explanatory feedback includes diagnosing the misconceptions that generated the error and setting new goals that remediate the error, and correct misconceptions. Error flagging and directive feedback can be presented concurrently, whereas explanatory feedback would require more time and resources from the driver and be more appropriately presented retrospectively. For example, if the driver's distraction level increases beyond a threshold, a real-time alert (i.e., error flagging) can be provided, and the explanations for the alert (e.g., too many off-the-road glances to a navigational display) can be provided to the driver immediately after the trip is over.

Kluger and DeNisi[32] state that participants who receive negative feedback are likely to exert more effort than those who receive positive feedback. In some situations, negative feedback may be necessary to educate the driver about risky driving patterns. For example, if drivers deviate from their lanes, providing a warning can help avoid a collision. Retrospectively pointing out the number of lane deviations during a trip can help the drivers understand how unsafe their driving is. However, too much negative feedback can undermine driver acceptance of feedback. Positive feedback, on the other hand, promotes acceptance. For example, workers accept an ergonomic intervention more, for which they have to learn to perform their jobs in a new way, if they are provided with positive feedback (e.g., providing feedback on increased productivity or on better posture).[33] Fogg and Nass[34] found that people who received random positive feedback during a computer game thought the interaction was more enjoyable and were more willing to continue working with the computer than people who did not receive any feedback. The results were the same even when participants were told that feedback was unreliable. Participants liked the computer better when it praised them even if they were told that the feedback was unreliable, compared with when it criticized them.[35]

Including positive feedback in addition to negative feedback can help change drivers' attitudes toward the technology. If feedback is provided over a longer timescale (e.g., retrospective and cumulative), then driver acceptance is critical. Otherwise once a trip is completed, drivers can leave their cars without receiving retrospective or cumulative feedback on their performance. Toledo and Lotan[26] investigated driving performance over a 5-month period as influenced by cumulative feedback presented on a personal web page. Using this page, drivers could access the information on all their previous trips and also receive information about the performance of other drivers. Initially, feedback improved safety, but this effect diminished over time as the drivers accessed their web pages less frequently. One approach to improve participation is to include positive feedback. For example, a driver's acceptance of retrospective feedback can be enhanced if feedback also praises the driver for safe driving maneuvers, such as abiding speeding limits and maintaining safe following distances, in addition to presenting any unsafe driving maneuvers.

29.5 CONCLUSION

Most of the research in the driving domain considers only the ability of immediate feedback to mitigate distraction and enhance driving performance. In addition to enhancing immediate performance, feedback can also promote safer long-term driver behavior. The effects of feedback on immediate driving performance can be observed in performance measures such as braking, speed variation, and time headway maintenance. The long-term changes in behavior that result from feedback may be better awareness of certain safety critical situations, greater responsiveness to the roadway environment, and diminished willingness to engage in various types of distracting activities. Concurrent and delayed feedback can have the greatest effect on immediate driving performance, whereas retrospective and cumulative feedback can have a greater effect on long-term behavior. The combination of concurrent feedback and feedback at longer timescales may have more powerful effects than either type of feedback alone. Owing to the very limited number of studies on different feedback timescales, many of these assertions are based on theory, not empirical evidence. Future research should compare different feedback timescales and assess their relative influence on short-term and long-term driving behavior. In addition to establishing the most appropriate timing of feedback for a given situation, research is needed to determine the pairing of the type of feedback with the timing of the feedback.

REFERENCES

1. Haigney, D. and Westerman, S. J., Mobile (cellular) phone use and driving: a critical review of research methodology, *Ergonomics*, 44(2), 132–143, 2001.
2. Sharon, T., Selker, T., Wagner, L., and Frank, A. J., CarCoach: a generalized layered architecture for educational car systems, *Proceedings of IEEE International Conference on Software—Science, Technology and Engineering*, Herzelia, Israel, pp. 13–22, 2005.
3. Lee, J. D., Caven, B., Haake, S., and Brown, T. L., Speech-based interaction with in-vehicle computers: The effect of speech-based e-mail on drivers' attention to the road, *Human Factors*, 43, 631–640, 2001.
4. Horrey, W. J. and Wickens, C. D., Examining the impact of cell phone conversations on driving using meta-analytic techniques, *Human Factors*, 48(1), 196–205, 2006.
5. Donmez, B., Boyle, L., Lee, J. D., and McGehee, D., Drivers' attitudes towards imperfect distraction mitigation strategies, *Transportation Research Part F: Psychology and Behaviour*, 9(6), 387–398, 2006.
6. Donmez, B., Boyle, L., and Lee, J. D., The impact of distraction mitigation strategies on driving performance, *Human Factors*, 48(4), 785–804, 2006.
7. Donmez, B., Boyle, L., and Lee, J. D., Taxonomy of mitigation strategies for driver distraction, *Proceedings of the Human Factors and Ergonomics Society 47th Annual Meeting*, Denver, CO, pp. 1865–1869, 2003.
8. Donmez, B., Boyle, L., and Lee, J. D., Safety implications of providing real-time feedback to distracted drivers, *Accident Analysis & Prevention*, 39(3), 581–590, 2007.

9. Evans, L., *Traffic Safety*, Science Serving Society, Bloomfield Hills, MI, 2004.
10. Sheridan, T., Driver distraction from a control theory perspective, *Human Factors*, 46(4), 587–599, 2004.
11. Lee, J. D., McGehee, D., Brown, T. L., and Reyes, M., Collision warning timing, driver distraction, and driver response to imminent rear end collision in a high fidelity driving simulator, *Human Factors*, 44(2), 314–334, 2002.
12. Campbell, J. L., Richard, C. M., Brown, J. L., and McCallum, M., *Crash Warning System Interfaces: Human Factors Insights and Lessons Learned*, DOT HS 810 697, National Highway Traffic Safety Administration, Washington, D.C., 2007.
13. Parasuraman, R., Hancock, P. A., and Olofinboba, O., Alarm effectiveness in driver centered collision warning systems, *Ergonomics*, 39, 390–399, 1997.
14. Kulmala, R., The potential of ITS to improve safety on rural roads, ICTCT Workshop, Budapest, 1998.
15. Donmez, B., Evaluating driver distraction mitigation strategies based on a novel taxonomy, Master's thesis, University of Iowa, Iowa City, IA, 2004.
16. Parasuraman, R. and Riley, V., Humans and automation: Use, misuse, disuse, abuse, *Human Factors*, 39(2), 230–253, 1997.
17. Burns, P. C. and Lansdown, T. C., E-distraction: The challenges for safe and usable internet services in vehicles, www-nrd.nhtsa.dot.gov/departments/nrd-13/driver distraction/Topics043100029.htm. Accessed August 2007.
18. Lees, M. N. and Lee, J. D., The influence of distraction and driving context on driver response to imperfect collision warning systems, *Ergonomics*, 50(8), 1264–1286, 2007.
19. Lee, J. D. and See, K. A., Trust in automation: Designing for appropriate reliance, *Human Factors*, 46(1), 50–80, 2004.
20. Munro, A., Fehling, M. R., and Towne, D. M., Instruction intrusiveness in dynamic simulation training, *Journal of Computer-Based Instruction*, 12, 50–53, 1985.
21. Arroyo, E., Sullivan, S., and Selker, T., CarCoach: a polite and effective driving coach, *Proceedings of the CHI: Conference on Human Factors in Computing Systems*, Montreal, Canada, pp. 357–362, 2006.
22. Monk, C. A., Boehm-Davis, D. A., and Trafton, J. G., Recovering from interruptions: implications for driver distraction research, *Human Factors*, 46(4), 650–663, 2004.
23. Ballas, J., Heitmeyer, C., and Perez, M., Evaluating two aspects of direct manipulation in advanced cockpits, *CHI'92: Human Factors in Computing Systems*, Bauersfeld, P., Bennett, J., and Lynch, G. (Eds.), ACM Press, New York, pp. 27–34, 1992.
24. Chapman, P. and Underwood, G., Forgetting near-accidents: the roles of severity, culpability and experience in the poor recall of dangerous driving situations, *Applied Cognitive Psychology*, 14, 31–44, 2000.
25. McGehee, D. V., Raby, M., Carney, C., Lee, J. D., and Reyes, M. L., Extending parental mentoring using and event-triggered video intervention in rural teen drivers, *Journal of Safety Research*, 38, 215–227, 2007.
26. Toledo, T. and Lotan, T., In-vehicle data recorder for evaluation of driving behavior and safety, *Transportation Research Record*, 1953, 112–119, 2006.
27. Vakil, S. S. and Hansman, R. J., Approaches to mitigating complexity-driven issues in commercial autoflight systems, *Reliability Engineering and System Safety*, 75, 133–145, 2002.
28. Graesser, A. C., Person, N. K., and Magliano, J. P., Collaborative dialogue patterns in naturalistic one-to-one tutoring, *Applied Cognitive Psychology*, 9, 495–522, 1995.
29. Lepper, M. R., Aspinwall, L. G., Mumme, D. L., and Chabay, R. W., Self-perception and social-perception processes in tutoring: subtle social control strategies of expert tutors, in *Self-Inference Process: The Ontario Symposium*, Olson, J. M., Zanna, M. P. (Eds.), Lawrence Erlbaum Associates, Inc., Hillsdale, NJ, 217–237, 1990.
30. Chi, M. T. H., Siler, S. A., Jeong, H., Yamauchi, T., and Hausmann, R. G., Learning from human tutoring, *Cognitive Science*, 25, 471–533, 2001.

31. Sanders, M., The effect of immediate feedback and after-action reviews (AARS) on learning, retention, and transfer, Master's thesis, University of Central Florida, Orlando, Florida, 2005.
32. Kluger, A. N. and DeNisi, A., The effects of feedback interventions on performance: a historical review, a meta-analysis, and a preliminary feedback intervention theory, *Psychological Bulletin*, 119(2), 254–284, 1996.
33. Branderburg, D. L. and Mirka, G. A., Assessing the effects of positive feedback and reinforcement in the introduction phase of an ergonomic intervention, *Human Factors*, 47(3), 526–535, 2005.
34. Fogg, B. J. and Nass, C., Silicon sycophants: the effects of computers that flatter, *International Journal of Human-Computer Studies*, 46(4), 551–561, 1997.
35. Reeves, B. and Nass, C., *The Media Equation: How People Treat Computers, Television, and New Media Like Real People and Places*, Cambridge University Press, New York, 1996.

30 Driver Distraction Injury Prevention Countermeasures—Part 1: Data Collection, Legislation and Enforcement, Vehicle Fleet Management, and Driver Licensing

Michael A. Regan, Kristie L. Young, and John D. Lee

CONTENTS

30.1 INTRODUCTION

Each year, around 1.2 million people worldwide die as a result of road crashes.[1] For every death, around 50 million people are injured and around 15 million injuries are severe enough to require hospitalization.[2] Findings from the analysis of police-reported crashes, reviewed in Chapter 16, suggest that driver distraction is a contributing factor in 10–12% of crashes. Converging data, from the 100-car Naturalistic Driving study in the United States,[3] suggest that distraction is a contributing factor in up to 23% of crashes and near-crashes. Globally, therefore, driver distraction is a significant cause of unintentional death. As such, the development, implementation, and evaluation of injury countermeasures to prevent and mitigate the effects of distraction are critical to reducing existing road trauma and preventing distraction from escalating into a bigger problem than it already is.

Injury prevention countermeasure development for distraction is in its infancy relative to other road safety issues, even in developed countries with relatively good road safety records. This is not surprising. Governments continue to rely heavily, often overly, on crash data to justify and stimulate countermeasure development. However, to date, distraction has been poorly defined, systems for accurately and reliably collecting and analyzing data on its role in crashes do not exist in many jurisdictions, and many policymakers are unaware of converging evidence, from epidemiological and other studies, that implicates distraction as a road safety problem. This has thwarted attempts by governments to strategically target key distraction problems using evidence-based strategies, and to justify adequate resources for meaningful implementation of effective countermeasures.

Noteworthy is a lack of published data on the effectiveness of existing distraction prevention and mitigation measures. The limited data that do exist, reviewed in this book, pertain to the impact of banning mobile phone use while driving. Vehicle manufacturers, to their credit, have been proactive in undertaking and commissioning research to understand distraction, and in developing methods, tools, guidelines, and standards for the design and evaluation of products to limit distraction. Even for these interventions, however, there is limited published data on their effectiveness in limiting distraction, let alone enhancing safety. Hedlund and Leaf[4] come to a similar conclusion. They discuss the efficacy of some countermeasures developed to prevent or mitigate the effects of distraction: mobile phone laws, graduated driver licensing restrictions on mobile phone use for novice drivers, general distraction laws, communications and outreach, and employer programs. They conclude that, with the exception of mobile phone laws, nothing is currently known about the effectiveness of any of these countermeasures in reducing distraction, and even for mobile phone laws, they conclude that the effectiveness data are "uncertain."

In this, and the following two chapters, we discuss countermeasure options (existing and proposed) for preventing and mitigating the effects of distraction. The recommendations derive from the body of material reviewed in this book, from the authors' collective understanding of issues relevant to the topic, and from specific sources (discussed in this chapter) that provide some initial guidance in the area. For the sake of simplicity and to package the recommendations in a manner familiar to governments, road transport authorities, and designers, the countermeasures have been assembled under the following headings: data collection, legislation and enforcement, vehicle fleet management, driver licensing, education and training, vehicle design, technology design, and road design. This chapter focuses on data collection, legislation and enforcement, vehicle fleet management, and driver licensing. The remaining areas are addressed in Chapters 31 and 32.

Given the paucity of data on countermeasure effectiveness and differences in the nature and extent of the distraction problem between jurisdictions, it is difficult to say which countermeasures are most likely to be effective in reducing distraction; even the best countermeasures will be ineffective unless they are properly designed, implemented, and routinely evaluated. Johnston[2] (Chapter 4, p. 16) asserts that current best practice in road safety countermeasure development has the following defining features:

- Routine surveillance of safety progress, using comprehensive, high-quality data systems, covering the gamut of road safety problems
- Strategic targeting of the key problems using evidence-based strategies and program options
- The provision of adequate resource for meaningful implementation
- Rigorous evaluation of the effectiveness of the interventions
- Continuous improvement in implementation based upon the evaluation results and maximum coordination among all relevant institutions

The countermeasures presented in this chapter and in Chapters 31 and 32, therefore, should be regarded as options rather than prescriptions for countermeasure development; although Chapters 16 through 18 provide quantitative data on the role of distraction in crashes and near-crashes that can be used to prioritize the choice of options.

In the writing of this and the two chapters that follow, several documents were reviewed that provide some initial thoughts on countermeasure development: the summary and proceedings of the National Highway Traffic Safety Administration (NHTSA) Driver Distraction Expert Working Group Meetings[5]; a discussion document, prepared by Transport Canada concerned with strategies for reducing driver distraction from in-vehicle telematics devices[6]; the preface to a special section on driver distraction in the journal *Human Factors*[7]; a U.S. NHTSA highway safety countermeasures guide for state highway safety offices[4]; three keynote papers presented at the First International Conference on Driver Distraction in Sydney, Australia, in June 2005[8–10]; a Monash University Accident Research Centre (MUARC) Submission to the Parliament of Victoria Road Safety Committee Inquiry into Driver Distraction[11]; the August 2006 report of the Parliament of Victoria Road Safety Committee Inquiry into Driver Distraction (henceforth, referred to as the "Australian Distraction Inquiry Report")[12]; the Summary of Proceedings and Recommendations deriving from the International Conference on Distracted Driving, held in Toronto, Canada, in October 2005[13]; and several key review articles and reports.[14–19]

The options for countermeasure development in this and subsequent chapters have been shaped and influenced by ideas, insights, and principles presented in previous chapters of this book. Notable is the overarching policy framework for the management of distraction advocated by Tingvall, in Chapter 33, which is consistent with the Swedish Vision Zero philosophy of traffic safety management.[20] Tingvall argues that a government policy that takes distraction seriously should have the following assumptions as its basis:

- Distraction is a serious problem and is often the initial event in a chain of events that leads to a serious health loss.
- While some distraction is not legally allowed, it is understood to exist now and in the future.
- Based on the above, distraction must be taken into account for all systems, products, and services that exist within the road transport system.
- Distraction should be reduced as well as prepared for in the integrated safety chain.
- In developing technology to reduce the consequences of distraction, consideration must be given to possible modifications in behavior arising from driver interaction with the technology that might diminish the intended safety benefits.

It is with these principles, ideas, and frameworks as a backdrop that we present the options for countermeasure development that follow.

Appended to this chapter are the recommendations deriving from the Australian State of Victoria's Inquiry into Driver Distraction, referred to earlier. These recommendations contain interesting insights into current political thinking on how to manage driver distraction in a jurisdiction—the State of Victoria, in Australia—that has, for over a decade, had one of the lowest rates of road trauma in the developed world.

30.2 DATA COLLECTION AND ANALYSIS

Currently, in many countries, there do not exist adequate data on the role of distraction as a contributing factor in crashes and near-crashes. This prevents an accurate assessment from being made of the frequency of such crashes, the number of people being killed and injured, and the factors that give rise to them. For most policymakers around the world, "hard" data of this kind is needed to justify and drive countermeasure development. Even where such hard data do exist, Gordon (in Chapter 16) points out that current estimates are highly likely to be underestimated, and there is considerable variation in the size of estimates across studies.

The recommendations made in the following sections aim to improve the quality of distraction-related data collection and analysis.

30.2.1 Defining Driver Distraction

Distraction is a poorly defined concept. Even within this book definitions of it vary widely. Reaching agreement on a commonly accepted definition of distraction is arguably the single most important activity to be undertaken in understanding and managing the problem. The lack of a consistent definition across studies makes the comparison of research findings difficult or impossible. Inconsistent definitions also lead to different interpretations of crash data and, ultimately, to different estimates of the role of distraction in crashes. The definition coined in this book—*distraction is the diversion of attention away from activities critical for safe driving toward a competing activity*—is presented as a first step in resolving these issues.

30.2.2 Sources of Distraction

Once there is agreement on a suitable definition of distraction, there are subsequent issues that need to be addressed in developing an agreed classification system that can be used to review existing coding structures in crash data systems and future crash studies (see Chapter 16): factors related to impairment, such as alcohol, fatigue and psychological states, need to be distinguished from distraction; inside- and outside-the-vehicle distractions should be distinguished from one another so that they are not included within the same factor or code; distraction needs to be distinguished from poorly allocated attention related to the primary task of driving; decisions must be made on what individual distractions should be grouped together; and it must be decided how to code distraction in terms of the object or scene or the behavior involved.

In Chapter 15, a taxonomy is presented for categorizing sources of distraction currently known to exist as contributing factors in crashes and near-crashes. The taxonomy, which derives from the definition of distraction coined in this book, addresses these issues and provides a suitable starting point for collecting and analyzing data in current and future crash and near-crash information systems. It will need to be refined as the driving task, and the sources of distraction associated with it, continue to evolve. In particular, as noted in Chapter 15, greater effort is needed to identify and classify sources of distraction deriving from outside the vehicle, which have been the subject of relatively less research than those deriving from inside the vehicle.

30.2.3 Crash Data Collection and Analysis Procedures

Improved processes for collecting and analyzing distraction-related crash data are required. In Chapter 16, three complimentary approaches are recommended for improving crash data collection and analysis processes.

The first is to improve the traditional way that Police-reported crash data are collected and analyzed:

- Improve data capture at the crash scene by, for example, improving the design of reporting forms, training investigators to collect distraction-related data, and making use of technology to capture and store crash scene information.
- Improve system processes by, for example, capturing the raw data, reviewing coding structures, and using trained coders.
- The sources of distraction listed in reporting forms should be derived from taxonomic descriptions, such as that presented in Chapter 15 of this book, and the listed categories should be as uniform as possible across territorial boundaries—for benchmarking purposes and for comparing crash data over time and across jurisdictions.

The second approach is to undertake specialized crash studies to collect information on distraction. In-depth studies, in which investigators and other experts attend crash scenes for a selected number of crashes and interview victims, provide more detailed insights into the contributory role of distraction in crashes. However, like traditional police-derived crash reports, they rely heavily on self-report (from drivers and witnesses). Although early in-depth studies, such as the Indiana Tri-level study in the United States,[21] provided important information on the role of distraction in crashes at that time, the driving task, the potential sources of distraction associated with it, and methods and tools available for identifying and classifying distraction as a contributing factor in crashes have evolved since then. Confidential accident and incident reporting systems, such as those operating in commercial and military aviation, provide another option for collecting and analyzing distraction-related data. However, in the relatively less regulated driving domain, they are likely to be more logistically difficult to implement and rely on the voluntary and possibly biased reporting of accidents and incidents by drivers.

The third approach is to build into vehicles technology that will record pre-crash, crash, and postcrash information about the role of distraction in crashes and near-crashes—information that cannot be reliably obtained from driver or eye witness accounts. Naturalistic driving studies, such as the 100-car study,[3] exemplify this approach. Event data recorders (EDRs) (sometimes referred to as "black box" recorders), which are now installed in many new production vehicles, also provide an opportunity to record distraction-related crash data, on a much wider scale. It is now possible to expand the functionality of these devices to record information about the use and status of vehicle systems being used by a driver around the time of a collision. The United States has been active in legislating, regulating, and standardizing the fitment of EDRs since 1997.[12,22] Aftermarket video camera-based incident recording devices of various kinds are also entering the market. These can record images and sounds, a few seconds before and after crashes, near-crashes, or incidents, that are detected by accelerometers and other sensors built into the unit. Typically, these units store digital video images of the driver, passengers, and the road scene ahead of the vehicle. The data recorded from these devices can be used to complement that obtained from EDRs in understanding the role of distraction in crashes, near-crashes, and incidents.

In collecting distraction-related data, it is desirable to take advantage of special driving populations, such as vehicle fleets.[5] Many police agencies, for example, keep detailed crash records and use EDRs and video-based technologies in their vehicles that can be used to capture data on the role of distraction in crashes and near-crashes for this driving population.

Ultimately, a combination of approaches is needed to build up a complete picture of the role of distraction in crashes.

30.2.4 Epidemiological Research

More epidemiological research is needed to quantify the increased crash risk associated with driver involvement in distracting activities. Few studies have been undertaken—and those that have, have focused primarily on the use of mobile phones and the carriage of passengers while driving. As discussed by McEvoy and Stevenson, in Chapter 17, until recently establishing risks for other types of distracting activities has been difficult because accurately measuring drivers' exposure to various distracting activities before a crash and during equivalent control intervals has been limited to self-report. With the development of the naturalistic driving study, which combines experimental techniques with epidemiological methods, this drawback can be overcome using video evidence and other sensor data. However, there are other limitations associated with naturalistic driving studies (see Chapter 6), which must be addressed in undertaking future studies of this kind.

30.2.5 Other Road Users

The study of distraction has been confined almost entirely to the road transport domain, although some related work has been going on in the computing and aviation domains under the guise of "interruptions." Even within the road transport domain, the focus of distraction efforts to date has been on drivers—distracted walking and distracted riding, whether on bicycles or motorcycles, are potential areas of concern that are totally unexplored and unresearched. Notable also is the paucity of research on driver distraction in the public and commercial transport sectors. The limited research undertaken, reviewed in Chapter 14, suggests that distraction is a problem in bus and heavy vehicle transport operations. Bus drivers, in particular, are required to take on multiple, and at times competing, roles while driving, which make them particularly vulnerable to the effects of distraction. This is exacerbated by the demands of bus driving itself, which is arguably a less "satisficing" task than ordinary driving, particularly in residential areas. Much further research is required to identify and classify the sources of distraction that exist in the public and commercial transport sectors and to quantify their impact on driving performance and safety. In the meantime, Chapter 14 provides initial guidance on preventing and mitigating the effects of distraction in bus operations.

In summary, although there is converging evidence that distraction is a significant road safety issue, further work is needed to improve data collection, analysis, and reporting systems to quantify the nature and extent of the problem more accurately—for all road users. This will, in turn, stimulate and support further countermeasure development.

30.3 LEGISLATION AND ENFORCEMENT

Well-designed legislation that is properly enforced and accompanied by ongoing publicity that is directly linked to enforcement has been shown to be effective in reducing road trauma in Australia and other countries (e.g., Ref. 23). Traffic law and its enforcement is a common tool for seeking to constrain road user behavior to its lowest risk forms[2] and applies to individuals (as in laws that prohibit the use of handheld mobile phones) as well as to institutions (as in laws that mandate that manufacturers position visual display units [VDUs] in locations that cannot be seen by drivers while driving).

Traffic law and its enforcement is an important tool for shaping behavior associated with distraction at all three levels of driving control (see Chapter 4). At the *strategic* level, laws that prohibit driver exposure to distracting objects, events, and activities can be effective in changing societal judgment of what constitutes acceptable risk and safe driving. As noted in Chapter 4, social norms that render as taboo driver engagement in distracting activities prohibited by law may be far more powerful than subtle design modifications in preventing and mitigating the effects of distraction. Traffic law and its enforcement can also be used to shape behavior at the *tactical* and *operational* levels of driving control. At the tactical level, laws could be used to prescribe, for example, minimum headways or maximum speeds at which drivers are allowed to operate vehicles when using handsfree cell phones or other devices permitted to be used under current legal regimes. Similarly, laws that prescribe the optimal location and design of systems that have the potential to distract drivers will directly impact on driving behavior at the operational level, by reducing workload.

Laws can also be used in different ways to limit distraction at different stages of the *integrated safety chain* (see Chapter 33). To the extent that they limit exposure to risk, laws can be effective at the beginning of the integrated safety chain in preventing distraction-related crashes—for example, by mandating that certain functions deemed to be distracting are locked out or in mandating the use of workload managers to support normal driving. Laws can also be used to mandate the fitting to vehicles of devices that mitigate, in real time, the effects of distraction at later stages of the integrated safety chain. They can be used, for example, to mandate the installation of distraction warning systems to support the driver when there is a deviation from normal driving and of driver support systems that sense that the driver is distracted and intervene earlier to mitigate the effects of crashes that are unavoidable.

Although there is variation in road safety laws across countries, those that relate to the mitigation of driver distraction tend to be limited in scope and quite similar. These are reviewed briefly, before discussing options and priorities for future countermeasure development.

30.3.1 EXISTING LAWS

There exist general and specific laws relating to driver distraction. In Australia, for example, police have discretion under their own State and Territory legislation to reprimand drivers who they think are driving "carelessly" or "dangerously." This includes careless or dangerous driving that arises from driver distraction. General laws, such as these, which target driving "without due care and attention," or similar behaviors, are in effect in all provinces and territories in Canada and states in the United States[24,25] (as cited in Ref. 13). In Australia, this general legislation, and in particular the careless driving provision, tends to be used in circumstances where a driver—for whatever reason—has been distracted and a crash occurs. Under this legislation, the charge is heard and determined by a court.[11]

There also exist more specific laws relating to distraction. In Australia, for example, Harmonized Australian Road Rules (ARR) were introduced nationally in 1999. One of these (ARR 300[26]) states that "the driver of a vehicle (except an emergency vehicle or police vehicle) must not use a handheld mobile phone while the vehicle is moving, or is stationary but not parked, unless the driver is exempt from this rule… ." Around 40 countries worldwide have similar bans or restrictions on the use of handheld mobile phones[27–29] (as cited in Ref. 13). Regan et al.[11] have highlighted a number of deficiencies associated with the Australian Rule in its current form. These stem mainly from the inability of the law to keep up with technological changes in the design and functionality of the mobile phone and are worth noting, given the similarity between this and other laws elsewhere:

- The rule relates only to handheld phones, even though handsfree phones carry similar increases in crash risk (see Chapter 11 of this book).
- The rule appears to allow drivers to use handsfree phones (such as a cradle-mounted phones) to send text messages, download video clips, and access other functions and services.
- Given that drivers of police and emergency vehicles are expected to drive, at times by themselves, at high speed and in demanding conditions that require complex maneuvering of their vehicles, the exemption accorded to them does not seem justified on road safety grounds (although it is acknowledged that their exposure to the technology would likely be less than that for ordinary drivers).
- The rule allows for the use while driving of Citizen's Band (CB) and other two-way radios. On the basis of the material reviewed in this book, these would be expected to induce levels of distraction comparable to that of the mobile phone.
- While the rule pertains to mobile phones only, there now exist other technologies, such as personal digital assistants (PDAs) that can be used to converse, send text messages, and perform other functions that can be performed using a mobile phone.
- The rule is difficult to enforce (e.g., in heavy traffic, at night, and in vehicles with heavily tinted windows), and around 30% of Australian drivers are known to regularly violate it.

A related Australian Road Rule (ARR 299[30]) states that, "A driver must not drive a motor vehicle that has a television receiver or visual display unit in or on the vehicle operating while the vehicle is moving, or is stationary but not parked, if any part of the image on the screen (a) is visible to the driver from the normal driving position or (b) is likely to distract another driver." Similar laws exist in other countries[25,27] (as cited in Ref. 13). Drivers are exempt from this rule if they are driving a bus, where the VDU is or displays a destination sign or other bus sign, or the VDU is part of a driver's aid (e.g., dispatch or navigation system). Regan et al.[11] have also highlighted deficiencies associated with this rule in its current form:

- There is no known published empirical evidence that a VDU in one vehicle is capable of distracting a driver driving another vehicle.
- The law pertains only to visually induced distraction. Auditory information displayed by TV, video, and DVD players may divert attention away from tasks critical for safe driving.
- It is not clear whether visual display screens which form part of portable devices, such as mobile phones, MP3 players, ipods, and PDAs, should be classified as "visual display units"
- Some driver's aids (dedicated or nomadic), if poorly designed, have potential to distract the driver.
- The law is difficult to enforce. It is difficult for police to know, for example, which features and services are exempt and whether a VDU in one vehicle is distracting the driver of another vehicle.

In addition to these two ARRs there is another (ARR 297)[31] that relates to a driver not having proper control of a vehicle. However, "proper control" has not been defined in the rules and nor have the courts ruled on what constitutes proper control (p. 121).[12]

Hedlund et al.[13] report that one Canadian province (Newfoundland and Labrador), three U.S. states (Connecticut, New Jersey, and New York), and the District of Columbia prohibit *all* drivers from using handheld mobile phones[24,25,28] (as cited in Ref. 13). Twelve U.S. states and the District of Columbia prohibit all mobile phone use by drivers with a learner's permit or provisional license or by drivers under 18.[4] Several U.S. states also prohibit all mobile phone use by school bus drivers; and in jurisdictions where no such laws exist, some 26 communities prohibit the use of handheld mobile phones.[13] No jurisdiction in the United States restricts handsfree phone use for all drivers, and at least 40 countries are known to prohibit handheld phone use.[4] In Europe, most EU member states have laws that ban the use of handheld phones, or plan to introduce them, and generally allow the use of handsfree phones on the condition that drivers do not endanger traffic.[12] Interestingly, Sweden, which has an excellent road safety record, has no law prohibiting the use of handheld mobile phones.

In the United States the emerging trend is to legislate against a multitude of behaviors.[12] In Washington, D.C., for example, there exists legislation that specifically targets the offense of distracted driving, bans talking on handheld phones while the vehicle is in use, and bans all phone use by school bus and learner drivers. Other activities covered by the legislation include "... reading, writing, performing personal grooming, interacting with pets or unsecured cargo, or engaging in any other activity, which causes distraction and results in inattentive driving."[12]

As noted previously, traffic laws can also apply to institutions. For example, there exist Australian Design Rules (ADRs), under the Motor Vehicles Act 1989, relating to the fitment and location of television and VDUs to new vehicles, which overlap to some extent with the requirements of ARR 299. Part 18 of ADR 42/04[32] states that, "All receivers or visual display units and their associated equipment must be securely mounted in a position, which does not obscure the driver's vision" and that, "... unless a driver's aid, all television receivers or visual display units must be installed so that no part of the screen is visible to the driver from the normal driving position." There are, however, systems entering the Australian market that could be interpreted as "drivers aids," which have potential to distract the driver if viewed from the normal driving position. Further, there is currently no regime in place in Australia to ensure that all television receivers and VDUs installed in Australian vehicles as aftermarket products are fitted in accordance with the requirements of the ADR.

30.3.2 Safety and Economic Impact of Existing Laws

Surprisingly few studies have evaluated the safety and economic impacts of existing laws relating to distraction.

Research suggests that banning the use of handheld mobile phones while driving initially lowers the rate of handheld mobile phone use (by up to 50%), before figures subsequently rise back up to prelegislation levels.[33–36] The reasoning behind this subsequent rise and return to prelegislation levels may be twofold: after a brief period of compliance, drivers may judge that the risk of getting caught is minimal, and return to using handheld mobile phones while driving; and after a reduction in publicity about the risks of mobile phone use while driving, drivers may forget or underestimate these risks.[36]

Very few studies have examined the economic implications of banning the use of mobile phones while driving. Two early studies (Hahn and Tetlock, 1999; Redelmeier and Weinstein, 1999; as cited in Ref. 37), attempted to quantify the monetary benefits associated with a ban on handheld mobile phones and the monetary costs associated with the loss of consumer convenience in being able to use the devices while driving. Both studies concluded that a ban on the use of hand-held mobile phones would not be economically efficient. The Hahn and Tetlock benefit–cost analysis estimated that a ban on mobile phones would result in a societal loss of US $23 billion annually. The cost-effectiveness analysis conducted by Redelmeier and Weinstein estimated that the cost per quality adjusted life year (QALY) saved would be US $300,000. However, in a more recent study,[37] the key assumptions for the two earlier studies were revised so that they were consistent, reflected the latest information available, and assumed a ban on the use of both handheld and handsfree mobile phones. It was concluded that the estimated net benefit of a ban on mobile phone use while driving (in the United States) was close to zero; that is, the value of preventing crashes caused by mobile phone use while driving is approximately equal to the value of the calls that would be eliminated by a ban.

Of course, this discussion raises the more general issue of what might be called the "distraction paradox"—the fact that, at times, it may be beneficial in safety terms for drivers to willingly and deliberately expose themselves to known sources of distraction. DVD players, for example, even if they can be heard by the driver, have potential to placate children for lengthy periods on long trips, thereby, reducing the intensity and duration of passenger-related distraction. Similarly, conversations, either with a passenger or with someone at the other end of a mobile phone, may help maintain alertness and delay the onset of drowsiness and fatigue. The net interactive effect of a given source of distraction on driving performance and safety is thus a balance between its behavioral benefits and its costs. There is, however, no known research on this topic.

30.3.3 Where to from Here?

Hedlund et al.[13] (p. 10) argue that, to be more effective, laws intended to reduce driver distraction should, "follow the same principles as all good traffic safety laws: They should

- be written well, without loopholes or unintended consequences;
- place minimal burden on law enforcement in observing and documenting the prohibited behaviour and in documenting and assisting in the prosecution of the offence; and
- have the full support of prosecutors and judges."

Some further specific recommendations for improving the effectiveness of existing laws are made below.

30.3.3.1 Technologies

Existing legislation that prohibits the use of handheld mobile phones and other information and communication technologies while driving needs to keep pace with the uncertain evolution of these technologies.[13] Exemptions, where these are provided, should be justified on road safety grounds. This applies to exemptions that apply to technologies (in Australia, e.g., the prohibition on the use of handheld phones does not include CB radios or other two-way radios; and television receivers and VDUs are allowed to be viewed by drivers if they are, or are part of, a driver's aid such as a navigation system) and exemptions that apply to specific users of the devices (e.g., the drivers of public transport, emergency, and police vehicles).[11] High-risk groups, such as learner and probationary drivers, should be prohibited from using handheld and handsfree mobile phones, and other technologies known to significantly increase crash risk while driving.[11,13] As for any road safety countermeasures, the long-term effects of legislation that prohibits the use of mobile phones and other technological devices while driving must be evaluated on a regular basis. The outcomes of such evaluations should inform the design of measures to optimize the effectiveness of the legislation.

Police enforcement of existing laws that prohibit driver use of handheld mobile phones appears to be inadequate in some countries—in Australia, for example, around 30% of drivers use handheld mobile phones while driving even though it is illegal to do so. In addition to determining why this is so and what can be done to improve the effectiveness of current enforcement practices, it is important to exploit emerging technological countermeasures that might obviate the need for police enforcement. If the technology exists, for example, to prevent mobile phones and other devices from being used in cinemas and hospitals, why not use technology in a similar way to block reception when a mobile phone or other device is being used illegally in a vehicle that is traveling above a certain maximum speed on the road network? Workload managers, discussed in Part 8 of this book, provide another means for selectively restricting driver access to mobile phone functions at times when driver workload is estimated to be high. The penalties for violation of laws that prohibit the use of mobile phones and other technological devices while driving should be commensurate with those pertaining to other deviant behaviors, such as speeding and drink driving, which carry comparable increases in crash risk.[11]

As discussed earlier in this chapter, a total ban on the use (not carriage) of all mobile phones (handheld and handsfree) while driving for work purposes has been implemented by some employers. Such bans appear to be justified on road safety grounds, at least when the devices are used to converse. The authors of Chapters 16 and 17, for example, report data from the 100-car naturalistic driving study[3] that show that increases in crash risk associated with dialing a handheld device and talking and listening on a handheld device are 2.8 and 1.3, respectively, implying that talking and listening are less risky than dialing. However, the population-attributable risk percentages derived for each of these two sets of activities (which take into account driver exposure) are the same—3.6%. Hence, for the population at large, dialing a handheld device and talking and listening on a handheld device are associated with approximately equal increases in crash risk (because drivers spend a greater percentage of time talking and listening to handheld devices than dialing). McEvoy and Stevenson, in Chapter 17, also cite converging evidence from epidemiological research in Australia, which demonstrates increases in crash risk associated with the use of handheld and handsfree mobile phones of 4.1 and 3.8 times, respectively.

Whether or not there should be a total societal ban on the use of mobile phones while driving is a matter for policymakers to decide. Logically, at least, it makes no sense to ban handheld phones and not handsfree phones if the increase in crash risk associated with the use of the device in these two modes is similar. Perhaps, this is why Sweden, with an excellent road safety record, has chosen not to ban the use of either handheld or handsfree phones. The potential gains in safety of implementing a total societal ban on the use of mobile phones would likely equal any consequent losses in economic productivity, based on the data reviewed in this chapter—at least when mobile phones are used for conversing. Such a ban would likely be difficult to enforce, and its initial impact in reducing mobile phone use while driving may not persist in the longer term. As noted previously, surveys of public opinion suggest that support for a total ban on phone usage while driving would be low. Given that the mobile phone is a flexible platform that is capable of hosting a range of relatively low-cost functions that have potential to support the driving task and enhance safety (e.g., satellite navigation, intelligent speed adaptation), and given that when it is used to converse it may have some safety benefits (e.g., in mitigating the effects of drowsiness and fatigue; see Chapter 21), it may be premature at this point in time to implement a total societal ban on its use while driving. Further research is needed to determine new ways of limiting levels of distraction associated with mobile phone use (e.g., through better design and by supporting use of it with real-time distraction prevention and mitigation countermeasures), for all functions that can be accessed when using the device while driving, and in exploiting the potential of the devices to host functions that have potential to assist the driver and enhance safety. Such activity might help bring together vehicle manufacturers, aftermarket suppliers, and nomadic device developers in achieving the common goal of optimizing driver safety.

Finally, transport authorities, in conjunction with automotive manufacturers and providers of aftermarket products, need to develop verification processes for the installation of new technologies so that vehicle owners and potential purchasers can be assured that the installation satisfies the design rules that apply in that jurisdiction.[11,12] There is also a need for the development of safety standards, ratings, and labels for aftermarket products.[13]

As discussed in Chapter 2, the whole issue of whether a distracted driver is blameworthy when circumstances act to displace the primacy of their social role as a driver—such as when a driver diverts attention away from activities critical for safe driving toward a screaming baby to fulfill their role as a parent—requires careful thought. This is an important issue, but it is bound to be one of the more difficult legal issues to resolve.

30.3.3.2 Other Sources of Distraction

Emerging laws in the United States recognize the role that other sources of distraction, such as grooming and carrying animals, can also play in degrading driving performance and increasing crash risk. However, specific laws such as these may be difficult to enforce and no known studies have evaluated the effectiveness of them. As for any road law, they should target those activities, which confer the greatest risk to safety (see Chapters 16 through 18).

More general distracted driving laws that prohibit drivers from engaging in any activity that diverts attention away from activities critical for safe driving also operate in some parts of the United States. These give publicity on the dangers of distracted driving more relevance and credibility.[12] No studies, however, have evaluated whether such general distracted driving laws have any effect, and it is unlikely that they will be effective unless they are vigorously publicized and enforced.[4] In addition, objective criteria for identifying driving behaviors that are indicative of distracted driving need to be developed, to enable police to detect and penalize drivers who engage in distracted driving, similar to those developed for sobriety testing of drunk drivers.

As part of graduated driver licensing regimes, there is justification for restricting the carriage of multiple passengers by learner and probationary drivers, at least for part of the probationary period. Such laws are already in force in some jurisdictions. In the United States, for example, laws in 35 states and the District of Columbia limit the number of passengers allowed with a driver with a provisional license, and there is evidence that these restrictions reduce teenage driver crashes and injuries.[4] However, the actual extent to which reduced distraction contributes to the effectiveness of such restrictions in reducing road trauma remains unclear.

In summary, there is scope for improving existing laws. They should be data-driven, justifiable on road safety grounds, enforceable, in pace with technological developments, evaluated, and recognize that drivers are in some situations biologically and socially primed to be distracted.

30.4 VEHICLE FLEET MANAGEMENT

In Australia, about a quarter of all vehicles involved in crashes are business vehicles,[38] and nearly 50% of all Australian workplace fatalities occur on roads if traveling to and from work is included.[12] The situation is similar in many other developed countries. Given the dangers associated with work-related driving, effort is needed to reduce this problem, or at least to ensure it does not grow.

Employers in Australia, and many other countries, are required to provide a duty of care to drivers of their vehicles as an occupational health and safety requirement. Fleet owners and managers, therefore, are in a powerful position to develop and implement policies that internally regulate driver exposure to distracting activities in vehicles driven for work purposes and to purchase and lease vehicles and equipment that are best designed to minimize driver distraction. In Australia, the companies purchase for their fleets around 60% of all new vehicles sold in the country. These vehicles, when sold, will filter rapidly through the rest of the community, further enhancing the safety of private motorists.

Many government agencies and corporations around the world have implemented specific policies on mobile phone use while driving by their employees. In Australia, for example, several large companies, including Shell, BP, ExxonMobil, BHP Billiton, and BOC Gases, have banned the use of handheld and handsfree phones in company vehicles.[12] Some employers have extended this ban to cover the use of mobile phones while walking around worksites.[12] Many corporations around the world also have more general "safe driving policies" that include advice on the management of driver distraction.

Employers are in a particularly powerful position to prevent and mitigate within society the effects of driver distraction. There are several reasons for this. First, they are able to influence driving behavior at all levels—at the strategic level (e.g., by limiting the availability of distracting technologies and devices to employees and reducing productivity pressures to use mobile phones on the job), at the tactical level (e.g., through on-the-job education and training in how to self-regulate driving behavior in response to distraction), and at the operational level (e.g., through the provision to drivers of vehicles equipped with technologies designed to minimize distraction). They also have at their disposal a captive audience to which they can apply a wide range of traffic safety strategies: exposure control (e.g., through company regulations that prohibit use of mobile phones); crash prevention (e.g., through the purchase or lease of vehicles equipped with real-time distraction mitigation systems); injury control (e.g., through the purchase or lease of vehicles equipped with passive safety features, such as airbags that protect the driver in the event that a distraction-related crash is unavoidable); behavior modification (through education, and enforcement of company regulations); and post-injury control (e.g., through the purchase or lease of vehicles equipped with automatic crash notification systems in the event of an unavoidable distraction-related crash). Finally, vehicle fleet managers have discretion in choosing, for the vehicles they purchase, a wide range of distraction prevention and mitigation technologies that are capable of addressing all stages of the integrated safety chain (Chapter 33)—from normal driving through to postcrash.

The following sections outline a range of initiatives that employers have at their disposal to prevent and mitigate distraction-related crashes.

30.4.1 Company Policies to Manage Distraction

30.4.1.1 Responsibilities

The following, general, recommendations are made to manage driver distraction within corporate vehicle fleet safety management programs:

- Road transport authorities, occupational health and safety authorities, and other stakeholders need to work together to encourage an occupational health and safety approach to driver distraction for people who drive as part of their work.[12]
- Governments should play a leading role in developing their own vehicle safety policies for the management of driver distraction and in encouraging the private sector to follow suit.
- Governments need to provide employers, government and private, with advice and guidance in developing vehicle safety policies: advice to employers on their legal responsibilities and potential liabilities in relation to driver distraction; guidance on strategies that could be adopted by them, and by those they contract to perform services for them, to limit the adverse effects of distraction; and product information that stimulates them to purchase vehicle makes and models and nomadic devices that minimize driver distraction.

30.4.1.2 Company Policies—General Issues

Company policies designed to manage distraction should address the following general issues:

- Employees should be made aware of the existence and contents of the company's existing policy, and versions of it.
- The policy should provide clear guidance on what the company believes are acceptable circumstances in which it is appropriate for drivers to willingly engage in distracting activities, and those which are prohibited.
- The policy should explain to employees their legal and company responsibilities, penalties for violation of the policy, and incentives for adherence to it.
- The policy should identify the range of distractions that can adversely affect driving performance and the relative risks involved in engaging in distracting activities while driving.
- The policy should contain guidance for employees in how to minimize the effects of driver distraction.

30.4.1.3 Company Policies—Content

The following issues should be considered in developing company policies, programs, and strategies for managing distraction.

Collection, monitoring, and analysis of crash data. Systems are needed to measure and quantify driver exposure to distractions while driving, for quantifying the extent to which distractions contribute to injury and noninjury crashes, and for determining whether company policies are effective in reducing injury and property damage attributable to distraction.

Exposure reduction. Effective enforcement of the company's distraction policy is the principle means by which driver exposure to prohibited sources of distraction can be controlled.

Enforcement. Penalties for failing to adhere to the company policy, and aspects of it, should be determined and documented, along with incentives for driver compliance with the policy.

Education. Education programs should cover the following basic issues:

- The company policy on distraction
- National and state legislation relevant to distraction
- Penalties for violating company policy and state legislation
- The definition and nature of driver distraction
- Sources of distraction
- The impact of distraction on driving performance and crash risk
- Relative risks involved in using mobile phones in different weather conditions, geographical areas (e.g., country areas), and traffic conditions
- Individual differences in vulnerability to distraction
- The role of passengers in managing distraction
- Strategies for minimizing distraction, including knowledge about features of technologies in the vehicle and the safest ways of using them to reduce distraction

Training. Company training programs should focus on developing the following knowledge and skills:

- Knowledge of mobile phone features (e.g., voice recognition) and features of other technologies that reduce distraction
- How to use vehicle technologies and nomadic devices in the safest manner to minimize distraction while driving
- Optimal modes of self-regulation to reduce the effects of distraction (e.g., slowing down, increasing following distance)
- Self-awareness of the relative effects of distraction on driving performance deriving from different phone tasks (e.g., handheld, handsfree, text messaging), conversation complexity, driving task demand, weather conditions, and so on

More detailed advice on training and education initiatives to limit distraction is provided in Chapter 31.

Technology design. The design and placement of technologies in vehicles used by employees while driving critically determines the extent to which they are vulnerable to distraction as follows:

- Employers should request, when purchasing vehicles, mobile phones, and other technologies, evidence that they comply with best practice human factors and ergonomic guidelines and standards for minimizing driver distraction.
- Employees should be provided with vehicles, technologies, and nomadic devices with the best features for minimizing distraction. Several helpful features to reduce distraction when using a handsfree phone exist: speed and voice dialing, a large speed and voice dial memory capacity, automatic radio muting, large display screen and control buttons, automatic answering facility, automatic brightness control, and long display illumination times.
- Ideally, nomadic devices that are allowed to be used while driving should be connected with the vehicle via Bluetooth or a simple physical connection (e.g., plug, port, or dock—as is possible with iPod devices) to enable the driver to operate the device through normal vehicle controls and displays (which, hopefully, are more compatible for use while driving).
- New and existing nomadic and retrofitted aftermarket devices should be installed and located in vehicles in accordance with design guidelines and best practice ergonomic and human factors guidelines and standards to minimize distraction.
- A driver distraction subcommittee should be formed—involving management and employees—that is responsible for developing criteria for the purchase of in-vehicle technologies, ensuring that they are properly located in vehicles, and assessing them for ergonomic design and usability before deployment in company vehicles.
- Technologies that enable in-vehicle device use to be restricted or locked out in circumstances when it is unsafe to use the device, such as beyond certain speeds, in certain locations, when performing certain maneuvers, when the windscreen wipers are activated, and so on, are preferred.

There is, in summary, much that can be done to minimize distraction in vehicles being driven for work purposes, and employees are in a powerful position to do so.

30.5 LICENSING

The licensing system provides an important mechanism for reducing the adverse effects of distraction.[4,11,19] It can be used to shape and modify driving behavior at all three levels of driving control (strategic, tactical, and operational), over variable timescales. However, little has been done to date to exploit it as a driver distraction countermeasure. The following, general, recommendations derive from earlier suggestions made by Young et al.[19] and Regan et al.[11]

Most existing licensing handbooks for Learner and Probationary drivers contain only limited reference to distraction as a potential risk for drivers. Such documents should include information about the range of distractions inside and outside the vehicle that can adversely impact on driver performance and safety, how they do so, the relative risks deriving from engagement in these activities, the factors that make young drivers more vulnerable to the effects of distraction, and practical strategies for avoiding and coping with distractions, including advice on technology features and modes of interaction with technologies that minimize distraction.

Knowledge tests, undertaken to obtain learner permits and probationary licenses, should include items that test driver knowledge of these issues. So-called "hazard perception tests," which test for the ability to detect, recognize, and respond appropriately to traffic hazards, should be designed to more closely simulate the demands of real driving by incorporating surrogate driving and competing tasks (such as radio-tuning) that allow road authorities to assess a driver's ability to perceive and effectively respond to potential and actual hazards when loaded or distracted. Practical driving tests, undertaken in real vehicles with license testers, should be designed to assess driver awareness of distractions, their willingness to engage in distracting activities, and their ability to safely compensate for the effects of distraction.

Graduated licensing schemes should be designed to systematically, and chronologically, expose learner and probationary drivers to potentially distracting activities (such as operating entertainment systems, using handsfree mobile phones, and carrying passengers) based on their level of driving experience and demonstrated competence in safely managing the effects of distraction. As noted earlier in this chapter there is justification, on road safety grounds, for banning mobile phone use and restricting the carriage of multiple passengers by learner and probationary drivers, at least for part of the probationary period. Such laws are already in force in some jurisdictions.

Testing for the presence of knowledge and skills acquired through education and training is important in ensuring that drivers are properly equipped to drive safely and in motivating learner drivers to undertake education and training programs that have been proven to be effective in reducing crash risk. Chapter 31 contains recommendations for the design and content of driver education and training initiatives for preventing and mitigating the effects of distraction. It identifies specific distraction-related knowledge and skills, which could be tested for within the driver licensing system.

30.6 CONCLUSIONS

Although there is converging evidence that distraction is a road safety problem, better data are needed to more accurately characterize and quantify the problem and to prioritize countermeasure development. Agreement on a suitable definition of distraction, from which can be extracted a taxonomy for classifying sources of distraction, is critical in bettering our understanding of the true nature and role of distraction in accidents and incidents, for all road users. The definition coined in this book—*distraction is the diversion of attention away from activities critical for safe driving toward a competing activity*—is presented as a first step in resolving this issue.

Legislation is common among jurisdictions and has potential to prevent and mitigate the effects of distraction if well written and enforceable. Evidence to suggest that existing legislation is effective in doing so, however, is limited. Legislation needs to be evidence-based, target high-risk groups, be evaluated on a regular basis, and keep pace with the evolution of technologies. Exemptions from laws, where provided, should be justified on road safety grounds and penalties for violations of laws that prohibit driver engagement in distracting activities should be comparable to those pertaining to other driving behaviors, which carry similar increases in crash risk. Developments in technology are making it possible to both improve and obviate the need for police enforcement and these should be pursued. Whether a distracted driver is blameworthy when circumstances act to displace the primacy of their social role as a driver is an important issue to resolve.

Employers have a significant role to play in limiting distraction in society and have at their disposal many options for doing so. The critical starting point is development of a company-wide endorsed policy for managing distraction. Governments can play a leading role in developing policies for managing distraction and in encouraging and supporting the private sector to follow suit.

Little has been done to date in exploiting the driver licensing system as a driver distraction countermeasure. There are many options for doing so. Graduated driver licensing systems that progressively delimit driver exposure to potentially distracting activities are likely to yield significant road safety benefits.

ACKNOWLEDGMENTS

The authors would like to thank Dr. Peter Burns, from Transport Canada, for reviewing and commenting on an earlier version of this chapter. We also thank Dr. Craig Gordon, from the New Zealand Ministry of Transport, and Dr. Suzanne McEvoy, from the George Institute for International Health, Australia, for their valuable comments on selected sections of an earlier version of this chapter.

A.30.1 APPENDIX A: RECOMMENDATIONS OF THE PARLIAMENT OF VICTORIA

A.30.1.1 Inquiry into Driver Distraction

A.30.1.1.1 Introduction

The Road Safety Committee of the Parliament of Victoria, Australia, handed down in August 2006 a 200-page report on its Inquiry into Driver Distraction (see Ref. 12). This Committee comprises seven members of Parliament, drawn from both houses and all political parties. To the knowledge of the authors, it is the first known report to be tabled on the topic of driver distraction by a bipartisan Parliamentary committee, and provides a comprehensive summary of Australian and international research and activities in the area. The Committee made 31 recommendations for addressing distraction as a road safety issue. These are reproduced below (with permission), and provide an interesting insight into recent thinking on this issue, in a jurisdiction (the Australian State of Victoria) that has one of the lowest rates of road trauma in the developed world.

A.30.1.1.2 Recommendations

1. That VicRoads adopt a clearer concise definition of driver distraction, consistent with the definition arising out of the 2005 Toronto conference on driver distraction, and establish a range of categories of distraction sources. Any definition and categorisation should distinguish distraction from other driver behaviours such as fatigue and inattention.
2. That VicRoads and Victoria Police develop methods to enable the future assessment of the role of distraction in crashes on Victorian roads, including a review of existing traffic crash reporting systems. Consultation should take place with other Australasian jurisdictions and the Australian Transport Safety Bureau on appropriate methods and classification of distraction.
3. That VicRoads undertake a comprehensive roadside observational study to determine the prevalence of both handheld and handsfree mobile phone use by drivers in Victoria that will provide a benchmark for future studies and a basis for measuring the effect of any countermeasures.
4. That VicRoads continue to monitor research on the effects of various aspects of mobile phone use on driving performance, with a particular emphasis on:
   - the context, duration and content of conversations;
   - experimental validity and repeatability;
   - age-related differences;
   - phone design and new technology; and
   - experience with using a mobile phone while driving.
5. That VicRoads and Victoria Police improve crash data systems on mobile phone use, including type of device and the context in which it was being used when the crash occurred.
6. That the state government work with the vehicle industry to encourage development of safer in-car mobile phone technology, including integrated speech-controlled phone communication systems.
7. That relevant state government agencies implement targeted publicity campaigns warning drivers of the dangers of mobile phone distraction, including
   - the use of hands-free phones in hazardous traffic conditions;
   - the dangers of text and video messaging; and
   - the greater risks associated with complex phone conversations

   In developing publicity campaigns, the Government should examine the recent '*Switch off before you drive off*' campaign undertaken in the United Kingdom.
8. That VicRoads review the results of the NSW Roads and Traffic Authority study of the distraction from in-vehicle videos and possible subsequent Australian Transport Safety Bureau investigations for their implications in addressing driver distraction in Victoria.
9. That VicRoads undertake a survey on the current use of video, audio and other electronic devices by drivers in Victoria to establish a benchmark for future usage surveys and a basis for measuring the effect of any countermeasures.

10. That VicRoads and Victoria Police improve crash data systems on video, audio, and other electronic device use, including the type of device and the context in which it was being used when the crash occurred.
11. That VicRoads and the Transport Accident Commission undertake a publicity campaign warning of the dangers of drivers being distracted by "everyday" activities and the need to remain alert to the driving task.
12. That VicRoads, in consultation with local councils, develop a set of guidelines to regulate the location, size, and content of all road authority and other signs within road reserves. Such guidelines will be designed to minimise potential driver distraction and will apply to individual signs as well as the total signscape along a road. That following the implementation of the above guidelines, VicRoads and local councils aim to remove superfluous and obsolete signs.
13. That VicRoads, the Department of Sustainability and Environment and municipalities develop a more consistent and stringent approach to the installation, use and content of scrolling, moving and video-style advertising within and adjacent to road reserves. Any installations should be monitored for their effect on road safety.
14. That VicRoads, the Department of Sustainability and Environment and municipalities develop more prescriptive regulations and guidelines controlling advertising in or near road reserves, including the need to control the content of advertisements.
15. That any future consideration of the laws dealing with mobile phone use while driving, take into consideration the potential safety and economic benefits to be gained from using handsfree mobile phones.
16. That VicRoads monitor, evaluate, and publish the results of the impact on road crashes and driver performance of a ban on all mobile phone use while driving by learner permit and first-year probationary licence drivers under Victoria's revised Graduated Licensing System.
17. That in relation to the road rule on the use of television and video-screen devices in vehicles, Victoria Police and VicRoads implement separate penalties for installations, which could distract the driver and those, which may distract drivers of other vehicles.
18. That VicRoads develop, in conjunction with the automotive manufacturer and aftermarket motor accessory industry, a verification process for the installation of video and TV screens in motor vehicles so that vehicle owners and potential purchasers can be assured that the installation satisfies Australian Design Rules.
19. That VicRoads review the intent of Australian Road Rule 299 (television receivers/visual display units) and Australian Road Rule 300 (use of handheld mobile phones) in view of emerging technologies and consider the appropriateness of having two separate rules.
20. That following the development of a clear definition and categorisations of driver distraction (see Recommendation 1), Victoria Police and VicRoads introduce an appropriate road rule to prohibit driving while undertaking activities, which could distract from safe driving.

21. That following the implementation and evaluation of the recently announced changes to the Graduated Licensing Scheme, the Government reconsider the issue of restricting the carriage of multiple passengers by novice drivers.
22. That VicRoads liaise with the Australian Transport Council with a view to further research and development into the potential benefits to be gained from various emerging driver assistance technologies including:
    - Electronic stability control
    - Driver workload managers
    - Speech recognition devices
23. That VicRoads liaise with the Australian Transport Council with a view to further research and development to ensure that driver assistance technologies minimise potential driver distraction through appropriate system integration, driver-machine interfaces, and the positioning of vehicle displays and controls.
24. That the Minister for Transport raise at the Australian Transport Council the need to undertake public and industry consultation leading to a memorandum of understanding between governments and industry to reduce driver distraction from in-vehicle electronic devices.
25. That the Government increase the profile of driver distraction as a road safety issue. This should include:
    - addressing the issue in the forthcoming Victorian road safety strategy;
    - school road safety programs; and
    - development of suitable publicity for use by the rental car industry.
26. That VicRoads develop a comprehensive and prioritised program of research and policy initiatives on driver distraction to improve road safety in Victoria.
27. That VicRoads and the driver training industry incorporate driver distraction material in driver training and licensing processes and publications.
28. That VicRoads and WorkSafe encourage an occupational health and safety approach to driver distraction for people who drive as part of their work.
29. That the state government implement vehicle safety policies to encourage government and vehicle fleet drivers, while driving, to:
    - minimise hands-free mobile phone use;
    - more safely use other electronic devices, such as navigation systems, and
    - avoid or minimise nonelectronic distractions.
30. That VicRoads and Victoria Police investigate how information from Event Data Recorders in modern motor vehicles can be used to provide new insights into the role of driver distraction in crashes and other information to improve road safety in Victoria. This should include data access, privacy, and resourcing issues.
31. That VicRoads investigate how video camera event recordings of driver behaviour and traffic conditions when collisions or near-crashes occur can be used to provide new insights into driver distraction and other aspects of road safety.

REFERENCES

1. World Health Organization, *World Report on Road Traffic Injury Prevention*, WHO, Geneva, Switzerland, 2004.
2. Johnston, I., Highway safety, *The Handbook of Highway Engineering*, Fwa, T.F. (Ed.), CRC Press, London, 2006, pp. 4-1–4-39.
3. Klauer S.G., Dingus T.A., Neale V.L., Sudweeks J.D., and Ramsey D.J., *The Impact of Driver Inattention on Near-Crash/Crash Risk: An Analysis Using the 100-Car Naturalistic Driving Study Data*, Technical Report No. DOT HS 810 594, National Highway Traffic Safety Administration, Washington, D.C., April 2006.
4. Hedlund, J.H. and Leaf, W.A., *Countermeasures That Work: A Highway Safety Countermeasures Guide for State Highway Safety Offices*, National Highway Traffic Safety Administration, Washington, D.C., 2007.
5. National Highway Traffic Safety Administration, NHTSA driver distraction expert working group meetings: summary and proceedings, September 28 and October 11, 2000, NHTSA, Washington, D.C., 2000.
6. Transport Canada, Strategies for reducing driver distraction from in-vehicle telematics devices: A discussion document (No. TP 14133 E), Transport Canada, Ottawa, Canada, 2003.
7. Lee, J.D and Strayer, D., Preface to the special section on driver distraction, *Human Factors*, 46(4), 583–586, 2004.
8. Burns, P.C., Driver distraction countermeasures, *First International Conference on Driver Distraction*, Sydney, Australia, June 2–3, 2005.
9. Regan, M., Driver distraction: Reflections on the past, present and future, *Journal of the Australasian College of Road Safety*, 16(2), 22–33, 2005.
10. Lee, J.D., Driver distraction: Breakdowns of a multi-level control process, *Journal of the Australasian College of Road Safety*, 16(2), 33–38, 2005.
11. Regan, M., Young, K., and Johnston, I., MUARC submission to the Parliamentary Road Safety Committee inquiry into driver distraction, Monash University Accident Research Centre, Melbourne, Australia, 2005.
12. Parliament of Victoria, Inquiry into driver distraction—Report of the Road Safety Committee on the inquiry into driver distraction, Parliamentary Paper No. 209, Session 2003–2006. Melbourne, Australia; Parliament of Victoria, 2006, available at http://www.parliament.vic.gov.au/rsc/Distraction/default.htm.
13. Hedlund, J., Simpson, H., and Mayhew, D., *International Conference on Distracted Driving—Summary of Proceedings and Recommendations*, Traffic Injury Research Foundation, Toronto, Canada, 2005.
14. Caird, J. and Dewar, R., Driver distraction, *Human Factors in Traffic Safety*, Dewar, R. and Olson, P. (Eds.), Chapter 10, pp. 195–229, Lawyers and Judges Publishing Company Inc., New York, 2006.
15. Haigney, D. and Westerman, S.J., Mobile (cellular) phone use and driving: A critical review of research methodology, *Ergonomics*, 44(2), 132–143, 2001.
16. Horrey, W.J. and Wickens C.D., Examining the impact of cell phone conversations on driving using meta-analytic techniques, *Human Factors*, 48(1), 196–205, 2006.
17. Kircher, K., *Driver Distraction— A Review of the Literature*, VTI Rapport 594A, VTI, Lindkoping, Sweden, 2007.
18. McCartt, A.T., Hellinga, L.A., and Bratiman, K.A., Cell phones and driving: Review of research, *Traffic Injury Prevention*, 7, 89–106, 2006.
19. Young, K., Regan, M., and Hammer, M., *Driver distraction: A review of the literature*, MUARC Report 206, Monash University Accident Research Centre, Melbourne, Australia, 2003.

20. Tingvall, C., The Swedish Vision Zero and how Parliamentary approval was obtained, *Proceedings of the Road Safety Research, Policing, and Education Conference*, November 16–17, Wellington, New Zealand, 1988.
21. Treat, J.R., Tumbas, N.S., McDonald, S.T., Shinar, D., Hume, R.D., Mayer, R.D., Stansifer, R.L., and Castallen, N.J., *Tri-level Study of the Causes of Traffic Accidents: Final Report—Executive Summary*, Report No. DOT-HS-034-3-535-79-TAC(S), National Highway Traffic Safety Administration, Washington, D.C., 1979.
22. National Highway Traffic Safety Administration, Event Data Recorders (EDRs), Final regulatory evaluation, U.S. Department of Transportation, NHTSA, Washington, D.C., 2006.
23. Cameron, M. and Newstead, S., Mass media publicity supporting police enforcement and its economic value, *28th Annual Conference Symposium on Mass Media Campaigns in Road Safety*, September 30, 1996.
24. Booth, R., Legislation, regulation and enforcement for dealing with distracted driving, *International Conference on Distracted Driving*, Toronto, Canada, October 2–5, 2005.
25. Wilson, J., Legislation and regulation in Canada with respect to driver distraction, *International Conference on Distracted Driving*, Toronto, Canada, October 2–5, 2005.
26. Australian Road Rule 300—Use of hand-held mobile phones. Part 18, Division 1, available at http://www.rta.nsw.gov.au/rulesregulations/downloads/arrc.pdf.
27. McCartt, A.T., Cell phone and other distractions, *International Conference on Distracted Driving*, Toronto, Canada, October 2–5, 2005.
28. Sundeen, M., Distracted driving legislation, *International Conference on Distracted Driving*, Toronto, Canada, October 2–5, 2005.
29. Vanlaar, W., Legislation, regulation and enforcement for dealing with distracted driving in Europe, *International Conference on Distracted Driving*, Toronto, Canada, October 2–5, 2005.
30. Australian Road Rule 299 – Television receivers and visual display units in motor vehicles. Part 18, Division 1, available at www.rta.nsw.gov.au/rulesregulations/downloads/p18.pdf.
31. Australian Road Rule 297—Driver to have proper control of a vehicle, etc. Part 18, Division 1, available at www.rta.nsw.gov.au/rulesregulations/downloads/p18.pdf.
32. Australian Design Rule ADR 42/04. Part 18: Television and visual display units, available at http://rvcs-prodweb.dot.gov.au/files/4204.pdf.
33. McCartt, A. and Geary, L., Longer term effects of New York State's law on driver's hand-held cell phone use, *Injury Prevention*, 10, 11–15, 2004.
34. Johal, S., Napier, F., Britt-Crompton, J., and Marshall, T., Mobile phones and driving, *Journal of Public Health*, doi:10.1093/pubmed/fdh213, 2005.
35. Rajalin, S., Summala, H., Poysti, L., Anteroinen, P., and Porter, B.E., In-car cell phone use and hazards following hands free legislation, *Traffic Injury Prevention*, 6, 225–229, 2005.
36. Hussain, K., Al-Shakarchi, J., Mahmoudi, A., Al-Mawlawi, A., and Marshall, T., Mobile phones and driving: A follow-up, *Journal of Public Health*, 28(4), 395–396, 2006.
37. Cohen, J.T. and Graham, J.D., A revised economic analysis of restrictions on the use of cell phones while driving, *Risk Analysis*, 23(1), 5–17, 2003.
38. Symmons, M. and Haworth, N., Safety attitudes and behaviours in work-related driving—Stage 1: Analyses of crash data, Report 232, Monash University Accident Research Centre, Melbourne, Australia, 2005.

31 Driver Distraction Injury Prevention Countermeasures—Part 2: Education and Training

Michael A. Regan, John D. Lee, and Kristie L. Young

CONTENTS

31.1 INTRODUCTION

In the previous chapter, options for countermeasure development were presented to prevent and mitigate the effects of driver distraction in the areas of data collection, legislation and enforcement, vehicle fleet management and driver licensing. In this chapter, options are presented for countermeasure development in the areas of driver education and training.

Distraction, as an issue, has been largely neglected in the design of driver education and training programs. The number of educational initiatives is limited and few have existed long enough to be evaluated.[1] Mayhew and Simpson[2] note the paucity of research data that exists on the level of public awareness and understanding of distraction. The data that do exist (e.g., Refs 3–7) seem to suggest that, as a whole, the driving public has little understanding of what activities are distracting, of the relative risks associated with different sources of distraction, of the impact of distraction on performance and the mechanisms that mediate its effects, and of the need to self-regulate in response to distractions other than the mobile phone. Also, there is a perception that the risk of being apprehended for violating distraction laws is low. Although there is some limited evidence, reviewed in this book, that the ability to combine competing tasks can be improved with practice, distraction as an issue has also been largely neglected in the design of driver training programs. The authors are unaware of any specific reference to distraction in the driver training literature, and there is only scant reference to distraction as an issue in training-related licensing materials issued by road authorities to learner and probationary drivers.

Driver education and training programs need to address, as a core competency, the ability of drivers to safely manage distraction. The terms "driver training" and "driver education" are poorly defined in the road safety literature. They are often used interchangeably to describe the same programs or initiatives. In attempting to separate the two, Senserrick and Haworth[8] distinguish between driver training, which focuses on the development of a specific set of skills, and driver education, which refers to the more "... contemplative and value-based instruction of knowledge and attitudes relating to safe driving behavior" and covers a broader range of topics than training, over a longer time period (p. 5). They view driver training, therefore, as a component of the broader field of driver education. Others have come to a similar conclusion (e.g., Refs 9 and 10). For the purposes of the present chapter, this would seem a reasonable distinction, although, as noted by Senserrick and Haworth,[8] it is often difficult in practice to distinguish between the two. Education programs may include some training, and training programs do not necessarily take place in isolation from driver education.

It may be artificial, and even counterproductive, to attempt to separate these two areas of countermeasure development. This is exemplified by the Goals for Driver Education (GDE) matrix,[11] which derives from the earlier work of Keskinen, 1996, cited in Ref. 12, and illustrates how inextricably intertwined are the various elements of driver education and training. The cells in the matrix (reproduced in Table 31.1) have been used previously to define detailed competencies that are needed to be a safe driver, that can be addressed through driver education and training.[9,13] The matrix is described in the following section and provides a useful conceptual framework for determining and structuring options for countermeasure development to prevent and mitigate the effects of distraction.

Data on the effectiveness of driver training and education in reducing crashes and crash risk is equivocal, which may be bound up in problems of definition and other methodological issues that are beyond the scope of this chapter to discuss. Recent reviews of literature on the evaluation of driver education and training initiatives yield little or no evidence that either improve safety,[2,8,9,14] although there is evidence that well-designed and evaluated training programs which target specific skills critical for safe driving are effective in improving those skills (e.g., Refs 15 and 16).

Driving is a multitask activity performed at different levels of control.[17] At the strategic level, tasks include journey planning, selection of transport mode, and route choice, at a timescale of minutes to weeks. At the tactical level, tasks include adhering to traffic rules, giving way to other road users, and overtaking. The timescale at this level of control is seconds to minutes. At the operational level, tasks include lateral and longitudinal control of the vehicle, at a timescale of milliseconds to seconds. In Chapter 4, distraction was described as a breakdown in multilevel control in activities critical for safe driving and competing activities at one or more of these levels. In this chapter we assume that the ability to maintain multilevel control of driving and competing tasks requires knowledge and skills that can be acquired through education and training. In the sections that follow, options are presented for countermeasure development to support educational and training initiatives.

31.2 THE GOALS FOR THE DRIVER EDUCATION MATRIX

The GDE model, or matrix (see Table 31.1), was developed to provide a theoretical framework for defining the competencies needed to be a safe driver and the goals of driver education and training.[18] It spans competencies for basic vehicle maneuvering (bottom left corner) all the way through to self-reflection of one's personal lifestyle (upper right corner). The idea is that skills lower in the hierarchy are exercised under the guidance of higher-level goals and motives. This means that, in addition to basic driving skills, driver training and education should take into account drivers' goals connected with driving and with life itself. According to this view, failure or success at higher levels in the hierarchy will affect the demands on skills at lower levels. The model reflects the fact that drivers are individuals and that their problems and skill deficits may lie in different boxes of the matrix[12] (p. 312).

Hernetkoski and Keskinen[13] provide a succinct overview of the matrix. First, they describe the vertical hierarchy. The bottom three levels of the matrix (see Table 31.1) derive from the earlier work of Michon[17] and relate, from bottom to top, to the operational, tactical, and strategic levels of driving control, respectively. The lowest level in the hierarchy, "vehicle maneuvering," relates to knowledge and skills relevant for

TABLE 31.1
Goals of Driver Education Matrix

	Content of Driver Education		
Hierarchical level of behavior (extent of generalization):	**Knowledge and skills the driver has to master**	**Risk-increasing factors the driver must be aware of and be able to avoid**	**Self-evaluation**
Goals for life and skills for living (global)	**10.** Knowledge about/control over how life goals and personal tendencies affect driving behavior: • Lifestyle • Group norms • Motives • Self-control • Personal values • etc.	**11.** Risky tendencies: • Acceptance of risks • Self-enhancement through driving • High level of sensation seeking • Complying to social pressure • Use of alcohol and drugs • Values, attitudes toward society • etc.	**12.** Self-evaluation/awareness of: • Personal skills for impulse control • Risky tendencies • Safety-negative motives • Personal risky motives • etc.
Goals and context of driving (specific trip)	**7.** Knowledge and skills on: • Effect of trip goals on planning • Planning and choosing routes • Evaluation of driving time • Effects of social pressure in car • Evaluation of necessity of trip • etc.	**8.** Risks connected with: • Driver's condition (mood, BAC, etc.) • Purpose of driving • Driving environment (e.g., urban/rural) • Social context and company • Extra motives (competing, etc.) • etc.	**9.** Self-evaluation/awareness of: • Personal planning skills • Typical goals of driving • Typical risky driving motives • etc.

Mastery of traffic situations (specific situation)	**4.** Knowledge and skills on: • Traffic rules • Observation/selection of signals • Anticipation of course or situation • Speed adjustment • Communication • Driving path • Driving order • Distance to others/safety margins	**5.** Risks caused by: • Wrong expectations • Risky driving style (e.g., aggressive) • Unsuitable speed adjustment • Vulnerable road users • Not obeying rules/unpredictable behavior • Information overload • Difficult conditions (e.g., darkness) • Insufficient automatism/skills	**6.** Self-evaluation/awareness of: • Strong and weak points of basic traffic skills • Personal driving style • Personal safety margins • Strong and weak points of skills for hazard situations • Realistic self-evaluation • etc.
Vehicle Maneuvering	**1.** Knowledge and skills on: • Control of direction/position • Tire grip and friction • Vehicle properties • Physical phenomena • etc.	**2.** Risks connected with: • Insufficient automatism/skills • Unsuitable speed adjustment • Difficult conditions, low friction, etc.	**3.** Self-evaluation/awareness of: • Strong and weak points of basic maneuvering skills • Strong and weak points of skills for hazard situation • Realistic self-evaluation

Source: Adapted from Hatakka, M., Keskinen, E., Hernetkoski, K., Gregerson, N. P., and Glad, A., In *Driver Behaviour and Training*, Dorn, L. (Ed.), Ashgate, England, United Kingdom, 2003, p. 313. With permission.

basic vehicle handling such as accelerating, braking, changing gears, steering, and so on. The second lowest level, "mastery of traffic situations," relates to knowledge and skills relevant for adaptation of driver behavior to the behavior of other road users and to the traffic environment. This includes perception and anticipation of the behavior of other road users, making the driver's own behavior predictable to others, and knowledge of and adherence to traffic rules. The third level relates to the "goals and context of driving." At this level, drivers decide for what purpose, where, with whom, with what, and at what time to drive. This level relates to decisions about planning and choosing of the driving route, driver state, and the company of passengers. The highest level in the hierarchy is labeled "goals for life and skills for living." This refers to the motives and goals of the person in a broad sense, and includes personal skills for handling different situations of life in general (Keskinen, 1996, cited in Ref. 13).

The horizontal dimensions of the model, going from left to right, describe the competencies that are needed to be a safe driver. The first column, "knowledge and skills," describes what a good driver needs to know and be able to do at each of the four vertical levels to drive a vehicle safely and cope with traffic. The lower half of the column encompasses competencies that are addressed in traditional driver training programs. The upper half encompasses competencies that are being increasingly addressed in postlicensing training programs.[13] The second column, "risk-increasing factors," relates to the first but emphasizes knowledge and skills related to factors that increase or decrease crash risk, ranging from those connected directly to driving conditions (e.g., the effects of ice and snow) through to risks deriving from social pressure and lifestyle (at the upper end of the column). The competencies in the lower half of this column are usually addressed in defensive driving courses and so-called insight learning courses[12] (p. 314). The third column, "self-evaluation" refers to the process of reflective thinking whereby an individual tries to get feedback on his or her personal actions "from within the self"[13] (p. 56). The idea here is to train self-evaluative skills, which "are a feature of experts and do not develop automatically"[12] (p. 314).

31.3 THE GOALS FOR DRIVER EDUCATION MATRIX, OTHER CONCEPTUAL FRAMEWORKS, AND DRIVER DISTRACTION

It is beyond the scope of this chapter to prescribe instruction theories, methods, and media that are appropriate for imparting, through education and training, the skills, knowledge, and attitudes necessary to prevent and mitigate the effects of distraction. Other reference sources provide guidance on this issue for driver education and training generally (e.g., Refs 8, 9, 19–22). It is possible, however, to use the GDE matrix to define, in a systematic way, some basic competencies that might be addressed in driver distraction education and training programs. To this end, the cells of the GDE matrix have been numbered in Table 31.1, from 1 to 12. In the following section of the chapter we consider, for each cell in the matrix, distraction management competencies that might be addressed in driver education and training programs.

The GDE matrix is only one conceptual framework that can be used for this purpose. The ideas presented in Chapter 4 provide another perspective from which to derive education and training needs. In that chapter, driving, as a control process, was described at three levels (operational, tactical, and strategic), each operating at different time horizons. The three levels were also used to describe the control of attention to competing activities. At the operational level drivers control resource *investment*; at the tactical level they control task *timing*; and at the strategic level they control *exposure* to potentially demanding situations. Distraction-related mishaps result from a breakdown of control at any one level, and from the accumulation of control problems that compound as they propagate across levels. In the following section, we also identify, within the relevant cells of the GDE matrix, distraction management competencies deriving from this perspective that might be addressed in driver education and training programs.

The GDE matrix defines a hierarchy of competencies needed to be a safe driver of current generation vehicles. However, as noted in previous chapters, the driving task is rapidly evolving as new technologies (e.g., in-vehicle navigation, intelligent speed adaptation, adaptive cruise control) make their way into the cockpit that automate, partly or fully, some elements of driving. This has important implications for education and training, particularly as it relates to the management of distraction. Relevant here is a conceptual framework proposed by Donmez et al.[23] (see Table 31.2). Donmez et al. distinguishes between driving- and nondriving-related mitigation strategies for driver distraction that are applicable for different levels of system automation. Table 31.2 provides a framework for considering the various mechanisms by which technology can be used to minimize distraction, by moderating the demands of both the driving task and competing tasks. The table shows, in the left column, three levels of automation—high (automation takes control and ignores human), moderate (automation executes action only if human approves) and low (human does it all)—that can be used in designing a distraction mitigation system. The table further distinguishes between driving-related (e.g., steering, braking) and nondriving-related tasks (e.g., tuning a radio, talking on a mobile phone). Strategies that address driving-related tasks focus on the roadway environment and directly support driver control of the vehicle, whereas strategies for nondriving-related tasks focus on modulating driver interaction with telematics devices.[23,24] Within each of these strategies (i.e., driving and nondriving related), the support provided by technology can be system or

TABLE 31.2
Mitigation Strategies for Driver Distraction

	Driving-Related Strategies		Non-Driving-Related Strategies	
Level of Automation	**System Initiated**	**Driver Initiated**	**System Initiated**	**Driver Initiated**
High	Intervening	Delegating	Locking and Interrupting	Controls-presetting
Moderate	Warning	Warning tailoring	Prioritizing and filtering	Place keeping
Low	Informing	Perception augmenting	Advising	Demand minimizing

Source: Adapted from Donmez, B., Boyle, L. N., and Lee, J. D., *Human Factors*, 48(4), 786, 2006. With permission.

driver initiated. Within this conceptual framework, drivers interact with technology to varying degrees, depending on the particular mitigation strategy employed. These technologies, in turn, introduce new training needs, which will vary according to whether the strategies employed are system initiated or driver initiated.

In the sections that follow we consider, using the GDE matrix as the principle organizing structure, distraction education and training needs from the perspectives of all three of these frameworks.

31.4 HIERARCHY OF COMPETENCIES NEEDED TO EDUCATE AND TRAIN DRIVERS TO MANAGE DISTRACTION

In Table 31.1, the cells of the GDE matrix have been numbered from 1 to 12. In this section we consider, for each cell in the matrix, distraction management competencies that might be addressed in driver education and training programs. These derive from the three conceptual frameworks described earlier.

31.4.1 Cells 1 to 3—Vehicle Maneuvering

31.4.1.1 Cell 1 (Vehicle Maneuvering—Knowledge and Skills)

This cell of the GDE matrix pertains to knowledge and skills the driver has to master relating to vehicle handling; that is, to control of the vehicle and knowledge of vehicle properties, tire grip and friction, physical phenomena, and so on.

From a distraction perspective this cell is relevant to driver operation of vehicle systems that have potential to distract them, whether they be factory fitted, aftermarket, or nomadic. It is critical within this cell of the matrix that drivers develop the knowledge and skills they need to use systems, design features, and functions in a manner that minimizes the potential for saturation effects at the operational level of control (see Chapter 4) and hence minimizes distraction. Several recommendations can be made here.

Drivers need to be able to operate vehicle systems in a way that minimizes distraction. Here, for example, it is critical to ensure that drivers be aware of design features in their vehicles (e.g., steering wheel-mounted controls, speed alerters and limiters) that have potential to limit distraction and that drivers be able to use these features in the manner that is least distracting.

Drivers need to be able to perform all *competing* tasks that are nondriving-related and legally allowed (e.g., tuning radios, adjusting climate control systems, using hands-free mobile phones) in a way that minimizes distraction. Here, for example, it is critical to ensure that drivers are aware of design features in their vehicles (e.g., radio presets, steering wheel controls, voice-activated dialing) that have the potential to limit distraction and that drivers are able to use these features in a manner that is least distracting.

Where vehicles are equipped with advanced driver assistance systems (e.g., in-vehicle navigation systems, adaptive cruise control), it is critical that drivers be aware of those system functions and features that have the potential to distract them and that drivers are able to operate the system in the manner that is least distracting.

Where vehicles are equipped with real-time distraction mitigation systems (e.g., workload managers and distraction warning systems), drivers must be able to operate the systems in the manner intended.

Drivers must be made aware of, and be prepared for, unexpected or unwanted system functions and operations that have potential to distract them (e.g., seatbelt reminder warnings, low fuel warnings, radio mute functions which suddenly deactivate and startle the driver).

Drivers need to be trained how and under what circumstances automation in a vehicle cockpit is system initiated (see Table 31.2). For example, if a system warns a driver (under "driving related strategies") to take a necessary action (e.g., slow down, increase headway) and if the driver fails to self-regulate early in a chain of events in response to the demands of a distracting task or is judged to be too distracted to react in a timely manner later in the chain, it is important that the driver understand the meaning and intent of the information displayed.

Where the system intervenes to take control of the vehicle (see Table 31.2), for example if the driver fails to self-regulate or is it too distracted to react in a timely manner, and this intervention is noticeable to the driver, training can be important in demonstrating system effectiveness and promoting trust in system operation. On the other hand, if a system locks out or interrupts (under "nondriving activities"; see Table 31.2) information displayed by a telematics device because the demands of the driving situation are judged to be too high, knowledge of system operation may not be necessary. If the system locks out access to control functions, however, knowledge of system operation may be desirable to ensure driver acceptance of the system.

Training is likely to be more important where the provision of automation is driver initiated (see Table 31.2). Drivers may, for example, delegate authority to a system, so that it warns them or takes control of the vehicle if it judges that they are failing to self-regulate or are too distracted to react in a timely manner when interacting with a competing task. Under such circumstances, drivers need to know when and under what circumstances it is appropriate to delegate authority to a system, and what form the level of intervention will take.

From an automation perspective, training is likely to be most important for "nondriving-related strategies" (see Table 31.2) that are driver initiated in relation to telematics devices with low levels of automation. In this case it is important for drivers to know, for example, which modes of operation for a particular telematics device are least distracting (i.e., demand minimizing).

31.4.1.2 Cell 2 (Vehicle Maneuvering—Risk Awareness)

Cell 2 covers vehicle maneuvering factors that increase risk and that the driver must be aware of and be able to avoid, such as vehicle properties, friction, and so on.

From a distraction perspective, education and training have several roles to play here:

- To ensure that drivers understand what distraction is
- To ensure that drivers understand why, when, and how vehicle systems, design features, and functions can distract them

- To ensure that drivers understand the limitations of real-time distraction mitigation systems, so that they do not become complacent and believe that they can divert attention away from activities critical for safe driving in circumstances that will not be detected by the system
- To ensure that drivers understand the adverse effects that distraction can have on their driving performance (e.g., impaired event detection, lateral and longitudinal control)
- To ensure that drivers understand the relative risks associated with interaction with vehicle systems, design features, and functions that are known to degrade safety
- To make drivers aware of how inclement road conditions can interact with distracting activities to undermine their performance

31.4.1.3 Cell 3 (Vehicle Maneuvering—Self-Evaluation)

Cell 3 focuses on awareness and self-evaluation of personal strengths and weaknesses in relation to driving skills, maneuvering in hazardous situations, and so on.

In this case, education and training have three main roles to play in relation to distraction:

- To make drivers reflect on their own strengths and weaknesses in their ability to limit distraction when using vehicle systems, design features, and functions
- To expose drivers, in a safe environment, to distractions deriving from inappropriate operation of vehicle systems, design features, and functions within vehicles they drive, enabling them to reflect and become self-aware of the impact of these distractions on their driving performance
- To support feedback control (see Chapter 4) by making drivers self-aware of the effects of distraction deriving from inappropriate operation of vehicle systems, design features, and functions that have no overt impact on performance in normal conditions (e.g., cognitive distraction when talking on a mobile phone) but which could compromise performance and safety in safety-critical situations (e.g., when a pedestrian unexpectedly steps out from behind a parked car)

31.4.2 Cells 4 to 6—Mastery of Traffic Situations

31.4.2.1 Cell 4 (Mastering Traffic—Knowledge and Skills)

Cell 4 pertains to knowledge and skills the driver has to master relating to road rules, speed adjustment, safety margins, signaling, and so on.

The perspective in this cell is that the driver is operating not in isolation but in conjunction with other road users. At this level of the GDE matrix, it is usually assumed that drivers are operating at the tactical level of control. From a distraction perspective, however, drivers must control both resource investment (operational control) and task timing (tactical control) in managing distraction in traffic situations. Hence, following from the discussion in Chapter 4, it is assumed that in managing distraction the driver is operating at this level of the matrix at both the tactical and operational levels of control. Drivers need to know how to control resource investment and task timing in a range of traffic situations. From a distraction perspective, education and training have several roles to play here.

Education is important in providing drivers with knowledge about road laws and regulations designed to limit distraction.

Training is necessary to support feedback control (see Chapter 4) and to enable drivers to make the best use of feedback that leads to immediate performance improvement in response to distraction—for example, to train them to interpret and respond appropriately to warnings and feedback from real-time distraction mitigation systems and from tactile edge lines on roads.

Similarly, training can support feed-forward control (see Chapter 4) by equipping drivers with the ability to better anticipate and respond to the demands of driving and of the other objects, events, and activities that compete for their attention. Training products and techniques that have been developed to improve young drivers' ability to detect, perceive, and respond to actual and potential traffic hazards (e.g., Refs. 16, 25–28) could also improve feed-forward control.

Training programs for enhancing situation awareness (e.g., Ref. 29) could support feed-forward control by helping the driver develop an accurate internal model of the future state of the traffic system (see Chapter 4).

Improving drivers' ability to combine driving and competing tasks can limit distraction. Relevant research in this area has been conducted in the aviation domain (to improve the attention-sharing ability of military pilots; Gopher, 1992) and in the driving domain.[26] A CD-ROM-based training product has been shown in a simulator evaluation study to significantly improve the attention-sharing skills of young novice drivers.[16] Products such as this have the potential to support adaptive control at both the operational and tactical levels of control (see Chapter 4).

Drivers need to be able to self-regulate at the tactical level of control—to know *when* it is appropriate to adjust their driving behaviors to compensate for the added or anticipated load imposed by a competing task (e.g., when approaching an intersection; when hearing an ambulance siren; when driving into a storm); and, conversely, to know when to moderate their interactions with competing tasks to minimize interference with the driving task (e.g., when a conversation with a passenger becomes too complex or emotional). Training can facilitate development of this ability.

Similarly, drivers need training in how to self-regulate at the operational level of control—that is, to know *how* to adjust their driving behavior to compensate for the added or anticipated load imposed by a competing task (e.g., by slowing down, increasing headway, taking an easy route, avoiding bad weather), and, conversely, to know how to moderate their interactions with competing tasks to minimize interference with the driving task (e.g., by using speed-dial on a mobile phone, avoiding complex or emotional conversations, avoiding reading and sending text messages).

Distraction can also be reduced by improving the ability of drivers to interrupt competing activities and return to activities that are critical for safe driving. Concurrent feedback (see Chapter 29) provides a suitable mechanism for enhancing this ability.

Finally, training can play an important role in enabling drivers to adopt a graded rather than a "brittle" resource allocation and performance tradeoff when interacting with competing tasks (see Chapter 4).

31.4.2.2 Cell 5 (Mastering Traffic—Risk Awareness)

Cell 5 covers risk-increasing factors relating to the selection of inappropriate speeds, narrow safety margins, neglect of road rules, difficult driving conditions, vulnerable road users, and so on.

From a distraction perspective, education and training have several roles to play here:

- To provide drivers with knowledge about the full range of activities that have potential to distract them (see Chapter 15)
- To provide drivers with knowledge about the adverse effects of different distracting activities on performance (see Chapters 11 through 14), and the mechanisms that moderate these effects (see Chapters 3, 4, and 19)
- To provide drivers with knowledge about the relative risks associated with different sources of distraction (see Chapters 15 through 17)

It is important that drivers understand that the distraction posed by a competing activity may persist even after it has ended, for example when they continue to think about aspects of the driving task or about what they have just seen on an advertising billboard.

31.4.2.3 Cell 6 (Mastering Traffic—Self-Evaluation)

Cell 6 focuses on driver awareness and self-evaluation of personal skills, driving style, hazard perception, and so on, from the viewpoint of strengths and weaknesses. Education and training have several roles to play here:

- To expose drivers in a safe environment to high risk distractions to enable them to become self-aware of the impact of these distractions on their driving performance.
- To make drivers self-aware of the effects of distractions deriving from different sources which have no overt impact on performance (e.g., cognitive distraction when talking on a mobile phone) but which can compromise performance and safety in safety-critical situations (e.g., when a child chasing a ball unexpectedly runs across the road into the path of a vehicle). This supports feedback control at the operational level of control (see Chapter 4).
- To make drivers self-aware of the effects of distraction on their driving performance when in different driving states (e.g., when fit to drive, when fatigued, drowsy, emotionally upset, inebriated, and so on).
- To make drivers self-aware of the differential effects of distraction on their driving performance when driving at different speeds, on different routes and in other situations that moderate their vulnerability to distraction.
- To make drivers self-aware of strengths and weaknesses in their ability to self-regulate in response to distraction.

- To make drivers self-aware of strengths and weaknesses in their ability to take into account future demands as well as current demands in deciding whether, when and how to attend to competing tasks.
- To make drivers self-aware of their strengths and weaknesses in being able to determine when their driving performance is being compromised by distraction to a dangerous degree.
- To calibrate drivers, so that they acquire the ability to match the joint demands of driving (with and without competing tasks) to their own driving capabilities[30,31]—and, in doing so, to support and enhance adaptive control (see Chapter 4).

31.4.3 Cells 7 to 9—Goals and Context of Driving

31.4.3.1 Cell 7 (Driving Goals and Context—Knowledge and Skills)

Cell 7 pertains to journey-related knowledge and skills the driver has to master such as the effect on safety of goals, environment choice, effects of social pressure, evaluation of necessity to drive, and so on.

Drivers at this level of the matrix are operating at the strategic level of control. At this level they control exposure to potentially demanding situations. Here, drivers require knowledge and skills to develop and optimize strategic control. From a distraction perspective, education and training have several roles to play here.

First, education is important in providing drivers with knowledge about driving routes, situations, and scenarios in which drivers are most vulnerable to the effects of distraction (see Chapters 16, 17, and 19).

Second, education and training can provide drivers with knowledge and skills that enable them to plan and manage their interactions with competing tasks that are known to be contributing factors in crashes (see Chapter 15):

- When interacting with *things brought into the vehicle* (e.g., to plan a route beforehand to minimize eyes off-road-time when navigating using a paper map; not to use the phone while driving; if it is necessary to use a mobile phone, to use a hands-free phone and keep calls short; to tell the person at other end of the mobile phone that they are driving)
- When interacting with *vehicle controls and devices* (e.g., using radio preset buttons to find a radio station rather than a more distracting mode of operation; choosing to dial a phone only during the day on straight roads rather than at night on curves)
- When interacting with *vehicle occupants* (e.g., looking at them infrequently rather than frequently when conversing with them; using them as copilots to assist with driving tasks such as navigation or to answer the car or cellular phone if it rings)
- When interacting with *objects that move or have potential to move within the vehicle* (e.g., properly restraining animals prior to departure; placing drink containers in locations where they won't fall over and spill)

- When engaging in *internalized activities* (e.g., to the extent that it is possible, choosing when and where to allow themselves to daydream)
- When engaging with *external objects and events* (e.g., choosing not to dwell visually on crash scenes; using optimal search strategies to safely navigate to a destination)

Third, education and training can support feed-forward control by developing in drivers the ability to anticipate the confluence of driving and competing task events (see Chapter 4).

Education and training can also play an important role in training drivers and passengers to cooperate as a team in the vehicle cockpit to limit distraction. Regan et al.,[32] for example, have advocated the use of team training techniques, such as Crew Resource Management (CRM), to equip young passengers with the skills, knowledge, and attitudes required to behave and perform as copilots rather than as backseat drivers when traveling with young drivers. This includes performing some secondary tasks for the driver to reduce exposure to risk (e.g., answering the phone), performing for the driver some aspects of the driving task itself (e.g., navigating, alerting the driver to unnoticed hazards) and in behaving in a way that minimizes distraction (e.g., moderating conversation with drivers when traveling through busy intersections).

Finally, following from Table 31.2, education and training have two additional roles to play:

- Where there is *system* initiation of technologies that automatically regulate driver exposure to distraction (e.g., workload managers; see Table 31.2), drivers need to understand the manner and circumstances under which this occurs. A lack of such understanding might generate distraction—for example, if a driver is expecting a phone call that does not arrive when expected because it is temporarily suppressed by a workload manager.
- Where there is *driver* initiation of technologies that regulate driver exposure to distraction (see Table 31.2), drivers need to know when, and under what circumstances, it is appropriate to initiate the technology, and what form the level of intervention will take.

31.4.3.2 Cell 8 (Driving Goals and Context—Risk Awareness)

Cell 8 covers risk-increasing factors the driver must be aware of and be able to avoid relating to trip goals, driving state, social pressure, purpose of driving, and so on. From a distraction perspective, this cell in the matrix is concerned with factors that increase the risk of having a distraction-related crash that relate to trip goals, driving state, social pressure, and so on.

Drivers require knowledge about factors at this level of the matrix that moderate distraction-related crash risk, such as driver state (e.g., fatigued, inebriated), productivity pressures (e.g., when driving for work purposes), circumstances that act to displace the primacy of their social role as driver (e.g., when tending to a screaming child; when using an iPod; when interacting socially with passengers; see Chapter 2), age, experience, driving routes, and travel patterns (e.g., day versus night driving).

The technologies that drivers interact with provide a mechanism for regulating their exposure to distraction, and hence their exposure to risk. The different distraction mitigation strategies shown in Table 31.2 involving the use of technology can affect driver exposure to distraction in different ways, some of which have implications for training. As noted in Chapter 2, for example, driving is, in its current form, a "satisficing" task—it does not require continuously perfect behavior and total attention, which leaves free attentional resources that drivers can use to do other things that might distract them. Mitigation strategies in Table 31.2 that involve high levels of automation, therefore, are likely to make the driving task more "satisficing" than it already is, perhaps encouraging drivers to take on other roles which expose them to other forms of distraction. In this context, the role of education and training should be to make drivers aware of the potential risks that may arise in using free attentional resources to willingly engage in distracting activities that have potential to compromise their safety.

31.4.3.3 Cell 9 (Driving Goals and Context—Self-Evaluation)

Cell 9 concerns awareness and self-evaluation of personal planning skills, typical driving goals, and driving motives. The role of education and training is to develop knowledge and skills that enable drivers to self-evaluate their ability to manage distraction.

Education and training would seem to have three main goals here:

- In making drivers self-aware of their strengths and weaknesses in planning trips in a manner that limits their exposure to distractions
- In making drivers self-aware of the different factors that influence their propensity to engage in distracting activities (e.g., their awareness of the demands associated with a distracting activity in a given environment; their appreciation of their own ability to handle the demands associated with the activity in that situation; the propensity of the individual to take risks; whether laws exist that permit engagement in the activity; productivity and other pressures; driving culture and societal norms)
- In making drivers self-aware of different risky driving motives, attitudes and emotions that influence their propensity to engage in distracting activities

31.4.4 Cells 10 to 12—Goals for Life and Skills for Living

31.4.4.1 Cell 10 (Life Goals and Skills—Knowledge and Skills)

Cell 10 relates to how general life goals and values, behavioral style, group norms, and other factors affect driving. Here, the role of education and training is to make drivers aware of these factors and their relationship with distraction.

31.4.4.2 Cell 11 (Life Goals and Skills—Risk Awareness)

Cell 11 is concerned with the driver's personal control over risks connected with life goals and values, behavioral style, social pressure, substance abuse, and so on. Here, the role of education and training is to provide drivers with knowledge and skills to manage distraction-related risk that may derive from these higher-order goals.

31.4.4.3 Cell 12 (Life Goals and Skills—Self-Evaluation)

Finally, Cell 12 pertains to the driver's impulse control, motives, lifestyle, values, and so on. Here, the role of education and training is to make drivers self-aware of how these personal factors moderate their propensity to be distracted.

At this level of the GDE matrix, the link between life goals and distraction is less straightforward, which makes it difficult to formulate concrete options for countermeasure development. Further research and thinking is needed to develop this level of the GDE matrix for the management of distraction.

31.5 SUMMARY AND CONCLUSIONS

The management of distraction by drivers can be regarded as an ability that can be developed and improved through education and training. In this chapter we have used three organizing frameworks to define a range of distraction management competencies that might be addressed in driver education and training programs.

The GDE matrix is a useful organizing structure for defining the competencies needed to manage distraction. It emphasizes the overlap between education and training and the role of higher-level goals and motives in guiding the development and deployment of skills lower in the hierarchy. The focus of this framework, however, is on the development of competencies needed to be a safe driver, regardless of the level of vehicle system automation. Donmez et al.[23] distinguish between driving- and nondriving-related mitigation strategies for driver distraction that are applicable for different levels of system automation. To this end, Table 31.2 provides a complementary framework for considering the various mechanisms by which technology can be used to minimize distraction, by moderating the demands of both the driving task and competing tasks, and for considering the implications of this for driver education and training. The theoretical framework presented in Chapter 4 provides another, complementary, perspective by considering distraction as a problem of control at three levels (operational, tactical, and strategic), each operating at different time horizons. According to this view, distraction-related mishaps result from a breakdown of control at any one level, and from the accumulation of control problems that compound as they propagate across levels. Understanding why and how each control type might fail at each time horizon prompts consideration and derivation of education and training initiatives that can be implemented to support and maintain control, and prevent such failures.

Given the paucity of empirical data that is currently available to guide the design and evaluation of education and training programs, the choice of competencies presented in this chapter is based largely on the material reviewed in this book and on the general knowledge and judgment of the authors. In the absence of any empirical data on the effectiveness of driver education and training initiatives directly concerned with the prevention and mitigation of distraction, it is difficult to know which of the competencies presented are most likely to be effective in preventing and mitigating the effects of distraction. Paradoxically, it is the competencies at the higher levels of the GDE matrix, which are the least developed and researched, that may have the greatest influence in doing so.

It is beyond the scope of this chapter to prescribe theories of learning, instructional methods, and media and delivery mechanisms that can be used to turn the learning objectives presented here into actual countermeasures; although, where relevant countermeasures and delivery mechanisms already exist, they have been identified. Noteworthy, in this context, however, are the ideas introduced by Donmez et al. in Chapter 29. They highlight the role that different types of feedback, delivered over different timescales, can have in preventing and mitigating the effects of distraction. They distinguish between four such timescales—concurrent (milliseconds), delayed (seconds), retrospective (minutes, hours), and cumulative (days, weeks, months)—which have implications both for the design of education and training programs. Retrospective and cumulative feedback appear, in particular, to be suited to the design of distraction education and training programs. Retrospective feedback, which is provided to the driver immediately after a trip, provides information to the driver about inappropriate and appropriate behavior during that trip (e.g., number of distracting tasks performed during dangerous situations, duration of eyes-off-road time, etc.). Such feedback can influence future driving behavior by informing drivers of what constitutes safe driving, when not to engage in distracting activities, and what speed to maintain in different driving conditions. Cumulative feedback, which integrates driving data over many trips spanning several weeks and months, can help drivers assess their overall level of driving performance by highlighting persistent distraction-related behaviors that compromise their safety. Donmez et al. argue that feedback at this timescale is more likely to lead to lasting behavioral change, whereas concurrent feedback can enhance driving performance. New technologies exist to support the provision of feedback at multiple timescales and, in doing so, to positively change driver performance and induce lasting behavioral change.

Young novice drivers, in their roles as drivers and passengers, would appear to be the group most important to target for education and training, as data reviewed in this book suggest that they are relatively more likely than more experienced drivers to be exposed to distraction and appear to be relatively more vulnerable to its effects. Commercial drivers, especially urban bus drivers, warrant particular attention. There is at present little empirical evidence to suggest that older drivers require special education and training, although this may be warranted in future as they become more tech savvy and drive in proportionately larger numbers.

Driver education and training are, in most countries, responsibilities shared between professional driving instructors, parents (in the case of young novice drivers), and other supervising drivers. Increasingly, automotive companies, point-of-sale staff, and even rental car companies are becoming involved in driver education and training. Responsibility for the design and development of driver training curricula and materials is also shared between multiple stakeholders—professional driving instructors, road authorities (e.g., in prescribing jurisdictional training and assessment regimes and standards; in developing materials to structure and guide the accumulation of on-road driving practice during the learner and probationary periods), educational authorities (for school-based driver training programs), training authorities (for accrediting driving instructors and the training they provide), and vehicle manufacturers, suppliers, and sales personnel. The recommendations made in this chapter are therefore relevant to a wide range of stakeholders. The roles and responsibilities of the various parties in providing training in distraction management need to be carefully defined and delineated.

Awareness campaigns involving mass media and pitched at the public at large can complement education and training initiatives. Such campaigns need to be designed to increase public understanding of the nature of driver distraction, its relative standing in relation to other traffic safety issues, the relative dangers associated with engaging in distracting activities (deriving from inside and outside the vehicle), its impact on driving performance and safety, the factors that increase driver vulnerability to distraction, strategies for minimizing the effects of distraction, and the penalties associated with engaging in distracting activities, where applicable.[1,33] Awareness campaigns should also seek to raise employer awareness of tools, available to them, which can be used to limit employee exposure to distraction while driving company vehicles (see Chapter 30).

An important issue to resolve is the point in training at which it is best to start exposing young novice drivers to distracting activities, such as carrying passengers or using mobile phones. Unfortunately, there is no known empirical research on this topic to guide countermeasure development, despite evidence that even apparently "automated" tasks, such as manual gear-shifting, significantly impair the sign detection performance of novice drivers using standard shift compared with novice drivers using automatic transmission.[34] How quickly drivers become invulnerable to the demands of secondary activities varies considerably. Epidemiological and crash data, reviewed in this book (see Chapters 16 and 17), have been used to support the introduction of licensing restrictions for young and novice drivers, as part of GDL systems. In several jurisdictions, young and novice drivers are banned from using mobile phones and from carrying passengers in the first year of solo driving. At some point in their driving careers, however, they will inevitably be exposed to one or more of these distractions. Clearly, research is needed to determine at what point in the learning curve it is appropriate to introduce these and other sources of distraction (over which some degree of exposure control can be exercised) and in what manner, instructionally, to do so.

Ideally, education and training in how to manage driver distraction should commence early in the life cycle of the road user, and be structured such that the knowledge, skills, and attitudes acquired (e.g., in how to manage distraction as a pedestrian or bicycle rider) transfer positively to driving later in life. Determining how such early programs should be designed and delivered to promote positive skill transfer is a challenging area for further research. As technologies continue to evolve, at an increasingly rapid rate, a more critical issue is how to continue, through education and training, to maintain this skill transfer and provide drivers with better support for both feedback and feed-forward control of distraction.

In the next chapter, options for countermeasure development are presented in the areas of vehicle, technology, and road design.

ACKNOWLEDGMENTS

The authors would like to thank Dr. Peter Burns, Transport Canada, and Phil Wallace, Learning Systems Analysis Pty Ltd, Australia, for their insightful comments on earlier versions of this chapter.

REFERENCES

1. Hedlund, J., Simpson, H., and Mayhew, D., International conference on distracted driving—Summary of proceedings and recommendations, *International Conference on Distracted Driving*, Traffic Injury Research Foundation, Toronto, Canada, 2006.
2. Mayhew, D. R. and Simpson, H. M., The safety value of driver education and training, *Injury Prevention,* 8(Suppl II), 3–8, 2002.
3. Baker, S. and Spina, K., Drivers' attitudes, awareness and knowledge about driver distractions: Research from two central Sydney communities, *First International Conference on Distracted Driving*, Sydney, Australia, 2005.
4. Barker, C., Key findings from focus group research on inside-the-vehicle distractions in New Zealand, *First International Conference on Distracted Driving*, Sydney, Australia, 2005.
5. Beirness, D. J., Distracted driving: The role of survey research, *International Conference on Distracted Driving*, Toronto, Canada, 2005.
6. Rudin-Brown, C. M., Driver distraction from in-vehicle telematics devices: The public opinion, *International Conference on Distracted Driving*, Toronto, Canada, 2005.
7. Sundeen, M., Distracted driving legislation, *International Conference on Distracted Driving*, Toronto, Canada, 2005.
8. Senserrick, T. and Haworth, N., *Review of Literature Regarding National and International Young Driver Training, Licensing and Regulatory Systems*, Report No. 239, Monash University Accident Research Centre, Melbourne, Australia, 2005.
9. Engström, I., Gregersen, N. P., Hernetkoski, K., Keskinen, E., and Nyberg, A., *Young Novice Drivers Education and Training: Literature Review*, Report No. VTI Report 491A, VTI, Linköping, Sweden, 2003.
10. Woolley, J., *In-Car Driver Training at High Schools: A Literature Review*, Report No. Transport SA Report 6/2000, Transport Systems Centre, University of South Australia, Adelaide, Australia, 2000.
11. Hatakka, M., Keskinen, E., Gregersen, N. P., Glad, A., and Hernetkoski, K., From control of the vehicle to personal self-control; broadening the perspectives to driver training, *Transportation Research, Part F*, 5, 201–215, 2002.
12. Hatakka, M., Keskinen, E., Hernetkoski, K., Gregerson, N. P., and Glad, A., Goals and contents of driver education, in *Driver Behaviour and Training*, Dorn, L. (Ed.), Ashgate, England, United Kingdom, 2003.
13. Hernetkoski, K. and Keskinen, E., Used methods and incentives to influence young drivers' attitudes and behaviour, in *Young Novice Drivers Education and Training: Literature Review*, Engström, I., Gregersen, N. P., Hernetkoski, K., Keskinen, E., and Nyberg, A. (Ed.), VTI, Linköping, Sweden, 2003.
14. Christie, R., *The Effectiveness of Driver Training as a Road Safety Measure*, Report No. 01/03, Royal Automobile Club of Victoria, Melbourne, Australia, 2001.
15. Fisher, D. L., Laurie, N. E., Glaser, R., Connerney, K., Pollatsek, A., Duffy, S., and Brock, J., The use of a fixed base driving simulator to evaluate the effects of experience and PC based risk awareness training on drivers' decisions, *Human Factors,* 44, 287–302, 2002.
16. Regan, M. A., Triggs, T., and Godley, S., Simulator-based evaluation of the DriveSmart novice driver CD ROM training product, *Road Safety Research, Policing and Education Conference*, Brisbane, Australia, 2000.
17. Michon, J. A., A critical review of driver behaviour models: What do we know, what should we do?, in *Human Behaviour and Traffic Safety*, Evans, L. and Schwing, R. (Eds.), Plenum Press, New York, 1985.

18. Falkmer, T. and Gregerson, N. P., The TRAINER project—The evaluation of a new simulator-based driver training methodology, in *Driver Behaviour and Training*, Dorn, L. (Ed.) Ashgate, England, United Kingdom, 2003.
19. Global Road Safety Partnership, Road safety publicity campaigns, GRSP, Geneva, Switzerland, 2002.
20. Simons-Morton, B. G., Hartos, J. L., Leaf, W. A., and Preusser, D., The persistence of effects of the checkpoints program on parental restrictions of adolescent driving privileges, *American Journal of Public Health*, 95, 447–452, 2005.
21. Tovey, M. D. and Lawlor, D. R., *Training in Australia: Design, Delivery, Evaluation, Management*, Pearson Prentice Hall, Australia, 2004.
22. Wallace, P. R. and Regan, M. A., Case study: Converting human factors research into design specifications for instructional simulation, *Third International SimTect Conference*, Adelaide, Australia, 1998.
23. Donmez, B., Boyle, L. N., and Lee, J. D., The impact of distraction mitigation strategies on driving performance, *Human Factors*, 48(4), 785–804, 2006.
24. Lee, J. D., Driver distraction: Breakdowns of a multi-level control process, *Journal of the Australasian College of Road Safety*, 16(2), 33–38, 2005.
25. Fisher, D. L., Using eye movements to evaluate a PC-based risk awareness and perception training program on a driving simulator, *Human Factors*, 48(3), 447–464, 2006.
26. Regan, M. A., Deery, H., and Triggs, T., Training for attentional control in novice car drivers, *42nd Annual Meeting of the Human Factors and Ergonomics Society*, Chicago, IL, 1998.
27. Regan, M. A., Deery, H. A., and Triggs, T. J., A technique for enhancing risk perception in novice car drivers, *Road Safety Research, Policing and Education Conference*, Wellington, New Zealand, 1998.
28. Triggs, T. J. and Regan, M. A., Development of a cognitive skills training product for novice drivers, *Road Safety Research, Policing and Education Conference*, Wellington, New Zealand, 1998.
29. Endsley, M. R. and Garland, D. J., Pilot situation awareness training in general aviation, *XIVth Triennial Congress of the International Ergonomics Association and 44th Annual Meeting of the Human Factors and Ergonomics Society*, San Diego, CA, 2000.
30. Mitsopoulos, E., Triggs, T., and Regan, M., Examining novice driver calibration through novel use of a driving simulator, *SimTect 2006* Melbourne, Australia, 2006.
31. Triggs, T. J., Human performance and driving: the role of simulation in improving young driver safety, *12th Triennial Congress of the International Ergonomics Association*, Toronto, Canada, 1994.
32. Regan, M., Salmon, P. M., Mitsopoulous, E., Anderson, J., and Edquist, J., Crew resource management and young driver safety, *Human Factors and Ergonomics Society 49th Annual Meeting*, Orlando, FL, 2005.
33. Regan, M., Young, K., and Johnston, I., MUARC Submission to the Parliamentary Road Safety Committee Inquiry into Driver Distraction, Monash University Accident Research Centre, Melbourne, Australia, 2005.
34. Shinar, D., Meir, M., and Ben-Shoham, I., How automatic is manual gear shifting? *Human Factors* 40(4), 647–654, 1998.

32 Driver Distraction Injury Prevention Countermeasures—Part 3: Vehicle, Technology, and Road Design

Michael A. Regan, Trent W. Victor, John D. Lee, and Kristie L. Young

CONTENTS

32.1 INTRODUCTION

A number of design challenges are created by the proliferation of systems and functions that interact with the driver in the vehicle and the road environment. These include technology-related design challenges (e.g., integration of many functions, life cycle gaps, false alarms) and safety-related design challenges (e.g., distraction, drowsiness, automation, and behavioral adaptation).

This chapter addresses the challenges that exist in designing systems and functions that minimize the impact of distraction on safety while maximizing the possibilities that are offered by new technology. It presents and discusses countermeasures that relate to the vehicle itself, the technology brought into or added to the vehicle, and the road environment in which the vehicle is traveling. The first section focuses on what designers of vehicle systems and functions can do to prevent distraction, both in the design phase before manufacturing and during driving (in real time). The second section deals with what designers of the road environment can do to limit distraction.

32.2 VEHICLE AND TECHNOLOGY DESIGN COUNTERMEASURES

32.2.1 Background

The number and types of technologies—vehicle-integrated, aftermarket, and nomadic (portable)—entering the vehicle cockpit that have the potential to distract the driver is increasing rapidly. There is evidence (see Chapter 16) that distraction deriving from driver interaction with technologies whilst driving accounts for around 15–20% of all distraction-related crashes. Hedlund et al.[1] highlight (p. vii) several trends in the rollout of vehicle and nomadic technologies, which they believe challenge traditional methods of ensuring the safety of vehicles and equipment through regulation:

- "Electronics and telematics devices are becoming multi-functional. For example, the device "previously known as a cell phone" now can send and receive email, take pictures, and provide location and route information.

- Devices are becoming increasingly portable, no longer attached to a telephone line or an automobile. Consumers can bring their communications and entertainment with them wherever they go.
- The industry is highly diverse, ranging from traditional suppliers of original and aftermarket automobile equipment to consumer electronics manufacturers. It does not fit well within the traditional automobile industry regulatory structure.
- New products are developed, introduced, and modified very rapidly. For example, a typical user replaces his or her cell phone every 18–24 months."

Two additional trends are also noteworthy in this respect. First, interfaces within the vehicle allow nomadic devices to communicate with other devices built into the vehicle. For example, nomadic devices (such as mobile phones) can now display information on dedicated screens within the vehicle. Second, information outside the vehicle, such as speed limits, destination information, and the location of speed cameras, can be duplicated inside the vehicle.[2,3]

As Hedlund et al.[1] point out, the challenge is "to assess the distracting potential of new technology and take proactive steps to prevent it from increasing crash risks, while preserving its potential benefits" (p. 7). These technology trends challenge traditional design methods to adapt to rapid innovation and development.

In the next two sections, a number of countermeasures addressing distraction-safety concerns are made with regard to driver use of in-vehicle technologies while driving. We focus first on what designers of systems and functions can do to "design-out" distraction in the design-phase, before vehicles and other technology are manufactured. Thereafter, we focus on what can be done by designers of real-time distraction prevention and mitigation systems.

32.2.2 Design-Phase Distraction Countermeasures

One main countermeasure is to *design systems so that they do not distract*. But *who* are the designers, and *how* do they know what is distracting? These are deceptively simple questions requiring consideration of many issues. The following section presents and discusses some of the main issues, which are taken up in greater detail elsewhere in this book (e.g., Chapters 22 through 25 and Chapter 33).

As noted by Stevens (Chapter 22), individuals at a number of organizations are typically involved in designing, producing, and providing elements of systems. These include vehicle manufacturers, aftermarket system producers, providers of nomadic device functionality, manufacturers of parts enabling the use of nomadic devices while driving (e.g., cradle, connectors), and information and service providers. Thus, in an answer to the question "who are the designers?" it can be concluded that organizations as well as individuals must be regarded as "designers" and countermeasures should be implemented in organizations as well as by individuals.

32.2.2.1 Guidelines and Standards as Countermeasures

Designers must balance safety with many competing priorities such as cost, packaging, complexity, esthetics, and so on (see Chapter 33) and are often faced with hard questions such as, "Do the safety benefits achieved by moving this display up 5 cm motivate the costs of redesign?" System design and construction includes overall design, installation, information presentation, interaction with displays and controls, system behavior, and information about the system.

Ergonomic guidelines and standards (see Part 7 of this book) are important tools that can be used at different stages in the user-centered design (UCD) process to support the ergonomic design and evaluation of vehicle-human-machine interfaces for driving-related systems. Various distraction-related interface design and evaluation guidelines and standards have been developed around the world, and problems associated with them are reviewed in Part 7 of this book. Guidelines and standards can be seen as the main distraction countermeasures in the design-phase. They describe *how* to know what is distracting and *what* to do about it.

There are three general types of standards (see Burns, Chapter 23):

- *Design standards* provide precise specifications for a vehicle or vehicle system in terms of, for example, physical attributes or geometry.
- *Performance-based standards* set out the minimum level of performance that a system must meet when tested in accordance with a prescribed test method.
- *Process-oriented standards* are concerned with the systems and procedures that an organization should establish and follow during its development and implementation cycle.

Ideally, a combination of the standards should be used because each type of standard has its limitations. Design standards are the most straightforward to implement, but limit innovation; whereas process-oriented standards support innovation, but require a great deal of organizational commitment to implement. The development of performance standards should be a cooperative effort, involving many stakeholders: governments, the technology industry, product manufacturers, service providers and consumer groups.[1]

Design and performance standards. Since design and performance guidelines regarding distraction are typically combined in the same documents, these two types of standards will be considered together here. A number of design and performance guidelines and standards exist, such as the European Statement of Principles (EsoP), the American Automobile Manufacturers (AAM), and the Japan Automobile Manufacturers Association (JAMA) guidelines, and the ISO standards, which aim to limit distraction deriving from in-vehicle information and communication systems. These are reviewed in Part 7 of this book, and the issues that remain to be resolved in ensuring that they achieve their intended purpose have been discussed. Here, a macro perspective is taken to elicit some more general conclusions.

Ideally, the approach to develop and implement mandatory performance standards for vehicle electronic devices, similar to the vehicle safety standards currently in effect in many countries,[1,4] would prescribe "practical, repeatable methods to measure the distracting effect of these devices and reliable benchmark levels of unacceptable performance" (Ref. 1, p. 7). These should be consistent for original, aftermarket, and portable devices, and should not stifle product innovation.[1] The state of the art has progressed, as can be witnessed in Part 3 of this book, but more research on this topic is needed to achieve consensus on many of the performance-related guidelines and standards.

The following main issues are identified with regard to the *content* of existing design and performance standards:

- One main concern is the current lack of scientific knowledge to provide unequivocal assessment and robust compliance criteria (see Chapters 22 through 25 and Chapter 33). Further research is needed to develop new principles and performance criteria for good interface design, and to confirm and refine those that already exist.
- There remains a research gap in how to combine individual methods for assessing the level of distraction deriving from interaction with in-vehicle information system (IVIS) technologies into an overall integrated methodology to make predictions about safety in use. Stevens (see Chapter 22) outlines the issues that would need to be addressed in doing so.
- There is a particular need for more detail among the elaborations supporting the principles, including more emphasis on concise, unambiguous, and traceable references (Chapter 23). Insufficient detail may lead to variations in design and a lack of reliability between experts. Guidelines and standards still range from those that are more precise and detailed to those that are less prescriptive and rely on expert judgment. When existing guidelines and standards are less prescriptive, it is usually because there exists insufficient scientific research and data. Some degree of expert judgment will always remain necessary.
- Guidelines and standards should themselves be designed from a user-centered development perspective, and properly evaluated to ensure they are valid, reliable, effective, user friendly, and acceptable to end users.[5]
- Current guidelines have been developed principally to reduce the amount of distraction deriving from driver interaction with vehicle information and communication systems. Consequently, the primary focus of these guidelines is on the ergonomic design of systems, unrelated or secondary to driving, rather than on the design of systems that are integral to driving itself; although many of the principles contained within them are relevant to the design of interfaces for driving-related systems. As noted in Chapter 3, driving is a multitask activity and, as such, some elements of one driving task (e.g., changing gears, reading a speedometer) may distract the driver from elements of another (e.g., monitoring for traffic hazards). It would seem appropriate, therefore, to expand the scope of existing guidelines and standards to incorporate design guidance that seeks to minimize distraction from driving-related systems that could distract the driver from other driving-related activities.

Akumatsu (Chapter 24) and Stevens (Chapter 22) discuss the role that checklists, in addition to guidelines and standards, can play in enabling experts to make rapid and structured assessments of key safety-related features of an IVIS in a standard way, to identify where more detailed assessments may be required, and to determine whether human interaction with the device complies with the principles contained within industry guidelines. As Akamutsu points out, however, automotive companies often have their own checklists for in-vehicle human machine interface (HMI) design, which are usually confidential and are not made available to the public. The Transport Research Laboratory checklist, described in Chapter 21, would seem to provide a good platform for the development of a more globally acceptable checklist that incorporates what vehicle manufacturers have learned. Another promising example is the SafeTE checklist, which was developed in Sweden by Volvo Technology and the Swedish National Road and Transport Research Institute (VTI) for the Swedish Road Administration.[6] Table 15.5 in Chapter 15, provides a framework for a checklist that would focus more specifically on the assessment of distraction.

In Chapter 33, we invited two representatives of the automotive industry to provide their views on driver distraction. The following, general recommendations are distilled from the contributions made by Eckstein (BMW) and Hammer (GM Holden) with regard to design-phase countermeasures and the *implementation* of guidelines:

- The availability and presentation of information and communication functionalities while driving is a responsibility shared by many stakeholders. Specialist suppliers must increasingly share responsibility for designing systems that minimize distraction.
- All stakeholders need to develop information and communication systems to the same standard, independent of their functionality and degree of integration.
- The development of HMI design guidelines, such as the ESoP, AAM, and JAMA guidelines, is not in itself sufficient to ensure that distraction is minimized. Accompanying measures are needed to ensure that all system types (vehicle integrated, aftermarket, and nomadic) are designed in accordance with the guidelines and that drivers install and use them in a responsible manner. The final report of the e-Safety Working Group on HMI[7] summarizes those measures deemed necessary in Europe.
- Automobile manufacturers and suppliers should actively contribute to HMI research, standardization, and guideline development. The automotive industry, in particular, needs to be proactive and to establish agreed-on, voluntary design guidelines where this has not been done.
- Cars are no longer designed for a single country, so HMI guidelines need to be made available as harmonized global industry guidelines to achieve real societal road safety benefits.
- Smaller countries need government partnerships to help defray the large capital investments required for test facilities. Sweden is considered an exemplar in this area.

Eckstein argues, in Chapter 33, that road authorities should in several ways play a decisive role in limiting distraction from vehicle technologies as follows:

- Ensuring that HMI design guidelines are effectively disseminated, known, and used by all responsible stakeholders

- Providing general information to drivers on safe use of in-vehicle information and communication systems
- Promoting self-commitment, by the manufacturers and suppliers of after-market and nomadic devices, to comply with automotive HMI design guidelines such as the ESoP
- Monitoring the impact of HMI design guidelines on the market for after-market and nomadic devices
- Evaluating the safety impact of in-vehicle information and communication systems by collecting and analyzing crash and naturalistic driving data
- Taking measures to ensure (a) secure fixing of devices in accordance with relevant guidelines; (b) hands-free use of nomadic devices, and (c) restricted access to drivers while playing movies, TV, and video games

Process-oriented standards. A process-oriented standard, or code of practice, outlines the iterative steps that should be undertaken in designing, developing, evaluating, and deploying a vehicle system, device, or product from a UCD perspective. Process-oriented standards already exist for the user-centered procurement, design, development, evaluation, and deployment of systems in other domains, such as the military.[8]

Like Ekstein and Hammer, Burns (Chapter 23) notes that existing voluntary design guidelines and standards are insufficient by themselves. Burns argues that a process-oriented safety management systems (SMS) approach is needed to (a) ensure that designers are aware of guidelines and standards and have the resources and expertize to apply them, (b) ensure that the risks of driver distraction are routinely considered within the product development process, and (c) ensure that safety and usability testing in the automotive industry is prioritized and implemented more effectively. An SMS approach resembles the automotive industry process standards except that the targets are product "safety" rather than "quality." SMS stems from James Reason's work on managing organizational risk and human error.[9]

As yet, such a standard does not exist in the automotive sector. Transport Canada is promoting an SMS approach, working with the automotive industry and other relevant stakeholders to ensure that systems entering the market meet certain minimum requirements (Chapter 23). The parliament of Victoria also recommended in its Report on the Road Safety Committee Inquiry into Driver Distraction[10] that Australia go down a similar path. In the absence of process-oriented standards, there will remain considerable variability in the ergonomic quality of vehicle cockpit system interfaces across makes and models of vehicles. The development of such a voluntary process-oriented standard, or code of practice, is certainly an important priority for countermeasure development. Key components of the SMS approach[11] include voluntary commitments to:

- Define clear accountabilities and responsibilities that commit senior management to safety policies, measurable safety objectives, and clear organizational responsibilities and accountabilities for safety
- Monitor safety performance to comply with best practice human factors and ergonomic guidelines, checklists, and standards
- Develop a safety assessment process

- Communicate the SMS to all relevant parties
- Hold safety training
- Perform periodic audits
- Document the SMS

32.2.2.2 Guidelines and Standards for Advanced Driver Assistance Systems

Advanced driver assistance systems (ADAS) that automate partly, or fully, some driving-related functions and tasks (e.g., forward collision warning (FCW), lane departure warning, intelligent speed adaptation (ISA), in-vehicle navigation systems) provide another mechanism by which driver workload and distraction can be reduced—provided, of course, the systems are well-designed and evaluated. Unfortunately, very few guidelines and standards currently exist for the ergonomic design and evaluation of such systems (but see Campbell et al.[12]).

ISA, for example, is a system for which there currently exists no human factors design and evaluation guidelines or standards. This system has been shown, in numerous studies, to be highly successful in reducing speed and in conferring a range of other safety benefits,[13–15] is available in several countries as a commercial aftermarket product, and is being strongly promoted by many governments around the world. Little is known, however, about the extent to which this system moderates driver workload and distraction and to what extent this varies among different interface design configurations. It is possible that such systems, even though they could reduce crash risk, might increase driver workload and distraction if poorly designed. Is it better, for example, to provide the driver inside the vehicle with a continuous display of the posted speed limit along the road network, or to warn the driver only when the posted speed limit has been exceeded? The first design strategy is more likely than the second to divert drivers' attention away from activities critical for safe driving (assuming the ISA display is visual), but in doing so may make drivers less likely to want to continue to monitor redundant information from the speedometer and external speed signs (everyday driving activities which, in themselves, could divert drivers' attention away from activities critical for safe driving). The second design strategy is less likely than the first to divert drivers' attention away from activities critical for safe driving, but it may encourage drivers to continue to look at the speedometer and at external speed signs, activities which, as noted, could divert their attention away from activities critical for safe driving. Although existing guidelines, reviewed in Part 7 of this book, provide some general guidance relevant to the design of interfaces for ADAS, the focus of these, as previously noted, is on information and communication systems. Clearly, more suitably tailored guidelines and standards are needed for the design of ADAS technologies, which are derived from a proper understanding of the manner and contexts in which they are used.

More generally, as noted in Chapter 3 (see also Refs. 2 and 3), technological developments are making it possible to display to the driver, inside the vehicle, driving-related information currently displayed on signs and other media outside the vehicle. ISA, noted earlier, is capable, for example, of displaying to the driver inside the vehicle the posted speed limit along all segments of the roadway. Systems also exist, which are capable of displaying inside the vehicle a host of other information currently found on guide, warning and regulatory signs outside the vehicle. The information can be displayed to the driver any time, in any sensory modality, and in any design configuration. This is an interesting development. If the display of information from outside the vehicle is well designed and integrated with information already inside the vehicle, it has potential to reduce overall driver workload and, in turn, vulnerability to distraction. If not, drivers may find themselves overwhelmed by spatially and temporally overlapping and redundant information from within and outside the vehicle, which has potential to distract, overload, and confuse them. Little, however, is known about the impact on driver workload and distraction of the in-vehicle echoing of external information, and very little guidance exists to inform design efforts (one exception being Saad and Dionisio[15]). Clearly there is a need for further research in this area to support guideline and standards development efforts.

32.2.2.3 Human-Centered Design Process as a Countermeasure

An important, overarching countermeasure for distraction is use of a human-centered design process. The ISO standard for the UCD process (ISO 13407)[17] aims to help those responsible for managing hardware and software design processes to identify and plan effective and timely UCD activities. It defines a general process for including human-centered activities throughout a development life cycle, but does not specify exact methods. The methods for human-centered design and evaluation of potentially distracting in-vehicle systems and functions must therefore be defined to fit with this more general process.

A critical part of the UCD process is evaluation, ideally through testing with actual users. Any development process, which claims to have met the recommendations in ISO 13407 shall specify (a) the procedures used, (b) the information collected, and (c) the use made of the results. Within the automotive domain, there has been a proliferation of methods and metrics, which have been used at different stages of the UCD process to inform and refine the design of systems and functions to limit the impact of distraction on driving performance (see Part 3 of this book). However, there is currently little consensus regarding which assessment methods and metrics should be used for the evaluation of particular activities with potential to distract (see Chapter 7), and at which stages of the UCD process particular methods are most appropriate to use.

The outputs of projects such as Human Machine Interface and the Safety of Traffic in Europe (HASTE),[13] Adaptive Integrated Driver-vehicle Interface (AIDE),[18] and Crash Avoidance Metric Partnership (CAMP)[19] provide some initial guidance in assessment. Rapid evaluation procedures are needed that are cost-effective, easy to use by designers, and do not require sophisticated equipment and months of time. The lane change test is such a procedure although, as is noted in Chapter 25, the notion that a single, low-cost, test can assess the interference with driving of a competing task, regardless of its visual, auditory, cognitive, and psychomotor demands, may be unrealistic. Table 15.5 in Chapter 15 of this book, identifies mechanisms that moderate the effects of distraction that might be considered in refining the sensitivity and scope of existing assessment procedures, including checklists. In future, rapid evaluation procedures might include computational models of driver performance such that preliminary design concepts can be evaluated more quickly and earlier in the design and development cycle. Some models of this kind already exist (see Chapter 25). There also remains a need to develop reference tasks that provide benchmarks against which the impact on driving performance of distracting activities (driving related and nondriving related) can be established. These must be unambiguously defined, repeatable across different test environments, and induce mechanisms of distraction identical to those induced by the distracting activity under investigation.

32.2.2.4 HMI Integration as a Countermeasure

As the number of systems and functions that interact with the driver increases, so to do the number and complexity of input/output (I/O) devices and associated driver behaviors. HMI integration needs to be achieved to realize the full potential of intelligent vehicle technologies and is now the subject of some major research initiatives, for example the AIDE project,[20] the Integrated Vehicle-Based Safety Systems (IVBSS) project,[21] and the Integrated Safety System (INSAFES) project.[22] HMI Integration leads to many technological challenges. For example, packaging problems, a many-to-many mapping between applications, I/O devices, reconfigurability, and scalability. There are many human factors-related reasons for integration. For example, for solving problems associated with presentation of multiple simultaneous warnings[21,22] and problems of driver interaction with multiple different systems. Perhaps the main benefit from integration is achieved by using real-time centralized HMI management to resolve presentation conflicts between applications, for example, conflict between a telephone call and a navigation message presented simultaneously.

32.2.2.5 Safe Integration of Nomadic (Portable) Devices as a Countermeasure

A special case of integration that needs specific attention because of the distraction problem is that of the safe integration of nomadic (portable) devices; that is, mobile devices such as smartphones, navigation systems, and music players. The safety problems associated with nomadic devices are well recognized (see, e.g., Part 5 of this book), and integration of cell phones is in part even regulated through legislation (see Chapters 11 and 30). However, the benefits of different integration implementations need to be better understood. The assumption that integrated systems are necessarily better than non-integrated systems may not always be true and warrants more research. For example, Transport Canada[11] has called for a limit to the implementation of open architectures for portable devices brought into the vehicle. It is important to note that nomadic device integration can be achieved at different levels of integration, ranging from partial to full integration: physical integration (crashworthiness, power), synchronization of information (e.g., using Bluetooth technology), integration with driver–vehicle-environment (DVE) state (see, e.g., Ref. 20), audio integration, text/menu integration, vehicle integration (e.g., car-area-network [CAN]), visual information integration of nondynamic images, and visual information integration of dynamic/moving images.

One main research initiative within this area is taking place within the European AIDE project.[20] Work in AIDE on nomadic device integration involves both the technical development of possible solutions for nomadic device integration as well as the establishment of the nomadic device forum (that brings together vehicle manufactures, electronic device manufacturers, telecom industries, public authorities, and other stakeholders). To date, technical results include the identification of key integration "use cases" and basic architectural requirements. Moreover, a potential integration solution, based on a Bluetooth protocol, has been identified. One current on-market example of nomadic device integration is Ford Sync.

32.2.2.6 Incentive Schemes as Countermeasures

Under a voluntary regime, even the most ergonomically designed vehicle cockpits and technologies will be ineffective in limiting the effects of distraction if there is no demand for them by consumers. There exist, however, many mechanisms for stimulating voluntary demand by consumers for vehicles and technologies that optimize safety.[23,24] One option, suggested by Hedlund et al.,[1] that is likely to be effective, is to make bonus safety points available to vehicle manufacturers for electronic systems that have been assessed as meeting minimum requirements for limiting distraction. Such incentive schemes already exist for other technologies (e.g., seatbelt reminders, air bags) under the European new car assessment program (EuroNCAP) and similar programs operating in other jurisdictions.

32.2.3 Real-Time Distraction Countermeasures—Preventing and Mitigating Distraction While Driving

This section focuses on distraction countermeasures that actively operate in *real-time* while driving, so-called "real-time distraction countermeasures" (RDC). Engström and Victor (Chapter 26) define two main classes of existing RDC functions—real-time distraction *prevention* (with the focus on workload management) and real-time distraction *mitigation* (with the focus on distraction warning/feedback).

32.2.3.1 Real-Time Distraction Prevention

Real-time distraction prevention countermeasures, also commonly known as *workload managers*, include functions that serve the main purpose of preventing mental overload and distraction from occurring in the first place; for example, by prioritizing and scheduling system-initiated information according to the current driving situation or driver state (see Chapters 26 and 27 for details). The most common real-time distraction prevention functions include the following:

- *Information scheduling*, with the general purpose of ensuring that the driver receives information only when it is needed and when the driver is available to receive it (Chapters 26 and 27)
- *Demand-based advisories*, which are issued to discourage the use of functions such as MP3 music players, manually dialing 7- or 10-digit phone numbers, reading navigation maps, and using turn-by-turn navigation systems (Chapter 27)
- *Function lockout*, which involves the entire disabling of a function or subfunction in certain conditions (Chapters 26 and 27)
- *Adaptation of information format*, which involves altering the way the information is presented (not just the timing) to the current context (i.e., the driving situation or the presence of concurrent messages) (Chapter 26)

Although the majority of work within this area has been on embedded in-vehicle systems, such as the on-market systems in Volvo and SAAB Cars, it is important to note that nomadic-device-based solutions can also implement distraction prevention functions. For example, a cell phone service provider can take messages for the driver when they detect the phone is being driven, or require the person to confirm they are not driving before they answer or make a call. Further, shared solutions, which involve embedded-in-vehicle and nomadic-device-based solutions, are also possible. Currently, the area of context-aware, vehicle-integrated nomadic devices is attracting considerable research interest.

32.2.3.2 Real-Time Distraction Mitigation

Real-time distraction mitigation countermeasures are functions that serve the main purpose of mitigating distraction once it occurs. For example, by redirecting drivers' attention to the relevant aspects of the driving task (see Chapter 26). These functions provide feedback to help the driver shift attention back to driving when she/he is judged as being "too distracted" according to predetermined criteria set by the system, the driver, or the owner. The idea is to help the driver realize that he or she is being "tricked" into a distractive behavior. Two main functions are under development (e.g., in Volvo, SAAB, and in the AIDE and Safety Vehicles using Adaptive Interface Technology [SAVE-IT; see Chapters 27 and 28] projects) as follows:

- The *visual distraction alert*. The basic idea here is to help the driver to realize that she/he is glancing away from the road for too long or too often, and to "train" the driver to recognize a limit. As such, it alerts the driver to inappropriate behavior, and does not necessarily have a direct coupling to driving performance deterioration.
- The *cognitive distraction alert*. Feedback is given to the driver in situations where the driver is cognitively distracted, that is, when excessive attention is directed to internal thoughts or auditory content. The feedback can be given to wake the driver up, giving a reminder to increase scanning behavior.

Although research into real-time distraction mitigation functions is quite active, the field is still rather immature. In general, more research is needed to determine the effectiveness of distraction mitigation functions in improving performance and safety, in the short and long term.

32.2.3.3 Driver–Vehicle-Environment Adaptive Collision Warning Functions as Distraction Countermeasures

Driver-vehicle-environment (DVE) adaptive collision warning functions adapt warnings to certain states of the driver, vehicle, or environment (see Chapters 26 and 28). By doing so, they primarily improve the warning functionality with regard to effectiveness and acceptance. DVE states that warrant adaptation of warnings include eyes-off-road, visual time-sharing, cognitive distraction, high driving demand, driver impairment, driver intent of maneuvering, high traffic risk, and driver characteristics. Possible types of adaptation include altering the timing, intensity, duration, complexity, and modality of warnings. The primary role of the warning adaptivity is not to counteract distraction *per se*, but rather to optimize collision warning functionality. For example, a static warning will likely warn a driver too late of an unexpected event that occurs when the driver is distracted because of the increased RT associated with unexpected events.

Most research has been performed on distraction-adaptive (DA) forward collision warning (FCW) (see Chapters 26 and 28). Such FCW systems generally adjust the warning timing based on where attention is allocated, although other types of adaptation (e.g., of warning intensity) are possible as well. Other DA functions have also been tested: DA lane departure warnings either cancel a warning when the driver is attentive or provide an earlier lane departure warning if distraction is detected; DA curve speed warnings involve issuing an earlier warning when distraction is detected; and DA adaptive cruise control involves automatically and gradually increasing the headway to the vehicle in front or changing the set speed if distraction is detected to allow for longer reaction time.

In summary, the RDC research field is very active, with some first-generation products already available on the market. However, the field is immature and to a large extent technology-driven. There is still little consensus regarding which RDC functions are the most efficient and useful. The full benefits of adaptive functions are not yet clear.

32.2.4 Summary—Vehicle and Technology Design Countermeasures

Tingvall's "integrated safety chain" (see Figure 33.1 in Chapter 33) provides a perspective on how best to proceed in addressing the distraction problem as it relates to the use of technology in vehicles. Tingvall differentiates in Chapter 33 between five stages in the sequence of events leading up to a crash: "normal driving," "deviation from normal driving," "emerging situation," "critical situation," and "crash unavoidable." The distraction countermeasures outlined here can be placed within each of these stages to provide a holistic view.

With regard to the material reviewed in this book, distraction countermeasures in the "normal driving" stage are achieved by the following:

- *Design-phase distraction countermeasures*, such as the use of guidelines and standards, to reduce distraction and workload from interactions with in-vehicle, aftermarket, and nomadic devices
- *HMI integration*, including nomadic device integration
- *Real-time distraction prevention functions* that serve the main purpose of preventing mental overload and distraction from occurring in the first place; for example, by prioritizing and scheduling system-initiated information according to the current driving situation or driver state

Various systems have been developed to support drivers who momentarily deviate from normal driving, which may occur because drivers are inattentive, distracted, or in some other state. Some systems were not originally designed *per se* as distraction mitigation systems (e.g., forward collision warning, lane departure warning, intelligent speed adaptation), but have the potential to mitigate the effects of distraction by warning drivers when they fail to self-regulate in response to distraction (e.g., if they exceed the speed limit; if they get too close to the vehicle ahead). Others, such as the real-time distraction mitigation functions (the visual distraction alert and the cognitive distraction alert; see Chapters 26 and 28) have been designed specifically to warn drivers when they are driving in a distracted state that warrants intervention.

Tingvall describes an "emerging situation" as one that is less transient than a deviation from normal driving. Here the driver may not be aware of a sudden event that has potential to threaten safety—the driver may be drifting off the road or may be rapidly approaching a vehicle ahead. As discussed in Chapter 26, more aggressive real-time distraction warnings could be issued at this point. However, DA collision warning functions can alter the timing, intensity, duration, complexity, and modality of warnings. For example, an FCW is delivered earlier if the driver is distracted.

A "critical situation" is one where, for whatever reason, previous defenses in the integrated safety chain have failed and, in response to an impending crash, there is loss of control of the vehicle by the driver. Here immediate correction is required. Active safety systems, such as electronic stability control, lane-keeping assistance with active steering, and emergency braking assist, are available to support the driver to regain control of the vehicle and avoid a crash that might be attributable fully, or in part, to distraction. The sensors that comprise these systems can also be used to prime the early activation of passive safety systems.

At the tail end of the integrated safety chain, when a "crash is unavoidable," the distracted driver can benefit from the crash mitigation effects (e.g., lower impact speed) of previous countermeasures such as electronic stability control, lane-keeping assistance with active steering, and emergency braking assist. At impact, the driver still has many passive vehicle safety features to rely on for protection. A forgiving road infrastructure can also provide additional occupant protection at this stage of a crash.

The integrated safety chain (and other similar models) is useful not only in helping to identify where and how in the crash sequence existing and emerging driver distraction countermeasure technologies can be effective in protecting the driver. It is also useful in creating conceptual links between the operation of these technologies and other vehicle active and passive safety technologies that exist to protect the driver at successive stages in the chain and in creating conceptual links at each stage of the chain between driver, vehicle, and road infrastructure-related countermeasures.

In summary, the automotive industry has been proactive in developing a wide range of countermeasures to prevent and mitigate the effects of distraction. However, as an industry, it cannot be expected to shoulder the burden of good design in limiting distraction. Driving is a complex, multitask activity, and elements of the driving task itself (relating to both vehicle control and roadway monitoring) have the potential to divert the attention of the driver away from activities critical for safe driving. Responsibility for mitigating, through driver-centered design, the propensity for competing tasks to divert attention away from activities critical to safe driving is therefore a joint responsibility shared by multiple stakeholders: vehicle manufacturers, aftermarket suppliers, nomadic device suppliers, and road authorities. In the final section of this chapter, we consider what can be done to design the road environment to limit distraction.

32.3 ROAD DESIGN COUNTERMEASURES

32.3.1 Background

Data from crash studies, reviewed in Chapter 16, suggest that ~30% of distraction-related crashes derive from the driver being distracted by sources outside the vehicle. Known sources of distraction external to the vehicle that have been identified as contributing factors in crashes were identified and categorized in Chapter 15. These include animals, architecture, advertising billboards, construction zones/equipment, crash scenes, incidents (e.g., road rage, near-misses), insects, landmarks, road signs, road users, scenery, vehicles, and weather (e.g., lightning). The potential impact of these sources of distraction on driving performance and safety can be moderated, to varying degrees, through road design.

Relative to the amount of research on sources of distraction deriving from inside the vehicle, far less has been done in relation to external sources of distraction. Most of the relevant work, reviewed in Chapter 13, has focused on billboards, advertising signs, and, to a lesser extent, traffic signs.

32.3.2 Options for Road-Design-Based Countermeasure Development

As noted in Chapter 13, there is currently a lack of research evidence on which to form guidelines or standards about how much distraction from outside the vehicle is "safe." There are, however, some general recommendations for countermeasure development that can be made at this point in time.

Methods are needed for identifying sources of distraction on or near roads that adversely affect driving performance and safety, or have the potential to do so. General approaches for identifying sources of distraction with potential to compromise safety were described in Chapters 15 through 17. More specific guidance, relating to sources deriving from outside the vehicle, is provided in Chapter 13. Road safety audits, for example, which are routinely undertaken in Australia and other parts of the world, could include criteria for the identification and assessment of roadway-related activities, objects, and events that could distract drivers and degrade driving performance and safety.

There is a need to develop a taxonomy of those objects, events, and activities on or near road reserves that have the potential to distract. The taxonomy proposed in Chapter 15 provides an initial attempt at categorizing sources of distraction outside the vehicle currently known to be contributing factors in crashes and near-crashes, including those on or near the road reserve.

Needed are methods and metrics for assessing how sources of distraction from outside the vehicle impact driving performance. Work in this area is still in its infancy relative to that undertaken in relation to distractions deriving from inside the vehicle. As for the assessment of in-vehicle distraction, there is a need to develop reference tasks, which induce "acceptable" levels of distraction against which the impact of external distractions on driving performance can be assessed. Some initial thoughts on this issue are contained in Chapter 13.

Data are needed on the effects on driving performance and safety of external sources of distraction, individually and in combination. Although limited data exists on the impact of static and dynamic advertising billboards on driving performance, the data is inconclusive (at least for static billboards) and there is a scarcity of published research on the impact on driving performance and safety of other sources of distraction external to the vehicle. Even for advertising billboards, the focus has been on those erected as dedicated structures at specific locations. Nothing is known about those that are placed on the sides of buildings and on the backs of buses, taxis, and other moving objects.

There is some limited evidence, reviewed in Chapter 13, that both younger and older drivers are more vulnerable to the effects of distractions deriving from outside the vehicle. Given that distraction is a joint property of the demands of driving and competing tasks (see Chapter 4), UCD of the road environment to support the driving task has potential to reduce driver workload and, in turn, reduce driver vulnerability to distraction—not just for young and elderly drivers, but for all road users. Reducing the demands of the roadway, however, may not be sufficient. The roadway needs to be designed in such a way that the demands can be anticipated so that drivers can devote attention to the road when needed. Although some guidelines for UCD of roadways for the elderly have been developed,[25] no known guidelines exist in relation to young novice drivers.

Little specific guidance exists for the UCD of the traffic management system to minimize driver workload for the population at large, although some general guidance on human factors issues relevant to the design, operation, and evaluation of the road environment can be found in the literature.[26–30] Self-explaining roads are a promising development.[30,31] These are designed to increase the likelihood that a driver will automatically adopt appropriate speed and steering profiles without depending on road signs. "The geometric features of the road encourage the desired driver behavior, and do not rely on the driver's ability or willingness to read and obey road signs. A perfect self-explaining road would not require speed limit signs and curve advisory signs" (p. xii).[30] Burns[32] suggests that self-explaining roads, although used mainly as a tool for speed management, may also play a role in managing distraction. Road traffic environments that are more or less tolerant of driver distraction could be designed to make this tolerance self-evident, although drivers may compensate for greater levels of tolerance by engaging more in distracting activities. Alternatively, roadways could be designed to convey an illusion to drivers that they must be vigilant, making the road look intolerant to distraction, when in reality it is very tolerant to frequent inattention. Such an illusion might make drivers generally attentive to the road.

The degree to which the traffic engineer can prevent and mitigate the effects of specific distractions deriving from outside the vehicle, which are known to contribute to crashes and near-misses (see Chapter 15), varies according to the source. There are, however, some sources of distraction known to be contributing factors in crashes that would appear to be amenable to intervention by traffic engineers.

Animals. On or near stretches of road where it is known that roadway incursions by animals are problematic, warning signs (e.g., "Kangaroos next 5 km") and even barriers could be used to minimize the likelihood of an interaction between drivers and animals that leads to distraction. Such countermeasures have already been implemented in many countries, to prevent physical contact between vehicles and animals.

Architecture. It is possible to visually mask (e.g., with trees) prominent architectural structures likely to distract drivers or to reroute traffic away from them, although such countermeasures may not always be practical to implement. Scenic routes are, by definition, set up to distract drivers. Not only do they contain architecture, prominent landmarks, and other things attractive to drivers, but they are often located in rural environments, with narrow, winding, roads that increase workload and hence increase driver vulnerability to distraction. If such routes are to be promoted, it is incumbent upon relevant stakeholders to ensure, as a duty of care to drivers, that drivers are aware of the risk of distraction and that road design measures are put in place to minimize distraction and its effects on driving performance and safety. Basic measures, such as reducing the speed limit along scenic routes, would reduce driver workload, provide drivers with more time to recover from the effects of distraction, and reduce impact forces in the event of a distraction-related crash.

Advertising billboards. Some guidelines and checklists assess whether a proposed advertising sign or billboard is likely to pose an unacceptable risk to drivers. A checklist developed for this purpose by the roads and traffic authority in the Australian state of Victoria (VicRoads), is presented as an example in the appendix to this chapter, along with their operational requirements for installing variable message signs, used for displaying advertisements. Other examples can be found in a report on the findings of an Australian parliamentary inquiry into driver distraction.[10] Given the paucity of research data on the effects of advertising signage (with respect to both design and location), on driving performance and safety (and during and after driver exposure to the advertising material), such checklists and guidelines are generally based on accumulated wisdom and past traffic engineering practice. Clearly, there is a need for more prescriptive guidelines, checklists, and regulations controlling the location, size, and content of advertising on or near road reserves. Further research is needed to inform and support the development of these.[33]

Crash scenes. So-called rubber necking is a commonly observed behavior in the vicinity of crashes. It is possible to route traffic away from crash scenes or to visually mask the scene in some way, although these countermeasures may not always be practical or easy to implement. Reducing traffic speed in the vicinity of crash scenes would reduce driver workload, provide drivers with more time to recover from the effects of distraction, and reduce impact forces in the event of a crash.

Landmarks. It is possible to visually mask (e.g., with trees) prominent landmarks most likely to distract drivers or to route traffic away from them; but, again, such countermeasures may not always be practical to implement. Indeed, it may be counterproductive to do so if certain landmarks are critical for route finding. Masking them may lead drivers to invest more effort in trying to locate the landmarks, or in choosing new ones, which may lead to greater distraction.

Road signs. Most jurisdictions have in place specific guidelines and standards for the location, design, and use of traffic signs on roads. Traffic signs, even though they directly support the driving task, can be sources of distraction. If poorly designed or located, traffic signs may prolong the diversion of attention away from activities critical for safe driving. If absent in locations where they should be (e.g., in the case of missing street signs and numbers), they may encourage drivers to adopt compensatory search strategies which divert attention away from activities critical for safe driving; and if poorly colocated, they may induce visual search competition or clutter, impairing visual search and diverting attention away from activities critical for safe driving. It is important that existing guidelines and standards, which regulate the location and design of road signs on road reserves are revised to minimize the potential for distraction from signs—together, and in combination. While there is very little specific research available on traffic signs and distraction to support this activity, there is available general guidance in the literature on USD of traffic signs[34] (Chapter 13).

Road users. Road users can distract drivers in different ways: by behaving unpredictably, erratically, or irresponsibly as pedestrians, riders, or drivers; by attracting attention by virtue of their appearance, size, or other defining feature; by attracting attention by virtue of the vehicle they are riding or driving; and by other means. Given that distraction is a property of the joint demands of driving and competing tasks, the road engineer basically has two options for limiting driver distraction deriving from other road users. First, as discussed earlier, UCD of the road traffic system has the potential to reduce driver workload and, hence, vulnerability to distraction from other road users. Another option is to prevent or minimize, where possible, driver interaction with other road users, especially in traffic situations in which drivers are known to be most vulnerable to the effects of distraction. The extant data suggests that distraction is largely associated with rear-end crashes, same travel-way/same direction crashes, single-vehicle crashes, and crashes occurring at night (see Chapter 16). High friction road surfaces (for rear-end crashes), median barriers (for same travel-way/same direction crashes), sealed road shoulders (for single-vehicle crashes), and improved lighting and delineation of roads at night (for crashes occurring at night) are examples of traditional traffic engineering countermeasures that are likely to mitigate the effects of distraction deriving from interaction between road users.

Weather. There is little the traffic engineer can do to prevent or reduce the distracting effects of weather phenomena. Minimizing sun glare is perhaps one exception. Appropriate routing of roads to minimize driver exposure to sun glare is a countermeasure that is likely to be effective, given that sun glare has been shown to be a contributing factor in distraction-related crashes (see Chapter 16).

32.3.3 A Distraction-Tolerant Road System

Ultimately, as underscored by Tingvall in Chapter 33, the aim should be to create a distraction-tolerant road system, at all stages of the "injury safety chain" (see Figure 33.1, Chapter 33), such that, in the event of a distraction-related crash (or, indeed, any other crash), no road user is killed or seriously injured. In the context of Figure 33.1, the countermeasures already described can be regarded as ones that aim to prevent or reduce the potential for distraction and, hence, promote "normal driving." At later stages of the crash sequence, however, the traffic engineer, like the vehicle engineer, has other countermeasures at his or her disposal to mitigate the effects of distraction. Tactile edge linings and real-time over-speed feedback warning signs, for example, can be used to provide feedback to distracted drivers who "deviate from normal driving," who are unaware that they are speeding or veering off the road. In "critical situations," if a crash is still avoidable, high friction road surfaces and sealed road shoulders can support the driver in avoiding a crash. At the terminal stage of the crash sequence (i.e., "crash unavoidable"), wire rope barriers and other road treatments can be used to minimize crash impact. As noted earlier, further research is needed to identify common crash and near-crash configurations and environmental conditions in which distraction is a contributing factor to enable the development of more targeted countermeasures at each stage of the integrated safety chain.

As noted in the previous section, technological developments are making it possible to display to the driver, inside the vehicle, traffic-related and nontraffic-related information currently displayed on signs and other media outside the vehicle. Little is known about the impact on driving performance and safety of this in-vehicle echoing of external information. Does the driver continue, for example, to search for speed limit signs outside the vehicle even though there is no longer any need to? Clearly there is a need for further research in this area.

Finally, there is a need for vehicle manufacturers to enter into a direct diallog with traffic engineers—to ensure that there are no incompatibilities in the spatial and temporal design of traffic messages and signals impinging on the driver from within and outside the vehicle that could increase the potential for distraction. Future roadways may even include infrastructure that communicates with the car, such that it can tell when a driver is distracted. If information regarding driver state can be communicated to the roadside, then road signs and traffic control devices could adapt to draw drivers' attention back to the road. As for vehicle cockpit design, there is a critical need for institutional arrangements and process-oriented design standards, which ensure that, at a macro level, there is mutual cooperation and cross-talk between the automotive industry, road authorities, local councils, suppliers, the aftermarket industry, and other relevant stakeholders to ensure that, through coordinated design, driver interaction with competing tasks deriving from inside and outside the vehicle (driving and nondriving related) does not lead to a breakdown in multilevel control processes at one or more levels of control (tactical, strategic, and operational).

32.4 DISCUSSION AND CONCLUSIONS

This chapter has provided an overview and discussion of numerous countermeasures that relate to the vehicle itself, the technology brought into, or added to, the vehicle, and the road environment in which the vehicle is traveling. The first section dealt with what designers of systems and functions can do to prevent distraction in the design-phase before manufacturing and during driving (in real time). The second section dealt with what designers of the road environment can do to counteract distraction.

Several developments in vehicle and technology design are likely to be effective as countermeasures in preventing and mitigating the effects of distraction:

1. Further development and refinement of distraction assessment methods and of design, performance, and process guidelines and standards
2. Increased commitment to the UCD process
3. Increased integration of the human-machine interface, including nomadic devices

4. Continued development and validation of RDC:
   - Real-time distraction prevention (workload management) functions such as information scheduling, demand-based advisories, function lockout, and adaptation for information format
   - Real-time distraction mitigation functions such as visual distraction alert and cognitive distraction alert
5. Continued development and validation of DA collision warning functions such as FCW, lane departure warning, curve speed warning, and adaptive cruise control

In this chapter we have also provided specific guidance for preventing and mitigating the effects of some specific distractions deriving from outside the vehicle that are known to contribute to crashes and which appear to be amenable to intervention by traffic engineers. There are, however, further countermeasures that are needed in this area: methods are needed for identifying sources of distraction external to the vehicle that do, or have the potential to, distract; a taxonomy of objects, events and activities that could distract must be developed (the taxonomy in Chapter 15 is only a starting point); methods and metrics for measuring the impact on driving performance and safety of external sources of distraction are needed; and traffic engineers need guidance on how to design, from a user-centered perspective, the traffic management system to reduce driver workload and limit distraction.

Effective USD of vehicles and technology is fundamental to the prevention and mitigation of distraction. There are, however, some possible downsides of improved design that must be considered. Ergonomically designed interfaces can, for example, encourage drivers to use them more often, thus increasing their exposure to risk (the so-called "usability paradox"; see Chapter 4); and the more automated is the driving task, the more "satisficing" (see Chapter 2) it will be, freeing up driver attention that can be used by drivers to take on other roles within the vehicle that may distract them. Effective USD requires continuous evaluation and feedback to identify and rectify such unintended side effects.

Vehicle cockpits are evolving rapidly, and the role of the driver, like that of the pilot, will change over time—from being the active controller of the vehicle to being more of a systems monitor.[27] Already there exist intelligent transport systems, which are capable of automating, partly or fully, all of the key components of driving: finding one's way; following the road; monitoring speed; avoiding collisions; following traffic rules; and controlling the vehicle.[35] These technologies will create new styles and modes of human-machine interaction. At the same time, they will make it possible to display inside the vehicle information currently displayed outside it; and in doing so, blur the distinction between what the driver sees and responds to inside and outside the vehicle. The implications of these developments from a distraction perspective are many and varied. Ultimately, they will provide an impetus for closer cooperation and shared responsibility between vehicle and road designers for ensuring that the design of the driving task to limit distraction is an integrated activity. Tingvall's integrated safety chain provides a conceptual roadmap for how that integration might proceed. Designing vehicles and technologies to limit the effects of distraction will remain a challenging area for countermeasure development.

ACKNOWLEDGMENT

The authors would like to thank Peter Burns, from Transport Canada, for reviewing and commenting on an earlier version of this manuscript.

REFERENCES

1. Hedlund, J., Simpsom, H., and Mayhew, D., International conference on distracted driving. Summary of proceedings and recommendations, *International Conference on Distracted Driving*, Toronto, Canada, 2006.
2. Regan, M. A., A sign of the future – 1: Intelligent transport systems, in *The Human Factors of Transport Signs*, Castro, C. and Horberry, T. (Eds.), CRC Press, Boca Raton, FL, 2004, pp. 213–224.
3. Regan, M. A., A sign of the future – 2: Human factors, in *The Human Factors of Transport Signs*, Castro, C. and Horberry, T. (Eds.), CRC Press, Boca Raton, FL, 2004, pp. 225–238.
4. Burns, P. C., Driver distraction countermeasures, *First International Conference on Driver Distraction*, Sydney, Australia, 2005.
5. Go, E., Morton, A., Famewo, J., and Angel, H., *Final Report: Evaluation of Industry Safety Principles for In-Vehicle Information and Communication Systems*, Report prepared for Transport Canada, Humansystems Inc., 2006.
6. Engström, J. and Mårdh, S., *SafeTE final report*, Report No. Report 2007:36, Swedish Road Agency (SRA), 2007.
7. Augello, D., Becker, S., Eckstein, L., Gelau, C., Hallen, A., König, W., Pauzie, A., and Stevens, A., *Recommendations from the eSafety-HMI Working Group – Final Report*, European Commission, 2005.
8. MIL-H-46855B, Human Engineering Requirements for Military Systems, Equipment and Facilities, U.S. Department of Defense, Washington, D.C.
9. Reason, J., *Managing the Risk of Organizational Accidents*, Ashgate, Hampshire, England, 1997.
10. Parliament of Victoria, *Inquiry into Driver Distraction—Report of the Road Safety Committee on the Inquiry into Driver Distraction*, Report No. Parliamentary Paper No. 209 Session 2003–2006, Parliament of Victoria, Melbourne, Australia, 2006.
11. Transport Canada, *Strategies for Reducing Driver Distraction from In-Vehicle Telematics Devices: A Discussion Document*, Report No. TP 14133 E, Transport Canada, Ottawa, Canada, 2003.
12. Campbell, J. L., Richard, C. M., Brown, J. L., and McCallum, M., *Crash Warning System Interfaces: Human Factors Insights and Lessons Learned*, Report No. DOT HS 810 697, National Highway Traffic Safety Administration, Washington, D.C., 2007.
13. Carsten, O., Merat, N., Janssen, W., Johansson, E., Fowkes, M., and Brookhuis, K., HASTE final report, human machine interface and the safety of traffic in Europe (HASTE) project, 2005.
14. Regan, M., Young, K., Triggs, T., Tomasevic, N., Mitsopoulus, E., Tierney, P., Healey, D., Tingvall, C., and Stephan, K., Impact on driving performance of intelligent speed adaptation, following distance warning and seatbelt reminder systems: Key findings from the TAC SafeCar project, *IEE Proceedings Intelligent Transport Systems* 53(1), 51–62, 2006.
15. Saad, F. and Dionisio, C., Pre-evaluation of the "mandatory active" LAVIA: Assessment of usability, utility and acceptance, *14th World Congress on Intelligent Transport Systems*, Beijing, China, 2007.

16. Campbell, J. L., Carney, C., and Kantowitz, B. H., *Human Factors Design Guidelines for Advanced Traveler Information Systems (ATIS) and Commercial Vehicle Operations (CVO)*, Report No. FHWA-RD-98-057, USDOT FHWA, 1998.
17. ISO 13407 Human-centered design processes for interactive systems, 1999.
18. Johansson, E., Engström, J., Cherri, C., Nodari, E., Toffetti, A., Schindhelm, R., and Gelau, C., Review of existing techniques and metrics for IVIS and ADAS assessment, Deliverable D2.2.1.,AIDE IST-1-507674-IP, 2004.
19. Angell, L., Auflick, J., Austria, P. A., Kochlar, D., Tijerina, L., Biever, W., Diptiman, J., Hogsett, J., and Kiger, S. (CAMP) *Driver Workload Metrics Project: Final Report*, Report No. DOT HS 810 635, National Highway Traffic Safety Administration, Washington, D.C., 2006.
20. Engström, J., Arfwidsson, J., Amditis, A., Andreone, L., Bengler, K., Cacciabue, P. C., Janssen, W., Kussman, H., and Nathan, F., Towards the automotive HMI of the future: Mid-term results of the AIDE project, in *Advanced Microsystems for Automotive Applications*, Valldorf, J. and Gessner, W. (Eds.), Springer, Berlin, 2006.
21. Resendes, R. and Ference, J., Integrated vehicle-based safety systems program plan, U.S. Department of Transportation, Washington, D.C., 2004.
22. Sjögren, A., D60.1 project presentation: Preventive and active safety applications (PReVENT) project, integrated safety system (INSAFES) sub-project, 2005.
23. Regan, M. A., Removing obstacles to the deployment of vehicle e-safety technologies, *Smart Demo 2005 Exhibition and Conference*, Victoria Park, Adelaide, Australia, 2005.
24. Tingvall, C., Paper presented in plenary session on roles of the industry, the government and the drivers, *20th Enhanced Safety of Vehicles Conference*, Lyon, France, 2007.
25. Fildes, B., Corben, B., Morris, A., Oxley, J., Pronk, N., Brown, L., and Fitharris, M., *Road Safety Environment and Design for Older Drivers*, Report No. AP-R169/00, Austroads, Sydney, Australia, 2000.
26. Castro, C. and Horberry, T., *The Human Factors of Transport Signs*, CRC Press, Boca Raton, 2004.
27. Fuller, R. and Santos, J. A., *Human Factors for Highway Engineers*, Pergamon, Amsterdam, 2002.
28. Ogden, K. W., Human factors in traffic engineering, In *Traffic Engineering and Management*, Young, W. (Ed.), Monash University Institute of Transport Studies, Melbourne, Australia, 2003.
29. Dewar, R. E., Roadway design, in *Human Factors in Traffic Safety*, Dewar, R. E. and Olson, P. L. (Eds.), Lawyers and Judges Publishing Company, Tucson, Arizona, 2002.
30. Keith, K., Trentacoste, M., Depue, L., Granda, T., Huckaby, E., Ibarguen, B., Kantowitz, B., Lum, W., and Wilson, T., *Roadway Human Factors and Behavioural Safety in Europe*, Report No. FHA Report FHWA-PL-05-005, Federal Highway Administration, Washington, D.C., 2005.
31. Theeuwes, J. and Godthelp, H., Self explaining roads: How people categorize roads outside the built-up area, *Road Safety in Europe and Strategic Highway Research Program*, Lille, France, 1995.
32. Burns, P. C., personal communication, 2007.
33. Smiley, A., Smahel, T., and Eizenman, M., Impact of video advertising on driver fixation patterns, Transportation Research Board, 2004.
34. Lansdown, T. C., Considerations in evaluation and design of roadway signage from the perspective of driver attentional allocation, in *The Human Factors of Transport Signs*, Castro, C. and Horberry, T. (Eds.), CRC Press, Boca Raton, 2004.
35. Falkmer, T. and Gregerson, N. P., The TRAINER project—The evaluation of a new simulator-based driver training methodology, in *Driver Behaviour and Training*, Dorn, L. (Ed.), Ashgate, England, UK, 2003, pp. 317–330.

APPENDIX 1: VICROADS' TEN-POINT ROAD SAFETY CHECKLIST TO ASSIST IN THE LOCATION OF NEW ADVERTISING SIGNS

An advertisement, or any structure, device, or hoarding for the exhibition of an advertisement, is considered to be a road safety hazard if it

1. Obstructs a driver's line of sight at an intersection, curve or point of egress from an adjacent property; or
2. Obstructs a driver's view of a traffic control device, or is likely to create a confusing or dominating background which might reduce the clarity or effectiveness of a traffic control device; or
3. Could dazzle or distract drivers because of its size, design, or coloring, or it being illuminated, reflective, animated or flashing; or
4. Is at a location where particular concentration is required (e.g., high pedestrian volume intersection); or
5. Is likely to be mistaken for a traffic control device, for example, because it contains red, green, or yellow lighting, or has red circles, octagons, crosses or triangles, or arrows; or
6. Requires close study from a moving or stationary vehicle in a location where the vehicle would be unprotected from passing traffic; or
7. Invites drivers to turn where there is fast moving traffic or the sign is so close to the turning point that there is no time to signal and turn safely; or
8. Is within 100 meters of a rural railway crossing; or
9. Has insufficient clearance from vehicles on the carriageway; or
10. Could mislead drivers or be mistaken as an instruction to drivers

VicRoads operational requirements for the installation of variable advertising message signs are that the sign

- Not display animated or moving images, or flashing or intermittent lights
- Not be brighter than 0.25 candela/m^2
- Remain unchanged for a minimum of 30 s
- Not be visible from a freeway
- Satisfy the ten point checklist

Source: From Parliament of Victoria, *Inquiry into Driver Distraction—Report of the Road Safety Committee on the Inquiry into Driver Distraction*, Report No. Parliamentary Paper No. 209 Session 2003–2006, Parliament of Victoria, Melbourne, Australia, 2006. With permission.

33 Government and Industry Perspectives on Driver Distraction

Claes Tingvall, Lutz Eckstein, and Mike Hammer

CONTENTS

33.1 INTRODUCTION

There are many potential sources of distraction in the driving environment. Many are neither new nor technical in origin. Although some cannot be avoided, drivers have, to a varying extent, some latitude in deciding where, when, and how to engage in potentially distracting activities. The development of effective countermeasures for preventing and mitigating the effects of distraction requires a concerted effort from multiple stakeholders in society, each of whom faces a different set of constraints in addressing the issue of distraction. It is important, in a book such as this, to understand the perspectives of these stakeholders. These perspectives represent an important complement to the research-oriented perspectives of most of the other chapters in the book.

The proliferation of entertainment and vehicle information and communication systems (VICS), entering the vehicle cockpit—Original Equipment Manufacturer (OEM), aftermarket, and nomadic devices—has served to highlight the driver's limited ability to perform competing tasks while driving and, hence, the potential for these technologies to distract the driver. Whether justified or not, these developments have focused particular attention on the automotive industry.

Vehicle manufacturers must balance many competing priorities when designing in-vehicle technologies: cost, consumer appeal, customer needs across different countries and cultures, and safety concerns. In doing so, they operate in a highly regulated environment, which has the potential to stifle innovation and their competitiveness within the global market.

Transport safety authorities must also balance many competing priorities in developing policies and programs to prevent and mitigate the effects of distraction. Their agenda is somewhat broader than that of vehicle manufacturers, and they have at their disposal a broader range of countermeasures to address the problem. Their policies must be evidence based and must balance the complex, and often conflicting, needs and desires of a broad range of stakeholders in society, including researchers, designers, suppliers, politicians, and their constituents.

In this chapter, we have brought together three complementary perspectives on the issue of driver distraction. In Section 33.2, Professor Claes Tingvall, Director of Traffic Safety at the Swedish Road Administration, articulates his vision for an integrated approach to the management of distraction as a safety problem that is based on the assumption that distraction cannot be eliminated and must, therefore, be regarded as a major design factor that requires consideration at every level of the road transport system. Following that, senior representatives from two prominent vehicle manufacturers—Mike Hammer, from General Motors Holden (Holden), based in Melbourne, Australia, and Dr Lutz Eckstein, responsible for Ergonomics and Human Machine Interaction within BMW Group, based in Munich, Germany—each provide their own unique perspectives on the issue of driver distraction.

33.2 DISTRACTION FROM THE VIEW OF GOVERNMENTAL POLICY MAKING

Claes Tingvall

The road transport system is complex, generally unsafe, and not energy efficient. It is also an open system, with a number of stakeholders involved as users and providers of services, infrastructure, and vehicles. The stakeholders are loosely connected to each other, mainly through standards and regulations, rather than through any overall structure or model. The legal requirements or standards imposed on each component of the system might be clear in some circumstances, but are mostly either lacking or fail to take relevant human behavior into account. Although it is clear in most countries what road users are expected to do, and especially, not do, the rules that apply to road users are seldom based on what can reasonably be expected from the human, or even enforced in a rational way. Laws that forbid drivers from using handheld phones, or from engaging in other potentially distracting activities, exist in most countries, and in the case of an accident, the road user can be brought to justice. Fines, temporary license withdrawal, imposition of demerit points, and even imprisonment, if an accident is severe, are punishments meted out for distracted driving. This "blaming the victim" approach,[1] which typifies the *status quo*, has unfortunately slowed down the introduction of effective countermeasures aimed at reducing and mitigating the consequences of distracted driving.

In more modern transport safety management policies, such as the European Traffic Safety Programme 2001–2010[2] or the Swedish Vision Zero,[3] the chains of responsibility, the safety philosophy, and the countermeasures are more closely linked to the limitations and capabilities of the road user. The endpoint for such a policy is to dramatically reduce, and even eliminate, serious consequences resulting from road crashes. The whole driving chain, from normal driving to crash, can in such a policy be treated as a process where intervention at any stage can take place. Although in the past, primary safety (sometimes misunderstood as active safety) and secondary safety (sometimes misunderstood as passive safety) were two more or less separate areas of prevention, safety developments are now taking place in what is sometimes called integrated safety, or the integrated safety chain.[4] The integrated safety chain is the sequence from "normal driving" to crash, broken down into stages of progression toward the crash (see Figure 33.1). The integrated safety chain can be used to identify possible interventions at all stages and combinations of stages. It is particularly useful in developing technologies in both vehicles and infrastructure and also for more general safety interventions like enforcement.

Before the integrated safety chain can be applied, however, it is essential that the requirements of the driver are well defined and understood by all involved in safety policy development. On the one hand, all drivers should be expected to fulfill the basic requirements: they should be sober, use fitted seat belts, and not drive faster than the posted speed limit. On the other, if these requirements are too demanding, we either end up in a situation where few humans can be allowed to drive, or we are back to the current situation where we turn a blind eye to reality and just blame the victim. These requirements seem easy to fulfill; yet there is still dramatic potential to fulfill them in most countries of the world. Extending these basic requirements to distraction—that is, to require all drivers not to be distracted at all times—is not only impossible, but is a major safety hazard for the road transport system. In doing so, it would give designers and those responsible for the operation of the road transport system a misleading safety principle on which they would base their actions. It would be tantamount to saying "Even if the thing you design is distracting, it doesn't matter—because, by law, people aren't allowed to be distracted."

It is time to rethink the basic design philosophies of the future road transport system. If the safety policy implies that all risk factors that cannot be eliminated should be taken into account in the design and functioning of the system, distraction must be one of these factors. The consequence of this policy would be a statement like "While trying to reduce driver distraction at all times, we must in all situations build our design and functionalities on the basis that the driver is distracted." Although this design philosophy seems to imply that all stakeholders have a role in reducing distraction, it has a major implication for both car manufacturers and

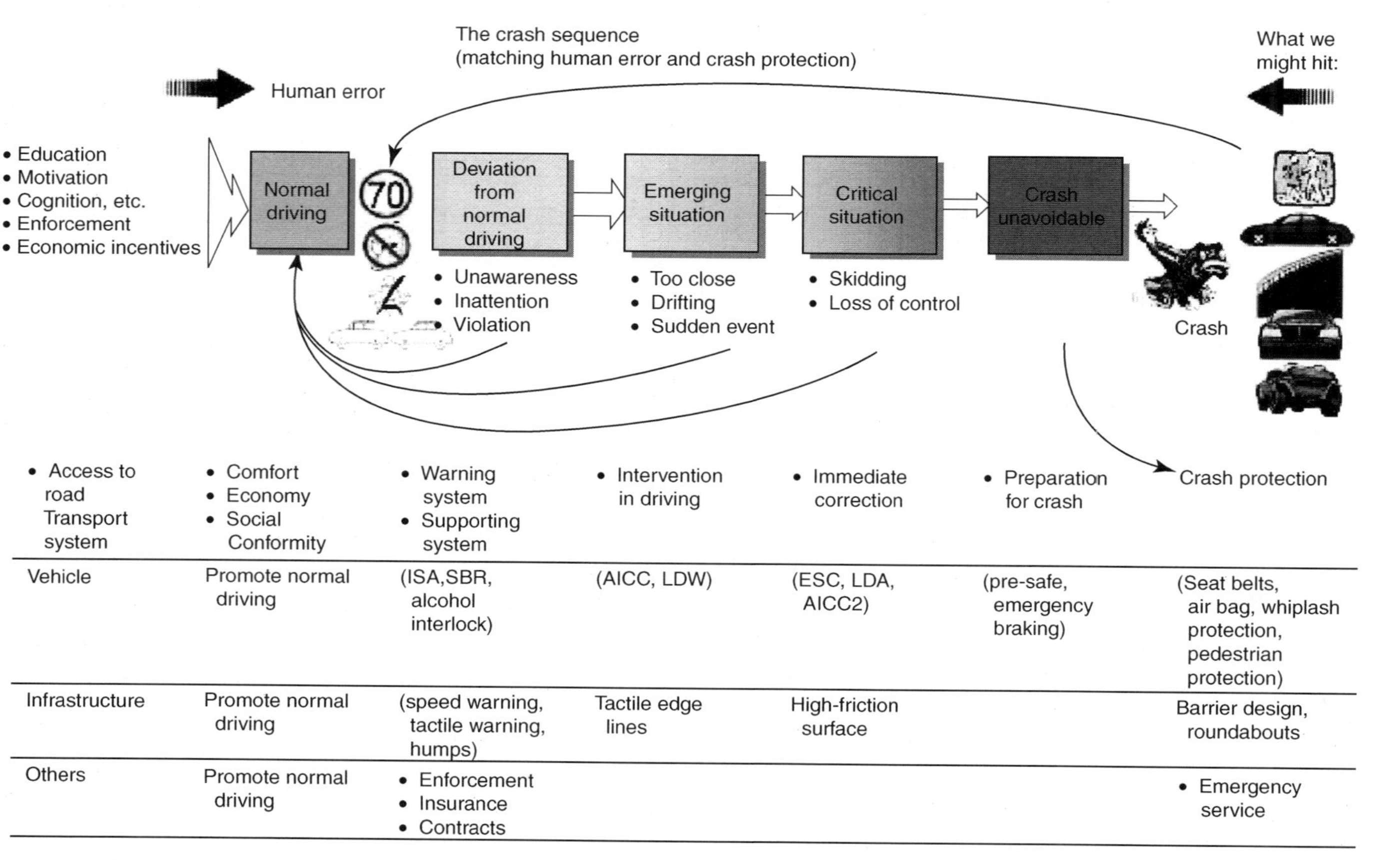

FIGURE 33.1 The integrated safety chain: matching human error and crash protection.

infrastructure designers. This task is not just limited to reducing the consequences of a crash for the distracted driver, but also to reducing the likelihood of crashes occurring as a result of distraction.

The starting point for the integrated safety chain (see Figure 33.1) is the support of "normal" driving. Supporting the driver in adhering to the posted speed limit, in driving with less than the prescribed legal blood alcohol concentration (BAC), and in using fitted seat belts are most important. Driving at safe following distance behind vehicles ahead is another important part of normal driving, which can also be supported with currently available systems.[5,6] The next step in the integrated safety chain is to help the driver to return to normal driving when he or she deviates from a safe driving mode. This could happen, for example, when a distracted driver drifts from the intended road path. Both the vehicle and the infrastructure could help the driver to regain control, and return to normal driving. Lane departure warning (LDW) systems or roadway tactile edge lining could do so via the provision of warnings. In the next phase of the integrated safety chain, the situation is more critical (e.g., the driver drives off the road), and the vehicle might partly take control from the driver if the driver is judged not to be in control any longer. Vehicle systems capable of doing so could include electronic stability control (ESC) and lane departure assist (LDA). Such systems could help the driver to regain control of the vehicle or prepare for the next stage. In the next stage, if the driver has not regained control, the vehicle can prepare for a crash—by, for example, automatically applying the brakes. In the final stage of the integrated safety chain—the crash—both the vehicle and the infrastructure can help to avoid serious injuries. In the event of a crash, both the speed of the vehicle and the posted speed limit must be in accordance with the capacity of the vehicle and the infrastructure to protect the vehicle occupants.

This integrated safety chain sits well with the problem of distraction, as it tries to stimulate the driver to regain control of the vehicle at various points along the chain. Used correctly, this approach will have dramatic safety effects in reducing and mitigating the effects of distraction; but it could also be misused by designers if the driver modifies his/her behavior and starts to depend on the systems. Therefore, it is imperative that new technologies designed to prevent, mitigate, and tolerate the effects of distraction are carefully trialed and evaluated before deployment. All stages of the integrated safety chain are of importance, not only the later stages. By avoiding distraction, or avoiding the early consequences of distraction at an early stage, we do not have to stress the more dramatic later stages where the vehicle gradually takes over control and we do not have to deploy crash-related protection systems other than in rare cases.

The road infrastructure has a major role to play in reducing and mitigating the consequences of distraction. Tactile edge lining (often referred to as "rumble strips"), as noted previously, is an example of a simple road treatment that can warn drivers if they drift off the roadway. Better intersection design can also play a role in reducing distraction-related crashes. Also, less prominent hazards, like roadworks, should be handled on the basis that drivers are distracted. It is not uncommon for drivers to drive into roadworks without taking any evasive action, which can have serious consequences for road workers and drivers. For those responsible for safety at roadworks, it should be natural to take precautions based on the assumption that drivers are distracted.

A governmental policy that takes distraction seriously could have the following content:

- Distraction is a serious problem and is often the initial event in a chain of events that leads to a serious health loss.
- Although distraction is not legally allowed, it is well understood to exist now and in the future.
- On the basis of the above-mentioned factors, distraction must be taken into account for all systems, products, and services that exist within the road transport system.
- Distraction should be both reduced and prepared for in the integrated safety chain.
- In developing technology to reduce the consequences of distraction, consideration must be given to possible modifications in behavior arising from driver interaction with the technology that might diminish the intended safety benefits.

In essence, policy development must be based on the existence of distraction as a safety problem and on the assumption that distraction cannot be eliminated. Distraction must, therefore, be regarded as a major design factor that requires consideration at every level of the road transport system. At the same time, efforts should be made to reduce distraction. The integrated safety approach, described here, is the best working tool available to handle distraction.

33.3 THE BMW PERSPECTIVE ON DRIVER DISTRACTION

Lutz Eckstein

33.3.1 Driver Distraction: A Societal Problem

Stutts et al. assert that "distraction occurs when a driver is delayed in the recognition of information needed to safely accomplish the driving task because some event, activity, object, or person within or outside the vehicle compels or induces the driver's attention shifting away from the driving task. The presence of a triggering event distinguishes a distracted driver from one who is simply inattentive, drowsy or lost in thought" (Ref. 7, p. 3).

Field studies analyzing driver behavior under everyday driving conditions show that many driver distractions are neither new nor technological in nature.[8] For example, of all secondary tasks that drivers engaged in during the National Highway and Traffic Safety Administration (NHTSA)-funded 100-Car Naturalistic Driving Study (2006),[9] "reaching for a moving object" was shown to have the highest impact on the likelihood of a crash or near crash, followed by "external distraction," "reading," "applying makeup," and "dialing a hand-held device." Driver distraction must be, therefore, regarded as a societal problem, not as a problem of a specific industry. Consequently, management of this problem requires concerted action by multiple stakeholders, including OEMs, system manufacturers, service providers, employers, drivers, and road authorities.

33.3.2 Importance of Human-Machine Interaction for BMW

For BMW, the human-machine interaction (HMI) plays a crucial role in achieving the brand's goal of providing sheer driving pleasure. Therefore, BMW engineers and psychologists have actively contributed for many years to national and international research standardization activities (e.g., through International Organization for Standardization Technical Committee 22, Subcommittee 13) and to the formulation of guidelines on HMI design for information and communication systems (e.g., the European Statement of Principles, ESoP[10]) and its revision in 2005. On the basis of these activities, BMW has introduced several innovations and is continuously improving its products. Features characterizing BMW's integrated information and communication system iDrive, for example, include a central information display positioned close to the driver's normal line of sight, display contrast and character sizes designed in accordance with International Organization for Standardization (ISO) 15008, and a central control element, which is easy to reach and operate without glancing at it. To further improve the HMI, BMW has concentrated its competence in a dedicated team of people responsible for the HMI of the entire vehicle functionality. This team applies state-of-the-art methods in designing the HMI as well as in evaluating its effectiveness, acceptance, and effect on driving performance.

33.3.3 Goal of Human-Machine Interaction Design

Accident probability ($p_{accident}$ per unit of time) depends on three major factors: the driver (e.g., age, experience, motivation, vigilance, travel patterns), the vehicle (e.g., type, age, maintenance, safety features, HMI), and the environment (e.g., road type, road curvature, road condition, time of day, weather). The driver acts as a workload manager; that is, he/she needs to be able to decide whether to start interacting with other persons, objects, or displays and controls. A prerequisite for a correct driving decision is the anticipation of his/her vehicle's behavior, the future development of the traffic situation, and the manual, visual, and cognitive workload associated with the interaction. Once he/she starts an interaction, the driver needs to be able to decide whether to continue or interrupt this interaction. This means that the design of interaction must allow for monitoring of the traffic situation, for example, by being interruptible at any time without subjectively perceived costs. Integrating accident probability along the duration of the trip results in accident risk, *R*. Vehicle or system design does not influence the integration constants but should aim at minimizing the intensity of interaction, which is reflected by $p_{accident}$.

In the 100-car study, drivers were engaged in tasks other than driving in 54% of all 20,000 baseline epochs of six seconds duration each. This may in some cases be due to a low workload associated with the driving task, but it may also suggest that people want to make use of the steadily increasing amount of time they spend in their vehicles. Since a vehicle manufacturer cannot influence driver behavior directly, the primary goal for system designers should be to provide an HMI that satisfies several criteria: (1) it should be compatible with the driving task (e.g., highly mounted displays, interruptible and driver-paced interaction); (2) it should enable the driver to anticipate the associated visual, auditory, manual, and cognitive workload of interaction; and (3) it should strive to call for minimal attention to enable the driver to monitor the traffic situation at every point during the interaction. Therefore, in accordance with the ESoP on HMI,[10] task-based criteria (e.g., total task time, total glance time per task) are not adopted by BMW in the context of safety, because they may exclude future functions with potential safety benefits. Furthermore, these simplistic criteria may impede the development of new types of driver-system interactions and even suggest that it is always safe to perform a task that complies to a task-based criterion like total glance time—it may not be regarded as safe to look away from the road scene even once, if the driving maneuver is very demanding (e.g., left turn with oncoming traffic).

33.3.4 Minimizing Driver Distraction: Proposed Actions

The availability and presentation of information and communication functionalities to drivers while driving is a responsibility shared by several stakeholders. The distribution of this responsibility depends on the type of system in question (integrated, aftermarket, or nomadic device) as well as on the specific aspect of HMI under consideration (installation, information presentation, interaction, system performance).

- Installation of the system, for example, may be undertaken by the OEM (integrated system), a dealer/supplier (aftermarket system), or a private person/driver (nomadic device).
- Information presentation is not only defined by the system manufacturer but is also influenced by service providers offering information/content (e.g., running text).
- System interaction is partly defined by the system hardware and partly by the system software. System hardware and software may be provided by one party (e.g., OEM) but may also come from different parties, for example, in the case of nomadic devices.

All stakeholders, therefore, need to develop information and communication systems to the same standard, independent of their functionality and degree of integration. The ESoP summarizes important principles for designing the HMI of information and communication systems. Until now, however, only vehicle manufacturers have committed themselves to developing their systems in accordance with the ESoP.

Since the driver ultimately decides when to perform which task and which system type (e.g., OEM-integrated or nomadic device) he/she prefers, the existence of an HMI design guideline may not be sufficient—it may be necessary to identify accompanying actions to ensure that all system types are designed in accordance with the ESoP and that drivers install and use them in a responsible manner. The final report of the European eSafety HMI Working Group (HMI WG) gives a very comprehensive summary of the measures deemed necessary.[11] Concerning nomadic devices (e.g., mobile telephones), the HMI WG recommends that the product-responsible organizations make sure that specific functions not compatible with the driving task are inaccessible to the driver while the vehicle is in motion. The product-responsible organization should also provide an installation kit in accordance with the principles given in the ESoP and appropriate consumer information. The responsibility for the installation itself will rest with the driver (or his employer in case of nonprivate use).

From the HMI WG's perspective, road authorities play a decisive role in actively ensuring that the ESoP is effectively disseminated, known, and used by all responsible stakeholders. Authorities should also provide general information to drivers on safe use of in-vehicle information and communication systems and promote self-commitment of ESoP compliance for manufacturers of aftermarket systems and nomadic devices. Authorities should also monitor the impact of the ESoP on the market for aftermarket and nomadic devices and evaluate the safety impact of in-vehicle information and communication systems by collecting and analyzing accident data. Moreover, they should take measures to ensure secure fixing of devices in accordance with ESoP guidelines, hands-free use of nomadic devices, and the inaccessibility to drivers while driving of movies, TV, and video games.

33.3.5 Summary

For BMW, the HMI plays a crucial role in achieving the brand's goal of providing sheer driving pleasure. BMW actively contributes to research, standardization, and guideline development on HMI, forming the basis for continuous improvement and innovation in the design of displays, controls, and HMIs. For BMW, the primary goal is to provide an HMI that (1) is compatible with the driving task (e.g., high-mounted displays, interruptible and driver-paced interaction); (2) enables the driver to anticipate the associated visual, auditory, manual, and cognitive workload of interaction; and (3) strives to call for minimal attention to enable the driver to monitor the traffic situation at every point during the interaction. BMW does not adopt task-based criteria (e.g., total task time) in the context of safety. Nevertheless, the efficiency of interaction is important for achieving high customer satisfaction. Based on its extensive experience with integrated navigation systems and continuous accident analysis, BMW regards the driver as an active workload manager.

From a traffic safety perspective, BMW is convinced that the concerted action of multiple stakeholders is needed to ensure (1) that all system types meant for use by the driver while driving are designed according to the same principles and standards and (2) that drivers use these systems in a responsible way.

33.4 THE HOLDEN PERSPECTIVE ON DRIVER DISTRACTION

Mike Hammer

33.4.1 Introduction

Driver inattention and driver distraction are increasingly important issues for the automotive industry. Our busy multitasking lifestyle places more demands on the driver's attention than ever before.

In-vehicle information systems are becoming more sophisticated and common in cars today. Many luxury cars now come standard with navigation systems, integrated telephones, and various driver information systems all controlled from either a large central color screen or a combination of steering wheel switches and instrument cluster displays. Many of these systems are designed to assist the driver with the task of driving and navigation, but unless carefully designed, they can become a source of distraction.

It is important to understand the nature of how and why drivers become distracted to develop effective countermeasures and designs for interfaces that can be used safely by the driver. Many automotive companies are doing extensive research into this and are forming partnerships with government and university research institutions to advance fundamental knowledge in this area. The Cooperative Research Centre for Advance Automotive Technology (AutoCRC) in Australia is an example of such a partnership and receives strong government support.

To minimize distraction, a system must have good basic ergonomics; that is, it must be easy to learn and intuitive to use. In addition, the system must have specific design features to reduce distraction, such as task chunk-ability, and allow the user to control the pace of interaction with the system (e.g., no time-outs). The system must also have design features to individually address the different types of distraction (e.g., visual and cognitive).[12]

If an in-vehicle information system is designed with these attributes, it will often provide a safer means for the driver to access information than other conventional alternatives. For example, there are many studies that show that a well-designed turn-by-turn navigation system is less distracting than using a paper map.[12–15]

Exposure also needs to be taken into account. If a system is easier to use, it may encourage more use, resulting in more eyes-off-road time. People's willingness to engage in distracting activity is an area in which little research has been done. Early evidence suggests that younger, less experienced drivers are more likely to engage in distracting activity and are less aware of the impact on driving than older, more experienced drivers.[16]

Driver inattention is likely to be a much larger factor in crashes than official data currently show. Many crashes are not correctly classified, often being attributed to speed, following too closely, and other factors. For example, the 100-Car Naturalistic Driving Study carried out by Virginia Tech Transport Institute (VTTI)[17] shows that most rear-end crashes are caused by driver inattention (rather than following too closely, as generally thought). The human factor will always be present, and needs to be studied and understood at a fundamental level if a "Vision Zero" road toll is to be achieved. Vision Zero is a road safety philosophy, adopted by the Swedish government, which states that, eventually, no one will be killed or seriously injured within the road transport system. Vision Zero explicitly states that the responsibility for road safety is shared by the transport system designers and the road user.

33.4.2 Future Trends

The next wave of technology features will come as vehicles become fully connected to the Internet via telematics. There is a strong consumer demand for location-based services such as traffic information, emergency callout, and motoring-related location-based information, such as the availability of fuel and parking.

It is estimated that 46 million vehicles were equipped with telematics by the end of 2006,[18] and by the end of 2011, subscriber revenues will total US$38 billion.[19] General Motors (GM) in the United States recently announced that, by 2007, telematics will be a standard fitment on all GM cars manufactured in the United States. In Australia, there are approximately 7000 passenger cars equipped with OEM telematics systems, with a growth rate of 118% per year.[20]

Electronic systems are becoming so complex that they are increasingly being designed by specialist electronics suppliers rather than the automotive manufacturers. Navigation systems, for example, are generally designed by large electronics suppliers, and it is these suppliers that implement the operational behavior of the system. Thus, specialist suppliers increasingly share responsibility for designing systems that minimize driver distraction.

33.4.3 Vehicle Design

Vehicle designers must balance many aspects of the vehicle interface when designing in-vehicle information and entertainment systems. The systems must appeal to customers in the showrooms and in use, address customers' needs, be intuitive to use, and minimize driver distraction. Such a balance is only achievable when the interface is designed from a position of fundamental knowledge of the human factors issues involved at many levels; that is, understanding of cognitive and perceptual processes as well as conventional ergonomic principles. The interface must first be built around sound ergonomic principles that enable ease of use, intuitive task flow, and good learnability in a time-shared environment. If done properly, this can be achieved somewhat independently of the system styling, which creates the initial customer appeal.

Once the basic ergonomic qualities of the interface are established, specific design features to manage driver distraction and workload can be built into the design. The design can then be further developed and validated using an inventory of testing techniques involving both laboratory and on-road testing. This process is typical of that used by many automotive companies.

The building blocks for an effective design are shown in Figure 33.2. Good ergonomic design forms the foundation from which a good interface is built (taking into

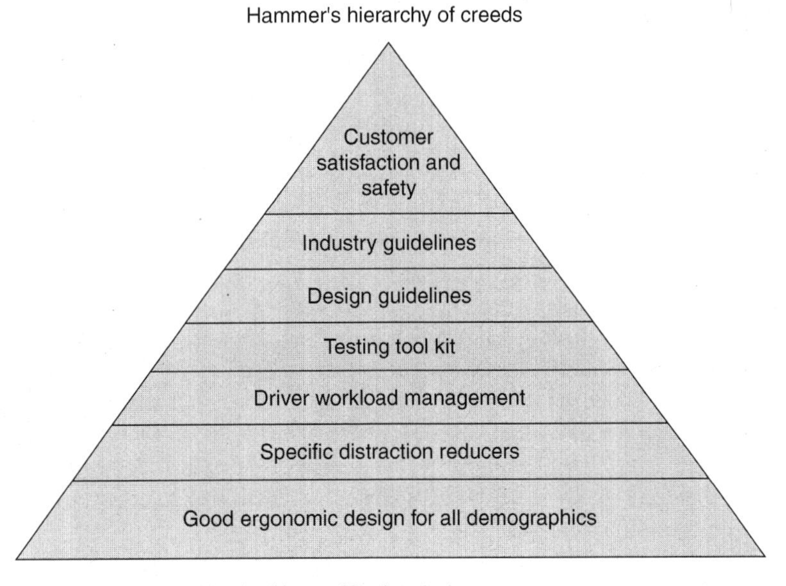

FIGURE 33.2 Building blocks for an effective design.

account the specific need to minimize visual and cognitive load). A well-designed interface will be intuitive and easy to learn and use, and suit both novices and experienced users. Special needs of different demographic groups should also be taken into account. For example, the interface should appeal to younger, more technologically literate, users as well as to those not familiar with the latest technology. Older users require more contrast, and their eyes take longer to change focus from the road to the interface and back again. It follows that an interface designed to take these limitations into account will have greater usability for all demographic groups. An interface designed using good ergonomic principles is likely to be less distracting than a poorly designed interface; for example, it has been shown that an interface with many similar-appearing buttons significantly increases the number and duration of glances required to complete a task.[21]

Basic ergonomic issues, such as the above-mentioned example, influence driver distraction and must be considered in this context. Specific techniques to reduce driver distraction are also applied to the design. These techniques include, but are not limited to, the following:

- Displays should be mounted high up on the instrument panel. This reduces eyes-off-road time and keeps the road scene within the driver's central field of view.
- Drivers must control the pace of interaction with the system. The driver's attention must be able to be time-shared between the important task of driving safely and operating the system. The driver can only attend to the system during periods of low workload and when it is safe to do so. Therefore, there may be long delays between the steps required to complete a task.
- Tasks must be chunkable. For the above-mentioned reason, tasks must be interruptible and resumable after a short delay without the operator losing track of where he/she is in the operation (to minimize short-term memory load).
- Information presented should be easily comprehended by a few short glances. Two seconds is generally considered the maximum allowable glance duration. In practice, average glance times vary from 0.6 to 1.7 s depending on the task complexity with a standard deviation of up to 0.8 s.[21] In the 100-Car Naturalistic Driving Study,[9] it was found that glance times less than 2 s were associated with negligible increased risk of a crash.

A variety of test techniques are then used to develop and validate the design. Testing is initially done in a static buck (generally a mock-up or prototype interface on a bench or a vehicle interior mock-up), and later on road or in a driving simulator. Testing is mostly focused on usability and visual distraction, but more emphasis is now being placed on developing techniques to measure cognitive distraction and manage driver workload. Saab, for example, recently introduced a "workload manager" into their cars, which suppresses mobile phone calls and some warnings during periods of high driver workload.

The automotive market is fiercely competitive. When designing an in-vehicle system and interface, the customer's requirements cannot be forgotten. A distraction-free vehicle will not be successful if the customer will not purchase it. The vehicle must reduce distraction but still address customers' needs.

In large industries, such as the automotive industry, individual job responsibilities often change, so it is important to codify this knowledge into design guidelines and procedures. Within an industry as vast and multinational as the automotive industry, cars are no longer designed for a single country. Guidelines need to be made available as harmonized global industry guidelines to achieve real societal benefits in road safety.

33.4.4 Countermeasures

33.4.4.1 Crash Avoidance Technology

One promising method to reduce the risk from driver inattention is crash avoidance or active safety technology. These systems can warn drivers when they are getting into a potentially dangerous situation or if the traffic condition suddenly changes when they are not paying attention to the driving task, allowing them to redirect their attention and take corrective action before it is too late. Crash avoidance technologies include

- Warning sign recognition
- Lane departure warning
- Curve speed warning
- Front collision warning
- Blind spot collision warning
- Pedestrian recognition
- Night vision
- Reversing aids
- Distraction warning

Cars are appearing on the market now with these technologies, and they will become more common over the next few years. The introduction of these vision- and radar-based technologies is expected to significantly reduce the risk of crashes due to driver inattention and distraction.[22]

33.4.4.2 Speech Recognition

Speech recognition is generally regarded as a means to reduce driver distraction associated with complex tasks. Voice is a mode that interferes less with the task of driving than manual control, and voice control does not generally require the diversion of visual attention away from the driving task. Some cognitive load is still required to operate secondary controls using speech, and this needs to be managed.

Although speech recognition systems are beginning to appear in some vehicles, they still have significant limitations: recognizing some accents (particularly the Australian accent), operating reliably in a high-noise environment, and recognizing natural speech. These systems are not yet available for all major languages. However, this is a rapidly developing area, both in the understanding of the nature of speech and in the availability of inexpensive and powerful in-vehicle computing hardware, and it is expected that these problems will be solved in the next few years.

33.4.5 Industry Standards and Guidelines

The increasing demand that in-vehicle information systems place on the driver's attention has been recognized internationally. In the past 3–4 years, the automotive industries in Europe, the United States, and Japan have all issued voluntary guidelines for the design of in-vehicle information systems. These guidelines are the result of extensive research in this area and provide a more up-to-date and flexible means of regulation than legislation, which, in this fast-moving area of technology, can quickly become out of date and may even become an impediment to innovation and, thus, the competitiveness of the country of enforcement. For example, new dual-view display screen technology allows the driver and passenger to see different images, but cars fitted with this new technology will not be able to be used in some countries (e.g., Canada) due to the actual wording of the legislation. This does not imply that the automotive industry does not support legislation where applicable; in fact, many important road safety outcomes are a result of properly enforced legislation. The mandatory wearing of seat belts is a case in point. Legislation, however, tends to be slow moving and to remain fixed once enacted. Therefore, careful consideration needs to be given before applying it to areas of fast-moving technology. For example, it is unlikely that corresponding legislation could be changed quickly enough to allow a country's high-technology industries to remain competitive. Industry, regulators, and researchers need to work together to field safe and satisfying systems. As noted previously, the automotive industry is one of the most global of large industries, and therefore, standards and guidelines also need to be harmonized between countries.

33.4.6 Next Steps

Some important steps need to be taken if timely progress is to be made in this important area of road safety:

- In countries that manufacture cars, the automotive industry needs to be proactive and establish agreed voluntary design guidelines, if this has not already been done.
- The supplier industry needs to become more involved in driver distraction research. In-vehicle information systems are increasingly becoming the product of specialist supplier companies. The principles outlined in this chapter need to be built into these devices at the basic design level.
- Nomadic devices are becoming very popular in cars. In many cases, these devices are not designed for in-vehicle use, but the same principles need to be applied to these devices if they are to be used by the driver. It might be possible to address these with docking stations along the lines of GM's iPod integration device.
- Driver awareness and training is an often overlooked area. Specific training to help drivers become aware of, and minimize, inattention induced by distraction can make an impact on road safety. Specific training can also

be designed for younger drivers as it is well known that young drivers often overestimate their abilities, are poor judges of risk, and are more likely to use in-vehicle devices.
- Smaller countries need government partnerships to help defray the large capital investments required for test facilities. Sweden is an exemplar in this area.

Driver distraction and inattention is a complex issue. It cannot be solved by hastily drafted legislation or other quick fixes. A thorough knowledge and understanding of driver behavior (through, e.g., naturalistic studies), distraction, ergonomics, and the effectiveness of countermeasures (through testing and research) will be required if we are to make a significant impact on this area of road safety.

REFERENCES

1. Haddon, W., Advances in the epidemiology of injuries as a basis for public policy, *Public Health Reports* 95(5), 411–421, 1980.
2. WHO, *World Report on Traffic Injury Prevention*, ISBN-10 9241562609, 2004.
3. Communication from the Commission, *Saving 20,000 Lives on Our Roads. A Shared Responsibility*, Report No. COM 2003/311, Luxembourg, 2003.
4. Kanianthra, J., Re-inventing safety: Do technologies offer opportunities for meeting future safety needs?, *ESV Conference*, Lyon, France, 2007.
5. Krafft, M., Kullgren, A., Lie, A., and Tingvall, C., The use of seat belts in cars with smart seat belt reminders—Results of an observational study, *Traffic Injury Prevention* 7, 125–129, 2006.
6. Regan, M., Young, K., Triggs, T., Tomasevic, N., Mitsopoulus, E., Tierney, P., Healey, D., Tingvall, C., and Stephan, K., Impact on driving performance of intelligent speed adaptation, following distance warning and seatbelt reminder systems: Key findings from the TAC SafeCar project, *IEE Proceedings Intelligent Transport Systems* 53(1), 51–62, 2006.
7. Stutts, J. C., Reinfurt, D. W., Staplin, L., and Rodgman, E. A., The role of driver distraction in traffic crashes, AAA Foundation for Traffic Safety, Washington, D.C., 2001.
8. Stutts, J., Feaganes, J., Rodgman, E., Hamlett, C., Meadows, T., and Reinfurt, D., Distractions in everyday driving, AAA Foundation for Traffic Safety, Washington, D.C., 2003.
9. Klauer, S. G., Dingus, T. A., Neale, V. L., Sudweeks, J. D., and Ramsey, D. J., *The Impact of Driver Inattention on Near-Crash/Crash Risk: An Analysis Using the 100-Car Naturalistic Driving Study Data*, Virginia Tech Transportation Institute, Blacksburg, VA, 2006.
10. European Commission (EC), *Recommendation on Safe and Efficient In-Vehicle Information and Communication Systems: A European Statement of Principles on Human Machine Interface*, Report No. OJ L19, 25.1.2000, 2000.
11. Augello, D., Becker, S., Eckstein, L., Gelau, C., Hallen, A., König, W., Pauzie, A., and Stevens, A., *Recommendations from the eSafety-HMI Working Group*—Final Report, Report prepared for the European Commission, 2005.
12. Perez, W. A., *The Safety Evaluation of Travtek, Safety Evaluation of Intelligent Transport Systems*, Workshop Proceedings, ITS America Safety & Human Factors Committee & National Highway Traffic Safety Administration, 1995.
13. Dingus, T., McGehee, D., Hulse, M., Manakkal, N., Mollenbauer, M., and Fleischman, R., *Travtek Evaluation Task C3—Camera Car Study*, Performance and Safety Sciences, Inc., Iowa City, IA, 1995.

14. Srinivasan, R. and Jovanis, P. P., Effect of in-vehicle route guidance systems on driver workload and choice of vehicle speed: Findings from a driving simulator experiment, in *Ergonomics and Safety of Intelligent Driver Interfaces*, Noy, Y. I. (Ed.), Lawrence Erlbaum Associates Inc., Mahwah, NJ, pp. 97–114, 1997.
15. Young, K., Regan, M., and Hammer, M., *Driver Distraction: A Review of the Literature,* MUARC Report No. 206, Monash University Accident Research Centre, Clayton, Victoria, 2003.
16. McEvoy, S. P., Stevenson, M. R., and Woodward, M., The impact of driver distraction on road safety: Results from a representative survey in two Australian states, *Injury Prevention* 12(4), 242, 2006.
17. Dingus, T. A., Klauer, S. G., Neale, V. L., Petersen, A., Lee, S. E., Sudweeks, J., Perez, M. A., Hankey, J., Ramsey, D., Gupta, S., Bucher, C., Doerzaph, Z. R., Jermeland, J., and Knipling, R. R., *The 100-Car Naturalistic Driving Study, Phase II: Results of the 100-Car Field Experiment*, Virginia Tech Transportation Institute, Blacksburg, VA, 2006.
18. Harel, K., The convergence conundrum, Telematics West, 2001.
19. ABI Research, Automotive Telematics: Global Consumer Telematics Markets and Technologies, 2006.
20. Holden Market Research, 2004—Telematics Sales History, 2004.
21. Wierwille, W. W., Visual and manual demands of in-car controls and displays, in *Automotive Ergonomics*, Peacock, B. and Karwowski, W. (Eds.), Taylor and Francis, London, pp. 299–320, 1993.
22. Lee, J. D., McGehee, D. V., Brown, T. L., and Reyes, M. L., Collision warning timing, driver distraction, and driver response to imminent rear-end collisions in a high-fidelity driving simulator, *Human Factors* 44(2), 314, 2002.

Part 9

Conclusions

34 Some Concluding Remarks

Michael A. Regan, Kristie L. Young, and John D. Lee

CONTENTS

In this book, we have reviewed much of the knowledge that exists on driver distraction—what it means, theories describing its mechanisms, its effects on driving performance and safety, and strategies for preventing or mitigating its effects. On the basis of the material reviewed, several conclusions can be drawn.

There is converging evidence that driver distraction is a significant road safety problem worldwide. Findings from the analysis of police-reported crashes, reviewed in Chapter 16, suggest that driver distraction is a contributing factor in 10 to 12% of crashes. Data from the 100-car naturalistic driving study in the United States, also reviewed in Chapter 16, suggest that distraction from secondary tasks may be a contributing factor in up to 23% of crashes and near-crashes. Although estimates vary due to differences in definitions, data collection methods, and classification schemes, there is good reason to believe that all of these estimates underestimate the true scale of the problem. About one-third of all distractions appear to derive from outside the vehicle, and between about 15% and 20% involve driver interaction with technology. Distraction appears to be largely associated with rear-end crashes, same travelway or same direction crashes, single-vehicle crashes, and crashes occurring at night.

Driver distraction is a complex, multidimensional problem. The impact of distraction on driving performance and safety depends on many interrelated factors, such as the concurrent demands of driving and nondriving tasks (see Chapter 4); moderating factors such as the state, age, level of experience, and personality of the driver (see Chapter 19); understanding what is distracting the driver (see Chapter 15); how often and for how long the driver is distracted (see Chapters 17 and 18); when and where the driver is distracted (see Chapters 16 and 17); the momentary configuration of physical circumstances that determine whether the driver fails to maintain an appropriate distribution of attention relative to the changing demands of the roadway (see Chapters 2 and 4); the degree to which the driver, the vehicle being driven, and the physical environment is tolerant of the consequences of distraction (see Chapter 33); and even a certain amount of luck. Therefore, the prevention, mitigation, and management of distraction is a complex undertaking.

It is unlikely that distraction will ever be eradicated as a road safety problem. There are various reasons for this: humans are fundamentally limited in their capacity to simultaneously attend to multiple activities; driving is a "satisficing" task, leaving free attention that can be used to accomplish other tasks; vehicles are likely to continue to be designed solely for single-person operation; drivers are biologically primed to be attracted, sometimes beyond their control, to certain objects, events, and activities that are salient or novel; different social roles motivate drivers, sometimes through necessity, to engage in certain activities that have potential to distract them; and new sources of distraction will continue to emerge as the driving task, and society itself, evolve. At best, driver distraction can be effectively managed.

Effective management of road traffic safety issues has a number of defining characteristics. Johnston[1] argues that the current "best practice" model for road traffic policy making, intervention programming, and effective implementation of integrated countermeasure programs has the following defining features (Chapter 4, p. 16):

- Routine surveillance of safety progress, using comprehensive, high-quality data systems, covering the gamut of road safety problems
- Strategic targeting of the key problems using evidence-based strategies and program options
- The provision of adequate resource for meaningful implementation
- Rigorous evaluation of the effectiveness of the interventions
- Continuous improvement in implementation based upon the evaluation results and maximum coordination among all relevant institutions

When judged against these criteria, it is clear from the material reviewed in this book that countermeasure development for preventing and mitigating the effects of distraction is still in its infancy, even in developed countries with relatively good road safety records. Perhaps this is not surprising. Governments continue to rely heavily, often overly, on crash data to justify and stimulate countermeasure development. However, to date, distraction has been poorly defined and systems for accurately and reliably collecting and analyzing data on its role in crashes do not exist in many jurisdictions. Technological change introduces new distractions at a great rate and makes crash data a lagging and ineffective indicator of the distraction problem. Many policymakers are also unaware of converging evidence, from epidemiological and other studies, that implicates distraction as a road safety problem. In turn, this has thwarted attempts by governments to strategically target key distraction problems using evidence-based strategies, and to justify adequate resources for meaningful implementation of effective countermeasures.

Rigorous evaluation of intervention effectiveness is also lacking. Noteworthy is a lack of published research on the effectiveness of existing distraction prevention and mitigation measures. The limited data that do exist pertain mainly to the impact of banning handheld mobile phone use while driving and suggest that, while rates of handheld mobile phone use initially decline after such bans are implemented, they drift back toward prelegislation levels at a rate contingent on the amount of ongoing enforcement of the ban and associated publicity. Vehicle manufacturers, to their credit, have been proactive in undertaking and commissioning research to understand distraction, and in developing methods, tools, guidelines, and standards for the design and evaluation of products to limit distraction. Even for these interventions, however, there is limited published data on their effectiveness in limiting distraction, let alone enhancing safety. Effective mechanisms for ensuring that the outputs of distraction countermeasure evaluation are fed back into the countermeasure development process do not currently exist, and are complicated by institutional and industry complexity.

Maximum coordination among all relevant institutions is a crucial element in effective countermeasure development. One of the defining features of road traffic safety, however, is its institutional complexity.[1] Many different stakeholders have a vested interest in managing the safety of the road system. The management of driver distraction is no different and is further complicated by technological developments that outpace the rate at which effective legislation can be written and adequately enforced. These developments make it possible for drivers to interact with functions available on nomadic and aftermarket devices, developed by industries not accustomed to considering how its products affect driving safety.[2] Other developments that enable drivers to see and hear inside the vehicle traffic and nontraffic-related information normally displayed to them visually outside the vehicle adds yet another layer of complexity. These developments appear to be occurring in isolation. Who, for example, is responsible for coordinating the simultaneous flow and integration of information to drivers from inside and outside the vehicle? Who is responsible for coordinating the simultaneous flow and integration of information to drivers from Original Equipment Manufacturer (OEM), aftermarket, and nomadic devices within the vehicle? Lacking in the management of distraction are institutional arrangements, which ensure that there is mutual cooperation and cross talk between the automotive industry, traffic engineers, aftermarket suppliers, and nomadic device suppliers to ensure that the total demands of driving are within the capacity of the driver. Without such cooperation, the best efforts of one sector in limiting the effects of distraction might be partly or completely undermined by another.

The management of distraction is therefore a fertile area for countermeasure development. The key to effectively tackling the driver distraction issue is to stop blaming drivers who deliberately or inadvertently fail to attend to activities critical for safe driving. We must start looking at the issue from a broader, system-wide perspective. To this end we have proposed in this book (Chapters 30 through 33) an integrated approach to the prevention and mitigation of distraction and have recommended specific countermeasures for addressing the problem across a broad range of areas.

The injury prevention countermeasures presented are framed in large part around two organizing frameworks: a conceptual framework, referred to in Chapter 33 as an "integrated safety chain" that stimulates consideration of options for limiting distraction pertaining to the driver, vehicle, and roadway environment at each stage leading to a crash; and a theoretical account of driver distraction, presented in Chapter 4 that describes the diversion of attention away from activities critical for safe driving toward a competing activity as a breakdown in multilevel control processes, with different timescales characterizing each level. These conceptual frameworks are useful in stimulating interdisciplinary thinking about the interactions that occur between the three road traffic entities—driver, vehicle, and environment—and between each stage in the crash sequence, and how these might guide countermeasure development. Many of the countermeasures presented in this book are derived not from "hard data," but from current understanding of the mechanisms that appear to characterize distraction. There is no guarantee that the countermeasures recommended, whether derived from "hard" or "soft" data, will be effective in preventing and mitigating the effects of distraction. This can only be determined through careful evaluation. The fact that efforts to manage distraction are still in their infancy may be a virtue from an evaluation perspective, as it may enable the effectiveness of new policies and programs that are rolled out to be evaluated in relative isolation.

Some concluding comments can be made about our current state of knowledge regarding driver distraction.

Distraction is a poorly defined concept. Even within this book definitions of it vary widely. The lack of a consistent definition across studies makes the comparison of research findings difficult or impossible. Inconsistent definitions also lead to different interpretations of crash data and, ultimately, to different estimates of the role of distraction in crashes. The definition of distraction coined in Chapters 1 and 3 of this book—*distraction is the diversion of attention away from activities critical for safe driving toward a competing activity*—is presented as a first step in resolving these issues. Deriving from this definition we have further proposed, in Chapter 15, a taxonomic description of those sources of distraction that have been identified as contributing to crashes and near-crashes. The taxonomy is intended to resolve confusion about what are, and are not, the sources of distraction; to provide a framework for classifying sources of distraction; and to support the development of more reliable and less variable methods for collecting and coding crash and epidemiological data. The taxonomy will, of course, require further refinement as driving and non-driving tasks performed while driving continue to evolve.

The impact of distraction on driving performance depends on many interrelated factors (see Chapters 4 and 19). Much of the distraction research to date has focused on the impact of sources of distraction deriving from within the vehicle related to technology use. Little is known about the impact on performance of other sources of distraction, identified in Chapter 15, deriving from inside or outside the vehicle. Of those studies that have investigated the distraction potential of technologies, surprisingly few have investigated the distraction potential of everyday driving-related tasks associated with driving (e.g., changing gears, monitoring speedometers, etc.). The lack of research on factors—the age, state (e.g., drowsy, drunk, etc.), level of experience, and personality of the driver—that moderate the effects of distraction, and the mechanisms through which this moderation occurs is also notable. The manner in which drivers self-regulate in response to distraction, and in response to other road users they perceive to be distracted is not well understood; and evidence for it is sparse and ambiguous[3] (see Chapter 19). As noted in Chapter 20, age may become a more salient moderating factor as the driving population continues to age. Owing to age-related functional declines, older drivers appear on first principles to be relatively more vulnerable to the effects of distraction. There exists, however, little research to confirm this. Even less is known about how older drivers self-regulate in response to distraction. Similarly, as noted in Chapter 21, further research is needed to establish whether fatigue effects moderate the effects of distraction and the potential role of distractions in offsetting the effects of driver fatigue.

Techniques are available to improve the quality of crash data, and these were discussed in Chapter 16. The solutions, however, are not simple. Unlike crashes that involve drugs, alcohol, or speed, in which there is a clear marker of a causal agent, crashes deriving from distraction leave no telltale trace.[4] In managing distraction, therefore, it is not appropriate to rely solely on crash data to prioritize countermeasure development; although, for governments, crash data is likely to remain important in justifying and stimulating countermeasure development. Approaches to data collection that involve collection of naturalistic driving data and data derived from on-board data loggers show promise, and help to provide an indication of the level of underestimation involved in traditional crash studies in quantifying the role of distraction in crashes. In the end, however, all methods have their limitations. As noted by Caird and Dewar,[3] the degree to which drivers are absorbed in thought, their allocation of attention to competing tasks, and strategic choices to self-regulate in response to distraction are behaviors not evident through observation alone. A combination of different methods will need to be used to investigate crashes to build up a complete crash picture.

With the advent of naturalistic driving studies, remarkable progress has been made in the capacity to assess drivers' exposure to many distracting activities while driving (see Chapter 17). However, much work remains to be done in this area. Establishing risk estimates for the full gamut of distracting activities that occur while driving remains an important area for research, particularly for activities unrelated to technology use and those deriving from driver interaction with objects and events outside the vehicle. Further investigation into the circumstances during which distracting activities present the greatest risk is also warranted. Further naturalistic driving studies, employing sound epidemiological methods and larger, more representative driving populations are therefore warranted.

The automotive industry has been proactive in developing countermeasures to prevent and mitigate the effects of distraction (see Parts 7 and 8 of this book). However, as an industry, it cannot be expected to shoulder the burden of good design in limiting distraction. Driving is a complex, multitask activity, and elements of the driving task itself (relating to both vehicle control and roadway monitoring) have the potential to divert the attention of the driver away from activities critical for safe driving. From a design perspective, the implication of this is that distraction can be limited by reducing the demands of driving tasks themselves (which, in turn, reduces vulnerability to distraction from competing tasks) and by directly limiting the distraction potential of competing tasks. The critical units of analysis for driver-centered design to limit distraction, therefore, should not be the physical aspects of the vehicle-driver interface, but rather the driving tasks, such as navigation, following the road, monitoring speed, avoiding collisions, following traffic rules, and controlling the vehicle (Brown, 1986, cited in Ref. 5) as well as the nondriving tasks that have potential to divert attention away from activities critical for safe driving. Responsibility for mitigating, through driver-centered design, the propensity for competing tasks to divert attention away from activities critical to safe driving is therefore a joint responsibility shared by multiple stakeholders—vehicle manufacturers, aftermarket suppliers, nomadic device suppliers, and traffic engineers. Also critical is the development of institutional arrangements which ensure that, at a macro level, there is mutual cooperation and cross talk between the relevant stakeholders to develop a coordinated design for driver interaction with the myriad of competing tasks (driving and nondriving) that compete for driver attention, so that it does not lead to a breakdown in multilevel control processes (tactical, strategic, or operational). As argued in Chapter 22, coordinated design of this kind can only be achieved through a formal requirement, in the form of a process-based code of practice that requires all stakeholders to adhere to a common safety management system—a systematic process that defines and prioritizes human factors and safety considerations that must be addressed throughout the design cycle.

At a more micro level, as argued in Chapter 25, the development of tools, methods, and metrics for designing products to limit distraction should be seen as a product in itself that is calibrated to the needs of its users and packaged in an appropriate manner. Development of a human factors and ergonomic "toolbox," which includes formative (design) methods, guidelines, and clear decision criteria is seen as an essential target for future research and development. Guidelines and standards (both design and performance standards) exist which aim, through good design, to limit distraction deriving from in-vehicle information and communication systems that are peripherally related, or unrelated, to driving. These are reviewed in Part 7 of this book, and the issues that remain to be resolved in ensuring that they achieve their intended purpose have been discussed. Noteworthy is that the principles contained in some guidelines (e.g., the European ESoP guidelines) currently apply solely to in-vehicle information systems. These will need to be reexamined and extended to advanced driver assistance systems, given that these are designed to assist the driver in performing driving tasks (see Chapter 22). The increasing integration of functions within in-vehicle systems will also necessitate revision or extension of existing guidelines (see Chapter 23), and as noted in Chapter 25, few current guidelines have concrete performance criteria. Guidelines by themselves are, however, insufficient. Mechanisms are needed to ensure that designers are aware of guidelines and standards, have the resources and skills to apply them effectively, and comply with them. The development of tools, methods, and metrics for designing products to limit distraction should not be confined to the automotive industry. Aftermarket suppliers, nomadic device suppliers, traffic engineers, and other relevant stakeholders are in need of guidance in how to ergonomically design the road environment and the nomadic devices brought into vehicles, to limit distraction. The establishment of institutional arrangements that ensure that this is done as a coordinated activity is critical.

There has been a proliferation of methods and metrics for measuring the impact of distraction on driving performance (see Part 3 of this book). These pertain, however, almost entirely to the measurement of distraction deriving from driver interaction with technologies within the vehicle. Certain challenges still remain. The repertoire of methods and metrics must be expanded to enable the assessment of the impact on performance of distractions deriving from outside the vehicle, and as for in-vehicle distractions, to develop reference tasks that provide a benchmark against which the impact on driving performance of external distractions (both driving and nondriving related) can be established. Some guidance on this issue is given in Chapter 13.

Collectively, assessment methods for all distractions must be cost-effective and easy to use by designers. Appropriate reference tasks must be developed, which are unambiguously defined, repeatable across different test environments, and induce mechanisms of distraction identical to those induced by the distracting activity under investigation.

There is currently little consensus regarding which assessment methods and metrics should be used for the evaluation of particular activities with potential to distract drivers (see Chapter 7). The outputs of projects such as HASTE, AIDE, and CAMP will provide some guidance in this area. Rapid evaluation procedures are needed that do not require sophisticated equipment and months of time. The lane change test is such a procedure although, as is noted in Chapter 25, the notion that a single, low cost test can assess the interference with driving of a competing task, regardless of its visual, auditory, cognitive, and psychomotor demands may be unrealistic. Table 15.5 in Chapter 15 of this book identifies mechanisms that moderate the effects of distraction that might be considered in refining the sensitivity and scope of existing assessment procedures, including checklists. In future, rapid evaluation procedures might include computational models of driver performance such that preliminary design concepts can be evaluated more quickly and earlier in the design and development cycle. Some models of this kind already exist (see Chapter 25).

Real-time distraction countermeasures (RDCs) have perhaps the greatest potential as a design countermeasure to prevent and mitigate the effects of distraction and save lives (see Chapters 26 through 28). These systems adaptively prevent or limit driver exposure to competing tasks when the concurrent demands of driving are estimated to be high (real-time distraction prevention) and mitigate the effects of distraction once it occurs, by providing feedback and warnings to drivers that redirects their attention to relevant aspects of the driving task (real-time distraction mitigation). These systems have many advantages over nonadaptive approaches to system design. First, these approaches are potentially capable of detecting whether a driver is distracted regardless of the competing activity (driving or nondriving related) and whether driver engagement in the competing activity is voluntary or involuntary, regardless of whether the impetus for the competing activity derives from inside or outside the vehicle. Second, the system can be optimized so that it is adaptive to factors that moderate the effects of distraction (e.g., driving demand, competing task demand, driver state, age, and experience; such as by issuing more conservative warnings if the driver is inebriated). Third, they can be used to prime and activate the operation of other active and passive safety systems at different stages of the integrated safety chain to optimize driver safety during all stages of the crash sequence. Finally, through the provision of real-time feedback to drivers, they have potential to provide long-term benefits in calibrating drivers to the dangers of distraction so they can better manage distraction, even when they drive vehicles not equipped with such systems. As discussed in Chapter 29, the design of feedback to mitigate distraction, whether it be in real time or delayed is a whole research field in itself. RDC is currently a very active research field, with some first generation products on the market. However, the field is still developing and largely technology driven. Further work is needed to identify the RDC functions that have the largest impact on driving safety, efficiency, and comfort; to improve the measurement and assessment of driving task demand and driver state; to optimize the systems for driver acceptance; and to develop suitable methods for evaluating their impact on driver acceptance, performance, and safety.

There are some possible drawbacks of improved vehicle and technology design that must be considered. Well ergonomically designed interfaces can, for example, encourage drivers to use them more often, thus increasing their exposure to risk (the so-called "usability paradox," as discussed in Chapter 3). The more automated is the driving task, the less demanding it may be, or appear to be (see Chapter 2), freeing up driver attention that can be used by drivers to take on other roles within the vehicle that may have potential to distract them. Emerging technology poses obvious distractions in the form of infotainment systems. More subtle threats, however, may lie in driver support systems, particularly as many of these systems are combined. Technology that automates elements of the driving task could surprise drivers and distract them as they try to figure out how to get it to do what they want. Effective driver-centered design requires continuous evaluation and feedback to identify and rectify these and other unintended side effects.

A fundamental issue, raised in several chapters, is how to bridge the gap between measurement of distraction and its link with real-life crash causation. As noted in Chapter 25, equations are needed that relate performance in various tests to fatalities and injuries likely at that level of performance. Distraction metrics are surrogate metrics intended to predict crash involvement. Data derived from naturalistic driving studies may provide a mechanism for bridging the performance and safety gap by identifying critical scenarios and events that characterize real-world crashes that can be used as assessment scenarios. However, the combinations of coincidences between tasks (e.g., navigating, dialing) and traffic conditions (e.g., intersection, merging) that give rise to crashes and near-misses may be difficult to test in experimental settings and to observe in naturalistic driving studies.[3] As noted in Chapter 22, a particular challenge is how to combine individual assessment methods into an overall integrated methodology to make predictions about safety in use. The control theoretic approach to understanding the processes that underlie distraction, described in Chapter 4, attempts to describe the relationship between distraction related performance degradation and crash risk, and provides a theoretical account of the critical links that may be useful in framing future research activities.

It is notable is that the study of distraction has been confined almost entirely to the road transport domain, although some related work has been going on in the computing and aviation domains under the guise of "interruptions."[6,7] Even within the road transport domain, the focus of distraction efforts to date has been on drivers, whereas distracted walking and distracted riding, whether on bicycles or motorcycles, are potential areas of concern that appear to be totally unexplored and researched. Notable also is the paucity of research on driver distraction in the public and commercial transport sectors. The limited research undertaken in this field, reviewed in Chapter 14, suggests that distraction is a problem in bus and heavy vehicle transport operations. Bus drivers, in particular, are required to take on multiple, and at times competing roles while driving, which make them particularly vulnerable to the effects of distraction. This is exacerbated by the demands of bus driving itself, which is arguably a less "satisficing" task than ordinary driving, particularly in residential areas. Further research is required to identify and classify the sources of distraction that exist in the public and commercial transport sectors and to quantify their impact on driving performance and safety. In the meantime, Chapter 14 provides initial guidance on preventing and mitigating the effects of distraction in bus operations.

Not all distraction is bad distraction. The driver distraction issue has a flip side too. Some potentially distracting activities may have safety benefits, such as combating the effects of drowsiness or fatigue (as in the case of a truck driver using a CB radio as discussed in Chapter 21). There are also situations in which the attention of the individual in charge of the vehicle is drawn to circumstances other than its momentary control that may be beneficial for the personal safety and even survival of the driver, for example when taking a hand off the wheel to parry the attack of a snake coiled on the passenger seat (see Chapter 2). The scientific, philosophical, legal, and moral issues concerning the nature and interpretation of driver behavior in these circumstances are important ones that remain to be explored.

The tragic incident that occurred on the morning of December 31, 2001, on Port Arlington Road, near Geelong, Australia—in which a 24-year-old female dentist preparing to send a mobile phone generated text message while driving crashed into, and killed, a 36-year-old mechanical engineer riding a bicycle—need not have happened. It is hoped that the knowledge provided in this book will help to prevent further tragedies of this kind from occurring.

REFERENCES

1. Johnston, I., Highway safety, in *The Handbook of Highway Engineering*, Fwa, T. F. (Ed.), CRC Press, Boca Raton, 2006.
2. Hedlund, J., Simpson, H., and Mayhew, D., International conference on distracted driving: Summary of proceedings and recommendations, *International Conference on Distracted Driving*, Toronto, Canada, 2006.
3. Caird, J. and Dewar, R., Driver distraction, in *Human Factors in Traffic Safety*, Dewar, R. and Olson, P. (Eds.), Lawyers and Judges Publishing Company Inc., New York, 2006.
4. Lee, J. D. and Strayer, D., Preface to the special section on driver distraction, *Human Factors* 46(4), 583–586, 2004.
5. Hatakka, M., Keskinen, E., Hernetkoski, K., Gregerson, N. P., and Glad, A., Goals and contents of driver education, in *Driver Behaviour and Training*, Dorn, L. (Ed.), Ashgate, England, UK, 2003.
6. McFarlane, D. C., Comparison of four primary methods for coordinating the interruption of people in human–computer interaction, *Human-Computer Interaction* 17(1), 63–139, 2002.
7. Monk, C. A., Boehm-Davis, D. A., and Trafton, J. G., Recovering from interruptions: Implications for driver distraction research, *Human Factors* 46(4), 650–663, 2004.

Index

A

B

C

E

F

G

H

M

N

O

P

Q

R

S

T

U

V

W

Y